Gene Therapy
of Cancer

Gene Therapy of Cancer

Translational Approaches from Preclinical Studies to Clinical Implementation

Edited by

Edmund C. Lattime, PhD
Department of Surgery
UMDNJ-Robert Wood Johnson Medical School
The Cancer Institute of New Jersey
New Brunswick, New Jersey

Stanton L. Gerson, MD
Division of Hematology, Oncology, and
Ireland Cancer Center at University Hospitals of Cleveland
and Case Western Reserve University
Cleveland, Ohio

Academic Press

San Diego   London   Boston   New York   Sydney   Tokyo   Toronto

Cover photograph: Low and high magnification of squamous cell carcinoma cells stained with hematoxylin and eosin showing adenovirus-induced cytopathic effects. For more details, see Chapter 15, color figure 4.

This book is printed on acid-free paper. ⊚

Academic Press
a division of Harcourt Brace & Company
525 B Street, Suite 1900, San Diego, California 92101-4495, USA
http://www.apnet.com

Academic Press Limited
24-28 Oval Road, London NW1 7DX, UK
http://www.hbuk.co.uk/ap/

Library of Congress Catalog Card Number: 98-85115

International Standard Book Number: 0-12-437190-6

PRINTED IN THE UNITED STATES OF AMERICA
98 99 00 01 02 03 MM 9 8 7 6 5 4 3 2 1

To Holly, Deb, and our children Sarah, Ruth, James, and David,
for their encouragement and tolerance
that led us to conceive and complete this book.

Contents

PART I Genetic and Immunologic Targets for Gene Therapy

CHAPTER 1 Tumor Suppressor Genes as Targets for Cancer Gene Therapy

RAYMOND D. MENG AND WAFIK S. EL-DEIRY

CHAPTER 2 Characterization of Specific Genetic Alterations in Cancer Cells

KAY HUEBNER, TERESA DRUCK, PIOTR HADACZEK, PETER A. MCCUE, AND HENRY C. MAGUIRE, JR.

CHAPTER 3 Immunologic Targets for the Gene Therapy of Cancer

SUZANNE OSTRAND-ROSENBERG, VICKY S. GUNTHER, TODD A. ARMSTRONG, BETH A. PULASKI, MATTHEW R. PIPELING, VIRGINIA K. CLEMENTS, AND NANA LAMOUSÉ-SMITH

CHAPTER 11   Transfer of Drug Resistance Genes into Hematopoietic Progenitors

OMER N. KOC, BRIAN M. DAVIS, JANE S. REESE, SARAH E. FRIEBERT, AND STANTON L. GERSON

CHAPTER 12   Cytosine Deaminase as a Suicide Gene in Cancer Gene Therapy

CRAIG A. MULLEN

CHAPTER 13   Preemptive and Therapeutic Uses of Suicide Genes for Cancer and Leukemia

FREDERICK L. MOOLTEN AND PAULA J. MROZ

PART IV   Targeting Oncogenes and Growth Factors for Gene Therapy

CHAPTER 14   Antisense Strategies in the Treatment of Leukemias

BRUNO CALABRETTA, TOMASZ SKORSKI, GERALD ZON, MARIUSZ Z. RATAJCZAK, AND ALAN M. GEWIRTZ

CHAPTER 15   Selectively Replicating Viruses as Therapeutic Agents against Cancer

DAVID KIRN

PART V   Molecular Vaccine Strategies for Cancer

CHAPTER 16   *ras* Oncogene Products as Tumor-Specific Antigens for Activation of T-Lymphocyte-Mediated Immunity

SCOTT I. ABRAMS, J. ANDREW BRISTOL, AND JEFFREY SCHLOM

CHAPTER 17   Polynucleotide-Mediated Immunization Therapy of Cancer

STEPHEN ANDREW WHITE, ROBERT MARTIN CONRY,
THERESA V. STRONG, DAVID TERRY CURIEL,
AND ALBERT FRANCES LoBUGLIO

CHAPTER 18   DNA and Dendritic Cell-Based Genetic Immunization against Cancer

LISA H. BUTTERFIELD, ANTONI RIBAS,
AND JAMES S. ECONOMOU

PART VI   Genetically Modified Cells for Immunization

CHAPTER 19   Engineering Cellular Cancer Vaccines: Gene and Protein Transfer Options

MARK TYKOCINSKI

Contributors

Numbers in parentheses indicate the pages on which the authors' contributions begin.

Scott Abrams, PhD **(251)**
National Institutes of Health, National Cancer Institute, Laboratory of Tumor Immunology and Biology, Bethesda, Maryland 20892

G. P. Adams, PhD **(113)**
Fox Chase Cancer Center, Medical Oncology, Philadelphia, Pennsylvania 19111

Todd Armstrong, PhD **(33)**
University of Maryland, Department of Biological Sciences, Baltimore, Maryland 21250

Angelo Baccala **(333)**
The Johns Hopkins University School of Medicine, The Johns Hopkins Oncology Center and Brady Urological Institute, Baltimore, Maryland 21287

Christopher Baum, PhD **(51)**
Heinrich-Pette-Institut, D-20251 Hamburg, Germany

Carmela Beger **(139)**
University of California, San Diego, Departments of Medicine and Biology, La Jolla, California 92093-0665

J. A. Bristol, PhD **(251)**
National Institutes of Health, National Cancer Institute, Laboratory of Tumor Immunology and Biology, Bethesda, Maryland 20892

Lisa Butterfield, PhD **(285)**
University of California, Los Angeles, Division of Surgical Oncology, Los Angeles, California 90095

Bruno Calabretta, MD, PhD **(223)**
Thomas Jefferson University, Kimmel Cancer Center, Department of Microbiology and Jefferson Cancer Institute, Director, Leukemia and Lymphoma Program, Philadelphia, Pennsylvania 19107

Alfred Chang, MD **(349)**
University of Michigan, Department of Surgery, Ann Arbor, Michigan 48109

Saswati Chatterjee, PhD **(95)**
City of Hope National Medical Center, Division of Pediatrics, Duarte, California 91010

Si.-Yi Chen, PhD **(389)**
Wake Forest University School of Medicine, Comprehensive Cancer Center, Department of Cancer Biology, Winston-Salem, North Carolina 27157

Virginia Clements **(33)**
University of Maryland, Department of Biological Sciences, Baltimore, Maryland 21250

Robert Conry, MD **(271)**
University of Alabama at Birmingham, Birmingham, Alabama 35294

Mark Cooper, MD **(77)**
Copernicus Gene Systems, Inc., Vice President of Science and Medical Affairs, Cleveland, Ohio 44106

David Curiel, MD **(271)**
University of Alabama at Birmingham, Birmingham, Alabama 35294

Brian Davis **(177)**
Case Western Reserve University, Graduate Student, Cleveland, Ohio 44106-4937

Teresa Druck **(21)**
Jefferson Medical College, Kimmel Cancer Center, Department of Microbiology-Immunology, Philadelphia, Pennsylvania 19103

James Economou, MD **(285)**
University of California, Los Angeles, Division of Surgical Oncology, Los Angeles, California 90095

Laurence Eisenlohr, VMD, PhD **(125)**
Thomas Jefferson University, Department of Microbiology and Immunology, Philadelphia, Pennsylvania 19107

Wafik El-Deiry, PhD **(3)**
University of Pennsylvania School of Medicine, Howard Hughes Medical Institute, Department of Medicine, Genetics, Cancer Center and The Institute for Human Gene Therapy, Laboratory of Molecular Oncology and Cell Cycle Regulation, Philadelphia, Pennsylvania 19104

Grace Fisher-Adams, PhD **(95)**
City of Hope National Medical Center, Division of Pediatrics, Duarte, California 91010

Jed Freeman **(155)**
Cancer Care, Cancer Care Consultants, Department of Oncology, Redding, California 96001

Scott Freeman, MD **(155)**
Tulane University Medical School, Department of Pathology, New Orleans, Louisiana 70122-2699

Sarah Friebert, MD **(177)**
Case Western Reserve University, Pediatric Hematology/Oncology Fellow, Cleveland, Ohio 44106-4937

Stanton Gerson, MD **(177)**
Case Western Reserve University, Professor of Medicine, Cleveland, Ohio 44106-4937

Alan Gewirtz, MD **(223)**
University of Pennsylvania School of Medicine, Department of Pathology and Internal Medicine, Philadelphia, Pennsylvania 19104

Leonard Gomella, MD **(125)**
Thomas Jefferson University, Department of Urology, Philadelphia, Pennsylvania 19107

Vicky Gunther **(33)**
University of Maryland, Department of Biological Sciences, Baltimore, Maryland 21250

Piotr Hadaczek, PhD **(21)**
Medical Academy, Department of Genetics and Pathology, 70-111 Szczecin, Poland

Evan Hersh, MD **(319)**
Arizona Cancer Center, Tucson, Arizona 85724

Kay Huebner, PhD **(21)**
Jefferson Medical College, Kimmel Cancer Center, Department of Microbiology-Immunology, Philadelphia, Pennsylvania 19103

David Kirn, MD **(235)**
Onyx Pharmaceutical, Director, Clinical Research, Richmond, California 94806

Omer Koc, MD **(177)**
Case Western Reserve University, Assistant Professor of Medicine, Cleveland, Ohio 44106-4937

Martin Krüger **(139)**
University of California, San Diego, Departments of Medicine and Biology, La Jolla, California 92093-0665

Nana Lamous-Smith **(33)**
University of Maryland, Department of Biological Sciences, Baltimore, Maryland 21250

Edmund Lattime, PhD **(125)**
The Cancer Institute of New Jersey, Associate Director, and UMDNJ–Robert Wood Johnson Medical School, Professor of Surgery, New Brunswick, New Jersey 08901

Ho Lim, MD **(333)**
The Johns Hopkins University School of Medicine, The Johns Hopkins Oncology Center and Brady Urological Institute, Baltimore, Maryland 21287

Albert LoBuglio, MD **(271)**
University of Alabama at Birmingham, Birmingham, Alabama 35294

Michael Lotze, MD **(359)**
University of Pittsburgh Medical Center, Molecular Genetics and Biochemistry, Pittsburgh, Pennsylvania 15261

Di Lu, MD, PhD **(95)**
University of Kansas, Department of Pathology, Lawrence, Kansas 66045

Henry Maguire, Jr., MD **(21)**
Jefferson Medical College, Kimmel Cancer Center and Department of Medicine, Philadelphia, Pennsylvania 19103

Aizen Marrogi, MB **(155)**
Louisiana State University Medical School, Department of Surgery and Gene Therapy Program, New Orleans, Louisiana 70112

Michael Mastrangelo, MD **(125)**
Thomas Jefferson University, Department of Medicine, Division of Medical Oncology, Philadelphia, Pennsylvania 19107

A. M. McCall, PhD **(113)**
Fox Chase Cancer Center, Medical Oncology, Philadelphia, Pennsylvania 19111

Peter McCue, MD **(21)**
Jefferson Medical College, Kimmel Cancer Center and Department of Pathology, Anatomy and Cell Biology, Philadelphia, Pennsylvania 19103

Raymond Meng (3)
University of Pennsylvania School of Medicine, Laboratory of Molecular Oncology and Cell Cycle Regulation, Cell and Molecular Biology Graduate Program, Philadelphia, Pennsylvania 19104

Frederick Moolten, MD (209)
Edith Nourse Rogers Memorial Veterans Hospital, Bedford, Massachusetts 01730, and Boston University School of Medicine, Department of Microbiology, Boston, Massachusetts 02215

Paula Mroz (209)
Edith Nourse Rogers Memorial Veterans Hospital, Bedford, Massachusetts 01730

James Mulé, PhD (373)
University of Michigan, Department of Surgery, Ann Arbor, Michigan 48109-0666

Craig Mullen, MD, PhD (201)
University of Texas M. D. Anderson Cancer Center, Departments of Experimental Pediatrics and Immunology, Houston, Texas 77030

Anupama Munshi, PhD (155)
Louisiana State University Medical School, Department of Surgery and Gene Therapy Program, New Orleans, Louisiana 70112

Wolfram Ostertag, PhD (51)
Heinrich-Pette-Institut, D-20251 Hamburg, Germany

Suzanne Ostrand-Rosenberg, PhD (33)
University of Maryland, Department of Biological Sciences, Baltimore, Maryland 21250

Jean-Marie Peron, MD (359)
Hopital Purpan, Service d'Hépatogastroentérologie, CHRU Toulouse, France

Matthew Pipeling (33)
University of Maryland, Department of Biological Sciences, Baltimore, Maryland 21250

Beth Pulaski, PhD (33)
University of Maryland, Department of Biological Sciences, Baltimore, Maryland 21250

Rajagopal Ramesh, PhD (155)
Louisiana State University Medical School, Department of Surgery and Gene Therapy Program, New Orleans, Louisiana 70112

Mariusz Ratajczak (223)
University of Pennsylvania, Department of Internal Medicine, Philadelphia, Pennsylvania 19104

Jane Reese (177)
Case Western Reserve University, Senior Research Assistant, Cleveland, Ohio 44106-4937

Antoni Ribas, MD (285)
University of California, Los Angeles, Division of Surgical Oncology, Los Angeles, California 90095

A. P. Salas, MD (349)
University of Michigan, Department of Surgery, Ann Arbor, Michigan 48109

Jeffrey Schlom, PhD (251)
National Institutes of Health, National Cancer Institute, Laboratory of Tumor Immunology and Biology, Bethesda, Maryland 20892

Elizabeth Shaughnessy, MD, PhD (95)
University of Cincinnati, Department of Surgery, Cincinnati, Ohio 45267-0772

Michael R. Shurin, MD, PhD (359)
University of Pittsburgh Medical Center, Department of Surgery, Pittsburgh, Pennsylvania 15261

Jonathan W. Simons, MD (333)
The Johns Hopkins Oncology Center, Baltimore, Maryland 21231

Tomasz Skorski, PhD (223)
Thomas Jefferson University, Kimmel Cancer Center, Department of Microbiology and Jefferson Cancer Institute, Director, Leukemia and Lymphoma Program, Philadelphia, Pennsylvania 19107

Carol Stocking, PhD (51)
Heinrich-Pette-Institut, D-20251 Hamburg, Germany

Theresa Strong (271)
University of Alabama at Birmingham, Birmingham, Alabama 35294

Mark Tykocinski, MD (301)
University of Pennsylvania, Professor and Chairman, Department of Pathology and Laboratory Medicine, Philadelphia, Pennsylvania 19104

Dorothee von Laer (51)
Heinrich-Pette-Institut, D-20251 Hamburg, Germany

L. M. Weiner, MD (113)
Fox Chase Cancer Center, Medical Oncology, Philadelphia, Pennsylvania 19111

Katharine Whartenby (155)
Brown University and Rhode Island Hospital, Department of Medicine, Division of Clinical Pharmacology, Providence, Rhode Island 02903

Stephen White (271)
University of Alabama at Birmingham, Birmingham, Alabama 35294

Lee Wilke (373)
University of Michigan, Department of Surgery, Ann Arbor, Michigan 48109-0666

K. K. Wong, Jr., MD (95)
City of Hope National Medical Center, Department of Hematology and Bone Marrow Transplantation, Duarte, California 91010

Flossie Wong-Staal, PhD (139)
University of California, San Diego, Departments of Medicine and Biology, La Jolla, California 92093-0665

Gerald Zon (223)
Lynx Therapeutics Inc., Hayward, California 34545

Preface

Our goal in compiling *Gene Therapy of Cancer* is to produce a comprehensive review of the scientific basis and current approaches to this rapidly changing field. The result is a surprisingly broadly based perspective which traverses basic and applied science, gene transfer therapeutic strategies, and translational therapeutic applications. Since the clinical benefits of the gene therapy of cancer remain an unrealized vision for the future, we view this text as a work in progress and invite you to participate in the excitement, dilemmas, and new directions that are presented within these pages.

The organization and emphasis of the chapters was influenced by extensive discussions held between the editors and past and current members of the Experimental Therapeutics Study Section, as well as other colleagues in the field. The discussions with our Experimental Therapeutics II Study Section colleagues focused on key issues explored in these pages. These include: opportunities for current and future research, including defining predictive markers of therapeutic benefit; using immunovaccines and other immunotherapeutics; determining how to target tumor cells with new genes when, at first impression, it would seem impossible to target every tumor cell; how to define therapeutic benefit in hematopoietic stem cell drug resistance gene transfer; and which of the numerous antioncogene strategies is likely to yield therapeutic benefit. Answers to some of these questions are in the enclosed chapters; others await the results of current research.

This has been an exciting project for us and each of the contributing authors. We think you will find each chapter to be both provocative and timely. We hope you enjoy it and welcome your comments.

STANTON L. GERSON, M.D.

EDMUND C. LATTIME, Ph.D.

PART I

Genetic and Immunologic Targets for Gene Therapy

CHAPTER 1

Tumor Suppressor Genes as Targets for Cancer Gene Therapy

RAYMOND D. MENG AND WAFIK S. EL-DEIRY

Laboratory of Molecular Oncology and Cell Cycle Regulation, Howard Hughes Medical Institute, Departments of Medicine and Genetics, Cancer Center, and The Institute for Human Gene Therapy, University of Pennsylvania School of Medicine, Philadelphia, Pennsylvania 19104

I. INTRODUCTION

Cancer cells accumulate numerous genetic alterations that contribute to tumorigenesis, tumor progression, and chemotherapeutic drug resistance. Most of these alterations affect the regulation of the cell cycle. In normal cells, a balance is achieved between proliferation and cell death by tightly regulating the progression through the cell cycle with cellular checkpoints [1]. Before the cell can enter the next cell cycle phase, it must pass through a checkpoint that decides if all the previous processes have been completed. The decision to enter the cell cycle is made during the G1 phase by cyclins and their regulatory units, the cyclin-dependent kinases (CDKs). Control of the CDKs is achieved by phosphorylation on different sites of the protein and by the activity of CDK inhibitors, comprised of two families [2]. The INK4 proteins, consisting of p15, p16, p18, and p19, bind to cdk4 or cdk6, whereas the CIP/KIP proteins, including p21, p27, and p57, bind to cyclin–CDK complexes.

Cell cycle progression is also affected by environmental stimuli. Thus, if a eukaryotic cell is deprived of nutrients or growth factors, the cell cycle machinery responds by activation of a checkpoint, leading to cell cycle arrest until conditions become favorable for cell division [3]. In mammalian cells, growth factors and mitogens regulate the expression level of cyclin D1, which is involved in driving foward the G1- to S-phase transition [4]. In mammalian cells, the growth-factor-dependent G1 to S checkpoint is regulated by the retinoblastoma protein [5]. Thus, once a cell has traversed this "restriction point" [6], it is committed to going through the S, G2, and M phases, leading to two daughter cells. One of the hallmarks of cancer cells is loss of checkpoint control. Both the viral oncoproteins [7] and mutations in human cancer [4] appear to specifically target two parallel yet interrelated cell-cycle-controlling pathways: the p16–cyclin D1–CDK4–Rb and the ATM–p53–p21 pathways. The p16–cyclin D1–CDK4–Rb pathway helps the cell progress through the G1 phase. Cyclin D1 binds to and positively regulates CDK4, which helps to phosphorylate Rb. In a hypophosphorylated state, Rb is a negative regulator of cell cycle progression because it binds to and inactivates transcription factors. However, when it is hyper-phosphorylated by cyclin E–CDK2, Rb becomes inactive, and the cell can continue into the S phase. p16 is a negative regulator of CDK4, which prevents it from phosphorylating Rb. Abnormalities of each of the components of this pathway have been identified in human cancer and are usually mutually exclusive [4].

The detection of DNA damage is governed by the tumor suppressor p53. Following DNA damage, p53 arrests the cell to allow time for repair, or if the damage is extensive enough, p53 initiates programmed cell death, or apoptosis. Loss of these various molecular checkpoints has been found to underlie the development of many tumors because cell cycle progression becomes dysregulated. The accumulation of genetic alterations also contributes to enhanced chemoresistance, resulting from the loss of the ability to respond to DNA damage (reviewed by El-Deiry [8]). Of the many alterations that have been identified, changes involving the tumor suppressors, such as p53, are the most common. With loss of growth suppression, progression through the cell cycle remains unchecked, and tumorigenesis results. Therefore, a major strategy in gene therapy for cancer has focused on replacing the tumor suppressors in cancer cells that have been lost through deletion or mutation. This review will focus on five tumor suppressors, p53, p21, p16, Rb, and p27, that have been studied

as potential targets for gene replacement in cancer (see Fig. 1). The following sections will discuss the background of each gene and its current role in gene therapy for cancer, including its use in any clinical trials.

II. LIMITATIONS AND PROGRESS TOWARD GENE THERAPY FOR CANCER

A. Vectors for Gene Delivery

The field of gene therapy, including research into cancer gene therapy, is currently subject to three major constraints: (1) gene delivery, (2) gene expression, and (3) efficacy and toxicity (reviewed by Verma and Somia [9]). For gene delivery, viruses have often been used as vectors because they can transfer genes efficiently and at high expression of the target gene. The early gene therapy experiments used retroviruses because they can integrate into the chromosomes of the host cell to provide prolonged gene expression and because they can carry large genes, up to 10 kb. Retroviruses, however, have two disadvantages. First, because retroviruses integrate into host chromosomes, they can only efficiently infect cells that are dividing; hence, quiescent or terminally differentiated cells, such as neurons, are not effectively infected. Second, it is difficult to prepare sufficiently high titers of retroviruses for *in vivo* gene therapy.

In contrast to retroviruses, adenoviruses infect a wide range of cells, including dividing and nondividing cells, and they can be prepared at extremely high titers, on the order of 10^{12} pfu/mL which is sufficiently concentrated for *in vivo* use (reviewed by Hitt *et al.* [10]). Although they offer high gene expression, adenoviruses do not undergo chromosomal integration. In cancer gene therapy, however, successful eradication of a tumor would obviate the need for prolonged gene expression, which would be a greater requirement for human gene therapy for somatic diseases. In fact, almost 75% of the patients enrolled in gene therapy clinical trials last year were being treated for malignancies, primarily solid tumors [11]. The most significant drawback to adenoviruses, however, is that they elicit a strong host immune response; thus, the preclinical *in vivo* experiments have been conducted in immunodeficient hosts, either by direct suppression of host immunity [12] or by modification of the adenovirus vector to lessen its inherent immunogenicity [13]. Although it may seem intuitive that a heightened immune response may be good in cancer gene therapy, it is less desirable on a practical scale because the immune response helps to eliminate the vector and to decrease the expression of the transduced gene.

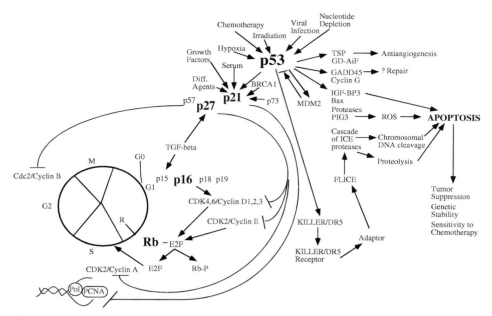

FIGURE 1   Tumor suppressors that are targets for gene therapy of cancer. This review discusses five tumor suppressors involved in cell cycle regulation that have been studied as potential targets for gene replacement in cancer. Loss of these tumor suppressors, most notably p53, results in tumor development and progression. p53 mediates the cellular response to DNA damage, resulting in growth arrest or in apoptosis. p21 is a main effector of p53 that mediates growth arrest and is a CDKI, along with p16 and p27, which help to regulate G1 transition. Rb helps to mediate cell cycle progression from G1 to S phase.

Recent research has focused on improving the gene delivery capability of the adenovirus vectors. The introduction of adenoviruses into cells was enhanced following its formation in a complex with DNA via poly-L-lysine [14]. The specific targeting of adenoviruses has been attempted at the transcriptional level by adding tissue-specific promoters to the adenovirus backbone (reviewed in Miller and Whelan, [15]). Recently, tetracycline-responsive recombinant adenoviruses have been developed to try to regulate gene expression following infection [16,17].

Some efforts have been directed toward the development of new virus vectors for gene therapy. First, adeno-associated viruses, derived from a nonpathogenic and defective human parvovirus, offer certain advantages. They are relatively safe, provide high titers, and can infect a wide range of cell types, including quiescent cells (reviewed in Flotte and Carter [18]). A p53-adeno-associated virus vector transduced wild-type p53 into a human cancer cell line, resulting in growth suppression and cytotoxicity [19]. Second, a novel chimeric vector was derived from both adenovirus and retrovirus genes [20]. These vectors were then used to infect human bladder cancer and ovarian cancer cells, causing them to act as transient producer cells, which release retroviruses that can then integrate into nearby cells. Third,

lentiviruses, such as HIV, are being used to infect nondividing and even growth-arrested cells. Vectors based on lentiviruses that have been deleted of their virulence genes are being created [21]. Fourth, replication-restricted herpes simplex viruses were effectively used to infect melanomas and may be used in gene therapy [22]. Fifth, anecdotally, p53 DNA has been injected directly into liver cancers with some success, although the mechanism for tumor growth inhibition remains unclear [23].

B. Route of Administration

With respect to delivery *in vivo*, infectivity represents the gateway into the target cells once the virus of choice reaches its destination. However, the ability of a particular agent to reach the tumor is influenced by its route of administration. The most direct route is intratumoral, which practically assures delivery to the target tissue. Furthermore, the existence of a bystander effect obviates the need to deliver the agent into every cell, and local delivery minimizes toxicity to the normal tissues, where the agent is not present.

Intratumoral delivery of adenoviruses that selectively replicate in p53 deficient tumor cells have been designed that may take full advantage of introducing a small amount of virus in some tumor cells followed by propagation and toxicity only within the tumor mass [24,25]. The adenovirus E1b gene region, which binds to and inactivates wild-type p53 in host cells, was deleted from an adenovirus backbone [24]. Consequently, because it can no longer inactivate p53, this vector can only replicate in p53-null cells, effectively targeting this virus to cells with p53 mutation, which are usually cancer cells. It was shown that normal cells are highly resistant to the E1b-deleted adenovirus [25]. Although animal studies have suggested low toxicity and excellent efficacy, these investigations have been limited by the use of immunodeficient mice and by the poor infectivity of normal rodent cells compared to the human cancer xenografts. Clinical trials are currently underway with the E1b-deleted adenovirus in cancer patients, and it remains to be seen if survival will be affected and what short- and long-term side effects may occur in human beings.

Delivery of virally expressed genes by intravascular or intracavitary injections also presents barriers to the delivery of the target genes. In intravascular administration, instillation into a peripheral vein dilutes the vehicle, so only a small portion may ultimately reach the tumor. Intravascular administration also elicits a powerful immune response [26]. Tropism for organs such as the liver, for example by adenovirus, can be a disadvantage if delivery is intended elsewhere or may be advantageous if the liver is the target [27]. Even with regional intravascular administration, the virus must traverse the endothelial wall and travel against pressures within an expanding tumor mass. In the case of intracavitary administration (i.e., intrapleural or intraperitoneal), the surface of the tumor mass is coated by virus, but intratumoral delivery within a solid mass represents an important barrier.

III. p53

A. Introduction

Among the tumor suppressors being considered for gene replacement in malignancies, p53 has been the focus of many groups for several reasons. First, p53 plays a pivotal role in the fate of a cell following DNA damage. It determines if the damaged cell will undergo growth arrest in order to repair itself [28] or if the cell will undergo programmed cell death or apoptosis because the damage is too extensive [29,30]. Therefore, loss of p53 or mutations in p53, some of which can act in a dominant-negative manner to inhibit residual wild-type p53, significantly contribute to tumor development, progression, and chemotherapeutic resistance (re-

viewed by Velculescu and El-Deiry, [31]). The largest study to date, involving a survey of the toxicity of hundreds of anticancer drugs towards over 60 human cancer and leukemia cell lines, has indicated that the vast majority of clinically useful drugs are most effective in cells that express wild-type p53 [32]. Second, mutations in p53 are, in fact, the most common genetic alterations in tumors, being mutated or deleted in over 50% of all human cancers [33]. Germline transmission of a mutant p53 allele predisposes individuals with Li–Fraumeni syndrome to a high risk of cancers [34]. Genetically engineered mice had both alleles of p53 deleted or "knocked-out"; 75% developed tumors by 6 months of age, and all died by 2 years [35]. Third, loss of p53 results in decreased apoptosis [36] and decreased susceptibility to radiotherapy or to chemotherapy [29,30].

With respect to gene replacement therapy, p53 is a potent inducer of cancer cell apoptosis and is effective despite the presence of multiple genetic changes in the cancer cells [37]. Because p53 offers a promising way to regulate the growth of cancers *in vivo,* much work has been focused on replacing or even overexpressing wild-type p53 in the hope that aberrant cell cycle control can once again be tightly regulated. It is important, however, to realize that despite its strengths, p53-gene therapy has important limitations that must be considered for its clinical development as an anticancer agent.

B. Gene Therapy with p53

The initial p53-gene therapy experiments used retroviruses to deliver the tumor suppressor gene into various cancer cell lines. Two groups introduced wild-type p53 via a retrovirus vector into a non-small-cell lung cancer line and suppressed the growth of the tumor both *in vitro* [38] and *in vivo* in a nude mouse model [39]. It was also shown in a phase I clinical trial that retrovirus-transferred p53 can be used to infect human non-small-cell lung cancers by intratumoral injection and that it may effectively limit growth in a small minority of these patients with advanced terminal cancer [40].

Currently, the majority of p53-directed gene replacement strategies have shifted toward using an adenovirus vector because the adenovirus can infect numerous cell types and because it can be produced in high titers. It has been previously reported that an adenovirus expressing β-galactosidase (Ad-LacZ) is capable of infecting tumor cells from a wide range of tissues, including brain, lung, breast, ovarian, colon, and prostate (see Table 1) [41]. Furthermore, the ability of the adenovirus to infect these tumor cells was independent of the p53 status of the host cell. Some cell lines, however, remain inherently resistant to adenovirus infection. For exam-

TABLE 1   Infection of Selected Cell Lines by Ad-p53

Cell type	Name of cell line	p53	MOI[a]	In Vitro	In Vivo	References
Bladder	UMUC3	mut	20	X		43
Breast	MDA-MB-435	mut	N.I.[b]	X		42
	MCF7	wt	10	X		45
	SKBr3	mut	<30	X		41
Choriocarcinoma	JEG3	wt	100	X		43
Colon	HCT116	wt	<30	X		41
	RKO	wt	<30	X		41
	SW480	mut	<30	X		41
Glioblastoma	Del4A	mut	<30	X		41
	U-87 MG	wt	100	X		46
	G122	mut	30	X	X	47
Head and Neck	Tu-177	mut	100	X	X	48
	MDA 886	wt	100	X		48
	JSQ-3	mut	20	X	X	49
Leukemia	ML-1	null	N.I.	X		41
	HL60	null	N.I.	X		41
Lung	H-358	null	10	X		45
	H460	wt	<30	X		41
	A549	wt	<30	X		41
Lymphoma	RAMOS	mut	N.I.	X		41
	CA46	mut	N.I.	X		41
Melanoma	A875	wt	100	X		43
	7336	wt	100	X		43
Nasopharyngeal	RPMI 2650	wt	10	X	X	50
	Fadu	mut	250	X		50
	Detroit 562	mut	250	X		50
Ovarian	SKOV3	null	<30	X		41
	OVCAR	mut	<30	X		41
Prostate	LNCaP	wt	<30	X		41
	DU-145	mut	<30	X		41

[a] MOI indicates >50% transduction.
[b] N.I., not infected by the adenovirus, even at the highest MOI tested.

ple, it was reported that two leukemia and two lymphoma cell lines showed less than 0.001% infectivity following infection with an Ad-LacZ at an MOI of 150 [41]. Other adenovirus-resistant cells include a breast cancer cell line MDA-MB-435 with mutant p53 [42] and a choriocarcinoma cell line JEG3 with wild-type p53 [43]. The explanation for cellular resistance to adenovirus infection remains unknown, primarily because the cellular receptor responsible for adenovirus binding is unknown. A recent report suggested that α-integrins are required for efficient internalization of adenoviruses [44], but FACS analysis did not show decreased integrin expression in a breast cancer cell line that was infected poorly with a p53 adenovirus (Ad-p53) [42]. Although the mechanism is unknown, inefficient adenovirus infection of cells remains an important cause of nonresponsiveness to p53 treatment.

To help determine successful responsiveness to Ad-p53, markers for productive p53 infection are currently being studied. Unfortunately, examining p53 levels following Ad-p53 infection is difficult because the majority of human cancer cells already overexpress p53 because of its mutant status. In these cases, assessment of p53 protein levels following Ad-p53 infection is not useful. The expression of the p21$^{WAF/CIP1}$ protein has instead been evaluated as a marker for efficient transduction of wild-type p53. p21, a CDK inhibitor, is a downstream target of p53 that is transcriptionally induced following p53 elevation in response to DNA damage [51]. p21 acts as a negative growth regulator during the G1 cell cycle checkpoint by binding and inhibiting cyclin/CDKs [51, 52]. It was hypothesized that p21 would be a good marker for p53 transduction because in tumor cells expressing mutant p53, p21 is expressed at low levels [53].

Consequently, following Ad-p53 infection in different cancer cell lines, high expression of p21 is induced both *in vitro* and *in vivo,* as detected by immunocytochemistry or by Western blot analysis, and this induction is independent of p53 status [41,50]. Recently, the responsiveness to p53 gene therapy has also been correlated with the phosphorylation state of the endogenous retinoblastoma (Rb) protein [43]. It was shown that persistence of the hypophosphorylated form of Rb predicted effective growth suppression of cancer cells following Ad-p53 infection. Although there are targets of p53 that are secreted by cells (i.e., the angiogenesis inhibitor thrombospondin-1 and IGF-BP3), there is some evidence that inhibition of angiogenesis may in part account for a p53-dependent bystander effect [54].

C. Cell Cycle Arrest and Apoptosis Induced by Ad-p53

Numerous studies have documented the efficacy of Ad-p53 infection in cancer cell growth inhibition ([40]; reviewed by Roth and Cristiano [55]). The cell cycle arrest is both dose- and time-dependent. One study reported that the degree of apoptosis is lessened in cells harboring wild-type p53 when compared to those with mutant p53, suggesting that some cells with wild-type p53 are resistant to Ad-p53 in the absence of DNA damage [41]. In contrast, growth inhibition by Ad-p53 did not seem to depend on p53 status in prostate cancer cells [56].

Although most studies have focused on the effect of Ad-p53 infection on cancer cells, the effect on normal cells has important implications regarding toxicity. Until a vector that specifically targets cancer cells is developed, inadvertent infection of normal cells by Ad-p53 could therefore lead to undesired side effects. One study compared the response of a human fibroblast cell line GM38 and two nasopharyngeal carcinoma cell lines to Ad-p53 infection and found that the fibroblasts had lower transfection efficiency and transgene expression following Ad-p53 infection. Consequently, these fibroblasts remained resistant to p53-mediated cytotoxicity, although some cell death was observed [57]. The cell cycle distribution of another human fibroblast cell line was also not affected by Ad-p53 [49]. Our group has recently submitted a paper showing that infection of normal human keratinocytes by Ad-p53 is also very poor [58]. It is therefore likely that normal cells may be more resistant to the effects of Ad-p53 because of their poor infectivity and possibly the fact that they have intact the mechanisms to survive despite wild-type p53 expression. For example, normal cells may induce MDM2, which may then lead to p53 degradation.

p53 differs from the other tumor suppressors in that it can induce apoptosis. Apoptosis is a highly desired effect in cancer therapy. What may prove useful in p53-gene therapy is a greater understanding of the apoptotic pathways regulated by p53 (reviewed by Canman and Kastan [59]). p53 seems to be involved in multiple apoptotic pathways. Although p53 can regulate the expression of genes involved in apoptosis, such as BAX or FAS/APO1, p53-dependent apoptosis can still occur independently of those target genes [60,61]. In a search for other effectors of p53-dependent apoptosis, we recently identified the *KILLER/DR5* gene [62]. *KILLER/DR5* is a novel member of the TNF-receptor family and is most identical to DR4 [63]. *KILLER/DR5* is strongly induced in cell lines with wild-type p53 following DNA damage initiated by ionizing radiation or by chemotherapy, and by Ad-p53 infection. Thus, *KILLER/DR5* may represent a new class of genes that may be targeted for gene replacement in malignancies, and its induction following Ad-p53 infection may correlate with apoptosis induction.

D. Applications of p53-Gene Therapy

1. MDR1-Overexpressing Cancer Cells

One area in which p53 gene therapy may prove useful is the infection of cancer cells that overexpress the multidrug resistance gene, MDR1 (reviewed by El-Deiry [64]). MDR1 is a cell surface membrane glycoprotein that decreases the intracellular concentration of chemotherapeutic agents. It has been reported that colon or breast cancer cell lines that are up to 1,000-fold more resistant to adriamycin because of overexpression of MDR1 are instead readily infected by Ad-p53 and undergo apoptosis, as assessed by detection of nuclear fragmentation [41]. Interestingly, it has been shown that p53 may be able to regulate MDR1, so loss of p53 would result in upregulation of MDR1 and consequently an increase in chemoresistance [65,66]. Therefore, in cell lines that are resistant to chemotherapy, Ad-p53 may effectively bypass the effect of MDR1.

2. p53 and Chemotherapy

Although the administration of Ad-p53 as a single agent may develop into an important anticancer treatment strategy, the combination of p53 gene therapy with traditional cancer treatments is a logical and practical future direction to pursue. p53 seems to play an important role in mediating DNA damage induced by various drugs because mutations in or loss of p53 are associated with decreased susceptibility to chemotherapy [30,32]. Therefore, replacing p53 in cell lines that lack functional p53 may render them more susceptible to chemotherapy. It was first reported that human non–small cell lung cancer lines with mutant p53 became more sensitive to cisplatin following transduction of wild-type p53 [67]. A recent study found that enhanced expression of wild-type p53 increased the susceptibility of a myeloid leukemia cell line to etoposide [68]. Our laboratory has also observed that the combination of DNA damaging agents and p53 infection universally results in synergy in cancer cell killing [58]. Thus, there is hope that combining Ad-p53 with DNA damaging chemotherapeutic agents may be effectively developed as a useful modality for the treatment of patients with cancer.

3. p53 AND RADIOTHERAPY

It was originally shown that thymocytes readily undergo apoptosis following exposure to γ-irradiation but that loss of wild-type p53 expression in thymocytes dramatically reduced the number of apoptotic cells [29]. In another experiment, transfection of the human papilloma virus (HPV) 16 E6 gene, which binds and inactivates p53, into human diploid fibroblasts rendered these cells more resistant to irradiation, presumably because of loss of p53 [69]. Therefore, loss of p53 may render cancer cells more resistant to radiotherapy. Efforts to combine p53-gene therapy with radiotherapy have focused on enhancing the radiosensitivity of cancer cells. Thus, p53 has been transduced into ovarian cancer cells that lack functional p53 in order to render them more radiosensitive [70]. p53 therapy has also been used to decrease the inherent radioresistance of cancer cells. The administration of Ad-p53 to a radioresistant colon cancer cell line with mutant p53 led to enhanced susceptibility to subsequent irradiation both *in vitro* and *in vivo* [71]. Irradiation and Ad-p53 infection also decreased the growth of a radioresistant head and neck cancer cell line *in vivo* in a mouse model more effectively than either treatment alone [49].

Unfortunately, it has become apparent that the presence of p53 is not always sufficient nor necessary for radiosensitization [72]. It has been shown that some human colorectal adenoma and carcinoma cell lines lacking wild-type p53 can still undergo γ-irradiation-induced apoptosis [73]. Furthermore, introduction of the HPV16 E6 gene, which binds and inactivates p53, into several tumor lines did not increase their radioresistance [74]. In conclusion, the role played by p53 in radiosensitivity may be a cell-type-specific phenomenon that needs to be established for each tumor.

IV. p21^{WAF1/CIP1}

A. Introduction

p21^{WAF1/CIP1} was originally cloned as a transcriptionally activated target of p53 [51] that was found to be a potent universal inhibitor of CDK's [52,75]. It was shown that following DNA damage, p21 is required for p53 mediated G1 arrest [76]. p21 also associates with PCNA, which results in inhibition of DNA polymerase δ processivity *in vitro* [77]. The negative growth regulatory effects of p21 are also observed during differentiation [78]. Because p21 is an important downstream target of p53 and because it helps to mediate the growth suppressive effects of p53, its effectiveness as an anticancer treatment in gene therapy replacement has been studied. In addition, p21 has growth regulatory effects independent of p53 [64], as it has been shown that expression of p21 effectively inhibits cancer cell growth *in vivo* [79].

B. Gene Therapy with p21

Most of the studies on p21-gene therapy have used adenovirus vectors. It has been shown that p21-expressing adenovirus (Ad-p21) can infect a variety of cancer cell types and can produce readily detectable p21 protein within 24 hours after infection (see Table 2) [80]. The induction of p21 was comparable to if not greater than that induced by Ad-p53 [43,45]. Infection with Ad-p21 was able to inhibit tumor growth both *in vitro* and *in vivo*, causing cell cycle arrest at G0/G1 and altering tumor morphology. Infection of normal tissues *in vivo* with Ad-p21 produced no adverse effects [80]. The exogenously transferred p21 was also shown to be functional, as histone H1 kinase assays showed that CDC2 activity and CDK2 activity were both decreased after Ad-p21 infection of glioma cells [81]. Although cell death was noted in some tumors after Ad-p21 infection, no evidence of massive apoptosis was observed [43,45,80,82].

C. Gene Therapy with a p21 Mutant Deficient in PCNA Interaction

It has been shown that the N-terminal (cyclin and CDK–interacting and –inhibitory) domain of p21 is sufficient for cancer cell growth inhibition [82]. p21-341 is a p21 mutant in which a premature stop codon has been inserted at nucleotide 341 of the human p21 sequence to delete the C-terminal domain that binds to PCNA. Transfection of p21-341 into human colon cancer cells inhibits their growth more than that of wild-type p21, and the inhibition is comparable to that obtained with p53 [82]. In addition, loss of the PCNA-interacting region of p21 contributes to a repair defect [83]. Thus, a strategy using p21 lacking the PCNA-interacting domain would be expected to inhibit growth but not stimulate DNA repair because of the absence of interaction

TABLE 2   Infection of Selected Cell Lines by Ad-p21

Cell type	Name of cell line	p53[a]	MOI[b]	In Vitro	In Vivo	References
Breast	MDA-MB-231	mut	10	X		45
	MCF-7	wt	10	X		45
	SKBr3	mut	20	X		82
Choriocarcinoma	JEG3	wt	100	X		43
Colon	SW480	mut	20	X		82
	HCT116	wt	20	X		43
Glioblastoma	U-373 MG	mut	10	X	X	81
	U-87 MG	wt	100	X		46
Head and Neck	Tu-138	mut	100	X		48
	Tu-177	mut	100	X		48
	MDA-686-LN	wt	100	X		48
	MDA 886	wt	100	X		48
Lung	H-358	null	10	X		45
	H460	wt	20	X		82
Melanoma	A875	wt	100	X		43
	7336	wt	100	X		43
Prostate	LNCaP	wt	20	X		56

[a] p53 mutations are being examined because p21 mutations are very rare in human cancers.
[b] MOI indicates >50% transduction.

with PCNA [82,83]. Therefore, an adenovirus containing p21-341 was constructed to evaluate whether it can play a role in the growth suppression of cancer cells. It was shown that the p21-341-adenovirus (Ad-p21-341) can infect colon cancer cells *in vitro* like Ad-p21, and it was confirmed that the truncated p21-341 protein associates with CDK2 but not with PCNA. Infection of colon cancer, breast cancer, and lung cancer cells with Ad-p21-341 produced a significant suppression of DNA synthesis that was comparable to that of Ad-p53 and that was independent of p53 status. Furthermore, DNA fragmentation was observed in lung and colon cancer cells following infection with Ad-p21-341. Therefore, Ad-p21-341 is an effective candidate for gene replacement in cancer, and studies are currently focusing on the mechanism of its growth inhibitory action and how it may be combined with agents that cause DNA damage that would require PCNA-dependent repair.

D. Gene Therapy with p21 as an Alternative to p53

In evaluating the potential for p21-gene therapy, most groups have compared it to Ad-p53. In contrast to Ad-p53, Ad-p21 causes little or no apoptosis following the infection of many cell lines, including those derived from head and neck cancer [84], prostate cancer [56], lung cancer [45], gliomas [85], and melanomas [43].

However, p21 is a potent suppressor of cancer cell growth. Thus, Ad-p21 may be an important alternative to Ad-p53 in specific situations in which p53 is inactivated (see Fig. 2). In some cell lines, p53 is not mutated but is still nonfunctional because it is targeted for degradation by the HPV type 16 or 18 E6 protein or because it is bound to inactivating cellular proteins such as the human homolog of the mouse double minute-2 (MDM2) oncoprotein. It is in those situations in which overexpression of these proteins may inactivate any transduced exogenous p53 that p21 may represent a more promising approach.

1. HPV16 E6 INACTIVATES p53

Infection by HPV type 16 or type 18 has been epidemiologically correlated with a greatly increased risk of cervical cancer worldwide [86]. It was discovered that the E6 protein of HPV targets human p53 for degradation through ubiquitin-mediated proteolysis [87]. In the presence of E6, a cellular protein called E6-associated protein (E6AP) binds to p53 and functions as an E3 ubiquitin ligase in mediating the degradation of p53.

In studies on the role of p53 in chemosensitivity, we created ovarian cancer cells that stably express HPV16 E6 protein, leading to endogenous p53 degradation [88]. Infection of these ovarian cancer cells with Ad-p53 produced only slight inhibition of DNA synthesis, and p21 expression was barely induced [82]. In contrast, infection of E6-overexpressing cells with Ad-p21-341 pro-

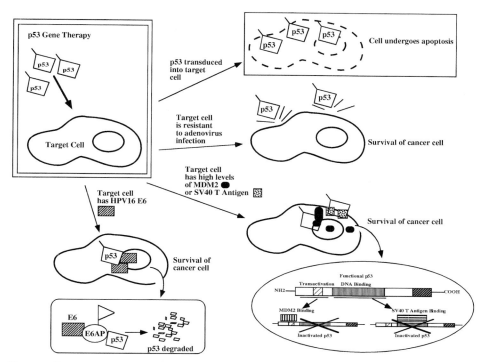

FIGURE 2 Mechanisms of resistance to p53-gene therapy. Although delivery of wild-type p53 by a virus vector suppresses the growth of and induces apoptosis in many human cancer cell lines, some cells are resistant to p53-gene therapy. An important cause of nonresponsiveness to Ad-p53 is target cell resistance to adenovirus infection, although the mechanism for this resistance is currently unknown. In other cell lines, overexpression of MDM2 or SV40 T antigen, two proteins that bind to and inactivate p53, can decrease the effectiveness of exogenously transduced p53. The expression of HPV16 E6 causes enhanced degradation of p53. In these cases, gene replacement with p21 may bypass this resistance.

duced a significant suppression of DNA synthesis at a much lower MOI. Whereas E6-overexpressing cells continued to proliferate at 3 days following infection with Ad-p53 at an MOI of 150, infection with Ad-p21-341 was cytotoxic to cells at the same time period, causing some DNA fragmentation indicative of apoptosis. Thus, in HPV-associated cancers in which E6 may be overexpressed, the inactivation of p53 may be bypassed by Ad-p21.

2. MDM2 OVEREXPRESSION INACTIVATES p53

p53 can be inactivated by binding to MDM2, which also targets p53 for degradation. The MDM2 oncogene is a target for transcriptional activation by p53 [89], but upon binding, MDM2 conceals the transactivation domain of p53 and inhibits p53-dependent transcriptional activation [90]. The importance of MDM2 in development was shown when its targeted disruption in mice led to embryonic lethality [91,92]. Thus, although p53 activates MDM2, it is MDM2 that downregulates p53 in a feedback loop that inhibits p53 function in both growth arrest and apoptosis. Although found in various

tumors, overexpression of MDM2 is most commonly reported in soft tissue sarcomas, occurring in over 30% of all cases [93]. Recently, in addition to inactivating p53, MDM2 was shown to promote the rapid degradation of p53 [94,95]. Therefore, in tumors where MDM2 is elevated, exogenous p53 gene replacement may not be the best therapy.

To test this hypothesis, we used Ad-p53 to infect several human cancer cell lines that have high expression levels of the MDM2 protein [43]. In comparison to cell lines with low levels of MDM2, the tumor cell lines with elevated MDM2 were resistant to the growth inhibitory effects of Ad-p53. Although cancer cells that overexpress MDM2 were still readily infected by Ad-p53 and induced high expression of p53 protein as determined by Western immunobloting, their rate of DNA synthesis was only slightly decreased as compared to mock-infected or Ad-LacZ-infected cells, and they displayed a blunted induction of p21. Because the inhibitory effect of the exogenous p53 was blunted in these cell lines, we tested if p21, a downstream target of p53, can bypass this MDM2-mediated inhibition of p53. Infection of MDM2-overexpressing cells with Ad-p21 resulted in overexpression of p21, a strong inhibition of cell cycle progression, and a decrease in cellular viability. Similar results were obtained with Ad-p21-341, suggesting that the cyclin-CDK-interacting domain is sufficient for bypassing p53 resistance in MDM2-overexpressing cells. Interestingly, persistence of the hyperphosphorylated form of the Rb protein correlated with tumor resistance to Ad-p53 infection, suggesting that in some situations, the phosphorylation state of Rb may be a good indicator of p53-mediated growth inhibition. A giant cell phenotype was also observed in cells infected with Ad-p21 but not with Ad-p53 [43]. Therefore, Ad-p21 may be able to effectively bypass MDM2 mediated inactivation of p53 in cancer therapy.

3. SV40 LARGE T ANTIGEN INACTIVATES p53

The p53 tumor suppressor was originally discovered in 1979 as a 53-kilodalton simian virus 40 (SV40) large T antigen-associated protein [96]. SV40 large T antigen (Tag) is known to bind to and to inactivate several tumor suppressor genes, including p53 [97]. It was recently reported that over 60% of human mesotheliomas express SV40-like sequences [98]. In several mesothelioma samples, it was shown that p53 coexpressed with Tag, and Tag coprecipitated with p53, suggesting that these sequences may bind to and inactivate p53 [99]. It was also shown that these SV40-like sequences from mesotheliomas can bind to the Rb family members as well [100]. Therefore, in some mesotheliomas that are refractory to standard therapy, experimental treatment with p21-gene therapy may be more appropriate than treatment with p53.

4. HEPATITIS B VIRUS "X PROTEIN" AND CYTOPLASMIC RETENTION OF p53

The hepatitis B virus X protein is negatively regulated by p53, and the X protein in turn inhibits p53 function [101]. The binding of X protein to p53 has been reported to exclude p53 from the nucleus [102]. Thus, in hepatitis-virus dependent liver cancer, p53 may be dysfunctional because of exclusion from the nucleus by X protein [103]. This would be another situation in which p21 or other tumor suppressors may be more effective than p53.

E. Combination of Ad-p21 and Chemotherapy or Radiotherapy

Like p53, p21 may also prove important in combination with radiotherapy or chemotherapy, either in the adjuvant setting or in progressive metastatic disease. Because p21 helps to induce cell cycle arrest following DNA damage, loss of p21 in cancer cells might cause a

deficiency in DNA repair leading to chemosensitivity. It has been reported that p21 $-/-$ cells have defective repair of damaged DNA in vitro and are more sensitive to ultraviolet radiation or to chemotherapeutic agents [83,104]. In circumstances in which the presence of wild-type p53 in a human tumor is not correlated with radiosensitivity, p21 may be involved in such radiosensitivity, and gene therapy with p21 may further enhance sensitivity. In a recent study, a retrovirus construct containing p21 was used to infect a rat glioma cell line [105]. In addition to tumor growth suppression, the introduction of p21 but not p53 rendered these cells more radiosensitive. In a colony formation assay following 8 Gy exposure, the number of p21-infected cells was decreased 93% compared to the controls and to the p53-infected cells. Second, in some tumor cells, p21 may help to mediate chemosensitivity. It has recently been shown that overexpression of p21 in a human sarcoma cell line that lacked both p53 and Rb resulted in enhanced sensitivity to several chemotherapeutic agents [106]. However, in a human colon cancer cell line with disruption of both alleles of p21, chemosensitivity to DNA crosslinking agents appears to be enhanced [83,104,107]. Recently, it was shown that p21 expression may contribute to sensitivity of cancer cells and apoptosis following exposure to antioxidants such as vitamin E [108].

V. p16^{INK4}

A. Introduction

p16$^{INK4/CDKN2}$ is a tumor suppressor gene that encodes a specific inhibitor of cyclin D–CDK4 and CDK6. By controlling the activity of CDK4, p16 helps to control the phosphorylation of Rb at late G1 [109]. p16 has been termed a tumor suppressor because it is frequently mutated or homozygously deleted in several types of cancers. p16 is also a major target for hypermethylation, leading to its inactivation in many cancers [110]. Homozygous p16 deletions have been found in over 50% of gliomas [111], but mutations in p16 are also found in other tumors, including esophageal, pancreatic, and non-small-cell lung cancer, and in familial melanomas (reviewed by Foulkes et al. [112]). Mice that had both alleles of p16 deleted became highly predisposed to the development of spontaneous tumors, and cells from these "knockout" mice were highly predisposed to transformation by the ras oncogene [113,114].

B. Gene Therapy with p16

A p16-expressing adenovirus (Ad-p16) was first used to infect non-small-cell lung cancer lines that had homozygous deletions of p16 (see Table 3) [115]. These cell lines were readily infected by Ad-p16 *in vitro* and *in vivo,* they expressed p16 at high levels, their growth rates were inhibited by up to 90%, and cell cycle arrest occurred at G0/G1. In contrast, infection of a normal mammary epithelial cell line with Ad-p16 did not cause growth inhibition or cell cycle arrest. Ad-p16 has also been used to successfully infect and inhibit tumor growth by up to 80% in several malignant glioma cell lines *in vitro,* with either wild-type p16 or homozygous p16 deletions [116]. Although Ad-p16 inhibited the growth of esophageal squamous cell cancers *in vitro,* no effect was observed for esophageal adenocarcinoma because these cells were poorly infected [117]. Because retroviruses infect hematopoietic cells more efficiently than adenoviruses, a p16 retrovirus was created and used to infect several leukemia cell lines *in vitro.* Strong growth inhibition was observed in three lines with homozygous deletions of p16, but no inhibition was observed for a leukemia cell line with mutant p16 [118].

The efficacy of p16 gene therapy has been compared to p21 and p53 therapy in causing growth inhibition in prostate cancer cells [56]. At comparable titers, p53 inhibited prostate cancer cell growth *in vitro* more significantly than either p21 or p16, which were comparable to each other. It was also shown that Ad-p53 induced a higher percentage of apoptosis among infected cells. In an *in vivo* model of prostate cancer in nude mice, all three viruses could inhibit tumor growth when initial tumor size was less than 200 mm³; however, only Ad-p53 was effective in larger tumors.

Unlike p53 or even p21, the role, played by p16 in radiosensitivity, or even if it plays a role, remains controversial. One group reported that the transfection of p16 into two human malignant melanoma cell lines, one of which had a homozygous deletion of p16, increased the

radiosensitivity of both cell lines [119]. In contrast, the levels of p16 in bladder cancer cells were not altered following irradiation [120], and the presence of p16 was not correlated with radiosensitivity or with the induction of p53 in several tumor cell lines [121]. In terms of chemosensitivity, one group reported that overexpression of p16 in an IPTG-p16 inducible cell line made them more resistant to methotrexate, vinblastine, and cisplatin [122]. In contrast, transfection of p16 into a glioma with a homozygous deletion of p16 did not increase its chemosensitivity to nitrogen mustards [123].

p16, however, has been combined with Ad-p53, which induces apoptosis, to infect a panel of cancer cells [124]. It was shown that infection of tumor cell lines *in vitro* with both viruses induced apoptosis, whereas neither virus alone, at the MOIs used, caused apoptosis. Such a strategy may prove useful for tumors that have mutations in different tumor suppressors, such as gliomas which often have deletions of p16 and mutations in p53. Thus, the combination of p16 and p53 may induce apoptotic cell death in cancer cells, although p16 alone has not been shown to induce apoptosis in any system. It is not entirely clear that *in vivo,* the combination of p53 and p16 offers anything that cannot be achieved by p53 alone, if used at a sufficiently high MOI. It has not been shown, for example, that the combination of p53 and p16 is either more tumor specific or less toxic to normal cells.

VI. Rb

A. Introduction

The retinoblastoma, or Rb, gene plays a role in the progression of the cell into the S phase. It was the first

TABLE 3   Infection of Selected Cell Lines by Ad-p16

Cell type	Name of cell line	p16	MOI[a]	*In vitro*	*In vivo*	References
Breast	MCF7	null	300	X		124
Cervical	C33A	wt	50	X		124
Colon	Lovo	null	30	X		124
Glioma	U-251 MG	null	125	X		116
	U-87 MG	null	125	X		116
Hepatocellular	HuH7	null	50	X	X	124
Leukemia	K562	null	NA[b]	X		118
	Jurkat	null	NA	X		118
	HL60	null	NA	X		118
Lung	H322	null	50	X		115
	H460	null	50	X	X	115

[a] MOI indicates >50% transduction.
[b] NA, not available.

tumor suppressor identified and has been shown to be a nuclear phosphoprotein that regulates cell cycle progression by binding to several transcription factors needed for DNA synthesis, most notably the E2F family of transcription factors [5]. The function of Rb depends on its phosphorylation state. If Rb is hypophosphorylated, it can bind to and inactivate E2F, halting progression through the cell cycle. Hyperphosphorylation of Rb, however, releases it from binding to E2F, and the cell enters the S phase. It has been hypothesized that Rb functions as a "guardian" at the R point, the point at which the cell commits itself in G1 to progress to the S phase [5]. Although Rb was originally identified as being deleted in the rare disease retinoblastoma, it can be found mutated in many other tumors, including osteosarcoma, breast cancers, hepatocellular carcinoma, and bladder cancer.

B. Gene Therapy with Rb

The initial studies on the role of Rb as a tumor suppressor focused on the replacement of Rb in various Rb-defective human cell lines *in vitro,* which suppressed the tumorigenicity of the cell lines (reviewed in Xu et al. [125]). An adenovirus construct encoding the full-length wild-type Rb was created and was used to infect several Rb $-/-$ cell lines, including a non-small-cell lung carcinoma, bladder cancer, breast cancer, and an osteosarcoma [126]. Following infection, the cell lines expressed high levels of exogenous Rb proteins, mostly in the hypophosphorylated or unphosphorylated forms, as determined by immunocytochemistry and by Western blot analyses. Infection of established bladder cancer tumors in mice with the Rb-adenovirus (Ad-Rb) slightly decreased the rate of growth of the tumors. Ad-Rb was also used to treat spontaneous pituitary melanotroph tumors in Rb $+/-$ mice *in vivo* [127]. Gene replacement with Ad-Rb decreased the growth of the tumors and increased the survival of the mice, compared to untreated controls.

C. Gene Therapy with Rb Mutants

Interestingly, an N-terminal truncated Rb mutant has been reported to cause enhanced tumor suppression *in vitro* compared to the full-length wild-type Rb [128]. When this Rb mutant was expressed in an adenovirus and used to infect human bladder cancer cells in an *in vivo* mouse model, complete growth inhibition of the tumors was observed, and tumor regression was noted in 50% of the tumors [126]. This Rb mutant may prove helpful in gene therapy of cancer, as some tumor cells,

even after the Rb gene has been replaced, are still resistant to growth inhibition, suggesting that the Rb pathway may be inactivated in these cells [125]. Another Rb mutant that may be useful for gene therapy is an Rb protein that is not phosphorylated and is therefore always present in its active cell cycle inhibitory form [129]. An adenovirus construct encoding this active Rb was used *in vivo* to control the proliferation of vascular endothelial cells in coronary artery disease.

VII. p27^{Kip1}

p27^{Kip1} is a universal cyclin dependent kinase inhibitor that was first identified as a downstream effector of TGF-β [130]. It belongs to the same family of CDK inhibitors as p21$^{WAF1/CIP1}$ and can also arrest cells in G1 [131]. p27^{Kip1} is believed to be a tumor suppressor because it maps to a chromosomal site often deleted in leukemias, because it functions as a CDK inhibitor as p21 does, and because it can be found mutated in some tumors [132]. Recent efforts have uncovered important information about frequent loss of p27 protein expression in colon, breast, and lung cancer through increased ubiquitin-mediated proteolysis of p27 in cancer [133–135]. To further study its possible role in growth suppression, two groups constructed adenoviruses encoding p27^{Kip1} [136,137]. p27 adenovirus efficiently infected a glioma and a squamous cell carcinoma cell line *in vitro,* producing profound growth suppression that was caused by cell cycle arrest at G0/G1 [137]. Tumor growth was also inhibited in an *in vivo* glioma model. When compared to Ad-p21 following infection of a breast cancer cell line, a p27 adenovirus produced greater cytotoxicity, caused G1/S arrest, and decreased CDK2 activity at a lower MOI, suggesting that in some breast cancer cell lines, p27 may be a better gene therapy agent than p21 [136]. The role of p27 in tumor suppression needs to be better refined, but p27 may be an interesting tumor suppressor for gene replacement because it seems to be regulated by TGF-β, suggesting possible combination therapy with TGF-β to induce apoptosis.

VIII. CLINICAL TRIALS

Of the tumor suppressors being considered for gene replacement in cancer, only p53 is being tested in clinical trials (see Table 4). Much of the pioneering work with p53-gene therapy was conducted by Jack Roth and colleagues at the M. D. Anderson Cancer Center. They reported the results of the first clinical trials using p53 delivered by a retrovirus vector to treat patients with non-small-cell lung cancer who had failed other treat-

TABLE 4 p53 Gene Therapy Clinical Trials in 1997

Cancer	Mode of treatment	Site of trial	Results
Bladder cancer	Adenovirus	Houston, TX	Phase I planned
Colorectal cancer	Adenovirus	Houston, TX	Phase I planned
Glioblastoma	Adenovirus	Houston, TX	Phase I planned
Head-neck squamous cell cancer	Adenovirus, multiple injections	Houston, TX	Phase I completed, 2 of 17 nonresectable had partial regression, 40% of resectable were disease-free for 6 months Phase II in progress
Head-neck squamous cell cancer, mutant p53	Adenovirus, intratumor	Pittsburgh, PA, London, England	Phase I planned, 6–18 patients
Liver cancer, primary and metastatic, mutant p53	Adenovirus, hepatic artery infusions	San Francisco, CA, Philadelphia, PA	Phase I planned, 21–42 patients
Liver cancer	Adenovirus	Houston, TX	Phase I planned
Malignant ascites	Adenovirus	Houston, TX	Phase I planned
Non–small cell lung cancer	Adenovirus and retrovirus, intratumor, combine with cisplatin	Houston, TX	Phase I completed, Adp53 slowed growth at high doses, combination with etoposide stabilized growth Phase II in progress
Non–small cell lung cancer, mutant p53	Adenovirus, intratumor	Mainz, Germany Basel, Switzerland	Phase I planned, 6–18 patients
Ovarian cancer	Adenovirus, combine with cisplatin	Houston, TX	Phase I planned
Ovarian, fallopian tube, or peritoneal cancer, mutant p53	Adenovirus, intraperitoneal	Karolinksa, Sweden Iowa City, IA	Phase I planned, 6–24 patients

Note. Data compiled from Roth and Cristiano [55], National Cancer Institute PDQ Clinical Trial Database, and *Genetic Engineering News,* June 15, 1997.

ments [40]. The virus was administered intratumorally and caused no toxic side effects up to 5 months later. Wild-type p53 was detected in lung biopsies by *in situ* hybridization and PCR amplification, and apoptosis, as determined by the TUNEL assay, was increased in post-treatment biopsy samples. Of the 9 patients in the study, 3 showed tumor growth stabilization, and 3 showed slight tumor regression. It was also reported by the same group that tumor growth inhibition was enhanced when p53 gene therapy was combined with systemic cisplatin [138]. Currently, phase II trials with p53 gene therapy for lung cancer and for head and neck cancer are being conducted. Another potentially interesting clinical trial is being conducted utilizing the E1b-deleted adenovirus, which appears to target mutant p53-expressing cells [24]. Preliminary results show no significant side effects from this vector [139].

IX. SUMMARY AND FUTURE WORK

The use of tumor suppressors in gene therapy represents an important strategy in the war on cancer.

p53-gene therapy remains the most important tumor suppressor strategy being developed, and its combination with chemotherapy or radiotherapy may prove to be even more beneficial. Currently, only p53-gene therapy has progressed to clinical trials. However, p53 may not represent the ideal choice for gene therapy in all cancers. In tumor cells that overexpress MDM2 or have HPV16 E6, other tumor suppressors such as p21 may be more desirable targets of gene therapy because they can bypass the inactivation of p53. In addition to p53 and p21, other tumor suppressors that have been studied for gene replacement, include p16, Rb, and p27. Although significant progress in gene therapy for cancer has been made within the past decade, several problems still need to be resolved. First, an efficient vector needs to be designed that can cause prolonged high expression of the transduced gene while only targeting cancer cells. Second, further criteria need to be established in scheduling the decision about which tumor suppressor to employ for gene therapy. The use of tumor suppressors represents a potentially important anticancer treatment that needs to be further investigated.

References

1. Hartwell, L. H., and Kastan, M. B. (1994). Cell cycle control and cancer. *Science* **266**, 1821–1828.
2. Morgan, D. O. (1995). Principles of CDK regulation. *Nature* **374**, 131–134.
3. Murray, A., and Hunt, T. (1993). "The Cell Cycle: An Introduction," Oxford University Press, Oxford, UK.
4. Sherr, C. J. (1996). Cancer cell cycles. *Science* **274**, 1672–1677.
5. Weinberg, R. A. (1995). The retinoblastoma protein and cell cycle control. *Cell* **81**, 323–330.
6. Pardee, A. B. (1989). G1 events and regulation of cell proliferation. *Science* **246**, 603–608.
7. Nevins, J. R. (1994). Cell cycle targets of the DNA tumor viruses. *Curr. Opin. Genet. Dev.* **4**, 130–134.
8. El-Deiry, W. S. (1997). Role of oncogenes in resistance and killing by cancer therapeutic agents. *Curr. Opin. Oncol.* **9**, 79–87.
9. Verma, I. M., and Somia, N. (1997). Gene therapy—promises, problems and prospects. *Nature* **389**, 239–242.
10. Hitt, M. M., Addison, C. L., and Graham, F. L. (1997). Human adenovirus vectors for gene transfer into mammalian cells. *Adv. Pharmacol.* **40**, 137–206.
11. Marcel, T., and Grausz, J. D. (1997). The TMC worldwide gene therapy enrollment report, end 1996. *Hum. Gene Ther.* **8**, 775–800.
12. Scaria, A., St. George, J. A., Gregory, R. J., Noelle, R. J., Wadsworth, S. C., Smith, A. E., and Kaplan, J. M. (1997). Antibody to CD40 ligand inhibits both humoral and cellular immune responses to adenoviral vectors and facilitates repeated administration to mouse airway. *Gene Ther.* **4**, 611–617.
13. Ilan, Y., Droguett, G., Chowdhury, N. R., Li, Y., Sengupta, K., Thummala, N. R., Davidson, A., Chowdhury, J. R., and Horwitz, M. S. (1997). Insertion of the adenoviral E3 region into a recombinant viral vector prevents antiviral humoral and cellular immune responses and permits long-term gene expression. *Proc. Natl. Acad. Sci. USA* **94**, 2587–2592.
14. Nguyen, D. M., Wiehle, S. A., Roth, J. A., and Cristiano, R. J. (1997). Gene delivery into malignant cells *in vivo* by a conjugated adenovirus/DNA complex. *Cancer Gene Ther.* **4**, 183–190.
15. Miller, N., and Whelan, J. (1997). Progress in transcriptionally targeted and regulatable vectors for genetic therapy. *Hum. Gene Ther.* **8**, 803–815.
16. Yoshida, Y., and Hamada, H. (1997). Adenovirus-mediated inducible gene expression through tetracycline-controllable transactivator with nuclear localization signal. *Biochem. Biophys. Res. Commun.* **230**, 426–430.
17. Hu S.-X., Ji, W., Zhou, Y., Logothetis, C., and Xu, H.-J. (1997). Development of an adenovirus vector with tetracycline-regulatable human tumor necrosis factor alpha gene expression. *Cancer Res.* **57**, 3339–3343.
18. Flotte, T. R., and Carter, B. J. (1997). *In vivo* gene therapy with adeno-associated virus vectors for cystic fibrosis. *Adv. Pharmacol.* **40**, 85–101.
19. Qazilbash, M. H., Xiao, X., Seth, P., Cowan, K. H., and Walsh, C. E. (1997). Cancer gene therapy using a novel adeno-associated virus vector expressing human wild-type p53. *Gene Ther.* **4**, 675–682.
20. Feng, M., Jackson, W. H. Jr., Goldman, C. K., Rancourt, C., Wang, M., Dusing, S. K., Siegal, G., and Curiel, D. T. (1997). Stable in vivo gene transduction via a novel adenoviral/retroviral chimeric vector. *Nat. Biotechnol.* **15**, 866–870.
21. Zufferey, R., Nagy, D., Mandel, R. J., Naldini, L., and Trono, D. (1997). Multiply attenuated lentiviral vector achieves efficient gene delivery *in vivo*. *Nat. Biotechnol.* **15**, 871–875.
22. Randazzo, B. P., Bhat, M. G., Kesari, S., Fraser, N. W., and Brown, S. M. (1997). Treatment of experimental subcutaneous human melanoma with a replication-restricted herpes simplex virus mutant. *J. Invest. Dermatol.* **108**, 933–937.
23. Habib, N. A., Ding, S. F., El-Masry, R., Mitry, R. R., Honda, K., Michail, N. E., Dalla Serra, G., Izzi, G., Greco, L., Bassyouni, M., El-Toukhy, M., and Abdel-Gaffar, Y. (1996). Preliminary report: The short-term effects of direct p53 DNA injection in primary hepatocellular carcinomas. *Cancer Detect. Prev.* **20**, 103–107.
24. Bischoff, J. R., Kirn, D. H., Williams, A., Heise, C., Horn, S., Muna, M., Ng, L., Nye, J. A., Sampson-Johannes, A., Fattaey, A., and McCormick, F. (1996). An adenovirus mutant that replicates selectively in p53-deficient human tumor cells. *Science* **274**, 373–376.
25. Heise, C., Sampson-Johannes, A., Williams, A., McCormick, F., von Hoff, D. D., and Kirn, D. H. (1997). ONYX-015, an E1b gene attenuated adenovirus, causes tumor-specific cytolysis and antitumoral efficacy that can be augmented by standard chemotherapeutic agents. *Nat. Med.* **3**, 639–645.
26. Peeters, M. J., Patijn, G. A., Lieber, A., Meuse, L., and Kay, M. A. (1996). Adenovirus-mediated hepatic gene transfer in mice: Comparison of intravascular and biliary administration. *Hum. Gene Ther.* **7**, 1693–1699.
27. Gao, G. P., Yang, Y., and Wilson, J. M. (1996). Biology of adenovirus vectors with E1 and E4 deletions for liver-directed gene therapy. *J. Virol.* **70**, 8934–8943.
28. Kastan, M. B., Onyekwere, O., Sidransky, D., Vogelstein, B., and Craig, R. W. (1991). Participation of p53 protein in the cellular response to DNA damage. *Cancer Res.* **51**, 6304–6311.
29. Lowe, S. W., Schmitt, E. M., Smith, S. W., Osborne, B. A., and Jacks, T. (1993). p53 is required for radiation-induced apoptosis in mouse thymocytes. *Nature* **362**, 847–849.
30. Lowe, S. W., Ruley, H. E., Jacks, T., and Housman, D. E. (1993). p53-dependent apoptosis modulates the cytotoxicity of anticancer agents. *Cell* **74**, 957–967.
31. Velculescu, V. E., and El-Deiry, W. S. (1996). Biological and clinical importance of the p53 tumor suppressor gene. *Clin. Chem.* **42**, 858–868.
32. Weinstein, J. N., Myers, T. G., O'Connor, P. M., Friend, S. H., Fornace, A. J., Jr., Kohn, K. W., Fojo, T., Bates, S. E., Rubinstein, L. V., Anderson, N. L., Buolamwini, J. K., van Osdol, W. W., Monks, A. P., Scudiero, D. A., Sausville, E. A., Zaharevitz, D. W., Bunow, B., Viswanadhan, V. N., Johnson, G. S., Wittes, R. E., and Paull, K. D. (1997). An information-intensive approach to the molecular pharmacology of cancer. *Science* **275**, 343–349.
33. Greenblatt, M. S., Bennett, W. P., Hollstein, M., and Harris, C. C. (1994). Mutations in the p53 tumor suppressor gene: Clues to cancer etiology and molecular pathogenesis. *Cancer Res.* **54**, 4855–4878.
34. Malkin, D., Li, F. P., Strong, L. C., Fraumeni, J. F. Jr., Nelson, C. E., Kim, D. H., Kassel, J., Gryka, M. A., Bischoff, F. Z., Tainsky, M. A., and Friend, S. H. (1990). Germ line p53 mutations in a familial syndrome of breast cancer, sarcomas, and other neoplasms. *Science* **250**, 1233–1238.
35. Donehower, L. A., Harvey, M., Slagle, B. L., McArthur, M. J., Montgomery, C. A., Jr., Butel, J. S., and Bradley, A. (1992). Mice deficient for p53 are developmentally normal but susceptible to spontaneous tumours. *Nature* **356**, 215–221.
36. Symonds, H., Krall, L., Remington, L., Saenz-Robles, M., Lowe, S., Jacks, T., and Van Dyke, T. (1994). p53-dependent apoptosis suppresses tumor growth and progression in vivo. *Cell* **78**, 703–711.

37. Baker, S. J., Markowitz, S., Fearon, E. R., Willson, J. K., and Vogelstein, B. (1990). Suppression of human colorectal carcinoma cell growth by wild-type p53. *Science* **249,** 912–915.

38. Cai, D. W., Mukhopadhyay, T., Liu, Y. J., Fujiwara, T., and Roth, J. A. (1993). Stable expression of the wild-type p53 gene in human lung cancer cells after retrovirus-mediated gene transfer. *Hum. Gene Ther.* **4,** 617–624.

39. Fujiwara, T., Cai, D. W., Georges, R. N., Mukhopadhyay, T., Grimm, E. A., and Roth, J. A. (1994). Therapeutic effect of a retroviral wild-type p53 expression vector in an orthotopic lung cancer model. *J. Natl. Cancer Inst.* **86,** 1458–1462.

40. Roth, J. A., Nguyen, D., Lawrence, D. D., Kemp, B. L., Carrasco, C. H., Ferson, D. Z., Hong, W. K., Komaki, R., Lee, J. J., Nesbitt, J. C., Pisters, K. M., Putnam, J. B., Schea, R., Shin, D. M., Walsh, G. L., Dolormente, M. M., Han, C. I., Martin, F. D., Yen, N., Xu, K., Stephens, L. C., McDonnell, T. J., Mukhopadhyay, T., and Cai, D. (1996). Retrovirus-mediated wild-type p53 gene transfer to tumors of patients with lung cancer. *Nat. Med.* **2,** 974–975.

41. Blagosklonny, M. V., and El-Deiry, W. S. (1996). In vitro evaluation of a p53-expressing adenovirus as an anti-cancer drug. *Int. J. Cancer* **67,** 386–392.

42. Nielsen, L. L., Dell, J., Maxwell, E., Armstrong, L., Maneval, D., and Catino, J. J. (1997). Efficacy of p53 adenovirus-mediated gene therapy against human breast cancer xenografts. *Cancer Gene Ther.* **4,** 129–138.

43. Meng, R. D., Shih, H., Prabhu, N. S., George, D. L., and El-Deiry, W. S. (1998). Bypass of abnormal MDM2 inhibition of p53-dependent growth suppression. *Clin Cancer Res.,* **4,** 251–259.

44. Wickham, T. J., Mathias, P., Cheresh, D. A., and Nemerow, G. R. (1993). Integrins alpha v beta 3 and alpha v beta 5 promote adenovirus internalization but not virus attachment. *Cell* **73,** 309–319.

45. Katayose, D., Wersto, R., Cowan, K. H., and Seth, P. (1995). Effects of a recombinant adenovirus expressing WAF1/CIP1 on cell growth, cell cycle, and apoptosis. *Cell Growth Differ.* **6,** 1207–1212.

46. Gomez-Manzano, C., Fueyo, J., Kyritsis, A. P., McDonnell, T. J., Steck, P. A., Levin, V. A., and Yung, W. K. A. (1997). Characterization of p53 and p21 functional interactions in glioma cells en route to apoptosis. *J. Natl. Cancer Inst.* **89,** 1036–1044.

47. Koch, H., Harris, M. P., Anderson, S. C., Machemer, T., Hancock, W., Sutjipto, S., Wills, K. N., Gregory, R. J., Shepard, H. M., Westphal, M., and Maneval, D. C. (1996). Adenovirus-mediated p53 gene transfer suppresses growth of human glioblastoma cells *in vitro* and *in vivo. Int. J. Cancer* **67,** 808–815.

48. Clayman, G. L., El-Naggar, A. K., Roth, J. A., Zhang, W.-W., Goepfert, H., Taylor, D. L., and Liu, T.-J. (1995). *In vivo* molecular therapy with p53 adenovirus for microscopic residual head and neck squamous carcinoma. *Cancer Res.* **55,** 1–6.

49. Chang, E. H., Jang, Y.-J., Hao, Z., Murphy, G., Rait, A., Fee, W. E., Jr., Sussman, H. H., Ryan, P., Chiang, Y., and Pirollo, K. F. (1997). Restoration of the G1 checkpoint and the apoptotic pathway mediated by wild-type p53 sensitizes squamous cell carcinoma of the head and neck to radiotherapy. *Arch. Otolaryngol. Head Neck Surg.* **123,** 507–512.

50. Zeng, Y.-X., Prabhu, N. S., Meng, R., and El-Deiry, W. S. (1997). Adenovirus-mediated p53 gene therapy in nasopharyngeal cancer. *Int. J. Oncol.* **11,** 221–226.

51. El-Deiry, W. S., Tokino, T., Velculescu, V. E., Levy, D. B., Parsons, R., Trent, J. M., Lin, D., Mercer, W. E., Kinzler, K. W., and Vogelstein, B. (1993). WAF1, a potential mediator of p53 tumor suppression. *Cell* **75,** 817–825.

52. Harper, J. W., Adami, G. R., Wei, N., Keyomarsi, K., and Elledge, S. J. (1993). The p21 Cdk-interacting protein Cip1 is a potent inhibitor of G1 cyclin-dependent kinases. *Cell* **75,** 805–816.

53. El-Deiry, W. S., Harper, J. W., O'Connor, P. M., Velculescu, V. E., Canman, C. E., Jackman, J., Pietenpol, Burrell, M., Hill, D. E., Wang, Y., Wiman, K. G., Mercer, W. E., Kastan, M. B., Kohn, K. W., Elledge, S. J., Kinzler, K. W., and Vogelstein, B. (1994). WAF1/CIP1 is induced in p-53-mediated G1 arrest and apoptosis. *Cancer Res.* **54,** 1169–1174.

54. Xu, M., Kumar, D., Srinivas, S., Detolla, L. J., Yu, S. F., Stass, S. A., and Mixson, A. J. (1997). Parenteral gene therapy with p53 inhibits human breast tumors in vivo through a bystander mechanism without evidence of toxicity. *Hum. Gene Ther.* **8,** 177–185.

55. Roth, J. A., and Cristiano, R. J. (1997). Gene therapy for cancer: What have we done and where are we going? *J. Natl. Cancer Inst.* **89,** 21–39.

56. Gotoh, A., Kao, C., Ko, S.-C., Hamada, K., Liu, T. J., and Chung, L. W. K. (1997). Cytotoxic effects of recombinant adenovirus p53 and cell cycle regulator genes (p21$^{WAF1/CIP1}$ and p16^{CDKN4}) in human prostate cancers. *J. Urol.* **158,** 636–641.

57. Li, J.-H., Li, P., Klamut, H., and Liu, F.-F. (1997). Cytotoxic effects of Ad5CMV-p53 expression in two human nasopharyngeal carcinoma cell lines. *Clin. Cancer Res.* **3,** 507–514.

58. Blagosklonny, M. V., and El-Deiry, W. S. (1998). Acute overexpression of wt p53 facilitates anticancer drug-induced death of cancer and normal cells. *Int. J. Cancer* **75,** 933–940.

59. Canman, C. E., and Kastan, M. B. (1997). Role of p53 in apoptosis. *Adv. Pharmacol.* **41,** 429–460.

60. Knudson, C. M., Tung, K. S., Tourtellotte, W. G., Brown, G. A., Korsmeyer, S. J. (1995). Bax-deficient mice with lymphoid hyperplasia and male germ cell death. *Science* **270,** 96–99.

61. Fuchs, E. J., McKenna, K. A., and Bedi, A. (1997). p53-dependent DNA damage-induced apoptosis requires Fas/Apo-1-independent activation of CPP32β. *Cancer Res.* **57,** 2550–2554.

62. Wu, G. S., Burns, T. F., McDonald III, E. R., Jiang, W., Meng, R., Krantz, I. D., Kao, G., Gan, D.-D., Zhou, J.-Y., Muschel, R., Hamilton, S. R., Spinner, N. B., Markowitz, S., Wu, G., and El-Deiry, W. S. (1997). KILLER/DR5 is a DNA damage-inducible p53-regulated death receptor gene. *Nat. Genet.* **17,** 141–143.

63. Pan, G., O'Rourke, K., Chinnaiyan, A. M., Gentz, R., Ebner, R., Ni, J., and Dixit, V. M. (1997). The receptor for the cytotoxic ligand TRAIL. *Science* **276,** 111–113.

64. El-Deiry, W. S. (1998). p21/p53 cellular growth control, and genomic integrity. *Curr. Top. Microbiol. Immunol.* **227,** 121–137.

65. Chin, K. V., Ueda, K., Pastan, I., and Gottesman, M. M. (1992). Modulation of activity of the promoter of the human MDR1 gene by Ras and p53. *Science* **255,** 459–462.

66. Thottassery, J. V., Zambetti, G. P., Arimori, K., Schuetz, E. G., and Schuetz, J. D. (1997). p53-dependent regulation of MDR1 gene expression causes selective resistance to chemotherapeutic agents. *Proc. Natl. Acad. Sci. USA* **94,** 11037–11042.

67. Fujiwara, T., Grimm, E. A., Mukhopadhyay, T., Zhang, W.-W., Owen-Schaub, L. B., and Roth, J. A. (1994). Induction of chemosensitivity in human lung cancer cells *in vivo* by adenovirus-mediated transfer of the wild-type p53 gene. *Cancer Res.* **54,** 2287–2291.

68. Skladanowski, A., and Larsen, A. K. (1997). Expression of wild-type p53 increases etoposide cytotoxicity in M1 myeloid leukemia cells by facilitated G2 to M transition: Implications for gene therapy. *Cancer Res.* **57,** 818–823.

69. Tsang, N.-M., Nagasawa, H., Li, C., and Little, J. B. (1995). Abrogation of p53 function by transfection of HPV16 E6 gene enhances the resistance of human diploid fibroblasts to ionizing radiation. *Oncogene* **10**, 2403–2408.

70. Gallardo, D., Drazan, K. E., and McBride, W. H. (1996). Adenovirus-based transfer of wild-type p53 gene increases ovarian tumor radiosensitivity. *Cancer Res.* **56**, 4891–4893.

71. Spitz, F. R., Nguyen, D., Skibber, J. M., Meyn, R. E., Cristiano, R. J., and Roth, J. A. (1996). Adenovirus-mediated wild-type p53 gene expression sensitizes colorectal cancer cells to ionizing radiation. *Clin. Cancer Res.* **2**, 1665–1671.

72. Bristow, R. G., Benchimol, S., and Hill, R. P. (1996). The p53 gene as a modifier of intrinsic radiosensitivity: Implications for radiotherapy. *Radiother. Oncol.* **40**, 197–223.

73. Bracey, T. S., Miller, J. C., Preece, A., and Paraskeva, C. (1995). γ-Radiation-induced apoptosis in human colorectal adenoma and carcinoma cell lines can occur in the absence of wild type p53. *Oncogene* **10**, 2391–2396.

74. Huang, H., Li, C. Y., and Little, J. B. (1996). Abrogation of p53 function by transfection of HPV16 E6 gene does not enhance resistance of human tumour cells to ionizing radiation. *Int. J. Radiat. Biol.* **70**, 151–160.

75. Xiong, Y., Hannon, G. J., Zhang, H., Casso, D., Kobayashi, R., and Beach, D. (1993). p21 is a universal inhibitor of cyclin kinases. *Nature* **366**, 701–704.

76. Deng, C., Zhang, P., Harper, J. W., Elledge, S. J., and Leder, P. (1995). Mice lacking p21$^{CIP1/WAF1}$ undergo normal development, but are defective in G1 checkpoint control. *Cell* **82**, 675–684.

77. Flores-Rozas, H., Kelman, Z., Dean, F. B., Pan, Z. Q., Harper, J. W., Elledge, S. J., O'Donnell, M., and Hurwitz, J. (1994). Cdk-interacting protein 1 directly binds with proliferating cell nuclear antigen and inhibits DNA replication catalyzed by the DNA polymerase delta holoenzyme. *Proc. Natl. Acad. Sci. USA* **91**, 8655–8659.

78. Zhang, W., Grasso, L., McClain, C. D., Gambel, A. M., Cha, Y., Travali, S., Deisseroth, A. B., and Mercer, W. E. (1995). p53-independent induction of WAF1/CIP1 in human leukemia cells is correlated with growth arrest accompanying monocyte/macrophage differentiation. *Cancer Res.* **55**, 668–674.

79. Wu, H., Wade, M., Krall, L., Grisham, J., Xiong, Y., and Van Dyke, T. (1996). Targeted in vivo expression of the cyclin-dependent kinase inhibitor p21 halts hepatocyte cell-cycle progression, postnatal liver development, and regeneration. *Genes Dev.* **10**, 245–260.

80. Yang, Z.-Y., Perkins, N. D., Ohno, T., Nabel, E. G., and Nabel, G. J. (1995). The p21 cyclin-dependent kinase inhibitor suppresses tumorigenicity in vivo. *Nat. Med.* **1**, 1052–1056.

81. Chen, J., Willingham, T., Shuford, M., Bruce, D., Rushing, E., Smith, Y., and Nisen, P. D. (1996). Effects of ectopic overexpression of p21$^{WAF1/CIP1}$ on aneuploidy and the malignant phenotype of human brain tumor cells. *Oncogene* **13**, 1395–1403.

82. Prabhu, N. S., Blagosklonny, M. W., Zeng, Y.-X., Wu, G. S., Waldman, T., and El-Deiry, W. S. (1996). Suppression of cancer cell growth by adenovirus expressing p21$^{WAF1/CIP1}$ deficient in PCNA interaction. *Clin. Cancer Res.* **2**, 1221–1229.

83. McDonald, E. R. III, Wu, G. S., Waldman, T., and El-Deiry, W. S. (1996). Repair defect in p21$^{WAF1/CIP1}$ −/− human cancer cells. *Cancer Res.* **56**, 2250–2255.

84. Clayman, G. L., Liu, T.-J., Overholt, S. M., Mobley, S. R., Wang, M., Janot, F., and Goepfert, H. (1996). Gene therapy for head and neck cancer: Comparing the tumor suppressor gene p53 and a cell cycle regulator WAF1/CIP1 (p21). *Arch. Otolaryngol. Head Neck Surg.* **122**, 489–493.

85. Gomez-Manzano, C., Fueyo, J., Kyritsis, A. P., Levin, V. A., and Yung, W. K. A. (1997). Crossroads of death and life signals in gliomas: p53 and p21 interactions. *Neurology* **48**, V3003.

86. Villa, L. L. (1997). Human papillomaviruses and cervical cancer. *Adv. Cancer Res.* **71**, 321–341.

87. Scheffner, M., Werness, B. A., Huibregtse, J. M., Levine, A. J., and Howley, P. M. (1990). The E6 oncoprotein encoded by human papillomavirus types 16 and 18 promotes the degradation of p53. *Cell* **63**, 1129–1136.

88. Wu, G. S., and El-Deiry, W. S. (1996). Apoptotic death of tumor cells correlates with chemosensitivity, independent of p53 or bcl-2. *Clin Cancer Res.* **2**, 623–633.

89. Barak, Y., Juven, T., Haffner, R., and Oren, M. (1993). mdm2 expression is induced by wild type p53 activity. *EMBO J.* **12**, 461–468.

90. Momand, J., Zambetti, G. P., Olson, D. C., George, D., and Levine, A. J. (1992). The mdm-2 oncogene product forms a complex with the p53 protein and inhibits p53-mediated transactivation. *Cell* **69**, 1237–1245.

91. Jones, S. N., Roe, A. E., Donehower, L. A., and Bradley, A. (1995). Rescue of embryonic lethality in Mdm2-deficient mice by absence of p53. *Nature* **378**, 206–208.

92. Montes de Oca Luna, R., Wagner, D. S., and Lozano, G. (1995). Rescue of early embryonic lethality in mdm2-deficient mice by deletion of p53. *Nature* **378**, 203–206.

93. Oliner, J. D., Kinzler, K. W., Meltzer, P. S., George, D. L., and Vogelstein, B. (1992). Amplification of the gene encoding a p53-associated protein in human sarcomas. *Nature* **358**, 80–83.

94. Haupt, Y., Maya, R., Kazaz, A., and Oren, M. (1997). Mdm2 promotes the rapid degradation of p53. *Nature* **387**, 296–299.

95. Kubbutat, M. H. G., Jones, S. N., and Vousden, K. H. (1997). Regulation of p53 stability by mdm2. *Nature* **387**, 299–303.

96. Linzer, D. I. H., and Levine, A. J. (1979). Characterization of a 54K dalton cellular SV40 tumor antigen present in SV40-transformed cells and uninfected embryonal carcinoma cells. *Cell* **17**, 43–52.

97. Tiemann, F., Zerrahn, J., and Deppert, W. (1995). Cooperation of simian virus 40 large and small T antigens in metabolic stabilization of tumor suppressor p53 during cellular transformation. *J. Virol.* **69**, 6115–6121.

98. Carbone, M., Pass, H. I., Rizzo, P., Marinetti, M., Di Muzio, M., Mew, D. J., Levine, A. S., and Procopio, A. (1994). Simian virus 40-like DNA sequences in human pleural mesothelioma. *Oncogene* **9**, 1781–1790.

99. Carbone, M., Rizzo, P., Grimley, P. M., Procopio, A., Mew, D. J. Y., Shridhar, V., de Bartolomeis, A., Esposito, V., Giuliano, M. T., Steinberg, S. M., Levine, A. S., Giordano, A., and Pass, H. I. (1997). Simian virus-40 large-T antigen binds p53 in human mesotheliomas. *Nat. Med.* **3**, 908–912.

100. DeLuca, A., Baldi, A., Esposito, V., Howard, C. M., Bagella, L., Rizzo, P., Caputi, M., Pass, H. I., Giordano, G. G., Baldi, F., Carbone, M., and Giordano, A. (1997). The retinoblastoma gene family pRb/p105, p107, pRb2/p130 and simian virus-40 large T-antigen in human mesotheliomas. *Nat. Med.* **3**, 913–916.

101. Wang, X. W., Forrester, K., Yeh, H., Feitelson, M. A., Gu, J., and Harris, C. C. (1994). Hepatitis B virus X protein inhibits p53 sequence-specific DNA binding, transcriptional activity, and association with transcription factor ERCC3. *Proc. Natl. Acad. Sci. USA* **91**, 2230–2234.

102. Ueda, H., Ullrich, S. J., Gangemi, J. D., Kappel, C. A., Ngo, L., Feitelson, M. A., and Jay, G. (1995). Functional inactivation but not structural mutation of p53 causes liver cancer. *Nat. Genet.* **9**, 41–47.

103. Takada, S., Kaneniwa, N., Tsuchida, N., and Koike, K. (1997). Cytoplasmic retention of the p53 tumor suppressor gene product is observed in the hepatitis B virus X gene-transfected cells. *Oncogene* **15,** 1895–1901.

104. Waldman, T., Lengauer, C., Kinzler, K. W., and Vogelstein, B. (1996). Uncoupling of S phase and mitosis induced by anticancer agents in cells lacking p21. *Nature* **381,** 713–716.

105. Hsiao, M., Tse, V., Carmel, J., Costanzi, E., Strauss, B., Haas, M., and Silverberg, G. D. (1997). Functional expression of human p21(WAF1/CIP1) gene in rat glioma cells suppresses tumor growth in vivo and induces radiosensitivity. *Biochem. Biophys. Res. Comm.* **233,** 329–335.

106. Li, W. W., Fan, J., Hochhauser, D., and Bertino, J. R. (1997). Overexpression of p21^{wafl} leads to increased inhibition of E2F-1 phosphorylation and sensitivity to anticancer drugs in retinoblastoma-negative human sarcoma cells. *Cancer Res.* **57,** 2193–2199.

107. Fan, S., Chang, J. K., Smith, M. L., Duba, D., Fornace, A. J., Jr., and O'Connor, P. M. (1997). Cells lacking CIP1/WAF1 genes exhibit preferential sensitivity to cisplatin and nitrogen mustard. *Oncogene* **14,** 2127–2136.

108. Chinery, R., Brockman, J. A., Peeler, M. O., Shyr, Y., Beauchamp, R. D., and Coffey, R. J. (1997). Antioxidants enhance the cytotoxicity of chemotherapeutic agents in colorectal cancer: A p53-independent induction of p21$^{WAF1/CIP1}$ via C/EBP delta. *Nat. Med.* **3,** 1233–1241.

109. Serrano, M., Hannon, G. J., and Beach, D. (1993). A new regulatory motif in cell-cycle control causing specific inhibition of cyclin D/CDK4. *Nature* **366,** 704–707.

110. Herman, J. G., Merlo, A., Mao, L., Lapidus, R. G., Issa, J. P., Davidson, N. E., Sidransky, D., and Baylin, S. B. (1995). Inactivation of the CDKN2/p16/MTS1 gene is frequently associated with aberrant DNA methylation in all common human cancers. *Cancer Res.* **55,** 4525–4530.

111. Nobori, T., Miura, K., Wu, D. J., Lois, A., Takabayashi, K., and Carson, D. A. (1994). Deletions of the cyclin-dependent kinase-4 inhibitor gene in multiple human cancers. *Nature* **368,** 753–756.

112. Foulkes, W. D., Flanders, T. Y., Pollock, P. M., and Hayward, N. K. (1997). The CDKN2A (p16) gene and human cancer. *Mol. Med.* **3,** 5–20.

113. Serrano, M., Lee, H., Chin, L., Cordon-Cardo, C., Beach, D., and DePinho, R. A. (1996). Role of the INK4a locus in tumor suppression and cell mortality. *Cell* **85,** 27–37.

114. Serrano, M., Lin, A. W., McCurrach, M. E., Beach, D., and Lowe, S. W. (1997). Oncogenic ras provokes premature cell senescence associated with accumulation of p53 and p16^{INK4a}. *Cell* **88,** 593–602.

115. Jin, X., Nguyen, D., Zhang, W.-W., Kyritsis, A. P., and Roth, J. A. (1995). Cell cycle arrest and inhibition of tumor cell proliferation by the p16^{INK4} gene mediated by an adenovirus vector. *Cancer Res.* **55,** 3250–3253.

116. Fueyo, J., Gomez-Manzano, C., Yung, W. K. A., Clayman, G. L., Liu, T.-J., Bruner, J., Levin, V. A., and Kyritsis, A. P. (1996). Adenovirus-mediated p16/CDKN2 gene transfer induces growth arrest and modifies the transformed phenotype of glioma cells. *Oncogene* **12,** 103–110.

117. Schrump, D. S., Chen, G. A., Consuli, U., Jin, X., and Roth, J. A. (1996). Inhibition of esophageal cancer proliferation by adenovirusly mediated delivery of p16^{INK4}. *Cancer Gene Ther.* **3,** 357–364.

118. Quesnel, B., Preudhomme, C., Lepelley, P., Hetuin, D., Vanrumbeke, M., Bauters, F., Velu, T., and Fenaux, P. (1996). Transfer of p16$^{INK4A/CDKN2}$ gene in leukaemic cell lines inhibits cell proliferation. *Br. J. Hematol.* **95,** 291–298.

119. Matsumura, Y., Yamagishi, N., Miyakoshi, J., Imamura, S., and Takebe, H. (1997). Increase in radiation sensitivity of human malignant melanoma cells by expression of wild-type p16 gene. *Cancer Lett.* **115,** 91–96.

120. Ribeiro, J. C., Hanley, J. R., and Russell, P. J. (1996). Studies of X-irradiated bladder cancer cell lines showing differences in p53 status: Absence of a p53-dependent cell cycle checkpoint pathway. *Oncogene* **13,** 1269–1278.

121. Valenzuela, M. T., Nunez, M. I., Villalobos, M., Siles, E., McMillan, T. J., Pedraza, V., and Ruiz de Almodovar, J. M. (1997). A comparison of p53 and p16 expression in human tumor cells treated with hyperthermia or ionizing radiation. *Int. J. Cancer* **72,** 307–312.

122. Stone, S., Dayananth, P., and Kamb, A. (1996). Reversible, p16-mediated cell cycle arrest as protection from chemotherapy. *Cancer Res.* **56,** 3199–3202.

123. Hama, S., Sadatomo, T., Yoshioka, H., Kurisu, K., Tahara, E., Naruse, I., Heike, Y., and Saijo, N. (1997). Transformation of human glioma cell lines with the p16 gene inhibits cell proliferation. *Anticancer Res.* **17,** 1933–1938.

124. Sandig, V., Brand, K., Herwig, S., Lukas, J., Bartek, J., and Strauss, M. (1997). Adenovirally transferred p16$^{INK4/CDKN2}$ and p53 genes cooperate to induce apoptotic tumor cell death. *Nat. Med.* **3,** 313–319.

125. Xu, H.-J. (1997). Strategies for approaching retinoblastoma tumor suppressor gene therapy. *Adv. Pharmacol.* **40,** 369–397.

126. Xu, H.-J., Zhou, Y. L., Seigne, J., Perng, G. S., Mixon, M., Zhang, C. Y., Li, J., Benedict, W. F., and Hu, S. X. (1996). Enhanced tumor suppressor gene therapy via replication-deficient adenovirus vectors expressing an N-terminal truncated retinoblastoma protein. *Cancer Res.* **56,** 2245–2249.

127. Riley, D. J., Nikitin, A. Y., and Lee, W. H. (1996). Adenovirus-mediated retinoblastoma gene therapy suppresses spontaneous pituitary melanotroph tumors in Rb +/− mice. *Nat. Med.* **2,** 1316–1321.

128. Xu, H.-J., Xu, K., Zhou, Y., Li, J., Benedict, W. F., and Hu, S. X. (1994). Enhanced tumor cell growth suppression by an N-terminal truncated retinoblastoma protein. *Proc. Natl. Acad. Sci. USA* **91,** 9837–9841.

129. Chang, M. W., Barr, E., Seltzer, J., Jiang, Y. O., Nabel, G. J., Nabel, E. G., Parmacek, M. S., and Leiden, J. M. (1995). Cytostatic gene therapy for vascular proliferative disorders with a constitutively active form of the retinoblastoma gene product. *Science* **267,** 518–522.

130. Koff, A., Ohtsuki, E. M., Polyak, K., Roberts, J., and Massague, J. (1993). Negative regulator of G1 in mammal cells: Inhibition of cyclin E-dependent kinase by TGF-β. *Science* **260,** 536–539.

131. Ponce-Casteneda, M., Lee, M., Latres, E., Polyak, K., Lacombe, L., Montgomery, K., Matthew, S., Krauter, K., Sheinfeld, J., Massague, J., and Cordon-Cardo, C. (1995). p27^{kip1} chromosomal mapping to 12p12.2–12p13.1 and absence of mutations in human tumors. *Cancer Res.* **55,** 1211–1214.

132. Spirin, K. S., Simpson, J. F., Takeuchi, S., Kawamata, N., Miller, C. W., and Koeffler, H. P. (1996). p27/Kip1 mutation found in breast cancer. *Cancer Res.* **56,** 2400–2404.

133. Esposito, V., Baldi, A., De Luca, A., Groger, A. M., Loda, M., Giordano, G. G., Caputi, M., Baldi, F., Pagano, M., and Giordano, A. (1997). Prognostic role of the cyclin-dependent kinase inhibitor p27 in non-small cell lung cancer. *Cancer Res.* **57,** 3381–3385.

134. Loda, M., Cukor, B., Tam, S. W., Lavin, P., Fiorentino, M., Draetta, G. F., Jessup, J. M., and Pagano, M. (1997). Increased proteasome-dependent degradation of the cyclin-dependent ki-

nase inhibitor p27 in aggressive colorectal carcinomas. *Nat. Med.* **3,** 231–234.

135. Tan, P., Cady, B., Wanner, M., Worland, P., Cukor, B., Magi-Galluzzi, C., Lavin, P., Draetta, G., Pagano, M., and Loda, M. (1997). The cell cycle inhibitor p27 is an independent prognostic marker in small (T1a,b) invasive breast carcinomas. *Cancer Res.* **57,** 1259–1263.

136. Craig, C., Wersto, R., Kim, M., Ohri, E., Li, Z., Katayose, D., Lee, S. J., Trepel, J., Cowan, K., and Seth, P. (1997). A recombinant adenovirus expressing p27^kip1 induces cell cycle arrest and loss of cyclin-Cdk activity in human breast cancer cells. *Oncogene* **14,** 2283–2289.

137. Chen, J., Willingham, T., Shuford, M., and Nissen, P. D. (1996). Tumor suppression and inhibition of aneuploid cell accumulation in human brain tumor cells by ectopic overexpression of the cyclin-dependent kinase inhibitor p27^kip1. *J. Clin. Invest.* **97,** 1983–1988.

138. Nguyen, D. M., Spitz, F. R., Yen, N., Cristiano, R. J., and Roth, J. A. (1996). Gene therapy for lung cancer: Enhancement of tumor suppression by a combination of sequential systemic cisplatin and adenovirus-mediated p53 gene transfer. *J. Thorac. Cardiov. Sur.* **112,** 1372–1376.

139. Kirn, D., Ganley, I., Nemunaitis, J., Otto, R., Soutar, D., Kuhn, J., Heise, C., Propst, M., Maack, C., Eckhardt, G., Kaye, S., and Von Hoff, D. (1997). A phase I clinical trial with ONYX-015 (a selectively replicating adenovirus) administered by intratumoral injection in patients with recurrent head and neck carcinoma. *Cancer Gene Ther.* **4,** 05.

CHAPTER 2

Characterization of Specific Genetic Alterations in Cancer Cells

KAY HUEBNER, TERESA DRUCK, PIOTR HADACZEK,[1] PETER A. MCCUE, AND HENRY C. MAGUIRE, JR.

Kimmel Cancer Center, Jefferson Medical College, Philadelphia, Pennsylvania 19107

I. INTRODUCTION

The genetic changes found in human neoplasms suggest that hematopoietic tumors frequently develop because of inappropriate expression (usually overexpression) of a growth promoting gene (oncogene). In contrast, the malignant progression of carcinomas frequently involves loss of expression of tumor suppressor genes. Gene therapy might be used to turn off an activated oncogene, e.g., by antisense treatment, whereas gene therapy to overcome tumor suppressor gene loss would necessarily focus on gene replacement in the tumor cell or pharmacologically substituting for lost gene function. On the other hand, the protein products of mutations that activate oncogenes or that inactivate tumor suppressor genes are both potential tumor antigens. Increasingly, characterization of the molecular changes that contribute to the malignant phenotype provides information impacting on tumor diagnosis and patient prognosis.

Insights into the genetic causes of specific cancer types have suggested fundamental differences between

---

[1] *Present address:* Department of Genetics and Pathology, Medical Academy, Al. Powstancow Wlkp. 72, 70-111 Szczecin, Poland.

hematopoietic and solid tumors. The nonrandom chromosome translocations typifying specific leukemias or lymphomas and involving only one homolog of a specific chromosome were the first markers for positional cloning of genes at the chromosome breakpoints.[1] Numerous translocation breakpoint genes have been cloned since the early 1980s and act, in general, as dominant growth promoting oncogenes through overexpression, aberrant expression or inappropriate expression, or through fusion with juxtaposed genes to give expression of chimeric gene products (for review, see Newell [1]). Activation of oncogenes or fusion genes seems to be an early step in the progression of the hematopoietic tumors and suggests possible therapies based on shut-off of the oncogene by antisense methods (see Chapter 15).

The cytogenetics of solid tumors is more complex, with most tumor types showing aneuploid karyotypes with numerous alterations; experts began to observe that a majority of kidney tumors, for example, had visible alterations, usually deletions of different sizes, on the short arm (or p arm, for petite) of one chromosome 3 [2,3]. Chromosomal deletions in tumors could be studied by molecular genetics beginning in the late 1980s after the discovery of restriction fragment length polymorphisms (RFLPs, illustrated in Fig. 1). This was a method whereby two alleles at a specific chromosome region could be recognized on a Southern blot because, after treatment of DNA with a particular restriction enzyme, they were different lengths; by comparing restriction enzyme digested DNA from normal cells of a patient with DNA from the patient's tumor cells, it was possible to observe the loss of alleles at specific chromosome regions, such as 3p in kidney tumors [4], 11p in Wilms tumors, or 13q in retinoblastomas (for review, see Knudson [5]).

For most solid tumors, which represent more than 85% of human cancers, complete loss of normal function of a series of growth regulating genes is the common mechanism of tumor progression (for review, see Knudson [5]). The genes involved are frequently inhibitors of cell growth or division; their loss or inactivation contributes to the malignant phenotype, and thus they have been called antioncogenes or tumor suppressor genes. The existence of tumor suppressor genes was suggested early on by suppression of the malignant phenotype in human–human hybrids of normal and tumor cells and was predicted by the consideration of familial embryonal tumors by Knudson and Strong (cited in Knudson [5]).

In fact, a good model for the loss of tumor suppressor genes as the limiting step in tumor formation is the classic human retinoblastoma model. Retinoblastoma arises by inactivation of both alleles of the RB1 gene [5]. Moreover, *in vitro* reintroduction of this gene into retinoblastoma cells reverses the malignant phenotype. Application of RFLP analysis to the genetics of solid malignant tumors has provided evidence for loss of alleles at specific chromosomal sites in many other tumors, and analyzing a collection of tumors of a specific type for loss of alleles (allelotyping) is often the first step in the molecular genetic characterization of specific classes of tumors.

Both oncogenes and tumor suppressor genes can code for proteins with a wide range of subcellular locations and functions, from cell surface molecules (receptors, cell adhesion molecules) to membrane–cytoskeletal interface proteins, to cytoplasmic proteins and nuclear proteins (transcription factors, zinc finger proteins, homeobox-containing proteins).

By looking for loss of alleles of many different genes and anonymous DNA fragments (markers) in a specific chromosome region in a large panel of tumors, a common deleted region can be identified. Within that common region, then, must be the gene that is the target of the deletion, and the retained copy of that target gene would be expected to be mutated in the tumor, thereby coding for a functionally inactive protein product. The tumor suppressor theory, coupled with the ability to pinpoint genes that are commonly lost in specific tumors and to clone them by positional cloning methods, has led to the identification of more than a dozen suppressor genes, which have been inactivated in specific cancers (e.g., the *VHL* gene in renal cell carcinomas) or in a broad spectrum of tumors (e.g., the p53 gene in colon, lung, esophageal, and other malignant tumors). Deleted regions are still being identified for many of the major types of carcinomas, and the more closely chromosomes are examined, the more complex are the patterns of deletions found. In most cases the identification and cloning of tumor suppressor genes have been aided by the occurrence of familial forms of specific tumor types, in which there is hereditary inactivation of one germline tumor suppressor gene allele.

Relatively little is yet known about specific genetic alterations occurring in many of the major types of human cancer such as melanoma and carcinomas of the cervix, ovary, prostate, and kidney, and searches for relevant oncogene and tumor suppressor gene alterations in these human tumors are under way in numerous laboratories.

II. SUPPRESSOR GENES

For several years our laboratories have been engaged in a search for a tumor suppressor gene involved in

---

[1]For explanations of basic nomenclature and concepts of molecular biology see Benjamin Lewin's excellent basic test, *Genes V,* Oxford University Press, New York, 1995.

development of clear cell renal carcinoma (RCC). Examples of results obtained during this search will be used to illustrate the methods involved in identification and characterization of tumor specific gene alterations.

A. Our First Candidate RCC Suppressor Gene

We became interested in the study of the short arm of chromosome 3 (3p) in kidney tumors when we mapped the receptor protein tyrosine phosphatase gamma gene, PTPγ, to chromosome region 3p14.2 centromeric to the position of a constitutional chromosome translocation break [6], observed in a large Italian-American family [7]. Family members who inherited the chromosome translocation developed multiple RCCs a decade earlier than the age of occurrence of most sporadic RCCs. It was postulated that the chromosome translocation interrupts a tumor suppressor gene, a gene that might also be important in sporadic RCCs, at chromosome region 3p14.2 in this family. Because the PTPγ gene had enzymatic properties suggestive of possible suppressor activity (as a receptor tyrosine phosphatase, the enzyme might send a growth inhibitory signal opposing a receptor tyrosine kinase gene growth promoting signal) and because it mapped near the translocation break, it was an attractive tumor suppressor gene candidate. Thus, we studied this gene in detail in kidney tumors: we have determined where it is relative to the 3p14.2 breakpoint, its gene structure, whether it is lost in sporadic RCCs, and how loss at 3p14.2 relates to the loss of other regions of 3p in kidney cancer.

The first step in determining the gene structure of the full length PTPγ gene was to collect DNA from a series of kidney tumors with matched normal DNA from the same patients and begin testing the DNAs for loss of alleles for various regions of 3p, because kidney tumors lose alleles at several independent regions of 3p [8]. Additionally, cytogenetic and molecular studies had shown previously that the short arm of chromosome 3 (3p) was frequently affected by chromosomal rearrangements and deletions in several malignant diseases. Thus, discovery of a tumor suppressor gene on 3p could be important in the study of a number of cancers.

1. CHROMOSOME 3p DELETIONS IN RCC

Several common regions of 3p deletion in specific tumor types had been delineated by loss of heterozygosity (LOH) studies, singularly or in combination, at chromosome regions 3p25, 3p21.3, and 3p14.3–p13 (individual references cited in Lubinski *et al.* [9]). The 3p14.3–p13 region included the 3p14.2 breakpoint of the constitutional chromosome translocation associated with hereditary RCC. Our preliminary study was under-

taken to define more narrowly the 3p14–12 LOH targets in 30 primary RCCs by assessing the status of nine simple sequence repeat polymorphisms (described in the legend to Fig. 1) in matched normal/tumor DNA pairs.

Partial losses in the 3p21–p12 region were observed, allowing determination of common regions of loss of heterozygosity in 15 RCCs. Results suggested that most RCCs exhibit loss in a region that includes the familial chromosome translocation point at 3p14.2 [9].

In some cases, regions of loss were delimited by retention of both alleles in neighboring loci, allowing identification of common regions of deletion, as illustrated in Fig. 1. Deletions are indicated by gaps in the vertical lines representing chromosome 3p in Fig. 1, which outlines a general strategy for locating tumor suppressor loci and isolating the suppressor gene within the common deleted regions.

To narrow the common region of loss further, more DNA markers between D3S1300 and D3S1312 (shown in Fig. 1) were isolated from cloned fragments of chromosomal (genomic) DNA that cover the region between D3S1300 and D3S1312. The kidney tumor DNAs were then retested for loss of these new markers. The result was that the common deletion was narrowed to about a megabase [10], a region of a million base pairs that included the 3p14.2 translocation breakpoint and could code for perhaps 10 genes. However, we knew that the PTPγ gene took up about half of this region of a megabase because portions of the cloned DNA between D3S1300 and D3S1312 contained portions of the PTPγ gene, as determined by sequencing of PTPγ DNA and cloned chromosomal DNA fragments from the region. Thus, as a possible target of loss of heterozygosity, the PTPγ gene was a candidate tumor suppressor gene and had to be assessed for mutations in the PTPγ allele remaining in the kidney tumors.

2. ASSESSMENT OF MUTATIONS IN CANDIDATE TUMOR SUPPRESSOR GENES

To look for mutations in a specific gene in a tumor cell, the DNA sequence of the gene from the normal or constitutional cells of an individual must be compared to the sequence of the same gene from the individual's tumor DNA. This could be done, perhaps most quickly, by determining the sequence of the complementary DNA (cDNA) of the gene from the normal and tumor cells. However, cDNA, a DNA copy of the messenger RNA (mRNA) transcript of the gene, must be made from RNA and, for many clinical samples, RNA is more difficult to obtain and store. Also, a mutated suppressor gene might not be expressed as an RNA transcript in the tumor cell. Thus, it is usually safest to compare the normal and tumor sequence of the candidate suppressor gene by sequencing the chromosomal copy of the gene,

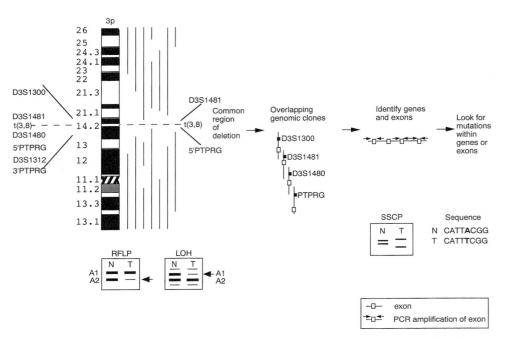

FIGURE 1   Identification of a candidate suppressor gene by loss of heterozygosity studies. Normal and tumor DNA from a large group of cases of a particular tumor type, in this illustration RCC, is tested for loss of one copy of specific chromosome regions by either RFLP or LOH analysis. **RFLP analysis** involves cutting of the normal and tumor DNA (1–10 mg) at specific 5- or 6-bp sequences with a restriction enzyme, separating the DNA fragments by size on a gel, transferring the size-fractionated DNA to a nylon membrane by a blotting technique and hybridization to a DNA probe (a cloned fragment of DNA from the region of interest, in this case the 5′ *PTPRG* probe in region 3p14.2). If, after enzyme digestion, the individual DNA involved had two different sized digestion products because of the presence of an enzyme restriction site on the *A2* allele that was not present on the *A1* allele, then the probe will see two bands on the membrane, as in lane N (for normal DNA). If the tumor (lane T) has lost one of these alleles, the tumor DNA (compare T to N for normal in RFLP figure, lower left) will show a missing or much diminished signal for one allele, *A2* in the example (*A2* shows a faint presence in lane T because the tumor from which the DNA was derived almost always has some normal tissue admixed that contributes the A2 allele in the T lane). **LOH analysis,** also at bottom left, gives similar results but is based upon a different kind of DNA polymorphism. Scattered throughout human chromosomes are small regions of repeated sequence such as CACACACA, flanked by unique sequences; the length of a specific repeated region can be highly variable, not only from individual to individual but from allele to allele. Such small repeated runs are called microsatellites and tend to exhibit frequencies of heterozygosity of 60% or more; i.e., 60% or more of individuals show two alleles of different size, so such polymorphisms are useful in looking for loss of alleles in tumors. The microsatellite polymorphisms are assessed by polymerase chain reaction (PCR) amplification of the specific microsatellite region of interest from very small amounts of DNA (10–20 ng) from the normal and the tumor tissues. **PCR amplification** is a method of making many copies of a specific DNA fragment *in vitro* by mixing heat stable DNA polymerase with template DNA and oligonucleotides (called primers) complementary to DNA sequences flanking the region to be amplified. Cycling the *in vitro* reaction repeatedly through temperatures that allow successive annealing of the primers to the template, synthesis of the single strand of DNA from each of the flanking primers, and melting of the complementary DNA duplexes allows a geometric increase in the number of template fragments. The amplified products are labeled during amplification, and then the normal and tumor labeled alleles are run on a gel that separates different sized amplified bands. For example, see the drawing of an LOH gel at the bottom left of the figure. The alleles are marked to the right; the fainter upper and lower bands represent artifactual "shadow bands" usually seen on these gels; allele *A1* has been lost in the tumor (T) as noted by the arrowhead; the faint *A1* allele is caused by some normal tissue in the tumor sample. Above the RFLP and LOH panels is a representation of the short arm of chromosome 3 with the interrupted vertical lines to the right representing the extent of loss (gaps in lines) of 3p regions from one copy of chromosome 3p in a number of tumors. In this example, all the tumors exhibited partial loss of one chromosome 3p in the region including the band 3p14.2; by determining the smallest common region of loss among this group of tumors we see that it includes the 3p14.2 break of the t(3;8) chromosome translocation and a region on either side of the break. This region of ~1 Mb should contain the tumor

the actual gene itself. The gene, unlike the mRNA transcript, consists of regulatory regions at the front followed by coding exons, the blocks of DNA carrying amino acid coding portions, interspersed with noncoding blocks, or introns, flanking each coding exon. To know if a candidate gene is mutated or normal, it is necessary to sequence, or otherwise assess the integrity of, each coding block or exon along with a bit of flanking intron on each side (because a mutated intron sequence can affect proper splicing together of the exons in the RNA transcript). Thus, the intron–exon structure of the PTPγ gene was determined and sequences flanking each exon were defined for use in amplification of each of the 30 PTPγ exons from each kidney tumor. The individual exons could then be examined for mutations by SSCP (single strand conformation polymorphism) analysis and sequencing, as illustrated in Fig. 1. SSCP analysis is a method invented in Japan [11] to detect differences in a specific fragment of DNA derived from different individuals or, as illustrated in Fig. 1, differences in the mobility of the exon from the normal tissue DNA compared to tumor DNA from the same individual. If there is a nucleotide difference, the two matched DNAs will show mobility differences on a polyacrylamide gel (see Fig. 1 for examples). Although we found many interindividual differences in mobility of various exons of the PTPγ gene, we did not find mobility differences when comparing normal to tumor DNA of individual cases; i.e., we did not find tumor-specific mutations of the PTPγ gene in kidney tumors [10]. Thus, the search for the presumptive target tumor suppressor gene in the remaining 0.5 Mb of the common allelic deletion region continued.

B. The Second Candidate RCC Suppressor Gene

If the PTPγ gene centromeric to the hereditary RCC translocation break at 3p14.2 (illustrated in Fig. 1) was not the RCC tumor suppressor gene, then it was likely that the target gene was on the other side, the telomeric side, of the translocation break, where we focused our continuing search. This search was aided by the work of a number of investigators who had isolated markers and genomic clones for this region of 3p14.2 (Boldog et al. [12] and cited in Kastury et al. [13,14]). The search was also aided by development of an ingenious new method called representational difference analysis (RDA) [15–17], a method based upon polymerase chain reaction (PCR) amplification of restriction-enzyme-digested normal and tumor DNA, which results in isolation of DNA fragments from chromosome regions that differ between constitutional and matched tumor DNA. DNA fragments or probes isolated by RDA are then used to screen panels of cancer-derived cell lines to see if they frequently show deletion or amplification of the specific probe. One such probe, BE758-6 or D3S3155, was homozygously deleted, i.e., entirely missing, in a number of colon-cancer-derived cell lines [16]. Complete loss of a DNA locus in cancer cells is a strong indication that the homozygously deleted region could contain a portion of a tumor suppressor gene, because homozygous deletion is one way of inactivating both alleles of a gene. We showed that the BE758-6 locus mapped closely telomeric to the t(3;8) translocation and determined the size of the homozygous deletions in a number of cancer-derived cell lines [14], as illustrated in Fig. 2. Because the common region of homozygous loss was only 50 to 200 kilobases (kb) in length, this was a region small enough to look for candidate genes, also as illustrated in Fig. 2. The 200-kb region of deletion was isolated from human DNA libraries as overlapping cosmid clones, and each cosmid clone of ~50 kb was tested for the presence of portions of a gene by a method called exon trapping (references cited in Ohta et al. [18]).

Several exons were isolated and tested by Northern blot hybridization, a method for determining if an exon is part of an mRNA expressed in normal cells of various

suppressor gene that is the target of the 3p14.2 allelic loss. **SSCP analysis:** To identify the suppressor gene, the 1 Mb region must be retrieved from human genomic libraries using the probes employed in defining the deletions, as illustrated by the overlapping genomic clones. All genes within the large overlapping genomic clones must then be identified by one of several methods, usually by first identifying exons (where the exon is the box and introns are flanking lines). Individual exons can then be amplified from normal and tumor DNA (where arrows flanking the exon represent the oligonucleotide primers for PCR amplification) and run on a slightly different type of gel that separates the two strands, to determine if the exons of candidate suppressor genes exhibit mutations in the RCCs; if there is even one nucleotide difference in the exon amplified from tumor DNA relative to normal DNA, the strands will usually run differently, as shown in the SSCP illustration at the lower right. The aberrant band can then be cut from the gel and sequenced, as shown at the lower right. If normal and tumor DNAs show a different sequence, as illustrated in the example, and if many of the tumor samples show some abnormality in an exon of the same gene, then this gene is the likely tumor suppressor gene. The gene function can then be studied to see if it has the properties of a tumor suppressor gene by transfecting the full-length gene into kidney tumor cells to see if it affects the phenotype of the tumor (does it grow more slowly, become less tumorigenic in nude mice, become less metastatic?) or by preparing antibodies to the normal and abnormal suppressor proteins to study expression of the protein during development and in specific tissues. Transgenic mice or mice in which both copies of the suppressor gene have been knocked out can also contribute to study of the function of the suppressor gene.

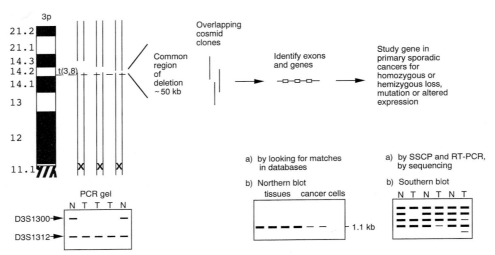

FIGURE 2   Identification of a candidate suppressor gene by homozygous loss of a chromosome region. Sometimes tumor-derived cell lines exhibit complete loss of a specific chromosome region; an example is complete loss of some portions of chromosome region 3p14.2 in various types of cancer cell lines, as illustrated at the upper left. The common region of homozygous loss was narrowed to less than 50 kb by assessing the status of many 3p14.2-linked DNA probes in tumor-derived cell lines, as illustrated in the drawing at upper and lower left; the position of DNA probes is shown in Fig. 1. In the illustration the D3S1300 microsatellite region was PCR amplified from three cancer cell lines (T) (middle three T lanes in drawing at lower left) and from DNA of two normal (N) cell lines; at the same time the D3S1312 microsatellite locus from within the PTPγ gene was amplified from all five DNAs. The D3S1300 locus was entirely absent from the middle three (T) lanes, indicating absence of the locus from the three tumor derived cell lines, whereas the D3S1312 locus was present in all, demonstrating the integrity of all DNA samples. The smallest common region of deletion identified was less than 50 kb, the size of about one gene. Overlapping cosmid clones covering the deleted region were selected and exons identified by "exon-trapping" as described in Ohta *et al.* [18]. Most methods for identifying exons can find false exons, so it is necessary to verify exons by determining if they are part of actual genes. Trapped exons were sequenced and compared to all sequences in various databases. Our trapped exon matched several sequences in the EST (expressed sequence tag) database, and the same exon detected a 1.1-kb RNA signal in various tissues on a Northern blot, as illustrated in the bottom central part of the figure. A **Northern blot** is prepared by electrophoresing a few micrograms of RNA extracted from tissues or cell lines in each lane of an agarose gel; after electrophoretic separation of the RNAs by size, the gel is covered with a nylon filter and absorbent paper and the RNA is transferred to the nylon filter by capillary motion of fluid from the gel to the paper. The nylon filter is then dried and hybridized to a labeled probe, in this case the putative exon, which in this example detects abundant mRNA in normal tissues and little or none in the cancer cells in the last three lanes. The Northern data and the matching clone in the EST database suggest that the trapped exon is a real exon. Further tests might be to determine the structure of the gene, i.e., where the other exons and introns are. In this example, the trapped exon turned out to be exon 5 of the *FHIT* gene and the complete structure is shown in Fig. 3. The next step was to determine if the gene, which encompassed the common deleted region, was altered in cancer cells. First each exon in a panel of cancer cells was tested for mutation by amplification followed by SSCP. For *FHIT,* no mutations were found. If the *FHIT* gene is usually inactivated by deletion of both alleles, then Southern blot might be a good way to find alterations, as shown at bottom right. For **Southern blot** analysis, ~10 µg of DNA from matched normal and tumor samples is electrophoresed on an agarose gel after each DNA sample is cut with the same restriction enzyme, as described in Fig. 1 for RFLP. The DNA is then transferred to a filter as described for Northern blots, and the filter is hybridized to the labeled *FHIT* cDNA. If the pattern of hybridization detected on the filter is different for the normal/tumor pairs, as illustrated at bottom right, then it can be determined if one or both alleles of the candidate gene are altered in the tumors.

tissue types, to determine if it could be part of a real gene. Only one potential exon trapped from the region passed tests suggesting it was part of a gene: it detected a 1.1 kb mRNA on Northern blots and there was an anonymous cDNA clone in a public database that par-

tially matched this exon. The full-length cDNA was isolated from a commercial library and sequenced, and its intron–exon structure was characterized, as described in the preceding section for PTPγ. With this information it was then possible to look for point mutations or other alterations in the candidate gene in tumor cells. The gene was designated *FHIT* because it resembled a family of genes called the *HIT* family and because it encompassed the region of 3p14.2 known as *FRA3B*, a chromosomal fragile region susceptible to breakage by agents that interfere with DNA replication. In determining the structure of the *FHIT* gene, we found that the 5′ end of the gene was on the centromeric side of the t(3;8) translocation break whereas the major part of the gene was on the telomeric side; thus, one copy of the *FHIT* gene was interrupted and inactivated in the hereditary RCC family members, and both copies of a portion of the *FHIT* gene were deleted in a large number of cancer-derived cell lines [18] from stomach, colon, cervical, breast (see Fig. 3), and other tumor types, strengthening *FHIT*'s suppressor gene candidacy.

To determine if a candidate gene is definitely a tumor suppressor gene it is necessary to show that both copies of the gene are inactivated and incapable of carrying out the normal function in specific cancer cells. This is usually done by (1) looking for loss of one allele combined with point mutation of the other allele in a panel of primary tumors of a particular type as described in Fig. 1; (2) looking for loss of both alleles in the tumors; and (3) looking for reduced or absent expression of the RNA and protein products of the candidate gene, as described in Fig. 2.

Loss of at least one allele of the *FHIT* gene occurs in a large fraction of primary tumors of many organs, including kidney [10], lung [19], pancreas [20], digestive tract [14], breast [21], and cervical tumors [22]. Several laboratories have studied the *FHIT* gene for the presence of inactivating mutations in tumor cells, and only

a few such mutations have been reported [19,23]. To determine if both alleles of a gene have been partially deleted in primary tumors is a much more difficult task because primary tumors almost always include from 10 to 50% nontumor cells, so DNA and RNA from these tumors will include nucleic acids from normal cells with two intact tumor suppressor alleles. Thus, simple PCR amplification of a *FHIT* exon from primary tumor DNA or RNA will usually reveal presence of the exon, although there are ways of quantitating PCR products from tumors. On the other hand, Northern and Southern blot analyses are well suited to reveal diminution of expression or presence of a gene in a mixture of tumor and normal cells, as illustrated in Fig. 2. PCR studies and Northern and Southern blot analysis of primary tumors and cancer-derived cell lines have suggested that the primary mechanism of inactivation of the *FHIT* gene is through independent loss of portions of both *FHIT* alleles in cultured and uncultured cancer cells [24,25], as illustrated for some examples in Fig. 3.

To confirm that the *FHIT* gene is indeed inactivated in tumors, antibodies prepared against the purified Fhit protein were used in immunocytochemical analysis of frozen sections of lung tumors [26] and in immunohistochemical staining of cervical tumors [27]; absence of Fhit protein was observed in more than 70% of the tumors.

Additionally, when the *FHIT* gene was reintroduced into cancer cells that had lost both copies of the *FHIT* gene, those cancer cells no longer formed tumors in nude mice [28], indicating that Fhit protein functions as a tumor suppressor gene in four different cancer cell lines. Thus, the *FHIT* gene is a strong candidate tumor suppressor gene (as summarized in Fig. 4) that is fre-

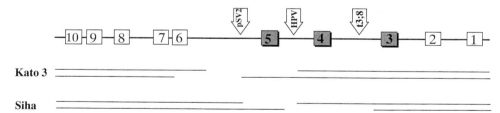

FIGURE 3 The *FHIT* gene. The 1-Mb *FHIT* gene is shown with the 10 small exons that make up the 1.1-kb cDNA. Exons 3, 4, and 5 are shadowed to emphasize their proximity to the familial kidney tumor associated t(3;8) translocation break, the HPV16 integration site identified in a cervical carcinoma, and the fragile sites in intron 5 identified by plasmid pSV2 integration in aphidicolin-treated hybrid cells (references cited in Huebner [25]). Arrows indicate approximate locations of the familial translocation and fragile regions. Note that the fragile sites flank the first *FHIT* coding exon, exon 5. The paired lines below the gene represent the status of the *FHIT* gene in cancer-derived cell lines. Kato 3 is a gastric carcinoma derived cell line exhibiting complete loss of several markers within *FHIT* intron 5, but careful analysis of the *FHIT* locus by Southern blot, Northern blot, and PCR amplification of Kato 3 DNA and RNA revealed that one Kato 3 *FHIT* allele is missing exon 5 and the other is missing at least exon 6 [24]. Similarly, the cervical-carcinoma-derived cell line Siha exhibits independent loss within both *FHIT* alleles. The overlapping portion of the two deletions would be observed by PCR amplification of appropriate intron 4 markers as a small homozygous deletion. Both of these cell lines express no Fhit protein [24].

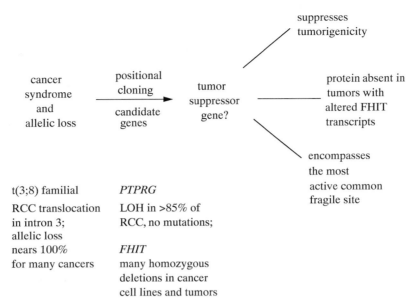

FIGURE 4 Anatomy of a probable tumor suppressor gene. Two candidate genes were isolated from the region of loss of heterozygosity at 3p14.2, a region that is bisected by the familial RCC translocation. The first candidate gene, *PTPRG* or PTPγ, does not show inactivation of both alleles in the cancer cells studied. The second gene, *FHIT,* is a stronger candidate tumor suppressor gene because it is interrupted by the familial translocation, numerous cancer-derived cell lines exhibit loss of portions of both *FHIT* alleles, and numerous primary tumors show reduced *FHIT* RNA expression and absence of Fhit protein expression. Replacement of the gene in Fhit-negative tumor cell lines abolishes their tumorigenicity. Thus, the second candidate gene is probably a tumor suppressor gene for a large fraction of many important human cancers.

quently inactivated in many important human cancers through carcinogen-induced breakage of the fragile region within the gene [29,30]. Breakage and deletion within this gene may be a very early marker of clonal expansion of carcinogen-damaged cells. Thus, assessment of the structure and expression of the suppressor gene, not only in kidney cancer but in cancers of multiple organs, relative to natural history of these neoplasias, may provide a useful marker for many important tumors.

III. ONCOGENES

Oncogenes are genes that encode proteins that contribute to the malignant phenotype. Usually an oncogene evolves from a native gene (called a proto-oncogene) in the cellular genome that, via one of several mechanisms listed in Table 1, becomes activated. An important feature of oncogenes is that they behave genetically as dominants. In contrast to tumor suppressor genes, activation of a single allele is sufficient to cause an increase in the malignant phenotype of a cell. A general discussion of oncogenes can be found in Cooper [31].

A. Activation by Point Mutation

Oncogenes in human tumors commonly become activated through point mutations. The Kirsten *ras* (K-*ras*) gene provides an example. The K-*ras* proto-oncogene, located on the short arm of chromosome 12 at 12p12.1, encodes a protein of molecular massapproximately 21 kDa that is attached to the cytoplasmic side of the cell membrane. The ras proteins function to transduce signals from receptors on the cell surface by binding to GTP in a pathway of cell signaling that ultimately results in promotion of gene transcription and cell division.

TABLE 1 Activation of Oncogenes in Human Cancer

1. Activating point mutations of protooncogenes
2. Overexpression of protooncogenes
   (a) Amplification of protooncogene
   (b) Promoter or enhancer insertion through chromosomal translocation
   (c) Unidentified changes in regulatory elements.
3. Chromosomal translocations resulting in gene fusion and fusion proteins that behave as oncogenes.
4. Infection with DNA viruses whose gene products inactivate tumor suppressor genes.

As a result of an activating point mutation, generally in codon 12, 13, or 61, the ras protein product is no longer regulated by normal means so that the growth stimulatory signals that it ordinarily transmits are self-generated and markedly increased. Usually, only one of the two K-*ras* alleles is activated.

K-*ras* is a member of a family of three similar *ras* genes. The other two *ras* genes, H-*ras* and N-*ras*, have a high homology with K-*ras*, and their expressed proteins are of similar molecular weight, structure, and function. If a given class of tumor has a high incidence of *ras* mutation, the mutation is almost always of one particular *ras* gene, e.g., N-*ras* in acute myeloid leukemia, and K-*ras* in pancreatic adenocarcinoma.

A classical way to identify point mutations that activate proto-oncogenes (or inactivate tumor suppressor genes) is by RNA–DNA RNase A mismatch cleavage analysis [32]. The DNA segment in question is hybridized with a labeled RNA probe representing the wild-type (unmutated) gene and the dimer is digested with RNase A, an enzyme that cleaves mismatched DNA-RNA hybrids. The digestion products of the dimer, consisting of the mutated DNA and the RNA probe, differ in size from those derived from a hybrid of the same RNA probe with wild-type (unmutated) DNA. The differences are easily visualized by electrophoresis of the labeled digested RNA samples and subsequent autoradiography. More recently, another technique to identify point mutations, SSCP analysis, has become popular [11] to distinguish mutated DNA from wild-type DNA, as illustrated previously in Fig. 1. Both of these methods identify variant tumor DNA fragments on gels; the variant bands are then sequenced to confirm the presence of the mutation and to identify the mutated codon.

K-*ras* mutations at codon 12 are found in more than 90% of pancreatic adenocarcinomas [33]. Identification of K-*ras* mutations in fine needle aspirates of pancreatic tumors has been proposed as a diagnostic tool for pancreatic adenocarcinoma [34]. The clustering of mutations at this site is probably a footprint of the particular carcinogen(s) causing pancreatic cancer, although these carcinogens remain to be identified. Activating K-*ras* mutations are frequently found in other tumors such as colon adenocarcinomas (40–50%) and lung cancers (25%).

B. Overexpression or Amplification

Another common way in which proto-oncogenes become activated is through overexpression, for example, of c-erbB-2 [35] in breast tumors. The c-erbB-2 gene product is a receptorlike molecule similar in structure to the epidermal growth factor receptor; it is overexpressed in a significant proportion of carcinomas of the breast, ovary, and pancreas, in adenocarcinomas of the lung, and in certain other carcinomas. For instance, increased c-erbB-2 protein was found in 15 of 23 pancreatic adenocarcinomas by immunohistochemistry [36]. In contrast, c-erbB-2 overexpression is rarely found in melanomas, gliomas, neuroblastomas, or lymphomas. Frequently, the overexpression correlates with the presence of increased numbers of copies of the c-erbB-2 gene. This amplification of copy number is readily demonstrated by probing, or hybridizing, tumor DNA immobilized on a nylon filter (as illustrated for Southern blotting in Fig. 2) with labeled c-erbB-2 DNA. If the tumor DNA carries amplified copies of the c-erbB-2 gene, the signal detected in the tumor DNA will be much stronger than the signal detected in normal DNA. The c-erbB-2 protein, through unknown mechanisms, can also be overexpressed in tumors with single genomic copies of the c-erbB-2 gene. Increased expression of the c-erbB-2 protein product can be recognized by immunohistochemistry using antibodies specific for the c-erbB-2 protein, and such overexpression has been proposed as an indicator of poor prognosis for breast, ovarian, and bladder cancer patients [35].

C. Translocations

As noted earlier, proto-oncogenes can be overexpressed through chromosomal translocation as is observed in a number of lymphomas and leukemias [1]. Burkitt's lymphoma is a classic example. Through chromosomal breaks and reciprocal translocations, the *myc* gene on chromosome 8 is brought close to an immunoglobulin gene on chromosome 14 (heavy chain gene), 2 (kappa chain gene), or 22 (lambda chain gene) with resultant overexpression of the *myc* gene product, presumably caused by promote/enhancer control by the immunoglobulin gene. Activation of the *myc* gene by translocation is observed in several other hematological neoplasms as well.

Another mechanism by which translocations can activate oncogenes is by the creation of chimeric genes that produce fusion proteins that contribute to the malignant phenotype. An example of this is the Philadelphia chromosome in chronic myelogenous leukemia (CML). In CML the reciprocal translocation results in the *abl* gene on chromosome 9 translocating to chromosome 22 adjacent to the *bcr* gene. The unique bcr/abl fusion protein that results initiates or adds to the malignant phenotype. Such chimeric proteins are unique tumor specific markers that can be targets for gene therapy.

D. Viral Oncogenes

Viral genes can also behave as oncogenes. Clinically relevant examples are some of the early genes of specific strains of human papilloma (wart) viruses; these are double-stranded DNA viruses. Over 60 different strains of human wart virus have been described and most have been related to benign lesions. However, strains 16 and 18 are associated with cancer of the uterine cervix in approximately 70% of cases. Two of the human papilloma virus gene products, designated E6 and E7, behave as oncogenes. Experimentally, it has been shown that they inactivate, respectively, the tumor suppressor genes p53 and Rb, thereby contributing to the malignant phenotype (reviewed in Howley [37]).

IV. SUMMARY

The genetic changes observed thus far in human neoplasms suggest that hematopoietic tumors originate because of inappropriate expression (usually overexpression) of a growth-promoting gene, which leads to extended growth or life span of hematopoietic clones. Thus, a strategy for a gene based therapy for hematopoietic or solid tumors with an activated oncogene would be to turn off expression of the activated oncogene, perhaps by antisense treatment (see Chapter 13). Such treatment would be tumor specific in the case of the unique chimeric oncogenes created by some tumor specific chromosome translocations.

Carcinomas, which are driven primarily by tumor suppressor gene loss, would need to have their inactivated suppressor genes replaced, either by transfection with a copy of the normal suppressor gene (Chapter 12) or perhaps by intervening pharmacologically in the signal pathway disrupted by loss of the suppressor activity.

Genetic changes in both hematopoietic and solid tumors are still being discovered, and no doubt hematopoietic tumors will be found to involve some tumor suppressor genes. Some solid tumors, on the other hand (most sarcomas in fact), exhibit specific chromosome translocations, a number of which have been cloned and some of which result in production of chimeric transcription factors. The understanding of the molecular genetic changes that drive the establishment and spread of specific human tumors is already contributing to diagnostic and prognostic characterization of cancers and to decisions concerning treatment. As described and discussed in a number of chapters of this volume, knowledge of the specific oncogene and suppressor gene mutations that characterize specific cancers will, in the near future, serve as the basis for novel therapeutic strategies.

Acknowledgments

This work was supported by NIH. Grant CA21124 and Kimmel Cancer Center support Grant CA56336, the U.S.–Poland Maria Sklo- dowska–Curie Joint Fund II, and a gift from Mr. R. R. M. Carpenter, III and Mrs. Mary K. Carpenter. Portions of this chapter are adapted from Huebner *et al.,* (1996). *Semin Oncol.* **23**, 22–30.

References

1. Nowell, P. C. (1994). Cytogenetic approaches to human cancer genes. *FASEB J.* **8**, 408–413.
2. Carroll, P. R., Murty, V. V. S., Reuter, V., Jhanwar, S., Fair, W. R., Whitmore, W. F., and Chaganti, R. S. K. (1987). Abnormalities at chromosome region 3p12–14 characterize clear cell renal carcinoma. *Cancer Genet Cytogenet.* **26**, 253–259.
3. Szucs, S., Muller-Brechlin, R., DeRiese, W., and Kovacs, O. (1987). Deletion 3p: The only chromosome loss in a primary renal cell carcinoma. *Cancer Genet Cytogenet.* **26**, 369–373.
4. Zbar, B., Brauch, H., Talmage, C., and Linehan, M. (1987). Loss of alleles of loci on the short arm of chromosome 3 in renal cell carcinomas. *Nature (Lond).* **327**, 721–724.
5. Knudson, A. G. (1993). Antioncogenes and human cancer. *Proc. Natl. Acad. Sci. USA.* **90**, 10914–10921.
6. LaForgia, S., Lasota, J., Latif, F., Boghosian-Sell, L., Kastury, K., Ohta, M., Druck, T., Atchison, L., Cannizzaro, L., Barnea, G., Schlessinger, J., Modi, W., Kuzmin, I., Tory, K., Zabar, B., Croce, C. M., Lerman, M., and Huebner, K. (1993). Detailed genetic and physical map of the 3p chromosome region surrounding the familial RCC chromosome translocation, t(3;8)(p14.2;q24.1). *Cancer Res.* **53**, 3118–3124.
7. Cohen, A. J., Li, F. P., Berg, S., *et al.* (1979). Hereditary renal-cell carcinoma associated with chromosomal translocation. *N. Engl. J. Med.* **301**, 592–595.
8. Hadaczek, P., Podolski, J., Toloczko, A., Kurzawski, G., Sikorski, A., Rabbitts, P., Huebner, K., and Lubinski, J. (1996). Accumulation of losses at 3p common deletion sites is characteristic of clear cell renal cell carcinoma. *Virchows Arch.* **429**, 37–42.
9. Lubinski, J., Hadaczek, P., Podolski, J., Toloczko, A., Sikorski, A., McCue, P., Druck, T., and Huebner, K. (1994). Common regions of deletion in chromosome regions 3p12 and 3p14.2 in primary clear cell renal carcinomas. *Cancer Res.* **54**, 3710–3713.
10. Druck, T., Kastury, K., Hadaczek, P., Podolski, J., Toloczko, A., Sikorski, A., Ohta, M., LaForgia, S., Lasota, J., McCue, P., Lubinski, J., and Huebner, K. (1995). Loss of heterozygosity at the familial RCC t(3;8) locus in most clear cell renal carcinomas. *Cancer Res.* **55**, 5348–5353.
11. Hayashi, K. (1992). PCR-SSCP: A method for detection of mutations. *GATA* **9**, 73–79.
12. Boldog, F. L., Gemmill, R. M., Wilke, C. M., Glover, T. W., Nilsson, A. S., Chandrasekharappa, S. C., Brown, R. S., Li, F. P., and Drabkin, H. A. (1993). Positional cloning of the hereditary renal carcinoma 3 : 8 chromosome translocation breakpoint. *Proc. Natl. Acad. Sci. USA* **90**, 8509–8513.
13. Kastury, K., Ohta, M., Lasota, J., Moir, D., Dorman, T., LaForgia, S., Druck, T., and Huebner, K. (1996). Structure of the human receptor tyrosine phosphatase gamma gene (*PtPRG*) and relation to the familial RCC t(3;8) chromosome translocation. *Genomics* **32**, 225–235.
14. Kastury, K., Baffa, R., Druck, T., Ohta, M., Cotticelli, M. G., Inoue, H., Negrini, M., Rugge, M., Huang, D., Croce, C. M., Palazzo, J., and Huebner, K. (1996). Potential gastrointestinal tumor suppressor locus at the 3p14.2 *FRA3B* site identified by homozygous deletions in tumor cell lines. *Cancer Res.* **56**, 978–983.
15. Lisitsyn, N. A., Lisitsina, N. M., and Wigler, M. (1993). Cloning the differences between two complex genomes. *Science* **259**, 946–951.

16. Lisitsyn, N. A., Lisitsina, N. M., Dalbagni, G., Barker, P., Sanchez, C. A., Gnarra, J., Linehan, W. M., Reid, B. J., and Wigler, M. H. (1995). Comparative genomic analysis of tumors: Detection of DNA losses and amplification. *Proc. Natl. Acad. Sci. USA* **92,** 151–155.

17. Lisitsyn, N. A. (1995). Representational difference analysis: Finding the differences between genomes. *Trends Genet.* **11,** 303–307.

18. Ohta, M., Inoue, H., Cotticelli, M.G., Kastury, K., Baffa, R., Palazzo, J., Siprashvili, Z., Mori, M., McCue, P., Druck, T., Croce, C. M., and Huebner, K. (1996). The *FHIT* gene, spanning the chromosome 3p14.2 fragile site and renal carcinoma-associated t(3;8) breakpoint, is abnormal in digestive tract cancers. *Cell* **84,** 587–597.

19. Sozzi, G. Veronese, M. L., Negrini, M., Baffa, R., Cotticelli, M. G., Inoue, H., Tornielli, S., Pilotti, S., De Gregorio, L., Pastorino, U., Pierotti, M. A., Ohta, M., Huebner, K., and Croce, C. M. (1996) The *FHIT* gene at 3p14.2 is abnormal in lung cancer. *Cell* **85,** 17–26.

20. Shridhar, R., Shridhar, V., Wang, X., Paradee, W., Dugan, M., Sarkar, F., Wilke, C., Glover, T. W., Vaitkevicius, V. K., and Smith, D. I. (1996). Frequent breakpoints in the 3p14.2 fragile site, *FRA3B*, in pancreatic tumors. *Cancer Res.* **56,** 4347–4350.

21. Ahmadian, M., Wistuba, I. I., Fong, K. M., Behrens, C., Kodagoda, D. R., Saboorian, M. H., Shay, J., Tomlinson, G. E., Blum, J., Minna, J. A., and Gazdar, A. F. (1997). Analysis of the *FHIT* gene and *FRA3B* region in sporadic breast cancer, preneoplastic lesions and familial breast cancer probands. *Cancer Res.* **57,** 3664–3668.

22. Larson, A. M., Kern, S., Curtiss, S., Gordon, R., Cavenee, W. K., and Hampton, G. M. (1997). High resolution analysis of chromosome 3p alterations in cervical carcinoma. *Cancer Res.* **57,** 4082–4090.

23. Gemma, A., Hagiwara, K., Ke, Y., Burke, L. M., Khan, M. A., Nagashima, M., Bennett, W. P., and Harris, C. C. (1997). *FHIT* mutations in human primary gastric cancer. *Cancer Res.* **57,** 1435–1437.

24. Druck, T., Hadaczek, P., Fu, T-B., Ohta, M., Siprashvili, Z., Baffa, R., Negrini, M., Kastury, K., Veronese, M. L., Rosen, D., Rothstein, J., McCue, P., Cotticelli, M. G., Inoue, H., Croce, C. M., and Huebner, K. (1997). Structure and expression of the human *FHIT* gene in normal and tumor cells. *Cancer Res.* **57,** 504–512.

25. Huebner, K., Hadaczek, P., Siprashvili, Z., Druck, T., and Croce, C. M. (1997). The *FHIT* gene, a multiple tumor suppressor gene encompassing the carcinogen sensitive chromosome fragile site, *FRA3B. Biochim. Biophys. Acta.* (Reviews on Cancer), **1332,** M65–M70.

26. Sozzi, G. Tornielli, S. Tagliabue, E., Sard, L., Pezzella, F., Pastorino, U., Minoletti, F., Pilotti, S., Ratcliffe, C., Veronese, M. L., Goldstraw, P., Huebner, K., Croce, C. M., Pierotti, M. A. (1997). Absence of Fhit protein in primary lung tumors and cell lines with *FHIT* gene abnormalities. *Cancer Res.* **57,** 5207–5212.

27. Greenspan, D. L., Connolly, D. C., Wu, R., Lei, R. Y., Vogelstein, J. T. C., Kim, Y.-T., Mok, J. E., Muñoz, N., Bosch, X., Shah, K., and Cho, K. R. (1997). Loss of *FHIT* expression in cervical carcinoma cell lines and primary tumors. *Cancer Res.* **57,** 4692–4698.

28. Siprashvili, Z., Sozzi, G., Barnes, L. D., McCue, P., Robinson, A. K., Eryomin, V., Sard, L., Tagliabue, E., Greco, A., Fusetti, L., Schwartz, G., Pierotti, M. A., Croce, C. M., and Huebner, K. (1997). Replacement of Fhit in cancer cells suppresses tumorigenicity. *Proc. Natl. Acad. Sci. USA* **94,** 13771–13776.

29. Zimonjic, D. B., Druck, T., Ohta, M., Kastury, K., Popescu, N. C., and Huebner, K. (1997). Positions of chromosome 3p14.2 fragile sites (*FRA3B*) within the *FHIT* gene. *Cancer Res.* **57,** 1166–1170.

30. Sozzi, G., Sard, L., Marchetti, A., De Gregorio, L., Musso, K., Tornielli, S., Veronese, M. L., Incarbone, M., Manenti, G., Pastorino, U., Huebner, K., Croce, C. M., and Pierotti, M. A. (1997). Association between cigarette smoking and *FHIT* gene alterations in lung cancer. *Cancer Res.* **57,** 2121–2123.

31. Cooper, G. M. (1995). Oncogenes, 2nd ed. Jones and Bartlett, Boston.

32. Winter, E., Yamamoto, F., Almoguera, C., and Perucho, M. (1985). A method to detect and characterize point mutations in transcribed genes: Amplification and overexpression of mutant c-Ki-*ras* alleles in human tumor cells. *Proc. Natl. Acad. Sci. USA* **82,** 7575–7579.

33. Almoguera, C., Shibata, D., Forrester, K., Martin, J., Arnheim, N., and Perucho, M. (1988). Most human carcinomas of the exocrine pancreas contain mutant c-K-*ras* genes. *Cell* **53,** 549–554.

34. Shibata, D., Almoguera, C., Forrester, K., Jordon, D., Martin, S. E., Cosgrove, M. M., Perucho, M., and Arnheim, N. (1990). Detection of c-K-*ras* mutations in fine needle aspirates from human pancreatic carcinomas. *Cancer Res.* **50,** 1279–1283.

35. Maguire, H. C., Jr., and Greene, M. I. (1990). Neu (c-erbB-2), a tumor marker in carcinoma of the female breast. *Pathobiology* **58,** 297–303.

36. Williams, T. M., Weiner, D. B., Greene, M. I., and Maguire, H. C. Jr. (1991). Expression of c-erbB-2 in human pancreatic adenocarcinomas. *Pathobiology* **59,** 46–52.

37. Howley, P. M. (1995). "Viral Carcinogenesis," pp. 38–58 in *The Molecular Basis of Cancer,* Ed. J. Mendelsohn *et al.* W. B. Saunders, Philadelphia.

Chapter 3

Immunologic Targets for the Gene Therapy of Cancer

Suzanne Ostrand-Rosenberg, Vicky S. Gunther, Todd A. Armstrong, Beth A. Pulaski, Matthew R. Pipeling, Virginia K. Clements, and Nana Lamousé-Smith

Department of Biological Sciences, University of Maryland, Baltimore, Maryland 21250

I. INTRODUCTION

For at least 100 years immunologists have proposed activating the immune system to specifically target and eradicate autologous tumor cells. Until the advent of molecular gene transfer techniques and increased knowl-

edge of the basic pathways of lymphocyte activation, however, effective methods for harnessing the immune system as a therapeutic agent were unsuccessful, despite numerous efforts and enthusiasm for the approaches. During the last 10 to 15 years, however, the field of immunotherapy of cancer has been inundated with novel strat-

33

egies for the treatment of malignancies. As a result, there is considerable optimism in both the basic science and clinical arenas for pursuing immunotherapy strategies. The renewed enthusiasm for immunotherapy is largely because of the recent and cumulative explosive expansion of knowledge in basic immunology and molecular biology, and the rapidly improved understanding of regulation of the immune response. Acquisition of this basic information has allowed tumor immunologists to design approaches to control immune responses against tumors, leading, in at least some cases, to impressive antitumor responses in experimental systems. Tumor immunologists are now at the cusp of applying to the treatment of human cancers what has been learned and successfully applied in animal models.

Many of the current immunotherapy/gene therapy strategies are presented in detail in subsequent chapters of this book. To fully appreciate and understand these strategies, it is necessary to understand how the immune system responds to antigen and to appreciate the characteristics of tumor cells that make them potential targets for an immune response. The goal of this chapter, therefore, is to provide a concise overview of the induction of immunity and to clarify how an immune response can be specifically enhanced and directed toward tumor cells.

II. CELL-MEDIATED VERSUS HUMORAL (ANTIBODY-MEDIATED) IMMUNE RESPONSES TO TUMOR CELLS

The immune system mediates two basic types of responses: Antibody-mediated (B cell) responses and cell-mediated (T-cell) responses. Because antibodies react with soluble molecules and are incapable of entering intact, live cells, they are very effective agents against extracellular and soluble pathogens, but they are not effective agents against intracellular pathogens or malignant cells whose tumor antigens are intracellular molecules and not expressed at the surface of the malignant cells. As a result, antibodies have been used in tumor settings when tumor antigens are extracellular molecules or in situations when tumor antigens are membrane-bound, integral membrane proteins. In these settings, monoclonal antibodies (mAbs) and recombinant antibodies have been tagged with markers such as radioisotopes or fluorescent probes and used as imaging reagents, or they have been coupled to toxic agents, such as radioisotopes or toxins, in an attempt to specifically deliver a "lethal hit" to targeted tumor cells. Most of the antibody approaches involve passive transfer of

in vitro generated reagents, which can mediate immediate antitumor responses, but which do not result in long-term immunologic memory. Some of these antibodies have shown treatment potential in patients with established tumors, and they are undergoing continual development to decrease their inherent immunogenicity and to increase their targeting specificity [1–4].

The utility of antibodies as therapeutic agents obviously depends on their specific binding to tumor antigens and on their binding affinity. Until recently, the identification of new tumor-specific or tumor-reactive antibodies has depended on conventional immunization techniques and the availability of tumor cells for screening the resulting monoclonal antibodies. The recently developed SEREX approach, however, may greatly increase the number and repertoire of serologically defined tumor antigens. SEREX, an acronym for *serological analysis of tumor antigens by recombinant expression cloning,* uses cancer patients' sera to screen autologous tumor cell cDNA expression libraries [5,6]. This newly developed technique has already identified several human tumor antigens that were previously identified via T-cell reactivity, as well as identifying previously uncharacterized antigens from a variety of human cancers [5–8]. Because SEREX uses patients' sera as the detection system, the tumor antigens characterized are immunogenic in tumor-bearing individuals and therefore may be useful targets for antibody-mediated immunotherapy.

In contrast, T lymphocytes, which mediate cell-mediated immunity, are effective agents against intracellular pathogens and malignant cells because they recognize antigens synthesized within target cells. Many therapeutic strategies involving T-cell activation are also aimed at inducing active antitumor immunity in the tumor-bearing patient, such that immunologic memory will be induced. Induction of immunologic memory to tumor antigens has several advantages in a therapy setting. For example, many metastatic lesions are refractory to conventional treatments. Induction of protective immunity against metastatic cells would have the obvious benefit of controlling malignant cell growth and providing long-term protection against distant metastases that are otherwise untreatable. Likewise, active antitumor immunity would provide long-term protection against recurrence of primary tumor. Many recent studies, therefore, have explored the harnessing and enhancing of T-lymphocyte-mediated immunity to malignant cells, and most of these approaches involve gene therapy. The remainder of this chapter will therefore focus on T-cell responses to tumor cells.

III. CD4⁺ AND CD8⁺ T LYMPHOCYTES RESPOND TO TUMOR ANTIGENS PRESENTED IN THE CONTEXT OF MOLECULES ENCODED BY THE MAJOR HISTOCOMPATIBILITY COMPLEX

T lymphocytes recognize target cells via their antigen specific T-cell receptor for antigen (TcR). The TcR is a membrane anchored heterodimeric molecule with specificity for antigen that is associated with additional protein molecules known as the CD3 complex. The TcR/CD3 complex of T cells interacts with antigen bound to a major histocompatibility complex (MHC) encoded molecule on the antigen presenting cell (APC) or target cell. CD8⁺ cytotoxic T lymphocytes (CTLs) typically recognize antigen bound to MHC class I molecules, whereas CD4⁺ T helper lymphocytes (Th cells) recognize antigen bound to MHC class II molecules. Although specificity of antigen recognition is accomplished by the TcR, additional accessory molecules expressed by the T cell, such as CD4, CD8, ICAMs (intracellular adhesion molecules), and LFAs (lymphocyte function associated antigens) are also involved in T-cell activation. By binding to their cognate receptors on the APC, these accessory molecules stabilize the binding of the T cell to its target cell and in some cases also deliver specific activation signals to the T lymphocyte [9,10].

A. CD8⁺ CTLs Recognize Endogenously Synthesized Antigens

The actual antigen bound by the MHC and TcR is a processed peptide fragment derived from a larger intact protein. MHC class I–bound peptides are typically derived from proteins synthesized within the target cell (endogenously synthesized proteins) and processed via the cell's proteosome into peptides that are transported into the endoplasmic reticulum (ER), where they are bound to newly synthesized MHC class I molecules. The resulting MHC class I/peptide complex then traffics via the default secretory pathway to the target cell membrane, where it is anchored via its hydrophobic region. Figure 1 shows a schematic diagram of the trafficking pathway of endogenously synthesized molecules.

Because of this pathway, virtually any endogenously synthesized self-protein can be presented by MHC class I molecules and hence be available as a potential target for CD8⁺ CTL. CD8⁺ T cells are therefore particularly good candidates for recognizing tumor cells, because many tumor antigens are endogenously synthesized molecules that are not expressed at the cell surface in intact form but are presented as peptides bound to endogenously synthesized MHC class I molecules [11,12].

B. CD4⁺ T Lymphocytes Recognize Exogenously Synthesized Antigens

Like CD8⁺ T lymphocytes, CD4⁺ T cells recognize processed peptide fragments of antigen. However, in the case of CD4⁺ T cells, processed peptide is bound by MHC class II instead of MHC class I molecules. In addition, the peptides seen by CD4⁺ T cells are from a different source than those seen by CD8⁺ T cells, and they associate with MHC class II molecules via a different pathway than peptides presented by MHC class I molecules. MHC class II molecules are only expressed on certain cells (B lymphocytes, dendritic cells [DCs], macrophages, and Langerhans cells), and only these cells serve as APC for CD4⁺ T lymphocytes. Typically, class II⁺ APC phagocytose or endocytose soluble antigen and process it to peptides in their endosomes, where the antigen then binds to class II molecules. The antigenic peptide/MHC class II complex is then transferred and inserted into the plasma membrane. Unlike class I molecules, class II molecules traffic to an endosomal compartment because they associate in the endoplasmic reticulum (ER) with a chaperon molecule called invariant (Ii) chain. Ii chain has two functions: (1) When tightly bound to class II in the ER, Ii prevents binding of endogenously synthesized peptides to class II molecules, and (2) Ii contains a trafficking motif that guides the class II/Ii complex to the endosomal compartment. Once in the endosomal compartment, the Ii chain dissociates from class II and is degraded, and the processed exogenous peptides that are also in the endosome bind to the class II molecule [11–13]. (See Fig. 1 for a schematic diagram of the trafficking pattern of MHC class II–bound peptides.)

The MHC class I and class II antigen processing and presentation pathways therefore yield very different repertoires of antigenic peptides presented to CD8⁺ vs. CD4⁺ T cells.

C. Activation of Tumor-Specific CD4⁺ and CD8⁺ T Lymphocytes Requires Two Signals

Extensive studies have demonstrated that activation of T lymphocytes requires two signals, as shown in Fig. 2A. The first signal is the antigen specific signal that is received by the responding CD4⁺ or CD8⁺ T cell when its TcR binds to the MHC/peptide complex on the APC.

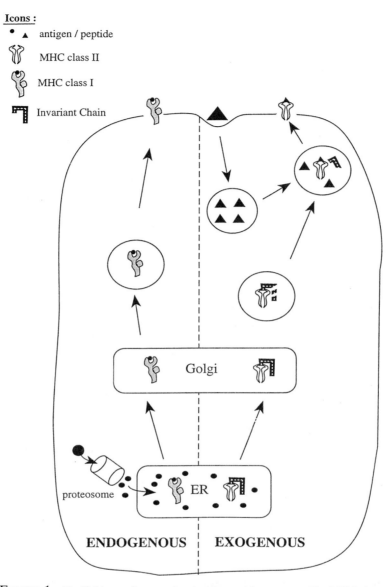

Icons :

• ▲ antigen / peptide

MHC class II

MHC class I

Invariant Chain

ENDOGENOUS | EXOGENOUS

FIGURE 1   Trafficking pathways of antigenic peptides presented by MHC class I and MHC class II molecules. Molecules synthesized within the APC (endogenously synthesized antigen) are digested in the proteosome into peptides and transported into the ER, where they bind to newly synthesized MHC class I molecules. The MHC class I/peptide complex then traffics via the default secretory pathway to the plasma membrane, where the hydrophobic region of the class I molecule is inserted into the membrane and the peptide/external domain of the class I molecule is displayed. Molecules synthesized outside of the APC (exogenously synthesized antigen) are internalized via endocytosis and digested within the endosomal compartment. Antigenic peptides are bound by MHC class II molecules that have been directed from the ER to the endosomal compartment by trafficking signals contained on the class II–associated Ii chain. The class II/peptide complex is inserted into the plasma membrane via the hydrophobic region of the class II molecule, and the antigenic peptide/external domain of the class II molecule is displayed.

The second or costimulatory signal is delivered to the T lymphocyte when a costimulatory molecule, such as B7.1 (CD80) or B7.2 (CD86), on the APC binds to its cognate receptor, CD28, on the responding T cell. Because only certain professional APCs express costimulatory molecules, only these APCs are capable of

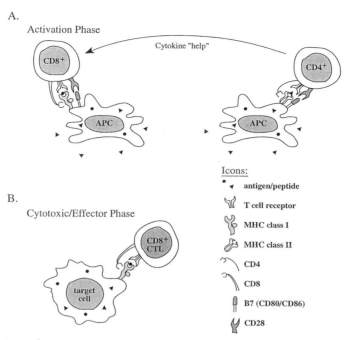

FIGURE 2 Roles of MHC class I, class II, and costimulatory molecules in the activation and effector function of CD8+ and CD4+ T lymphocytes. (A) CD8+ and CD4+ T lymphocytes require two signals for activation. The first signal is an antigen specific signal that is delivered when the peptide/MHC class I or peptide/MHC class II complex binds to the TcR of CD8+ or CD4+ T cells, respectively. The second signal is delivered when a costimulatory molecule on the APC binds to its counterreceptor, CD28, on the responding CD8+ or CD4+ T lymphocyte. (B) Activated CD8+ CTLs only need to recognize and bind to the appropriate peptide/MHC class I complex on the target cell to initiate cytolysis. (*Note:* Other accessory molecules, such as ICAM and LFA, which stabilize T-cell–APC or T-cell–target cell binding, are not shown.)

activating T cells. B7.1 and B7.2 are constitutively expressed on dendritic cells and are inducible on B lymphocytes and macrophages, whereas B7.2 is constitutively expressed on resting T cells. On B cells, B7.2 levels reach their maximum expression between 24 and 96 hours after B cell activation, whereas B7.1 levels peak between 48 and 72 hours after B cell activation. Binding of B cell CD40 to T-cell CD40L (B cell costimulatory molecules) also causes up-regulation of B cell encoded B7.1 and B7.2. Expression of T-cell costimulatory molecules, therefore, is fluid and depends on the activation state of the APC [14,15].

If the antigen specific signal is delivered in the absence of the costimulatory signal, then the responding T lymphocytes are anergized rather than activated. Although different APCs can deliver the antigen specific and the costimulatory signals, T cells are most efficiently activated if both signals are delivered by the same APC [16,17]. As a result, somatic cells that present MHC class I molecules with self-peptides and do not express costimulatory ligands may anergize T cells because they are unable to deliver the requisite costimulatory signal. Indeed, it has been hypothesized that the CD8+ T cells of tumor-bearing individuals are "anergized" or "tolerized" to their tumors because MHC class I+ tumor cells present tumor antigens but do not deliver a costimulatory signal [18,19]. Tolerization of tumor-specific CD4+ T lymphocytes is less likely to be a problem than tolerization of CD8+ T cells because professional APC that activate CD4+ T cells usually coexpress MHC class II molecules and costimulatory molecules.

Although costimulatory signals are required to activate naive T lymphocytes, they are not needed to reactivate memory T cells, nor are they needed as target molecules for T-cell-mediated cytolysis [20], as shown in Fig. 2B. Cell-based vaccines consisting of autologous tumor cells transfected/transduced with B7.1 and/or B7.2 genes have therefore been generated. These vaccines presumably activate naive antitumor T cells, and the activated T cells subsequently destroy unmodified, wild-type tumor cells in situ [18,19,21].

In addition to interacting with CD28, B7.1 and B7.2 have a second receptor on T cells, cytotoxic T-lymphocyte antigen 4 (CTLA4). In contrast to B7/CD28 interactions, which are stimulatory, B7/CTLA4 interactions block activation and promote T-cell death or apoptosis. The success of T-cell activation therefore hinges on signaling via CD28 instead of via CTLA4. A recently reported antitumor strategy capitalized on the differential function of CD28 versus CTLA4 and demonstrated that mice given anti-CTLA4 mAbs develop a potent antitumor immunity. Presumably, the anti-CTLA4 antibodies sterically inhibited signaling through CTLA4, thus favoring signaling and subsequent T-cell activation through CD28 [22].

D. Effective Antitumor Immunity Probably Requires Activation of Both CD4$^+$ and CD8$^+$ T Lymphocytes

Although CD8$^+$ T cells can be activated in the absence of CD4$^+$ T helper lymphocytes, optimal CD8$^+$ T-cell activation and long-term immunologic memory probably require coactivation of CD4$^+$ T lymphocytes. The added benefit of CD4$^+$ T-cell help is particularly apparent in tumor immunity, where numerous studies have demonstrated that improved CD4$^+$ T-cell help facilitates tumor rejection [23–26]. Induction of optimal antitumor immunity should therefore involve activation of both CD4$^+$ and CD8$^+$ tumor-specific T cells.

IV. ARE TUMOR-BEARING INDIVIDUALS ABLE TO RESPOND TO TUMOR ANTIGENS?

The active immunization therapies currently under development depend on the tumor-bearing patient's capacity to generate an immune response against autologous tumor. The immunocompetence of the host is therefore very important in assessing if an individual is a reasonable candidate for immunotherapy. Both anecdotal and experimental data indicate that tumor-bearing individuals, especially ones with advanced disease, have significantly reduced immunocompetence. The extent and onset of the immunosuppression varies depending on the type of tumor and tumor burden. For example, in some cases, patients or animals with experimental tumors are fully immunocompetent for all antigens except antigens present on the autologous tumor [27,28],

whereas in other cases, tumor bearers are generally immunosuppressed, as measured by lack of alloreactivity and the inability to generate CTL responses [29].

A. Defects in Intracellular Signaling Pathways May Inhibit Tumor-Specific T Cells from Responding to Tumor Cells

Although some individuals with large tumor burdens are significantly immunosuppressed, there is no consensus about the mechanism of suppression. Earlier studies indicated that some advanced tumor bearers contained suppressor T cells that down-regulate effector T cells [30,31]. In more recent experiments advanced tumor bearers were shown to lack certain components in the T-cell signaling pathway, such as the T-cell receptor ζ chain, p56^{lck}, and p59^{fyn}, and in some cases displayed abnormal intracellular signaling. A wide variety of signaling molecules and T-cell response parameters have been examined, including levels of CD3ζ, p56^{lck}, p59^{fyn}, and ZAP-70 proteins, NF-κB expression and/or translocation, tyrosine phosphorylation patterns, protein tyrosine kinase activity, and T-cell proliferative capacity. In many cases, reductions or defects were found in some tumor-bearing individuals [32–39]. The implication of these studies was that the immunodeficiency in tumor bearers was caused by aberrant or insufficient T-cell signaling during T-cell activation [32–34,40]. The physiological relevance and universality of these signaling abnormalities is unclear, however, because other studies using different tumors in both mice and patients did not show consistent defects in these signal transduction molecules despite decreased protein tyrosine phosphorylation or reduced T-cell activation or CTL activity [29,37,41].

Some of the discrepancies between these studies may be caused by differences among tumors. In addition, there appear to be technical explanations. In some mice with experimental tumors, loss of the CD3ζ chain and reduction of p56^{lck} was shown to result from proteases that were released from contaminating granulocytes in the T-cell preparations [42]. However, in the same experiments loss of the CD3ζ chain in peripheral blood lymphocytes (PBL) of tumor-bearing patients was not caused by proteases released by contaminating cells. As Levey and Srivastava point out in their review [43], the techniques used to isolate the various T-cell signaling molecules are highly error-prone, and hence different results in the different studies may be caused by technical differences. The same authors also point out that there is a wide range in expression of CD3ζ molecules among normal, non-tumor-bearing patients [36] so that differences in CD3ζ expression in tumor bearers may reflect genetic differences, rather than differences caused by the presence of tumor.

The various studies assessing p56[lck] and/or CD3ζ expression have been performed on T cells isolated from PBL of tumor-bearing individuals and/or tumor infiltrating lymphocytes (TILs). TILs were generally found to be more deficient in these signaling molecules than were T cells isolated from PBL, suggesting that T cells constantly in close contact with tumor cells may be more severely impacted than systemic, circulating T lymphocytes. However, as pointed out by Levey and Srivastava [43], the differences may be caused by differences in purity of the T-cell populations examined. For example, it is technically easier to prepare highly purified T cells from PBL, whereas isolation of TILs typically gives a more diverse cell mixture. If the potentially contaminating cells in the TIL isolates release proteases, then the TIL T cells could appear to have reduced expression of signaling molecules. Although TILs were found to be more defective in signaling molecule expression, PBL from tumor bearers also showed significant defects in some studies, suggesting that tumors also exert systemic effects on non-tumor-specific T cells.

As the preceding paragraphs demonstrate, the status of T-cell signaling molecules and T-cell responses in tumor-bearing individuals is not consistent from study to study. These differences may reflect technical differences used in the various studies, genuine heterogeneity between different tumors and/or individuals, and/or differences in T-cell populations. Regardless of the reasons for the discrepancies between the individual studies, it seems likely that at least some tumor-bearing individuals have reduced levels of proteins involved in signal transduction in T lymphocytes. Whether these reductions contribute to tumor growth by preventing T-cell activation and the development of effective antitumor immunity remains unclear.

As suggested by Zier and co-workers [35], introduction of cytokine genes into tumors and their subsequent inoculation into tumor-bearing individuals may restore T-cell signaling and promote the generation of antitumor immunity. Indeed, tumor cells transfected with IL-2, IL-4, GM-CSF, or IFN-γ (reviewed in Blankenstein *et al.* [44]) can cause regression of small established tumors in mice. However, other explanations for the effects of cytokine gene therapy are also possible. For example, as has been suggested by Pardoll and colleagues, GM-CSF may enhance tumor rejection by recruiting antigen presenting cells to the site and by facilitating their function [45]. Additional studies are clearly necessary to determine if the mechanism of cytokine gene therapy is via restoration of T-cell signaling or by other mechanisms.

B. A Number of Tumor-Associated Peptides Have Been Identified That Are Candidate Targets for Tumor-Specific T Lymphocytes

If tumor cells are to be targets for tumor-specific T lymphocytes, then the tumor cells must express target antigens that are recognized by T cells. Recent studies have identified a variety of such tumor antigens. The antigens fall into several categories, including (1) normal (i.e., unmutated) self-antigens expressed by tumor cells and minimally expressed by some normal cells, (2) tissue-specific differentiation antigens expressed by both tumor cells and normal cells that may be overexpressed by malignant cells, (3) mutated self-proteins unique to tumor cells, and (4) oncogenic proteins inappropriately expressed by tumor cells. Although the antigens in each of these categories are potential targets for activating tumor-specific T cells, it is not yet clear if CTLs to these epitopes can cause tumor rejection.

1. NORMAL (UNMUTATED) SELF-ANTIGENS EXPRESSED BY TUMOR CELLS AND MINIMALLY EXPRESSED BY SOME NORMAL CELLS

Typically, this class of tumor antigen is a nonmutated self-molecule that is expressed on tumor cells but also on a very limited range of normal, nonmalignant cells. The first antigens characterized were from human melanoma cells. Human CTLs from melanoma bearing patients were used to identify antigens from melanoma antigen loss variants that had been transfected with genomic melanoma DNA [46]. These and other studies (reviewed in Van den Eynde and Van der Bruggen [47], Van den Eynde and Boon [48], and Robbins and Kawakami [49]) identified self-antigens that are shared among many melanomas, as well as being expressed on other types of tumors. Subsequent studies also using CTLs have identified additional normal antigens from other tumor types. Because all of these studies used CTLs as the readout for antigenicity, these antigens are MHC class I restricted. Table 1 lists the antigens identified to date and shows their distribution on normal tissue and tumor cells.

2. TISSUE-SPECIFIC DIFFERENTIATION ANTIGENS EXPRESSED BY BOTH TUMOR CELLS AND NORMAL CELLS THAT MAY BE OVEREXPRESSED BY MALIGNANT CELLS

This group of antigens is shared between tumor cells and their normal cellular counterparts, and it was identified because CTLs reacted with both the tumor cells and normal cells. Representative antigens of this class are listed in Table 2 and include antigens such as carcinoembryonic antigen (CEA), which is shared between

TABLE 1   Self-Antigens Shared among Tumors and Expressed at Minimal Levels
on Some Normal Cells[a]

Antigen	Primary tumor[b]	Other tumors[b,c]	Normal cells	References
MAGE-1	Melanoma[c]	A,B,F,G,H,L,O,Q	Testis,[d] trophoblast	96,97
MAGE-3	Melanoma	A,B,C,H,M,Q	Testis,[d]	98–100
BAGE	Melanoma	C	Testis,[d] trophoblast	101
GAGE 1,2	Melanoma	A,B,D,L,Q	Testis,[e] trophoblast	102
RAGE-1	Renal carcinoma	D,R,S	Retina[d]	103
MUC-1[e]	Breast	J,L,P	Lactating breast[f]	104–106

[a] Table based on data from Van den Eynde and Van der Bruggen [47], Van den Eynde and Boon [48], and Restifo [107].

[b] ≥25% of tumors tested express antigen.

[c] A, NSCLC; B, Head and neck cancers; C, bladder; D, sarcoma; E, mammary carcinoma; F, gastric carcinoma; G, hepatocellular carcinoma; H, esophogeal carcinoma; I, prostatic carcinoma; J, colon carcinoma; K, renal cell carcinoma; L, ovarian carcinoma; M, neuroblastoma; N, leukemias and lymphomas; O, T cell leukemias; P, pancreas; Q, testicular seminoma; R, melanoma; S, mesothelioma; T, small cell lung cancer.

[d] These tissues do not express MHC class I molecules, so antigen is not presented.

[e] Underglycosylated mucin.

[f] Lactating breast is not accessible to T cells, so antigen may not be seen by the immune system.

carcinomas of the gut and normal colon, and immunoglobulin idiotypes, which are shared between malignant B cell lymphoma and normal B lymphocytes. Several melanocyte differentiation antigens (e.g., tyrosinase, MART 100) are also in this category.

Because these molecules are common and expressed at high levels by normal tissue, any immunotherapy strategies targeting these antigens must take into account the possibility of reactivity to normal tissues. Theoretically it would be undesirable if tumor rejection were accompanied by autoimmunity against normal tissue. However, it has been reported that prolonged survival and tumor regression in melanoma patients treated with immunotherapy targeting a common antigen is accompanied by transient vitiligo, an autoimmune response to normal melanocytes [50]. The potential danger of concomitant autoimmunity, therefore, is unclear, and may depend on the distribution and function of the targeted normal tissue.

3. MUTATED SELF-PROTEINS UNIQUE TO TUMOR CELLS

This class of tumor antigens includes mutated self-proteins, some of which are unique to individual tumors, and some of which have been found on several tumors. Some of these molecules, such as mutated β-catenin and CASP-8 are likely to be involved in the transformation process either as stimulators of cell proliferation or as inhibitors of apoptosis [51–54].

The Ras oncogene and the p53 tumor suppressor gene are also frequently mutated in many cancers, and several mutations of these genes are shared among tu-

TABLE 2   Tumor Antigens That Are Differentiation Antigens[a]

Antigen	Primary tumor[b]	Other tumors[b,c]	Normal cells	References
Tyrosinase	Melanoma		Melanocytes	108–111
CEA	Colon, GI malignancies	E, T	Colon	112
Ig idiotype	B cell lymphoma		B lymphocytes	113,114
gp 100	Melanoma		Melanocytes, retina[d]	115,116
melan A/Mart1	Melanoma		Melanocytes, retina[d]	117–119
gp75/TRP-1	Melanoma			120
PSA	Prostate		Prostate	121,122

[a] Table based on data of Van den Eynde and Van der Bruggen [47] and Restifo [107].

[b,c,d] See footnotes for Table 1.

mors. For some of these mutations CTLs can be generated that are cytotoxic for wild-type tumor cells expressing the corresponding antigen [55–57]; however, in other cases the CTLs are not cytotoxic for wild-type tumor cells [58,59]. Although some of the p53 mutations are widely expressed on human tumors, many mutated tumor antigens are restricted in their expression to individual or a small subset of tumors. Immunotherapeutic strategies targeting these antigens would therefore have to be customized to individual patients. hence, the utility of these antigens as generalized immunotherapeutic targets is unclear and may be limited.

4. ONCOGENIC PROTEINS INAPPROPRIATELY EXPRESSED AND OVEREXPRESSED BY TUMOR CELLS

This class of tumor antigens includes nonmutated oncogenic molecules whose expression is at least partially responsible for the tumor's malignant phenotype. These proteins are frequently growth factors or growth factor receptors, and they can be expressed at low levels by normal cells, sometimes only during certain stages of differentiation, but are constitutively expressed, usually at high levels, by particular malignant cells. Classical examples of these molecules include Her-2/neu, a growth factor receptor overexpressed in approximately 30% of breast and ovarian cancers as well as numerous other adenocarcinomas [60], and epidermal growth factor receptor (EGFR), a cell surface molecule also expressed by breast carcinoma cells [61]. Spontaneous anti-Her-2/neu T-cell responses have been detected in some patients with Her-2/neu-expressing cancers [60], and animal studies have demonstrated that immunity to Her-2/neu can be induced by immunization with selected peptides derived from the Her-2/neu protein [62]. Antibody responses to Her-2/neu have also been noted in patients [60], and animal studies using adoptive transfer of anti-Her-2/neu mAbs into Her-2/neu transgenic mice significantly reduced the onset of breast cancer [63].

As target antigens for immunotherapy, nonmutated oncogene products have a major advantage: Immunotherapy frequently selects for tumor antigen loss variants that are resistant to the host's immune response. If an overexpressed oncogenic protein is the target antigen, its loss would produce a less malignant or nonmalignant tumor cell [64]. Immunity to these self-proteins is therefore inducible, and because of their relationship to the malignant phenotype, such molecules may be desirable target antigens for immunotherapy.

5. VIRAL ANTIGENS

Viruses are clearly associated with some human tumors, and antigens expressed by the viruses may serve as target molecules for immunotherapy. The most prom-

inent association is between three of the human papilloma virus strains (HPV 16, 18, and 45), which are found in more than 90% of human uterine and cervical tumors. Because the E6 and E7 proteins are essential to maintain the transformed state, these molecules are favored as targets for immunotherapy [65]. Vaccines are currently under development for prophylactic administration prior to HPV infection, as well as for therapeutic treatment after HPV infection and viral integration [65,66].

V. IMMUNOTHERAPEUTIC STRATEGIES FOR THE TREATMENT OF CANCER

New immunotherapy strategies using gene therapy are rapidly being developed based on an improved understanding of the immune response and the availability of sophisticated gene manipulation techniques. Immunotherapy is unlikely to replace existing treatments that are curative (e.g., successful surgical resections), although it may significantly contribute to the treatment of cancers that are refractile to conventional treatments. Disseminated, metastatic lesions, which are largely unresponsive to existing treatments, may be particularly amenable to immunotherapy because of the systemic scope of the immune response. As novel strategies are designed and implemented, however, various questions must be considered.

A. Questions to Consider When Designing Immunotherapy Strategies for the Treatment of Cancer

1. *What type of immunity is desired?* Immunotherapy can be directed at activating antibody responses and/or cell-mediated responses. Because the different arms of the immune response require selective activation of different lymphocyte subsets, any immunotherapy must be targeted to the appropriate cell population. If antibody-mediated immunity is desired, then activation of Th2 helper T lymphocytes should be targeted, whereas if cell-mediated immunity is desired, then activation of Th1 helper T cells should be targeted [67].

2. *Is the patient immunosuppressed or tolerant to autologous tumor?* Some tumor-bearing patients are systemically immunosuppressed because of their tumors, whereas other patients may be immunocompromised because of chemotherapy and/or radiation therapy treatments. If tumor burden is the reason for immunosuppression, then

immunotherapy may only be warranted once tumor burden is reduced by conventional treatments.

Alternatively, immunocompetent patients may be actively tolerized to their tumors. Any decrease in immunocompetence or increase in tolerance will obviously complicate immunotherapy and initiation of immunotherapy treatments may need to be phased with respect to other therapies (i.e., radiation and chemotherapy) that potentially compromise the induction of immunity.

3. *Is the goal of the immunotherapy short-term tumor rejection or prolonged antitumor immunity?* The goal of the immunotherapy may dictate the type of immunity to be induced. If the role of immunotherapy is to eliminate residual primary or metastatic tumor that is not immediately accessible to conventional treatments, then a potent short-term response should be targeted. Alternatively, if the immunotherapy goal is to protect against outgrowth of micrometastases or recurrence of primary tumor, then long-term immunologic memory should be induced.

4. *Is the tumor antigen known?* If a particular tumor antigen or peptide has been previously implicated in tumor rejection, then the antigen or peptide could be the immunizing agent. However, relatively few tumor antigens have been identified, and none of these have definitively been shown to be efficacious targets for immunotherapy. The use of multiple tumor antigens/peptides, therefore, might improve the chances of successful immunotherapy. Immunization with multiple epitopes may also overcome the *in vivo* selection of antigen loss variants because it is less likely that single mutations could result in loss of all targeted tumor antigens.

5. *Will optimal immunotherapy require "customized" therapy for each patient, or will a generic, tumor type of immunotherapy be used?* If shared tumor antigens like those described in Tables 1, 2, and 3 are effective immunotherapy targets, then generic antitumor vaccines may be feasible. Patients could be typed for their MHC alleles and the same allele-specific tumor peptides used for patients sharing the same type of tumor and HLA alleles. However, if immunotherapy is most effective when the target antigens are mutated self-proteins (antigens in Table 4) or other tumor antigens unique to individuals, then therapy may need to be customized for each patient.

6. *How will the therapeutic genes be delivered in the immunotherapy setting?* A major problem for gene therapy is the delivery method for the therapeutic

genes. At present effective gene delivery methods for long-term, stable gene expression are very limited. Likewise, methods for delivering genes to selected locales in the body are limited. Although they are a problem for gene therapy of genetic diseases, these limitations may not be problematic for immunotherapy of cancer. If the immunotherapy goal is activation of the host's immune response, then transient rather than long-term, sustained gene expression may be sufficient.

B. A Variety of Novel Immunotherapeutic Strategies for the Treatment of Cancer Are Currently Being Studied and Developed

This final section briefly reviews some of the novel immunotherapeutic strategies currently in development. Many of these approaches and additional strategies are discussed in detail in the following chapters of this book.

1. *Immunization with tumor-cell-based vaccines.* Both autologous (syngeneic) and allogeneic tumor cells have been transfected/transduced with a variety of different genes and used as cell-based vaccines. This approach is based on the hypothesis that the genetically modified tumor cell can function as an APC and directly activate CD8[+] and/or CD4[+] T lymphocytes if it expresses a tumor regression antigen and the requisite MHC class I and/or class II molecules. Because the transfected/transduced genes are not the target molecules for the induced immune response, vaccination with the genetically modified tumor cells should induce immunity against the wild-type, unmodified tumor. Genes encoding the following molecules have been used: (1) Costimulatory molecules, such as B7.1 and B7.2, whose expression provides the "second

TABLE 3 Tumor Antigens That Are Oncopeptides Overexpressed on Tumors[a]

Antigen	Initial tumor[b]	Additional tumors[b,c]	References
her2/neu	Mammary carcinoma, ovary	L	123,124
p53 (wild type)	Squamous cell carcinoma	Many	125

[a] Table based on data from Van den Eynde and Van der Bruggen [47] and Restifo [107].
[b,c] See footnotes for Table 1.

TABLE 4   Tumor Antigens That Are Mutated Self-Proteins[a]

Antigen	Initial tumor[b]	Additional tumors[b,c]	References
CDK4	Melanoma		126
β-catenin	Melanoma		52,127
CASP-8	Squamous cell carcinoma		53
p53	Many	Many	57
ras	Colon, lung, pancreas	~30% all malignancies	55,56
MUM-1	Melanoma		128
bcr/abl[d]	Chronic myelogenous leukemia		129

[a] Table based on data from Van den Eynde and Van der Bruggen [47] and Restifo [107].
[b,c] See footnotes for Table 1.
[d] Chimeric protein from a chromosomal translocation.

signal" for T-cell activation (reviewed by Blankenstein *et al.* [44] and Musiani *et al.* [68]); (2) cytokines, such as IL-2, IL-3, IL-4, IL-6, IL-7, IL-10, IL-12, GM-CSF, TNF, and IFN-γ, whose expression should facilitate differentiation/ activation of effector cells requiring specific cytokines for maturation [44,68,69]; (3) allogeneic MHC class I molecules, such as HLA-B7, whose expression induces immunity to the alloantigen that is cross-reactive on the wild-type tumor cells [70]. (4) Syngeneic MHC class II molecules, whose expression enables the tumor cell to directly present antigen to CD4+ T lymphocytes and stimulate Th lymphocyte activation [17,71]. Although this approach is based on the hypothesis that the genetically modified tumor cell functions as the APC, mechanistic studies show that in some situations host cells are the APC [45,72–74], whereas in other situations the genetically modified cells are APCs [71,73,75].

2. *Immunization with tumor peptides.* As listed in Tables 1 to 4, a variety of MHC class I–restricted tumor antigens/peptides that are putative regression antigens have been identified. Experiments in both animal systems and patients are in progress to determine if these molecules are tumor regression antigens. The overall strategy for these experiments is to immunize with the antigen/ peptide and determine if antitumor immunity is induced and/or tumor regression occurs [76]. Two immunization strategies are being tested: (1) Immunization with antigen/peptide either alone or in adjuvant [77]; (2) immunization with peptide-pulsed professional APCs such as DCs [78–83]. Many of these studies have shown tumor regression in animal systems, but in one study peptide vaccination led to enhanced tumor growth

by induction of tumor-specific immunologic tolerance [84].

3. *Immunization with DNA or RNA encoding tumor antigen/peptide.* As an alternative to immunization with protein, DNA encoding tumor antigen/ peptide has been used. The DNA can be introduced either via direct injection into the target tissue [85,86] or following immunization with DCs that are stably transfected with tumor antigen encoding genes [87]. Several types of viruses, including poxvirus [88,89], adenovirus [90,91], and retrovirus [92] have been used to insert tumor antigen genes into DC. In another strategy DCs have been pulsed with tumor cell RNA and used as immunogens. This approach has the advantages that it is not necessary to characterize a relevant tumor antigen(s) and that only a small number of autologous tumor cells is necessary to provide the required RNA [93].

4. *Enhancement of antitumor immunity by antibody treatment.* Improved understanding of T-cell activation has led to two novel antibody treatment approaches for enhancing antitumor immunity in the tumor-bearing host. One approach involves blocking the inhibitory receptor, CTLA4, on T cells. T lymphocytes express two counterreceptors for B7.1 and/or B7.2: CD28 and CTLA4. Interaction of B7.1/B7.2 with CTLA4 leads to T-cell anergy or apoptosis, whereas interaction with CD28 results in T-cell activation. Administration of anti-CTLA4 mAbs to tumor-bearing mice induces a potent antitumor immunity, presumably by blocking the inhibitory pathway, and favoring the T-cell activation pathway [22]. In a second approach, mAbs have been used to directly activate tumor-reactive T cells. 4-1BB is the receptor on T lymphocytes for the costimulatory

molecule 4-1BB ligand (4-1BBL). Tumor-bearing mice given anti-4-1BB mAbs reject their tumors, presumably by providing the second signal necessary for T-cell activation [94].

5. *Adoptive immunotherapy with tumor-specific T lymphocytes.* Adoptive transfer of *in vitro* activated tumor-specific T lymphocytes into tumor-bearing individuals has been shown to mediate tumor rejection in animal systems [23]. With the characterization of human tumor antigens, it has become feasible to antigen stimulate autologous human T cells *in vitro* and reinfuse them into the tumor-bearing patient [76]. Because various studies have demonstrated that maximal antitumor responses occur when both CD4$^+$ and CD8$^+$ T cells are generated, optimal immunity will probably involve adoptive transfer of both lymphocyte populations [95].

VI. CONCLUSIONS

Recent advances in understanding T-cell activation and in defining tumor antigens suggest a variety of novel approaches for using immunotherapy in the treatment of cancer. Technical advances in molecular biology and gene therapy have made these approaches feasible. Many of these novel immunotherapeutic strategies have shown efficacy in animal studies, but their ultimate clinical use awaits their testing in clinical situations. Patients participating in phase I and phase II clinical trials typically have advanced disease and high tumor burden, and it is unlikely that immunotherapy will be effective in this setting. A more accurate assessment of the efficacy of immunotherapy, therefore, may only be made when patients with more moderate tumor burden or less advanced disease are treated.

References

1. Pai, L., and Pastan, I. (1997). "Immunotoxin Therapy." In Cancer: Principles & Practice of Oncology (V. DeVita, S. Hellman, and S. Rosenberg, Eds.), 5th ed., pp. 3045–3057. Lippincott–Raven, Philadelphia.
2. Scott, A., and Welt, S. (1997). Antibody-based immunological therapies. *Curr. Opin. Immunol.* **9,** 717–722.
3. Scott, A., and Cebon, J. (1997). Clinical promise of tumor immunology. *Lancet* **349,** 19–22.
4. Van de Winkel, J., Bast, B., and de Gast, G. (1997). Immunotherapeutic potential of bispecific antibodies. *Immunol. Today* **18,** 562–564.
5. Sahin, U., Tureci, O., Schmitt, H., Cochlovius, B., Johannes, T., Schmits, R., Stenner, F., Luo, G., Schobert, I., and Pfreundschuh, M. (1995). Human neoplasms elicit multiple immune responses in the autologous host. *Proc. Natl. Acad. Sci. USA* **92,** 11810–11813.
6. Sahin, U., Tureci, O., and Pfreundschuh, M. (1997). Serological identification of human tumor antigens. *Curr. Opin. Immunol.* **9,** 709–716.
7. Brass, N., Heckel, D., Sahin, U., Pfreundschuh, M., Sybrecht, G., and Meese, E. (1997). Translation initiation factor eIF-4gamma is encoded by an amplified gene and induces an immune response in squamous cell lung carcinoma. *Hum. Mol. Genet.* **6,** 33–39.
8. Tureci, O., Sahin, U., Schobert, I., Koslowski, M., Schmitt, H., Schild, H., Stenner, F., Seitz, G., and Rammensee, H. (1996). The SSX2 gene, which is involved in the t(X,18) translocation of synovial sarcomas, codes for the human tumor antigen HOM-Mel-40. *Cancer Res.* **56,** 4766–4772.
9. Bentley, G., and Mariuzza, R. (1996). The structure of the T cell antigen receptor. *Annu. Rev. Immunol.* **14,** 591–618.
10. Cantrell, D. (1996). T cell antigen receptor signal transduction pathways. *Annu. Rev. Immunol.* **14,** 259–274.
11. York, I., and Rock, K. (1996). Antigen processing and presentation by the class I major histocompatibility complex. *Annu. Rev. Immunol.* **14,** 369–396.
12. Lanzavecchia, A. (1996). Mechanisms of antigen uptake for presentation. *Curr. Opin. Immunol.* **8,** 348–354.
13. Watts, C. (1997). Capture and processing of exogenous antigens for presentation on MHC molecules. *Annu. Rev. Immunol.* **15,** 821–850.
14. Chambers, C., and Allison, J. (1997). Co-stimulation in T cell responses. *Curr. Opin. Immunol.* **9,** 396–404.
15. Lenschow, D., Walunas, T., and Bluestone, J. (1996). CD28/B7 system of T cell costimulation. *Annu. Rev. Immunol.* **14,** 233–258.
16. Liu, Y., and Janeway, C. (1992). Cells that present both specific ligand and costimulatory activity are the most efficient inducers of clonal expansion of normal CD4 T cells. *Proc. Natl. Acad. Sci. USA* **89,** 3845–3849.
17. Baskar, S., Glimcher, L., Nabavi, N., Jones, R. T., and Ostrand-Rosenberg, S. (1995). Major histocompatibility complex class II$^+$B7-1$^+$ tumor cells are potent vaccines for stimulating tumor rejection in tumor-bearing mice. *J. Exp. Med.* **181,** 619–629.
18. Chen, L., Ashe, S., Brady, W. A., Hellstrom, I., Hellstrom, K. E., Ledbetter, J. A., McGowan, P., and Linsley, P. S. (1992). Costimulation of antitumor immunity by the B7 counterreceptor for the T lymphocyte molecules CD28 and CTLA-4. *Cell* **71,** 1093–1102.
19. Townsend, S. E., and Allison, J. P. (1993). Tumor rejection after direct costimulation of CD8+ T cells by B7-transfected melanoma cells. *Science* **259,** 368–370.
20. Harding, F. A., and Allison, J. P. (1993). CD28-B7 interactions allow the induction of CD8+ cytotoxic T lymphocytes in the absence of exogenous help. *J. Exp. Med.* **177,** 1791–1796.
21. Baskar, S., Ostrand-Rosenberg, S., Nabavi, N., Nadler, L. M., Freeman, G. J., and Glimcher, L. H. (1993). Constitutive expression of B7 restores immunogenicity of tumor cells expressing truncated major histocompatibility complex class II molecules. *Proc. Natl. Acad. Sci. USA* **90,** 5687–5690.
22. Leach, D., Krummel, M., and Allison, J. (1996). Enhancement of antitumor immunity by CTLA-4 blockade. *Science* **271,** 1734–1736.
23. Greenberg, P. (1991). Adoptive T cell therapy of tumors: mechanisms operative in the recognition and elimination of tumor cells. *Adv. Immunol.* **49,** 281–355.
24. Kern, D., Klarnet, J., Jensen, M., and Greenberg, P. (1986). Requirement for recognition of class II molecules and processed tumor antigen for optimal generation of syngeneic tumor-specific class I-restricted CTL. *J. Immunol.* **136,** 4303–4310.
25. Schultz, K., Klarnet, J., Gieni, R., Hayglass, K., and Greenberg, P. (1990). The role of B cells for *in vivo* T cell responses to a friend virus-induced leukemia. *Science* **249,** 921–923.

26. Ostrand-Rosenberg, S., Thakur, A., and Clements, V. (1990). Rejection of mouse sarcoma cells after transfection of MHC class II genes. *J. Immunol.* **144**, 4068–4071.

27. Perdrizet, G., Ross, S., Stauss, H., Singh, S., Koeppen, H., and Schreiber, H. (1990). Animals bearing malignant grafts reject normal grafts that express through gene transfer the same antigen. *J. Exp. Med.* **171**, 1205–1220.

28. Wick, M., Dubey, P., Koeppen, H., Siegel, C., Fields, P., Chen, L., Bluestone, J., and Schreiber, H. (1997). Antigenic cancer cells grow progressively in immune hosts without evidence for T cell exhaustion or systemic anergy. *J. Exp. Med.* **186**, 229–238.

29. Levey, D., and Srivastava, P. (1995). T cells from late tumor-bearing mice express normal levels of p56^{lck}, p59^{fyn}, ZAP-70, and CD3-zeta despite suppressed cytolytic activity. *J. Exp. Med.* **182**, 1029–1036.

30. Bursuker, I., and North, R. (1984). Generation and decay of the immune response to a progressive fibrosarcoma. II. Failure to demonstrate postexcision immunity after the onset of T cell–mediated suppression of immunity. *J. Exp. Med.* **159**, 1312–1321.

31. North, R., and Bursuker, I. (1984). Generation of decay of the immune response to a progressive fibrosarcoma. I. Ly-1^{+}2^{-} suppressor T cells down-regulate the generation of Ly-1^{-}2^{+} effector cells. *J. Exp. Med.* **159**, 1295–1231.

32. Mizoguchi, H., O'Shea, J., Longo, D., Loeffler, C., McVicar, D., and Ochoa, A. (1992). Alterations in signal transduction molecules in T lymphocytes from tumor-bearing mice. *Science* **258**, 1795–1798.

33. Nakagomi, H., Peterson, M., Magnusson, I., Jublin, C., Matsuda, M., Mellstedt, H., Taupin, J., Vivier, E., Anderson, P., and Kiessling, R. (1993). Decreased expression of the signal transducing zeta chains in tumor-infiltrating T cells and NK cells of patients with colorectal carcinoma. *Cancer Res.* **53**, 5613–5616.

34. Finke, J., Zea, A., Stanley, J., Longo, D., Mizoguchi, H., Tubbs, R., Wiltrout, R., O'Shea, J., Kudoh, S., Klein, E., and Ochoa, A. (1993). Loss of T cell receptor zeta chain and p56^{lck} in T cells infiltrating human renal cell carcinoma. *Cancer Res.* **53**, 5613–5616.

35. Zier, K., Gansbacher, B., and Salvadori, S. (1996). Preventing abnormalities in signal transduction of T cells in cancer: the promise of cytokine gene therapy. *Immunol. Today* **39**, 39–45.

36. Zea, A., Curti, B., and Longo, D. (1995). *Clin. Canc. Res.* **1**, 1327–1335.

37. Tartour, E., Latour, S., and Mathiot, C. (1995). *Int. J. Cancer* **63**, 205–212.

38. Aoe, T., Okamoto, Y., and Saito, T. (1995). Activated macrophages induce structural abnormalities of the T cell receptor-CD3 complex. *J. Exp. Med.* **181**, 1881–1886.

39. Matusuda, N., Petersson, M., and Lenkei, R. (1995). *Int. J. Cancer* **61**, 765–772.

40. Salvadori, S., Gansbacher, B., Pizzimenti, A., and Zier, K. (1994). Abnormal signal transduction by T cells of mice with parental tumors is not seen in mice bearing IL-2-secreting tumors. *J. Immunol.* **153**, 5176–5182.

41. Wang, Q., Stanley, J., Kudoh, S., Myles, J., Kolenko, V., Yi, T., Tubbs, R., Bukowski, R., and Finke, J. (1995). T cells infiltrating non-Hodgkin's B cell lymphomas show altered tyrosine phosphorylation pattern even though T cell receptor/CD3-associated kinases are present. *J. Immunol.* **155**, 1382–1392.

42. Franco, J., Gosh, P., and Wiltrout, R. (1995). *Cancer Res.* **55**, 3840–3846.

43. Levey, D., and Srivastava, P. (1996). Alterations in T cells of cancer-bearers: whence specificity? *Immunol. Today* **17**, 365–368.

44. Blankenstein, T., Cayeux, S., and Qin, Z. (1996). Genetic approaches to cancer immunotherapy. *Rev. Phys. Biochem. Pharm.* **129**, 1–49.

45. Huang, A., Golumbek, P., Ahmadzadeh, M., Jaffee, E., Pardoll, D., and Levitsky, H. (1994). Role of bone marrow–derived cells in presenting MHC class I–restricted tumor antigens. *Science* **264**, 961–965.

46. Van der Bruggen, P., Traversari, C., Chomez, P., Lurquin, C., De Plaen, E., Van den Eynde, B., Knuth, A., and Boon, T. (1991). A gene encoding an antigen recognized by cytolytic T lymphocytes on a human melanoma. *Science* **254**, 1643–1647.

47. Van den Eynde, B., and Van der Bruggen, P. (1997). T cell–defined tumor antigens. *Curr. Opin. Immunol.* **9**, 684–693.

48. Van den Eynde, B., and Boon, T. (1997). Tumor antigens recognized by T lymphocytes. *Int. J. Clin. Lab. Res.* **27**, 81–86.

49. Robbins, P., and Kawakami, Y. (1996). Human tumor antigens recognized by T cells. *Curr. Opin. Immunol.* **8**, 628–636.

50. Rosenberg, S., and White, D. (1996). Vitiligo in patients with melanoma: normal tissue antigens can be targets for cancer immunotherapy. *J. Immunotherapy* **19**, 81–84.

51. Peifer, M. (1997). Beta-catenin as oncogene: the smoking gun. *Science* **275**, 1752–1753.

52. Rubinfeld, B., Robbins, P., El-Gamil, M., Albert, I., Porfiri, E., and Polakis, P. (1997). Stabilization of beta-catenin by genetic defects in melanoma cell lines. *Science* **275**, 1790–1792.

53. Mandruzzato, S., Brasseur, F., Andry, G., Boon, T., and Van der Bruggen, P. (1997). A CASP-8 mutation recognized by cytolytic T lymphocytes on a human head and neck carcinoma. *J. Exp. Med.* **186**, 785–793.

54. Boon, T., and Old, L. (1997). Tumor antigens. *Curr. Opin. Immunol* **9**, 681–683.

55. Skipper, J., and Stauss, H. (1993). Identification of two cytotoxic T lymphocyte–recognized epitopes in the Ras protein. *J. Exp. Med.* **177**, 1493–1498.

56. Peace, D., Smith, J., Chen, W., You, S., Cosand, W., Blake, J., and Cheever, M. (1994). Lysis of ras oncogene–transformed cells by specific cytotoxic T lymphocytes elicited by primary *in vitro* immunization with mutated ras peptide. *J. Exp. Med.* **179**, 473–479.

57. Noguchi, Y., Chen, Y., and Old, L. (1994). A mouse mutant p53 product recognized by CD4^{+} and CD8^{+} T cells. *Proc. Natl. Acad. Sci. USA* **91**, 3171–3175.

58. Houbiers, J., Nijman, H., Van Der Burg, S., Drijfhout, J., Kenemans, P., Van De Velde, C., Brand, A., Momberg, F., Kast, W., and Melief, C. (1993). *In vitro* induction of human cytotoxic T lymphocyte responses against peptides of mutant and wild-type p53. *Eur. J. Immunol.* **23**, 2072–2077.

59. Elas, A., Nijman, H., Van Der Minne, C., Mourer, J., Kast, M., Melief, C., and Schrier, P. (1995). Induction and characterization of cytotoxic T lymphocytes recognizing a mutated p21 RAS peptide presented by HLA-A2010. *Int. J. Cancer* 389–396.

60. Cheever, M., Disis, M., Bernhard, H., Gralow, J., Hand, S., Huseby, E., Qin, H., Takahashi, M., and Chen, W. (1995). Immunity to oncogenic proteins. *Immunol. Rev.* **145**, 33–59.

61. Baselga, J., and Mendelsohn, J. (1994). The epidermal growth factor receptor as a target for therapy in breast carcinoma. *Breast Canc. Res. Treat.* **29**, 127–138.

62. Disis, M., Gralow, J., Bernhard, H., Hand, S., Rubin, W., and Cheever, M. (1996). Peptide-based, but now whole protein, vaccines elicit immunity to HER-2/neu, an oncogenic self-protein. *J. Immunol.* **156**, 3151–3158.

63. Katsumata, M., Okudaira, T., Samanta, A., Clark, D., Drebin, J., Jolicoeur, P., and Greene, M. (1995). Prevention of breast

tumour development *in vivo* by downregulation of the p185neu receptor. *Nature Med.* **1,** 644–648.

64. Disis, M., and Cheever, M. (1996). Oncogenic proteins as tumor antigens. *Curr. Opin. Immunol.* **8,** 637–642.

65. Tindle, R. (1997). Human papillomavirus vaccines for cervical cancer. *Curr. Opin. Immunol.* **8,** 643–650.

66. Christensen, N., Cladel, N., and Reed, C. (1995). Post-attachment neutralisation of papillomaviruses by monoclonal and polyclonal antibodies. *Virology* **201,** 136–142.

67. Constant, S., and Bottomly, K. (1997). Induction of the TH1 and TH2 CD4⁺ T cell responses: alternative approaches. *Annu. Rev. Immunol.* **15,** 297–322.

68. Musiani, P., Modesti, A., Giovarelli, M., Cavallo, F., Colombo, M., Lollini, P., and Forni, G. (1997). Cytokines, tumour-cell death and immunogenicity: a question of choice. *Immunol. Today* **18,** 32–26.

69. Dranoff, G., Jaffee, E., Lazenby, A., Golumbek, P., Levitsky, H., Brose, K., Jackson, V., Hamada, H., Pardoll, D., and Mulligan, R. (1993). Vaccination with irradiated tumor cells engineered to secrete murine granulocyte-macrophage colony-stimulating factor stimulates potent, specific, and long-lasting anti-tumor immunity. *Proc. Natl. Acad. Sci. USA* **90,** 3539–3543.

70. Nabel, G., Gordon, D., Bishop, D., Nickoloff, B., Yang, Z., Aruga, A., Cameron, M., Nabel, E., and Chang, A. (1996). Immune response in human melanoma after transfer of an allogeneic class I major histocompatibility complex gene with DNA-liposome complexes. *Proc. Natl. Acad. Sci. USA* **93,** 15388–15393.

71. Armstrong, T., Pulaski, B., and Ostrand-Rosenberg, S. (1997). Tumor antigen presentation: changing the rules. *Canc. Immunol. Immunother.* **46,** 70–74.

72. Huang, A., Bruce, A., Pardoll, D., and Levitsky, H. (1996). Does B7-1 expression confer antigen-presenting cell capacity to tumors *in vivo*? *J. Exp. Med.* **183,** 769–776.

73. Cayeux, S., Richter, G., Noffz, G., Dorken, B., and Blankenstein, T. (1997). Influence of gene-modified (IL-7, IL-4, and B7) tumor cell vaccines on tumor antigen presentation. *J. Immunol.* **158,** 2834–2841.

74. Pulaski, B., Yeh, K., Shastri, N., Maltby, K., Penney, D., Lord, E., and Frelinger, J. (1996). IL-3 enhances CTL development and class I MHC presentation of exogenous antigen by tumor-infiltrating macrophages. *Proc. Natl. Acad. Sci. USA* **93,** 3669–3674.

75. Armstrong, T., Clements, V., and Ostrand-Rosenberg, S. (1998). MHC class II-transfected tumor cells directly present antigen to tumor-specific CD4⁺ T lymphocytes. *J. Immunol.* in press.

76. Rosenberg, S. (1997). Cancer vaccines based on the identification of genes encoding cancer regression antigens. *Immunol. Today* **18,** 175–182.

77. Melief, C., Offringa, R., Toes, R., and Kast, M. (1996). Peptide-based cancer vaccines. *Curr. Opin. Immunol.* **8,** 651–657.

78. Mukherji, B., Chakraborty, N., Yamasaki, S., Okino, T., Yamase, H., Sporn, J., Kurtzman, S., Ergin, M., Ozols, J., Meehan, J., and Mauri, F. (1995). Induction of antigen-specific cytolytic T cells in situ in human melanoma by immunization with synthetic peptide-pulsed autologous antigen presenting cells. *Proc. Natl. Acad. Sci. USA* **92,** 8078–8082.

79. Celluzzi, C., Mayordoma, J., Storkus, W., Lotze, M., and Falo Jr., L. (1996). Peptide-pulsed dendritic cells induce antigen-specific, CTL-mediated protective tumor immunity. *J. Exp. Med.* **183,** 283–287.

80. Zitvogel, L., Mayordomo, J., Tjandrawan, T., DeLeo, A., Clarke, M., Lotze, M., and Storkus, W. (1996). Therapy of murine tumors with tumor-peptide-pulsed dendritic cells: dependence on T cells,

B7 costimulation, and T helper cell 1-associated cytokines. *J. Exp. Med.* **183,** 87–97.

81. Paglia, P., Chiodoni, C., Rodolfo, M., and Colombo, M. (1996). Murine dendritic cells loaded *in vitro* with soluble protein prime cytotoxic T lymphocytes against tumor antigen *in vivo*. *J. Exp. Med.* **183,** 317–322.

82. Porgador, A., Snyder, D., and Gilboa, E. (1996). Induction of anti-tumor immunity using bone marrow-generated dendritic cells. *J. Immunol.* **156,** 2918–2926.

83. Nair, S., Snyder, D., Rouse, B., and Gilboa, E. (1996). Regression of tumors in mice vaccinated with professional antigen-presenting cells pulsed with tumor extracts. *Int. J. Cancer* **70,** 706–715.

84. Toes, R., Offringa, R., Blom, R., Melief, C., and Kast, M. (1996). Peptide vaccination can lead to enhanced tumor growth through specific T-cell tolerance induction. *Proc. Natl. Acad. Sci. USA* **93,** 7855–7860.

85. Irvine, K., Rao, R., Rosenberg, S., and Restifo, N. (1996). Cytokine enhancement of DNA immunization leads to effective treatment of established pulmonary metastases. *J. Immunol.* **156,** 238–245.

86. Doe, B., Selby, M., Barnett, S., Baenziger, J., and Walker, C. (1996). Induction of cytotoxic T lymphocytes by intramuscular immunization with plasmid DNA is facilitated by bone marrow–derived cells. *Proc. Natl. Acad. Sci. USA* **93,** 8578–8583.

87. Alijagic, S., Moller, P., Artuc, M., Jurgovsky, K., Czarnetzki, B., and Schadendorf, D. (1995). Dendritic cells generated from peripheral blood transfected with human tyrosinase induce specific T cell activation. *Eur. J. Immunol.* **25,** 3100–3107.

88. Restifo, N. (1996). The new vaccines: building viruses that elicit antitumor immunity. *Curr. Opin. Immunol.* **8,** 658–663.

89. Bronte, V., Charroll, M., Goletz, T., Wang, M., Overwijk, W., Marincola, F., Rosenberg, S., Moss, B., and Restifo, N. (1997). Antigen expression by dendritic cells correlates with the therapeutic effectiveness of a model recombinant poxvirus tumor vaccine. *Proc. Natl. Acad. Sci. USA* **94,** 3183–3188.

90. Arthur, J., Butterfield, L., Kiertscher, S., Roth, M., Bui, L., Lau, R., Dubinett, S., Glaspy, J., and Economou, J. (1997). A comparison of gene transfer methods in human dendritic cells. *Cancer Gene Ther.* **4,** 17–25.

91. Ribas, A., Butterfield, L., McBride, W., Jilani, S., Bui, L., Vollmer, C., Lau, R., Dissette, V., Hu, B., Chen, A., Glaspy, J., and Economou, J. (1997). Genetic immunization for the melanoma antigen MART-1/Melan-A using recombinant adenovirus-transduced murine dendritic cells. *Cancer Res.* **57,** 2865–2869.

92. Reeves, M., Royalo, R., Lam, J., Rosenberg, S., and Hwu, P. (1996). Retroviral transduction of human dendritic cells with a tumor-associated antigen gene. *Cancer Res.* **56,** 5672–5677.

93. Boczkowski, D., Nair, S., Snyder, D., and Gilboa, E. (1996). Dendritic cells pulsed with RNA are potent antigen-presenting cells *in vitro* and *in vivo*. *J. Exp. Med.* **184,** 465–472.

94. Melero, I., Shuford, W., Newby, S., Aruffo, A., Ledbetter, J., Hellstrom, K., Mittler, R., and Chen, L. (1997). Monoclonal antibodies against the 4-1BB T-cell activation molecule eradicate established tumors. *Nature Medicine* **3,** 682–685.

95. Yee, C., Gilbert, M., Riddell, S., Brichard, V., Fefer, A., Thompsom, J., Boon, T., and Greenberg, P. (1997). *J. Immunol.* **157,** 4079–4086.

96. Traversari, C., Van der Bruggen, P., Luescher, I., Lurquin, C., Chomez, P., Van Pel, A., De Plaen, E., Amar-Costesec, A., and Boon, T. (1992). A nonapeptide encoded by human gene MAGE-1 is recognized on HLA-A1 by cytolytic T lymphocytes directed against tumor antigen MZ2-E. *J. Exp. Med.* **176,** 1453–1457.

97. Van der Bruggen, P., Szikora, J., Boel, P., Wildmann, C., Somville, M., Sensi, M., and Boon, T. (1994). Autologous cytolytic T lymphocytes recognize a MAGE-1 nonapeptide on melanomas expressing HLA-Cw*1601. *Eur. J. Immunol.* **24**, 2134–2140.

98. Gaugler, B., Van den Eynde, B., Van der Bruggen, P., Romero, P., Gaforio, J., De Plaen, E., Lethe, B., Brasseur, F., and Boon, T. (1994). Human gene MAGE-3 codes for an antigen recognized on a melanoma by autologous cytolytic T lymphocytes. *J. Exp. Med.* **179**, 921–930.

99. Herman, J., Van der Bruggen, P., Luescher, I., Mandruzzato, S., Romero, P., Thonnard, J., Fleischhauer, K., Boon, T., and Coulie, P. (1996). A peptide encoded by human gene MAGE-3 and presented by HLA-B44 induces cytolytic T lymphocytes that recognize tumor cells expressing MAGE-3. *Immunogenet.* **43**, 377–383.

100. Van der Bruggen, P., Bastin, J., Gajewski, T., Coulie, P., Boel, P., De Smet, C., Traversari, C., Townsend, A., and Boon, T. (1994). A peptide encoded by human gene MAGE-3 and presented by HLA-A2 induces cytolytic T lymphocytes that recognize tumor cells expressing MAGE-3. *Eur. J. Immunol.* **24**, 3038–3043.

101. Boel, P., Wildmann, C., Sensi, M., Brasseur, R., Renauld, J., Coulie, P., Boon, T., and Van der Bruggen, P. (1995). BAGE, a new gene encoding an antigen recognized on human melanomas by cytolytic T lymphocytes. *Immunity* **2**, 167–175.

102. Van den Eynde, B., Peeters, O., De Backer, O., Gaugler, B., Lucas, S., and Boon, T. (1995). A new family of genes coding for an antigen recognized by autologous cytolytic T lymphocytes on a human melanoma. *J. Exp. Med.* **182**, 689–698.

103. Gaugler, B., Brouwenstijn, N., Vantomme, V., Szikora, J.-P., Van der Spek, C., Patard, J., Boon, T., Schrier, P., and Van den Eynde, B. (1996). A new gene coding for an antigen recognized by autologous cytolytic T lymphocytes on a human renal carcinoma. *Immunogenet.* **44**, 323–330.

104. Jerome, K., Wakefield, D., and Watkins, S. (1988). Tumor-specific cytotoxic T cell clones from patients with breast and pancreatic adenocarcinoma recognize EBV-immortalized B cells transfected with polymorphic epithelial mucin cDNA. *J. Immunol.* **151**, 1654–1662.

105. Barratt-Boyes, S. (1996). Making the most of mucin: a novel target for tumor immunotherapy. *Cancer Immunol. Immunother.* **43**, 142–151.

106. Finn, O., Jerome, K., Henderson, A., Pecher, G., Domenech, N., Magarian-Blander, J., and Barratt-Boyes, S. (1995). MUC1 epithelial tumor mucin-based immunity and cancer vaccines. *Immunol. Rev.* **87**, 982–990.

107. Restifo, N. (1997). "Cancer Vaccines." In Cancer: Principles & Practice of Oncology (V. DeVita, S. Hellman, and S. Rosenberg, Eds.), 5th ed., pp. 3023–3043. Lippincott–Raven, Philadelphia.

108. Topalian, S., Gonzales, M., Parkhurst, M., Li, Y., Southwood, S., Sette, A., Rosenberg, S., and Robbins, P. (1996). Melanoma-specific CD4$^+$ cells recognize nonmutated HLA-DR restricted tyrosinase epitopes. *J. Exp. Med.* **183**, 1965–1971.

109. Wolfel, T., Van Pel, A., Brichard, V., Schneider, J., Seliger, B., Meyer zum Buschenfelde, K., and Boon, T. (1994). Two tyrosinase nonapeptides recognized on HLA-A2 melanomas by autologous cytolytic T lymphocytes. *Eur. J. Immunol.* **24**, 759–764.

110. Kang, X., Kawakami, Y., El-Gamil, M., Wang, R., Sakaguchi, K., Yannelli, J., Appella, E., Rosenberg, S., and Robbins, P. (1995). Identification of a tyrosinase epitope recognized by HLA-A24-restricted, tumor-infiltrating lymphocytes. *J. Immunol.* **155**, 1343–1348.

111. Brichard, V., Herman, J., Van Pel, A., Wildmann, C., Gaugler, B., Wolfel, T., Boon, T., and Lethe, B. (1996). A tyrosinase nonapeptide presented by HLA-B44 is recognized on a human melanoma by autologous cytolytic T lymphocytes. *Eur. J. Immunol.* **26**, 224–230.

112. Gold, P., and Freeman, S. (1965). Specific carcinomembryonic antigens of the human digestive system. *J. Exp. Med.* **122**, 467.

113. Hsu, F., Benike, C., Fagnoni, F., Liles, T., Czerwinski, D., Taidi, B., Engelman, E., and Levy, R. (1996). Vaccination of patients with B-cell lymphoma using autologous antigen-pulsed dendritic cells. *Nature Med.* **2**, 52–58.

114. Hsu, F., Casper, C., Czerwinski, D., Kwak, L., Liles, T., Syrengelas, A., Taidi-Laskowski, B., and Levy, R. (1997). Tumor-specific idiotype vaccines in the treatment of patients with B cell lymphoma–long-term results of a clinical trial. *Blood* **89**, 3129–3135.

115. Bakker, A., Schreurs, M., Tafazzul, G., De Boer, A., Kawakami, Y., Adema, G., and Figdor, C. (1995). Identification of a novel peptide derived from the melanocyte-specific gp100 antigen as the dominant epitope recognized by an HLA-A2.1-restricted anti-melanoma CTL line. *Int. J. Cancer* **62**, 97–102.

116. Kawakami, Y., Eliyahu, S., Jennings, C., Sakaguchi, K., Kang, X., Southwood, S., Robbins, P., Sette, A., Appella, E., and Rosenberg, S. (1995). Recognition of multiple epitopes in the human melanoma antigen gp100 by tumor-infiltrating T lymphocytes associated with *in vivo* tumor regression. *J. Immunol.* **154**, 3961–3968.

117. Kawakami, Y., Eliyahu, S., Sakaguchi, K., Robbins, P., Rivoltini, L., Yannelli, J., Appella, E., and Rosenberg, S. (1994). Identification of the immunodominant peptides of the MART-1 human melanoma antigen recognized by the majority of HLA-A2-restricted tumor infiltrating lymphocytes. *J. Exp. Med.* **180**, 347–352.

118. Castelli, C., Storkus, W., Maeurer, M., Martin, D., Huang, E., Pramanik, B., Nagabhushan, T., Parmiani, G., and Lotze, M. (1995). Mass spectrometric identification of a naturally processed melanoma peptide recognized by CD8$^+$ cytotoxic T lymphocytes. *J. Exp. Med.* **181**, 363–368.

119. Coulie, P., Brichard, V., Van Pel, A., Wolfel, T., Schneider, J., Traversari, C., Mattei, S., De Plaen, E., Lurquin, C., Szikora, J., Renauld, J., and Bood, T. (1994). A new gene coding for a differentiation antigen recognized by autologous cytolytic T lymphocytes on HLA-A2 melanomas. *J. Exp. Med.* **180**, 35–42.

120. Wang, R., Parkhurst, M., Kawakami, Y., Robbins, P., and Rosenberg, S. (1996). Utilization of an alternative open reading frame of a normal gene in generating a novel human cancer antigen. *J. Exp. Med.* **183**, 1131–1140.

121. Correale, P., Walmsley, K., Nieroda, C., Zaremba, S., Zhu, M., Schlom, J., and Tsang, K. (1997). *In vitro* generation of human cytotoxic T lymphocytes specific for peptides derived from prostate-specific antigen. *J. Natl. Cancer Inst.* **89**, 293–300.

122. Alexander, R., Brady, F., Leffell, M., Tsai, V., and Celis, E. (1998). Specific T cell recognition of peptides derived from prostate specific antigen in patients with prostatic cancer. *Urology* **51**, 150–157.

123. Peoples, G., Goedegebuure, P., Smith, R., Linehan, D., Yoshino, I., and Eberlein, T. (1995). Breast and ovarian cancer-specific cytotoxic T lymphocytes recognize the same HER-2/neu-derived peptide. *Proc. Natl. Acad. Sci. USA* **92**, 432–436.

124. Fisk, B., Blevins, T., Wharton, J., and Ionnides, C. (1995). Identification of an immunodominant peptide of HER2/enu protooncogene recognized by ovarian tumor-specific cytotoxic T lymphocyte lines. *J. Exp. Med.* **181**, 2109–2117.

125. Ropke, M., Hald, J., Guldberg, P., Zeuthen, J., Norgaard, L. F., Svejgaard, A., Van Der Burg, S., Nijman, M., Melief, C., and Claesson, M. (1996). Spontaneous human squamous cell carcinomas are killed by a human cytotoxic T lymphocyte clone recognizing a wild type p53-derived peptide. *Proc. Natl. Acad. Sci. USA* **93,** 14704–14707.

126. Wolfel, T., Hauer, M., Schneider, J., Serrano, M., Wolfel, C., Klehmann-Hieb, E., De Plaen, E., Hankeln, T., Meyer zum Buschenfelde, K., and Beach, D. (1995). A p16^{INK4a} insensitive CDK4 mutant targeted by cytolytic T lymphocytes in a human melanoma. *Science* **269,** 1281–1284.

127. Robbins, P., El-Gamil, M., Li, Y., Kawakami, Y., Loftus, D., Appella, E., and Rosenberg, S. (1996). A mutated beta-catenin gene encodes a melanoma-specific antigen recognized by tumor infiltrating lymphocytes. *J. Exp. Med.* **183,** 1185–1192.

128. Coulie, P., Lehmann, F., Lethe, B., Herman, J., Lurquin, C. A., M., and Boon, T. (1995). A mutated intron sequence codes for an antigenic peptide recognized by cytolytic T lymphocytes on a human melanoma. *Proc. Natl. Acad. Sci. USA* **92,** 7976–7980.

129. Ten Bosch, G., Joosten, A., Kessler, J., Melief, C., and Leeksma, O. (1996). Recognition of BCR-ABL positive leukemic blasts by human CD4$^+$ T cells elicited by primary *in vitro* immunization with a BCR-ABL breakpoint peptide. *Blood* **88,** 3522–3527.

PART II

Vectors and Engineered Products for Gene Therapy

CHAPTER 4

Retroviral Vector Design for Cancer Gene Therapy

CHRISTOPHER BAUM, WOLFRAM OSTERTAG, CAROL STOCKING, AND DOROTHEE VON LAER

Heinrich-Pette-Institut für Experimentelle Virologie und Immunologie an der Universität Hamburg, 20251 Hamburg, Germany

I. INTRODUCTION

In the past years, oncology was the center of gene therapy research [1]. However, despite generous support by, e.g., the National Institutes of Health and related institutions in Europe, there is still a wide gap between the hopes raised and the results achieved. Most of the failures of gene therapy trials can be attributed to a discordant combination of overinterpreted clinical concepts and immature technology, including poor vector design [2]. Nevertheless, many former skeptics were turned to true believers, partly because of the enormous public and economic interest [3]. Thus, strong international competition in the field was generated. Fortunately, many researchers have maintained their scientific integrity, following valuable long-term concepts and improving basic vector technology.

An ideal vector should (1) allow efficient and selective transduction of the target cell of interest, (2) be maintained and (3) expressed at levels necessary for achieving therapeutic effects, and, last but not least, (4) be safe in terms of avoiding unwanted side effects in the host. Viruses are a perfect tool for gene transfer as they have evolved to deliver their genome efficiently to target cells with subsequent high-level gene expression. Vector systems for therapeutic gene transfer have been developed from different virus groups, each system having specific advantages and drawbacks. Retroviruses have several unique features that render them highly suitable for vector development. Retroviral vectors are therefore the prevalent system for gene transfer in human cells. Retroviruses integrate and express their genome in a stable manner, thus allowing long-term manipulation with transferred genes. This is a prerequisite for many gene therapy applications, including some approaches in cancer gene therapy. Integration usually does not alter host cell functions and is well tolerated. In the retroviral genome, *cis*-acting elements, responsible for reverse transcription, integration, and packaging, can be well separated from coding sequences. Such a genome structure facilitates the design of safe vectors and packaging cell lines.

However, with the transition to applications in human gene therapy, severe limitations of conventional retroviral vector systems have become apparent. These include low and variable particle titers, lack of appropriate vector targeting to specific cell types and genomic loci, failure to transduce quiescent cells, and relatively inefficient, position-dependent transcription. Fortunately, substantial progress in vector development has been made, based on deeper understanding of the biology of retroviruses and target cells. Here, we review some of the work relevant to cancer gene therapy.

We start with a short overview of potential applications of retroviral vectors in oncology. Then, we describe aspects of retrovirus biology relevant to gene therapy, which creates the basis for discussing principles of and specific recent advances in retroviral vector design.

II. APPLICATIONS FOR RETROVIRAL VECTORS IN ONCOLOGY

In oncology, several different strategies involving somatic gene transfer are currently considered (Table 1). We can distinguish between diagnostic and therapeutic approaches; in either case, both healthy tissues and tumor cells may be targeted. Each strategy has special implications for vector design.

Gene marking uses stable retroviral transduction of heterologous genetic sequences to analyze the biological (stem cell function, antiviral effects) or pathogenic (tumor cell contamination, graft-versus-host reaction) capacity of blood cell transplants [4–6]. Here, efficient transduction of long-lived hematopoietic cells is required. Moreover, long-term transgene expression is necessary for follow-up analyses involving phenotyping and preparative sorting of transduced cells, based on cell surface or cytoplasmic markers encoded by the vector [7–9].

Besides this entirely diagnostic approach, there are several therapeutic strategies targeting healthy tissues. These strategies are also relevant to gene therapy of some inborn genetic disorders or acquired viral infections, because of the use of marker genes that allow selection of transduced cells *in vivo*. Positive selection is established in the context of drug resistance gene transfer, negative selection in adoptive immunotherapy.

Drug resistance gene transfer in nontumor tissues such as bone marrow is aimed at augmenting the therapeutic index of anticancer chemotherapy [10,11] (see also Chapter 11). Protection at the level of hematopoietic progenitor cells reduces short-term toxicity, protection at the stem cell level might even prevent long-term toxicity and the mutagenicity of chemotherapy. The benefit for the patient will depend on the numbers of protected cells obtainable. These are expected to increase with each cycle of chemotherapy, because cells acquire a selective advantage upon expression of the drug resistance gene. Thus, this approach sets a paradigm for forced expansion of transduced cells *in vivo*. Like gene marking, this approach requires a number of technological improvements: First, helper functions of the vector systems and transduction conditions have to mediate efficient uptake (see Section V.A) and nuclear translocation (see Section V.B) in primitive hematopoietic cells (reviewed in Baum [12]). Second, the vector needs to be equipped with *cis* regulatory elements mediating dominant gene expression levels, and thus strong penetrance of the phenotype (see Section V.C) [13,14]. Third, coexpression of a second gene (see Section V.C.6) is important in this approach, because coordinate transfer of two complementary drug resistance genes greatly widens the therapeutic flexibility [15,16]. Finally, malignant cells must be excluded from productive transduction by vector targeting (see Sections V.A.1.b and V.C.1).

A paradigm for negative selection of transduced cells *in vivo* is established in *adoptive immunotherapy*. Here, ex vivo selected populations of allogenic donor lymphocytes are used to elicit an antiviral or antileukemic effect [17]. Gene transfer in lymphocytes serves for both positive and negative selection. After transduction, positive selection of gene-modified lymphocytes is performed ex vivo using cell surface markers. After reinfusion, concomitant expression of a negative selection marker (a suicide gene) is instrumental for treating eventually occurring graft versus host disease. Here, the key issue is to design vectors with reliable and persisting coexpression of two genes (see Sections V.C.2., V.C.4., and V.C.6.).

Positive or negative selection and monitoring of transduced cells *in vivo* is crucial for the development of artificial *mini-organs* (derived from genetically manipulated cells [18]). In oncology, these are of interest for systemic delivery of tumor antagonistic factors such as immunotoxins or inhibitors of angiogenesis. Further applications extend to genetic or acquired disorders, which can be treated by delivery with enzymes, hormones or ligands. Equipping mini-organs with regulatable promoters might allow the adjustment of supply according to individual clinical requirements (see Section V.C.5) [19].

TABLE 1   Somatic Gene Transfer in Oncology and Implications for Vector Design

Approach	Aim	Target cells	Vector requirements	Vector system
Gene marking	Diagnostic	Healthy hematopoietic or lymphocytic cells, tumor cells (both *ex vivo*)	Transduction of long-lived stem cells, stable gene expression	Retroviral vectors
Drug resistance gene transfer	Therapeutic (paradigm for positive selection of transduced cells *in vivo*)	Healthy hematopoietic cells (*ex vivo*)	Transduction of repopulating cells, stable and high gene expression	Retroviral vectors
Adoptive immunotherapy	Therapeutic (paradigm for negative selection of transduced cells *in vivo*)	Donor lymphocytes (*ex vivo*)	Transduction of lymphocytes, stable and high gene expression	Retroviral vectors
Mini-organs	Therapeutic	Healthy autologous or xenogenic cells (*ex vivo*)	Stable or inducible gene expression	Retroviral vectors
Suicide gene transfer	Therapeutic (but not systemic)	Tumor cells (usually *in vivo*)	Applicability *in vivo*, targeting to tumor cells; strong, but not necessarily stable gene expression	Retroviral vectors; alternatively herpes virus vectors or adenoviral vectors
Oncogene antagonism	Therapeutic (but not systemic)	Tumor cells (usually *in vivo*)	Applicability *in vivo*, targeting to tumor cells; strong, but not necessarily stable gene expression	Retroviral vectors; alternatively herpes virus vectors or adenoviral vectors
Tumor vaccination	Therapeutic	Tumor cells, antigen-presenting cells (*ex vivo* or *in vivo*)	Applicability *in vivo*; moderate, but not necessarily stable gene expression	Retroviral vectors; alternatively herpes virus vectors, adenoviral vectors, or physicochemically

Other therapeutic concepts rely on direct genetic manipulation of tumor cells. Some of these suffer from poor predictability, mostly because of the tremendous variability of tumor evolution among and within individual patients. Moreover, not all of these strategies acknowledge the systemic character of tumor diseases. Nevertheless, in selected patients, these strategies might offer interesting perspectives. Usually, vectors have to be applied *in vivo* to become effective, and sometimes even replication-competent vectors will be needed. Here, nonretroviral systems may offer important alternatives, given that the problem of instability of persistence or expression is of minor importance.

Transfer of prodrug converting enzyme genes (*suicide gene transfer*) is performed to render tumor cells susceptible to cytotoxic compounds requiring activation by a heterologous enzyme [20,21] (see also Chapters 10 and 12). Another approach to control tumor cells by gene transfer is *oncogene antagonism* [22,23]. Here, one attempts to counteract tumor promoting mutations of cellular genes. This is achieved by transducing tumor cells with wild-type copies of tumor suppressor genes or dominant negative proteins, antisense nucleotides,

or ribozymes directed against oncogenes and their products. In compact tumor masses, both suicide gene transfer and oncogene antagonism may profit from the "bystander effect." This refers to cytostatic effects observed in nontransduced cells, which result from delivery of proteins or activated cytotoxic drugs through direct intercellular exchange. However, this exchange might also dilute the effects in transduced cells [24]. Suicide gene transfer, in contrast to oncogene antagonism, should exclude healthy tissues.

Application of retroviral vectors *in vivo* requires production of complement-resistant particles at high titers. Selectivity can be achieved at the level of transduction, taking advantage of preferential infection of dividing cells by vectors based on murine retroviruses [25]. Specific targeting using engineered envelope proteins may, however, be superior (see Section V.A.1.b). At the level of transcriptional regulation, selectivity can be accomplished by insertion of promoters preferentially activated in tumor cells (see Section V.C.3). Depending on the tumor type, herpes viruses or adenoviruses, some of the latter specifically replicating in p53-negative cells, may represent alternative vectors [26]. Key aspects of targeting using specific receptors or promoters also apply to these nonretroviral systems.

Finally, *tumor vaccination* is performed to evoke a systemic immune response to tumor specific antigens [27,28]. This is accomplished by transfer and expression of genes that increase antigen presentation or improve effector cell functions. Target cells for transduction are tumor cells, antigen presenting cells, or tumor infiltrating T cells. Thus, tumor vaccination strategies are highly variable with respect to the target cell population and to the type and numbers of activating genes to be transferred. The main requirements for vector design are applicability *in vivo* (after injection in tumor masses) as well as moderate but not necessarily sustained gene expression levels in target cells. For this approach, alternative vector systems (such as adenoviral or herpes vectors) or physicochemical methods such as biolistics might find more widespread use than retroviral vectors.

In all these different strategies, important variables to account for in vector design are the route of gene transfer, the target cell population, the efficiency and specificity of transduction, and the level, duration, and specificity of transgene expression (Table 1). Therefore, appropriate components of the vector system have to be identified for each application ("tailored vectors"). Fortunately, substantial progress has been made toward all aspects of vector design relevant to cancer gene therapy. These vector improvements are based on detailed insights into the biology of retrovirus–host interactions.

III. BIOLOGY OF RETROVIRUSES

A. Classification

Sequence data and genome structure are the basis for the classification of retroviruses (Table 2) [29]. Each group contains several virus strains that differ in biological properties, such as receptor utilization and pathogenicity. Until recently, most retroviral vector systems discussed for human gene therapy were based on murine leukemia viruses (MLVs). These belong to the mammalian C-type retroviruses and are further classified according to the species distribution of their receptors. Ecotropic MLVs replicate only in rodent cells and xenotropic MLVs only in nonmurine cells, whereas polytropic and amphotropic MLVs can infect murine and nonmurine cells. The 10A1 strain has an overlapping but distinct host range because of the use of the receptor for the gibbon ape leukemia virus (GALV) in addition to the amphotropic receptor [30]. Except for ecotropic viruses, gene transfer into human cells is possible with all groups of viruses mentioned. For historical reasons,

TABLE 2   Retrovirus Genera

Genus	Example viruses
Avian leukosis sarcoma	Rous sarcoma virus (RSV)
Mammalian C type	Murine leukemia virus (MLV), several strains: such as Moloney-, Harvey-, Abelson-, 407A-MLV.
	Feline leukemia virus (FeLV)
	Gibbon ape leukemia virus (GALV)
D-type viruses	Mason-Pfizer monkey virus (MPMV)
B-type viruses	Mouse mammary tumor virus (MMTV)
HTLV-BLV group	Human T cell leukemia virus (HTLV)-1 and 2
Lentivirus	Human immunodeficiency virus (HIV)-1 and -2
Spumavirus	Human foamy virus (HFV)

retroviral vectors applied thus far in human gene therapy utilized the amphotropic receptor.

B. Retroviral Genes and Their Products

Retroviruses within a group share a very similar proviral structure. In the first three groups (see Table 2), including mammalian C-type retroviruses, the genome codes only for the virion structural proteins Gag, Pol, and env (Fig. 1) [29]. The *gag* gene products constitute the viral matrix and package the two retroviral RNA genomes into a viral nucleocapsid. Encoded by the *pol* gene, the virion also includes several enzymes necessary for virus replication. These are the reverse transcriptase, the integrase, and the viral protease, which cleaves the Gag and Pol precursors into the individual proteins. Receptor utilization is determined mainly by the glycosylated *env* gene product SU, which is anchored in the viral envelope by the transmembrane protein TM. The viruses of the HTLV-BLV group, the lentiviruses, and the spumaviruses are more complex and also encode specific nonvirion proteins with different regulatory functions [31,32]. Examples are the viral transcriptional activators of gene expression, such as tax in HTLV, *tat* in HIV, and *bel*-1 in foamy viruses.

C. Retroviral *cis* Elements

The *cis*-acting elements that regulate viral gene expression, reverse transcription, and integration of the provirus into the cellular DNA are organized very similarly in all retroviruses [29]. The provirus is flanked by the long terminal repeats (LTRs), carrying the terminal *att* sites, which are recognized by the integration machin-

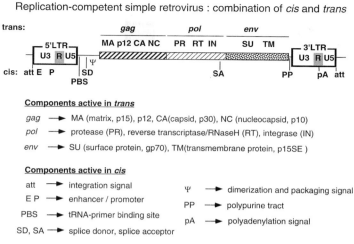

FIGURE 1   Scheme of the proviral form of a replication-competent simple C-type retrovirus. Sequences coding for *trans*-acting proteins are indicated above the drawing, *cis*-acting sequences below.

ery. The LTR is further divided into the three sections U3, R, and U5 (Fig. 1). The U3 region carries the viral enhancer and promoter elements. In the 3′ LTR, initiation of transcription is suppressed, possibly because of interference with the 5′ LTR. The polyadenylation signal resides in the R or U3 region. Transcription of viral genomic RNA begins at the R region in the 5′ LTR and ends with R in the 3′ LTR. The RNA genome is thus flanked by identical redundant regions (R), which play an important role during reverse transcription (see Section II.D). The U5 region contains sequences necessary for reverse transcription and terminates with the *att* site.

The untranslated leader comprises R and U5 regions of the 5′ LTR and sequences upstream of *gag*, including 18 nucleotides that form the primer binding site (PBS). The PBS is perfectly complementary to the 3′ terminus of the tRNA primer that initiates reverse transcription of the RNA genome into the minus strand of proviral DNA. Leader sequences downstream of the PBS contain the splice donor site for generating the subgenomic RNAs, as well as the packaging and dimerization signal, which directs incorporation of the two viral RNA genomes into virions. Optimal packaging can also require additional sequences, such as the first 400 nucleotides of *gag* in MLV-based vectors (called *gag*+ vectors) [33]. In HIV, the exact sequences sufficient for packaging have not yet been defined [34,35]; in addition to part of the leader, a region in the 5′ end of the *gag* gene, and sequences within *env* encompassing the Rev responsive element (RRE) appear to improve packaging [36].

The untranslated region between *env* and the 3′ LTR contains the polypurine tract (PP), a run of at least nine

A and G residues. Synthesis of the plus strand of proviral DNA is initiated here.

D. Retroviral Life Cycle

The retroviral life cycle is illustrated in Fig. 2. Initially, retroviruses bind through the Env protein SU to

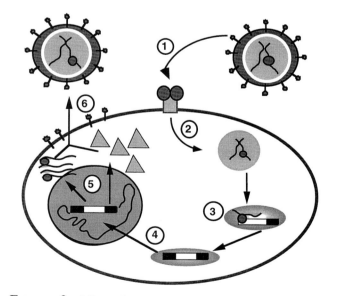

FIGURE 2   Life cycle of a replication-competent retrovirus: (1) virion binding; (2) virion penetration and uncoating; (3) reverse transcription of RNA genome into proviral DNA; (4) nuclear transport of preintegration complex and integration of provirus; (5) transcription of genomic and subgenomic mRNA and translation of viral gene products; (6) nucleocapsid assembly, budding, and maturation of virion.

a specific viral receptor on the cell surface. All known retroviral receptors are membrane proteins, and several have been cloned [37]. The receptors for amphotropic MLV (Pit-2) and for GALV (Pit-1) are phosphate transporters found on most human cells. Interaction of viral SU with the receptor exposes a fusion peptide in TM and triggers fusion of the viral and cellular membranes with subsequent release of the nucleocapsid into the cytoplasm [38]. Here the viral RNA genome is reverse transcribed into the proviral DNA by the viral reverse transcriptase (RT) [29]. The nucleocapsid protein NC is also required for this process. Reverse transcription is initiated at the PBS. RT synthesizes the negative strand complementary to the U5 and R region of the 5′ LTR, and the RNAse H activity of RT degrades the genomic RNA. The nascent DNA strand is transferred to the 3′ end of the RNA genome, where, starting with the U5 region, the negative strand is completed with concomitant degrading of the RNA genome. The PP tract escapes digestion and serves as a primer for plus strand DNA synthesis, which proceeds through U3 and R. The plus strand is then transferred to the 5′ end and transcription is completed. In brief, reverse transcription of the two RNA genomes generates a single provirus with two complete LTRs by duplicating the U3 and U5 regions of the RNA genome; importantly, the 3′ LTR and the 5′ LTR serve as templates for U3 and U5, respectively (Fig. 3). The infidelity of reverse transcription and recombinations occurring as a consequence of switching between the two RNA templates during reverse transcription lead to a high degree of variability

of retroviruses. This represents a potential drawback for vector design, because unpredictable errors may be introduced during vector infection [39]. However, other viral or nonviral methods of DNA transfer may be associated with even higher rates of recombination. In many instances, this results from the transfer of multiple copies of homologous DNA, which is excluded in retroviral systems.

After reverse transcription, the nucleocapsid proteins remain tightly associated with the proviral DNA. This complex carries all factors necessary for the integration of the viral DNA into the genome of the host cell. In MLVs this complex cannot pass through the nuclear pores, and nuclear transport requires mitosis with breakdown of the nuclear membrane [40,41]. In lentiviruses, such as HIV, mitosis is not required and the preintegration complex is targeted to the intact nucleus with the help of the matrix, integrase, and possibly the vpr protein [42]. Integration of the provirus is random with regard to position in the genome, with some preference for open chromatin [43]. Local structural features of host DNA rather than specific sequences influence the susceptibility to integration. Rarely, integration can produce alteration of the phenotype of an infected cell by activation or disruption of cellular genes [44]. Infection with actively replicating virus is accompanied by repetitive integration events in different cells and thus increases the possibility of proto-oncogene activation and the development of neoplasias [45]. In therapeutic retroviral gene transfer the probability of such insertional mutagenesis has been minimized by the use of replica-

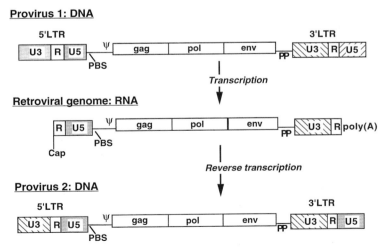

FIGURE 3 Structure of provirus and viral genome of MLV. Provirus 1 is transcribed into the retroviral genome flanked by the R sequences. Two RNA genomes are packaged into a virion and released from the cell. After infection of the target cell the genomic RNA is reverse transcribed into provirus 2. This again is flanked by two LTRs that contain the U5 region of the 5′ LTR and the U3 region of the 3′ LTR of provirus 1. Abbreviations are explained in Fig. 1.

tion incompetent retroviruses as vectors, which integrate at low copy numbers (usually one or two per cell). For a single integration event, the risk of inducing tumor promoting mutations is estimated to be in the range of 10^{-6} or lower [46]. Immune responses to altered cellular gene products further reduce the likelihood of inducing tumors by retroviral vector integration.

The integrated provirus is transcribed by the cellular RNA polymerase II. In simple retroviruses such as MLV, this process is solely dependent on the cellular transcription machinery. Between viral strains, binding sites for cellular transcription factors in the U3 region differ. Because the expression of many transcription factors is developmentally regulated, cell tropism and pathogenicity of retroviruses is influenced by the composition of *cis* elements in the LTR [47,48]. Complex retroviruses have a number of transcriptional and post-transcriptional transactivators that, in cooperation with cellular factors, influence viral transcription levels, nuclear export of RNAs, and splicing patterns [32].

Viral transcripts are modified by cellular capping enzymes at the 5' end, and a poly(A) tail is added to the 3' end at the R-U5 border following specific polyadenylation signals. Retroviral transcripts enter one of three pathways: (1) The RNA is translated into the Gag or Gag-Pro-Pol precursor proteins. (2) The RNA is spliced into subgenomic RNA. (3) The full-length viral RNA is packaged as a viral RNA genome into the virion and released from the cell. All subgenomic mRNAs are spliced from the same splice donor, generally located in the leader. In simple retroviruses only the *env* transcript is spliced. Complex retroviruses have several splice acceptors in the 3' half of the genome. Hereby, several different smaller spliced transcripts are generated that code for regulatory proteins [29,32].

The Gag precursor protein is always translated from the full-length viral RNA. Translation is continued past the stop codon to generate a Gag-Pro or Gag-Pro-Pol precursor protein at a low frequency (5–10%). The viruses of the avian sarcoma-leukosis virus complex (ASLV) are an exception as *gag* has no stop codon and is always translated as a Gag-Pro polyprotein [49]. After virus assembly, the viral protease (PR) is activated by autocatalytic release from the precursor. PR then cleaves the Gag precursors into the matrix (MA), capsid (CA), and nucleocapsid (NC) proteins, and several smaller peptides. The Pol precursor is cleaved by PR to yield the reverse transcriptase, which has an RNA- and DNA-directed polymerase and a ribonuclease H activity, and the smaller carboxyl terminal viral integrase. The Env protein is cleaved, most likely during transport to the cell surface, by a cellular protease into the viral surface protein (SU) and a transmembrane protein TM.

Viral genomic RNA is assembled into the nucleocapsid through specific interaction of the NC portion of the Gag precursor and *cis*-active viral packaging sequences [34]. B- and D-type retroviruses preassemble in the cytoplasm, whereas C-type viruses, such as MLV, assemble at the cytoplasmic membrane. The MA domain interacts with the inner surface of the cytoplasmic membrane and mediates budding of the virus [50]. The virus acquires the viral envelope by budding through membrane areas that contain Env proteins. However, viral glycoproteins are not necessary for virion formation and in the absence of Env, noninfectious, bald enveloped particles are released. Proteolytic cleavage of viral polyproteins begins during the budding process and is completed in the released particles, thereby generating the mature infectious virions [51].

IV. PRINCIPLES OF RETROVIRAL VECTOR SYSTEMS

First, we will discuss general rules for designing packaging cell lines and vectors. In Section V, we approach specific aspects related to vector entry, integration, and expression.

A. Packaging Cells

Retroviral vector systems are designed to mimic the infectious properties of retroviruses (stable transduction of target cells without inducing rearrangements and relatively stable gene expression) in replication-incompetent vector particles. The latter are produced from specialized packaging cell lines. In these cells, viral coding regions are physically and functionally separated from the vector genome. The strict separation of *cis*-active and *trans*-active components serves purposes of safety, efficiency, and increased flexibility (Fig. 4).

A modern, safety-modified packaging cell contains at least two expression constructs for viral genes, one encoding for *gag-pol* and one for *env* genes [52]. To prevent mobilization in retroviral particles, these mRNAs lack the packaging signal. In the case of MLV, this is easy to achieve, because retroviral coding regions are not sufficient for packaging. In HIV, the extended and less well defined packaging signal might represent a drawback to the development of helper cells (and vectors).

High expression of viral proteins in packaging cells is best accomplished by allowing direct selection for the promoter driving the retroviral genes [53]. This can be managed by linkage with a dominant selectable marker, with coexpression obtained by reinitiation of transcrip-

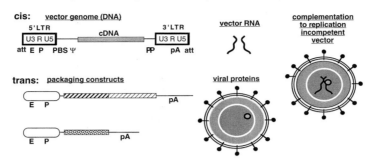

FIGURE 4 Separation of *cis*-acting sequences and *trans*-acting retroviral coding regions to generate safe packaging cell lines for release of replication-incompetent vectors. Abbreviations are explained in Fig. 1.

tion from the 3′ untranslated region, internal ribosomal entry sites for reinitiation of translation, or alternative splice signals. Alternatively, inducible promoters can be used, representing the method of choice when particle components have cytopathic effects (as in the case of some proteins used for pseudotyping, see Section 5.A.1.a). Encoded from such "packaging constructs," the packaging cell provides all *trans*-acting viral elements required for particle assembly, release, maturation, and transduction of the target cell.

The vector RNA is also generated inside the packaging cell and contains those *cis*-active elements required for a single transduction: cap site and poly(A) signal for the genomic message, packaging signal for incorporation into particles, PBS, PP, and R-U5 sequences for reverse transcription, and *att* sites for integration. Moreover, it contains the transgene cassette(s) of interest including enhancer/promoter sequences for initiation of transcription (Fig. 4).

Packaging cells not expressing RNAs with suitable packaging signals should not release infectious particles. However, depending on the cellular background, there is the potential risk of packaging endogenously expressed retroviral or retroviruslike sequences. In the case of MLV-based systems, this risk is highest in a rodent background. Here, packaging and transfer of VL30 sequences is observed as a frequent event, especially when vector titers are low [54,55], implying that this risk can be reduced when high-titer producer clones are selected for vector production. This is important because packaging of and recombination with endogenous retroviral elements is the most important event leading to the generation of replication-competent retroviruses from safety-modified packaging cells [56,57].

In immunocompromised primates and permissive mouse strains, replication-competent amphotropic MLV can induce lymphoma [45]. Also, amphotropic retrovirus induced spongiform encephalomyelopathy

has been observed after inoculation in newborn mice [58] (for a thorough and thoughtful discussion on safety aspects of nonhuman, xenogeneic viruses, see also Isacson and Brakefield [59]). More recently, many packaging cell lines have been developed or are under construction in a nonrodent background, such as human or canine, not known to express retroviral sequences packaged in MLV particles.

In the human host, most xenogeneic retroviruses are complement sensitive, precluding administration *in vivo*. Complement sensitivity is defined at two levels: first, by specific Env sequences; second, by protein modifications characteristic to the species background of the producer cell [60,61]. Complete complement resistance is achieved by producing vector particles with alternative envelopes (e.g., derived from feline leukemia virus) in human packaging cells. However, repetitive administration is likely to be hindered by the immunogenicity of retroviral proteins. Moreover, fundamental restrictions to successful transduction *in vivo* are present at the physicochemical level (such as particle concentration and motility). Ex vivo, these can be overcome by suitable transduction protocols (reviewed by Palsson and Andreadis [62]).

To produce replication-incompetent vectors, vector genomes are introduced into packaging cells either by transfection or by retroviral transduction. Packaging systems have also been developed to release high vector titers after transient transfection, or semipermanently from episomally replicating plasmids [63]. For clinical applications, stably transfected clonal packaging cell lines still represent the ultimate choice, because these allow vigorous preclinical testing of safety (especially the absence of replication-competent retroviruses [64]) and efficiency, as well as large-scale production of vector stocks from defined cell banks under conditions of good manufacturing practice (GMP). Titers released from retroviral packaging cell lines usually do not exceed 10^6

to 10^7 per milliliter of cell-free supernatant. Currently, this is the minimum required for ex vivo approaches, such as transduction of hematopoietic cells. When all components are improved, titers might be as high as 10^8. Concentration of the fragile retroviral particles is alleviated when the membrane is stabilized with nonretroviral components (see Section V.A.1.a).

B. Vector Architecture

The flexibility of the retroviral genome offers a great degree of freedom for the insertion of transgene cassettes. Still, retroviral vectors mimicking the basic architecture of their replication competent ancestors (LTR–leader–gene or genes–LTR) have found the most widespread use (Fig. 5A and B) [65]. These vectors are usually very stable and also mediate reasonable transgene expression in the cellular system of interest, provided that appropriate enhancers are employed (see Section V.B). The gene is expressed either from within the *gag* region [33,66], or from the *env* position, which can lead to higher translation efficacy (MFG vector [67]).

Efficiency as well as safety of retroviral vectors might be further improved when the leader is functionally inactivated or physically deleted in the target cell (Fig.

5C to F). Increased efficiency results from removing the long untranslated leader from the transcript, which might contain repressory elements for transcription or translation. This also excludes the packaging region from vector transcripts in the target cell, thus precluding transmission of the vector in the hypothetical case of accidental superinfection with replication-competent retroviruses.

There are several options for leader exclusion. First, the transgene plus its enhancer-promoter can be placed in the U3 or R region of the LTR, resulting in a "double copy" vector after completion of reverse transcription [68,69]. Duplication of the transgene cassette is expected to result in higher expression levels. Some sequences, however, are not compatible with this strategy, resulting in a high incidence of recombination. Second, "self-inactivating" or "suicide" vectors can be generated by deleting the enhancer-promoter or the promoter only in the U3 region of the LTR and placing the transgene of interest under control of an internal promoter, either in sense or in antisense orientation to the LTRs [70,71]. Third, in LTR-controlled vectors, sequences between PBS and the start codon of the transgene can be flanked by, e.g., loxP sites, allowing conditional deletion upon expression of the bacteriophage recombinase cre [72]. Finally, reversion of double copy vectors to a monocopy vector is possible with a self-contained loxP/cre vector

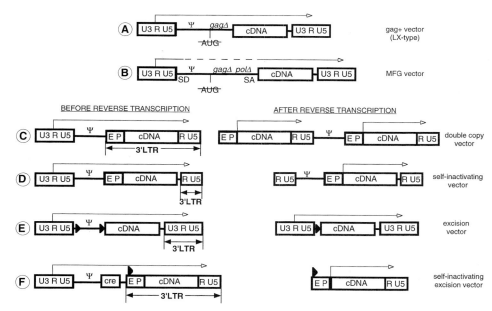

FIGURE 5 Flexibility in basic vector architecture. (A, B) LTR-controlled vectors including the packaging region (ψ) in the genomic transcript in transduced cells. *gag*Δ and *pol*Δ, residual fragments of viral genes *gag* and *pol*, respectively; ~~AUG~~, destroyed start codon of *gag*. (C–F) Different forms vectors excluding ψ from transcripts in transduced cells. Plasmid constructions are represented on the left (before reverse transcription), the status of the proviral form after reverse transcription is shown on the right. In (E) and (F), the status in transduced cells after site specific recombination is shown. The black filled triangles represent loxP sites, which are recognized by the site specific recombinase, Cre.

[73]. Given that stability, titer, and expression character-istics of these more sophisticated constructions are better defined, they represent valuable alternatives to conventional, LTR-controlled vectors.

V. ADVANCES IN RETROVIRAL VECTOR TAILORING

Besides the more general aspects of packaging cell line design and vector construction discussed earlier, specific advances can be noted with respect to distinct stages of the retroviral life cycle. Of special importance are those related to vector entry, integration (both defined by *trans*-active vector components), and expression (defined by *cis*-active elements) relevant to cancer gene therapy. This work should lead to vectors specifically tailored for clinical applications.

A. Components Active in *trans*

1. THE RETROVIRAL ENVELOPE

For many reasons, the amphotropic Env, hitherto used in all gene therapy trials involving retroviruses, is not a perfect choice for mediating vector entry. The cognate receptor, Pit-2, is too widely expressed to allow specific cell targeting, with the ironical exception of primitive hematopoietic cells, where expression is too poor to allow efficient transduction [74–77]. Moreover, the amphotropic Env is involved in an unexpected pathogenicity of replication-competent retroviruses, which is induction of spongiform encephalomyelopathy [58]. Alternative Env proteins such as that of the 10A1 strain are also associated with these potential drawbacks [78]. Moreover, the low stability of retroviral particles with conventional retroviral envelopes does not allow vector concentration for *in vivo* delivery. Importantly, both vector stability and targeting can be improved by altering the retroviral envelope. The two major approaches are discussed next.

a. Pseudotyped Retroviral Vectors   Coinfection with two viruses generates hybrid virions, which contain the genome and core proteins of one virus and mixed envelope glycoproteins of both viruses. The host range of these "pseudotypes" is determined by both envelope proteins [37,79]. Pseudotyping can be used to alter the host range of retroviral vectors. Pseudotyped MLV-derived vectors thus can transduce cells that are normally resistant to MLV because they lack a functional amphotropic receptor (reviewed in Friedmann and Yee [80]).

The mechanisms that determine whether a foreign viral envelope protein can be incorporated into the viral envelope are not well understood. The cytoplasmic anchor of the transmembrane (TM) Env protein was shown to guide MLV glycoproteins to the envelope of budding virus particles [81]. Thus, homologous Env proteins are efficiently incorporated into the MLV envelope, whereas heterologous proteins must be expressed at high densities in the cell membrane to allow pseudotype formation [82].

Pseudotyped retroviral vectors are generated by coexpression of vector RNA containing the packaging signal, with retroviral Gag and Pol, and the unrelated glycoprotein. Packaging systems for several pseudotypes of MLV have been developed. Pseudotypes that incorporate the glycoprotein of vesicular stomatitis virus (VSV-G) have an extremely broad host range [83]. The VSV-G protein enters the cell by interacting with an ubiquitous phospholipid component of cell membranes (Fig. 6) [84]. Mammalian, fish, and insect cells can be transduced [85–87]. CD34+ hematopoietic progenitors were shown to be up to 10-fold more susceptible to a VSV-G than to an amphotropic pseudotype [88]. In a recent study, we found that VSV-G pseudotypes can infect hematopoietic stem cell lines and fibroblasts equally well, but transduction of stem cells with amphotropic vectors was at least 100-fold less efficient [77]. This important observation reveals that the receptor deficiency of primitive hematopoietic cells to retroviral transduction [75,76] can be completely overcome by vector pseudotyping [77].

Additional advantages are that VSV-G confers great stability on the retroviral particle and that pseudotypes can be concentrated to high titers by ultracentrifugation. This is also of interest for *in vivo* applications. However, the host range of VSV-G is too broad to allow specific cell targeting, and the high immunogenicity of VSV-G is expected to preclude repetitive administrations *in vivo*. Another major drawback has been that the VSV-G protein is toxic for the cell. Pseudotypes are thus only produced for a limited period from already dying packaging cells [85]. Recently, stable packaging cell lines have been generated by placing the VSV-G gene behind an inducible promoter. In these lines the VSV-G gene is repressed but can be induced for vector production. However, here, too, vector production is accompanied by cell death [89].

Alternatively, pseudotypes of MLV vectors that incorporate Env proteins of the gibbon ape leukemia virus (GALV) transduce hemopoietic progenitors and lymphocytes more efficiently than do amphotropic pseudotypes [90,91]. Other retroviral Env proteins have also been utilized. Examples are the glycoproteins of the 10A1 and MCF MLV strains, HTLV-I, HIV-1, and human foamy virus (HFV) [30,92–95]. The tropism of these pseudotypes generally has not been properly evaluated to show advantage over amphotropic vectors.

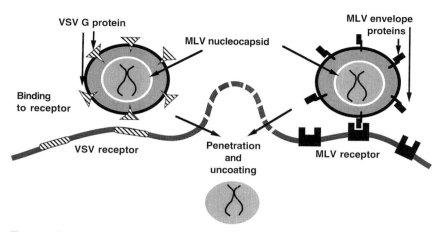

FIGURE 6   VSV-G protein pseudotyped retrovirus. Pseudotyping affects particle stability, receptor targeting, and mode of entry. After uncoating, different envelope pseudotypes follow the same pathway.

HFV pseudotypes are especially promising because it is assumed that all mammalian cell types are infectable. However, we have recently observed that the hematopoietic progenitor cell lines FDC-Pmix and FDC-P1 are partially resistant not only to amphotropic MLV but also to HFV infection [77,96].

New perspectives in pseudotype development are opened by generating chimeric envelopes. For instance, efficient pseudotyping of MLV with HFV surface proteins is only possible for chimeric envelope proteins containing an unprocessed cytoplasmic tail of MLV TM fused to a truncated HFV envelope protein [92]. Using similar chimeric envelope proteins it may be possible to generate a large panel of different pseudotypes with unrelated viral or nonviral membrane proteins which are normally not incorporated into retroviral envelopes efficiently.

b. Ligand-Directed Targeting   Ideally, vectors should be designed to selectively transduce specific target cells of interest present within mixed cell populations ex vivo or even intact organs *in vivo*. In oncogene antagonism and suicide gene transfer (see Section II), specific or at least preferential targeting to tumor cells has to be achieved *in vivo*. Here, binding of virus to nontarget cells will lead to considerable loss of the effective virus titer and also increases unwanted side effects. In contrast, drug resistance gene transfer to hematopoietic cells (see Section II) is performed ex vivo, with strict exclusion of malignant cells. Targeting retroviral transduction can be achieved at two levels: first, by colocalization of cells and viruses on a specific matrix (Fig. 7A), and second, by equipping retroviral particles with cell specific ligands (Fig. 7B to F).

Colocalization of cells and viruses on a biochemical matrix can only be used ex vivo, alleviating vector–cell

interactions at the physicochemical level. Colocalization can lead to higher transduction efficiency in cells with poor receptor representation, paradigmatically shown in fibronectin-assisted transduction of hematopoietic progenitor cells or lymphocytes with amphotropic vectors [9,97,98]. However, it remains to be seen whether this approach can be elaborated for target-specific virus uptake, e.g., by displaying ligands on the matrix that are selectively recognized by the target cell population of interest (Fig. 7A).

The alternative approach is to retarget retroviral entry via specific cell surface molecules by manipulating the viral envelope. To this end, several strategies have been followed (reviewed in Cosset and Russell [99]). Although specific binding is relatively easy to achieve, virus uptake with engineered envelopes often is much less efficient. One initial approach was to direct specific virus binding by creating a molecular bridge between the virion and the cell surface (Fig. 7B). Here, virus particles are coated with specific antibodies for the surface subunit of Env (SU) and the cells are incubated with an antibody specific for a membrane protein such as the epidermal growth factor receptor or the insulin receptor. Both antibodies are then linked by secondary antibodies or by biotin/streptavidin [100]. In other studies, ecotropic or avian retroviral envelope proteins were modified. Wild-type Env proteins of these viruses do not allow infection of human cells. Therefore, incorporation of specific binding epitopes can selectively retarget the virus to human cells with the complementary membrane protein of choice. Three general strategies have been followed:

1. Small peptides that specifically bind to cellular receptors are introduced into binding domains of the SU protein without affecting natural receptor

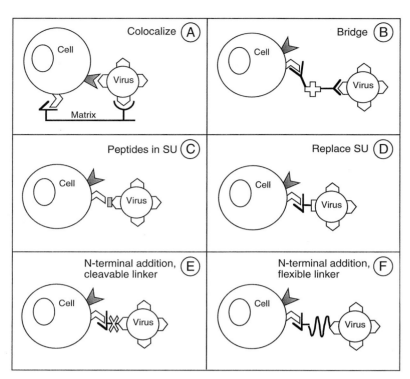

FIGURE 7   Targeting transduction via cellular receptors. The strategies discussed
in Section V.A.1.b are schematically represented. Entry either occurs via the differentiation specific cellular receptor (open symbol on cell surface) or still needs the natural
retrovirus receptor (shaded symbol). (A) Colocalization of virus and cell via matrix
proteins; (B) molecular bridge between virus and cell (here, cross-linked antibodies);
(C) a peptide in the binding domain of SU alters its tropism; (D) the binding domain
of SU is replaced with a targeting ligand; (E) an N-terminal addition is linked to SU
via a protease-cleavable linker, or, as shown in (F) via a flexible linker.

recognition [101,102]. Human breast cancer cell lines, overexpressing human epidermal growth factor receptors (HER-2 and HER-4) could be specifically targeted by insertion of the heregulin peptide, a ligand for HER-2 and HER-4, into ecotropic SU of MLV [103]. However, titers and efficiency of transduction were too low for *in vivo* applications (Fig. 7C).

2. The complete binding domains of SU are replaced by alternative ligands for cellular receptors (Fig. 7D). Erythroid progenitor cells have been targeted by replacing SU binding domains with erythropoietin [101]. Chimeric SU proteins that contain single chain antibodies (scAs) attached to the truncated retroviral Env proteins are a versatile system with the potential of targeting cells via specific epitopes of many different membrane proteins. An example is the scA B6.2, which binds to an antigen on breast and colon cancer cells [104]. This strategy works well with the Env proteins of the avian spleen necrosis virus,

whereas transduction with MLV-derived chimeric Env proteins is more inefficient and associated with low titers. Differences in the flexibility of Env proteins and in the pathways involved in virus internalization might be responsible for this discrepancy. In wild-type Env, virion binding causes a conformational change in SU, thereby exposing fusion domains of the Env transmembrane unit (TM), finally leading to viral penetration. In the chimeric envelope proteins described so far, conformation is generally altered and fusion processes not triggered efficiently. Therefore, most chimeric Env proteins support efficient binding of virions, but postbinding events are impeded or even completely blocked. This indicates that additional alterations in TM might be required to improve uptake. Alternatively, virion infectivity can be increased when also incorporating wild-type Env proteins that mediate fusion [105]. Such a receptor cooperation is also of central importance for the third targeting strategy:

3. With the aim of improving virus penetration, specific binding domains have been added to the N-terminus of the complete amphotropic Env proteins [106–109] (Fig. 7E and F). Amphotropic Env mediates infection of many human cell types, but N-terminal additions can block binding to the amphotropic receptor. Instead, these viruses bind to another membrane receptor of choice, but penetration still might require the amphotropic receptor. Two types of linkers between the retroviral Env and the N-terminal targeting domain allow such a two-step entry mechanism. In one approach, the ligand is fused to the amphotropic Env via a protease-cleavable linker [106] (Fig. 7E). Particles displaying these proteins bind to but do not infect cells. After protease cleavage, the N-terminal extension is released and the amphotropic binding domain exposed. Bound virus can then enter the cell efficiently by the amphotropic receptor. An unsolved problem *in vivo* is the systemic application of protease. An alternative would be to characterize cellular membrane proteases with their specific target sequences. In another approach, cooperation between two receptors is mediated by a flexible, proline-rich linker between amphotropic Env and the additional binding domain (Fig. 7F). Here, it is assumed that binding of the added specific ligand to its receptor triggers a conformational change that exposes amphotropic binding domains, which then mediate efficient entry via the amphotropic receptor [110]. A problem with both approaches may be that the amphotropic receptor is not expressed at sufficient levels on all cell types, as is evident from studies with early hematopoietic cells [77].

Although most investigators concentrate on positive targeting, negative targeting of selected cells can also be desirable in cancer gene therapy. An example is drug resistance gene transfer (see Section II), where the transduction of malignant cell is potentially hazardous because clones resistant to chemotherapy might be generated. Here the specific blockade to retrovirus entry found with many engineered Env proteins can potentially be exploited to increase safety of vectors.

B. Nuclear Transport and Integration

1. VECTORS DERIVED FROM COMPLEX RETROVIRUSES

Nuclear transport of the preintegration complex is restricted in those retroviruses (including MLVs) that require mitosis and breakdown of the nuclear membrane for integration into the host cell genome. Unlike MLV vectors, lentiviral vectors can transduce nondividing, yet postmitotic, cells, such as neurons and terminally differentiated macrophages [111–114]. Malignant cells, especially in larger tumors, where blood supply becomes limiting, are also often quiescent. Similarly, hematopoietic stem cells rarely cycle. Both cell types thus might be more efficiently transduced with lentiviral vectors.

Several lentiviral vectors have been derived from HIV-1 and HIV-2, but progress with packaging systems has been slow [115–117]. On the one hand, several gene products of HIV, such as the protease, the envelope proteins, and vpr, have proven to be toxic [118]. Therefore, and for other yet unknown reasons, vector titers have been low, although recently, a stable packaging cell line that produces titers as high as 10^5 per milliliter has been described [119]. Moreover, packaging sequences are not clearly separated from coding regions in the HIV genome and are dispersed throughout the genome [35,36,120]. Therefore, vectors and packaging constructs share common sequences with the potential to generate replication competent virus by homologous recombinations. Considering the high pathogenic potential of HIV, this risk must be eliminated. Currently, packaging of HIV-2 vectors in an HIV-1 packaging cell line is investigated. Homologies are minimal in this system. Hence, the risk of recombination is considerably reduced. Another alternative could be a vector derived from animal lentiviruses such as Maedi-visna virus. Such vectors would have the ability to transduce quiescent cells, but not the pathogenic potential of HIV [32].

Vectors derived from foamy viruses could have several advantages over lentiviral vectors. Foamy viruses are now generally considered to be apathogenic in humans, although this issue has been controversial in the past [121]. Foamy viruses have an increased packaging capacity (12 kb instead of 9 to 10 kb in MLVs). They infect many mammalian cell types, so the host range is generally considered to be broad [122,123]. However, we were not able to infect hematopoietic stem and progenitor cell lines from the mouse (FDC-Pmix and FDC-P1) with human or simian foamy viruses (HFV, SFV-5) [96]. This indicates that the viral host range may be more restricted than is generally believed. It has been postulated that HFV vectors transduce stationary cells more efficiently than MLV vectors. However, this issue is still controversial [124,125]. A major obstacle for the development of foamy virus packaging systems is that the packaging sequences in the viral genome have not been defined. Therefore, all vectors described so far to obtain infectivity in quiescent cells still contain viral genes. Knowledge of the biology of foamy viruses and other complex retroviruses is still limited, and extensive studies will be necessary before the value of these vectors for human gene therapy can be assessed.

2. TARGETING INTEGRASE

Targeting integrase to selected genomic loci is desirable to completely avoid insertional mutagenesis, to select integration sites supporting long-term expression, and to reduce clonal variability of gene expression. In the context of cancer gene therapy, these considerations are of relevance for gene marking, drug resistance gene transfer, and adoptive immunotherapy (see Section II).

Among the factors influencing site selection are overall DNA confirmation (open chromatin is a better target than heterochromatin), DNA sequence (in terms of local chemical or structural features rather than concrete motifs), DNA bending, and associated nuclear proteins (transcription factors, topoisomerases, replication proteins, matrix proteins) [126–128]. Integrases from different retroviruses differ with respect to target site selection, depending on the central core domain [129]. Systems for targeting retroviral integration to specific sequences are based on fusion of the IN protein with DNA-binding domains of well-characterized transcription factors, resulting in preferred, but not specific, integration to cognate sites [130–132]. A more efficient alternative might be to exploit the specificity of some yeast retrotransposons (Ty1, Ty3) for genes transcribed by RNA-polymerase III, which exist in multiple copies and where integration of a transgene is not expected to be hazardous [133,134]. Also, some human LINE elements and related retrotransposable sequences from other species encode endonucleases characterized by targeted DNA interaction [135]. It seems attractive to exploit such endonucleases, which are functionally distinct from integrases, for vector packaging systems.

A completely different approach is to block the viral integration process by eliminating *att* sites from the vector. Then selection can be made for integration via homologous recombination; this process, however, is limited by the cloning capacity of retroviral vectors (9 to 10 kb) and by the extremely low frequency of gene targeting in somatic cells [136].

C. *cis*-Active Elements

With the exception of gene marking, which theoretically can be performed without introducing active transcription units, all other applications for somatic gene transfer in oncology require a certain strength and duration of vector transcription (Table 1). Choosing appropriate *cis*-acting elements guarantees full penetrance of the phenotype of interest and thus influences the safety and efficiency of the gene transfer.

As opposed to physicochemical transfection methods, retroviruses are characterized by only moderate integration site dependence of gene expression. This implies that integration occurs at permissive loci or that retroviruses transfer genetic elements that can actively induce conformational or functional changes in their environment [137]. Such elements may reside in the enhancer region and involve poorly understood mechanisms, including secondary DNA structures [138]. Residual modulatory influences by the integration site usually lead to about 50-fold variation of gene expression levels among independent clones. However, depending on specific vector sequences and the genetic environment, complete extinction (silencing) of retroviral gene expression can also occur (reviewed by Baum [12] and Lund *et al.* [139]). This is most evident in embryonal stem cells [140] but has also been observed in hematopoietic stem cells [141] and more mature tissues, such as fibroblasts *in vivo* [142] and several somatic cell lines *in vitro* (reviewed by Lund *et al.* [139]). Therefore, modifying *cis*-acting elements in retroviral vectors can affect all aspects mentioned: differentiation-dependent gene expression levels, integration-site-dependent modifications, and incidence as well as kinetics of silencing. Thus, it is crucial to equip vectors with enhancer sequences that fit the host's transcriptional setting.

In simple retroviruses such as MLV and derived vectors, two major targets for transcriptional control have been identified: the dominant enhancer-promoter is located in the U3 region of the LTR, but sequences of the nontranslated leader (especially PBS) also contribute [143,144] (Fig. 8). Retroviral enhancers display recognition sites for a variety of transcription factors intimately involved in the differentiation processes of their natural target cell population. Precise consensus sequences, their numbers, and their relative orientation are crucial for enhancer strength and specificity [145]. Most retroviral enhancers are poorly expressed in more primitive, uncommitted cells such as embryonic and hematopoietic stem cells [144], mainly because these cells are not fully equipped with transcriptional activators or even express active repressors recognizing the retroviral enhancer. In permissive environments, such as in more mature hematopoietic cells, retroviral *cis* elements generally act quite autonomously and in a dominant manner, resulting in efficient transcription levels. Here, up to 0.1% of cellular transcripts can be generated from single-copy integrations, but even in more mature cells, differences in crucial enhancer elements drastically influence tissue tropism.

1. EARLY HEMATOPOIETIC CELLS

These represent a mixed cell population of primitive and uncommitted cells, with a latent, yet enormous po-

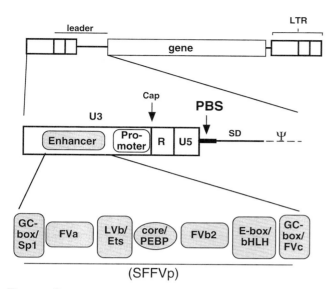

FIGURE 8 Dominant *cis*-acting elements of a murine leukemia virus reside in the U3 region of the LTR (specified in more detail for the strain SFFVp) and in the primer binding site (PBS) of the untranslated leader. SD, splice donor; ψ, packaging signal. For abbreviations of the enhancer boxes (gray) shown, refer to the text and Baum *et al.* [138].

tential for proliferation and stepwise differentiation following predefined genetic programs (reviewed by Baum [12] and Morrison *et al.* [146]). This is the target cell population for drug resistance gene transfer (see Section II), in which high levels of transgene expression are crucial for protection from chemotherapeutic side effects. Many vectors utilize control elements of the Moloney MLV (MoMLV) or the related Harvey murine sarcoma virus. These elements are strongly recruited in activated T cells but are only moderately active in more mature myeloid and erythroid precursor cells and are repressed to low levels in stem cells. Further repression of MoMLV-based vectors results from inhibitory elements targeting the PBS, both in embryonic stem cells and in early hematopoietic cells [13]. Based on systematic studies of transcription control of murine retroviruses in embryonic and early hematopoietic cells, we developed a series of vectors better adapted to the needs of these cells. The complex genealogy of these vectors is illustrated in Fig. 9.

cis-Active elements of MPSV, which differ from MoMLV by mutations in putative repressor sites and in one binding site for the transcription factor Sp1, perform better in hematopoietic progenitor and in embryonic stem cells. PCMV is an MPSV variant that has lost one copy of the enhancer's direct repeat. It arose by forced passage in embryonic carcinoma cells and contains the first retroviral enhancer known to be active in primitive embryonic stem cells [147]. When combined with the

leader of an endogenous retrovirus displaying an alternative PBS sequence, a vector resulted, allowing LTR-driven gene expression in undifferentiated embryonic stem cells. This chimeric virus is known as murine embryonic stem cell virus (MESV) [143]. The MESV backbone has been modified to include features of the MoMLV-based LX vectors [66] in the 5′ untranslated region (UTR) (packaging signal and untranslated *gag* sequences) and in the 3′ UTR (complete deletion of *env*). These modifications were incorporated to increase packaging efficiency and vector safety, but they did not improve gene expression as compared to MESV (MSCV, murine stem cell vector [148]). Vectors based on MESV (including MSCV) have found widespread use in experimental hematology, being associated with moderate, yet reliable transgene expression in myelo-erythroid progenitor cells [7].

In the MPSV-MESV hybrid vector (MPEV), the enhancer of MESV was replaced with the corresponding sequences of MPSV, roughly doubling gene expression levels because of the presence of the second copy of the direct repeat. Enhancers of Friend-MCF viruses like SFFVp (spleen focus-forming virus) were found to allow further increased gene expression levels in myeloerythroid cells [13]. An SFFVp-based vector can mediate sustained multilineage gene expression through serial transplantations in mice [149]. When the Friend-MCF-related U3 regions are combined with the nonrestrictive leader of MESV, novel vectors result, which we have named FMEV (Friend-MCF–MESV hybrid). These currently represent the best backbone for strong transgene expression in hematopoietic cells (Fig. 10A) [13,14].

The importance of improving enhancer strength became evident from comparative vector studies in the context of drug resistance gene transfer. Only MPEV and, even better, FMEV mediated high-dose drug resistance. Background-free selection of primary hematopoietic cells was thus possible when the human multidrug resistance 1 (MDR1) gene was expressed [13,14] (Fig. 10B). Moreover, intact proliferation and differentiation of transduced hematopoietic progenitor cells was observed in the presence of myeloablative doses of chemotherapeutic agents, indicating complete detoxification [14] (Fig. 10C). FMEV also allows dominant selection with MDR1 when a second gene is coexpressed. This is remarkable because coexpression of a second gene leads to reduced MDR1 expression when compared with the monocistronic counterpart [16]. Strong gene expression from FMEV vectors can also be instrumental for studies employing cell surface markers [150] or cytoplasmic proteins such as green fluorescent protein [151].

To further increase the transcriptional strength and specificity of FMEV, we are performing a molecular

**Evolution of vectors permissive for
potent gene expression in early hematopoietic and embryonic stem cells**

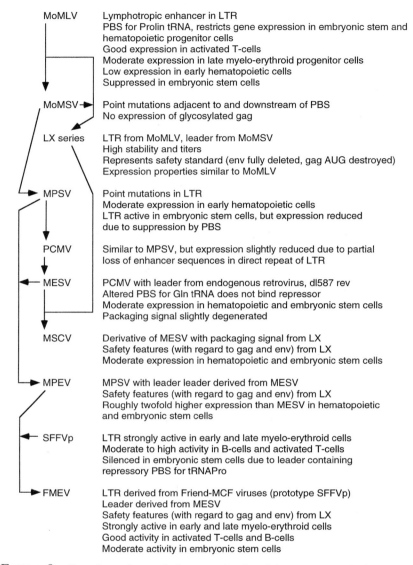

MoMLV — Lymphotropic enhancer in LTR
PBS for Prolin tRNA, restricts gene expression in embryonic stem and
hematopoietic progenitor cells
Good expression in activated T-cells
Moderate expression in late myelo-erythroid progenitor cells
Low expression in early hematopoietic cells
Suppressed in embryonic stem cells

MoMSV — Point mutations adjacent to and downstream of PBS
No expression of glycosylated gag

LX series — LTR from MoMLV, leader from MoMSV
High stability and titers
Represents safety standard (env fully deleted, gag AUG destroyed)
Expression properties similar to MoMLV

MPSV — Point mutations in LTR
Moderate expression in early hematopoietic cells
LTR active in embryonic stem cells, but expression reduced
due to suppression by PBS

PCMV — Similar to MPSV, but expression slightly reduced due to partial
loss of enhancer sequences in direct repeat of LTR

MESV — PCMV with leader from endogenous retrovirus, dl587 rev
Altered PBS for Gln tRNA does not bind repressor
Moderate expression in hematopoietic and embryonic stem cells
Packaging signal slightly degenerated

MSCV — Derivative of MESV with packaging signal from LX
Safety features (with regard to gag and env) from LX
Moderate expression in hematopoietic and embryonic stem cells

MPEV — MPSV with leader leader derived from MESV
Safety features (with regard to gag and env) from LX
Roughly twofold higher expression than MESV in hematopoietic
and embryonic stem cells

SFFVp — LTR strongly active in early and late myelo-erythroid cells
Moderate to high activity in B-cells and activated T-cells
Silenced in embryonic stem cells due to leader containing
repressory PBS for tRNAPro

FMEV — LTR derived from Friend-MCF viruses (prototype SFFVp)
Leader derived from MESV
Safety features (with regard to gag and env) from LX
Strongly active in early and late myelo-erythroid cells
Good activity in activated T-cells and B-cells
Moderate activity in embryonic stem cells

FIGURE 9   Genealogy of retroviral vectors developed for strong constitutive gene
expression in early hematopoietic cells and embryonic stem cells.

analysis of Friend-MCF types of enhancers. At least three crucial motifs contributing to strong and relatively lineage-independent activity in hematopoietic cells were identified: recognition sites for the ubiquitous transactivator, Sp1, ETS family members, and AML1/PEBP [138] (Fig. 8). As was expected, these are all important transcriptional regulators in hematopoietic cells [152]. Additional activation may result from E-Box binding basic helix–loop–helix factors [153], and Myb [154]. Similar recognition sites are represented in a number of endogenous promoters controlling differentiation-dependent cellular genes. Such cellular motifs can be successfully incorporated in retroviral vectors [155]. Variations in enhancer assembly, e.g., by developing hybrid enhancers composed of distinct modules of retroviral or endogenous enhancers, are expected to result in even higher gene expression levels. Other alterations may lead to more specific and lineage-restricted activity within the hematopoietic system. Thus, it seems possible to develop novel enhancers that are strongly recognized

A **Retroviral vectors**

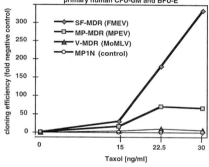

B **Selective advantage**

C **Colony morphology**

SF-MDR (FMEV) high proliferative potential
complete differentiation
low residual toxicity

MP-MDR (MPEV) medium proliferative potential
complete differentiation
only mderate residual toxicity

V-MDR (MoMLV) low proliferative potential
incomplete differentiation
dying cells due to pronounced
residual toxicity

FIGURE 10   Vector design determines phenotype, here shown for myeloprotection by drug resistance gene transfer. (A) Different types of retroviral vectors evaluated in context of transfer of the multidrug resistance gene (MDR1). (B) Relative selective advantage conferred to primary human hematopoietic colony forming units (CFUs) kept under selection with the chemotherapeutic agent Taxol, recognized by the MDR1-encoded efflux pump, P-glycoprotein. Data are calculated from Eckert *et al.* [14] and expressed as cloning efficiency fold negative control, i.e., cells transduced with MP1N. PC-MDR (MESV type) is only slightly better than V-MDR and therefore not shown. (C) Average colony morphology at selection with 15 ng Taxol/mL reveals importance of complete detoxification. This can only be achieved with vector backbones of strong transcriptional activity.

in hematopoietic progenitors, but have low activity in tumor cells, e.g., those of epithelial origin. With such hematopoiesis-specific enhancers, transduction of

nonhematopoietic tumor cells would have no significant consequences in terms of inducing drug resistance.

2. T LYMPHOCYTES

Although they represent a mature blood cell population, T cells can be very long-lived and have the capacity for limited clonal activation and expansion. In the switch between the resting and the activated status, chromosomal organization and transcription factor equipment is reordered. Thus, stably integrating retroviral vectors are a perfect tool for genetic manipulation of T cells, but vector expression may vary depending on the cellular activation status. All MLV-based vectors described in Fig. 10A mediate sufficient expression in activated T lymphocytes for application in adoptive immunotherapy or tumor vaccination [156]. The enhancer of SL3-3, a highly lymphotropic MLV, is an interesting alternative [157]. High expression levels in T cells have also been achieved using control elements of HIV-1, but only in the presence of the HIV-encoded transactivator, Tat [158]. As discussed for early hematopoietic cells, insights into the molecular mechanisms defining T lymphotropism of retroviral or endogenous enhancers is expected to create the basis for developing artificial transgene enhancers with increased T cell specificity. Interestingly, as with some endogenous T lymphocytic promoters, *reversible* down-regulation of retroviral gene expression was observed in resting T cells. This might be prevented by including scaffold attachment regions in the vector [159]. These *cis*-active sequences separate transcriptional domains and are known to increase autonomy, that is, decrease site specificity, of physicochemically transfected transgenes [160].

3. TUMOR CELLS

Mechanisms of tumor specific transcriptional controls are of interest for targeting of tumor cells in suicide gene transfer and oncogene antagonism, as outlined in Section II. Generally, the specificity of heterologous promoters in retroviral vectors is increased when more promiscuous retroviral enhancer sequences are deleted. Transcriptional targeting of tumors can be achieved using control elements of genes that are "tumor specific" or overexpressed in tumors. When targeting metastases, control elements of genes specific to the parental tissue of the tumor might also be sufficient. Also, hypoxia-responsive promoters have been proposed for tumor targeting [161]. A more indirect approach is the targeting of endothelial cells involved in tumor angiogenesis using "endotheliotropic" control regions. Thus, an ever increasing number of candidate promoters is being proposed (reviewed by Sikora [20] and Miller and Whelan

[162]). However, for most of these, evidence for tumor specificity *in vivo* is yet to be confirmed.

4. SILENCING

Silencing not only reduces the efficiency, it can also compromise the safety of gene transfer strategies. This is of special importance for negative selection of transduced cells (as required in suicide gene transfer, adoptive immunotherapy, or mini-organs, see Section II). Here, cells having silenced the vector will escape exogenous control. Silencing results from dominant negative influences of the integration site, or it may be directly triggered by the integrated vector. Silencing involves functional reorganizations within the chromosome. As a result, vector sequences can be methylated in CpG islands, which may play a role for fixation of downregulation [163,164]. The speed and incidence of silencing depend on the cellular background, on the genomic integration site, and (not well defined) on specific vector sequences, including transgene cDNAs (reviewed by Baum [12] and Lund *et al.* [139]). This opens perspectives for active prevention of silencing by vector improvements. Studies with housekeeping promoters indicate that Sp1 binding sites can counteract silencing to some extent [165]. The retroviral enhancers of MoMLVs, MPSVs, PCMVs, and Friend-MCF viruses differ with respect to number, affinity, and positioning of Sp1 binding sites [138,166]. The relevance to long-term expression remains to be shown. Furthermore, MESV-derived leader sequences or vectors containing other, even artificial primer binding sites avoiding transcriptional repression in embryonic and hematopoietic cells (see Section V.C.1.) might support long-term expression. However, silencing of MESV-leader-based vectors is also observed upon differentiation of embryonic stem cells permissive to vector expression in the undifferentiated state [140]. Inclusion of scaffold attachment regions, eventually shielding retroviral control regions from negative influences of the integration site, might prevent differentiation-dependent silencing to some extent. Our experiences with such vectors are promising, providing evidence for reduced integration-site dependence of vector expression [167]. Systematic analyses in appropriate primary cell systems and results from comparative clinical studies are still awaited to clarify the significance of this issue.

5. REGULATABLE PROMOTERS

Regulatable promoters are of interest for generating artificial mini-organs and also for drug resistance gene transfer (see Section II). Progress in regulatable promoter systems has recently been described [162]. Best documented in retroviral vectors is the tetracycline-regulated system, available both for conditional repression and for induction of transgene expression [168,169]. Moreover, a number of alternative artificial systems for conditional promoter induction or repression are already described [162]. The applicability of synthetic inducer/promoter systems has been demonstrated *in vivo* using retroviral vectors expressing erythropoietin from a tetracycline-regulated cassette [19]. Further advances in regulated vectors are expected to address potential limitations of the systems: side effects of the drugs administered for regulation, immunogenicity of the synthetic transactivators or repressors employed, toxic squelching effects eventually occurring from overexpressed synthetic transcription factors, clonal variabilities in inducibility related to the retroviral integration site, differentiation dependence of regulation, and maintenance of regulation over time.

6. COEXPRESSION STRATEGIES

Vectors expressing more than one transgene greatly widen the perspectives of most cancer gene therapy approaches. Depending on the specific application, coexpression is used to combine two selectable marker genes, a selectable marker gene with a nonselectable gene, or two nonselectable genes (Table 3). There are several options for simultaneously expressing different biological functions from a single vector (Fig. 11). In general, type and positioning of transgenes as well as cellular background and specific experimental conditions (especially the stringency of selection applied) greatly influence the efficacy of the coexpression strategy. Therefore, systematic comparative studies appear desirable for each coexpression vector developed for a specific clinical use.

a. Internal Promoters   To express genes from retroviral vectors, promoters can not only be placed in the LTR, they can also be inserted in either orientation in the sequences between the leader and the 3' LTR (Fig. 11A). These internal promoters can be used in vectors when the U3 promoter has been deleted, or in addition to an LTR-controlled transcription unit. However, when two promoters are located close to each other, there is the potential of promoter interference, leading to shutdown of one promoter to the advantage of its neighbor [170–172]. The stronger promoter (or the promoter selected for) either tends to exploit enhancer sequences of the neighboring promoter or inhibits formation of the Pol II initiation complex at the internal promoter. Here, separation by transcriptional termination signals would be a possible solution [172,173]. However, this is inappropriate in retroviral vectors because it would lead to premature termination of genomic messages in packaging cells. Placing the internal promoter in antisense orientation to the LTR might reduce interference at the

TABLE 3   Reasons for Expressing Two or More Genes from a Single Vector

Combination	Approach (see Section II)	Example
Selectable marker gene only.	Drug resistance gene transfer	Complementary drug resistance genes (to widen spectrum of resistance).
		Drug resistance gene(s) plus suicide gene (to remove transduced cells in case of pathogenicity).
	Adoptive immunotherapy	Surface marker plus suicide gene (to select transduced cells before reinfusion).
	Suicide gene transfer	Two suicide genes (improves efficacy).
Selectable marker gene plus nonselectable gene.	Mini-organs	Suicide gene plus therapeutic gene of interest (to remove transduced cells in case of pathogenicity).
	Oncogene antagonism	Suicide gene plus anti-oncogene (improves efficacy).
Nonselectable genes only.	Oncogene antagonism	Complementary anti-oncogenes (improves efficacy).
	Tumor vaccination	Cooperating immunostimulatory genes (improves efficacy).

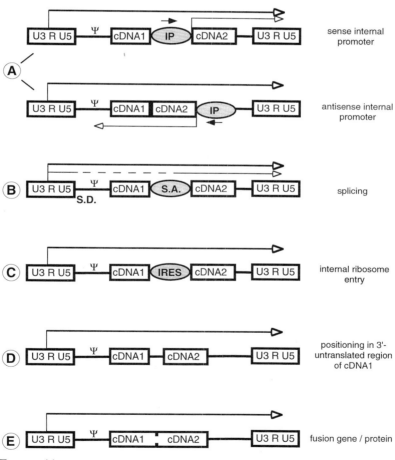

FIGURE 11   Strategies for coexpression of two genes from a retroviral vector (see Section V.B.6). Open arrows indicate mRNAs, the bold arrow represents the genomic message of the vector. ψ, packaging signal; IP, internal promoter (orientation indicated by the filled arrow); S.D. and S.A., splice donor and splice acceptor, respectively; IRES, internal ribosomal entry site.

transcriptional level but necessarily generates antisense RNA, which is expected to disturb translation of the cotransferred gene. Therefore, vectors containing internal promoters might generate unwanted effects, especially under conditions of dominant selection for only one promoter, as in adoptive immunotherapy.

b. Alternative Splicing   For reasons not entirely understood, retroviral splice donor and splice acceptor sequences are only partly recognized in host cells. This leads to a defined ratio of genomic and subgenomic messages and can be exploited for constructing splicing vectors, which sometimes, but not always, yield good results [16,174]. Generating a spliced, subgenomic message can be associated with improved nuclear export, increased half-life of cytoplasmic RNA, or improved translation efficacy. Importantly, type and positioning of the transgenes will affect the efficacy of alternative splicing. Finally, cDNAs inserted in splice vectors must be free of cryptic splice signals (Fig. 11B).

c. Internal Ribosome Entry   The internal ribosomal entry site (IRES) was originally described in picorna viruses. The IRES is a complex domain of the RNA (in the size of a few hundred base pairs), generating a specific structure allowing cap-independent initiation of translation. When introduced in front of the start codon of the transgene, bi- or even oligocistronic vectors can be generated [175]. Compared to internal promoters and alternative splicing, IRES control has the advantage of exploiting a single mRNA for translation of two (or more) proteins (Fig. 11C). However, not every cDNA is fully compatible with translation via an IRES, and sometimes alternative and mutually exclusive rather than simultaneous initiation of translation might predominate. Many reports state that IRES-dependent initiation of translation occurs as efficiently as that from capped RNAs, with capped RNAs referring to transgenes expressed from within the *gag* region of vector mRNAs [171,176,177]. However, recently it was demonstrated that MLVs also use an IRES related mechanism for translating Gag proteins [178], and there is accumulating evidence that initiation of translation from within the *gag* region is suboptimal [67]. Also, we observed that some large IRES-containing RNAs might be associated with reduced nuclear export or cytoplasmic half-life [150]. Thus, it is still unclear whether IRES control occurs without loss of efficiency. Moreover, to what extent IRES translation is subject to differentiation-dependent control remains to be elucidated.

d. Positioning in Untranslated Vector Regions   Sometimes it is sufficient to express RNA without translation, as in approaches utilizing antisense RNA or ribozymes for oncogene antagonism (see Section II). These therapeutic RNAs can be located in untranslated vector regions, preferably in the 3′ untranslated region of another gene coexpressed from the vector (Figure 11D). The same strategy cannot be recommended for open reading frames: Spontaneous reinitiation of translation from the 3′ untranslated region of a gene occurs at greatly reduced efficiency [53].

e. Fusion Proteins   Multifunctional fusion proteins are a good choice for coexpression provided that the domains of interest are active in similar subcellular localizations (Fig. 11E). Some cytosolic proteins might also function when expressed as cytoplasmic tail of a membrane-anchored fusion protein [179]. It remains to be determined whether the efficacy of the fusion protein is comparable to those of the individual components. A potential risk of this approach is that the fusion site might give rise to an immunogenic peptide.

7. *cis*-ACTING ELEMENTS IN cDNAs

Even cDNAs can contain *cis*-acting elements, active at either the transcriptional [180] or the posttranscriptional level. This aspect of vector design is often neglected, but it can have profound influence on overall vector performance. The retroviral life cycle implies that aberrant signals for splicing, termination and polyadenylation, primer binding, or cryptic PP tracts will reduce vector titers or give rise to rearranged vector copies, with unpredictable immunologic or toxicological consequences. Examples relevant to cancer gene therapy are the drug resistance genes MDR1 [16,181] and ALDH1 [182]. A stable, selectable marker gene coexpressed with the unstable sequence can serve as a tool to tag hot spots of recombination, providing the basis for cDNA improvement [16]. Also, cDNAs can harbor silencer elements or might contain enhancers influencing levels as well as tissue tropism of vector expression. Thus, a lot of fine-tuning work is required to develop stable and, hence, safe vectors suitable for actual clinical use. Evolution has done that work for retroviral genes. Vector designers usually follow empirical approaches, not always the most elegant and effective way to success.

VI. OUTLOOK

Retroviral vector systems have dominated cancer gene therapy research in the past years, and they will certainly continue to play an important role. However, in future clinical trials it will be of outstanding importance to use specifically tailored and highly effective vectors. Only then can the perspectives of gene therapy concepts be evaluated. Based on a deeper understanding of the biology of retroviruses and their target cells, improved vector systems have already been created that now await clinical testing to assess efficacy and safety. Key developments include the advent of complex retrovirus-based systems for transduction of nondividing cells, pseudotyping and envelope engineering to widen or specify the host range at the level of transduction, and higher diversity in enhancer choice based on deeper insights into the transcriptional control of retroviral transgenes. Especially, further progress in the field of transductional and transcriptional targeting will have substantial impact for the therapeutic quality of cancer gene therapy approaches. So far, vector design has been dominated and also limited by deductive analyses of virus–host interactions. Future vector design should also follow a more evolutionary approach, taking advantage of the inherent genetic variability of viruses. Therefore, we need to establish intelligent systems for selecting and screening improved mutants. For widely applicable oncological strategies, tailoring can be performed as an international, multicenter effort. Unfortunately, for more specialized applications with small patient numbers this will be unaffordable. Here, concentration in specific centers of expertise might represent a solution. Importantly, many aspects of vector tailoring worked out using retroviral systems will also be applicable to nonretroviral systems (based on adenoviruses, adeno-associated viruses, herpes viruses, or physicochemical methods), which are emerging as an important alternative for some approaches in cancer gene therapy and will substantially widen the perspectives of the field.

References

1. Mulligan, R. C. (1993). The basic science of gene therapy. *Science* **260**, 926–932.
2. Friedmann, T. (1996). Human gene therapy–an immature genie, but certainly out of the bottle. *Nat. Med.* **2**, 144–145.
3. Dickman, S. (1997). Richard Mulligan: from skeptic to true believer. *Curr. Biol.* **7**, R601–R602.
4. Brenner, M. K. (1994). Genetic marking and manipulation of hematopoietic progenitor cells using retroviral vectors. *Immunomethods* **5**, 204–210.
5. Dunbar, C. E. (1996). Gene transfer to hematopoietic stem cells: implications for gene therapy of human disease. *Annu. Rev. Med.* **47**, 11–20.
6. Rooney, C. M., Smith, C. A., Ng, C. Y., Loftin, S., Li, C., Krance, R. A., Brenner, M. K., and Heslop, H. E. (1995). Use of gene-modified virus-specific T lymphocytes to control Epstein-Barr-virus-related lymphoproliferation. *Lancet* **345**, 9–13.
7. Pawliuk, R., Eaves, C. J., and Humphries, K. R. (1997). Sustained high-level reconstitution of the hematopoietic system by preselected hematopoietic cells expressing a transduced cell-surface antigen. *Hum. Gene Ther.* **8**, 1595–1604.
8. Phillips, K., Gentry, T., McCowage, G., Gilboa, E., and Smith, C. (1996). Cell-surface markers for assessing gene transfer into human hematopoietic cells. *Nat. Med.* **2**, 1154–1156.
9. Fehse, B., Uhde, A., Fehse, N. Eckert, H. G., Clausen, J., Rüger, R., Koch, S., Ostertag, W., Zander, A. R., and Stockschläder, M. (1997). Selective immunoaffinity-based enrichment of CD34+ cells transduced with retroviral vectors containing an intracytoplasmatically truncated version of the human low-affinity nerve growth factor receptor (ΔLNGFR) gene. *Hum. Gene Ther.* **8**, 1815–1827.
10. Gottesman, M. M., Germann, U. A., Aksentijevich, I., Sugimoto, Y., Cardarelli, C. O., and Pastan, I. (1994). Gene transfer of drug resistance genes. Implications for cancer therapy. *Ann. N.Y. Acad. Sci.* **716**, 126–138.
11. Baum, C., Margison, G. P., Eckert, H.-G., Fairbairn, L., Ostertag, W., and Rafferty, J. A. (1996). Gene transfer to augment the therapeutic index of anticancer chemotherapy. *Gene Ther.* **3**, 1–3.
12. Baum, C. (1997). Gene transfer and transgene expression in hematopoietic cells. In: *Concepts in Gene Therapy* (M. Strauss and J. A. Barranger, eds.), pp. 233–265. DeGruyter, Berlin.
13. Baum, C., Hegewisch-Becker, S., Eckert, H.-G., Stocking, C., and Ostertag, W. (1995). Novel retroviral vectors for efficient expression of the multidrug resistance (*mdr*-1) gene in early hematopoietic cells. *J. Virol.* **69**, 7541–7547.
14. Eckert, H.-G., Stockschläder, M., Just, U., Hegewisch-Becker, S., Grez, M., Uhde, A., Zander, A., Ostertag, W., and Baum, C. (1996). High-dose multidrug resistance in primary human hematopoietic progenitor cells transduced with optimized retroviral vectors. *Blood* **88**, 3407–3415.
15. Galipeau, J., Benaim, E., Spencer, H. T., Blakley, R., and Sorrentino, B. P. (1997). A bicistronic retroviral vector for protecting hematopoietic cells against antifolates and P-glycoprotein effluxed drugs. *Hum. Gene Ther.* **8**, 1773–1783.
16. Hildinger, M., Fehse, B., Hegewisch-Becker, S., John, J., Rafferty, J. R., Ostertag, W., and Baum, C. (1998). Dominant selection of hematopoietic progenitor cells with retroviral MDR1 co-expression vectors. *Hum. Gene Ther.* **9**, 33–42.
17. Bonini, C., *et al.* (1997). HSV-TK gene transfer into donor lymphocytes for control of allogeneic graft-versus-leukemia. *Science* **176**, 1719–1724.
18. Bohl, D., and Heard, J.-M. (1997). *In vivo* secretion of therapeutic proteins from neo-organs. In: *Concepts in Gene Therapy* (M. Strauss and J. A. Barranger, eds.), pp. 297–314. De Gruyter, Berlin.
19. Bohl, D., and Heard, J. M. (1997). Modulation of erythropoietin delivery from engineered muscles in mice. *Hum. Gene Ther.* **8**, 195–204.
20. Sikora, K. (1994). Genetic approaches to cancer therapy. *Gene Ther.* **1**, 149–151.
21. Harris, J. D., Gutierrez, A. A., Hurst, H. C., Sikora, K., and Lemoine, N. R. (1994). Gene therapy for cancer using tumour-specific prodrug activation. *Gene Ther.* **1**, 170–175.
22. Roth, J. A., Nquyen, D., Lawrence, D. D., Kemp, B. L., Carrasco, C. H., *et al.* (1996). Retrovirus-mediated wild-type p53 gene transfer to tumors of patients with lung cancer. *Nat. Med.* **2**, 985–991.
23. Yang, Z. Y., Perkins, N. D., Ohno, T., Nabel, E. G., and Nabel, G. J. (1995). The p21 cyclin-dependent kinase inhibitor suppresses tumorigenicity *in vivo*. *Nat. Med.* **1**, 1052–1056.
24. Wygoda, M. R., Wilson, M. R., Davis, M. A., Trosko, J. E., Rehemtulla, A., and Lawrence, T. S. (1997). Protection of herpes simplex virus thymidine kinase-transduced cells from ganciclovir-mediated cytotoxicity by bystander cells: the Good Samaritan effect. *Cancer Res.* **57**, 1699–1703.
25. Hurford, R. J., Dranoff, G., Mulligan, R. C., and Tepper, R. I. (1995). Gene therapy of metastatic cancer by *in vivo* retroviral gene targeting. *Nat. Genet.* **10**, 430–435.

26. Bischoff, J. R., Kirn, D. H., Willimas, A., Heise, C., Horn, S., Muna, M., Ng, L., Sampson-Johannes, A., Fattaey, A., and McCormick, F. (1996). An adenovirus mutant that replicates selectively in p53-deficient human tumor cells. *Science* **274**, 373–376.

27. Blankenstein, T., Cayeux, S., and Qin, Z. (1996). Genetic approaches to cancer immunotherapy. *Rev. Physiol. Biochem. Pharmacol.* **129**, 1–49.

28. Rosenthal, F. M., Zier, K. S., and Gansbacher, B. (1994). Human tumor vaccines and genetic engineering of tumors with cytokine and histocompatibility genes to enhance immunogenicity. *Curr. Opin. Oncol.* **6**, 611–615.

29. Coffin, J. M. (1996). Retroviridae: the viruses and their replication. In: *Field's Virology* (B. N. Fields, D. M. Knipe, and P. M. Howley, eds.), pp. 763–844. Lippincott–Raven, Philadelphia.

30. Miller, A. D., and Chen, F. (1996). Retrovirus packaging cells based on 10A1 murine leukemia virus for production of vectors that use multiple receptors for cell entry. *J. Virol.* **70**, 5564–5571.

31. Mergia, A., Shaw, K. E., Lowe, E., Barry, P. A., and Luciw, P. A. (1990). Simian foamy virus type 1 is a retrovirus which encodes a transcriptional transactivator. *J. Virol.* **64**, 3598–3604.

32. Luciw, P. A. (1996). Human immunodeficiency viruses and their replication. In: *Field's Virology* (B. N. Fields, D. M. Knipe, and P. M. Howley, eds.), pp. 1881–1975. Lippincott–Raven, Philadelphia.

33. Bender, M. A., Palmer, T. D., Gelinas, R. E., and Miller, A. D. (1987). Evidence that the packaging signal of Moloney murine leukemia virus extends into the gag region. *J. Virol.* **61**, 1639–1646.

34. Berkowitz, R., Fisher, J., and Goff, S. P. (1996). RNA packaging. *Curr. Top. Microbiol. Immunol.* **214**, 177–218.

35. Berkowitz, R., Hammarskjold, M.-L., Helga-Maria, C., Rekosh, D., and Goff, S. (1995). 5' regions of HIV-1 RNAs are not sufficient for encapsidation: implications for the HIV-1 packaging signal. *Virology* **212**, 718–723.

36. Richardson, J. H., Child, L. A., and Lever, A. M. (1993). Packaging of human immunodeficiency virus type 1 RNA requires *cis*-acting sequences outside the 5' leader region. *J. Virol.* **67**, 3997–4005.

37. Weiss, R. A. (1993). Pseudotyped viruses and envelope composition. In: *The Retroviridae* (J. A. Levy, ed.), pp. 5–8. Plenum Press, New York.

38. Hunter, E., and Swanstrom, R. (1990). Retrovirus envelope glycoproteins. *Curr. Top. Microbiol. Immunol.* **157**, 187–253.

39. Temin, H. M. (1993). Retrovirus variation and reverse transcription: abnormal strand transfers result in retrovirus genetic variation. *Proc. Natl. Acad. Sci. USA* **90**, 6900–6903.

40. Roe, T., Reynolds, T., Yu, G., and Brown, P. O. (1993). Integration of murine leukemia virus DNA depends on mitosis. *EMBO J.* **12**, 2099–2108.

41. Miller, D. G., Adam, M. A., and Miller, A. D. (1990). Gene transfer by retrovirus vectors occurs only in cells that are actively replicating at the time of infection. *Mol. Cell. Biol.* **10**, 4239–4242.

42. Bukrinsky, M. I., Haggerty, S., Dempsey, P., Sharova, N., Adzhubei, A., Spitz, L., Lewis, P., Goldfarb, D., Emerman, M., and Stevenson, M. (1993). A nuclear localization signal within HIV-1 matrix protein that governs infection of non-dividing cells. *Nature* **365**, 666–669.

43. Rohdewohld, H., Weiher, H., Reik, W., Jaenisch, R., and Breindl, M. (1987). Retrovirus integration and chromatin structure: Moloney murine leukemia proviral integration sites map near DNase I-hypersensitive sites. *J. Virol.* **61**, 336–343.

44. Jonkers, J., and Berns, A. (1996). Retroviral insertional mutagenesis as a strategy to identify cancer genes. *Biochim. Biophys. Acta* **1287**, 29–57.

45. Donahue, R. E., Kessler, S. W., Bodine, D., McDonagh, K., Dunbar, C., Goodman, S., Agricola, B., Byrne, E., Raffeld, M., Moen, R., *et al.* (1992). Helper virus induced T cell lymphoma in nonhuman primates after retroviral mediated gene transfer. *J. Exp. Med.* **176**, 1125–1135.

46. Stocking, C., Bergholz, U., Friel, J., Klingler, K., Wagener, T., Starke, C., Kitamura, T., Miyajima, A., and Ostertag, W. (1993). Distinct classes of factor-independent mutants can be isolated after retroviral mutagenesis of a human myeloid stem cell line. *Growth Factors* **8**, 197–209.

47. Tsichlis, P. N., and Lazo, P. A. (1991). Virus-host interactions and the pathogenesis of murine and human oncogenic retroviruses. In: *Retroviral Insertion and Oncogene Activation* (H. G. Kung and P. K. Vogt, eds.), pp. 95–173. Springer-Verlag, Berlin.

48. Ostertag, W., Stocking, C., Johnson, G. R., Kluge, N., Kollek, R., Franz, T., and Hess, N. (1987). Transforming genes and target cells of murine spleen focus-forming viruses. *Adv. Cancer Res.* **48**, 193–355.

49. Luciw, P. A., and Leung, N. J. (1992). Mechanisms of retrovirus replication. In: *The Retroviridae 1* (J. A. Levy, ed.), pp. 159–298. Plenum Press, New York.

50. Kräusslich, H.-G., and Welker, R. (1996). Intracellular transport of capsid components. In: Morphogenesis and Maturation of Retroviruses (H. G. Kräusslich, ed.), pp. 25–64. Springer-Verlag, Berlin.

51. Einfeld, D. (1996). Maturation and assembly of retroviral glycoproteins. In: Morphogenesis and Maturation of Retroviruses (H. G. Kräusslich, ed.), pp. 133–176. Springer-Verlag, Berlin.

52. Miller, A. D. (1990). Retrovirus packaging cells. *Hum. Gene Ther.* **1**, 5–14.

53. Cosset, F.-L., Takeuchi, Y., Battini, J.-L., Weiss, R. A., and Collins, M. K. L. (1995). High-titer packaging cells producing recombinant retrovirus resistant to human serum. *J. Virol.* **69**, 7430–7436.

54. Wagener, T., Stocking, C., and Ostertag, W. (1995). unpublished data.

55. Hatzoglou, M., Hodgson, C. P., Mularo, F., and Hanson, R. W. (1990). Efficient packaging of a specific VL30 retroelement by psi 2 cells which produce MoMLV recombinant retroviruses. *Hum. Gene Ther.* **1**, 385–397.

56. Chong, H., and Vile, R. G. (1996). Replication-competent retrovirus produced by a "split-function" third generation amphotropic packaging cell line. *Gene Ther.* **3**, 624–629.

57. Vanin, E. F., Kaloss, M., Broscius, C., and Nienhuis, A. W. (1994). Characterization of replication-competent retroviruses from nonhuman primates with virus-induced T-cell lymphomas and observations regarding the mechanism of oncogenesis. *J. Virol.* **68**, 4241–4250.

58. Münk, C., Lohler, J., Prassolov, V., Just, U., Stockschlader, M., and Stocking, C. (1997). Amphotropic murine leukemia viruses induce spongiform encephalomyelopathy. *Proc. Natl. Acad. Sci. USA* **94**, 5837–5842.

59. Isacson, O., and Brakefield, X. O. (1997). Benefits and risks of hosting animal cells in the human brain. *Nat. Med.* **3**, 964–969.

60. Takeuchi, Y., Cosset, F. L., Lachmann, P. J., Okada, H., Weiss, R. A., and Collins, M. K. (1994). Type C retrovirus inactivation by human complement is determined by both the viral genome and the producer cell. *J. Virol.* **68**, 8001–8007.

61. Takeuchi, Y., Porter, C. D., Strahan, K. M., Preece, A. F., Gustafsson, K., Cosset, F. L., Weiss, R. A., and Collins, M. K. (1996). Sensitization of cells and retroviruses to human serum by (alpha 1-3) galactosyltransferase. *Nature* **379**, 85–88.

62. Palsson, B., and Andreadis, S. (1997). The physico-chemical factors that govern retrovirus-mediated gene transfer. *Exp. Hematol.* **25**, 94–102.

63. Kinsella, T. M., and Nolan, G. P. (1996). Epsiomal vectors rapidly and stably produce high-titer recombinant retrovirus. *Hum. Gene Ther.* **7**, 1405–1413.

64. Wilson, C. A., Ng, T. H., and Miller, A. E. (1997). Evaluation of recommendations for replication-competent retrovirus testing associated with use of retroviral vectors. *Hum. Gene Ther.* **8**, 869–874.

65. Correll, P. H., Colilla, S., and Karlsson, S. (1994). Retroviral vector design for long-term expression in murine hematopoietic cells *in vivo. Blood* **84**, 1812–1822.

66. Miller, A. D., and Rosman, G. J. (1989). Improved retroviral vectors for gene transfer and expression. *Biotechniques* **7**, 980–982.

67. Krall, W. J., Skelton, D. C., Yu, X.-J., Riviere, I., Lehn, P., Mulligan, R. C., and Kohn, D. B. (1996). Increased levels of spliced RNA account for augmented expression from the MFG retroviral vector in hematopoietic cells. *Gene Ther.* **3**, 37–48.

68. Hantzopulos, P. A., Sullenger, B. A., Ungers, G., and Gilboa, E. (1989). Improved gene expression upon transfer of the adenosine deaminase minigene outside the transcriptional unit of a retroviral vector. *Proc. Natl. Acad. Sci. USA* **86**, 3519–3523.

69. Adam, M. A., Osborne, W. R., and Miller, A. D. (1995). R-region cDNA inserts in retroviral vectors are compatible with virus replication and high-level protein synthesis from the insert. *Hum. Gene Ther.* **6**, 1169–1176.

70. Yu, S. F., von Ruden, T., Kantoff, P. W., Garber, C., Seiberg, M., Ruther, U., Anderson, W. F., Wagner, E. F., and Gilboa, E. (1986). Self-inactivating retroviral vectors designed for transfer of whole genes into mammalian cells. *Proc. Natl. Acad. Sci. USA* **83**, 3194–3198.

71. Olson, P., Nelson, S., and Dornburg, R. (1994). Improved self-inactivating retroviral vectors derived from spleen necrosis virus. *J. Virol.* **68**, 7060–7066.

72. Bergemann, J., Kuhlcke, K., Fehse, B., Ratz, I., Ostertag, W., and Lother, H. (1995). Excision of specific DNA-sequences from integrated retroviral vectors via site-specific recombination. *Nucleic Acids Res.* **23**, 4451–4456.

73. Russ, A. P., Friedel, C., Grez, M., and von Melchner, H. (1996). Self-deleting retrovirus vectors for gene therapy. *J. Virol.* **70**, 4927–4932.

74. Beck-Engeser, G., Stocking, C., Just, U., Albritton, L., Dexter, M., Spooncer, E., and Ostertag, W. (1991). Retroviral vectors related to the myeloproliferative sarcoma virus allow efficient expression in hematopoietic stem and precursor cell lines, but retroviral infection is reduced in more primitive cells. *Hum. Gene Ther.* **2**, 61–70.

75. Crooks, G. M., and Kohn, D. B. (1993). Growth factors increase amphotropic retrovirus binding to human CD34+ bone marrow progenitor cells. *Blood* **82**, 3290–3297.

76. Orlic, D., Girard, L. J., Jordan, C. T., Anderson, S. M., Cline, A. P., and Bodine, D. M. (1996). The level of mRNA encoding the amphotropic retrovirus receptor in mouse and human hematopoietic stem cells is low and correlates with the efficiency of retrovirus transduction. *Proc. Natl. Acad. Sci. USA* **93**, 11097–11102.

77. von Laer, D., Thomsen, S., Vogt, B., Donath, M., Kruppa, J., Rein, A., Ostertag, W., and Stocking, C. (1998). Entry of amphotropic and 10A1 pseudotyped murine retroviruses is restricted in hematopoietic stem cell lines. *J. Virol.* **72**, 424–430.

78. Munk, C., and Stocking, C. (1997). unpublished data.

79. Rubin, H. (1965). Genetic control and cellular susceptibility to pseudotypes of Rous sarcoma virus. *Virology* **26**, 270–282.

80. Friedmann, T., and Yee, J.-K. (1995). Pseudotyped retroviral vectors for studies of human gene therapy. *Nat. Med.* **1**, 275–277.

81. Januszeski, M. M., Cannon, P. M., Chen, D., Rozenberg, Y., and Anderson, W. F. (1997). Functional analysis of the cytoplasmic tail of Moloney murine leukemia virus envelope protein. *J. Virol.* **71**, 3613–3619.

82. Suomalainen, M., and Garoff, H. (1994). Incorporation of homologous and heterologous proteins into the envelope of Moloney murine leukemia virus. *J. Virol.* **68**, 4879–4889.

83. Emi, N., Friedmann, T., and Yee, J.-K. (1991). Pseudotype formation of murine leukemia virus with the G protein of vesicular stomatitis virus. *J. Virol.* **65**, 1202–1207.

84. Conti, C., Mastromarino, P., and Orsi, P. (1991). Role of membrane phospholipids and glycolipids in cell-to-cell fusion of VSV. *Comp. Immun. Microbiol. Infect. Dis.* **14**, 303–313.

85. Burns, J. C., Friedmann, T., Driever, W., Burrascano, M., and Yee, J.-K. (1993). Vesicular stomatitis virus G glycoprotein pseudotyped retroviral vectors: concentration to very high titer and efficient gene transfer into mammalian and nonmammalian cells. *Proc. Natl. Acad. Sci. USA* **90**, 8033–8037.

86. Matsubara, T., Beeman, R. W., Shike, H., Besansky, N. J., Mukabayire, O., Higgs, S., James, A. A., and Burns, J. C. (1996). Pantropic retroviral vectors integrate and express in cells of the malaria mosquito, *Anopheles gambiae. Proc. Natl. Acad. Sci. USA* **93**, 6181–6185.

87. Yee, J.-K., Miyanohara, A., LaPorte, P., Bouic, K., Burns, J. C., and Friedmann, T. (1994). A general method for the generation of high-titer, pantropic retroviral vectors: Highly efficient infection of primary hepatocytes. *Proc. Natl. Acad. Sci. USA* **91**, 9564–9568.

88. Akkina, R. K., Walton, R. M., Chen, M. L., Li, Q.-X., Planelles, V., and Chen, I. S. Y. (1996). High-efficiency gene transfer into CD34+ cells with a human immunodeficiency virus type 1-based retroviral vector pseudotyped with vesicular stomatitis virus envelope glycoprotein G. *J. Virol.* **70**, 2581–2585.

89. Yang, Y. P., Vanin, E. F., Whitt, M. A., Fornerod, M., Zwart, R., Schneiderman, R. D., Grosveld, G., and Nienhuis, A. W. (1995). Inducible, high-level production of infectious murine leukemia retroviral vector particles pseudotyped with vesicular stomatitis virus G envelope protein. *Hum. Gene Ther.* **6**, 1203–1213.

90. Bunnell, B. A., Mesler Muul, L., Donahue, R. E., Blaese, R. M., and Morgan, R. A. (1995). High-efficiency retroviral-mediated gene transfer into human and nonhuman primate peripheral blood lymphocytes. *Proc. Natl. Acad. Sci. USA* **92**, 7739–7743.

91. von Kalle, C., Kiem, H.-P., Goehle, S., Darovsky, B., Heimfeld, S., Torok-Storb, B., Storb, R., and Schuening, F. G. (1994). Increased gene transfer into human hematopoietic progenitor cells by extended *in vitro* exposure to a pseudotyped retroviral vector. *Blood* **84**, 2890–2897.

92. Lindemann, D., Bock, M., Schweizer, M., and Rethwilm, A. (1997). Efficient Pseudotyping of murine leukemia virus particles with chimeric human foamy virus envelope proteins. *J. Virol.* **71**, 4815–4820.

93. Wilson, C., Reitz, M. S., Okayama, H., and Eiden, M. V. (1989). Formation of infectious hybrid virions with gibbon ape leukemia virus and human T-cell leukemia virus retroviral envelope glycoproteins and the gag and pol proteins of Moloney murine leukemia virus. *J. Virol.* **63**, 2374–2378.

94. Mammano, F., Salvatori, F., Indraccolo, S., De Rossi, A., Chieco-Bianchi, L., and Göttlinger, H. G. (1997). Truncation of the human immunodeficiency virus type 1 envelope glycoprotein allows efficient pseudotyping of Moloney murine leukemia virus particles and gene transfer into CD4+ cells. *J. Virol.* **71**, 3341–3345.

95. Loiler, S. A., DiFronzo, N. L., and Holland, C. A. (1997). Gene transfer to human cells using retrovirus vectors produced by a new polytropic packaging cell line. *J. Virol.* **71**, 4825–4828.

96. von Laer, D. (1997). unpublished data.

97. Hanenberg, H., Xiao, L. X., Dilloo, D., Hashino, K., Kato, I., and Williams, D. A. (1996). Colocalization of retrovirus and target cells on specific fibronectin fragments increases genetic transduction of mammalian cells. *Nat. Med.* **2**, 876–882.

98. Moritz, T., Patel, V. P., and Williams, D. A. (1994). Bone marrow extracellular matrix molecules improve gene transfer into human hematopoietic cells via retroviral vectors. *J. Clin. Invest.* **93**, 1451–1457.

99. Cosset, F. L., and Russell, S. J. (1996). Targeting retrovirus entry. *Gene Ther.* **3**, 946–956.

100. Etienne-Julan, M., Roux, P., Carillo, S., Jeanteur, P., and Piechaczyk, M. (1992). The efficiency of cell targeting by recombinant retroviruses depends on the nature of the receptor and the composition of the artificial cell-virus linker. *J. Gen. Virol.* **73**, 3251–3255.

101. Kasahara, N., Dozy, A. M., and Kan, Y. W. (1994). Tissue-specific targeting of retroviral vectors through ligand-receptor interactions. *Science* **266**, 1373–1376.

102. Valsesia, W. S., Drynda, A., Deleage, G., Aumailley, M., Heard, J. M., Danos, O., Verdier, G., and Cosset, F. L. (1994). Modifications in the binding domain of avian retrovirus envelope protein to redirect the host range of retroviral vectors. *J. Virol.* **68**, 4609–4619.

103. Xiaoliang, H., Kasahara, N., and Wai Kan, Y. (1995). Ligand-directed retroviral targeting of human breast cancer cells. *Proc. Natl. Acad. Sci. USA* **92**, 9747–9751.

104. Chu, T. H., and Dornburg, R. (1995). Retroviral vector particles displaying the antigen-binding site of an antibody enable cell-type-specific gene transfer. *J. Virol.* **69**, 2659–2663.

105. Chu, T. H., and Dornburg, R. (1997). Toward highly efficient cell-type-specific gene transfer with retroviral vectors displaying single-chain antibodies. *J. Virol.* **71**, 720–725.

106. Nilson, B. H., Morling, F. J., Cosset, F. L., and Russell, S. J. (1996). Targeting of retroviral vectors through protease-substrate interactions. *Gene Ther.* **3**, 280–286.

107. Somia, N. V., Zoppe, M., and Verma, I. M. (1995). Generation of targeted retroviral vectors by using single-chain variable fragment: an approach to *in vivo* gene delivery. *Proc. Natl. Acad. Sci. USA* **92**, 7570–7574.

108. Marin, M., Noel, D., Valsesia, W. S., Brockly, F., Etienne, J. M., Russell, S., Cosset, F. L., and Piechaczyk, M. (1996). Targeted infection of human cells via major histocompatibility complex class I molecules by Moloney murine leukemia virus-derived viruses displaying single-chain antibody fragment-envelope fusion proteins. *J. Virol.* **70**, 2957–2962.

109. Russell, S. J., Hawkins, R. E., and Winter, G. (1993). Retroviral vectors displaying functional antibody fragments. *Nucleic Acids Res.* **21**, 1081–1085.

110. Valsesia-Wittmann, S., Morling, F. J., Hatziioannou, T., Russell, S. J., and Cosset, F. L. (1997). Receptor co-operation in retrovirus entry: recruitment of an auxiliary entry mechanism after retargeted binding. *EMBO J.* **16**, 1214–1223.

111. Naldini, L., Blömer, U., Gallay, P., Ory, D., Mulligan, R., Gage, F. H., Verma, I. M., and Trono, D. (1996). *In vivo* gene delivery and stable transduction of nondividing cells by a lentiviral vector. *Science* **272**, 263–267.

112. Naldini, L., Blömer, U., Gage, F. H., Trono, D., and Verma, I. M. (1996). Efficient transfer, integration, and sustained long-term expression of the transgene in adult rat brains injected with a lentiviral vector. *Proc. Natl. Acad. Sci. USA* **93**, 1382–11388.

113. Reiser, J., Harmison, G., Kluepfel-Stahl, S., Brady, R. O., Karlsson, S., and Schubert, M. (1996). Transduction of nondividing cells using pseudotyped defective high-titer HIV type particles. *Proc. Natl. Acad. Sci. USA* **93**, 15266–15271.

114. Blömer, U., Naldini, L., Kafri, T., Trono, D., Verma, I. M., and Gage, F. H. (1997). Highly efficient and sustained gene transfer in adult neurons with a lentivirus vector. *J. Virol.* **71**, 6641–6649.

115. Carroll, R., Lin, J.-T., Dacquel, E. J., Mosca, J. D., Burke, D. S., and St. Louis, D. C. (1994). A human immunodeficiency virus type 1 (HIV-1)-based retroviral vector system utilizing stable HIV-1 packaging cell lines. *J. Virol.* **68**, 6047–6051.

116. Shimada, T., Fujii, H., Mitsuya, H., and Neinhuis, A. (1991). Targeted and highly efficient gene transfer into CD4+ cells by a recombinant human immunodeficiency virus retroviral vector. *J. Clin. Invest.* **88**, 1043–1047.

117. Richardson, J. H., Kaye, J. F., Child, L. A., and Lever, A. M. L. (1995). Helper virus-free transfer of human immunodeficiency virus type 1 vectors. *J. Gen. Virol.* **76**, 691–696.

118. Konvalinka, J., Litterst, M. A., Welker, R., Kottler, H., Rippmann, F., Heuser, A. M., and Krausslich, H. G. (1995). An active-site mutation in the human immunodeficiency virus type 1 proteinase (PR) causes reduced PR activity and loss of PR-mediated cytotoxicity without apparent effect on virus maturation and infectivity. *J. Virol.* **69**, 7180–7186.

119. Corbeau, P., Kraus, G., and Wong-Staal, F. (1996). Efficient gene transfer by a human immunodeficiency virus type 1 (HIV-1)-derived vector using a stable HIV packaging cell line. *Proc. Natl. Acad. Sci. USA* **93**, 14070–14075.

120. Parolin, C., Dorfman, T., Palu, G., Gottlinger, H., and Sodroski, J. (1994). Analysis in human immunodeficiency virus type 1 vectors of *cis*-acting sequences that affect gene transfer into human lymphocytes. *J. Virol.* **68**, 3888–3895.

121. Schweizer, M., Turek, R., Hahn, H., Schliephake, A., Netzer, K.-O., Eder, G., Reinhardt, M., Rethwilm, A., and Neumann-Haefelin, D. (1995). Markers of foamy virus infections in monkeys, apes and accidentally infected humans: appropriate testing fails to confirm suspected foamy prevalence in humans. *AIDS Res. Hum. Retroviruses* **11**, 161–170.

122. Hooks, J. J., and Gibbs, C. J. J. (1995). The foamy viruses. *Bacteriol. Rev.* **39**, 169–185.

123. Mikovits, J. A., Hoffman, P. M., Rethwilm, A., and Ruscetti, F. W. (1996). *In vitro* infection of primary and retrovirus-infected human leukocytes by human foamy virus. *J. Virol.* **70**, 2774–2780.

124. Bieniasz, P. D., Weiss, R. A., and McClure, M. O. (1995). Cell cycle dependence of foamy retrovirus infection. *J. Virol.* **69**, 7295–7299.

125. Russell, D. W., and Miller, A. D. (1996). Foamy virus vectors. *J. Virol.* **70**, 217–222.

126. Sandmeyer, S. B., Hansen, L. J., and Chalker, D. L. (1990). Integration specificity of retrotransposons and retroviruses. *Annu. Rev. Genet.* **24**, 491–518.

127. Withers-Ward, E. S., Kitamura, Y., Barnes, J. P., and Coffin, J. M. (1994). Distribution of targets for avian retrovirus DNA integration *in vivo*. *Genes Dev.* **8**, 1473–1487.

128. Muller, H. P., and Varmus, H. E. (1994). DNA bending creates favored sites for retroviral integration: an explanation for preferred insertion sites in nucleosomes. *EMBO J.* **13**, 4704–4714.

129. Shibagaki, Y., and Chow, S. A. (1997). Central core domain of retroviral integrase is responsible for target site selection. *J. Biol. Chem.* **272**, 8361–8369.

130. Katz, R. A., Merkel, G., and Skalk, A. M. (1996). Targeting of retroviral integrase by fusion to a heterologous DNA binding domain: *in vitro* activities and incorporation of a fusion protein into viral particles. *Virology* **217**, 178–190.

131. Goulavic, H., and Chow, S. A. (1996). Directed integration of viral DNA mediated by fusion proteins consisting of human

immunodeficiency virus type 1 integrase and *Escherichia coli* LexA protein. *J. Virol.* **70**, 37–46.

132. Bushman, F. D., and Miller, M. D. (1997). Tethering human immunodeficiency virus type 1 preintegration complexes to target DNA promotes integration at nearby sites. *J.Virol.* **71**, 458–464.

133. Dildine, S. L., and Sandmeyer, S. B. (1997). Integration of the yeast retrovirus-like element Ty3 upstream of a human tRNA gene expressed in yeast. *Gene* **194**, 227–233.

134. Devine, S. E., and Boeke, J. D. (1996). Integration of the yeast retrotransposon Ty1 is targeted to regions upstream of genes transcribed by RNA polymerase III. *Genes Dev.* **10**, 620–633.

135. Feng, Q., Moran, J., Kazazian, H., and Boeke, J. D. (1996). Human L1 retrotransposon encodes a conserved endonuclease required for retrotransposition. *Cell* **87**, 905–916.

136. Ellis, J., and Bernstein, A. (1989). Gene targeting with retroviral vectors: recombination by gene conversion into regions of nonhomology. *Mol. Cell. Biol.* **9**, 1621–1627.

137. Pazin, M. J., Sheridan, P. L., Cannon, K., Cao, Z., Keck, J. G., Kadonga, J. T., and Jones, K. A. (1996). NF-kappa B-mediated chromatin reconfiguration and transcriptional activation of the HIV-1 enhancer *in vitro*. *Genes Dev.* **10**, 37–49.

138. Baum, C., Itoh, K., Meyer, J., Laker, C., Ito, Y., and Ostertag, W. (1997). The potent enhancer activity of the polycythemic strain of spleen focus-forming virus in hematopoietic cells is governed by a binding site for Sp1 in the upstream control region and by a unique enhancer core motif, creating an exclusive target for PEBP/CBF. *J. Virol.* **71**, 6323–6331.

139. Lund, A. H., Duch, M., and Pedersen, F. S. (1996). Transcriptional silencing of retroviral vectors. *J. Biomed. Sci.* **3**, 365–378.

140. Laker, C., Meyer, J., Schoopen, A., Friel, J., Ostertag, W., and Stocking, C. (1998). Host *cis*-mediated extinction of a retrovirus permissive for expression in embryonal stem cell. *J. Virol.* **72**, in press.

141. Challita, P.-M., and Kohn, D. B. (1994). Lack of expression from a retroviral vector after transduction of murine hematopoietic stem cells is associated with methylation *in vivo*. *Proc. Natl. Acad. Sci. USA* **91**, 2567–2571.

142. Scharfmann, R., Axelrod, J. H., and Verma, I. (1991). Long-term *in vivo* expression of retrovirus-mediated gene transfer in mouse fibroblast implants. *Proc. Natl. Acad. Sci. USA* **88**, 4626–2630.

143. Grez, M., Akgün, E., Hilberg, F., and Ostertag, W. (1990). Embryonic stem cell virus, a recombinant murine retrovirus with expression in embryonic stem cells. *Proc. Natl. Acad. Sci. USA* **87**, 9202–9206.

144. Stocking, C., Grez, M., and Ostertag, W. (1993). Regulation of retrovirus infection and expression in embryonic and hematopoietic stem cells. In: *Virus Strategies. Molecular Biology and Pathogenesis* (W. Doerfler and P. Böhm, eds), pp. 433–455. VCH Verlagsgesellschaft, Weinheim.

145. Speck, N. A., Renjifo, B. V., Golemis, E., Fredrickson, T. N., Hartley, J. W., and Hopkins, N. (1990). Mutations of the core or adjacent LVb elements of the Moloney leukemia virus enhancer alters disease specificity. *Genes Dev.* **4**, 223–242.

146. Morrison, S. J., Uchida, N., and Weissman, I. L. (1995). The biology of hematopoietic stem cells. *Annu. Rev. Cell Dev. Biol.* **11**, 35–71.

147. Hilberg, F., Stocking, C., Ostertag, W., and Grez, M. (1987). Functional analysis of a retroviral host range mutant: altered long terminal repeat sequences allow expression in embryonal carcinoma cells. *Proc. Natl. Acad. Sci. USA* **84**, 5232–5236.

148. Hawley, R. G., Lieu, F. H., Fong, A. Z., and Hawley, T. S. (1994). Versatile retroviral vectors for potential use in gene therapy. *Gene Ther.* **1**, 136–138.

149. Tumas, D. B., Spangrude, G. J., Brooks, D. M., Williams, C. D., and Chesebro, B. (1996). High-frequency cell-surface expression of a foreign protein in murine hematopoietic stem cells using a new retroviral vector. *Blood* **87**, 509–517.

150. Hildinger, M., and Baum, C. (1997). unpublished data.

151. Limon, A., Briones, J., Puig, T., Carmona, M., Fornas, O., Cancelas, J. A., Nadal, M., Garcia, J., Rueda, F., and Barquinero, J. (1997). High-titer retroviral vectors containing the enhanced green fluorescent protein gene for efficient expression in hematopoietic cells. *Blood* **90**, 3316–3321.

152. Shivdasani, R. A., and Orkin, S. (1996). The transcriptional control of hematopoiesis. *Blood* **87**, 4025–4039.

153. Nielsen, A. L., Pallisgaard, N., Pedersen, F. S., and Jorgensen, P. (1994). Basic helix-loop-helix proteins in murine type C retrovirus transcriptional regulation. *J. Virol.* **68**, 5638–5647.

154. Zaiman, A. L., and Lenz, J. (1996). Transcriptional activation of a retrovirus enhancer by CBF (AML1) requires a second factor: evidence for cooperativity with c-Myb. *J. Virol.* **70**, 5618–5629.

155. Malik, P., Krall, W. J., Yu, X. J., Zhou, C., and Kohn, D. B. (1995). Retroviral-mediated gene expression in human myelomonocytic cells: a comparison of hematopoietic cell promoters to viral promoters. *Blood* **86**, 2993–3005.

156. Plavec, I., Voyovich, A., Moss, K., Webster, D., Hanley, M. B., Escaich, S., Ho, K. E., Boehnlein, E., and DiGiusto, D. L. (1996). Sustained retroviral gene marking and expression in lymphoid and myeloid cells derived from transduced hematopoietic progenitor cells. *Gene Ther.* **3**, 717–724.

157. Couture, L. A., Mullen, C. A., and Morgan, R. A. (1994). Retroviral vectors containing chimeric promoter/enhancer elements exhibit cell-type-specific gene expression. *Hum. Gene Ther.* **5**, 667–677.

158. Parolin, C., Taddeo, B., Palu, G., and Sodroski, J. (1996). Use of *cis*- and *trans*-acting viral regulatory sequences to improve expression of human immunodeficiency virus vectors in human lymphocytes. *Virology* **222**, 415–422.

159. Agarwal, M., and Böhnlein, E. (1997). Oral presentation. In: "Euroconference on Stem Cells for Gene Therapy" (Chaired by J. M. Heard and W. Ostertag) Sitges, Spain: 27–30 April 1997.

160. Phi-Van, L., von Kries, J. P., Ostertag, W., and Strätling, W. H. (1990). The chicken lysozyme matrix attachment region increases transcription from a heterologous promoter in heterologous cells and dampens position effects on the expression of transfected cells. *Mol. Cell. Biol.* **10**, 2302–2307.

161. Dachs, G. U., Patterson, A. V., Firth, J. D., Ratcliffe, P. J., Townsend, K. M., Stratford, I. J., and Harris, A. L. (1997). Targeting gene expression to hypoxic tumor cells. *Nat. Med.* **3**, 515–520.

162. Miller, N., and Whelan, J. (1997). Progress in transcriptionally targeted and regulatable vectors for genetic therapy. *Hum. Gene Ther.* **8**, 803–815.

163. Gautsch, J. W., and Wilson, M. C. (1983). Delayed de novo methylation in teratocarcinoma suggests additional tissue-specific mechanisms for controlling gene expression. *Nature* **301**, 32–37.

164. Bird, A. P. (1986). CpG-rich islands and the function of DNA methylation. *Nature* **321**, 209–213.

165. Macleod, D., Charlton, J., Mullins, J., and Bird, A. P. (1994). Sp1 sites in the mouse aprt gene promoter are required to prevent methylation of the CpG island. *Genes Dev.* **8**, 2282–2292.

166. Grez, M., Zörnig, M., Nowock, J., and Ziegler, M. (1991). A single point mutation activates the Moloney murine leukemia

virus long terminal repeat in embryonal stem cells. *J. Virol.* **65,** 4691–4698.

167. Zaehres, H., and Baum, C. (1997). unpublished data.

168. Gossen, M., Bonin, A. L., and Bujard, H. (1993). Control of gene activity in higher eukaryotic cells by prokaryotic regulatory elements. *Trends Biochem. Sci.* **18,** 471–475.

169. Gossen, M., Freundllieb, S., Bender, G., Muller, G., Hillen, W., and Bujard, H. (1995). Transcriptional activation by tetracyclines in mammalian cells. *Science* **268,** 1766–1769.

170. Emerman, M., and Temin, H. M. (1984). Genes with promoters in retrovirus vectors can be independently suppressed by an epigenetic mechanism. *Cell* **39,** 449–467.

171. Ghattas, I. R., Sanes, J. R., and Majors, J. E. (1991). The encephalomyocarditis virus internal ribosome entry site allows efficient coexpression of two genes from a recombinant provirus in cultures cells and in embryos. *Mol. Cell. Biol.* **11,** 5848–5849.

172. Eggermont, J., and Proudfoot, N. J. (1993). Poly(A) signals and transcriptional pause sites combine to prevent interference between RNA polymerase II promoters. *EMBO J.* **12,** 2539–2548.

173. Proudfoot, N. J. (1986). Transcriptional interference and termination between duplicated alpha-globin gene constructs suggests a novel mechanism for gene regulation. *Nature* **322,** 562–565.

174. Ahlers, N., Hunt, N., Just, U., Laker, C., Ostertag, W., and Nowock, J. (1994). Selectable retrovirus vectors encoding Friend virus gp55 or erythropoietin induce polycythemia with different phenotypic expression and disease progression. *J. Virol.* **68,** 7235–7243.

175. Adam, M. A., Ramesh, N., Miller, A. D., and Osborne, W. R. (1991). Internal initiation of translation in retroviral vectors carrying picornavirus 5′ nontranslated regions. *J. Virol.* **65,** 4985–4990.

176. Boris-Lawrie, K. A., and Temin, H. M. (1993). Recent advances in retrovirus vector technology. *Curr. Opin. Genet. Dev.* **3,** 102–109.

177. Morgan, R. A., Couture, L., Elroy-Stein, O., Ragheb, J., Moss, B., and Anderson, W. F. (1992). Retroviral vectors containing putative internal ribosome entry sites: development of a polycistronic gene transfer system and applications to human gene therapy. *Nucleic Acids Res.* **20,** 1293–1299.

178. Berlioz, C., and Darlix, J. L. (1995). An internal ribosomal entry mechanism promotes translation of murine leukemia virus gag polyprotein precursors. *J. Virol.* **69,** 2214–2222.

179. Germann, U. A., Chin, K.-V., Pastan, I., and Gottesman, M. M. (1990). Retroviral transfer of a chimeric multidrug resistance-adenosine deaminase gene. *FASEB J.* **4,** 1501–1506.

180. Artelt, P., Grannemann, R., Stocking, C., Friel, J., Bartsch, J., and Hauser, H. (1991). The prokaryotic neomycin-resistance-encoding gene acts as a transcriptional silencer in eukaryotic cells. *Gene* **99,** 249–254.

181. Sorrentino, B. P., McDonagh, K. T., Woods, D., and Orlic, D. (1995). Expression of retroviral vectors containing the human multidrug resistance 1 cDNA in hematopoietic cells of transplanted mice. *Blood* **86,** 491–501.

182. Bunting, K. D., Webb, M., Giorgianni, G., Galipeau, J., Blakley, R. L., Townsend, A., and Sorrentino, B. P. (1997). Coding region-specific destabilization of mRNA transcripts attenuates expression from retroviral vectors containing class 1 aldehyde dehydrogenase cDNAs. *Hum. Gene Ther.* **8,** 1531–1543.

CHAPTER 5

Non-Infectious Gene Transfer and Expression Systems for Cancer Gene Therapy

MARK J. COOPER

Case Western Reserve University School of Medicine, and Copernicus Gene Systems, Inc., Cleveland, Ohio 44106

I. INTRODUCTION

Gene therapy provides a significant opportunity to devise novel strategies for the control or cure of cancer. Current approaches to cancer gene therapy typically employ viral-based vectors to express suitable target genes in human cancer cells either *ex vivo* or *in vivo* [1–4]. Therapeutic gene targets currently being evaluated include susceptibility genes, such as herpes simplex thymidine kinase followed by ganciclovir treatment [5–15]; genes that target the immune system to eliminate cancer cells, such as cytokines [16–35], co-stimulatory molecules [36], foreign histocompatibility genes [37,38], antisense constructs to insulin-like growth factor I [39,40], and polynucleotide vaccines [41–43]; replacement of wild-type tumor suppressor genes, such as p53 [44–49]; and antisense blockade of oncogenes, such as K-*ras* [50–52]. To move gene therapy into the mainstream of cancer therapeutics, however, it will ultimately be necessary to devise strategies to administer a gene therapy reagent to a patient in the familiar context of a pharmaceutical and to perform gene transfer

77

in vivo. Currently utilized viral-based gene therapy vectors, including retroviral and adenoviral vectors, fail to realize this potential because of limitations in their expression characteristics, lack of specificity in targeting tumor cells for gene transfer, and safety concerns regarding induction of secondary malignancies and recombination to form replication-competent virus. These limitations have refocused efforts to develop noninfectious gene transfer technologies for *in vivo* gene delivery of plasmid-based expression vectors. These vectors exist as extrachromosomal elements in populations of transiently transfected tumor cells. As discussed later, incorporation of transcription control sequences, including tissue-specific enhancers and inducible promoters, and elements permitting controlled vector replication in tumor cells have the potential to yield cancer gene therapy vectors that are both safe and effective for direct *in vivo* gene transfer.

II. ADVANTAGES AND DISADVANTAGES OF INFECTIOUS, VIRAL-BASED VECTORS FOR HUMAN GENE THERAPY

A number of viruses that infect humans, including retroviruses, adenoviruses, and adeno-associated viruses, have been modified to generate efficient expression vectors. These vectors either integrate into genomic DNA or persist as extrachromosomal elements, and they have distinct expression characteristics, as summarized in Table 1. The primary advantage of these vectors is the ability to infect a high percentage of target cells *in vitro,* and probably *in vivo* [7,53,54]. Whereas retroviral vectors yield one or several integrated proviral copies per cell, other vectors can introduce higher copy num-

bers of transcriptional cassettes, thereby enhancing transient levels of gene expression. Some viral-based vectors, such as those derived from recombinant adenoviruses, may replicate in transduced cells at a low level, although this feature has usually been interpreted as an undesired feature raising safety concerns regarding unregulated, systemic gene transfer [4,55–57]. Although viral-based vectors may be particularly useful for gene transfer *ex vivo,* this approach requires costly manipulations of tumor biopsies to yield either transient [58] or stably selected and characterized transfectants [19]. The latter approach may prove to be a particularly poor choice for gene targets that stimulate the immune system to eliminate tumor cells, because representation of tumor heterogeneity is likely lost prior to gene transfer.

Although high-level infectivity of viral-based vectors remains an attractive feature, there are multiple safety concerns and technical features that limit their applications, including (1) safety concerns regarding integration of vector DNA into host cell genomic DNA, which may induce secondary malignancies by activation of proto-oncogenes or inactivation of tumor suppressor genes [59]; (2) potential for recombination events to produce infectious virus able to replicate *in vivo* (recombination could occur either *in vitro* during vector preparation, or possibly *in vivo,* particularly when using vectors derived from pathogenic human viruses, such as adenoviruses) [2,3,55–57,60,61]; (3) presentation of viral antigens on the surface of infected human cells, resulting in T-cell recognition and destruction of transduced cells [62], (4) lack of specificity of cell types recognized by endogenous viral coat proteins, resulting in unintended transduction of nontargeted cell types *in vivo*; (5) heterogeneity of expression of viral coat protein receptors by tumor cell targets, thereby limiting the tumor cell population that can be transduced (viral-receptor-

TABLE 1   Infectious, Viral-Based Vectors for Cancer Gene Therapy

Vector	Integration or extrachromosomal distribution	Expression limited to cells undergoing replication at time of infection	References
Retrovirus	I	Yes	63
Adenovirus	E	No	55–57, 189
Adeno-associated virus	I[a]	Yes[b]	190
Herpes simplex virus	E	No	191
Vaccinia virus	E	No	192
Autonomous parvovirus (LuIII)	E	Yes	193

Note. Abbreviations: I, integration; E, extrachromosomal.
[a] Integration in replicating cells, transient extrachromosomal persistence in stationary phase cells.
[b] 90% of expression limited to cells traversing S phase.

negative cells may be selected for during treatment); (6) the fact that retroviral vectors will not express target genes in nonreplicating tumor cells [63]; (7) technical limitations regarding strategies to produce higher levels of gene expression in an infected cell; (8) difficulties in reproducibly producing, concentrating, delivering, and storing high-titer viral vectors for clinical use; (9) complement-mediated mechanisms of inactivation, which may limit use of some viral-based vectors in vivo [64]; (10) the potential for some virally-encoded proteins to yield undesired toxic effects in addition to immune recognition, leading to altered cell functions or transformation [2,4]; and (11) the immunogenicity of viral-based vectors, resulting in incrementally decreased effectiveness during repeated treatments in vivo [2,4]. These safety concerns and limitations in the ability of some infectious, viral-based vectors to yield maintained, high-level gene expression in transiently transfected tumor cells have led to the development of alternative, noninfectious gene expression and gene transfer technologies, as reviewed later.

III. RATIONALE FOR CONSIDERING NONINFECTIOUS, PLASMID-BASED EXPRESSION SYSTEMS

Initial assumptions regarding requirements for effective cancer gene therapy have changed since the demonstration of a significant "innocent bystander" effect using gene targets that confer antibiotic susceptibility, such as herpes simplex virus thymidine kinase followed by ganciclovir treatment [8], or genes that activate the immune system to recognize and kill tumor cells [16–43]. It may therefore not be necessary to transfect 50 to 100% of tumor cells in order to produce cure. These recent findings provide an important rationale to consider non-viral-based vectors for gene therapy applications, particularly constructs that yield higher levels of gene expression per transfected cell than is likely possible when using a viral-based vector. Moreover, new technical advances in receptor-mediated gene delivery of plasmid-based vectors now yield transient transfection efficiencies in vivo that approximate those observed using viral-based vectors [65–67].

IV. GENE TRANSFER TECHNOLOGIES FOR PLASMID-BASED VECTORS: PRECLINICAL MODELS AND CLINICAL CANCER GENE THERAPY TRIALS

Several gene transfer methods yield efficient transient transfection efficiencies following either in vitro or in vivo applications, as listed in Table 2. Although some of these methods are limited by the target cell type transfected or by the specificity of gene transfer, receptor-mediated gene transfer technologies have the potential to yield efficient and specific gene delivery to targeted tumor cells in vivo and therefore may have widespread utility.

A. Direct Injection of DNA

Perhaps the simplest formulation for in vivo gene transfer of plasmid vectors into cells is by direct administration of supercoiled DNA into tissues. Early studies demonstrated that DNA can be directly introduced into cells in vivo by simply injecting target organs with viral DNA. For example, when polyoma virus [68,69] or ground squirrel hepatitis virus [70] DNA was directly injected into mice or ground squirrels, respectively, the animals developed systemic infection and active virus particles were recovered. In these studies, however, very inefficient initial levels of in vivo gene transfer of purified virion DNA could be detected because of amplification of the gene transfer mechanism via systemic virus infection. In related studies, gene expression was observed in the liver and spleen of newborn rats 2 days following intraperitoneal injection of calcium-phosphate-precipitated plasmid DNA encoding the chloramphenicol acetyltransferase reporter gene [71]. More recently, direct injection of naked plasmid DNA was shown to yield significant levels of gene expression in rat skeletal and cardiac muscle, but not in kidney, lung, liver, or brain [72,73]. For example, direct injection of 25 μg of p-CMVint-lux plasmid DNA encoding the luciferase marker gene driven by the CMV immediate-early promoter into the rectus femoris muscle of mice yielded peak gene expression at day 14, and expression was detectable for up to 120 days [74]. The mechanism by which plasmid DNA is taken up by muscle cells is unclear but does not seem to be related to direct cell

TABLE 2   Gene Transfer Technologies for Plasmid-Based Vectors

Gene transfer method	Gene transfer limited to specific tissues	Ability to target tumor cells
Direct injection of naked DNA	Yes	No
Particle bombardment	Yes	No
Calcium phosphate	No	No
Liposome/DNA complexes	No	No
Ligand/DNA conjugates	No	Yes

injury to the sarcolemmal membrane [75]. In more recent studies, significant gene expression has also been observed following direct injection of naked plasmid DNA into rat or cat liver [76] and rabbit thyroid follicular cells [77], expanding the tissue types that can be transfected using this method.

Gene expression in transfected muscle cells is sufficient to produce antiviral immunity. For example, mice having their quadriceps muscle injected with a plasmid encoding influenza A nucleoprotein developed humoral and cytotoxic T cell responses to this antigen and were protected from subsequent challenge with influenza A virus [78]. In a similar fashion, direct intramuscular gene transfer of plasmid DNA encoding HIV envelope protein (gp160) in mice confers humoral and cell mediated immunity against recombinant envelope protein, and sera from these animals neutralizes HIV infectivity *in vitro* [79]. Direct injection of plasmid DNA also results in efficient gene delivery to subcutaneous tissues, including keratinocytes, fibroblasts, and dendritic cells [80]. This later approach may be superior to direct muscle injection for the development of cytotoxic T cell immunity, perhaps because of antigen presentation by macrophages and dendritic cells in the subcutaneous tissues [80].

In a similar fashion, intramuscular or intradermal gene transfer of plasmid vectors encoding tumor-associated antigens may yield effective cancer vaccines. This approach requires prior knowledge of potential tumor-associated antigens in a given patient's tumor that have presumably not yet been adequately presented to the host immune system. A variety of tumor-associated antigens have been identified that have the potential to stimulate a cytotoxic T cell response and are therefore candidate antigens for tumor vaccines. In human melanoma, such tumor-associated antigens include p97, MAGE-1, MAGE-2, MAGE-3, Melan-A, MART-1, gp100, and tyrosinase [81–90]. Cytotoxic T cell responses also have been demonstrated against mucin products of the MUC-1 gene in patients with pancreatic and breast carcinomas, and antigenicity appears to be related to underglycosylated forms of the protein found in tumor cells [91]. In ideal circumstances, tumor-associated antigens would only be expressed by the tumor and not by normal tissues, and a cancer vaccine would generate a tumor-specific immune response. Because peptide fragments of cellular proteins are displayed on the cell surface in conjunction with major histocompatibility antigens by TAP transporter proteins [92], the immune system is able to survey for the presence of gene mutations that can result in generation of novel peptide antigens. Tumor-specific cytotoxic T cell immunity has been demonstrated against peptide fragments from oncogenes or tumor suppressor genes that

are mutated during the generation of the malignancy. These vaccines include peptides encoding point mutations in *ras* genes [93–96] and p53 [97], and the unique breakpoint in the *bcr-abl* fusion gene [98].

One example of a successful cancer vaccine model is development of antitumor immunity to tumor cells expressing human carcinoembryonic antigen (CEA). CEA is expressed at high levels in several types of human adenocarcinomas, including colon, breast, gastric, pancreatic, and non-small-cell lung carcinomas [41,42,99]. CEA is also expressed at high levels in human fetal gut and at low levels in normal colonic mucosal cells, but it is not expressed in murine tissues [41]. Therefore, mice immunized with CEA protein would be expected to develop antitumor immunity to syngeneic tumor cells expressing human CEA. This result has been demonstrated by using a recombinant vaccinia virus vector encoding human CEA cDNA to immunize mice [41]. These studies used a murine colon carcinoma cell line, MC38, that had been transduced with a retroviral vector encoding CEA cDNA, generating the modified MC38-CEA-2 cell line. Vaccinia-vector-immunized syngeneic C57BL/6 mice developed humoral and cell-mediated immunity to CEA, and MC38-CEA-2 cells injected in immunized animals were rejected [41].

To extend these studies, Curiel and colleagues have demonstrated that C57BL/6 mice can develop antitumor immunity to MC38-CEA-2 cells by directly injecting plasmid DNA encoding CEA cDNA into striated muscle [43]. In these studies, the tongues of C57BL/6 mice were injected weekly with 100 μg of plasmid DNA encoding CEA. After 4 doses, these animals produced anti-CEA antibodies and developed cell-mediated immunity to MC38-CEA-2 cells. Importantly, these immunized mice rejected MC38-CEA-2 cells that were subcutaneously inoculated in the animals 1 week following the last immunization. These results demonstrate the ability to generate an effective cancer vaccine by expressing a tumor-associated antigen following direct *in vivo* gene transfer of plasmid DNA.

Further issues that need to be addressed by the use of polynucleotide vaccines include the choice of specific tumor-associated antigens likely to produce antitumor immunity in cancer patients of a given tumor type and the clinical setting in which this approach is likely to be effective. For example, administration of a tumor vaccine in an adjuvant setting following initial surgical removal of the primary mass may improve conditions for success by selecting a population of patients who have not yet received immune suppressive cytotoxic chemotherapy and whose tumor burden is small. In addition, analysis of tumor tissue for expression of relevant tumor antigens may be quite important, because multiple tumor-associated antigens may need to be targeted to address clonal evolution of heterogeneous populations of tumor cells.

B. Particle-Mediated Gene Delivery

An alternative approach for delivery of plasmid constructs into human cells *in vivo* is to coat metallic particles with a DNA vector and then introduce the particles directly into tissues using a "gene gun" to accelerate the particles to a high velocity [100]. Subcutaneous tissues can be directly transfected *in vivo* because the particles can penetrate to this depth. Visceral tissues have also been transfected *in vivo* in animals, although this approach requires an operative procedure to bring the tissue of interest in close approximation to the gene gun instrument. Nevertheless, particle-mediated gene transfer of a plasmid vector encoding influenza virus hemagglutinin subtype 1 has been demonstrated to immunize mice against challenge with a lethal inoculum of influenza virus [101]. This approach has significant potential for development of cancer vaccines, because efficient gene transfer of polynucleotide vaccines into subcutaneous tissues may be particularly effective in presenting antigens to the immune system [80]. Cancer preclinical models using particle-mediated gene transfer into subcutaneous tumor explants have also demonstrated improved survival of tumor bearing mice using a variety of cytokine targets, including IL-2, IL-6, and interferon-gamma [102].

C. Gene Transfer of DNA Precipitated with Calcium Phosphate

Plasmid DNA precipitated with calcium phosphate can efficiently transfect cells in tissue culture, as reported by Graham and Van der Eb in 1973 [103]. More then a decade ago, this technique was also used for *in vivo* gene transfer of viral and plasmid DNA into liver and spleen by either direct inoculation into the tissue bed or intraperitoneal instillation [68–71]. Despite these initial promising results and the ease of preparing these DNA precipitates, this method has largely been supplanted by alternative approaches that are thought to yield superior *in vivo* transfection efficiencies. Nevertheless, this method has recently been employed in preclinical cancer gene therapy studies evaluating introduction of HSV-TK into melanoma explants [104]. In these studies, plasmid DNA encoding HSV-TK was precipitated with calcium phosphate and directly injected into established B16 melanoma tumor explants in syngeneic C57/BL mice. After administration of intraperitoneal gan-

ciclovir, treated animals achieved a partial tumor regression.

D. Liposome-Mediated Gene Delivery

Polycationic lipids can be mixed with plasmid DNA to form liposome structures that are thought to fuse with the target cell membrane and thereby mediate gene delivery [105]. Several lipid preparations have been formulated for this application, including mixtures of dioleoyl phosphatidylethanolamine (DOPE) with DOTMA (lipofectin), DOSPA (lipofectamine), DDAB (lipofectace), DOGS (transfectam), DOTAP, DMRIE, and DC cholesterol (reviewed in Felgner *et al.* [106]). This approach can yield very high transfection efficiencies *in vitro* and can also be used for direct *in vivo* gene transfer. Plasmid DNA has been delivered to tumor explants in syngeneic mice by injecting the tumor nodule with liposome/DNA complexes, achieving a transient transfection efficiency of approximately 1 to 10% [37]. A particular advantage of this approach is the ease of preparing DNA/liposome complexes, the stability of the individual components, and the versatility to transfect a variety of tumor types. The liposome/DNA complex can be directly injected into a palpable tumor nodule [38]. Alternatively, visceral tumor masses can be directly instilled with liposome/DNA complexes by employing radiological procedures, such as CAT scans, to identify the location of the tumor and assist in percutaneous tumor injection [107,108]. Alternative approaches include using bronchoscopy, cystoscopy, or endoscopy to directly inject liposome/DNA complexes into visualized tumor masses. Liposome/DNA complexes are nonimmunogenic and have minimal systemic toxicities [109,110], and they can be administered repeatedly to the same patient with expectations of equivalent efficiencies of gene transfer.

Liposome/DNA complexes administered intravenously also can deliver plasmid vectors into multiple tissue types. In 1983, Nicolau *et al.* injected rats intravenously with a plasmid vector encoding rat preproinsulin I complexed with liposomes composed of phosphatidylcholine, phosphatidylserine, and cholesterol [111]. In these studies, radioactive labeled liposomes were shown to be taken up specifically by liver and spleen, and 6 hours after injection treated animals experienced a fall in serum glucose and an increase in serum, liver, and splenic insulin levels relative to control animals. More recently, Zhu and Debs demonstrated gene expression in diverse tissue types, including liver, spleen, kidney, lung, heart, lymph nodes, and bone marrow, following intravenous administration of chloramphenicol acetyltransferase reporter plasmids complexed with liposomes composed of DOTMA and DOPE lipids [112]. Gene expression was detected for up to 9 weeks following gene transfer. This widespread gene delivery raises the possibility of using intravenous administration of liposome/DNA complexes to introduce target genes in multiple foci of metastatic disease. For example, a study employing a p53 mutant human breast cancer xenogeneic model suggests that intravenous administration of liposome/DNA complexes encoding wild-type p53 may reduce the size of primary tumor explants and decrease the development of metastatic disease to lungs [113]. Although the liposome formulations described earlier do not specifically target tumor cells, ongoing studies suggest that it may be possible to increase the specificity of liposome-mediated gene transfer by conjugating ligands for cell surface receptors to lipid moieties. In recent studies, receptor-mediated gene transfer *in vitro* has been demonstrated for liposome preparations targeting the folate and erbB-2 receptors [114–116]. Additionally, "stealth" liposomes have been developed to avoid rapid clearance by reticuloendothelial cells following an intravenous injection [116–120]. Stealth liposomes pool in tissues, such as tumors, that have increased vascular permeability. In preclinical models, tumor-bearing animals treated with chemotherapeutic agents encapsulated in stealth liposomes had improved survival compared to control groups treated with free drug alone [120]. An active area of current research is to develop targeted and stealthy liposome preparations suitable for intravenous delivery of plasmid constructs for cancer gene therapy.

Several preclinical models have demonstrated antitumor responses when plasmid vectors have been directly transferred into established tumor explants in syngeneic mice. For example, plasmids encoding the murine class I H-2K^s gene have been complexed with liposomes and injected into established CT26 colon carcinoma (H-2K^d) and MCA 106 fibrosarcoma (H-2K^b) cells. As reported by Plautz and Nabel, a cytotoxic T cell response to H-2K^s antigen was induced, and animals preimmunized to H-2K^s antigen demonstrated significant antitumor activity, with some animals achieving long-term survival [37]. In addition, this antitumor activity was cell line specific, because animals bearing MCA 106 tumors previously cured following injection with H-2K^s plasmid rejected secondary tumor challenges with parental MCA 106 cells but not challenges with syngeneic B16BL/6 melanoma cells. These findings suggest that expression of foreign class I histocompatibility antigens by these tumor cells resulted in recognition of heretofore unrecognized tumor-associated antigens by cytotoxic T cells. This hypothesis would account for the observed efficient tumor elimination and prolonged survival despite the fact that only a modest percentage of

the tumor cells were transiently transfected following direct tumor inoculation by DNA/liposome complexes.

In a pilot study at the University of Michigan, Nabel and colleagues have extended their preclinical model to a clinical cancer gene therapy protocol by evaluating liposome-mediated gene transfer of plasmids encoding HLA-B7 in patients with metastatic melanoma. In these studies, liposome/DNA complexes were directly injected into subcutaneous, nodal, and visceral masses, and 1 out of the first 5 patients evaluated demonstrated a significant response [38]. These encouraging findings have led to several active trials evaluating the expression of a bicistronic plasmid encoding HLA-B7 and beta$_2$-microglobulin in patients with metastatic colon cancer, renal cancer, and melanoma [121]. In these trials, plasmid DNA complexed with liposomes composed of DIMRIE and DOPE lipids is directly injected into tumor masses. In initial reports, HLA-B7 gene expression has been shown in tumor biopsies after gene transfer [107,108] and antitumor immunity has been observed in local tumor-infiltrating lymphocytes [122]. These clinical trials are currently in progress, and additional data regarding generation of T cell immunity to HLA-B7 target cells, tumor responses, survival, and toxicities of the treatment are currently pending. Using a similar strategy, plasmids encoding human IL-2 cDNA are being directly transferred to tumor cells *in vivo* in patients with advanced cancer [121] or small cell lung cancer [121] in an attempt to stimulate T-cell-mediated antitumor immunity.

E. Ligand/DNA Conjugates

Negatively charged plasmid DNA molecules and polycations, such as poly(L-lysine), can form complex structures consisting of unimolecular and multimolecular complexes (with respect to the DNA) [65,66]. To enable efficient and cell-specific gene transfer, the poly(L-lysine) polymer can be modified by covalently attaching ligands that can subsequently bind to specific cellular receptors [123]. If the DNA/poly(L-lysine) complex contains a suitable ligand, then the DNA/poly(L-lysine) complex can be internalized in the cell when the receptor undergoes endocytosis. Most of these early DNA/poly(L-lysine) formulations were multimolecular complexes, approximately 100 to 200 nm in diameter [65], which may have limited their ability to enter cells via receptor-mediated endocytosis. Additionally, efficient expression of the internalized plasmid requires several additional steps, including exit from the endosome prior to destruction of the DNA by fusion of the endosome with lysosomes and transfer of the plasmid DNA to the nucleus [65].

Initial formulations of poly(L-lysine)/DNA complexes for *in vivo* gene transfer targeted the liver asialoglycoprotein receptor for gene delivery using asialoorosomucoid covalently linked to poly(L-lysine) [123]. Gene expression was transient, although preferential gene transfer to the liver was observed. In later studies, gene expression was improved by performing a partial hepatectomy in association with receptor-mediated gene transfer [124]. Further improvements in gene expression were achieved by using endosomolytic agents, such as defective adenovirus particles or peptides derived from the N-terminal region of influenza virus hemagglutinin HA-2 protein, to enable transferrin-conjugated poly(L-lysine)/DNA complexes to exit the endosome and enter the cytoplasm, for eventual transfer to the nucleus [125–127]. This modification has been shown to achieve transient gene expression in lung tissue following direct instillation of ligand/DNA complexes into the airway of rats [128]. Gene transfer *in vitro* has been demonstrated in primary intestinal mucosal cells and the transformed Caco$_2$ colon adenocarcinoma cell line [129], suggesting an approach for gene delivery into tumor cells.

Recent studies have focused on formulations of condensed, unimolecular DNA/poly(L-lysine) complexes that efficiently enter the cell via receptor-mediated endocytosis [65,66]. Such complexes are toroids of approximately 10 to 15 nm in diameter and achieve efficient and specific gene transfer following intravenous gene delivery. For example, condensed, unimolecular galactosylated DNA/poly(L-lysine) complexes encoding human factor IX cDNA efficiently target the hepatic asialoglycoprotein receptor, and transfected rats have detectable human factor IX in their serum for up to 140 days [66]. This result was achieved without the need for partial hepatectomy. Condensed DNA/poly(L-lysine) complexes have also been prepared by coupling the FAB fragment of an antibody recognizing the polymeric immunoglobulin receptor [67]. These complexes have a diameter of approximately 25 nm and yield efficient gene transfer into target rat lung epithelial cells following intravenous administration. Approximately 18% of tracheal epithelial cells were transfected as monitored by expression of the beta-galactosidase marker gene following a single intravenous injection of 300 μg of plasmid DNA formulated in these condensed complexes [67]. Expression was specific for tissues expressing the polymeric immunoglobulin receptor. In related studies, the mannose receptor on macrophages has been targeted for *in vivo* gene delivery by formulating condensed mannosylated DNA/poly(L-lysine) complexes [130]. In these studies, efficient and specific gene transfer was shown to correlate with the formulation of unimolecular, condensed DNA/poly(L-lysine) complexes. In

recent studies, the serpine enzyme complex receptor (SEC-R) also has been targeted for gene transfer using condensed DNA/poly(L-lysine) particles [131]. Gene transfer *in vitro* correlated closely with the level of cell surface SEC-R expression. Together, these studies suggest that coupling poly(L-lysine) to ligands that recognize cellular receptors preferentially expressed by tumor cells may provide an efficient and specific approach for *in vivo* gene transfer of plasmid vectors into cancer cells.

V. PLASMID EXPRESSION VECTORS

Unlike viral-based infectious vectors, plasmid vectors must be introduced into cells by specific gene transfer technologies, as reviewed earlier. Once introduced into a cell, however, plasmids have specific advantages compared to viral vectors, including (1) no potential to be infectious; (2) levels of gene expression per cell that are equivalent to other viral vectors that persist as extra-chromosomal elements (see Table 1); (3) lack of immunogenicity (allowing for multiple treatments) [109]; (4) lack of toxicity following intravenous injection [110]; (5) low probability of integration during transient periods of expression, thereby reducing potential for insertional mutagenesis; (6) easy coupling to liposome or receptor-mediated gene delivery systems; and (7) long-term stability, requiring no special preparation or storage requirements. Modifications in vector design, including tissue-specific promoters, inducible promoters, and elements enabling the plasmid to replicate extra-chromosomally in tumor cells, further enhance the safety of plasmid vectors and significantly augment the level of expression observed in transiently transfected tumor cells.

A. Tissue-Specific Promoters

The cytomegalovirus immediate-early promoter is often utilized in gene therapy studies because of its high level of activity in diverse tissue types [132,133]. Although it is desirable to express target genes at high levels in tumor cells, transcriptionally active promoters, such as CMV, will also direct high-level expression in unintentionally transfected normal cells following *in vivo* gene transfer. To approach current limitations in the ability to specifically target a tumor cell for gene transfer, tissue specific promoters can be employed that limit expression of the therapeutic gene to tumor cells and normal cells of a specific lineage. Many tissue-specific promoters have been developed [134,135], and a short list includes the insulin promoter (β islet cells of the pancreas) [136]; elastase promoter (acinar cells of the pancreas) [137]; whey acidic protein promoter (breast) [138]; tyrosinase promoter (melanocytes) [139]; tyrosine hydroxylase promoter (sympathetic nervous system) [140]; neurofilament protein promoter (brain neurons) [141]; glial fibrillary acidic protein promoter (brain astrocytes) [142]; Ren-2 promoter (kidney) [143]; collagen promoter (connective tissues) [144]; α-actin promoter (muscle) [145]; von Willebrand factor promoter (endothelial cells) [146]; α-fetoprotein promoter (hepatoma) [147]; albumin promoter (liver) [147]; surfactant promoter (lung) [148]; CEA promoter (gastrointestinal tract, tumors of colon, breast, lung) [149]; uroplakin II promoter (bladder) [150]; T-cell-receptor promoter (T lymphocytes) [151]; immunoglobulin heavy chain promoter (B lymphocytes) [152]; prostatic-specific-antigen promoter (prostate) [153]; and protamine promoter (testes) [154].

Tissue-specific promoters have been utilized in gene therapy studies to evaluate tumor-specific killing mediated by expression of the herpes simplex thymidine kinase gene followed by exposure to ganciclovir. For example, use of the albumin and α-fetoprotein promoter in retroviral constructs encoding HSV-TK specifically killed hepatoma cell lines but had marginal activity in other tumor cells derived from breast, colon, or skin [147]. In other studies, Vile and Hart recently reported use of plasmid DNA encoding HSV-TK transcriptionally regulated by the murine tyrosinase promoter to treat B16 melanoma tumors growing as subcutaneous explants in syngeneic mice [104]. Established tumors, approximately 4 mm in diameter, were directly injected with 20 μg of calcium phosphate precipitated plasmid DNA, and 2 days later mice were administered daily injections of intraperitoneal ganciclovir for 5 days. A statistically significant reduction in tumor size was observed compared to animals not receiving ganciclovir. No local toxicity was observed in the tissues adjacent to the tumor explant, as was expected based on the tissue specificity of the tyrosinase promoter. In similar studies, the CEA promoter also has been utilized to control transcription of HSV-TK [149]. CEA-expressing lung cancer cell lines were highly sensitive to ganciclovir *in vitro* and *in vivo* following gene transfer of these constructs, whereas CEA-nonexpressing lung cell lines were resistant to ganciclovir following gene transfer.

Another opportunity to specifically target tumor cells for gene expression is to utilize as promoters elements that become activated in chemotherapy resistant tumor cells. Based on the observation that the metallothionein promoter becomes activated in cisplatin-resistant ovarian carcinoma cells, plasmid DNA encoding the HSV-TK gene transcriptionally controlled by the metallothionein promoter has been introduced into cisplatin-sensitive and cisplatin-resistant ovarian carcinoma cell lines followed by treatment with ganciclovir [155]. No cytotoxicity was apparent in cisplatin-sensitive, parental 1A9 ovarian carcinoma cells, whereas a cisplatin-resistant subclone was efficiently killed by this treatment. These results suggest a specific approach for gene therapy of cisplatin-resistant ovarian carcinoma cells and underscore the potential of using tumor-specific promoter elements.

B. Inducible Promoters

In addition to using tissue-specific promoters to minimize target gene expression in unintentionally transfected cells, the timing and duration of gene expression also can be modulated by employing inducible promoters that can be externally controlled. Several inducible systems have been developed, and a few appear to be appropriate for use in clinical gene therapy trials because of their lack of apparent toxicity and demonstrated effectiveness *in vivo*. For example, a tetracycline-controlled expression system has been developed by Gossen and Bujard [156]. A novel hybrid transcriptional transactivation protein was constructed by ligating the ligand and DNA binding domains of the bacterial tetracycline repressor gene to the C-terminal region of the herpes virus VP16 transcriptional regulator protein containing its transactivation domain. In conjunction with reporter genes containing a heptad repeat of the consensus binding domain of the tetracycline repressor upstream of a minimal core element of the cytomegalovirus immediate-early promoter, tetracycline-controlled expression has been demonstrated *in vitro* and *in vivo* in transgenic mice [156–158]. The hybrid transcriptional transactivator binds to the tet operon in the absence of tetracycline, whereas tetracycline efficiently dissociates the transcriptional factor from its binding site. Hence, efficient reporter gene expression was observed in the absence of tetracycline, whereas transcription is virtually eliminated in the presence of 0.1 to 1 μg/mL of tetracycline, a concentration readily attainable in humans. This system has also been used to transiently express target genes following direct *in vivo* gene transfer of these plasmid constructs in rat myocardium [159]. More recently, a tetracycline-on system has been developed utilizing specific point mutations in the tetracycline repressor component of the hybrid transcriptional transactivator [160]. In other studies, tetracycline-controlled transcriptional repressors have been constructed by linking the KRAB transcriptional repressor downstream from the DNA binding domain of the tetracycline repressor [161].

O'Malley and colleagues have also described a novel, regulated transcriptional activator that consists of a truncated ligand binding domain of the human progesterone receptor (which binds tightly to the synthetic progesterone antagonist RU 486 but binds very poorly to progesterone), the DNA binding domain of the yeast transcriptional activator GAL4, and a C-terminal fragment of the herpes simplex VP16 transcriptional regulator protein [162]. In conjunction with a target gene containing four copies of the consensus GAL4 binding site, gene expression was activated only in the presence of RU 486 and regulation was achieved both *in vitro* and *in vivo* [162,163]. A similar gene switch has been developed by Delort and Capecchi that utilizes different domains of the progesterone receptor and GAL4 binding protein [164]. Wang *et al.* also have developed an inducible repressor system by substituting the KRAB transcriptional repressor domain for the VP16 transactivation domain [165]. The specificity of these inducible systems depends on the presence of the GAL4 consensus sequence upstream of the target gene of interest. Because GAL4-activated genes are not currently known to be present in the human genome, induction of gene expression *in vivo* is predicted to activate only the therapeutic target gene. In addition, the presence of endogenous progesterone receptors in tumor cells would not be expected to interfere with this expression system.

Lastly, other inducible transcriptional activation systems have been developed to control gene expression. These include the Drosophila ecdysone receptor gene switch and the rapamycin-controlled transactivation system [166,167]. The latter system utilizes two transcription factor fusion proteins that share a high-affinity binding site for rapamycin. The first element consists of the rapamycin binding protein, FKBP12, fused to the ZFHD1 DNA binding protein. The second element consists of a rapamycin binding protein, FRB, fused to the carboxy terminal portion of the NF-κB transcriptional activator protein. In the presence of rapamycin, these two fusion proteins bind to one another and reconstitute an active transcription factor for target reporter genes placed upstream of a minimal CMV promoter region carrying 12 binding sites for ZFHD1. Highly efficient and specific rapamycin-controlled gene expression has been demonstrated both *in vitro* and in *in vivo* preclinical models [167].

Several issues need to be addressed when considering any of these systems for cancer gene therapy. Although predicted to specifically inactivate or activate the transcription of target genes downstream from their respective consensus binding sequences in the presence of drug, further experimental testing is required to confirm that endogenous cellular genes, such as tumor suppressor genes and proto-oncogenes, are not unexpectedly regulated by these hybrid transcriptional repressors and transactivators, respectively. In addition, these hybrid transcriptional control proteins may very well generate antigenic peptide sequences derived from the bacterial tetracycline repressor, the yeast Gal4 protein, and the herpes simplex virus VP16 protein. An immune response may therefore be generated against tumor and normal cells following *in vivo* gene transfer. Although the toxicity of this immune response may be minimal, it may conceivably limit the duration of target gene expression in tumor cells following repetitive treatments.

Another example of an inducible promoter system utilizes transcriptional control elements that become active following radiation-induced injury. As developed by Weichselbaum and colleagues, the radiation-responsive consensus sequence from the early growth response (EGR-1) gene promoter was ligated upstream from a gene known to significantly enhance radiation injury, TNF-α [168]. This plasmid construct was electroporated into a hematopoietic cell line, HL525, known to be deficient in radiation-induced expression of TNF-α. These gene modified HL525 cells were injected into established radiation-resistant human squamous carcinoma xenografts in nude mice. Following radiation exposure to the tumor explant, the squamous carcinomas regressed and most of the animals were apparently cured. In contrast, control animals bearing squamous tumor explants that received radiation therapy alone, radiation plus HL525 cells transfected with the neomycin resistance gene, or TNF-α transfected HL525 cells without radiation all developed progressive tumor growth. These studies demonstrate the ability to induce gene expression *in vivo* in specific areas known to be involved by tumor by focused application of radiation and gene therapy.

C. Replicating Plasmid Vectors: Episomes

Expression of genes encoded by plasmids is generally transient unless specific modifications are made to enable the plasmid to replicate in human cells. In dividing tumor cells, plasmid-mediated gene expression falls to very low levels by several days after gene transfer. This decline in gene expression is mediated by several factors, including a logarithmic decline in the percentage of transfected cells during replication of the target population (because the plasmid does not replicate in human cells) [169], and potential loss of the transgene by nuclease destruction or by partitioning to nonnuclear compartments.

One approach to maintain plasmid copy numbers in transfected tumor cells is to incorporate sequences from human DNA that enable the plasmid to replicate extrachromosomally. Although sequence specific human

DNA origins have not been characterized, Calos and colleagues have identified DNA fragments that replicate semiconservatively during the S phase of the cell cycle when incorporated into plasmid vectors [170,171]. These vectors replicate once per cell cycle, and the plasmid copy number per cell therefore depends on the initial transfection conditions. In these studies, the size of the DNA fragment is an important factor in conferring replication competence, with random human DNA fragments over 10 to 15 kb in length having significant activity [172]. Similar sizes of randomly chosen yeast DNA also are replication-competent in human 293 cells [173], and large fragments of bacterial DNA have detectable although minor activity [170]. Plasmids containing these DNA fragments will replicate for several months in human cells if the vector additionally includes a portion of the Epstein–Barr virus (EBV) DNA origin (including a tandem array of repeated sequences), and if the transfected cells express the EBV early gene product, EBNA-1. EBNA-1 binds to these tandem repeat sequences and retains plasmid DNA in the nucleus of dividing cells, thereby conferring stable maintenance of the episomal plasmid [174]. In short-term assays, however, these DNA fragments alone enable plasmids to replicate transiently in human cells over several generations, although the copy number of these vectors is low [170]. The expression characteristics of plasmids containing such autonomously replicating human sequences and the potential role of these vectors for cancer gene therapy are currently undefined.

In other studies, a human artificial chromosome has been assembled *in vivo* by transfecting cells with specific fragments of telomeric and centromeric DNA [175]. These separate DNA fragments have recombined within the cell to form mini-chromosomes approximately 6 to 10 Mb in size, and stable vertical transfer of these extrachromosomal elements has been demonstrated over multiple generations *in vitro*. These properties make them well-suited for introduction into human stem cells, including ex vivo gene transfer into hematopoietic progenitor cells. However, the ability to isolate large quantities of homogeneous, unrearranged artificial chromosomes, transfect them into human cells, and then achieve transfer into the nucleus remains to be demonstrated.

Another approach to increase both the peak level and duration of gene expression mediated by plasmid vectors is to include sequences from DNA viruses that enable the plasmid to replicate in human cells. Two elements are required: (1) a viral DNA origin of replication and (2) a viral early gene product. The viral DNA origin alone is not functional in human cells. During the life cycle of DNA viruses, including Epstein–Barr virus and BK virus, an early gene product is synthesized

that directly binds to the viral DNA origin [176,177]. This protein/DNA complex is recognized by the infected human cell as a functional DNA origin, and the virus is able to replicate its DNA. In a similar fashion, plasmids encoding a viral DNA origin and its corresponding early gene product can replicate in human cells. Replicating episomal plasmid vectors have two predicted advantages compared to standard plasmid vectors for cancer gene therapy applications: (1) high-level gene expression caused by vector amplification and (2) maintenance of gene expression in transiently transfected cells caused by efficient vertical transfer of the episome during tumor cell division. These principles are summarized in Table 3 and illustrated in Fig. 1.

Plasmid vectors that replicate in human cells have been constructed from several viruses, including Epstein–Barr virus, BK virus, and SV40 [176–178]. For example, Epstein–Barr virus episomes replicate in lymphoid cells, achieving a steady-state copy number of approximately 10 to 50 copies [176]. Constructs derived from BK virus replicate in a wide range of cell types [177,179–181], and stable bladder cell transfectants have been characterized that have approximately 150 copies per cell [181]. In these studies, gene expression was proportional to the episomal plasmid copy number. Additionally, gene expression was maintained in a population of unselected, transiently transfected cells for at least 1 week following gene transfer of replicating episomes, whereas plasmid-based gene expression fell exponentially at a rate predicted by the doubling time of these cells. Hence, the predicted advantages of high-level, maintained gene expression of replicating episomal vectors compared to standard plasmids was observed in the BK-virus-derived system.

Despite the clear advantages of replicating plasmid vectors, a significant obstacle to their development is the transformation properties associated with suitable viral early genes that possess replication transactivator function. For example, the Epstein–Barr virus replication transactivator, EBNA-1, has transformation properties in transgenic mice [182]. In addition, papovavirus early gene products, including the large T antigens from BK virus and SV40 virus, have transformation proper-

TABLE 3   Features of Standard Plasmid and Replication-Competent Episomal Vectors

Expression vector	Peak level of gene expression	Sustained expression in dividing tumor cells
Standard plasmid	Low	No
Replication-competent episome	High	Yes

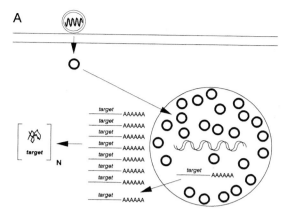

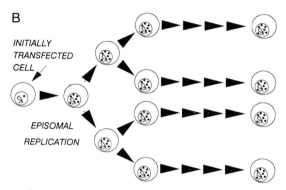

FIGURE 1   (A) Replicating episomal plasmids yield high levels of target gene expression caused by vector amplification. Depicted are multiple copies of the episomal plasmid in the nucleus of the transfected cell that have accumulated because of vector replication. An increased copy number of the expression vector produces high levels of target gene mRNA and consequently high levels of target gene protein. (B) High-level gene expression is maintained in transiently transfected tumor cells because of efficient vertical transfer of the episome (●) as these cells divide.

ties thought to be primarily mediated by binding to host tumor suppressor gene products, including p53, RB, and RB-related proteins, such as p107 and p130 [183–186].

To develop replicating episomal vectors for human gene therapy, our laboratory has recently developed a safety-modified SV40 large T antigen (107/402-T) that lacks detectable binding to human tumor suppressor gene products yet preserves replication competence (Fig. 2A) [187]. This large T antigen mutant has specific point mutations in codons 107 and 402, and it lacks detectable binding to p53, RB, and p107 proteins (Fig. 2B and C; Table 4). Episomal vectors incorporating the 107/402-T replicon amplify in a wide range of human and simian cell lines, but not dog or rodent cell lines (Table 5). In addition to gene transfer *in vitro,* we have observed that 107/402-T based episomal vectors replicate in human tumor cells following direct *in vivo* gene

transfer into human tumor xenografts in nude mice. An example of the replication activity of 107/402-T episomes in human hepatoma (Hep G2) and bladder (HT-1376) cell lines is shown in Fig. 2D and E. Based on transient gene transfer efficiencies and content of genomic DNA per cell, 107/402-T episomes achieve peak copy numbers of approximately 1,400 in HT-1376 cells and 25,000 in Hep G2 cells (Table 5). As a consequence of vector amplification, we observe significantly enhanced levels (>100-fold) of reporter gene expression when comparing this replicating episomal expression system to analogous, nonreplicating expression vectors (Fig. 3). When transferred into tumor cells maintained in log-phase growth, high levels of reporter gene expression were maintained for at least 1 to 2 weeks because of efficient vertical transfer of the replicating episomal expression vectors. In ongoing studies, we are developing systems to externally control vector amplification and to restrict vector replication to targeted tumor cells [188].

VI. FUTURE DIRECTIONS

Clinical cancer gene therapies have only recently been initiated, and results are currently very preliminary. At present, the optimal delivery system, expression vector, and target genes for a given tumor type are entirely unknown. Success of this modality will ultimately depend upon being able to express the therapeutic gene of interest at high levels, and being able to target the tumor cell for gene delivery will minimize toxicities. Incorporation of tissue-specific and inducible promoters in the vector design will likely permit appropriate control of vector expression and assist in limiting tumor cells for target gene expression. Coupling receptor-mediated gene transfer technologies with safety-modified, replicating episomal vectors may have the potential to target tumor cells for *in vivo* gene transfer of vectors capable of yielding sustained, high-level gene expression.

Gene therapy may have its most significant impact on patient survival when administered in an adjuvant setting, when tumor burden is at a minimal level. This may be particularly important for gene therapy approaches that attempt to stimulate the immune system to eliminate tumor cells. Appropriately, current clinical trials are administering gene therapy reagents to patients with either advanced, metastatic cancer or tumors having a poor prognosis based on local tumor growth, as in glioblastoma multiforme. Current trials will need careful analysis to develop second generation studies targeting high-risk groups having a lower tumor burden, such as patients with stage II breast cancer having more

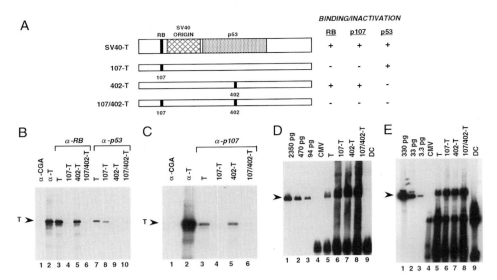

FIGURE 2   107/402-T lacks binding to human tumor suppressor genes and is replication competent. (A) Point mutations in replication-competent, safety-modified SV40 large T antigen mutants. Highlighted are domains of T antigen that bind to RB, p53, and the SV40 DNA origin. The codon 107 mutation substitutes lysine for glutamic acid, and the codon 402 mutation substitutes glutamic acid for aspartic acid [187]. (B and C) Co-immunoprecipitation analysis of binding of wild-type and mutant T antigens to human tumor suppressor gene products. 2×10^5 dpm of in vitro translated T antigens were mixed with CV-1 extracts overproducing human RB protein and anti-RB monoclonal antibody G3-245 (B, lanes 3–6), p53 and anti-p53 monoclonal antibody 1801 (B, lanes 7–10), and p107 and anti-p107 monoclonal antibody SD9 (C, lanes 3–6). As controls, wild-type T antigen is immunoprecipitated with either anti–chromogranin A monoclonal antibody LKH210 (lane 1) or anti-T-antigen monoclonal antibody 416 (lane 2). (D and E) 107/402-T is replication competent. Hep G2 hepatoma cells (D) were transfected with wild-type and mutant T antigen expression vectors, and total cellular DNA was harvested 2 days after transfection. DNA samples were sequentially digested with ApaI to linearize vector DNA and then with DpnI to distinguish amplified DNA from the input DNA used to transfect these cells. Because human cells lack adenine methylase activity, newly replicated DNA is resistant to digestion by DpnI. Hence, the presence of unit-length, linearized plasmid DNA, as indicated by the arrow, demonstrates newly replicated episome. Hybridization probe: pRC/CMV.107/402-T. (E) To evaluate amplification of a cotransfected plasmid in concert with T antigen episomes, HT-1376 bladder carcinoma cells were transfected with T antigen expression vectors and a reporter replication plasmid containing the SV40 DNA origin, pSV2CAT. DNA harvested from cells 4 days after gene transfer was sequentially digested with BamHI to linearize pSV2CAT and then with DpnI. Hybridization probe: BamHI-HindIII CAT fragment. CMV, pRC/CMV transfectants (no T antigen); DC, DpnI digestion control consisting of 5 µg of genomic DNA and 2 ng of either pRC/CMV.107/402-T (D, lane 9) or pSV2CAT (E, lane 9). (Reprinted with permission [187].)

than 10 positive axillary lymph nodes, or patients with colon, bladder, and lung cancer having local positive lymph node involvement.

The optimal gene targets for a given type of malignancy are unknown. At present, cancer gene therapies are focused on introducing genes in tumor cells to modulate the immune system to achieve antitumor immunity, induce susceptibility to exogenously administered drugs, block oncogene expression, or express wild-type tumor suppressor gene products. It is anticipated that the success of any of these target genes will be critically dependent upon the type of gene transfer and expression technologies employed. Nevertheless, the potential exists for selection of patient-specific target genes based upon

TABLE 4   Binding of Wild-Type and Mutant SV40 Large T Antigens to RB, p107, and p53 Tumor Suppressor Gene Products

Tumor suppressor gene	T[a]	107-T	402-T	107/402-T
RB	100	0.03	67	0.07
p107	100	0	79	0
p53	100	36.2	0	0

[a] Shown is the percentage of binding of T antigen mutants compared to wild-type T antigen. (Reprinted with permission [187].)

TABLE 5 Replication Activity of 107/402-T Based Episomes in Human and Animal Cell Lines

Species	Cell line	Type	Copy number/cell[a]
Human	HT-1376	Bladder	1,400
	5637	Bladder	100,000
	MCF-7	Breast	8,600
	T98G	Brain	25,000
	SW480	Colon	78
	Hs68	Fibroblast	82
	Hep G2	Hepatoma	25,000
	NCI-H69	Lung	9,000
	NCI-H82	Lung	1,200
	NCI-H146	Lung	2,200
	RAJI	Lymphoma	7,000
Simian	CV-1	Kidney	11,000
Dog	MDCK-2	Kidney	<1
	D17	Osteosarcoma	<1
Hamster	BHK	Kidney	<1
	V79	Lung	35
Rat	PC12	Pheochromocytoma	<1
Mouse	F9	Embryonal carcinoma	<1
	3T3	Fibroblast	<1

[a] Peak copy number achieved between days 2-6. (Reprinted with permission [187].)

a molecular genetic characterization of gene mutations and an evaluation of the determinants of immunogenicity. The promise of cancer gene therapy will be achieved when such tumor-specific analysis is incorporated in the design of clinical trials, and highly efficient and specific gene transfer and expression systems are developed.

Acknowledgments

This work was supported in part by National Institutes of Health grants R55CA/OD66780, RO1CA72737, RO1CA59646, and R43CA73376, and by Copernicus Gene Systems, Inc.

References

1. Friedmann, T.: Progress toward human gene therapy. *Science* **244**, 1275–1281, 1989.
2. Miller, A. D.: Human gene therapy comes of age. *Nature* **357**, 455–460, 1992.
3. Anderson, W. F.: Human gene therapy. *Science* **256**, 808–813, 1992.
4. Mulligan, R. C.: The basic science of gene therapy. *Science* **260**, 926–932, 1993.
5. Moolten, F. L.: Tumor chemosensitivity conferred by inserted herpes thymidine kinase genes: paradigm for a prospective cancer control strategy. *Cancer Res.*. **46**, 5276–5281, 1986.
6. Borrelli, E., Heyman, R., Hsi, M., *et al.*: Targeting of an inducible toxic phenotype in animal cells. *Proc. Natl. Acad. Sci. USA* **85**, 7572–7576, 1988.
7. Culver, K. W., Ram, Z., Wallbridge, S., *et al.*: In vivo gene transfer with retroviral vector-producer cells for treatment of experimental brain tumors. *Science* **256**, 1550–1552, 1992.
8. Freeman, S. M., Abboud, C. N., Whartenby, K. A., *et al.*: The "Bystander Effect": tumor regression when a fraction of the tumor mass is genetically modified. *Cancer Res.* **53**, 5274–5283, 1993.
9. Barba, D., Hardin, J., Sadelain, M., *et al.*: Development of anti-tumor immunity following thymidine kinase-mediated killing of experimental brain tumors. *Proc. Natl. Acad. Sci. USA* **91**, 4348–4352, 1994.
10. Smythe, W. R., Hwang, H. C., Amin, K. M., *et al.*: Use of recombinant adenovirus to transfer the herpes simplex virus thymidine kinase (HSVtk) gene to thoracic neoplasms: an effective *in vitro* drug sensitization system. *Cancer Res.* **54**, 2055–2059, 1994.
11. Chen, S. H., Shine, H. D., Goodman, J. C., *et al.*: Gene therapy for brain tumors: regression of experimental gliomas by adenovirus-mediated gene transfer *in vivo*. *Proc. Natl. Acad. Sci. USA* **91**, 3054–3057, 1994.

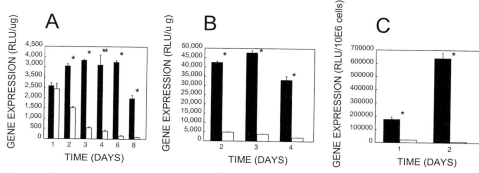

FIGURE 3 Episome-based gene expression in HT-1376 (A), Hep G2 (B), and RAJI (C). Cells were cotransfected with pRSVlacZII and either pRC/CMV.107/402-T (solid bars) or pRC/CMV (open bars). Shown are representative results from at least 2 separate experiments. Significance was determined using an unpaired, one-tailed Student's *t* test; *, P < 0.0001; **, P = 0.0001. (Reprinted with permission [187].)

12. Mullen, C. A., Kilstrup, M., Blaese, R. M.: Transfer of the bacterial gene for cytosine deaminase to mammalian cells confers lethal sensitivity to 5' fluorocytosine: A negative selection system. *Proc. Natl. Acad. Sci. USA* **89**, 33–37, 1992.

13. Mullen, C. A., Coale, M. M., Lowe, R., *et al.*: Tumors expressing the cytosine deaminase suicide gene can be eliminated *in vivo* with 5-fluorocytosine and induce protective immunity to wild type tumor. *Cancer Res.* **54**, 1503–1506, 1994.

14. Huber, R. E., Austin, E. A., Richards, C. A., *et al.*: Metabolism of 5-fluorocytosine to 5-fluorouracil in human colorectal tumor cells transduced with the cytosine deaminase gene: significant antitumor effects when only a small percentage of tumor cells express cytosine deaminase. *Proc. Natl. Acad. Sci. USA* **91**, 8302–8306, 1994.

15. Mroz, P. J., Moolten, F. L.: Retrovirally transduced *Escherichia coli* gpt genes combine selectability with chemosensitivity capable of mediating tumor eradication. *Hum. Gene Ther.* **4**, 589–595, 1993.

16. Tepper, R. I., Pattengale, P. K., Leder, P.: Murine interleukin-4 displays potent anti-tumor activity *in vivo*. *Cell* **57**, 503–512, 1989.

17. Watanabe, Y., Kuribayashi, K., Miyatake, S., *et al.*: Exogenous expression of mouse interferon gamma cDNA in mouse neuroblastoma C1300 cells results in reduced tumorigenicity by augmented anti-tumor immunity. *Proc. Natl. Acad. Sci. USA* **86**, 9456–9460, 1989.

18. Fearon, E. R., Pardoll, D. M., Itaya, T., *et al.*: Interleukin-2 production by tumor cells bypasses T helper function in the generation of an antitumor response. *Cell* **60**, 397–403, 1990.

19. Gansbacher, B., Zier, K., Daniels, B., *et al.*: Interleukin 2 gene transfer into tumor cells abrogates tumorigenicity and induces protective immunity. *J. Exp. Med.* **172**, 1217–1224, 1990.

20. Gansbacher, B., Bannerji, R., Daniels, B., *et al.*: Retroviral vector-mediated gamma-interferon gene transfer into tumor cells generates potent and long lasting antitumor immunity. *Cancer Res.* **50**, 7820–7825, 1990.

21. Colombo, M. P., Ferrari, G., Stoppacciaro, A., *et al.*: Granulocyte colony-stimulating factor gene transfer suppresses tumorigenicity of a murine adenocarcinoma *in vivo*. *J. Exp. Med.* **173**, 889–897, 1991.

22. Golumbek, P. T., Lazenby, A. J., Levitsky, H. I., *et al.*: Treatment of established renal cancer by tumor cells engineered to secrete interleukin-4. *Science* **254**, 713–716, 1991.

23. Esumi, N., Hunt, B., Itaya, T., *et al.*: Reduced tumorigenicity of murine tumor cells secreting gamma-interferon is due to nonspecific host responses and is unrelated to Class I major histocompatibility complex expression. *Cancer Res.* **51**, 1185–1189, 1991.

24. Hock, H., Dorsch, M., Diamantstein, T., *et al.*: Interleukin 7 induces CD4+ T cell-dependent tumor rejection. *J. Exp. Med.* **174**, 1291–1298, 1991.

25. Ley, V., Langlade-Demoyen, P., Kourilsky, P., *et al.*: Interleukin 2-dependent activation of tumor-specific cytotoxic T lymphocytes *in vivo*. *Eur. J. Immunol.* **21**, 851–854, 1991.

26. Asher, A. L., Mule, J. J., Kasid, A., *et al.*: Murine tumor cells transduced with the gene for tumor necrosis factor-α. *J. Immunol.* **146**, 3227–3234, 1991.

27. Blankenstein, T., Qin, Z., Uberla, K., *et al.*: Tumor suppression after tumor cell-targeted tumor necrosis factor α gene transfer. *J. Exp. Med.* **173**, 1047–1052, 1991.

28. Pardoll, D.: Immunotherapy with cytokine gene-transduced tumor cells: the next wave in gene therapy for cancer. *Curr. Opin. Oncol.* **4**, 1124–1129, 1992.

29. Porgador, A., Tzehoval, E., Katz, A., *et al.*: Interleukin 6 gene transfection into Lewis lung carcinoma tumor cells suppresses the malignant phenotype and confers immunotherapeutic competence against parental metastatic cells. *Cancer Res.* **52**, 3679–3686, 1992.

30. Aoki, T., Tashiro, K., Miyatake, S. I., *et al.*: Expression of murine interleukin 7 in a murine glioma cell line results in reduced tumorigenicity *in vivo*. *Proc. Natl. Acad. Sci. USA* **89**, 3850–3854, 1992.

31. Restifo, N. P., Spiess, P. J., Karp, S. E., *et al.*: A nonimmunogenic sarcoma transduced with the cDNA for interferon gamma elicits CD8+ T cells against the wild-type tumor: correlation with antigen presentation capability. *J. Exp. Med.* **175**, 1423–1431, 1992.

32. Tepper, R. I., Coffman, R. L., Leder, P.: An eosinophil-dependent mechanism for the antitumor effect of interleukin-4. *Science* **257**, 548–551, 1992.

33. Dranoff, G., Jaffee, E., Lazenby, A., *et al.*: Vaccination with irradiated tumor cells engineered to secrete murine granulocyte-macrophage colony-stimulating factor stimulates potent, specific, and long-lasting anti-tumor immunity. *Proc. Natl. Acad. Sci. USA* **90**, 3539–3543, 1993.

34. Porgador, A., Bannerji, R., Watanabe, Y., *et al.*: Antimetastatic vaccination of tumor-bearing mice with two types of IFN-gamma gene-inserted tumor cells. *J. Immunol.* **150**, 1458–1470, 1993.

35. Rosenthal, F. M., Cronin, K., Bannerji, R., *et al.*: Augmentation of antitumor immunity by tumor cells transduced with a retroviral vector carrying the interleukin-IL2 and interferon-gamma cDNAs. *Blood* **83**, 1289–1298, 1994.

36. Townsend, S. E., Allison, J. P.: Tumor rejection after direct costimulation of CD8+ T cells by B7-transfected melanoma cells. *Science* **259**, 368–370, 1993.

37. Plautz, G. E., Yang, Z. Y., Wu, B. Y., *et al.*: Immunotherapy of malignancy by *in vivo* gene transfer into tumors. *Proc. Natl. Acad. Sci. USA* **90**, 4645–4649, 1993.

38. Nabel, G. J., Nabel, E. G., Yang, Z. Y., *et al.*: Direct gene transfer with DNA-liposome complexes in melanoma: expression, biologic activity, and lack of toxicity in human. *Proc. Natl. Acad. Sci. USA* **90**, 11307–11311, 1993.

39. Trojan, J., Blossey, B. K., Johnson, T. R., *et al.*: Loss of tumorigenicity of rat glioblastoma directed by episome-based antisense cDNA transcription of insulin-like growth factor I. *Proc. Natl. Acad. Sci. USA* **89**, 4874–4878, 1992.

40. Trojan, J., Johnson, T. R., Rudin, S. D., *et al.*: Treatment and prevention of rat glioblastoma by immunogenic C6 cells expressing antisense insulin-like growth factor I RNA. *Science* **259**, 94–98, 1993.

41. Kantor, J., Irvine, K., Abrams, S., *et al.*: Antitumor activity and immune responses induced by a recombinant carcinoembryonic antigen-vaccinia virus vaccine. *J. Natl. Cancer Inst.* **84**, 1084–1091, 1992.

42. Kantor, J., Irvine, K., Abrams, S., *et al.*: Immunogenicity and safety of a recombinant vaccinia virus vaccine expressing the carcinoembryonic antigen gene in a nonhuman primate. *Cancer Res.* **52**, 6917–6925, 1992.

43. Conry, R. M., LoBuglio, A. F., Loechel, F., *et al.*: A carcinoembryonic antigen polynucleotide vaccine for human clinical use. *Cancer Gene Ther.* **2**, 33–38, 1995.

44. Huang, H. J. S., Yee, J. K., Shew, J. Y., *et al.*: Suppression of the neoplastic phenotype by replacement of the RB gene in human cancer cells. *Science* **242**, 1563–1566, 1988.

45. Chen, P. L., Chen, Y., Bookstein, R., *et al.*: Genetic mechanisms of tumor suppression by the human p53 gene. *Science* **250**, 1576–1580, 1990.

46. Baker, S. J., Markowitz, S., Fearon, E. R., *et al.*: Suppression of human colorectal carcinoma cell growth by wild-type p53. *Science* **249**, 912–915, 1990.

47. Cai, D. W., Mukhopadhyay, T., Liu, Y., *et al.*: Stable expression of the wild-type p53 gene in human lung cancer cells after retrovirus-mediated gene transfer. *Hum. Gene Ther.* **4,** 617–624, 1993.

48. Fujiwara, T., Grimm, E. A., Mukhopadhyay, T., *et al.*: A retroviral wild-type p53 expression vector penetrates human lung cancer spheroids and inhibits growth by inducing apoptosis. *Cancer Res.* **53,** 4129–4133, 1993.

49. Wills, K. N., Maneval, D. C., Menzel, P., *et al.*: Development and characterization of recombinant adenovirus encoding human p53 for gene therapy of cancer. *Hum. Gene Ther.* **5,** 1079–1088, 1994.

50. Mukhopadhyay, T., Tainsky, M., Cavender, A. C., *et al.*: Specific inhibition of K-*ras* expression and tumorigenicity of lung cancer cells by antisense RNA. *Cancer Res.* **51,** 1744–1748, 1991.

51. Zhang, Y., Mukhopadhyay, T., Donehower, L. A., *et al.*: Retroviral vector-mediated transduction of K-*ras* antisense RNA into human lung cancer cells inhibits expression of the malignant phenotype. *Hum. Gene Ther.* **4,** 451–460, 1993.

52. Gray, G. D., Hernandez, O. M., Hebel, D., *et al.*: Antisense DNA inhibition of tumor growth induced by c-Ha-ras oncogene in nude mice. *Cancer Res.* **53,** 577–580, 1993.

53. Cardoso, J. E., Branchereau, S., Jeyaraj, P. R., *et al.*: In situ retrovirus-mediated gene transfer into dog liver. *Hum. Gene Ther.* **4,** 411–418, 1993.

54. Li, Q., Kay, M. A., Finegold, M., *et al.*: Assessment of recombinant adenoviral vectors for hepatic gene therapy. *Hum. Gene Ther.* **4,** 403–409, 1993.

55. Stratford-Perricaudet, L. D., Makeh, I., Perricaudet, M., *et al.*: Widespread long-term gene transfer to mouse skeletal muscles and heart. *J. Clin. Invest.* **90,** 626–630, 1992.

56. LaSalle, G. L., Robert, J. J., Berrard, S., *et al.*: An adenovirus vector for gene transfer into neurons and glia in the brain. *Science* **259,** 988–990, 1993.

57. Mitani, K., Graham, F. L., Caskey, T.: Transduction of human bone marrow by adenoviral vector. *Hum. Gene Ther.* **5,** 941–948, 1994.

58. Jaffee, E. M., Dranoff, G., Cohen, L. K., *et al.*: High efficiency gene transfer into primary human tumor explants without cell selection. *Cancer Res.* **53,** 2221–2226, 1993.

59. Gunter, K. C., Khan, A. S., Noguchi, P. D.: The safety of retroviral vectors. *Hum. Gene Ther.* **4,** 643–645, 1993.

60. Cornetta, K. C., Morgan, R. A., Anderson, W. F.: Safety issues related to retroviral-mediated gene transfer in humans. *Hum. Gene Ther.* **2,** 5–20, 1991.

61. Donahue, R. E., Kessler, S. W., Bodine, D., *et al.*: Helper virus induced T cell lymphoma in nonhuman primates after retroviral mediated gene transfer. *J. Exp. Med.* **176,** 1125–1135, 1992.

62. Yang, R., Nunes, F. A., Berencsi, K., *et al.*: Cellular immunity to viral antigens limits E1-deleted adenoviruses for gene therapy. *Proc. Natl. Acad. Sci. USA* **91,** 4407–4411, 1994.

63. Miller, D. G., Adam, M. A., Miller, A. D.: Gene transfer by retrovirus vectors occurs only in cells that are actively replicating at the time of infection. *Mol. Cell Biol.* **10,** 4239–4242, 1990.

64. Cornetta, K. C., Moen, R. C., Culver, K., *et al.*: Amphotropic murine leukemia retrovirus is not an acute pathogen for primates. *Hum. Gene Ther.* **1,** 14–30, 1990.

65. Perales, J. C., Ferkol, T., Molas, M., *et al.*: An evaluation of receptor-mediated approaches for the introduction of genes in somatic cells. *Eur. J. Biochem.* **226,** 255–266, 1994.

66. Perales, J. C., Ferkol, T., Beegen, H., *et al.*: Gene transfer *in vivo*: sustained expression and regulation of genes introduced into the liver by receptor-targeted uptake. *Proc. Natl. Acad. Sci. USA* **91,** 4084–4090, 1994.

67. Ferkol, T., Perales, J. C., Eckman, E., *et al.*: Gene transfer into the airway epithelium of animals by targeting the polymeric immunoglobulin receptor. *J. Clin. Invest.* **95,** 493–502, 1995.

68. Israel, M. A., Chan, H. W., Hourihan, S. L., *et al.*: Biological activity of polyoma viral DNA in mice and hamsters. *J. Virol.* **29,** 990–996, 1979.

69. Dubensky, T. W., Campbell, B. A., Villarreal, L. P.: Direct transfection of viral and plasmid DNA into the liver or spleen of mice. *Proc. Natl. Acad. Sci. USA* **81,** 7529–7533, 1984.

70. Seeger, C., Ganem, D., Varmus, H. E.: The cloned genome of ground squirrel hepatitis virus is infectious in the animal. *Proc. Natl. Acad. Sci. USA* **81,** 5849–5852, 1984.

71. Benvenisty, N., Reshef, L.: Direct introduction of genes into rats and expression of the genes. *Proc. Natl. Acad. Sci. USA* **83,** 9551–9555, 1986.

72. Wolff, J. A., Malone, R. W., Williams, P., *et al.*: Direct gene transfer into mouse muscle *in vivo*. *Science* **247,** 1465–1468, 1990.

73. Acsadi, G., Jiao, S., Jani, A., *et al.*: Direct gene transfer and expression into rat heart *in vivo*. *New Biol.* **3,** 71–81, 1991.

74. Manthorpe, M., Cornefert-Jensen, F., Hartikka, J., *et al.*: Gene therapy by intramuscular injection of plasmid DNA: studies on firefly luciferase gene expression in mice. *Hum. Gene Ther.* **4,** 419–431, 1993.

75. Acsadi, G., Dickson, G., Lover, D. R., *et al.*: Human dystrophin expression in mdx mice after intramuscular injection of DNA constructs. *Nature* **352,** 815–818, 1991.

76. Hickman, M. A., Malone, R. W., Lehmann-Bruinsma, K.: Gene expression following direct injection of DNA into liver. *Hum. Gene Ther.* **5,** 1477–1483, 1994.

77. Sikes, M. L., O'Malley, B. W., Finegold, M. J.: *In vivo* gene transfer into rabbit thyroid follicular cells by direct DNA injection. *Hum. Gene Ther.* **5,** 837–844, 1994.

78. Ulmer, J. B., Donnelly, J. J., Parker, S. E., *et al.*: Heterologous protection against influenza by injection of DNA encoding a viral protein. *Science* **259,** 1745–1733, 1993.

79. Wang, B., Ugen, K. E., Srikantan, V., *et al.*: Gene inoculation generates immune responses against human immunodeficiency virus type 1. *Proc. Natl. Acad. Sci. USA* **90,** 4156–4160, 1993.

80. Raz, E., Carson, D. A., Parker, S. E., *et al.*: Intradermal gene immunization: the possible role of DNA uptake in the induction of cellular immunity to viruses. *Proc. Natl. Acad. Sci. USA* **91,** 9519–9523, 1994.

81. Estin, C. D., Stevenson, U. S., Plowman, G. D., *et al.*: Recombinant vaccinia virus vaccine against the human melanoma antigen p97 for use in immunotherapy. *Proc. Natl. Acad. Sci. USA* **85,** 1052–1056, 1988.

82. Van der Bruggen, P., Traversati, C., Chomez, P., *et al.*: A gene encoding an antigen recognized by cytolytic T lymphocytes on a human melanoma. *Science* **254,** 1643–1647, 1991.

83. Chen, Y. T., Stockert, E., Chen, Y., *et al.*: Identification of the MAGE-1 gene product by monoclonal and polyclonal antibodies. *Proc. Natl. Acad. Sci. USA* **91,** 1004–1008, 1994.

84. Coulie, P. G., Brichard, V., Van Pel, A., *et al.*: A new gene coding for a differentiation antigen recognized by autologous cytolytic T lymphocytes on HLA-A2 melanomas. *J. Exp. Med.* **180,** 35–42, 1994.

85. Kawakami, Y., Eliyahu, S., Sakaguchi, K., *et al.*: Identification of the immunodominant peptides of the MART-1 human melanoma antigen recognized by the majority of HLA-A2-restricted tumor infiltrating lymphocytes. *J. Exp. Med.* **180,** 347–352, 1994.

86. Cox, A. L., Skipper, J., Chen, Y., *et al.*: Identification of a peptide recognized by five melanoma-specific human cytotoxic T cell lines. *Science* **264,** 716–719, 1994.

87. Celis, E., Tsai, V., Crimi, C., *et al.*: Induction of anti-tumor cytotoxic T lymphocytes in normal humans using primary cultures and synthetic peptide epitopes. *Proc. Natl. Acad. Sci. USA* **91**, 2105–2109, 1994.

88. Brichard, V., Van Pel, A., Wolfel, T., *et al.*: The tyrosinase gene codes for an antigen recognized by autologous cytolytic T lymphocytes on HLA-A2 melanomas. *J. Exp. Med.* **178**, 489–495, 1993.

89. Bakker, A. B. H., Schreurs, M. W. J., de Boer, A. J., *et al.*: Melanocyte lineage-specific antigen gp100 is recognized by melanoma-derived tumor-infiltrating lymphocytes. *J. Exp. Med.* **179**, 1005–1009, 1994.

90. Boon, T., Cerottini, J. C., Van den Eynde, B., *et al.*: Tumor antigens recognized by T lymphocytes. *Annu. Rev. Immunol.* **12**, 337–365, 1994.

91. Jerome, K. R., Domenech, N., Finn, O. J.: Tumor-specific cytotoxic T cell clones from patients with breast and pancreatic adenocarcinoma recognize EBV-immortalized B cells transfected with polymorphic epithelial mucin complementary DNA. *J. Immunol.* **151**, 1654–1662, 1993.

92. Hill, A., Ploegh, H.: Getting the inside out: the transporter associated with antigen processing (TAP) and the presentation of viral antigen. *Proc. Natl. Acad. Sci. USA* **92**, 341–343, 1995.

93. Peace, D. J., Chen, W., Nelson, H., *et al.*: T cell recognition of transforming proteins encoded by mutated ras proto-oncogenes. *J. Immunol.* **146**, 2059–2065, 1991.

94. Jung, S., Schluesener, H. J.: Human T lymphocytes recognize a peptide of single point-mutated, oncogenic ras proteins. *J. Exp. Med.* **173**, 273–276, 1991.

95. Skipper, J., Stauss, H. J.: Identification of two cytotoxic T lymphocyte-recognized epitopes in the Ras protein. *J. Exp. Med.* **177**, 1493–1498, 1993.

96. Gedde-Dahl, T., Fossum, B., Eriksen, J. A., *et al.*: T cell clones specific for p21 ras-derived peptides: characterization of their fine specificity and HLA restriction. *Eur. J. Immunol.* **23**, 754–760, 1993.

97. Houbiers, J. G. A., Nijman, H. W., Van der Burg, S. H., *et al.*: In vitro induction of human cytotoxic T lymphocyte responses against peptides of mutant and wild-type p53. *Eur. J. Immunol.* **23**, 2072–2077, 1993.

98. Chen, W., Peace, D. J., Rovira, D. K., *et al.*: T-cell immunity to the joining region of p210$^{BCR-ABL}$ protein. *Proc. Natl. Acad. Sci. USA* **89**, 1468–1472, 1992.

99. Muraro, R., Wunderlich, D., Thor, A., *et al.*: Definition by monoclonal antibodies of a repertoire of epitopes on carcinoembryonic antigen differentially expressed in human colon carcinomas versus normal adult tissues. *Cancer Res.* **45**, 5769–5780, 1985.

100. Yang, N. S., Burkholder, J., Roberts, B., *et al.*: In vivo and in vitro gene transfer to mammalian somatic cells by particle bombardment. *Proc. Natl. Acad. Sci. USA* **87**, 9568–9572, 1990.

101. Fynan, E. F., Webster, R., Fuller, D. H., *et al.*: DNA vaccines: protective immunizations by parenteral, mucosal, and gene-gun inoculations. *Proc. Natl. Acad. Sci. USA* **90**, 11478–11782, 1993.

102. Sun, W. H., Burkholder, J. K., Sun, J., *et al.*: In vivo cytokine gene transfer by gene gun reduces tumor growth in mice. *Proc. Natl. Acad. Sci. USA* **92**, 2889–2893, 1995.

103. Graham, F. L., Van der Eb, A. J.: A new technique for the assay of infectivity of human adenovirus-5 DNA. *Virology* **52**, 456–467, 1973.

104. Vile, R. G., Hart, I. R.: Use of tissue-specific expression of the herpes simplex virus thymidine kinase gene to inhibit growth of established murine melanomas following direct intratumoral injection of DNA. *Cancer Res.* **53**, 3860–3864, 1993.

105. Felgner, P. L., Ringold, G. M.: Cationic liposome-mediated transfection. *Nature* **337**, 387–388, 1989.

106. Felgner, P. L., Zaugg, R. H., Norman, J. A.: Synthetic recombinant DNA delivery for cancer therapeutics. *Cancer Gene Ther.* **2**, 61–65, 1995.

107. Stopeck, A. T., Hersh, E. M., Akporiaye, E. T., *et al.*: Phase I study of direct gene transfer of an allogeneic histocompatibility antigen, HLA-B7, in patients with metastatic melanoma. *J. Clin. Oncol.* **15**, 341–349, 1997.

108. Rubin, J., Galanis, E., Pitot, H. C., *et al.*: Phase I study of immunotherapy of hepatic metastases of colorectal carcinoma by direct gene transfer of an allogeneic histocompatibility antigen, HLA-B7. *Gene Ther.* **4**, 419,425, 1997.

109. Nabel, E. G., Gordon, D., Yang, Z. Y., *et al.*: Gene transfer *in vivo* with DNA-liposome complexes: lack of autoimmunity and gonadal localization. *Hum. Gene Ther.* **3**, 649–656, 1992.

110. Stewart, M. J., Plautz, G. E., delBuono, L., *et al.*: Gene transfer *in vivo* with DNA-liposome complexes: safety and acute toxicity in mice. *Hum. Gene Ther.* **3**, 267–275, 1992.

111. Nicolau, C., Le Pape, A., Soriano, P., *et al.*: *In vivo* expression of rat insulin after intravenous administration of the liposome-entrapped gene for rat insulin I. *Proc. Natl. Acad. Sci. USA* **80**, 1068–1072, 1983.

112. Zhu, N., Liggitt, D., Liu, Y., *et al.*: Systemic gene expression after intravenous DNA delivery into adult mice. *Science* **261**, 209–211, 1993.

113. Lesoon-Wood, L. A., Kim, W. H., Kleinman, H. K., *et al.*: Systemic gene therapy with p53 reduces growth and metastases of a malignant human breast cancer in nude mice. *Hum. Gene Ther.* **6**, 395–405, 1995.

114. Wang, S., Lee, R. J., Cauchon, G.: Delivery of antisense oligodeoxyribonucleotides against the human epidermal growth factor receptor into cultured KB cells with liposomes conjugated to folate via polyethylene glycol. *Proc. Natl. Acad. Sci. USA* **92**, 3318–3322, 1995.

115. Lee, R. J., Huang, L.: Folate-targeted, anionic liposome-entrapped polylysine-condensed DNA for tumor cell-specific gene transfer. *J. Biol. Chem.* **271**, 8481–8487, 1996.

116. Goren, D., Horowitz, A. T., Zalipsky, S., *et al.*: Targeting of stealth liposomes to erbB-2 (Her/2) receptor: *in vitro* and *in vivo* studies. *Br. J. Cancer* **74**, 1749–1756, 1996.

117. Allen, T. M., Hanson, C.: Pharmacokinetics of stealth versus conventional liposomes: effect of dose. *Biochim. Biophys. Acta.* **1068**, 133–141, 1991.

118. Mayhew, E. G., Lasic, D., Babbar, S., *et al.*: Pharmacokinetics and antitumor activity of epirubicin encapsulated in long-circulating liposomes incorporating a polyethylene glycol-derivatized phospholipid. *Int. J. Cancer* **51**, 302–309, 1992.

119. Wu, N. Z., Da, D., Rudoll, T. L., *et al.*: Increased microvascular permeability contributes to preferential accumulation of Stealth liposomes in tumor tissue. *Cancer Res.* **53**, 3765–3770, 1993.

120. Yuan, F., Leunig, M., Huang, S. K., *et al.*: Microvascular permeability and interstitial penetration of sterically stabilized (stealth) liposomes in a human tumor xenograft. *Cancer Res.* **54**, 3352–3356, 1994.

121. Clinical Protocols. *Cancer Gene Ther.* **2**, 67–74, 1995.

122. Nabel, G. J., Gordon, D., Bishop, D. K., *et al.*: Immune response in human melanoma after transfer of an allogeneic class I major histocompatibility complex gene with DNA-liposome complexes. *Proc. Natl. Acad. Sci. USA* **93**, 15388–15393, 1996.

123. Wu, G. Y., Wu, C. H.: Receptor-mediated gene delivery and expression *in vivo*. *J. Biol. Chem.* **263**, 14621–14624, 1988.

124. Wu, G. Y., Wilson, J. M., Shalaby, F., *et al.*: Receptor-mediated gene delivery *in vivo*: partial correction of genetic analbuminemia in Nagase rats. *J. Biol. Chem.* **266**, 14338–14342, 1991.

125. Cotten, M., Wagner, W., Zatloukal, K., et al.: High-efficiency receptor-mediated delivery of small and large (48) kilobase gene constructs using the endosome-disruption activity of defective or chemically inactivated adenovirus particles. Proc. Natl. Acad. Sci. USA 89, 6094–6098, 1992.

126. Wagner, E., Zatloukal, K., Cotten, M., et al.: Coupling of adenovirus to transferrin-polylysine/DNA complexes greatly enhances receptor-mediated gene delivery and expression of transfected genes. Proc. Natl. Acad. USA 89, 6099–6103, 1992.

127. Wagner, E., Plank, C., Zatloukal, K., et al.: Influenza virus hemagglutinin HA-2 N-terminal fusogenic peptides augment gene transfer by transferrin-polylysine-DNA complexes: toward a synthetic virus-like gene-transfer vehicle. Proc. Natl. Acad. Sci. USA 89, 7934–7938, 1992.

128. Gao, L., Wagner, E., Cotten, M., et al.: Direct in vivo gene transfer to airway epithelium employing adenovirus-polylysine-DNA complexes. Human Gene Ther. 4, 17–24, 1993.

129. Batra, R. K., Berschneider, H., Curiel, D. T.: Molecular conjugate vectors mediate efficient gene transfer into gastrointestinal epithelial cells. Cancer Gene Ther. 1, 185–192, 1994.

130. Ferkol, T., Perales, J. C., Mularo, F., et al.: Receptor-mediated gene transfer into macrophages. Proc. Natl. Acad. Sci. USA 93, 101–105, 1996.

131. Ziady, A. G., Perales, J. C., Ferkol, T., et al.: Gene transfer into hepatoma cell lines via the serpin enzyme complex receptor. Am. J. Physiol. 273, G545–G552, 1997.

132. Furth, P. A., Hennighausen, L., Baker, C., et al.: The variability in activity of the universally expressed human cytomegalovirus immediate early gene 1 enhancer/promoter in transgenic mice. Nucl. Acids Res. 19, 6205–6208, 1991.

133. Cheng, L., Ziegelhoffer, P. R., Yang, N. S.: In vivo promoter activity and transgene expression in mammalian somatic tissues evaluated by using particle bombardment. Proc. Natl. Acad. Sci. USA 90, 4455–4459, 1993.

134. Jaenisch, R.: Transgenic animals. Science 240, 1468–1474, 1988.

135. Hanahan, D.: Transgenic mice as probes into complex systems. Science 246, 1265–1275, 1989.

136. Hanahan, D.: Heritable formation of pancreatic β-cell tumours in transgenic mice expressing recombinant insulin/simian virus 40 oncogenes. Nature 315, 115–122, 1985.

137. Ornitz, D. M., Hammer, R. E., Messing, A., et al.: Pancreatic neoplasia induced by SV40 T-antigen expression in acinar cells of transgenic mice. Science 238, 188–193, 1987.

138. Schoenenberger, C. A., Andres, A. C., Groner, B., et al.: Targeted c-myc gene expression in mammary glands of transgenic mice induces mammary tumours with constitutive milk protein gene transcription. EMBO J. 7, 169–175, 1988.

139. Vile, R. G., Hart, I. R.: In vitro and in vivo targeting of gene expression to melanoma cells. Cancer Res. 53, 962–967, 1993.

140. Sasaoka, T., Kobayashi, K., Nagatsu, I., et al.: Analysis of the human tyrosine hydroxylase promoter-chloramphenicol acetyltransferase chimeric gene expression in transgenic mice. Mol. Brain Res. 16, 274–286, 1992.

141. Julien, J. P., Tretjakoff, I., Beaudet, L., et al.: Expression and assembly of a human neurofilament protein in transgenic mice provide a novel neuronal marking system. Genes Dev. 1, 1085–1095, 1987.

142. Brenner, M., Kisselberth, W. C., Su, Y., et al.: GFAP promoter directs astrocyte-specific expression in transgenic mice. J. Neurosci. 14, 1030–1037, 1994.

143. Tronik, D., Dreyfus, M., Babinet, C., et al.: Regulated expression of the Ren-2 gene in transgenic mice derived from parental strains carrying only the Ren-1 gene. EMBO J. 6, 983–987, 1987.

144. Stacey, A., Bateman, J., Choi, T., et al.: Perinatal lethal osteogenesis imperfecta in transgenic mice bearing an engineered mutant pro-α1(I) collagen gene. Nature 332, 131–136, 1988.

145. Shani, M.: Tissue-specific and developmentally regulated expression of a chimeric actin-globin gene in transgenic mice. Mol. Cell Biol. 6, 2624–2631, 1986.

146. Jahroudi, N., Lynch, D. C.: Endothelial-cell-specific regulation of von Willebrand factor gene expression. Mol. Cell Biol. 14, 999–1008, 1994.

147. Huber, B. E., Richard, C. A., Krenitsky, T. A.: Retroviral-mediated gene therapy for the treatment of hepatocellular carcinoma: an innovative approach for cancer therapy. Proc. Natl. Acad. Sci. USA 88, 8039–8043, 1991.

148. Glasser, S. W., Korfhagen, T. R., Bruno, M. D., et al.: Structure and expression of the pulmonary surfactant protein SP-C gene in the mouse. J. Biol. Chem. 265, 21986–21991, 1990.

149. Osaki, T., Tanio, Y., Tachibana, I., et al.: Gene therapy for carcinoembryonic antigen-producing human lung cancer cells by cell type-specific expression of herpes simplex virus thymidine kinase gene. Cancer Res. 54, 5258–5261, 1994.

150. Lin, J. H., Zhao, H., Sun, T. T.: A tissue-specific promoter that can drive a foreign gene to express in the suprabasal urothelial cells of transgenic mice. Proc. Natl. Acad. Sci. USA 92, 679–683, 1995.

151. Krimpenfort, P., de Jong, R., Uematsu, Y., et al.: Transcription of T cell receptor β-chain genes is controlled by a downstream regulatory element. EMBO J. 7, 745–750, 1988.

152. Alexander, W. S., Schrader, J. W., Adams, J. M.: Expression of the c-myc oncogene under control of an immunoglobulin enhancer in Eμ-myc transgenic mice. Mol. Cell Biol. 7, 1436–1444, 1987.

153. Murtha, P., Tindall, D. J., Young, C. Y. F.: Androgen induction of a human prostate-specific kallikrein, hKLK2: characterization of an androgen response element in the 5′ promoter region of the gene. Biochemistry 32, 6459–6464, 1993.

154. Peschon, J. J., Behringer, R. R., Brinster, R. L., et al.: Spermatid-specific expression of protamine 1 in transgenic mice. Proc. Natl. Acad. Sci. USA 84, 5316–5319, 1987.

155. Rixe, O., Calvez, V., Mouawad, R., et al.: Trans-activation of the metallothionein promoter in cisplatin (CP) resistant cell lines: potential application for a specific gene therapy. Proc. Am. Assoc. Cancer Res. 36, 220, 1995.

156. Gossen, M., Bujard, H.: Tight control of gene expression in mammalian cells by tetracycline-responsive promoters. Proc. Natl. Acad. Sci. USA 89, 5547–5551, 1992.

157. Furth, P. A., St Onge, L., Boger, H., et al.: Temporal control of gene expression in transgenic mice by a tetracycline-responsive promoter. Proc. Natl. Acad. Sci. USA 91, 9302–9306, 1994.

158. Passman, R. S., Fishman, G. L.: Regulated expression of foreign genes in vivo after germline transfer. J. Clin. Invest. 94, 2421–2425, 1994.

159. Fishman, G. I., Kaplan, M. L., Buttrick, P. M.: Tetracycline-regulated cardiac gene expression in vivo. J. Clin. Invest. 93, 1864–1868, 1994.

160. Gossen, M., Freundlieb, S., Bender, G., et al.: Transcriptional activation by tetracyclines in mammalian cells. Science 268, 1766–1769, 1995.

161. Deuschle, U., Meyer, W. K., Thiesen, H. J.: Tetracycline-reversible silencing of eukaryotic promoters. Mol. Cell Biol. 15, 1907–1914, 1995.

162. Wang, Y., O'Malley, B. W., Tsai, S. Y., et al.: A regulatory system for use in gene transfer. Proc. Natl. Acad. Sci. USA 91, 8180–8184, 1994.

163. Wang, Y., DeMayo, F. J., Tsai, S. Y., *et al.*: Ligand-inducible and liver-specific target gene expression in transgenic mice. *Nature Biotechnology* **15**, 239–343, 1997.

164. Delort, J. P., Capecchi, M. R.: TAXI/UAS: a molecular switch to control expression of genes *in vivo*. *Hum. Gene Ther.* **7**, 809–820, 1996.

165. Wang, Y., Xu, J., Pierson, T., *et al.*: Positive and negative regulation of gene expression in eukaryotic cells with an inducible transcriptional regulator. *Gene Ther.* **4**, 432–441, 1997.

166. No, D., Yao, T. P., Evans, R. M.: Ecdysone-inducible gene expression in mammalian cells and transgenic mice. *Proc. Natl. Acad. Sci. USA* **93**, 3346–3351, 1996.

167. Rivera, V. M., Clackson, T., Natesan, S., *et al.*: A humanized system for pharmacologic control of gene expression. *Nature Med.* **2**, 1028–1032, 1996.

168. Weichselbaum, R. R., Hallahan, D. E., Beckett, M. A., *et al.*: Gene therapy targeted by radiation preferentially radiosensitizes tumor cells. *Cancer Res.* **54**, 4266–4269, 1994.

169. Biamonti, G., Della Valle, G., Talarico, D., *et al.*: Fate of exogenous recombinant plasmids introduced into mouse and human cells. *Nucleic Acids Res.* **13**, 5545–5561, 1985.

170. Krysan, P. J., Haase, S. B., Calos, M. P.: Isolation of human sequences that replicate autonomously in human cells. *Mol. Cell Biol.* **9**, 1026–1033, 1989.

171. Haase, S. B., Calos, M. P.: Replication control of autonomously replicating human sequences. *Nucl. Acids Res.* **19**, 5053–5058, 1991.

172. Heinzel, S. S., Krysan, P. J., Tran, C. T., *et al.*: Autonomous DNA replication in human cells is affected by the size and the source of the DNA. *Mol. Cell Biol.* **11**, 2263–2272, 1991.

173. Tran, C. T., Caddle, M. S., Calos, M. P.: The replication behavior of *Saccharomyces cerevisiae* DNA in human cells. *Chromosoma* **102**, 129–136, 1993.

174. Middleton, T., Sugden, B.: Retention of plasmid DNA in mammalian cells is enhanced by binding of the Epstein–Barr virus replication protein EBNA1. *J. Virol.* **68**, 4067–4071, 1994.

175. Harrington, J. J., Van Bokkelen, G., Mays, R. W., *et al.*: Formation of de novo centromeres and construction of first-generation human artificial microchromosomes. *Nat. Genet.* **15**, 345–355, 1997.

176. Yates, J. L., Warren, N., Sugden, B.: Stable replication of plasmids derived from Epstein–Barr virus in various mammalian cells. *Nature* **313**, 812–815, 1985.

177. Milanesi, G., Barbanti-Brodano, G., Negrini, M., *et al.*: BK virus-plasmid expression vector that persists episomally in human cells and shuttles into Escherichia coli. *Mol. Cell Biol.* **4**, 1551–1560, 1984.

178. Tsui, L. C., Breitman, M. L., Siminovitch, L., *et al.*: Persistence of freely replicating SV40 recombinant molecules carrying a selectable marker in permissive simian cells. *Cell* **30**, 499–508, 1982.

179. Grossi, M. P., Caputo, A., Rimessi, P., *et al.*: New BK virus episomal vector for complementary DNA expression in human cells. *Arch .Virol.* **102**, 275–283, 1988.

180. Grossi, M. P., Caputo, A., Paolini, L., *et al.*: Factors affecting amplification of BK virus episomal vectors in human cells. *Arch. Virol.* **99**, 249–259, 1988.

181. Cooper, M. J., Miron, S. M.: Efficient episomal expression vector for human transitional carcinoma cells. *Hum. Gene Ther.* **4**, 557–566, 1993.

182. Wilson, J. B., Bell, J. L., Levine, A. J.: Expression of Epstein–Barr virus nuclear antigen-1 induces B cell neoplasia in transgenic mice. *EMBO J.* **15**, 3117–3126, 1996.

183. Linzer, D. H., Levine, A. J.: Characterization of a 54K dalton cellular SV40 tumor antigen present in SV40-transformed cells and uninfected embryonal carcinoma cells. *Cell* **17**, 43–52, 1979.

184. DeCaprio, J. A., Ludlow, J. W., Figge, J., *et al.*: SV40 large tumor antigen forms a specific complex with the product of the retinoblastoma susceptibility gene. *Cell* **54**, 275–283, 1988.

185. Ewen, M. E., Xing, Y., Lawrence, J. B., *et al.*: Molecular cloning, chromosomal mapping, and expression of the cDNA for p107, a retinoblastoma gene product-related protein. *Cell* **66**, 1155–1164, 1991.

186. Claudio, P. P., Howard, C. M., Baldi, A., *et al.*: p130/pRb2 has growth suppressive properties similar to yet distinctive from those of retinoblastoma family members pRb and p107. *Cancer Res.* **54**, 5556–5560, 1994.

187. Cooper, M. J., Lippa, M., Payne, J. M., *et al.*: Safety-modified episomal vectors for human gene therapy. *Proc. Natl. Acad. Sci. USA* **94**, 6450–6455, 1997.

188. Brunovskis, P., Payne, J. M., Cooper, M. J.: Unpublished data, September 1997.

189. Quentin, B., Perricaudet, L. D., Tajbakhsh, S., *et al.*: Adenovirus as an expression vector in muscle cells *in vivo*. *Proc. Natl. Acad. Sci. USA* **89**, 2581–2584, 1992.

190. Russell, D. W., Miller, A. D., Alexander, I. E.: Adeno-associated virus vectors preferentially transduce cells in S-phase. *Proc. Natl. Acad. Sci. USA* **91**, 8915–8919, 1994.

191. Fink, D. J., Sternberg, L. R., Weber, P. C., *et al.*: In vivo expression of β-galactosidase in hippocampal neurons by HSV-mediated gene transfer. *Hum. Gene Ther.* **3**, 11–19, 1992.

192. Moss, B., Flexner, C.: Vaccinia virus expression vectors. *Annu. Rev. Immunol.* **5**, 305–324, 1987.

193. Maxwell, I. H., Maxwell, F., Rhode, S. L., *et al.*: Recombinant LuIII autonomous parvovirus as a transient transducing vector for human cells. *Hum. Gene Ther.* **4**, 411–450, 1993.

CHAPTER 6

Parvovirus Vectors for the Gene Therapy of Cancer

K. K. WONG, JR.,[1] ELIZABETH SHAUGHNESSY,[2,3] DI LU,[4,5]
GRACE FISHER-ADAMS,[4] AND SASWATI CHATTERJEE[4]

[1]Department of Hematology and Bone Marrow Transplantation,
[2]Department of General and Oncologic Surgery, and [4]Division of Pediatrics,
City of Hope National Medical Center, Duarte, California 91010

I. INTRODUCTION

Intensive investigation has demonstrated that a wide variety of clinically important diseases including viral infections and cancer arise as a direct result of aberrant gene expression, and thus may be considered "genetic" disorders [1]. Furthermore, the rapid development of recombinant DNA technology, in conjunction with expanding knowledge of these underlying pathogenic mechanisms, has led to strategies designed to actually correct disease at the molecular level. Gene transfer approaches have been developed to correct primary inherited diseases, such as cystic fibrosis or adenosine deaminase deficiency, as well as acquired disorders, such as cancer [2]. Although a variety of techniques are available for the introduction of genes into cells, few approach the efficiency necessary for actual gene therapy trials. To address this issue, viral vectors have been developed that exploit the natural ability of viruses to efficiently infect and transfer their genetic material into cells. To date, the most commonly used viral vector system is based upon murine retroviruses. Amphotropic, nonlentivirus retroviral vectors have a wide host range and efficiently integrate into

---

[3] *Present address:* Department of Surgery, University of Cincinnati
[5] *Present address:* Department of Pathology, University of Kansas

cellular DNA. However, they require proliferating target cells for efficient transduction, which potentially limits their effectiveness [3,4]. Thus, important nonproliferating populations such as primary neural and hematopoietic stem cells may be refractory to retroviral gene transfer. Furthermore, replication competent retroviruses (RCRs), which may be generated during the vector packaging process, thereby contaminating vector stocks, have been causally linked to the development of T-cell lymphomas in a primate model of hematopoietic progenitor transplantation with genetically modified cells [5]. Thus, safe and efficient gene transfer vectors that can circumvent these problems are continually being sought.

In this chapter, we will briefly review the basic biology of parvoviruses, specifically as it relates to the development of gene transfer vectors. We will focus upon potential advantages and disadvantages of parvovirus vectors, in addition to applications for the treatment of cancer that are being actively pursued. We refer the reader to several excellent recently published reviews describing the field of cancer gene therapy for more detailed discussions of the rationale and strategies currently undergoing evaluation (see Chapters 4, 9, and 10) [6,7].

II. BIOLOGY OF PARVOVIRIDAE AND VECTOR DEVELOPMENT

Parvoviruses are small, nonenveloped icosahedral viruses containing a 5-kb, single-stranded DNA genome flanked by palindromic inverted terminal repeats (ITRs). The ITRs serve as *cis*-active elements necessary for viral DNA replication. The family Parvoviridae is subdivided into three groups: the densoviruses, which are exclusive to arthropods, the autonomous parvoviruses, which are lytic and replicate in proliferating target cells, and the dependo- or adeno-associated viruses (AAVs), which require helper virus coinfection for productive infection [8]. The latter two groups infect a broad spectrum of vertebrates, including humans.

A. Adeno-Associated Virus (AAV)

Productive AAV infection requires defined "helper" functions provided by coinfection with adeno- or herpes viruses [9,10]. In the absence of helper virus coninfection, AAV genomically integrates into the host chromosome. For wild-type AAV, this integration appears to be site specific, inserting in human chromosome 19q13.2-13.4-qter [11–13]. Latent AAV infections have been maintained in tissue culture for more than 100 serial passages in the absence of selective pressure, at-

testing to the stability of viral integration [14]. AAV exhibits a wide host (from avians to primates) and tissue range. It is likely that cellular infection involves a receptor, as in the case of the autonomous parvovirus B19, although this has not yet been identified [15]. Importantly, since its original discovery, AAV has not been identified as a cause of disease in animals or humans [9,10]. On the contrary, infection with wild-type AAV may exert an oncoprotective effect [16,17]. AAV infection inhibits growth of human melanoma and cervical carcinoma cells, and transformation by the activated H-*ras* oncogene and bovine and human papillomaviruses *in vitro* [18–20]. These anti-oncogenic properties have been mapped to the p5 product (*Rep*78), a pleiotropic protein necessary for transcription regulation, integration, and viral *ori*-dependent replication, in conjunction with the ITRs [16].

Genetic analysis of the AAV genome has defined two major open reading frames (ORFs). The left ORF is comprised of two promoters at map positions 5 (p5) and 19 (p19), and encodes functions necessary for AAV *ori*-dependent replication (*rep*) and regulatory control of AAV promoters. The right ORF is under control of a promoter at map position 40 (p40) and encodes structural proteins necessary for virion encapsidation (*cap*). The AAV genome is flanked by palindromic ITRs, which, along with a recently described non-palindromic "D" sequence, are essential for viral DNA replication (*ori*), virion encapsidation, and host chromosomal integration [9,10,21,22]. In addition, AAV ITRs themselves possess weak intrinsic promoter activity [23]. Recombinant AAV-based vectors have been developed using several strategies [24,25]. In one strategy, the transgene of interest has been placed under control of an endogenous AAV promoter, typically p40. Another strategy has been to remove all endogenous AAV transcriptional units and replace them with strong, heterologous viral or cellular promoters (Fig. 1). This is currently the most common form, as removal of AAV sequences minimizes the opportunity to generate wild-type AAV by homologous recombination during the encapsidation process. For particularly large transgenes such as the cystic fibrosis transmembrane conductance reporter (CFTR), the ITRs themselves have been used to promote gene expression [23]. Finally, vectors have been developed that incorporate features of each of these strategies, e.g., mixing endogenous AAV with heterologous promoters, etc. In most instances, the p5 ORF has been removed to provide additional space for transgene insertion and because the p5 product (*rep 78/68*) has been demonstrated to inhibit either heterologous promoters or transformation efficiency [26].

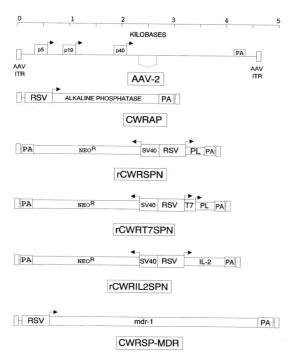

FIGURE 1   Comparison of wild-type AAV-2 sequence with recombinant constructs. All vectors are flanked by AAV-2 inverted terminal repeats (ITRs), which are necessary for DNA replication, encapsidation, and host cell integration. CWRAP encodes the thermostable placental alkaline phosphatase (PLAP) gene, and CWRSP-MDR encodes the human MDR1 cDNA under Rous sarcoma virus (RSV) promoter control. Other vectors are modifications of CWRSPN, which contains two transcriptional cassettes, one conferring G418 resistance and the other consisting of an RSV promoter, a multisite polylinker (PL), and an SV40 polyadenylation signal. CWRT7SPN is a modification of CWRSPN, containing a bacteriophage T7 promoter downstream of the RSV promoter, used for *in vitro* ribozyme expression, and rCWRIL2SPN encodes the human IL-2 cDNA under RSV promoter control.

The wild-type AAV genome is 4.8 nucleotides in length. Vectors greater than 115–120% of the wild-type size are generally not packaged efficiently [24,27,28]. However, it should be emphasized that most cDNAs are sufficiently small to meet this size constraint. AAV vectors are generally doubly defective; disruption of a requisite AAV-encoded function by insertion of an expression cassette necessitates provision of AAV functions *in trans* in addition to "helper" virus functions for productive replication. Thus, AAV vectors are often encapsidated in helper-virus-infected cells by cotransfection of vectors with a "helper" plasmid encoding AAV *rep* and/or *cap* functions but lacking ITRs so they cannot be packaged [24]. Robust expression of *cap* and appropriately regulated expression of *rep* appear important for efficient encapsidation [29,30]. Sequence homology between the vector and plasmid encoding AAV *rep* or *cap* functions is minimized or eliminated to prevent

generation of wild-type AAV by homologous recombination. Encapsidation efficiency may depend not only upon the size of the vector, but also upon the encoded transgene [24]. Generally, recombinant AAV vector titers of 10^5–10^7 functional transducing units/mL are obtained employing this system, with minimal wild type contamination [31]. However, this encapsidation strategy remains cumbersome and labor intensive, and it can be highly variable. Thus, other strategies including more efficient transfection protocols [32,33], better vector purification systems [34], and packaging cell lines analogous to those for retroviral vectors are under development in several laboratories [35–38].

What are the potential advantages of AAV vectors? AAV vectors retain the stability, wide host range, and lack of pathogenicity of the parental virus. Pretreatment of cells with inhibitors of proliferation including aphidicolin, fluorodeoxyuridine, methotrexate, or nocodazole does not affect vector transduction, implying that, unlike retroviral vectors, mitosis is not essential for AAV transduction [39,40]. These findings are supported by demonstration of efficient AAV vector transduction of nonproliferating respiratory epithelial cells [41] and cellular populations that are normally nonproliferating, including postmitotic neurons [42–44], retinal [45–46], cochlear [47], muscle (smooth [48,49], cardiac [50], and skeletal [51–54]), non-cytokine-stimulated hematopoietic progenitors [55,56], and primary human peripheral blood monocyte-macrophages [57,58] (reviewed in McKeon and Samulski [59]). Thus, important cellular populations of either slowly proliferating or quiescent cells may now be amenable to genetic manipulation. These and other independent studies demonstrate that transduction of primary cells is feasible [60–63], although one group has described difficulty with transducing primary cells [64]. Short, defined transcripts can readily be expressed from AAV vectors, a highly desirable feature when expressing either antisense RNA or ribozymes whose function could be significantly compromised by the addition of adventitious sequences (discussed later) [31]. All virus-encoded genes have been removed from most currently available rAAV vectors, potentially making them less immunogenic than vectors, such as adenovirus, that express multiple immunogenic viral proteins after transduction [54]. Additionally, AAV vectors frequently integrate as multicopy tandem repeats [54,65,66], potentially enhancing transgene expression.

Does removal of AAV-encoded genes alter the biology of recombinant vectors relative to the wild-type virus? For example, the p5 product (rep) mediates site-specific integration of wild-type AAV into human chromosomal DNA [67]. However, the *rep* gene product down-regulates expression from AAV and a variety of heterologous promoters, and has been removed from almost all rAAV vectors. Thus, p5-deleted AAV vectors potentially could display random or reduced integration efficiency. Indeed, nonintegrated episomal forms have been described following AAV vector transduction [41,68]. Using Southern blot analyses, we demonstrated that AAV vectors lacking the p5 ORF do indeed integrate into both cell lines and primary human hematopoietic cells [56]. However, evidence is accumulating that *rep* minus rAAV vectors do not integrate site specifically into human AAVS1 [69–71] (Chatterjee, Fisher-Adams, and Wong, submitted), and significant efforts are now underway to reinstate this desirable property to rAAV vectors. AAV is a single-stranded DNA virus, and a block in second-strand DNA synthesis occurring after vector entry which is complemented by the adenovirus E4 open reading frame (ORF) 6 gene product, has been demonstrated by some investigators [72,73]. Augmentation of second-strand synthesis might also contribute to enhanced transgene expression or transduction following treatment of rAAV-transduced cells with "DNA-damaging" or chemotherapeutic agents [24,73,74], findings that mirror reports from the late 1980s, which document limited wild-type AAV replication in cells exposed to various "genotoxic" stress [75–78]. Whether these findings are limited to specific vectors (the requirement for second-strand synthesis as done with vectors encoding β-galactosidase) or cell types, or are universally applicable, has yet to be determined.

B. Autonomous Parvoviruses

Unlike the AAVs, autonomous parvoviruses productively replicate in target cells, utilizing factors provided by the host cell. The genomic organization of autonomous parvoviruses is similar to AAV with two major ORFs. The left ORF is comprised of a promoter at map position 4 (p4), and encodes nonstructural proteins (NSs) necessary for DNA replication and transactivation of the promoter for the right ORF. The right ORF is under control of a promoter at map position 38 (p38) and encodes capsid proteins. Replication is limited to cells that enter S phase following virus entry. Consequently, autonomous parvoviruses are not known to develop latent infection; genomic integration has not been detected [79]. Like AAV, the two autonomous parvoviruses, the prototypic minute virus of mice (MVMp) and the rodent parvovirus H-1, have wide host ranges, which include human cells. H-1 is known to infect humans without apparent clinical sequelae [80]. Among the parvoviridae infecting humans, only the autonomous parvovirus B19 is known to be pathogenic.

The antineoplastic activity of parvoviruses was originally described in 1968 by Toolan and subsequently resulted in a clinical trial of H-1 in patients with osteosarcoma [81,82]. Extensive studies have subsequently demonstrated that MVMp and H-1 exert a preferential killing of transformed cells and thus may be uniquely suited for the gene therapy of cancer [82,83]. MVMp is oncotropic, binding to a cell-surface receptor known to be sensitive to neuraminidase and trypsin [84], and is cytolytic to transformed but not to normal cells [83]. This cytotoxic effect has been mapped to both the amino- and carboxyl terminal regions of the major nonstructural protein (NS-1), which is homologous to the AAV *rep78* gene product, and may be modulated by NS-2 [85,86]. *In vivo* experiments demonstrated MVMp-mediated suppression of Ehrlich ascites tumor growth in mice when virus was injected at a site distant from the tumor inoculum [87]. Similar studies have demonstrated selective inhibition of tumor growth following infection with H-1 both *in vitro* [82] and *in vivo* [88]. Experiments with mutants of the p53 tumor suppressor gene suggest that a functional wild-type p53 product contributes to cellular resistance to H-1-mediated cytotoxicity [89]. Thus, the autonomous parvoviruses MVMp and H-1 are well suited as vectors for *in situ* tumor injection as they are both oncotropic and specifically lytic to neoplastic cells.

Recombinant vectors based upon MVMp and LuIII, another autonomous parvovirus that efficiently infects human cells, have been described [90,91] (reviewed in Corsini *et al.* [92]). In these constructs, transgenes have been placed under control of viral p4 or p38 promoters or, more recently, under control of tissue-specific or inducible promoters [93]. Vectors are encapsidated by providing NS and/or capsid functions *in trans* as described earlier for AAV vectors [94,95].

III. APPLICATIONS OF RECOMBINANT PARVOVIRUS VECTORS TO CANCER GENE THERAPY

A. Genetic Marking of Hematopoietic Cells

The concept of marking cells to study their biological behavior *in vivo* has evolved from tracking cells with specific chromosomal markers to the current practice of genetically tagging cells using readily identifiable gene transfer vectors. This technique has demonstrated its utility in improving our understanding of the dynamic behavior of biological systems occurring during devel-

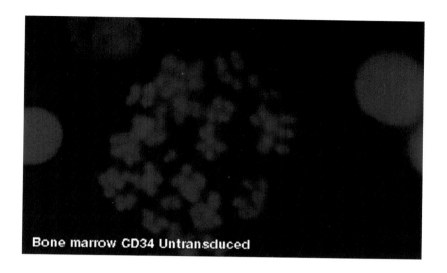

Bone marrow CD34 Untransduced

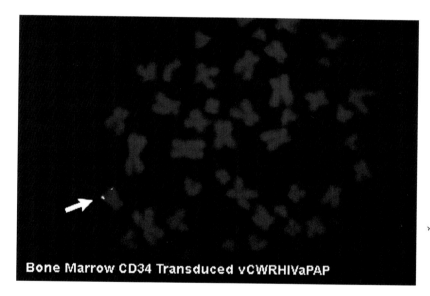

Bone Marrow CD34 Transduced vCWRHIVaPAP

Figure 2 Metaphase FISH analysis of rAAV-transduced primary human marrow-derived CD34 cells. The arrow denotes a vector-specfic signal.

opment, tumor growth, tumor metastases, and hematopoiesis [96,97].

Autologous bone marrow transplantation (auto BMT) is often used to rescue patients from otherwise myelotoxic doses of chemotherapy for various malignancies. Genetic marking of the transplant has been used to answer two critical questions: (1) Do contaminating tumor cells in the transplant contribute to tumor relapse, and (2) is the transplant responsible for long-term hematopoietic reconstitution? In patients with acute myeloid leukemia (AML), there is a 65% relapse rate following intensive chemotherapy and auto BMT. Brenner *et al.* harvested marrow from 12 patients with AML and 8 patients with neuroblastoma who had completed consolidation chemotherapy and had no evidence of residual disease. Approximately 30% of the harvested marrow was genetically marked with a retroviral vector. The patients were treated with myeloablative therapy, and marrow was infused [98]. Two patients with AML relapsed, both with genetically marked cells, implying that the "remission" marrow did contain tumor cells, which contributed to relapse. In addition, the marker gene was detectable in all hematopoietic lineages and remained detectable for up to 18 months posttransplant, the length of the study [99]. In a similar fashion, Deisseroth *et al.* determined that autologous marrow used for transplantation in patients with chronic myelogenous leukemia following intensive therapy also contained cells, which contributed to relapse [100]. Recently, Dunbar *et al.* described 11 patients who were transplanted with retroviral-vector-marked marrow and peripheral blood CD34$^+$ cells, a population enriched for hematopoietic progenitors. The marker gene was detectable in multiple lineages and persisted in 3 of 9 patients for greater than 18 months posttransplantation. Marked cells originated from both peripheral blood and marrow progenitors. However, the steady state levels of marking were low, with only 1:1,000 to 1:10,000 cells marker gene positive [101]. These findings confirm the utility of genetic marking and that a more accurate method for detecting minimal residual disease or for removal ("purging") of tumor cells from the transplant is necessary to improve this treatment approach.

Although these studies are very encouraging, it becomes apparent that the overall marking efficiency of hematopoietic progenitors using retroviral vectors is relatively low. This may result from the strict requirement of retroviral vectors for actively proliferating cellular targets for efficient gene transfer and proviral integration [3,4] combined with the nonproliferating state of true hematopoietic stem cells [102]. As a result, attempts to enhance retroviral vector transduction of hematopoietic progenitors have incorporated either coculture with retroviral packaging cell lines or stimulation of target

cells with a variety of cytokines to promote proliferation [103]. However, cytokine stimulation may shift stem cells along differentiation pathways, resulting in transduction of lineage-committed cells with limited self-renewal capacity [104].

Previous studies from our laboratory demonstrated that cellular proliferation was not requisite for efficient AAV transduction [39]. We have subsequently demonstrated transduction efficiencies of up to 70% in marrow-derived CD34$^+$-enriched cells isolated from over 20 individuals using a variety of AAV vectors. Furthermore, efficient gene transfer occurred with or without prior cytokine stimulation of target cells and resulted in stable vector integration by both Southern and FISH (fluorescent *in situ* hybridization) analyses (Fig. 2, see color insert) [56]. Similar findings have been obtained from other laboratories (Table 1, reviewed in Chatterjee *et al.* [105]). Miller *et al.* reported transgene expression in 20 to 40% of hematopoietic colonies following transduction with a recombinant AAV vector encoding human γ-globin [106]. Zhou and colleagues described transduction efficiencies of 33–75% and 50–80% to human umbilical cord blood and murine marrow hematopoietic progenitors, respectively, utilizing an AAV vector encoding neomycin phosphotransferase (Neo^R, conferring cellular resistance to G418) [55,107]. Again, efficient transduction of human umbilical cord progenitors occurred in the absence of cytokine prestimulation [55]. Finally, Walsh described the use of an AAV vector encoding the Fanconi anemia complementation group C (FACC) gene to phenotypically correct the Fanconi anemia defect *in vitro* within the peripheral blood CD34$^+$ cells isolated from a patient with this disorder [61]. In this study transduction efficiency was up to 60%, and transduced cells could be engrafted into a SCID-hu murine model [108]. Luhovy *et al.* demonstrated that rAAV vectors mediated transgene expression in 25% of human hematopoietic cells in long term culture (LTC-IC), a system that may assay more primitive progenitors, findings that we have noted as well [109,110]. However, it should be emphasized that not all laboratories have reported such encouraging results. These disparities may reflect multiple factors, including differing vector constructs, production and purification strategies, and source, method of purification, and culture conditions (amount and type of cytokines) of hematopoietic progenitors. For example, we have noted individual variability in transduction efficiencies, with cells from some individuals being poorly transduced. Furthermore, we have noted that transduction is better with CD34 cells isolated from the marrow than with those from mobilized peripheral blood stem cells, which may be reflective of the proliferating status of the cells (unpublished, Chatterjee and Wong).

TABLE 1   Transduction of Hematopoietic Cells with rAAV Vectors

Target cell	Transduction efficiency	Transgene	Reference
Murine marrow mononuclear cells	50–80%	neoR[a]	107
Human umbilical cord blood CD34+ cells	33–75%	neoR[a]	55
Human peripheral blood CD34+ cells	Up to 60%	FACC[b]	61
Human peripheral blood CD34+ cells	20–30%	Human γ-globin	106
Human marrow-derived CD34+ cells	>70%	PLAP[c]	56
Human marrow-derived CD34+ Lin⁻Thy⁺d cells	25%	β-galactosidase	109

[a] Neomycin phosphotransferase (NPT).
[b] Fanconi anemia, complementation group C gene.
[c] Placental alkaline phosphatase (PLAP).
[d] Denotes lineage negative (Lin⁻), Thy antigen positive cells.

To determine the frequency of potentially confounding wild-type AAV in hematopoietic tissues, Anderson *et al.* analyzed 106 human bone marrow samples using polymerase chain reaction (PCR) analysis, but were unable to detect wild-type AAV in CD34 positive progenitors. This suggests that vector rescue in transduced cells might be less likely *in vivo* [111]. Thus, for a variety of reasons, including high transduction efficiencies even in the absence of cytokine stimulation, AAV vectors appear promising for gene transfer to hematopoietic progenitors and are currently undergoing intensive evaluation for *in vivo* efficacy [63].

B. Chemoprotection of Hematopoietic Cells

High-dose chemotherapy coupled with stem cell rescue has shown promise in the treatment of a variety of advanced tumors [112]. However, complications directly attributable to myelosuppression, including infection or hemorrhage, still pose potentially life-threatening risks to this form of therapy. Conferring resistance of hematopoietic progenitors to the toxic effects of dose-intensified chemotherapy is one approach to this problem. The human multidrug resistance (MDR1) gene encodes a 170-kDa glycoprotein, P-glycoprotein, that actively exports a variety of toxic substances from the cell, including drugs commonly used in this setting such as anthracyclines, epipodophyllotoxins, and taxol [113]. Although MDR1 expression can be detected among primitive human hematopoietic progenitors [114], expression wanes as cells mature, and P-glycoprotein is seldom found among mature myeloid cells. Thus, granulocytes, myeloid cells that form the first line of defense

against bacterial infection, are exquisitely sensitive to the effects of chemotherapy. Therefore, augmented MDR1 expression among hematopoietic progenitors and their progeny may abrogate the myelosuppressive effects of dose-intensified chemotherapy and permit more intensive and effective regimens.

The feasibility of such an approach was established through the development of murine models. Mice transgenic for the human MDR1 gene were protected from the myelosuppressive effects of chemotherapeutic agents effluxed by the P-glycoprotein transporter. Myeloprotection correlated with specific, increased P-glycoprotein surface expression on hematopoietic cells [115] and could be transplanted to nontransgenic mice [116]. Using a more clinically relevant approach, transplantation of murine marrow mononuclear cells transduced with a retroviral vector encoding human MDR1 conferred myeloprotection against the leukopenic effects of taxol and permitted selection of transduced cells *in vivo* [117]. The myeloprotective effect of MDR1 could be serially transplanted to syngeneic animals [118], suggesting the genetic modification of true hematopoietic stem cells. Recently, Ward *et al.* transduced primary CD34+ human marrow progenitors prestimulated with interleukin-3 (IL-3), IL-6, and stem cell factor with a retroviral vector encoding MDR1 *in vitro*. Vector sequences could be identified by PCR analysis in 18–70% of erythroid (BFU-E) and 30–60% of granulocyte–monocyte (CFU-GM) colonies grown in methylcellulose, and augmented surface P-glycoprotein expression was demonstrated in 4–11.2% by flow cytometric analysis [119] (see also Chapter 11).

Encouraged by our previous studies with AAV-mediated gene transfer to hematopoietic progenitors, we constructed an AAV-based vector encoding the minimal ORF of the MDR1 cDNA. This vector, CWRSP-MDR, was approximately 200 bases larger than wild-type AAV yet demonstrated titers comparable to those of smaller AAV vectors. Flow cytometric analyses of CWRSP-MDR transduced CD34⁺-enriched normal human marrow demonstrated increased specific P-glycoprotein expression as determined by both antigenic (MRK-16 or UIC2 anti-MDR1 monoclonal antibodies) and functional (rhodamine 123 dye exclusion) assays (Fig. 3) [120]. Transduction efficiencies of CD34⁺ marrow-derived cells from 6 different donors ranged from 20 to 100%, values consistent with transduction of this cell population with other AAV-based vectors (Table 1). Southern hybridization analyses of high molecular weight cellular DNA isolated from CD34⁺ cells, digested with a restriction enzyme that cleaves once within the vector and hybridized with a vector-specific probe, demonstrated bands larger than the unit-length vector and were consistent with vector integration [56]. Importantly, no cellular toxicity was detected following vCWRSP-MDR transduction. Additional studies are underway to determine the utility of this vector both *in vitro* and *in vivo* in a murine model.

Baudard *et al.* recently reported upon the development of a liposome-encapsidated AAV bicistronic plasmid encoding the MDR1 gene and glucocerebrosidase. Administration of this construct to mice resulted in delivery of MDR1 and GC cDNAs in all organs tested [121].

C. Delivery of Transdominant Molecules

The discovery of oncogenes, tumor suppressor and mutator genes, has opened new areas of research in oncology aimed at discovering agents that could selectively inhibit the biological effects of oncogene products or restore the function of tumor suppressors and DNA-repair genes. Thus, a new field of research has arisen focusing upon the development of transdominant molecules designed to selectively inhibit gene expression. Transdominants can be easily classified into two main categories: **RNA-based** (antisense, sense decoys, and catalytic ribozymes), and **protein-based** (transdominant proteins, and single-chain antibodies). We will focus primarily on antitumor transdominants that have been expressed from parvovirus vectors.

1. ANTISENSE TRANSCRIPTS

During the last decade, synthetic antisense oligonucleotides designed to bind selectively to a complementary RNA sequence have been administered to interrupt targeted gene expression both in tissue culture and in

animal models *in vivo*. The mechanism of action of these antisense molecules appear to differ in different systems and include accelerated degradation of sense:antisense hybrids, and interference with trafficking, processing, or intra- or intermolecular interactions [122,123].

Numerous studies demonstrate the antioncogenic effects of antisense molecules targeting specific oncogenes as assessed by changes in morphology, growth rate, ability to form colonies in soft agar, and *in vivo* tumorigenicity. Molecular targets have included cellular oncogenes including *bcl2* [124], *p210^{bcr-abl}* [125], *erbB-2* [126], *fos* [127], and *myc* [128], as well as virus-encoded transforming genes, for example, E6/E7 ORF from human papillomavirus (HPV) [129].

The overall utility of antisense RNA as a therapeutic tool depends not only upon the optimal design and intracellular expression of this RNA but also upon the availability of safe and highly efficient delivery systems. Previous studies have demonstrated that AAV vectors can function as efficient vehicles for antisense RNA delivery. AAV vectors can be constructed that express short, distinct transcripts, a property that is useful for RNA-mediated inhibition of gene expression [31]. As a gene transfer approach to abrogate herpes simplex virus (HSV) infection, Wong and Chatterjee developed an AAV-based vector that encoded both neomycin phosphotransferase and an antisense transcript complementary to a portion of the 5′ noncoding leader and first coding "AUG" of the HSV-1 ICP4 transcript [130]. Vector-transduced clonal murine cells expressing this transcript were protected from cytopathogenicity, restricted HSV-1 production by 1,000 to 10,000-fold (99–99.9%), and demonstrated prolonged survival (75–80% viability by trypan blue dye exclusion versus 0% by day 4 postinfection) in comparison to control cells following viral challenge [multiplicity of infection (MOI) 0.1]. Similarly, an AAV-based vector encoding an antisense transcript complementary to both the HIV-1 *TAR* region and polyadenylation signal, common to all HIV strains, and conferring G418 resistance was constructed [131]. Transduced, clonally derived, G418-resistant 293 cells expressing this antisense transcript specifically restricted chloramphenicol acetyltransferase (CAT) expression directed from the HIV-1 LTR, as well as HIV replication following transfection with HIV_{IIIB}, an infectious molecular clone. Importantly, transduced, nonclonal but G418-resistant human CD4 T-cell lines (H9 and A3.01) expressing the antisense RNA demonstrated prolonged protection with a 1,000-fold reduction of titratable virus production after challenge with HIV_{IIIB} (MOI 1). Finally, Ponnazhagan and colleagues utilized an AAV vector encoding an antisense transcript complementary to the human α-globin gene to inhibit gene expression by up to 91% [132].

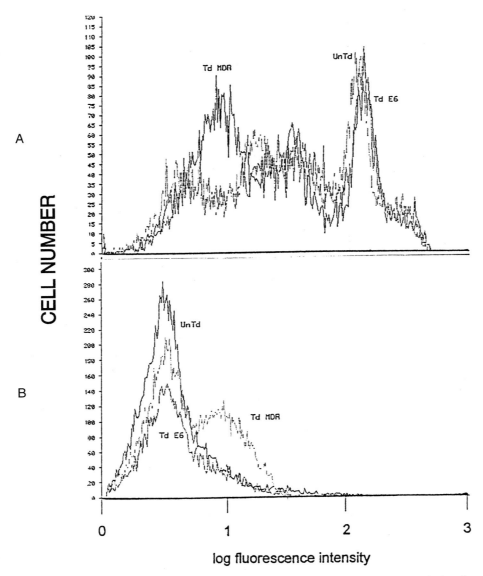

FIGURE 3   MDR1 expression following transduction of CD34[+]-enriched human hematopoietic cells with CWRSP-MDR. CD34[+]-enriched hematopoietic cells isolated from healthy donors were selected in flasks coated with anti-CD34 antibody. UnTd, untransduced; Td E6, control vector transduced; Td MDR, CWRSP-MDR-transduced. (A) Representative flow cytometry histogram of MDR1 antigen expression, depicting specific increased MDR1 antigen expression on CWRSP-MDR transduced cells. (B) Representative histogram of rhodamine 123 (Rho[123]) fluorescence. Rho[123] is actively effluxed by P-glycoprotein so that cells with functional P-glycoprotein appear Rho[123] dull (lower fluorescence).

2. RIBOZYMES

Ribozymes are a class of RNA discovered in the early 1980s that catalytically cleave RNA molecules in a sequence-specific fashion [133]. A variety of ribozyme motifs such as hammerhead, hairpin, axhead, group I intron, and RNase P are available for use as *trans*-acting catalysts, among which the hammerhead ribozyme is the most commonly used. The catalytic ribozyme core is targeted to specific regions of the target transcript by flanking anti-

sense sequences. Ribozymes (Rz) have been designed to efficiently cleave their targets *in trans* and inhibit gene expression from either cellular oncogenes or virus-encoded genes requisite for productive infection. Ribozymes may be more effective than antisense molecules as inhibitors of gene expression; they not only cleave the target transcript, but, because they act catalytically, each ribozyme can theoretically destroy multiple targets [133].

Several studies have demonstrated the efficacy of ribozyme-mediated inhibition of oncogene expression, with subsequent reversal of the transformed phenotype. Targets have included cellular oncogenes such as activated H-*ras* both *in vitro* [134] and *in vivo* [135], $p210^{bcr-abl}$ [136], and c-*fos* [137], and virus-encoded oncogenes (discussed later). In addition, hammerhead ribozymes targeting c-*fos* have reversed cisplatin resistance [138] and MDR1-mediated resistance to chemotherapeutic agents [139].

We constructed an AAV vector, termed CWRT7-SPN, that contains a highly expressed T7 bacteriophage promoter flanked by the Rous sarcoma virus (RSV) promoter and a multisite cloning polylinker (MCP) (Fig. 1). Oligonucleotides coding for specific ribozymes are inserted within the MCP, and ribozymes are expressed to a high level *in vitro* using T7 RNA polymerase. Two hammerhead ribozymes were designed to cleave all transcripts arising from the major transforming genes (E6 and E7 ORFs) from HPV type 16 and inserted into CWRT7SPN. Cleavage was designed to occur at nucleotides 110 and 558 relative to the transcriptional start site. Evidence of HPV 16 or 18 infection is demonstrable in approximately 85% of human cervical cancers and is believed to be an important cofactor in tumorigenesis [140]. Both ribozymes efficiently cleaved their cognate target transcripts, including the full-length 793-base transcript, in a cell-free system under a variety of conditions, including physiological temperature (Fig. 4). In addition, target cleavage occurred in the presence of an excess of total cellular RNA without preceding denaturation, and both ribozymes simultaneously cleaved the same target *in vitro,* indicating the feasibility of intracellular studies, which are currently ongoing [141]. Finally, hammerhead ribozymes targeting the activated H-*ras* [134] and $p210^{bcr-abl}$ [136] oncogenes have been cloned into AAV vectors and have demonstrated efficient target cleavage *in vitro.* Importantly, expression of the $p210^{bcr-abl}$ ribozyme encoded within an AAV vec-

tor inhibited proliferation of $p210^{bcr-abl}$-transformed (EM2) cells *in vitro* (Snyder, Forman, Wu, Chatterjee, and Wong, unpublished).

D. Modulation of Antitumor Immunity

The development and progression of malignancy is often considered a failure of the host immune system to recognize or to effectively respond to cancer antigens. A rationale for an immunological approach to cancer treatment can be found in reports of spontaneous regressions of advanced tumors including melanoma and renal cell carcinoma [142], both immunogenic tumors. Furthermore, lymphocytic infiltrates within tumors, implying an antitumor cellular imune response, often signify a better prognosis [143]. Antigenic determinants must be solely or selectively expressed on the tumor as compared to normal cells for an antitumor immune response to be mounted. These antigens must then be presented in the context of the major histocompatibility complex (MHC) in conjunction with costimulatory signals, such as B7.1 or B7.2, so as to be recognized by T cells and eliminated [144,145].

A variety of gene transfer strategies have been developed to potentiate antitumor immune responses. Rosenberg and colleagues have pioneered the use of genetically modified tumor-infiltrating lymphocytes (TILs) for specific delivery to malignant cells. In this strategy, autologous TILs are harvested from the tumor, genetically modified to express a cytokine gene [tumor necrosis factor (TNF), for example], expanded *ex vivo,* and readministered to the patient. Theoretically, modified TILs home to tumor-bearing areas and serve as sources for localized cytokine production for either immunomodulation or tumor toxicity, thus circumventing the significant toxicities seen after systemic cytokine administration [146]. Roberts recently described retroviral-vector-mediated transfer of a universal (MHC-unrestricted) chimeric T-cell receptor (TCR) gene composed of the invariant zeta chain of the TCR coupled to either the extracellular domain of the human CD4 receptor (CD4-UR), the primary receptor for HIV, or a single-chain antibody against the gp41 region of the HIV envelope (*env*) (SAb-UR) [147]. This strategy was employed to generate a rapid cytotoxic response directed against HIV-infected cells by circumventing MHC-restricted antigen presentation. Primary human CD8+ lymphocytes transduced with retroviral vectors encoding either CD4-UR or SAb-UR displayed both specific proliferative responses after exposure to HIV *env* and enhanced cytolysis of HIV-infected cells. A similar strategy could be employed to target cells bearing tumor antigens. Education of professional antigen-

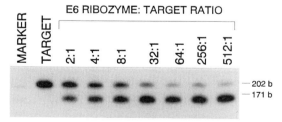

FIGURE 4 Hammerhead ribozyme-mediated cleavage of HPV16 E6 transcripts *in vitro*. The target sequence was at position 110 relative to the transcriptional start site. Target transcripts were internally labeled with ^{32}P and incubated with ribozyme at varying molar ratios at 37°C for 3 hours. Resulting products were resolved using a denaturing 5% polyacrylamide gel and analyzed by autoradiography. About 50% of the target was cleaved at a ratio of 4:1, and 90% was cleaved at a ratio of 64:1.

presenting cells to induce a specific antitumor response is also being investigated [148]. Finally, tumor cells themselves may be genetically modified to express cytokine genes [interleukin-2 (IL-2), IL-4, interferon-gamma (IFN-γ), TNF-α, granulocyte–macrophage colony-stimulating factor (GM-CSF) [149,150], costimulatory signals (B7-CD28) [151], or alloreactive antigens [152] to augment their immunogenicity. As with the case of TILs bearing cytokine genes, introduction of cytokine genes directly into tumor cells results in augmented localized cytokine expression in the area of disease, potentially enhancing tumor immunogenicity while limiting systemic toxicities.

Interleukin-2 (IL-2) is a pleiotropic cytokine with a central role in the immune response, inducing the proliferation of antigen-stimulated T cells, natural killer (NK) and lymphokine-activated killer (LAK) cells [153]. In murine models, expression of IL-2 within otherwise nonimmunogenic malignant cells has resulted in induction of cytotoxic antitumor responses, accompanied by protection against subsequent tumor challenge. This has been demonstrated using CT26, a murine colon cancer cell line [154], MBT-2, a murine bladder cancer cell line [155], CMS-5, a sarcoma line [156], and P815 mastocytoma cells [157], and has been useful in the prevention of tumor spread in a murine breast cancer metastasis model [158].

We constructed an AAV vector encoding the human IL-2 cDNA under Rous sarcoma virus promoter control to determine the potential role of AAV vectors for intratumor cytokine expression (Fig. 1). This vector, termed CWRIL2SPN, also confers resistance to G418. G418-selected, CWRIL2SPN-transduced human 293 cell clones constitutively secreted IL-2 as determined by both antigenic (ELISA) and functional assays (CTLL-2 proliferation bioassay) (Fig. 5). One clone (clone 9) secreted greater than 60,000 pg/mL/24 h/1 × 10^6 cells by ELISA and 10,000 U/10^6 cells/24 h by CTLL bioassay.

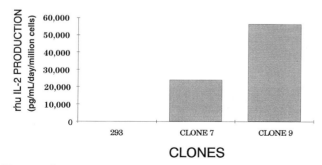

FIGURE 5   Recombinant human IL-2 production from 293-based clonal cells derived following CWRIL2SPN transduction, rhuIL-2 production was measured by ELISA. Expression from parental cells (293) and two G418-resistant clones (7 and 9) are depicted.

IL-2 expression was maintained after irradiation to prevent cell proliferation. This is important because the tumor cells, which would serve as a source of both IL-2 and tumor antigens, would be irradiated prior to autologous implantation *in vivo*. CWRIL2SPN, has also been used to transduce both human (MCF-7 and MDA) and murine (MOD) breast carcinoma cell lines. Again, expression of human IL-2 among clones was detectable, with actively greater than 100 U/10^6 cells/24 h. The ability of IL-2–expressing breast cancer cells to stimulate antitumor immunity will be examined *in vivo* using a murine model.

Philip and colleagues demonstrated that AAV vector transduction is not required for sustained transgene expression. They described expression of rhu-IL-2 for greater than 30 days in certain cells following transfection with an AAV plasmid–liposome complex. Transfection efficiencies ranged from 10 to 50% and included a wide variety of target cells including primary tumors (breast, ovarian, and lung), and T lymphocytes [159]. Although specific mechanisms are not yet known, prolonged transgene expression may result from increased plasmid stability resulting from the GC-rich AAV ITRs. This approach to enhancing antitumor immunity is about to undergo clinical trials. However, stable integrants using this system have not yet been described.

Okada *et al.* recently reported upon the use of a bicistronic rAAV vector encoding both the herpes simplex thymidine kinase and interleukin 2 (IL-2) genes (AAV-tk-IRES-IL2). Transduction of U-251SP cells, a human glioma cell line, with AAV-tk-IRES-IL2 rendered them sensitive to ganciclovir and resulted in expression of IL-2 in a dose-dependent fashion. Stereotactic delivery of 6 × 10^{10} AAV-tk-IRES-IL2 particles into day 7 tumors in nude mice followed by administration of GCV for 6 days resulted in a 35-fold reduction in the mean volume of tumors compared with controls [160].

Zhang *et al.* constructed an rAAV vector encoding a synthetic human type I interferon gene in addition to neomycin phosphotransferase. A variety of human tumor cell lines (293, HeLa, K562, Eskol) were transduced, and G418 was selected. All cell lines expressed interferon *in vitro*. In contrast to untransduced controls, transduced tumor cells failed to induce tumor formation in nude mice. Furthermore, transduction of established Eskol cells in nude mice resulted in tumor regression [161]. Clary *et al.* constructed an AAV plasmid encoding the murine gamma interferon gene (pMP6A-mIFN-gamma), and used liposome-mediated delivery to transfer it to the murine lung cancer cell line, D122. Gamma interferon and augmented major histocompatibility class I expression were documented in transformed cells, which were irradiated and used to vaccinate animals challenged with D122 cells. Animals vaccinated with gamma interferon expressing cells demonstrated a delay in primary (footpad) tumor growth and displayed a 57% reduction in pulmonary metastases versus a 0, 7, and 15% reduction in untreated controls or animals vaccinated with cells containing an empty AAV vector or a retroviral encoding gamma interferon, respectively [162].

Both the human IL-2 and murine interleukin-4 (IL-4) genes have been inserted into the oncotropic strain of MVM under control of the endogenous p38 promoter [163]. The tumorcidal NS genes were retained in these vectors. Intratumor IL-4 expression induces a potent antitumor eosinophilic response in mice [164]. Consistent with the biology of MVM, approximately 200 times more IL-2 was expressed in two transformed compared to untransformed parental fibroblasts following transduction. These vectors have the advantages of being oncotropic, directly tumorcidal due to expression of NS genes, and capable of enhancing antitumor immune responses by virtue of IL-2 and IL-4 expression.

Tumor cells may evade immune recognition because they lack accessory or costimulatory signals (B7.1 or B7.2) for efficient antigen presentation [165]. Chiorini et al. constructed an rAAV vector encoding the B7.2 costimulatory molecule. Transduction of the nonadherent human lymphoid cell line LP-1 resulted in increased B7.2 expression (from 6.8 to 78.0%). The ability to express costimulatory cells on tumor cells may permit the development of specific tumor vaccines [166].

Finally, Miyamura et al. engineered expression of hen egg white lysozyme (HEL) to the surface of empty capsids of the B19 parvovirus produced in a baculovirus expression system. Surface HEL was detectable from purified recombinant capsids by ELISA, immunoprecipitation, and immune electron microscopy, and enzymatically. Furthermore, rabbits inoculated with the recombinant capsids generated an anti-HEL immune response, implying that B19 empty capsids may be useful as platforms to induce immunity against specific proteins, potentially including tumor antigens [167]. Sedlik et al. subsequently demonstrated that porcine parvovirus particles could be used in a similar fashion to present epitopes of poliovirus [168].

1. Suicide Genes

The herpes simplex virus thymidine kinase (HSV-tk) gene sensitizes cells to the cytotoxic effects of specific drugs including ganciclovir and acyclovir [169]. (See Chapter 10 for a more detailed description.) Culver and colleagues exploited the transduction requirement of retroviral vectors for proliferating target cells to specifically introduce HSV-tk into rapidly proliferating brain tumors (gliomas) in rats [170]. Gliomas regressed completely in ganciclovir-treated animals. Interestingly, untransduced tumor cells in close proximity to transduced cells were also killed following ganciclovir treatment

("bystander effect"), which may result from a diffusible metabolite of ganciclovir [171]. As a prelude to the development of parvovirus vectors encoding suicide genes, Koering and colleagues inserted the HSV-tk gene within MVMp under control of the p38 promoter [172]. Acyclovir treatment of TK-minus cells stably transfected with this construct followed by infection with MVMp reduced cell survival by 3.5- to 5-fold compared to controls. Su et al. constructed an rAAV vector encoding the herpes simplex thymidine kinase gene under control of the liver specific albumin promoter coupled with the human alpha fetoprotein (AFP) enhancer. This construct also contained the neo^R gene. Only liver cells expressing AFP and albumin were sensitive to the effects of ganciclovir [173].

2. Interruption of Tumor Vascular Supply

Tumors secrete a number of "angiogenesis" factors, including vascular endothelial growth factor (VEGF), transforming growth factor (TGF)-β1, pleiotrophin, fibroblast growth factor (FGF), placental growth factor, and platelet-derived endothelial cell growth factor, which function to induce formation of supporting vasculature [174]. Provision of a suitable tumor blood supply is necessary for sustained tumor growth beyond several millimeters in size. Thus, interruption of this sometimes tenuous vascular supply could provide a therapeutic target to enhance tumor killing. AAV vectors have recently been shown to efficiently transduce endothelial and vascular smooth muscle cells from rodents, and nonhuman primates both in vitro and in vivo, and from humans in vitro [48,49]. This property could be exploited for expression of tumor angiogenesis inhibitors or thymidine kinase within tumor vessels [175,176].

3. Oncotropic Vectors

The natural tropism for MVM and H-1 could be exploited to develop oncotropic vectors. To this end, Dupont et al. recently described the construction of an MVM-based vector (MVM/p38cat) retaining the oncolytic NS proteins and encoding the chloramphenicol acetyltransferase (CAT) reporter gene under p38 promoter control. Encapsidated MMV/p38cat was used to transduce both primary nonmalignant and transformed cells. Vector DNA replication and CAT expression were limited to transformed cells (including human fibroblasts, epithelial cells, T lymphocytes and macrophages), although a direct correlation between DNA replication and transgene expression was not always apparent [177].

IV. PERSPECTIVES, PROBLEMS, AND FUTURE CONSIDERATIONS

Given the recognized limitations of current modalities, novel approaches are needed for more effective and nontoxic treatments for cancer. As outlined in the preceding discussion, parvovirus vectors offer a variety of advantages for the development of gene therapy strategies for cancer treatment. Vectors can be developed that transiently express transgene (MVM, H-1) or stably integrate into cellular DNA (AAV), that have an extremely wide host range (AAV), or that have specific oncotropic and tumorcidal activity (MVM, H-1).

However, several important issues remain. *In vivo* studies of parvovirus-vector-mediated gene transfer and stability of transgene expression are limited. Vector safety issues including potential risks of horizontal or vertical vector transmission, vector shedding, vector distribution within the host, and vector rescue need to be addressed. The potential effects of pre-existing antiparvovirus immunity upon vector transduction or survival of vector-transduced cells must be addressed. Finally, efficient, simple, reproducible encapsidation strategies need to be developed, particularly for the production of large quantities of wild-type virus–free, clinical-grade vector. This has become a major focus of several laboratories including our own.

Additional *in vivo* animal studies and actual human clinical trials will help to resolve many of these issues. The NIH Recombinant DNA Advisory Committee approved the first clinical protocol using an AAV vector in September 1994. In this study, an AAV vector encoding the CFTR gene will be used to determine transduction efficiencies, stability of gene expression, and safety issues following vector introduction to subsegmental regions of the lung in patients with mild cystic fibrosis [178]. Preliminary studies in a primate model revealed no evidence of toxicity [179]. We and others are actively studying AAV-mediated gene transfer in several animal models, with the anticipation of advancing to human clinical trials.

Newer applications for parvovirus vectors are continually being described. Recently, rAAV vectors have been shown to efficiently transduce murine skeletal muscle, with evidence of transgene expression for 12–18 months [52–54]. Of note is that transgenes were often under cytomegalovirus (CMV) immediate early (IE) promoter control, a highly active viral promoter that is frequently silenced in murine *in vivo* transduction models. In addition, immune responses to rAAV in these studies was low when compared to adenoviral vectors [53,54]. Finally, several potentially clinically relevant genes, including erythropoietin [52], and Factor IX [180], can be expressed at therapeutic levels following muscle transduction with rAAV vectors. Whether this important application can be translated to primates and adapted for potential cancer therapies remains to be determined.

It is perhaps of note that almost exactly 30 years ago, Helen Toolan at Sloan–Kettering administered H-1 parvovirus intravenously to two patients with incurable, disseminated osteosarcoma in an attempt to eradicate their disease [80]. Although only a modest therapeutic effect was noted, this important study laid the groundwork for the study and development of parvovirus vectors in our continuing struggle against cancer.

Acknowledgments

We thank Stephen Forman, John Zaia, John Kovach, Christine Wright, Susan Kane, and the City of Hope Bone Marrow Transplantation team, whose support made this work possible. We also thank members of our laboratory for their support, review, and critique. This work was supported in part by Grants CA59308, CA33572, AI25959, AI40001, CA75186, and CA71947 from the National Institutes of Health.

References

1. Bishop, J. M. Cancer: the rise of the genetic paradigm. *Genes Dev.* **9**, 1309–1315, 1995.
2. Mulligan, R. C. The basic science of gene therapy. *Science* **260**, 926–932, 1993.
3. Miller, D. G., Adam, M. A., Miller, A. D. Gene transfer by retrovirus vectors occurs only in cells that are actively replicating at the time of infection. *Mol. Cell. Biol.* **10**, 4239–4242, 1990.
4. Roe, T. Y., Reynolds, T. C., Yu, G., *et al.* Integration of murine leukemia virus DNA depends on mitosis. *EMBO J.* **12**, 2099–2108, 1993.
5. Donahue, R. E., Kessler, S. W., Bodine, D., *et al.* Helper virus induced T cell lymphoma in nonhuman primates after retroviral mediated gene transfer. *J. Exp. Med.* **176**, 1125–1135, 1992.
6. Dranoff, G., Mulligan, R. C. Gene transfer as cancer therapy. *Adv. Immunol.* **58**, 417–454, 1995.
7. Zhang, W.-W., Fujiwara, T., Grimm, E. A., *et al.* Advances in cancer gene therapy. *Adv. Pharm.* **32**, 289–341, 1995.
8. Siegl, G., Bates, R. C., Berns, K. I., *et al.* Characterization and taxonomy of parvoviridae. *Intervirology* **23**, 61–73, 1985.
9. Berns, K. I., Bohenzky, R. A. Adeno-associated viruses: an update. *Adv. Virus Res.* **32**, 243–306, 1987.
10. Berns, K. I., Giraud, C. Biology of adeno-associated virus. *Curr. Top. Microbiol. Immunol.* **218**, 1–23, 1996.
11. Kotin, R. M., Siniscalco, M., Samulski, R. J., *et al.* Site-specific integration by adeno-associated virus. *Proc. Natl. Acad. Sci. USA.* **87**, 2211–2215, 1990.
12. Kotin, R. M., Linden, R. M., Berns, K. I. Characterization of a preferred site on human chromosome 19q for integration of adeno-associated virus DNA by non-homologous recombination. *EMBO J.* **11**, 5071–5078, 1992.
13. Samulski, R. J., Zhu, X., Xiao, X., *et al.* Targeted integration of adeno-associated virus (AAV) into human chromosome 19. *EMBO J.* **10**, 3941–3950, 1991.
14. Berns, K. I., Pinkerton, T. C., Thomas, G. F., *et al.* Detection of adeno-associated virus (AAV)-specific nucleotide sequences in DNA isolated from latently infected Detroit 6 cells. *Virology* **68**, 556–560, 1975.
15. Brown, K. E., Anderson, S. M., Young, N. S. Erythrocyte P antigen: Cellular receptor for B19 parvovirus. *Science* **262**, 114–117, 1993.

16. Schlehofer, J. R. The tumor suppressive properties of adeno-associated viruses. *Mutat. Res.* **305**, 303–313, 1994.

17. Mayor, H. D. Defective parvoviruses may be good for your health! *Prog. Med. Virol.* **40**, 193–205, 1993.

18. Bantel-Schaal, U. Adeno-associated parvoviruses inhibit growth of cells derived from malignant human tumors. *Int. J. Cancer* **45**, 190–194, 1990.

19. Hermonat, P. L. Down-regulation of the human c-*fos* and c-*myc* proto-oncogene promoters by adeno-associated virus Rep78. *Cancer Lett.* **81**, 129–136, 1994.

20. Hermonat, P. L. Adeno-associated virus inhibits human papillomavirus type 16: a viral interaction implicated in cervical cancer. *Cancer Res.* **54**, 2278–2281, 1994.

21. Wang, X. S., Srivastava, A., Ponnazhagan, S., *et al.* Adeno-associated virus type 2 DNA replication *in vivo*: mutation analyses of the D sequence in viral inverted terminal repeats. *J. Virol.* **71**, 3077–3082, 1997.

22. Xiao, X., Samulski, R. J., Li, J., *et al.* A novel 165-base-pair terminal repeat sequence is the sole *cis* requirement for the adeno-associated virus life cycle. *J. Virol.* **71**, 941–948, 1997.

23. Flotte, T. R., Afione, S. A., Solow, R., *et al.* Expression of the cystic fibrosis transmembrane conductance regulator from a novel adeno-associated virus promoter. *J. Biol. Chem.* **268**, 3781–3790, 1993.

24. Muzyczka, N. Use of AAV as a general transduction vector for mammalian cells. *Curr. Top. Micro. Immunol.* **158**, 97–129, 1992.

25. Kotin, R. M. Prospects for the use of adeno-associated virus as a vector for human gene therapy. *Hum. Gene Ther.* **5**, 793–801, 1994.

26. Labow, M. A., Graf, L. H., Berns, K. I. Adeno-associated virus gene expression inhibits cellular transformation by heterologous genes. *Mol. Cell Biol.* **7**, 1320–1325, 1987.

27. Dong, J. Y., Frizzell, R. A., Fan, P. D. Quantitative analysis of the packaging capacity of recombinant adeno-associated virus. *Hum. Gene Ther.* **7**, 2101–2112, 1996.

28. Hermonat, P. L., Han, L., Bishop, B. M., *et al.* The packaging capacity of adeno-associated virus (AAV) and the potential for wild-type-plus AAV gene therapy vectors. *FEBS Lett.* **407**, 78–84, 1997.

29. Fan, P. D., Dong, J. Y. Replication of rep-cap genes is essential for the high-efficiency production of recombinant AAV. *Hum. Gene Ther.* **8**, 87–98, 1997.

30. Li, J., Xiao, X., Samulski, R. J. Role for highly regulated rep gene expression in adeno-associated virus vector production. *J. Virol.* **71**, 5236–5243, 1997.

31. Chatterjee, S., Wong, K. K., Jr. Adeno-associated viral vectors for the delivery of antisense RNA. Methods: A Companion to Methods Enzymol **5**, 51, 1993.

32. Mamounas, M., Leavitt, M., Yu, M., Wong-Staal, F. Increased titer of recombinant AAV vectors by gene transfer with adenovirus coupled to DNA-polylysine complexes. *Gene Ther.* **2**, 429–432, 1995.

33. Maxwell, F., Maxwell, I. H., Harrison, G. S. Improved production of recombinant AAV by transient transfection of NB324K cells using electroporation. *J. Virol. Methods* **63**, 129–136, 1997.

34. Tamayose, K., Shimada, T., Hirai Y. A new strategy for large-scale preparation of high-titer recombinant adeno-associated virus vectors by using packaging cell lines and sulfonated cellulose column chromatography. *Hum. Gene Ther.* **7**, 507–513, 1996.

35. Flotte, T. R., Barraza-Ortiz, X., Solow, R., *et al.* An improved system for packaging recombinant adeno-associated virus vector capable of *in vivo* transduction. *Gene Ther.* **2**, 29–37, 1995.

36. Clark, K. R., Johnson, P. R., Fraley, D. M., Voulgaropoulou, F. Cell lines for the production of recombinant adeno-associated virus. *Hum. Gene Ther.* **6**, 1329–1341, 1995.

37. Wong, K. K., Jr., Rosborough, E., Prasad, K.-M. R., *et al.* Development of cell lines that encapsidate adeno-associated virus vectors, submitted.

38. Trempe, J. P. Packaging systems for adeno-associated virus vectors. *Curr. Top. Microbiol. Immunol.* **218**, 35–50, 1996.

39. Podsakoff, G., Wong, K. K., Jr., Chatterjee, S. Stable and efficient gene transfer into non-dividing cells by adeno-associated virus (AAV)-based vectors. *J. Virol.* **68**, 5656–5666, 1994.

40. Alexander, I. E., Russell, D. W., Miller, A. D. DNA-damaging agents greatly increase the transduction of nondividing cells by adeno-associated virus vectors. *J. Virol.* **68**, 8282–8287, 1994.

41. Flotte, T. R., Afione, S. A., Zeitlin, P. L. Adeno-associated virus vector gene expression occurs in nondividing cells in the absence of vector DNA integration. *Am. J. Respir. Cell. Mol. Biol.* **11**, 517–521, 1994.

42. Kaplitt, M. G., Leone, P., Samulski, R. J., *et al.* Long-term gene expression and phenotypic correction using adeno-associated virus vectors in the mammalian brain. *Nature Genet.* **8**, 148–153, 1994.

43. Du, B., Terwilliger, E. F., Boldt-Houle, D. M., *et al.* Efficient transduction of human neurons with an adeno-associated virus vector. *Gene Ther.* **3**, 254–261, 1996.

44. Peel, A. L., Reier, P. J., Muzyczka, N., *et al.* Efficient transduction of green fluorescent protein in spinal cord neurons using adeno-associated virus vectors containing cell type-specific promoters. *Gene Ther.* **4**, 16–24, 1997.

45. Ali, R. R., Bhattacharya, S. S., Hunt, D. M., *et al.* Gene transfer into the mouse retina mediated by an adeno-associated viral vector. *Hum. Mol. Genet.* **5**, 591–594, 1996.

46. Flannery, J. G., Hauswirth, W. W., Muzyczka, N., *et al.* Efficient photoreceptor-targeted gene expression *in vivo* by recombinant adeno-associated virus. *Proc. Natl. Acad. Sci. USA* **94**, 6916–6921, 1997.

47. Lalwani, A. K., Mhatre, A. N., Muzyczka, N., *et al.* Development of *in vivo* gene therapy for hearing disorders: introduction of adeno-associated virus into the cochlea of the guinea pig. *Gene Ther.* **3**, 588–592, 1996.

48. Lynch, C. M., Geary, R. L., Dean, R. H., *et al.* Adeno-associated virus vectors for vascular gene delivery. *Circ. Res.* **80**, 497–505, 1997.

49. Arnold, T. E., Bahou, W. F., Gnatenko, D. *In vivo* gene transfer into rat arterial walls with novel adeno-associated virus vectors. *J. Vasc. Surg.* **25**, 347–355, 1997.

50. Kaplitt, M. G., Diethrich, E. B., Strumpf, R. K., *et al.* Long-term gene transfer in porcine myocardium after coronary infusion of an adeno-associated virus vector. *Ann. Thorac. Surg.* **62**, 1669–1676, 1996.

51. Bartlett, R. J., Ricordi, C., Sharma, K., *et al.* Long-term expression of a fluorescent reporter gene via direct injection of plasmid vector into mouse skeletal muscle: comparison of human creatine kinase and CMV promoter expression levels in vivo. *Cell Transplant* **5**, 411–419, 1996.

52. Kessler, P. D., Byrne, B. J., Kurtzman, G. J., *et al.* Gene delivery to skeletal muscle results in sustained expression and systemic delivery of a therapeutic protein. *Proc. Natl. Acad. Sci. USA* **93**, 14082–14087, 1996.

53. Xiao, X., Samulski, R. J., Li, J. Efficient long-term gene transfer into muscle tissue of immunocompetent mice by adeno-associated virus vector. *J. Virol.* **70**, 8098–8108, 1996.

54. Fisher, K. J., Wilson, J. M., Raper, S. E., *et al.* Recombinant adeno-associated virus for muscle directed gene therapy. *Nat. Med.* **3**, 306–312, 1997.

55. Zhou, S. Z., Cooper, S., Kang, L. Y., *et al.* Adeno-associated virus 2-mediated high efficiency gene transfer into immature

and mature subsets of hematopoietic progenitors cells in human umbilical cord blood. *J. Exp. Med.* **179,** 1867–1875, 1994.

56. Fisher-Adams, G., Wong, K. K., Jr., Forman, S., *et al.* Integration of adeno-associated virus vector genomes in Human CD34 cells following transduction. *Blood* **88,** 492–504, 1996.

57. Chatterjee, S., Podsakoff, G., Wong, K. K., Jr. Gene transfer into terminally differentiated primary human peripheral blood-derived mononuclear cells by adeno-associated virus. *Blood* **84,** 360a, 1994.

58. Inouye, R. T., Terwilliger, E. F., Pomerantz, R. J., *et al.* Potent inhibition of human immunodeficiency virus type 1 in primary T cells and alveolar macrophages by a combination anti-Rev strategy delivered in an adeno-associated virus vector. *J. Virol.* **71,** 4071–4078, 1997.

59. McKeon, C., Samulski, R. J. NIDDK workshop on AAV vectors: gene transfer into quiescent cells. *Hum. Gene Ther.* **7,** 1615–1619, 1996.

60. Muro-Cacho, C. A., Samulski, R. J., Kaplan, D. Gene transfer in human lymphocytes using a vector based on adeno-associated virus. *J. Immunother.* **11,** 231–237, 1992.

61. Walsh, C. E., Nienhuis, A. W., Samulski, R. J., *et al.* Phenotypic correction of Fanconi anemia in human hematopoietic cells with a recombinant adeno-associated virus vector. *J. Clin. Invest.* **94,** 1440–1448, 1994.

62. Goodman, S., Xiao, X., Donahue, R. E., *et al.* Recombinant adeno-associated virus-mediated gene transfer into hematopoietic progenitor cells. *Blood* **84,** 1492–1500, 1994.

63. Podsakoff, G., Shaughnessy, E. A., Lu, D., *et al.* Long term *in vivo* reconstitution with murine marrow cells transduced with an adeno-associated virus vector. *Blood* **84,** 256a, 1994.

64. Halbert, C. L., Alexander, I. E., Wolgamot, G. M., *et al.* Adeno-associated virus vectors transduce primary cells much less efficiently than immortalized cells. *J. Virol.* **69,** 1473–1479, 1995.

65. McLaughlin, S. K., Collis, P., Hermonat, P. L., *et al.* Adeno-associated virus general transduction vectors: analysis of proviral structures. *J. Virol.* **62,** 1963–1973, 1988.

66. Hargrove, P. W., Nienhuis, A. W., Kurtzman, G. J., *et al.* High-level globin gene expression mediated by a recombinant adeno-associated virus genome that contains the 3′ gamma globin gene regulatory element and integrates as tandem copies in erythroid cells. *Blood* **89,** 2167–2175, 1997.

67. Urcelay, E., Ward, P., Wiener, S. M., *et al.* Asymmetric replication *in vitro* from a human sequence element is dependent on adeno-associated virus rep protein. *J. Virol.* **69,** 2038–2046, 1995.

68. Malik, P., Kohn, D. B., Kurtzman, G. J., Podsakoff, G. M., *et al.* Recombinant adeno-associated virus mediates a high level of gene transfer but less efficient integration in the K562 human hematopoietic cell line. *J. Virol.* **71,** 1776–1783, 1997.

69. Shelling, A. N., Smith, M. G. Targeted integration of transfected and infected adeno-associated virus vectors containing the neomycin resistance gene. *Gene Ther.* **1,** 165–169, 1994.

70. Kearns, W. G., Cutting, G. R., Flotte, T. R., *et al.* Recombinant adeno-associated virus (AAV-CFTR) vectors do not integrate in a site-specific fashion in an immortalized epithelial cell line. *Gene Ther.* **3,** 748–755, 1996.

71. Sun, X. L., Antony, A. C., Srivastava, A., *et al.* Transduction of folate receptor cDNA into cervical carcinoma cells using recombinant adeno-associated virions delays cell proliferation in vitro and in vivo. *J. Clin. Invest.* **96,** 1535–1547, 1995.

72. Fisher, K. J., Gao, G. P., Weitzman, M. D., *et al.* Transduction with recombinant adeno-associated virus for gene therapy is limited by leading-strand synthesis. *J. Virol.* **70,** 520–532, 1996.

73. Ferrari, F. K., Samulski, R. J., Shenk, T., *et al.* Second-strand synthesis is a rate-limiting step for efficient transduction by re-

combinant adeno-associated virus vectors. *J. Virol.* **70,** 3227–3234, 1996.

74. Russell, D. W., Miller, A. D., Alexander, I. E. DNA synthesis and topoisomerase inhibitors increase transduction by adeno-associated virus vectors. *Proc. Natl. Acad. Sci. USA.* **92,** 5719–5723, 1995.

75. Yakobson, B., Koch, T., Winocour, E. Replication of adeno-associated virus in synchronized cells without the addition of a helper virus. *J. Virol.* **61,** 972–981, 1987.

76. Yakobson, B., Hrynko, T. A., Peak, M. J., *et al.* Replication of adeno-associated virus in cells irradiated with UV light at 254 nm. *J. Virol.* **63,** 1023–1030, 1989.

77. Yalkinoglu, A. Ö, Heilbronn, R., Bürkle, A., *et al.* DNA amplification of adeno-associated virus as a response to cellular genotoxic stress. *Cancer Res.* **48,** 3123–3129, 1988.

78. Yalkinoglu, A. O., Zentgraf, H., Hubscher, U. Origin of adeno-associated virus DNA replication is a target of carcinogen-inducible DNA replication. *J. Virol.* **65,** 3175–3184, 1991.

79. Cotmore, S. F., Tattersall, P. The autonomously replicating parvoviruses of vertebrates. *Adv. Virus Res.* **33,** 91–174, 1987.

80. Toolan, H. W., Saunders, E. L., Southam, C. M., *et al.* H-1 virus viremia in the human. *Proc. Soc. Exp. Biol. Med.* **119,** 711–715, 1965.

81. Toolan, H. W., Ledinko, N. Inhibition by H-1 virus on the incidence of tumors produced by adenovirus 12 in hamsters. *Virology* **35,** 475–478, 1968.

82. Van Pachterbeke, C., Tuynder, M., Cosyn, J. P., *et al.* Parvovirus H-1 inhibits growth of short-term tumor-derived but not normal mammary tissue cultures. *Int. J. Cancer* **55,** 672–677, 1993.

83. Rommelaere, J., Cornelis, J. J. Anti-neoplastic activity of parvoviruses. *J. Virol. Methods* **33,** 233–251, 1991.

84. Linser, P., Bruning, H., Armentrout, R. W. Specific binding sites for a parvovirus minute virus of mice on cultured mouse cells. *J. Virol.* **24,** 211–221, 1977.

85. Legendre, D., Rommelaere, J. Terminal regions of the NS-1 protein of the parvovirus minute virus of mice are involved in cytotoxicity and promoter *trans* inhibition. *J. Virol.* **66,** 5705–5713, 1992.

86. Legrand, C., Rommelaere, J., Caillet-Fauquet, P. MVM(p) NS-2 protein expression is required with NS-1 for maximal cytotoxicity in human transformed cells. *Virology* **195,** 149–155, 1993.

87. Guetta, E., Graziani, Y., Tal, J. Suppression of Ehrlich ascites tumors in mice by minute virus of mice. *J. Natl. Cancer Inst.* **76,** 1177–1780, 1986.

88. Dupressoir, T., Vanacker, J. M., Cornelis, J., *et al.* Inhibition by parvovirus H-1 of the formation of tumors in nude mice and colonies *in vitro* by transformed human mammary epithelial cells. *Cancer Res.* **49,** 3203–3208, 1989.

89. Telerman, A., Tuynder, M., Dupressoir, T., *et al.* A model for tumor suppression using H-1 parvovirus. *Proc. Natl. Acad. Sci. USA.* **90,** 8702–8706, 1993.

90. Russell, S. J., Brandenburger, A., Flemming, C. L., *et al.* Transformation-dependent expression of interleukin genes delivered by a recombinant parvovirus. *J. Virol.* **66,** 2821–2828, 1992.

91. Maxwell, I. H., Maxwell, F., Rhode, S. L., 3d, *et al.* Recombinant LuIII autonomous parvovirus as a transient transducing vector for human cells. *Hum. Gene Ther.* **4,** 441–450, 1993.

92. Corsini, J., Carlson, J. O., Maxwell, I. H., *et al.* Autonomous parvovirus and densovirus gene vectors. *Adv. Virus Res.* **47,** 303–351, 1996.

93. Maxwell, I. H., Maxwell, F., Long, C. J., *et al.* Autonomous parvovirus transduction of a gene under control of tissue-specific or inducible promoters. *Gene Ther.* **3,** 28–36, 1996.

94. Brandenburger, A., Russell, S. A novel packaging system for the generation of helper-free oncolytic MVM vector stocks. *Gene Ther.* **3**, 927–931, 1996.

95. Avalosse, B., Burny, A., Mine, N., *et al.* Method for concentrating and purifying recombinant autonomous parvovirus vectors designed for tumour-cell-targeted gene therapy. *J. Virol. Methods* **62**, 179–183, 1996.

96. Fekete, D. M., Perez-Miguelsanz, J., Ryder, E. F., *et al.* Clonal analysis in the chicken retina reveals tangential dispersion of clonally related cells. *Dev. Biol.* **166**, 666–682, 1994.

97. Lemischka, I. R. What we have learned from retroviral marking of hematopoietic stem cells. *Curr. Top. Micro. Immunol.* **177**, 59–71, 1992.

98. Brenner, M. K., Rill, D. R., Holladay, M. S., *et al.* Gene marking to determine whether autologous marrow infusion restores long-term haemopoiesis in cancer patients. *Lancet* **342**, 1134–1137, 1993.

99. Brenner, M. K., Rill, D. R., Moen, R. C., *et al.* Gene-marking to trace origin of relapse after autologous bone-marrow transplantation. *Lancet* **341**, 85–86, 1993.

100. Deisseroth, A. B., Zu, Z., Claxton, D., *et al.* Genetic marking shows that Ph+ cells present in autologous transplants of chronic myelogenous leukemia (CML) contribute to relapse after autologous bone marrow in CML. *Blood* **83**, 3068–3076, 1994.

101. Dunbar, C. E., Cottler-Fox, M., O'Shaughnessy, J. A., *et al.* Retrovirally marked CD34-enriched peripheral blood and bone marrow cells contribute to long-term engraftment after autologous transplantation. *Blood* **85**, 3048–3057, 1995.

102. Spangrude, G. J., Johnson, G. R. Resting and activated subsets of mouse multipotent hematopoietic stem cells. *Proc. Natl. Acad. Sci. USA.* **87**, 7433–7437, 1990.

103. Luskey, B. D., Rosenblatt, M., Zsebo, K., *et al.* Stem cell factor, interleukin-3, and interleukin-6 promote retroviral-mediated gene transfer into murine hematopoietic stem cells. *Blood* **80**, 396–402, 1992.

104. Williams, D. A. *Ex vivo* expansion of hematopoietic stem and progenitor cells—robbing Peter to pay Paul? *Blood* **81**, 3169–3172, 1993.

105. Chatterjee, S., Wong, K. K., Jr. Adeno-associated virus vectors for gene therapy of the hematopoietic system. *Curr. Top. Microbiol. Immunol.* **218**, 61–73, 1996.

106. Miller, J. L., Donahue, R. E., Sellers, S. E., *et al.* Recombinant adeno-associated virus (rAAV)-mediated expression of a human gamma-globin gene in human progenitor-derived erythroid cells. *Proc. Natl. Acad. Sci. USA.* **91**, 10183–10187, 1994.

107. Zhou, S. Z., Broxmeyer, H. E., Cooper, S. *et al.* Adeno-associated virus 2-mediated gene transfer in murine hematopoietic progenitor cells. *Exp. Hematology* **21**, 928–933, 1993.

108. Walsh, C. E., Liu, I. M., Wang, S., *et al. In vivo* gene transfer with a novel adeno-associated virus vector to human hematopoietic cells engrafted in SCID-hu mice. *Blood* **84**, 256a, 1994.

109. Luhovy, M., Prchal, J. T., Townes, T. M., *et al.* Stable transduction of recombinant adeno-associated virus into hematopoietic stem cells from normal and sickle cell patients. *Biol. Blood Marrow Transplant* **2**, 24–30, 1996.

110. Lu, D., Fisher-Adams, G., Forman, S. J., *et al.* Gene transfer to primitive human marrow and cord blood hematopoietic progenitor cells in long term culture with adeno-associated virus vectors, submitted.

111. Anderson, R. J., Prentice, H. G., Corbett, T. J., *et al.* Detection of adeno-associated virus type 2 in sorted human bone marrow progenitor cells. *Exp. Hematol.* **25**, 256–262, 1997.

112. Antman, K. H., Elias, A., Fine, H. A., Dose-intensive therapy with autologous bone marrow transplantation in solid tumors, in Forman, S. J., Blume, K. G., Thomas, E. D., (eds), Bone Marrow Transplantation. Cambridge, MA, Blackwell Scientific Publications, 1994, pp. 767–788.

113. Gottesman, M. M., Pastan, I. Biochemistry of multidrug resistance mediated by the multidrug transporter. *Annu. Rev. Biochem.* **62**, 385–427, 1993.

114. Chaudhary, P. M., Roninson, I. B. Expression and activity of P-glycoprotein, a multidrug efflux pump, in human hematopoietic stem cells. *Cell* 85–94, 1991.

115. Mickisch, G. H., Merlino, G. T., Galski, H., *et al.* Transgenic mice that express the human multidrug-resistance gene in bone marrow enable a rapid identification of agents that reverse drug resistance. *Proc. Natl. Acad. Sci. USA.* **88**, 547–551, 1991.

116. Mickisch, G. H., Aksentijevich, I., Schoenlein, P. V., *et al.* Transplantation of bone marrow cells from transgenic mice expressing the human MDR1 gene results in long-term protection against the myelosuppressive effect of chemotherapy in mice. *Blood* **79**, 1087–1093, 1992.

117. Sorrentino, B. P., Brandt, S. J., Bodine, D., *et al.* Selection of drug-resistant bone marrow cells in vivo after retroviral transfer of human MDR1. *Science* **257**, 99–103, 1992.

118. Hanania, E. G., Deisseroth, A. B. Serial transplantation show that early hematopoietic precursor cells are transduced by MDR-1 retroviral vector in a mouse gene therapy model. *Cancer Gene Ther.* **1**, 21–25, 1994.

119. Ward, M., Richardson, C., Pioli, P., *et al.* Transfer and expression of the human multiple drug resistance gene in human CD34+ cells. *Blood* **84**, 1408–1414, 1994.

120. Shaughnessy, E., Chatterjee, S., Podsakoff, G., *et al.* Efficient *MDR1* transgene expression in primary human CD34+ hematopoietic cells following transduction with an adeno-associated virus vector, submitted.

121. Baudard, M., Gottesman, M. M., Kearns, W. G., *et al.* Expression of the human multidrug resistance and glucocerebrosidase cDNAs from adeno-associated vectors: efficient promoter activity of AAV sequences and *in vivo* delivery via liposomes. *Hum. Gene Ther.* **7**, 1309–1322, 1996.

122. Neckers, L., Whitesell, L., Rosolen, A., *et al.* Antisense inhibition of oncogene expression. *Crit. Rev. Oncog.* **3**, 175–231, 1992.

123. Hélène, C. Control of oncogene expression by antisense nucleic acids. *Eur. J. Cancer* **30A**, 1721–1726, 1994.

124. Reed, J. C., Cuddy, M., Haldar, S., Croce, C., *et al.* BCL2-mediated tumorigenicity of a human T-lymphoid cell line: synergy with MYC and inhibition by BCL2 antisense. *Proc. Natl. Acad. Sci. USA.* **87**, 3660–3664, 1990.

125. Skorski, T., Nieborowska-Skorska, M., Nicolaides, N. C., *et al.* Suppression of Philadelphia leukemia cell growth in mice by BCR-ABL antisense oligodeoxynucleotide. *Proc. Natl. Acad. Sci. USA.* **91**, 4504–4508, 1994.

126. Colomer, R., Lupu, R., Bacus S. S., Gelmann, E. P.: erb B-2 antisense oligonucleotides inhibit the proliferation of breast carcinoma cells with erbB-2 oncogene amplification. *Br. J. Cancer* **70**, 819–825, 1994.

127. Mercola, D., Rundell, A., Westwick, J., *et al.* Antisense RNA to the c-*fos* gene; restoration of density dependent growth arrest in a transformed cell line. *Biochem. Biophys. Res. Commun.* **147**, 288–294, 1987.

128. Yokoyama, K., Imamoto, F. Transcriptional control of endogenous *myc* protooncogene by antisense RNA. *Proc. Natl. Acad. Sci. USA.* **84**, 7363–7367, 1987.

129. Steele, C., Cowsert, L. M., Shillitoe, E. J. Effects of human papillomavirus type 18-specific antisense oligonucleotides on the transformed phenotype of human carcinoma cell lines. *Cancer Res.* **53**, 2330–2337, 1993.

130. Wong, K. K., Jr., Rose, J. A., Chatterjee, S. Restriction of HSV-1 production in cell lines transduced with an antisense viral vector targeting the HSV-1 ICP4 gene, in Brown, F., Chanock, R., Ginsberg, H., Lerner, R. (eds), Vaccine 91. New York, Cold Spring Harbor, 1991, pp. 183–189.

131. Chatterjee, S., Johnson, P. R., Wong, K. K., Jr. Dual target inhibition of HIV-1 *in vitro* by means of an adeno-associated virus antisense vector. *Science* **258**, 1485–1488, 1992.

132. Ponnazhagan, S., Nallari, M. L., Srivastava, A. Suppression of human alpha-globin gene expression mediated by the recombinant adeno-associated virus 2-based antisense vectors. *J. Exp. Med.* **179**, 733–738, 1994.

133. Castanotto, D., Rossi, J. J., Sarver, N. Antisense catalytic RNAs as therapeutic agents. *Adv. Pharm.* **25**, 289–317, 1994.

134. Kashani, S. M., Funato, T., Tone, T., *et al.* Reversal of the malignant phenotype by an anti-ras ribozyme. *Antisense Res. Dev.* **2**, 3–15, 1992.

135. Tone, T., Kashani-Sabet, M., Funato, T., Shitara, T., *et al.* Suppression of EJ cells tumorigenicity. *In Vivo* **7**, 471–476, 1993.

136. Snyder, D. S., Wu, Y., Wang, J. L., *et al.* Ribozyme-mediated inhibition of *bcr-abl* gene expression in Philadelphia chromosome-positive cell line. *Blood* **82**, 600–605, 1993.

137. Scanlon, K. J., Jiao, L., Funato, T., *et al.* Ribozyme-mediated cleavage of c-*fos* mRNA reduces gene expression of DNA synthesis enzymes and metallothionein. *Proc. Natl. Acad. Sci. USA.* **88**, 10591–10595, 1991.

138. Funato, T., Yoshida, E., Jiao, L., Tone, T., *et al.* The utility of an anti-fos ribozyme in reversing cisplatin resistance in human carcinomas. *Adv. Enzyme Regul.* **32**, 195–209, 1992.

139. Holm, P. S., Scanlon, K. J., Dietal, M. Reversion of multidrug resistance in the P glycoprotein-positive human pancreatic cell line (EPP85-181RDB) by introduction of a hammerhead ribozyme. *Br. J. Cancer* **70**, 239–243, 1994.

140. Lowy, D. R., Kirnbauer, R., Schiller, J. T. Genital human papillomavirus infection. *Proc. Natl. Acad. Sci. USA.* **91**, 2436–2440, 1994.

141. Lu, D., Chatterjee, S., Brar, D., *et al.* High efficiency *in vitro* cleavage of transcripts arising from the major transforming genes of human papillomavirus type 16 mediated by ribozymes transcribed from an adeno-associated virus-based vector. *Cancer Gene Therapy* **1**, 267–277, 1994.

142. Challis, G. B., Stam, H. J. The spontaneous regression of cancer: A review of cases 1900–1987. *Acta Oncol.* **29**, 545–550, 1990.

143. Medeiros, L. J., Picker, L. J., Gelb, A. B., *et al.* Number of 'host' helper T cells and proliferating cells predict survival in diffuse small-cell lymphomas. *J. Clin. Oncol.* **7**, 1009–1017, 1989.

144. Zinkernagel, R. M., Doherty, P. C. MHC-restricted cytotoxic T cells: studies of the biological role of polymorphic major transplantation antigen determining T cell restriction specificity, function and responsiveness. *Adv. Immunol.* **27**, 51–177, 1979.

145. June, C., Bluestone, J., Nadler, L. M., *et al.* The B7 and CD28 receptor families. *Immunol. Today* **15**, 321–331, 1994.

146. Hwu, P., Rosenberg, S. A. The use of gene-modified tumor-infiltrating lymphocytes for cancer therapy. *Ann NY Acad. Sci.* **716**, 188–197, 1994.

147. Roberts, M. R., Qin, L., Zhang, D., *et al.* Targeting of human immunodeficiency virus-infected cells by CD8+ T lymphocytes armed with universal T-cell receptors. *Blood* **84**, 2878–2889, 1994.

148. Flamand, V., Sornasse, T., Thielemans, K., *et al.* Murine dendritic cells pulsed *in vitro* with tumor antigen induce tumor resistance *in vivo*. *Eur. J. Immunol.* **24**, 605–610, 1994.

149. Pardoll, D. M. New strategies for enhancing the immunogenicity of tumors. *Curr. Opin. Immunol.* **5**, 719–725, 1993.

150. Tepper, R. I., Mule, J. Experimental and clinical studies of cytokine gene-modified tumor cells. *Hum. Gene Ther.* **5**, 153–164, 1994.

151. Townsend, S., Allison, J. Tumor rejection after direct costimulation of CD8+ T cells by B7 transfected melanoma cells. *Science* **259**, 368–372, 1993.

152. Nabel, G. J., Nabel, E. G., Yang, Z.-Y., *et al.* Direct gene transfer with DNA-liposome complexes in melanoma: expression, biologic activity, and lack of toxicity in humans. *Proc. Natl. Acad. Sci. USA* **90**, 11307–11311, 1993.

153. Bruton, J. K., Koeller, J. M. Recombinant interleukin-2. *Pharmacotherapy* **14**, 635–656, 1994.

154. Fearon, E. R., Pardoll, D. M., Itaya, T., *et al.* Interleukin-2 production by tumor cells bypasses T helper function in the generation of an antitumor response. *Cell* **60**, 397–403, 1990.

155. Connor, J., Bannerji, R., Saito, S., *et al.* Regression of bladder tumors in mice treated with interleukin 2 gene-modified tumor cells. *J. Exp. Med.* **177**, 1127–1134, 1993.

156. Gansbacher, B., Zier, K., Daniels, B., *et al.* Interleukin 2 gene transfer into tumor cells abrogates tumorigenicity and induces protective immunity. *J. Exp. Med.* **172**, 1217–1224, 1990.

157. Haddada, H., Ragot, T., Cordier, L., *et al.* Adenoviral interleukin-2 gene transfer into P815 tumor cells abrogates tumorigenicity and induces antitumoral immunity in mice. *Hum. Gene Ther.* **4**, 703–711, 1993.

158. Coveney, E., Clary, B., DiMaio, J. M., *et al.* Inhibition of breast cancer metastasis by cytokine gene-modified tumor vaccination in tumor-bearing mice. *Surg. Forum* **XV**, 540–542, 1994.

159. Philip, R., Brunette, E., Kilinski, L., *et al.* Efficient and sustained gene expression in primary T lymphocytes and primary and cultured tumor cells mediated by adeno-associated virus plasmid DNA complexed to cationic liposomes. *Mol. Cell Biol.* **14**, 2411–2418, 1994.

160. Okada, H., Yoshida, J., Kurtzman, G., *et al.* Gene therapy against an experimental glioma using adeno-associated virus vectors. *Gene Ther.* **3**, 957–964, 1996.

161. Zhang, J. F., Taylor, M. W., Blatt, L. M., *et al.* Gene therapy with an adeno-associated virus carrying an interferon gene results in tumor growth suppression and regression. *Cancer Gene Ther.* **3**, 31–38, 1996.

162. Clary, B. M., Lyerly, H. K., Gilboa, E., *et al.* Active immunization with tumor cells transduced by a novel AAV plasmid-based gene delivery system. *J. Immunother.* **20**, 26–37, 1997.

163. Russell, S. J., Brandenburger, A., Flemming, C. L., *et al.* Transformation-dependent expression of interleukin genes delivered by a recombinant parvovirus. *J. Virol.* **66**, 2821–2828, 1992.

164. Tepper, R. I., Pattengale, P. K., Leder, P. Murine interleukin-4 displays potent anti-tumor activity *in vivo*. *Cell* **57**, 503–512, 1989.

165. Chen, L., Ashe, S., Brady, W. A., *et al.* Costimulation of antitumor immunity by the B7 counterreceptor for the T lymphocyte molecules CD28 and CTLA-4. *Cell* **71**, 1093–1102, 1992.

166. Chiorini, J. A., Kotin, R. M., Hallek, M., *et al.* High-efficiency transfer of the T cell co-stimulatory molecule B7-2 to lymphoid cells using high-titer recombinant adeno-associated virus vectors. *Hum. Gene Ther.* **6**, 1531–1541, 1995.

167. Miyamura, K., Kajigaya, S., Momoeda, M., *et al.* Parvovirus particles as platforms for protein presentation. *Proc. Natl. Acad. Sci. USA.* **91**, 8507–8511, 1994.

168. Sedlik, C., Casal, I., Leclerc, C., *et al.* Immunogenicity of poliovirus B and T cell epitopes presented by hybrid porcine parvovirus particles. *J. Gen. Virol.* **76**, 2361–2368, 1995.

169. Mullen, C. A. Metabolic suicide genes in gene therapy. *Pharmacol. Ther.* **63**, 199–207, 1994.

170. Culver, K. W., Ram, Z., Wallbridge, S., *et al. In vivo* gene transfer with retroviral vector-producer cells for treatment of experimental brain tumors. *Science* **256**, 1550–1552, 1992.

171. Bi, W. L., Parysek, L. M., Warnick, R., *et al. In vitro* evidence that metabolic cooperation is responsible for the bystander effect observed with HSV tk retroviral gene therapy. *Hum. Gene Ther.* **4**, 725–731, 1993.

172. Koering, C. E., Dupressoir, T., Plaza, S., *et al.* Induced expression of the conditionally cytotoxic herpes simplex virus thymidine kinase gene by means of a parvoviral regulatory circuit. *Hum. Gene Ther.* **5**, 457–463, 1994.

173. Su, H., Kan, Y. W., Xu, S. M. *et al.* Selective killing of AFP-positive hepatocellular carcinoma cells by adeno-associated virus transfer of the herpes simplex virus thymidine kinase gene. *Hum. Gene Ther.* **7**, 463–470, 1996.

174. Pluda, J. M. Tumor-associated angiogenesis: mechanisms, clinical implications, and therapeutic strategies. *Semin. Oncol.* **24**, 203–218, 1997.

175. Saleh, M., Wilks, A. F., Stacker, S. A. Inhibition of growth of C6 glioma cells *in vivo* by expression of antisense vascular endothelial growth factor sequence. *Cancer Res.* **56**, 393–401, 1996.

176. Ozaki, K., Terada, M., Sugimura, T., *et al.* Use of von Willebrand factor promoter to transduce suicidal gene to human endothelial cells, HUVEC. *Hum. Gene Ther.* **7**, 1483–1490, 1996.

177. Dupont, F., Tenenbaum, L., Guo, L. P., *et al.* Use of an autonomous parvovirus vector for selective transfer of a foreign gene into transformed human cells of different tissue origins and its expression therein. *J. Virol.* **68**, 1397–1406, 1994.

178. Flotte, T., Wetzel, R., Walden, S., *et al.* A phase I study of an adeno-associated virus-CFTR gene vector in adult CF patients with mild lung disease. *Hum. Gene Ther.* **7**, 1145–1159, 1996.

179. Conrad, C. K., Flotte, T. R., Guggino, W. B., *et al.* Safety of single-dose administration of an adeno-associated virus (AAV)-CFTR vector in the primate lung. *Gene Ther.* **3**, 658–668, 1996.

180. Herzog, R. W., High, K. A., Fisher, K. J., *et al.* Stable gene transfer and expression of human blood coagulation factor IX after intramuscular injection of recombinant adeno-associated virus. *Proc. Natl. Acad. Sci. USA* **94**, 5804–5809, 1997.

CHAPTER 7

Antibody-Targeted Gene Therapy

A. M. McCall, G. P. Adams, and L. M. Weiner

Fox Chase Cancer Center, Philadelphia, Pennsylvania 19111

I. INTRODUCTION

Two major obstacles must be addressed if the gene therapy of cancer is to be successful. The first involves the delivery of enough copies of a gene to the tumor site to guarantee efficacy. The efficient transfer of a gene to a particular tumor cell represents the second challenge. Currently, gene delivery systems are able to introduce genes into neoplastic cells *in vivo* but are nonselective in their targeting. By incorporating antibodies or antibody fragments into the different gene delivery systems, cell-specific targeting has been achieved *in vitro* (Fig. 1). Specific DNA transfer to tumor cells has been achieved using antibody DNA/poly-L-lysine complexes, a GAL4 DNA binding protein/scFv fusion protein and retroviral vectors expressing scFv-envelope fusion proteins. However, antibody-mediated gene transfer is very inefficient. New constructs are needed to improve the current gene delivery systems.

II. BACKGROUND: MONOCLONAL ANTIBODIES AND CANCER THERAPY

A. Monoclonal Antibodies and Antibody Fragments

Antibodies have long been popular targeting vehicles for the treatment of cancer. The unique structure of an antibody makes it ideally suited to targeting neoplastic cells. Each molecule consists of two light and two heavy chains, which are each divided into variable and constant regions. Variable regions (V_H and V_L) contain three complementarity-determining regions (CDRs) flanked on either side by framework regions. The six CDRs (three from the light chain and three from the heavy chain) make up the binding site of the antibody. This binding site is considered to be the antibody's idiotype. The heavy-chain-constant regions contain a minimum of three domains, C_H1, C_H2, and C_H3, whereas the light-chain-constant regions consist of only one do-

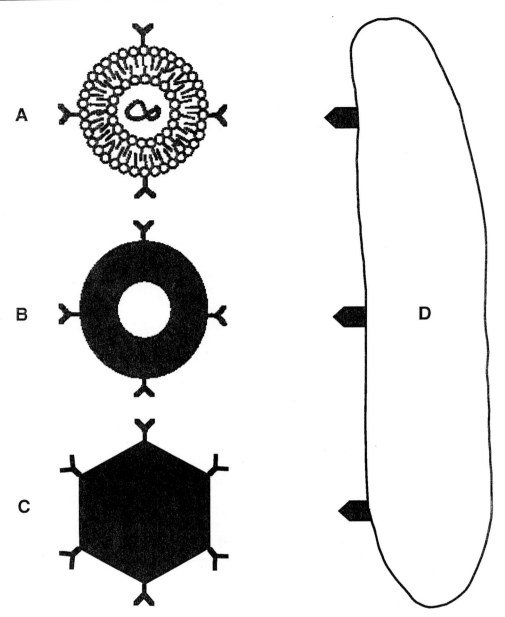

FIGURE 1   Different delivery systems in antibody-targeted gene therapy. (A) Immunoliposomes-DNA complex (a delivery system with potential use in antibody-targeted gene therapy). (B) Antibody–DNA/poly-L-lysine complex. (C) Antibody-coated retrovirus vector. (D) Target tumor cell bearing receptors recognized by the antibodies.

main (C_κ or C_λ). Domains C_H2 and C_H3 form the Fc region of the antibody, which determines its effector functions. For example, complement is activated by the Fc region and leukocytes recognize other epitopes of this region through their Fc receptors to mediate phagocytosis and antibody-dependent cellular cytotoxicity (ADCC).

For a long time, antibodies were of little practical use in the treatment of cancer. This was due in part to their lack of purity. With the advent of hybridoma technology by Kohler and Milstein [1], large quantities of antibodies could be generated in a monoclonal form. However, the search for tumor-specific targets has resulted in only a handful of potential antigens. Nevertheless, selective cell markers have been utilized in several tumor types. Monoclonal antibodies (mAbs) to these cell markers have been used successfully in the treatment of lymphomas and leukemias in patients [2–4].

However, only limited responses have been achieved with solid tumors [5]. The problems associated with solid tumors include a limited vascular supply [6], elevated interstital pressure, and inhomogenous tumor antigen expression. With the first two factors, antibodies find it difficult to penetrate the tumor. Moreover, heterogeneous antigen expression by tumor cells can prevent antibody binding to a large fraction of the cells. Tumor antigens can be either shed or internalized [7]. Shed antigen can inhibit the binding of antibodies to neoplastic cells, and antibodies that mediate their effects at the cell surface may be rendered ineffective by rapid internalization of the antigen. Until recently, most antibodies used in the treatment of patients were derived from mice and often led to the production of human anti-mouse antibodies (HAMAs) [8]. This can limit the number of treatments that a patient can receive.

Antibody fragments also have been used to target cancer. Whole antibodies can be digested with the enzyme pepsin to produce $F(ab)'_2$ fragments or the enzyme papain to produce Fab fragments. More recently, genetic engineering has been used to generate Fab fragments and single-chain Fv (scFv) molecules [9]. These contain V_H and V_L domains connected by a flexible polypeptide linker. All three smaller-based antibody fragments exhibit a rapid systemic clearance and have better tumor penetration than is possible with intact mAb [10]. This leads to an improved specificity of targeting. However, due to a rapid systemic clearance, these smaller molecules exhibit a tumor retention that is less than that of intact antibodies [11,12]. Moreover, the absence of a Fc region in these antibody fragments leads to a loss of Fc-dependent effector functions [13].

B. Unconjugated Antibodies

Antibodies can directly initiate killing of tumor cells either through binding alone or through Fc-mediated effector functions. Unconjugated mAbs directed against B-cell determinants have been found to induce objective clinical responses in a significant percentage of patients with chemotherapy-resistant B-cell lymphomas [14]. The precise mechanisms underlying the observed responses have not been fully unraveled, but these mAbs appear to act by interfering with ligand-receptor interactions or by triggering apoptosis. mAbs against ganglioside antigens in patients with melanoma and neuroblastoma appear to initiate lysis of tumor cells through complement-mediate cytotoxicity [15–17]. Treatment with unconjugated murine mAb 17-1A resulted in improved outcomes in patients with Dukes' C colonic neoplasms. In a randomized Phase III clinical trial, treat-

ment led to a 30% reduction in death ($P = 0.04$) and 27% reduction in recurrence ($P = 0.03$) of the cancer compared to the untreated control group [18].

One way of harnessing the effector mechanisms of the immune system is by directly conjugating the effector cells and tumor cells using bispecific mAbs. Bispecific mAbs can be generated by chemical conjugation, by gene fusion, or by the fusion of two hybridomas to create quadromas that secrete bispecific mAb. A bispecific mAb targeting HER2/neu and human CD16 induced potent cytokine release by natural killer cells and mature macrophages, and exhibited some clinical activity in a Phase I trial [19]. Additionally, a bispecific mAb targeting HER2/neu and human CD64 expressed by monocytes and activated neutrophils has been shown to be immunologically active and induces tumor inflammation [20].

Antibodies can also work through the idiotypic networks to exert their influence by acting as tumor vaccines [21]. In this case, an antibody (Ab1) specific for a particular tumor antigen is injected into mice to generate Ab2-secreting B cells. Ab2 recognizes the idiotype or binding site of Ab1. When patients are immunized with Ab2, they produce anti-anti-idiotype antibodies (Ab3), which recognize the original tumor antigen. Herlyn and colleagues treated 30 patients with advanced colorectal carcinoma with serial injections of polyclonal goat antibodies induced by immunizing the animals with murine 17-1A mAb [22]. Six patients experienced brief clinical responses, and all 30 developed antibodies directed against the immunizing goat antibody. Mittelman and colleagues treated 15 patients with metastatic melanoma using a murine anti-idiotype mAb directed against an antibody recognizing a high-molecular-weight human melanoma-associated antigen [23]. In this trial Ab3 was identified in seven of the patients, three of which showed a reduction in the size of metastases in the skin or lungs. Similar findings have been observed using antibodies directed against the carcinoembryonic antigen (CEA) [24].

C. Conjugated Antibodies

Cytotoxic agents can be coupled to antibodies in order to mediate killing of target tumor cells. Such agents include toxins, drugs, and radionuclides. Toxins that have been used in the treatment of cancer include the plant toxin ricin, *Pseudomonas* exotoxin, and diphtheria toxin. A toxin is usually composed of two chains: One chain (B chain) facilitates binding and intracellular transport; the second chain (A chain) is responsible for the catalytic activity of the toxin. In an immunotoxin, the mAb or antibody fragment is either chemically conjugated or gene fused to the A chain, replacing the cell binding and intracellular transport role of the B chain [25]. Vitetta and colleagues treated 14 patients with B-cell lymphoma using an anti-CD22 MAb conjugated to the ricin A chain and observed partial responses in 5 patients [26].

Chemotherapy agents can be conjugated to antibodies to confer specific tumor targeting. Doxorubicin has been conjugated to a human/mouse chimeric anticarcinoma antibody (BR96), which is reactive with a Lewis^y blood-group related antigen. The drug is liberated in the lysosomes due to an acid-labile hydrazone bond between doxorubicin and BR96. High cure rates in mice bearing advanced human, lung, breast, and colon tumor xenografts have been achieved [27].

Immunotoxins and drug immunoconjugates exert cytotoxicity through intracellular mechanisms, but radioimmunoconjugates have the ability to kill surrounding cells as a consequence of the long path lengths of the radionuclides. Radioimmunoconjugates have had the most success in the treatment of hematologic malignancies. Kaminski and colleagues [28] used iodine-131-labeled anti-CD20 mAb to treat 10 patients with chemotherapy-refractory B-cell lymphomas. Four patients experienced complete responses; 6 patients experienced partial responses. Press and colleagues [29] employed a similar strategy and also achieved promising results. In a group of 42 patients with chemotherapy-refractory lymphomas, 24 favorable biodistributions and 19 received high-dose radioimmunotherapy. A remarkable 84% of these patients experienced complete remissions and an additional 11% had partial remissions.

Antibodies have recently found new roles in the gene therapy of cancer. Elaborate gene delivery systems have utilized antibodies in the cell-specific transfer of genes, and intracellular antibodies or intrabodies have been used to block the expression of proteins related to the oncogenic process. Intrabodies will be reviewed in Chapter 26. The rest of this chapter will examine novel applications of antibodies in the delivery of genes to tumor cells.

III. RECENT ADVANCES: mAb-MEDIATED TARGETING AND CANCER GENE THERAPY

A. The Role of Antibodies in Nonviral Gene Delivery

1. DNA–poly-L-lysine COMPLEXES

DNA–poly-L-lysine complexes deliver DNA into cells by taking advantage of receptor-mediated endocy-

tosis. One particular receptor that has been useful for the delivery of DNA–poly-L-lysine complexes is the asialoglycoprotein receptor. The asialoglycoprotein receptor is found exclusively on the surface of hepatocytes [30]. Wu and Wu [31] covalently coupled galactose-terminal (asialo-) glycoprotein, asialoorosomucoid (AsOR), to poly-L-lysine. The conjugate was then complexed in a 2 : 1 molar ratio to the plasmid, pSV2CAT, containing the gene for the bacterial enzyme chloramphenicol acetyltransferase (CAT). When the DNA carrier system was labeled with ^{32}P and injected intravenously into rats, it was determined that 85% of the DNA–polylysine–AsOR was selectively taken up the liver. Using this gene delivery system, Wu and colleagues [32] attempted to correct a genetic metabolic disorder in the Nagase analbuminemic rat. This strain has virtually undetectable levels of circulating serum albumin as a result of a splicing defect in serum albumin mRNA. A plasmid containing the gene for human serum albumin was complexed with the asialoorosomucoid–poly-L-lysine conjugate and used to target hepatocytes in this model. Two weeks postinjection, human serum albumin was detected at levels of 34 μg/mL for up to 4 weeks. Clearly, the transient nature of this response is representative of one of the major hurdles facing the field.

The mannose receptor has been an effective target for receptor-mediated gene transfer. This receptor recognizes glycoproteins with mannose, glucose, fucose, and *N*-acetylglucosamine residues in exposed, nonreducing positions and is expressed on the surface of tissue macrophages [33,34]. A molecular conjugate, consisting of mannosylated polylysine complexed with an expression plasmid containing the *Photinus pyralis* luciferase receptor gene, was used to transfect macrophages in the liver and spleen of adult mice [35]. Luciferase activity was detected for up to 16 days posttransfection.

In order to improve the efficiency of uptake of the DNA/ligand–poly-L-lysine complexes, Perales and colleagues [36] devised a method for producing unimolecular complexes (i.e., complexes that contain a single molecule of DNA). By making the DNA/ligand–poly-L-lysine complexes smaller, it was felt that there would be a greater uptake by endocytic pathways. Briefly, a plasmid containing the human factor IX gene was condensed with galactosylated poly-L-lysine and titrated with NaCl to form complexes of defined size (10–12 nm in diameter) and shape. The molecular conjugate was injected into adult rats and was found to specifically target to the liver. Treatment with these unimolecular complexes led to prolonged expression of human factor IX, which could be detected up to 140 days after administration.

One of the problems associated with receptor-mediated endocytosis is that endocytosed DNA is trapped in intracellular vesicles and largely destroyed by lysosomal action. This was highlighted in a report by Cristiano and colleagues [37], who found that only 0.1% of hepatocytes were transfected by a *Escherichia coli* β-galactosidase gene condensed with AsOR–poly-L-lysine. One solution to this problem has been to exploit the properties of adenoviruses, which disrupt endosomes, allowing DNA to enter the cytoplasm. Incubating the replication-defective adenovirus with the AsOR–β-galactosidase gene/poly-L-lysine complex led to 100% of the hepatocytes being transfected and a 1,000-fold enhancement of β-galactosidase activity. The adenovirus was either covalently or noncovalently coupled to the ligand–DNA/poly-L-lysine complex. However, as the adenovirus has its own ligands for the binding of cellular receptors, this often leads to the loss of cell specific targeting by the ternary complex [38]. Michael and colleagues [39] were able to block the binding of serotype 5 adenovirus using mAb specific for the fiber protein and hence restore specific targeting.

In order to avoid using whole virus in the ternary complex, investigators have identified the part of the virus that is responsible for the disruption of endosomes. The 20 N-terminal residues of the influenza virus hemagglutinin subunit HA-2 were synthesized by Wagner and colleagues [40], and chemically coupled to poly-L-lysine. At a peptide/poly-L-lysine molar ratio of 8:1, the peptide–transferrin–DNA/poly-L-lysine complex produced up to a 1,000-fold increase in luciferase gene expression in the murine hepatocytic cell line BNL CL.2 compared to transferrin–DNA/poly-L-lysine complex alone. A dimeric derivative of the N-terminal peptide was found by Plank and colleagues [41] to lead to a 5,000-fold increase in luciferase gene expression in BNL CL.2 cells. Zauner and colleagues [42] observed a similar endosome-disruptive capacity in peptides derived from human rhinovirus serotype 2 (HRV2). At 24-amino-acid-long peptide derived from the N-terminus of VP1 of HRV2 was ionically bound to poly-L-lysine and conjugated to a transferrin–DNA/poly-L-lysine complex. When 150 micrograms of peptide was used in the ternary complex, there was a 500-fold increase in the activity of luciferase within the transfected NIH 3T3 cells.

2. ANTIBODIES AND DNA–poly-L-lysine COMPLEXES

Ligand-directed DNA–poly-L-lysine complexes provide a limited amount of flexibility in targeting the DNA–poly-L-lysine complexes to specific cancer cell types. For example, the transferrin receptor can be found on any cell, so transferrin is of little use in the targeting of particular neoplastic cells. For many cell types, no cell-specific ligand is known. Thus, the coupling of antibodies to DNA–poly-L-lysine complexes could provide a means of selectively targeting a wider range of cell types. This was demonstrated with the chimeric mouse–human antibody chCE7, which is specific for the CEA antigen. This antibody was covalently linked to poly-L-lysine, and the conjugate was condensed with a plasmid-encoding gamma interferon and transfected into neuroblastoma cells in the presence of chloroquine [43]. Like the previously discussed adenovirus and viral peptides, chloroquine will also lyse endosomes. Successful transfection was indicated by HLA-ABC expression on the neuroblastoma cells induced by the gamma interferon, thus enabling activation of autologous cytotoxic T lymphocytes *in vitro*.

Antibodies also have been used in the receptor-mediated transfer of genes into normal cells. The airway epithelial cells of rats have been the targets of receptor-mediated endocytosis using the polymeric immunoglobulin receptor (pIgR) [44]. Fab fragments of polyclonal antibodies, raised against the rat secretory component (SC) of the extracellular portion of pIgR,, were covalently-linked to poly-L-lysine. The anti-SC Fab–poly-L-lysine was condensed with plasmid pGL2 containing the luciferase gene and injected into the caudal vena cava of a rat. Luciferase enzyme activity in protein extracts from the liver and lung produced maximum values of approximately 14,000 and 350,000 integrated light units (ILU) per milligram of protein extract, respectively. As the lung and liver express pIgR, this demonstrated tissue-specific delivery of the luciferase gene by the anti-SC Fab–poly-L-lysine/DNA complex. Using an expression plasmid encoding the β-galactosidase gene, it was determined that the genes were being transferred to the surface epithelium of the airways and the submucosal glands. A follow-up study by Ferkol and colleagues [45] looked at the effect of multiple injections of anti-SC Fab-poly-L-lysine/DNA complex within the lungs of mice. Animals that received one injection of ternary complex exhibited luciferase activity of approximately 17,000 ILU/mg, whereas those that received three injections of ternary complex produced luciferase activity of about 3,800 ILU/mg. It was found that in the mice that had received three injections, a humoral immune response was mounted against the rabbit anti-SC Fab, which resulted in decreased luciferase gene transfer. Thus, the immune response may reduce the efficiency of gene transfer when antibody–DNA/poly-L-lysine complexes are injected more than once.

Human T lymphocytes have also been transfected by receptor-mediated gene transfer using anti-CD3 antibodies [46]. Biotinylated adenovirus d1312 and streptavidin–poly-L-lysine were used to enhance the endosomal release of the anti-CD3 mAb–poly-L-lysine/DNA complex. Following prestimulation with IL-2 and phytohemagglutinin, primary peripheral blood lymphocytes (>95% CD2+ cells) were treated with the ternary complex, which resulted in 5% of the cells expressing the β-galactosidase gene. It was found that up to 50% of Jurkat E6 cells (human acute T-cell leukemia line) were transfected when the synthetic influenza peptide IFN5 [41] was used as endosome-disruptive agent. Thus, like the neuroblastoma and airway epithelial cells, human T lymphocytes can be selectively targeted by antibodies delivering transgenes.

3. Antibodies and DNA–GAL4 Complexes

A novel approach was adopted by Fominaya and Wels [47] in the design of a nonviral gene delivery system. Instead of employing the endosome-disruptive function as a separate component of ternary complex, Fominaya and Wels [47] incorporated the *Pseudomonas* exotoxin A translocation domain in a chimeric multidomain protein. Target cell specificity of the chimeric protein was conferred by the FRP5 anti-HER2/*neu* scFv. The second unique feature of the nonviral gene delivery system was the inclusion of a DNA binding domain derived from the yeast GAL4 protein in the chimeric protein. A plasmid containing the luciferase receptor gene was designed with the GAL4-specific recognition sequence 5′-CGGN$_3$(T/A)N$_5$CCG-3′ [48]. In order to use the complex for the transfection of HER2/*neu* positive cells, the excess negative charge of the DNA was neutralized with poly-L-lysine. Transfection of HER2/*neu*-positive COS-1 cells with 240 ng of chimeric protein and 4 μg of pSV2G4LUC luciferase reporter plasmid DNA resulted in luciferase activity of nearly 5×10^6 relative light units/mg protein. This indicated the potential of a multidomain protein for the cell-specific delivery of genes.

4. Immunoliposome–DNA Complexes

An alternative method for targeted gene delivery includes liposomes (self-assembling colloidal particles in which a lipid bilayer encapsulates a fraction of the surrounding aqueous medium) that have been complexed with plasmids to deliver selected genes to targeted tumor cells [49]. The advantages of liposome–DNA complexes include a lack of immunogenicity, increased stability of the gene and minimal systemic toxicity *in vivo* [50,51]. However, cationic liposomes in liposome–DNA complexes tend to be nonspecific in their targeting due to their ability to directly fuse to cell surface membranes. Antibodies provide a means of specifically directing liposome–DNA complexes to target neoplastic cells. Immunoliposomes (liposomes complexed with antibodies) have demonstrated specific

targeting to HER2/*neu*-overexpressing breast cancer cells using the humanized mouse anti-HER2/*neu* antibody 4D5 (rhuMAbHER2) [52]. Nonspecific uptake of anti-HER2/*neu* immunoliposomes by negative control breast cancer cells was less than 0.2% of the uptake of HER2/*neu*-overexpressing breast cancer cells. However, one disadvantage associated with the use of immunoliposomes is that they are rapidly removed by cells of the reticuloendothelial system *in vivo*. As this is thought to be possibly mediated by Fc or C3b receptors interacting with the constant regions of the antibody molecules [53], the use of antibody fragments should solve this problem. Sterically stabilized liposomes, which are synthesized by conjugation of poly(ethylene glycol) (PEG) to the liposome surface, have lower reticuloendothelial uptake [54–56]. The anti-HER2/*neu* immunoliposomes used by Kirpotin and colleagues [52] were sterically stabilized and had rhuMAbHER2 covalently linked to the termini of the PEG molecule.

Another means of coupling antibodies or antibody fragments to liposomes has been developed by de Kruif and colleagues [57]. By fusing bacterial lipoprotein (LPP) nucleotide sequences to the DNA encoding the scFv, the scFv can be fatty acylated and hence incorporated into liposomes. de Kruif and colleagues [57] lipid-tagged human anti-CD22 scFv and fused them to liposomes. The resulting anti-CD22 immunoliposomes bound specifically to and were internalized by CD22+ cell lines and CD22+ peripheral blood B lymphocytes.

B. The Role of Antibodies in Retroviral Gene Delivery

The largest number of cancer gene therapy studies have utilized retroviral vectors [58]. Retroviral vectors are highly efficient at attaching to and internalizing into cells. Thus, retroviral vectors represent an excellent means for gene transfer. One of the drawbacks of retroviral vectors in gene therapy is a lack of target specificity. As with DNA–poly-L-lysine complexes, antibodies may represent tools for guiding retroviral vectors to deliver certain genes to cells.

To accomplish this objective, retroviral vectors carrying the trangene of interest are constructed in packaging cell lines [59]. The packaging cell lines provide all the proteins required to assemble a retrovirus (e.g., the products of the *gag, pol,* and *env* genes). A vector containing a retroviral packaging signal bounded by long terminal repeats enables the transgene to be incorporated into a retrovirus. The binding specificity or tropism of the retrovirus is dictated by the product of the *env* gene. An ecotropic envelope glycoprotein restricts infection of the retrovirus to rodent cells, whereas an amphotropic envelope glycoprotein permits infection of most mammalian cells. The ecotropic envelope glycoprotein of the Moloney murine leukemia virus (MoMLV) can be made as a precursor (Pr80env). This molecule is proteolytically cleaved to yield the mature surface (SU) (gp70) and transmembrane (TM) (p15E) proteins [60]. On the virion surface, three SU proteins and three TM proteins associate to form a homotrimeric complex [61]. The tropism of the ecotropic envelope glycoprotein lies within the SU protein.

One of the first uses of antibodies in the cell-specific targeting of retroviral vectors involved a bispecific antibody complex [62]. Anti-gp70 monoclonal antibodies were biotinylated and allowed to bind to the surface of ecotropic murine retroviruses, and human HeLa cells were coated with biotinylated B.9.12.1 anti-human MHC class I antibodies. Using the ψ2 packaging cell line, a plasmid containing a neomycin resistance gene was inserted into the retroviruses. Streptavidin was then used to bridge the ectropic murine retroviruses to the MHC class I molecules on the HeLa cells. From 2×10^5 cells, 50–80 G418-resistant clones were isolated. When the anti-class-I antibodies were substituted for anti-class-II antibodies in the bispecific complexes, 100–350 G418-resistant clones were isolated. Although this approach of redirecting retroviruses was successful, the efficiency of infection was low and the system is impractical for *in vivo* applications.

The genetic manipulation of the ecotropic envelope glycoprotein was pursued as a means of changing the tropism of the retroviral vectors while maintaining a high level of infection. However, altering the ecotropic envelope glycoprotein without the loss of infectively has proven to be difficult. No viral titer was obtained when Schnierle and colleagues [63] attempted to infect SKBR-3 and MDA-MB-453 cells with MoMLV retroviruses expressing chimeric envelope proteins. In this study, the cloning of the FRP5 anti-HER2/*neu* scFv between amino acids 6 and 7 of the MoMLV SU protein had completely ablated the infectivity of the SU protein (Table 1). FACS analysis of the CB25 packaging cells revealed that the chimeric envelope proteins had not been inserted into the cell membranes and therefore were unavailable for incorporation into the retroviral particles.

Marin and colleagues [64] encountered similar difficulties, producing very low retroviral titers from human TE671 cells when the B.9.12.1 anti-human MHC Class I scFv gene was inserted at position 6 of the MoMLV SU gene in the plasmid pMB34. The plasmid pMB34, which carries an antibiotic phleomycin selection marker, was transfected into the *env*-deficient packaging cell line TelCeb6 along with an nlsLACZ reporter-gene-carrying retroviral vector. Retroviruses derived from the TelCeb6 cells produced viral titers of 3–112 CFU/mL on human TE671 cells (Table 1). Interestingly, the eco-

TABLE 1　Antibody-Mediated Retroviral Infection of Nonmurine Cell Lines

Antibody fragment	Envelope protein fusion partner	Wild-Type envelope protein	Cell line	Retroviral titer (CFU/mL) [ref]
FRP5 anti-HER2 scFv	Residues 6+ of Moloney murine leukemia virus (MoMLV) SU protein		SKBR-3, MDA-MB-453 (both human)	0 [63]
OKT3 anti-CD3 scFv; A10, B3, C215 and F1 scFv	Residues 1+ or 7+ of MoMLV SU protein		Jurkat T cells, Colo 205 cells (both human)	0 [65]
B.9.12.1 anti-human MHC Class I scFv	Residues 6+ of MoMLV SU protein		TE671 (human)	3–112 [64]
Anti-DNP scFv	292–398 of Spleen necrosis virus (SNV) SU protein		CHO	30 [66]
	SNV TM proteins		CHO	20 [66]
B6.2 scFv (specific for antigen expression on human colon carcinoma cells)	Two thirds SNV SU protein		HeLa	10 [68]
	One third SNV SU protein		HeLa	5 [68]
	SNV TM protein		HeLa	20 [68]
	Two thirds SNV SU protein	+	HeLa	9×10^2 [68]
	One third SNV SU protein	+	HeLa	1.3×10^3 [68]
	SNV TM protein	+	HeLa	9.4×10^2 [68]
	SNV TM protein	+	HeLa	7×10^2 [69]
	SNV TM protein	(lower levels)	HeLa	7×10^3 [69]
Anti–low-density-lipoprotein receptor scFv	Residues 6+ MoMLV SU protein	+	HeLa	1×10^4 [71]

tropic retroviruses were still able to infect murine NIH 3T3 cells, having produced viral titers of 2×10^3 CFU/mL. This suggests that although the inserted scFv functions poorly in mediating the internalization of viral particles, the SU part of the chimeric envelope protein is still capable of recognizing its murine receptor and initiating infection.

In an attempt to improve the efficiency of infection, Ager and colleagues [65] fused the scFv to different positions of the SU envelope glycoprotein. In addition, linker sequences of different lengths were inserted between the scFv and the SU envelope glycoprotein. Despite these alterations, the ecotropic retroviruses expressing the chimeric envelope glycoprotein variants failed to infect human cells (Table 1). Nevertheless, Ager and colleagues [65] did demonstrate that varying the position of insertion of the scFv and different length linker sequences can have an effect on the functionality of the SU envelope glycoprotein as reflected by the infection of mouse cells. Five different scFv proteins were tested in the chimeric envelope glycoprotein construct. The first scFv (OKT3) was specific for T-cell surface marker CD3; the remaining scFv proteins (A10, B3, C215, and F1) were specific for distinct antigens expressed on colonic cancer cell lines. Four different constructs of the OKT3 scFv–SU fusion protein were made. The scFv proteins were fused to residues 1 and 7 of SU, employing either a noncleavable three-residue linker or the above linker with an additional factor Xa cleavage linker tetrapeptide. Eight different constructs of the anti–colonic cancer cell scFv–SU fusion proteins were made. Retroviral particles expressing scFv–SU chimeric constructs were harvested from TELCeB.6 packaging cells and were assayed for infectivity on murine NIH-3T3 fibroblasts. The investigators found a direct correlation between both the insertion position and the linker on the infectivity of the vectors. The greatest degree of infectivity was observed when the scFv was inserted at position +1 employing the heptapeptide linker. Thus, the insertion position in the SU envelope glycoprotein and the choice of linker can impact on the ability of a virus to infect cells.

Chu and colleagues [66] have adopted similar approaches to optimizing the cell-specific targeting of retroviral vectors, focusing on the avian spleen necrosis virus (SNV), which produces a 70-kDa SU and 20-kDa TM from a 90-kDa precursor (PR90env) [67]. Two chimeric constructs were made using an anti-DNP scFv. In the first construct, the anti-DNP scFv was fused to residues 292–398 of SU; in the second construct the anti-DNP scFv was fused directly to TM. scFv–SU$_{292-398}$ and scFv–TM chimeric envelope glycoproteins were expressed on the surface of retroviral particles harvested from the D17 dog osteosarcoma cell line. Upon the infection of DNP-conjugated CHO cells, the scFv–

SU$_{292-398}$ and scFV–TM chimeric envelope glycoproteins produced retroviral titers of 30 and 20 CFU/mL, respectively (Table 1). As the retroviral titers are low, the remaining parts of the retroviral envelope glycoprotein have little activity.

Using the spleen necrosis virus, Chu and Dornburg [68] were able to improve the efficiency of infection by coexpressing wild-type envelope glycoprotein with chimeric envelope glycoprotein in the homotrimeric complex on the viral surface. For cell-specific targeting, an scFv was used that recognizes an antigen expressed on human colon carcinoma cells. Constructs were designed with one third and two thirds, respectively, of the SU envelope glycoprotein removed and replaced with the B6.2 scFv. A third construct was made in which the scFv was directly fused to TM envelope glycoprotein. The constructs were cloned into the vectors pTC26 (one third SU removed), pTC24 (two thirds SU removed) and pTC25 (complete SU removal) and transfected into the packaging cell lines DSgp13 and DSH-cxl. Retroviruses produced by DSH-cxl cells expressed both wild-type and chimeric envelope glycoproteins, whereas retroviruses produced by DSgp13 cells expressed only chimeric envelope glycoproteins. The harvested viral particles were used to infect DLD-1 (human colon carcinoma cell line), HeLa, and HOS (human osteosarcoma cell line) cells. DSgp13-derived retroviruses expressing the three different constructs exhibited similar viral titers on HeLa cells (pTC26, 1×10^1; pTC24, 0.5×10^1; and pTC25, 2×10^1 CFU/mL) (Table 1). Thus, in terms of viral infectivity, the SU envelope glycoprotein had little impact. Compared to the DSgp13-derived retroviruses, the DSH-cxl-derived retroviruses expressing the three different constructs produced higher viral titers on HeLa cells (pTC26, 9×10^2; pTC24, 1.3×10^3; and pTC25, 9.4×10^2 CFU/mL (Table 1). The retroviral titers of the DLD-1 and HOS cells reflected the findings of the HeLa cells. From these increases in retroviral titer, it would appear that unmodified wild-type envelope glycoproteins have an important role to play in viral infectivity. Wild-type envelope glycoproteins may mediate their effects by allowing the viral membrane to efficiently fuse with the membrane of the target cell. Any changes in the envelope glycoprotein structure may nullify this effect.

Further investigations by Chu and Dornburg [69] reconfirmed that only a fully functional wild-type envelope glycoprotein can assist the chimeric envelope glycoprotein in efficient virus penetration. Moreover, the ratio of wild-type envelope glycoprotein to chimeric envelope glycoprotein in the viral membrane appeared to determine the efficiency of infection. These investigators were able to attain different levels of wild-type envelope glycoprotein by using SNV envelope glycoprotein-expressing constructs that either contained (pRD134) or lacked (pIM29) the adenovirus tripartite leader sequence. The adenovirus tripartite leader sequence enhances envelope expression in D17 cells about 10-fold [70]. The above plasmids were cotransfected with the plasmid pTC25 (described earlier) into DSH cells. When used to infect HeLa cells, retroviruses expressing pRD134-derived envelope glycoprotein produced viral titers that were 10-fold lower than those achieved with retroviruses expressing pIM29-derived envelope glycoprotein (7×10^2 and 7×10^3 CFU/mL, respectively) (Table 1). This demonstrated that lower levels of wild-type envelope glycoprotein allowed higher levels of infection (Table 1). To test the effects of nonfunctional wild-type envelope glycoprotein on infection, two mutant constructs were designed. The first construct (pTC12) contained two point mutations ($arg_{398} \rightarrow trp$, $ala_{399} \rightarrow pro$) at the SU/TM cleavage site; the second construct (pTC76) carried a point mutation ($asp_{192} \rightarrow arg$) in the middle of SU. The latter change reduced the efficiency of wild-type SNV infection by about 4,000-fold. These mutations lead to negligible or dramatically reduced retroviral titers on HeLa cells. Retroviruses that expressed the pTC12-derived envelope glycoproteins produced viral titers of <1 CFU/mL, whereas retroviruses that expressed the pTC76-derived envelope glycoproteins exhibited viral titers of 8×10^1 CFU/mL. Interestingly, when the retroviral stocks were concentrated 25-fold, a 200-fold rise in viral titers was observed, suggesting that the efficiency of infection increases at higher viral concentrations.

Wild-type envelope glycoproteins have also been coexpressed with chimeric envelope glycoproteins in MoMLV [71]. An anti–low-density-lipoprotein receptor scFv was inserted between residues 3 and 4 of the ecotropic SU envelope protein. $\psi2$ packaging cells were cotransfected with the plasmid pC7Env, which contains the gene for the scFv-envelope fusion protein, and a hygromycin B phosphotransferase expression vector. Hygromycin-B resistant clones were transfected with the gene for β-galactosidase. Retroviruses were produced with chimeric and wild-type enveloped glycoproteins on their surface and neomycin and β-galactosidase genes packaged inside. These viruses were harvested and used to infect HeLa cells, which resulted in a viral titer of 1×10^4 CFU/mL (Table 1). Thus, like Chu and Dornburg [68], Somia and colleagues [71] found the wild-type envelope glycoproteins help to stabilize that chimeric envelope glycoproteins and increase the efficiency of infection.

IV. FUTURE DIRECTIONS

Future advances are still required for efficient delivery of genes to targeted tumors *in vivo*. In the areas of endosome-mediated transfer and retroviral gene delivery, there is a need for the further development and refinement of the current strategies. Antibodies represent a means of achieving this end through their innate ability to bind specifically to a myriad of different antigens.

Of the different mechanisms of gene transfer, retroviral gene delivery holds the most potential in the treatment of cancer. However, as discussed earlier, attempts to redirect the retroviral envelope glycoproteins toward specific cell receptors have yielded poor viral infectivity. Reasonable infection of targeted cells has been achieved only in the presence of wild-type retroviral glycoproteins. As it has been difficult to modify the retroviral envelope glycoprotein without the loss of activity, a viable alternative would be to use an adapter molecule that contains both the retroviral receptor and a domain capable of specifically binding to the target cell population. Such an approach has been utilized by Ohno and colleagues [72]. A fusion protein containing the extracellular domain 3 of modified human H13 protein and transforming growth factor-α was used to cross-link an ecotropic AKR virus with the epidermal growth factor receptor (EGFR) expressed on the target cell surface. In principal, the modified human H13 protein would be equivalent to the murine ecotropic retrovirus receptor. As the adapter molecule contains the retroviral receptor, there is no need to modify the retroviral envelope glycoprotein. Thus, the high efficiency of viral transduction could be retained. However, in this study, the ability of the adaptor molecule to promote the infection of EGFR-positive human A431 cells by ecotropic AKR virus was not investigated.

One of the factors that has been overlooked during the development of cell-specific retroviral targeting is the choice of cell receptor. From a number of reports, it is apparent that cell receptors determine how many retroviruses can enter a cell. Some cell receptors are more efficient than others in allowing retroviral entry (Table 2). For example, MHC Class I molecules have permitted viral infection within target cells, albeit at low levels (Table 1) [62,64]. In contrast, no retroviral infection occurred when the transferrin receptor on Hep G2 cells was targeted by anti-transferrin receptor antibodies bridged to anti-gp70 antibodies [73]. The efficiency of viral infection could be optimized by selecting the right cell receptor. One of the problems that may occur with certain cell receptors is that the retrovirus may become trapped in the endosome following receptor internalization. The trapped retroviruses then are rapidly routed to the lysosomes for degradation. This was observed by Cosset [74] with ecotropic MLV expressing epidermal growth factor–SU fusion proteins. As would be expected, treatment with chloroquine reduced retroviral degradation in lysosomes and signifi-

TABLE 2   Ligand-Mediated Retroviral Infection of Human Cell Lines

Receptor	Targeting ligand	Cell line	Titer
High-density-lipoprotein receptor (HDL-R)	ApoA1-Protein A	HepG2	0^a
Galactose receptor	Biotinylated asialofetuin	HepG2	0^a
Insulin receptor	Biotinylated insulin	HeLa	8^a
Epidermal growth factor receptor (EGF-R)	Biotinylated EGF	A431	$8–23^a$
Erythropoietin receptor (EPO-R)	EPO	HEL	$\sim50^b$
HER3, HER4	Heregulin	SKBR-3	$<1^c$
		MDA-MB-453	$<1^c$
EGF-R	53 amino acids of EGF	A431	225^d
		TE671	46^d

[a] G418-resistant cells, [73].
[b] foci/mL [77].
[c] CFU/mL [63].
[d] CFU/mL [74].

cantly increased infection in human A431 and TF671 cells. It has been suggested by Weiss and Tailor [75] that in certain cell types the SU envelope glycoprotein of ecotropic MLV requires cleavage by pH-dependent cathepsins for endosomal escape. If these pH-dependent cathepsins were absent, then the ecotropic MLV would become trapped and degraded, as described earlier. This problem may be overcome by incorporating a translocation domain into the antibody–SU fusion proteins. The *Pseudomonas* exotoxin. A translocation domain could be ideally suited for this purpose [47].

Improvements can also be made to the antibody–DNA/poly-L-lysine complexes in order to improve the efficiency of gene transfer. One such improvement could include the utilization of nuclear localization signals, which could direct a higher degree of transcription of the transgene. Anti-HIV-1 Tat scFv proteins containing carboxy-terminal SV40 nuclear localization signals have been shown to direct heterologous proteins into the nucleus [76]. Similarly, SV40 nuclear localization signals could be fused to the C-termini of antibodies or antibody fragments directed against cell receptors.

Many of the barriers that antibodies have encountered in penetrating tumors will also have to be faced by gene delivery systems. These include limited vasculature, heterogeneous tumor antigen expression, and elevated interstitial pressure. For many gene delivery systems, these obstacles will be more difficult to overcome due to the far greater size of the DNA complexes and virus particles compared to antibodies. These issues will have to be addressed to exploit the high efficient gene transfer systems that can be anticipated in the near future.

References

1. Kohler, G., and Milstein, C. (1975). Continuous cultures of fused cells secreting antibody of predefined specificity. *Nature* **256**, 495–497.

2. Ghetie, M. A., Ghetie, V., and Vitetta, E. S. (1997). Immunotoxins for the treatment of B-cell lymphoma. *Mol. Med.* **3**, 420–427.

3. Caron, P. C., and Schienberg, D. A. (1993). Anti-CD33 monoclonal antibody M195 for the therapy of myeloid leukemia. *Leuk. Lymphoma* **11 (suppl 2)**, 1–6.

4. Kaminski, M. S., Zasadny, K. R., Francis, I. R., Milik, A. W., Ross, C. W., Moon, S. D., Crawford, S. M., Burgess, J. M., Petry, N. A., Butchko, G. M., Glenn, S. D., and Wahl, R. L. (1993). *N. Engl. J. Med.* **329**, 459–465.

5. Bodey, B., Siegel, S. E., and Kaiser, H. E. (1996). Human cancer detection and immunotherapy with conjugated and nonconjugated monoclonal antibodies. *Anticancer Res.* **16**, 661–674.

6. Jain, R. K., and Baxter, L. T. (1988). Mechanisms of heterogeneous distribution of monoclonal antibodies and other macromolecules in tumors: significance of elevated interstitial pressure. *Cancer Res.* **48**, 7022–7032.

7. Miller, R. A., Oseroff, A. R., Stratte, P. T., and Levy, R. (1983). Monoclonal antibody therapeutic trials in seven patients with T cell lymphoma. *Blood* **62**, 988–995.

8. Khazaeli, M. B., Conry, R. M., and LoBuglio, A. F. (1994). Human immune responses to monoclonal antibodies. *J. Immunother.* **15**, 42–52.

9. Huston, J. S., Levinson, D., Mudgett-Hunter, M., Tai, M., Novotny, J., Margolies, M. N., Ridge, R. J., Bruccoleri, R. E., Haber, E., Crea, R., and Oppermann, H. (1988). Protein engineering of antibody binding sites: recovery of specific activity in an anti-digoxin single-chain Fv analogue produced in *Escherichia coli*. *Proc. Natl. Acad. Sci. USA* **85**, 5879–5883.

10. Yokota, T., Milenic, D. E., Whitlow, M., and Scholm, J. (1992). Rapid tumor penetration of a single-chain Fv and comparison with other immunoglobulin forms. *Cancer Res.* **52**, 3402–3408.

11. Milenic, D. E., Yokota, T., Filupa, D. R., Finkelman, M. A. J., Dodd, S. W., Wood, J. F., Whitlow, M., Snoy, P., and Schlom, J. (1991). Construction, binding properties, metabolism, and tumor

targeting of a single-chain Fv derived from the pancarcinoma monoclonal antibody CC49. *Cancer Res.* **51**, 6363–6371.

12. Adams, G. P., McCartney, J. E., Tai, M., Oppermann, H., Huston, J. S., Stafford III, W. F., Bookman, M. A., Fand, I., Houston, L. L., and Weiner, L. M. (1993). Highly specific *in vivo* tumor targeting by monovalent and divalent forms of 741F8 anti-c-erB-2 single-chain Fv. *Cancer Res.* **53**, 4026–4034.

13. LoBuglio, A. F., and Saleh, M. N. (1992). Advances in monoclonal antibody therapy of cancer. *Am. J. Med. Sci.* **304**, 214–224.

14. Maloney, D. G., Liles, T. M., Czerwinski, D. K., Waldichuk, C., Rosenberg, J., Grillo-Lopez, A., and Levy, R. (1994). Phase I clinical trial using escalating single-dose infusion of chimeric anti-CD20 monoclonal antibody (IDEC C2B8) in patients with recurrent B-cell lymphoma. *Blood* **84**, 2457–2466.

15. Vadhan-Raj, S., Cordon-Cardo, C., Carswell, E., Mintzer, D., Dantis, L., Duteau, C., Templeton, M. A., Oettgen, H. F., Old, L. J., and Houghton, A. N. (1988). Phase I trial of a mouse monoclonal antibody against GD3 ganglioside in patients with melanoma: induction of inflammatory responses at tumor sites. *J. Clin. Oncol.* **6**, 1636–1648.

16. Cheung, N. V., Burch, L., Kushner, B. H., and Munn, D. H. (1991). Monoclonal antibody 3F8 can effect durable remissions in neuroblastoma patients refractory to chemotherapy: a phase II trial. *Prog. Clin. Biol. Res.* **366**, 395–400.

17. Handgretinger, R., Baader, P., Dopfer, R., Klingebiel, T., Reuland, P., Treuner, J., Reisfeld, R. A., and Neithammer, D. (1992). A phase I study of neuroblastoma with the anti-ganglioside GD2 antibody 14.G2a. *Cancer Immunol. Immunother.* **35**, 199–204.

18. Reithmüller, G., Schneider-Gädicke, E., Schlimok, G., Schmiegel, W., Raab, R., Höffken, K., Gruber, R., Pichlmaier, H., Hirche, H., Pichlmayr, R., Buggisch, P., and Witte, J. (1994). Randomised trial of monoclonal antibody for adjuvant therapy of resected Dukes' C colorectal carcinoma. *Lancet* **343**, 1172–1174.

19. Weiner, L. M., Clark, J. I., Davey, M., Li, W. S., Garcia de Palazzo, I., Ring, D. B., and Alpaugh, R. K. (1995). Phase I trial of 2B1, a bispecific monoclonal antibody targeting c-erbB-2 and Fc gamma RIII. *Cancer Res.* **55**, 4586–4593.

20. Valone, F. H., Kaufman, P. A., Guyre, P. M., Lewis, L. D., Memoli, V., Deo, Y., Graziano, R., Fisher, J. L., Meyer, L., Mrozek-Orlowski, M., Wardwell, K., Guyre, V., Morley, T. L., Arvizu, C., and Fanger, M. W. (1995). Phase Ia/Ib trial of bispecific antibody MDX-210 in patients with advanced breast or ovarian cancer that overexpresses the proto-oncogene HER-2/*neu*. *J. Clin. Oncol.* **13**, 2281–2292.

21. Jerne, N. K. (1974). Towards a network theory of the immune system. *Ann. Immunol.* **125**, 373–389.

22. Herlyn, D., Wettendorff, M., Schmoll, E., Iliopoulos, D., Schedel, I., Dreikhausen, U., Raab, R., Ross, A. H., Jaksche, H., Scriba, M., and Koprowski, H. (1987). Anti-idiotype immunization of cancer patients: modulation of the immune response. *Proc. Natl. Acad. Sci. USA* **84**, 8055–8059.

23. Mittelman, A., Chen, Z. J., Kageshita, T., Yang, H., Yamada, M., Baskind, P., Goldberg, N., Puccio, C., Ahmed, T., Arlin, Z., and Ferrone, S. (1990). Active specific immunotherapy in patients with melanoma. A clinical trial with mouse antiidiotypic monoclonal antibodies elicited with syngeneic anti-high-molecular-weight melanoma-associated antigen monoclonal antibodies. *J. Clin. Invest.* **86**, 2136–2144.

24. Foon, K. A., Chakraborty, M., John, W. J., Sherratt, A., Köhler, H., and Bhattacharya-Chatterjee, M. (1995). Immune response to the carcinoembryonic antigen in patients treated with an anti-idiotype antibody vaccine. *J. Clin. Invest.* **96**, 334–342.

25. Fitzgerald, D., and Pastan, I. (1989). Targeted toxin therapy for the treatment of cancer. *J. Natl. Cancer Inst.* **81**, 1455–1463.

26. Vitetta, E. S., Stone, M., Amlot, P., Fay, J., May, R., Till, M., Newman, J., Clark, P., Collins, R., Cunningham, D., Ghetie, V., Uhr, J. W., and Thorpe, P. E. (1991). Phase I immunotoxin trial in patients with B-cell lymphoma. *Cancer Res.* **51**, 4052–4058.

27. Trail, P. A., Willner, D., Lasch, S. J., Henderson, A. J., Hofstead, S., Casazza, A. M., Firestone, R. A., Hellström, I., and Hellström, K. E. (1993). Cure of xenografted human carcinomas by 96-doxorubicin immunoconjugates. *Science* **261**, 212–215.

28. Kaminski, M. S., Zasadny, K. R., Francis, I. R., Milik, A. W., Ross, C. W., Moon, S. D., Crawford, S. M., Burgess, J. M., Petry, N. A., Butchko, G. M., Glenn, S. D., and Wahl, R. L. (1993). Radioimmunotherapy of B-cell lymphoma with [131I]anti-B1(anti-CD20) antibody. *N. Engl. J. Med.* **329**, 459–465.

29. Press, O. W., Eary, J. F., Appelbaum, F. R., Martin, P. J., Badger, C. C., Nelp, W. B., Glenn, S., Butchko, G., Fisher, D., Porter, B., Matthews, D. C., Fisher, L. D., and Bernstein, I. D. (1993). Radiolabeled-antibody therapy of B-cell lymphoma with autologous bone marrow support. *N. Engl. J. Med.* **329**, 1219–1224.

30. Ashwell, G., and Morell, A. G. (1974). The role of surface carbohydrates in the hepatic recognition and transport of circulating glycoproteins. *Adv. Enzymol. Relat. Areas Mol. Biol.* **41**, 99–128.

31. Wu, G. Y., and Wu, C. H. (1988). Receptor-mediated gene delivery and expression *in vivo*. *J. Biol. Chem.* **263**, 14621–14624.

32. Wu, G. Y., Wilson, J. M., Shalaby, F., Grossman, M., Shafritz, D. A., and Wu, C. H. (1991). Receptor-mediated gene delivery *in vivo*. Partial correction of genetic analbuminemia in Nagase rats. *J. Biol. Chem.* **266**, 14338–14342.

33. Achord, D., Brot, F., and Sly, W. (1977). Inhibition of the rat clearance system for agalacto-orosomucoid by yeast mannans and by mannose. *Biochem. Biophys. Res. Commun.* **77**, 409–415.

34. Wileman, T., Boshans, R., and Stahl, P. (1985). Uptake and transport of mannosylated ligands by alveolar macrophages. Studies on ATP-dependent receptor-ligand dissociation. *J. Biol. Chem.* **260**, 7387–7393.

35. Ferkol, T., Perales, J. C., Mularo, F., and Hanson, R. W. (1996). Receptor-mediated gene transfer into macrophages. *Proc. Natl. Acad. Sci. USA* **93**, 101–105.

36. Perales, J. C., Ferkol, T., Beegen, H., Ratnoff, O. D., and Hanson, R. W. (1994). Gene transfer *in vivo*: sustained expression and regulation of genes introduction into the liver by receptor-targeted uptake. *Proc. Natl. Acad. Sci. USA* **91**, 4086–4090.

37. Cristiano, R. J., Smith, L. C., and Woo, S. L. C. (1993). Hepatic gene therapy: Adenovirus enhancement of receptor-mediated gene delivery and expression in primary hepatocytes. *Proc. Natl. Acad. Sci. USA* **90**, 2122–2126.

38. Wagner, E., Zatloukal, K., Cotten, M., Kirlappos, H., Mechtler, K., Curiel, D. T., and Birnsteil, M. L. (1992). Coupling of adenovirus to transferrin-polylysine/DNA complexes greatly enhances receptor-mediated gene delivery and expression of transfected genes. *Proc. Natl. Acad. Sci. USA* **89**, 6099–6103.

39. Michael, S. I., Huang, C., Rømer, M. U., Wagner, E., Hu, P., and Curiel, D. T. (1993). Binding-incompetent adenovirus facilitates molecular conjugate-mediated gene transfer by the receptor-mediated endocytosis pathway. *J. Biol. Chem.* **268**, 6866–6869.

40. Wagner, E., Plank, C., Zatloukal, K., Cotten, M., and Birnstiel, M. L. (1992). Influenza virus hemagglutinin HA-2 N-terminal fusogenic peptides augment gene transfer by transferrin-polylysine-DNA complexes: toward a synthetic virus-like gene transfer vehicle. *Proc. Natl. Acad. Sci. USA* **89**, 7934–7938.

41. Plank, C., Oberhauser, B., Mechtler, K., Koch, C., and Wagner, E. (1994). The influence of endosome-disruptive peptides on gene transfer using synthetic virus-like gene transfer systems. *J. Biol. Chem.* **269**, 12918–12924.

42. Zauner, W., Blaas, D., Kuechler, E., and Wagner, E. (1995). Rhinovirus-mediated endosomal release of transfection complexes. *J. Virol.* **69,** 1085–1092.

43. Coll, J.-L., Wagner, E., Combaret, V., Metchler, K., Amstutz, H., Iacono-Di-Cacito, I., Simon, N., and Favrot, M. C. (1997). *In vitro* targeting and specific transfection of human neuroblastoma cells by chCE7 antibody-mediated gene transfer. *Gene Ther.* **4,** 156–161.

44. Ferkol, T., Perales, J. C., Eckman, E., Kaetzel, C. S., Hanson, R. W., and Davis, P. B. (1995). Gene transfer into the airway epithelium of animals by targeting the polymeric immunoglobulin receptor. *J. Clin. Invest.* **95,** 493–502.

45. Ferkol, T., Pellicena-Palle, A., Eckman, E., Perales, J. C., Trzaska, T., Tosi, M., Redline, R., and Davis, P. B. (1996). Immunologic responses to gene transfer into mice via the polymeric immunoglobulin receptor. *Gene Ther.* **3,** 669–678.

46. Buschle, M., Cotten, M., Kirlappos, H., Mechtler, K., Schaffner, G., Zauner, W., Birnstiel, M. L., and Wagner, E. (1995). Receptor-mediated gene transfer into human T lymphocytes via binding of DNA/CD3 antibody particles to the CD3 T cell receptor complex. *Hum. Gene Ther.* **6,** 753–761.

47. Fominaya, J., and Wels, W. (1996). Target cell-specific DNA transfer mediated by a chimeric multidomain protein. Novel non-viral gene delivery system. *J. Biol. Chem.* **271,** 10560–10568.

48. Kodadek, T. (1993). How does the GAL4 transcription factor recognize the appropriate DNA binding sites in vivo? *Cell. Mol. Biol. Res.* **39,** 355–360.

49. Yoshida, J., and Mizuno, M. (1994). Simple method to prepare cationic multilamellar liposomes for efficient transfection of human interferon-β gene to human glioma cells. *J. Neurooncol.* **19,** 269–274.

50. Nabel, E. G., Gordon, D., Yang, Z. Y., Xu, L., San, H., Plautz, G. E., Wu, B., Gao, X., Huang, L., and Nabel, G. J. (1992). Gene transfer *in vivo* with DNA-liposome complexes: lack of autoimmunity and gonadal localization. *Hum. Gene Ther.* **3,** 649–656.

51. Stewart, M. J., Plautz, G. E., del Buono, L., Yang, Z. Y., Xu, L., Gao, X., Huang, L., Nabel, E. G., and Nabel, G. J. (1992). Gene transfer *in vivo* with DNA-liposome complexes: safety of autoimmunity and gonadal localization. *Hum. Gene Ther.* **3,** 267–275.

52. Kirpotin, D., Park, J. W., Hong, K., Zalipsky, S., Li, W., Carter, P., Benz, C. C., and Papahadjopoulos, D. (1997). Sterically stabilized anti-HER2 immunoliposomes: Design and targeting to human breast cancer cells *in vitro*. *Biochemistry* **36,** 66–75.

53. Debs, R. J., Heath, T. D., and Papahadjopoulos, D. (1987). Targeting of anti-Thy 1.1 monoclonal antibody conjugated liposomes in Thy 1.1 mice after intravenous administration. *Biochim. Biophys. Acta* **901,** 183–190.

54. Papahadjopoulos, D., Allen, T. M., Gabizon, A., Mayhew, E., Matthay, K., Huang, S. K., Lee, K.-D., Woodle, M. C., Lasic, D. D., and Redemann, C. (1991). Sterically stabilized liposomes: improvements in pharmacokinetics and antitumor therapeutic efficacy. *Proc. Natl. Acad. Sci. USA* **88,** 11460–11464.

55. Woodle, M. C., and Lasic, D. D. (1992). Sterically stabilized liposomes. *Biochim. Biophys. Acta* **1113,** 171–199.

56. Lasic, D., and Papahadjopoulos, D. (1995). Liposomes. *Science* **267,** 1275–1276.

57. de Kruif, J., Storm, G., van Bloois, L., and Logtenberg, T. (1996). Biosynthetically lipid-modified human scFv fragments from phage display libraries as targeting molecules for immunoliposomes. *FEBS Lett.* **399,** 232–236.

58. Cirielli, C., Capogrossi, M. C., and Passaniti, A. (1997). Antitumor gene therapy. *J. Neurooncol.* **31,** 217–223.

59. Vile, R. G., and Russell, S. J. (1995). Retroviruses as vectors. *Br. Med. Bull.* **51,** 12–30.

60. Hunter, E., and Swanstrom, D. (1990). Retrovirus envelope glycoproteins. *Curr. Top. Microbiol. Immunol.* **157,** 187–253.

61. Fass, D., Harrison, S. C., and Kim, P. S. (1996). Retrovirus envelope domain at 1.7 Å resolution. *Nature Struct. Biol.* **3,** 465–469.

62. Roux, P., Jeanteur, P., and Piechaczyk, M. (1989). A versatile and potentially general approach to the targeting of specific types by retroviruses: application to the infection of human cells by means of major histocompatibility complex class I and class II antigens by mouse ectropic murine leukemia virus-derived viruses. *Proc. Natl. Acad. Sci. USA* **86,** 9079–9083.

63. Schnierle, B. S., Moritz, D., Jeschke, M., and Groner, B. (1996). Expression of chimeric envelope proteins in helper cell lines and integration into Moloney murine leukemia virus particles. *Gene Ther.* **3,** 334–342.

64. Marin, M., Noël, D., Valseia-Wittman, S., Brockly, F., Etiennne-Julan, M., Russell, S., Cosset, F., and Piechaczyk, M. (1996). Targeted infection of human cells via major histocompatibility complex class I molecules by moloney murine leukemia virus-derived viruses displaying single-chain antibody fragment-envelope fusion proteins. *J. Virol.* **70,** 2957–2962.

65. Ager, S., Nilson, B. H. K., Morling, F. J., Peng, K. W., Cosset, F.-L., and Russell (1996). Retroviral display of antibody fragments: interdomain spacing strongly influences vector infectivity. *Hum. Gene Ther.* **7,** 2157–2164.

66. Chu, T. T., Martinez, I., Sheay, W. C., and Dornburg, R. (1994). Cell targeting with retroviral vector particles containing antibody-envelope fusion proteins. *Gene Ther.* **1,** 292–299.

67. Kewalramani, V. N., Panganiban, A. T., and Emerman, M. (1992). Spleen necrosis virus, an avian immunosuppressive retrovirus, shares a receptor with the type D simian retrovirus. *J. Virol.* **66,** 3026–3031.

68. Chu, T. T., and Dornburg, R. (1995). Retroviral vector particles displaying the antigen-binding site of an antibody enable cell-type specific gene transfer. *J. Virol.* **69,** 2659–2663.

69. Chu, T. T., and Dornburg, R. (1997). Toward highly efficient cell-type-specific gene transfer with retroviral vectors displaying single-chain antibodies. *J. Virology* **71,** 720–725.

70. Martinez, I., and Dornburg, R. (1995). Improved retroviral packaging lines derived from spleen necrosis virus. *Virology* **208,** 234–241.

71. Somia, N. V., Zoppé, M., and Verma, I. M. (1995). Generation of targeted retroviral vectors by using single-chain variable fragment: An approach to *in vivo* gene delivery. *Proc. Natl. Acad. Sci. USA* **92,** 7570–7574.

72. Ohno, K., Brown, G. D., and Meruelo, D. (1995). Cell targeting for gene therapy: Use of fusion protein containing the modified human receptor for ectropic murine leukemia virus. *Biochem. Mol. Med.* **56,** 172–175.

73. Etienne-Julan, M., Roux, P., Carillo, S., Jeanteur, P., and Piechaczyk, M. (1992). The efficiency of cell targeting by recombinant retroviruses depends on the nature of the receptor and the composition of the artificial cell-virus linker. *J. Gen. Virol.* **73,** 3251–3255.

74. Cosset, F., Morling, F. J., Takeuchi, Y., Weiss, R. A., Collins, M. K. L., and Russell, S. J. (1995). Retroviral retargeting by envelopes expressing an N-terminal binding domain. *J. Virol.* **69,** 6314–6322.

75. Weiss, R. A., and Tailor, C. S. (1995). Retrovirus receptors. *Cell* **82,** 531–533.

76. Yoneda, Y., Semba, T., Kaneda, Y., Noble, R. L., Matsuoka, Y., Kurihara, T., Okada, Y., and Imamoto, N. (1992). A long synthetic peptide containing a nuclear localization signal and its flanking sequences of SV40 T-antigen directs the transport of IgM into the nucleus efficiently. *Experimental Cell Research* **201,** 313–320.

77. Kasahara, N., Dozy, A. M., and Kan, Y. W. (1994). Tissue-specific targeting of retroviral vectors through ligand-receptor interactions. *Science* **266,** 1373–1376.

CHAPTER 8

The Use of Vaccinia Virus Vectors for Immunotherapy via *in Situ* Tumor Transfection

EDMUND C. LATTIME,[1],* LAURENCE C. EISENLOHR,[2] LEONARD G. GOMELLA,[3] AND MICHAEL J. MASTRANGELO[1]

[1]The Division of Medical Oncology, Department of Medicine, and Departments of [2]Microbiology and Immunology, and [3]Urology, Thomas Jefferson University, Philadelphia, Pennsylvania 19107

I. INTRODUCTION

The concept that the immune response may be manipulated so as to eliminate established neoplasms is an appealing one that has been under study for decades. This concept stemmed from the early suggestion of Lewis Thomas in 1959 that the immune response might be useful in ridding the body of aberrant cells [1] and was later refined into the immune surveillance hypothesis of Burnet in 1970, which in its simplest form hypothesized that the immune system would recognize incipient tumors as foreign and reject them, and that only those tumors that evaded this surveillance mechanism would persist and grow [2]. Whereas the broader interpretation of the immune surveillance hypothesis and its role in incipient tumors was disproved primarily in studies by Stutman, who showed that severely immunocompromised mice failed to display an increased incidence of most tumors [3], investigators continue to explore the role of induced antitumor immunity in

---

*Current address: The Cancer Institute of New Jersey and UMDNJ-RWJMS, New Brunswick, New Jersey 08901

mediating tumor regression. The immunogenicity of chemically induced tumors is supported by studies of Prehn and Main, who demonstrated in mice that immunization of a syngeneic (inbred mouse) with a given tumor protected against a subsequent challenge with the same, but not other tumors similarly derived [4]. Clinical support for the existence of tumor antigens is based on the reports that a number of human tumors, especially melanoma and renal cell, spontaneously regress presumably by the development of an antitumor immune response [5,6]. Based on these and other findings as well as the rapidly evolving understanding of basic immune regulation, studies by numerous investigators in preclinical and clinical settings have continued to focus on the harnessing of the immune response as a therapeutic for malignancy.

To date, investigators primarily have approached the immunotherapy of tumors in three ways. Local therapy with immune-active adjuvants has been shown to be highly effective in the case of localized tumors of the skin and bladder [7–9]. To the extent that such adjuvants lead to the generation of a systemic cell-mediated response via recruitment of immune effector cells or production of cytokines, these studies have led to the use of cytokine-gene-transfected tumor vaccines and our approach of *in situ* tumor transfection with cytokine genes. Second, studies from a number of investigators have focused on the generation of tumor vaccines. In their earliest manifestations, these included the use of whole tumor cells given either unmodified or following modification with viral antigens or haptens (reviewed in Mastrangelo *et al.* [10,11]). These vaccines, often given with adjuvants such as Bacille Calmette-Guerin (BCG) to provide an enhanced immune environment, are still used clinically. Success in inducing regression of clinically evident disease using these first-generation vaccines has been quite limited; however, in the adjuvant setting, where one would expect minimal residual disease, prolongation of disease-free survival has been reported using these first-generation approaches. More recently, these have given rise to studies of tumor extracts and, most recently, defined protein antigens and peptides [12] (see Chapter 3). There have as yet not been sufficient studies utilizing tumor-derived peptides for conclusions as to their efficacy to be made.

A third approach, which has led to numerous clinical trials in a variety of tumor types, involves the adoptive transfer of tumor reactive lymphocyte populations. This approach was made possible by the identification and characterization of cytokines such as interleukin 2 (IL-2), which allow the propagation of effector populations such as lymphokine activated killer (LAK) cells [13,14] and tumor-infiltrating T lymphocytes (TILs) [15–18]. Most recently, this approach has been modified to use cytokine-transfected tumor as a vaccine followed by removal and expansion of draining lymph node T cells, which are subsequently used adoptively. These studies are more fully described in Chapter 22.

The use of genetic means for traditional immunization strategies has centered on three approaches: First, molecular means have been used to produce protein antigens for immunization [19,20]. Second, viral vectors encoding tumor antigens have been used as immunogens (see Chapters 16 and 18). Third, naked DNA or plasmid vectors that give rise to tumor or allogeneic antigen expression have also been used (see Chapter 20).

II. GENERATION OF CELL-MEDIATED IMMUNE RESPONSES

Given the systemic or disseminated nature of most tumors, the goal of immunotherapeutic approaches must be the generation of a tumor-specific immune response in the host. Prior to discussing the current approaches to immunologically based gene therapy for cancer and our intralesional/intravesical approach, it would be helpful to discuss in some detail the cells and mechanisms involved in the generation of cell-mediated, and predominantly T-cell-mediated, immunity. For this discussion, we will focus almost exclusively on the cellular immune response with little discussion of the humoral (antibody)–based antitumor responses. Although a number of laboratories are focused on the generation of antitumor antibody responses [21] and monoclonal-antibody-based therapies continue to be studied [22] and reviewed [23,24], the preponderance of immunologically based gene therapy approaches are aimed at generating antitumor T-cell responses.

For a comprehensive discussion of immune mechanisms and targets for gene therapy see Chapter 3. A more restricted discussion follows here and will serve as a more specific background to our studies. Although there are certain nuances regarding tumors, it is important to point out that the underlying cells and mechanisms are the same whether one is generating an immune response to virus-infected cells, parasite-infected cells, or tumor cells.

Central to the generation of any T-cell-mediated response are (1) tumor antigens presented by antigen-presenting cells (APCs) in the context of major histocompatibility complex (MHC) antigens and costimulatory molecules such as B7 [25], (2) responder T cells with the appropriate T-cell receptor or recognition structures, and (3) resultant cytokine production required to both regulate the nature of the response and drive the expansion of the resultant effector T cells. In understanding the approaches to therapy to be discussed, it will be of value to highlight aspects of these three facets of the immune response with particular relevance to the generation of antitumor immunity.

A fundamental underlying requirement for the success of any form of immunotherapy is the expression of either unique or shared tumor antigens, which may include antigens that are shared with normal tissue but, by virtue of their overexpression or modified expression, may function as targets. Antigens currently under study include mutated or overexpressed oncogene products (see Chapters 16 and 17 and references [26–29]) as well as antigens also present on normal cells such as melanin-pigment-associated antigens in melanoma (see references [30–32] and Chapter 3). In the case of antigens shared with normal tissues, a logical concern is the generation of a concomitant autoimmune response. Although this concern may seem inconsequential when one is attempting to eradicate a life-threatening malignancy, autoimmune complications must be considered.

Targetable tumor antigens must be presented in an appropriate fashion, most efficiently by professional antigen-presenting cells such as dendritic cells and macrophages. In this light, it is important to note that responder T cells recognize antigen not in isolation but in the complex of antigenic peptides and self-MHC antigens (MHC Class II in the case of CD4+ helper cells, and MHC Class I in the case of CD8+ cytotoxic cells). Thus, expression of a shared tumor antigen among a number of patients may not elicit cross-reacting T-cell responses due to differences in MHC antigen expression. This MHC restriction phenomenon limits the application of immunotherapy strategies that utilize allogeneic tumor vaccines, that is, the use of tumor cells from one individual to immunize another. Although allogeneic tumors may share some antigens, and thus one might think that such an approach would be feasible, unless the donor and recipient are MHC matched, these shared antigens will not be recognized because they are expressed in the context of dissimilar MHC molecules. Based on this limitation, these restrictions favor approaches that utilize the patient's own tumor for immunization. For a more comprehensive review of MHC restriction, see Chapter 3 and Janeway *et al.* [33]. In addition to the MHC-antigen complex, optimal T-cell activation requires expression of costimulatory molecules, the most studied being B7, which is the ligand for CD28 on the responder T cell. In the absence of B7 expression, the antigen-presenting cell fails to stimulate a T-cell response and may even stimulate a state of tolerance or anergy to the expressed antigen (reviewed in references [34–36]).

Fundamental to stimulating an antitumor response is the need for T cells capable of recognizing the tumor antigen. In the case of antigens shared between normal and malignant tissues, one is faced with the necessity of overcoming or "breaking" tolerance to these self-antigens. A discussion of tolerance mechanisms can be found elsewhere (see Janeway *et al.* [33] and Chapter 3). Here it will suffice to point out that when one isolates lymphocytes from tumor masses, and in a some cases from peripheral blood, and stimulates them *in vitro* under the appropriate conditions, one is able to demonstrate the existence of T lymphocytes that recognize autologous tumor. Such antigen-reactive T cells have been identified in patients with a variety of tumor types [15–17,37,38] and have been shown to recognize tumor-specific antigens such as oncogene-encoded proteins [26–28] as well as antigens also found on normal cells such as the melanin-associated pigment antigens [30–32] alluded to earlier. Demonstrating a similar response *in vivo* has been harder to do and may be the result of tolerance and/or suppression mechanisms not found *in vitro*. In sum, however, it is clear that, based on *in vitro* analyses, tumor-reactive T cells are present in patients.

With regard to the requirement for the production of appropriate cytokine-based help for the development of an antitumor response, studies have focused on approaches that enhance proimmune cytokine production in the context of tumor, either adjuncts to vaccine strategies or, in the case of our studies, the *in situ* introduction of cytokine genes into tumors. In fact, the underlying hypothesis of the majority of ongoing gene therapy approaches is that by modulating the immune milieu at the local tumor or tumor vaccination sites using cytokine manipulation one may induce systemic antitumor immunity. Toward this end, multiple strategies are being employed, which include (1) the use of recombinant viral vectors encoding the genes for tumor-specific antigens plus immune-active cytokines (see Chapter 16), (2) immunization using vaccines that incorporate cytokine-gene-transfected tumor cells (see Chapter 21) or fibroblasts (see Chapter 23), and (3) as we have proposed, *in situ* introduction of cytokine genes or cytokine-producing cells at the tumor site *in vivo* using viral vectors. Studies have examined a panel of cytokines that target multiple regulatory mechanisms, including (1) the up-regulation of MHC and perhaps tumor-associated antigens [tumor necrosis factor (TNF), interferon gamma (IFN-γ)], (2) macrophage and antigen-presenting cell recruitment and activation with subsequent enhanced antigen processing and presentation [granulocyte macrophage colony stimulating factor (GM-CSF), IFN-γ], (3) cytokines that direct the T-cell helper response toward a TH$_1$ response and thus enhance the generation of delayed type hypersensitivity (DTH) and cytotoxic T-lymphocyte (CTL) responses (IFN-γ and IL-12) further described later, and (4) cytokines that drive the expansion of activated T cells (IL-2, IL-4).

Regarding the cytokine direction of cell-mediated responses, studies have shown that the profile of cytokines produced in an immune response correlates with and may direct the response to either a cellular response expressing DTH, termed TH_1, or a humoral response resulting in antibody production, termed TH_2. In studies of experimental and clinical parasite infections (leprosy and leishmania, respectively), the presence of a TH_1 response is associated with DTH and a good clinical course, whereas a TH_2 response is associated with an antibody response and a less favorable course [39,40]. Of particular interest therapeutically, TH_1 and TH_2 responses have been shown to cross-regulate one another with TH_1-associated cytokines (IFN-γ) inhibiting TH_2 immunity and TH_2-associated cytokines (IL-4, IL-10) suppressing the generation of cellular responses. For this reason, therapeutic strategies are being developed that maximize expression of TH_1 cytokines at the tumor site, which it is hoped will drive the antitumor responses toward the cell-mediated arm.

In addition to the cytokine regulatory networks in play during the generation of an immune response, the tumor itself can play a role in directing the nature of the ongoing response. In the first case, expression of tumor antigens in the context of MHC but in the absence of the costimulatory activity conferred by B7 expression may not only fail to stimulate a response but, as has been shown in a number of other systems, may actually confer tolerance or anergy [25,34–36,41,42]. A second characteristic of a number of tumors may drive the response toward TH_2, with resultant suppression of cell-mediated immunity (DTH and cytotoxic T-cell generation). Studies from our and other laboratories have reported the production of the cytokine IL-10 by a number of human [43–48] and murine [49,50] tumor types. Initial studies by Hersey et al. [45] found IL-10 production by human melanoma cell lines, and we and others subsequently found that biopsies from metastatic melanoma lesions expressed IL-10 mRNA and that tumor cell lines derived from these biopsies produced IL-10 protein [43,44]. Subsequently, we have also found IL-10 mRNA expression in biopsies of human transitional cell carcinoma of the bladder [46] and by bladder tumor cell lines [51]. The production of IL-10 by tumor and/or tumor-associated cells would be expected to drive any immune response toward TH_2 and operationally suppress cell-mediated immunity to the tumor. In fact, we have described just such an IL-10-driven inhibition of DTH to a tumor-associated antigen in our murine studies [52]. At least one mechanism of the IL-10 suppressive response may also include the down-regulation of the expression of the costimulatory antigen B7 and MHC antigen, thus tying together these two putative tumor-induced suppressive mechanisms [53].

In sum, the development of a productive cellular immune response toward tumor relies on the activation of a cascade of cells and cytokines. During the initiation of such a response, production of suppressive factors by tumor and/or immune cells can result in the inhibition of antitumor cellular immunity. Elucidation of these mechanisms has led a number of investigators including ourselves to put forth strategies for enhancing the generation of cellular responses directly or via overcoming suppressive influences. A discussion of a number of such strategies follows and is also presented by additional contributors to this volume.

III. CYTOKINE GENE TRANSFER STUDIES IN ANTITUMOR IMMUNITY

In an attempt to overcome a hypothesized lack of immune-stimulated cytokine production and/or further stimulate antitumor responses, a number of laboratories [54–60] have stably transfected murine and more recently human tumor cell lines with a variety of cytokine genes for use as vaccines. Inoculation of syngeneic mice with experimental tumors transfected with the genes for TNF [55], IL-2 [54,58], IFN-γ [61], IL-4 [62], and GM-CSF [59,60] has resulted in rejection of the injected tumor. In some cases, mice were shown to generate a measurable systemic antitumor response based on rejection of subsequent challenge with the nontransfected tumor [54,55,61–63]. In a limited number of cases, "vaccination" with such cells resulted in the elimination or reduced growth of preexisting tumor [58,59,62]. Although these studies have been less than overwhelming in their effects on existing tumors, they do show that localized cytokine/lymphokine production can enhance the generation of tumor-specific immunity. Most recently, this approach has been translated to clinical trials with positive immunologic findings (see Chapter 21).

In addition to systemic T-cell-dependent antitumor responses resulting from such treatment, local T-cell-independent antitumor activity has also been demonstrated [57]. IL-4-producing tumors and admixed normal tumor cells, when injected into T-lymphocyte-deficient nude (nu/nu) mice have been shown to regress as a result of a localized inflammatory response characterized by infiltration by eosinophils and macrophages [57]. Thus, although the goal of such genetic intervention is to optimize for the production of tumor-specific T-cell immunity, local antitumor effects may also contribute to a positive clinical outcome.

We would point out two limitations of the in vitro cytokine gene transfection approach to clinical translation. First, current cytokine studies have focused on in vitro transfection of tumor cell lines or fibroblasts prior to injection into mice or patients. Although this is a reasonable approach in preclinical studies, extension of this system to clinical trials will severely limit its availability. The requirement that autologous tumor, based on the need for proper antigen and MHC expression, be available, removed, transfected, cloned, and so on severely limits the number of suitable patients. A modified approach using a cytokine-transfected HLA-A2-matched allogeneic tumor vaccine is currently under study at a number of centers, but this approach would require a shared tumor antigen restricted at HLA-A2. Gene-transduced fibroblasts as a source of cytokine [64,65] have been injected with nontransduced tumor into mice with mixed results, and at least one preclinical study showed a marked advantage when the tumor itself produced the cytokine. A similar strategy of coinjecting tumor and cytokine-transfected autologous fibroblasts is currently in clinical trials by the Lotze group [64] (see Chapter 23).

The second possible limitation to vaccination using cytokine-gene-transfected autologous tumor cells is based on the need to reinject viable tumor. Preclinical studies would suggest that the transfected tumor will be rejected; however, loss of the cytokine gene or putative suicide genes could easily result in further tumor spread in the patient, a particular hazard in healthy patients with minimal tumor burden and in whom immunotherapeutic approaches have proven most efficacious. This concern has partially been addressed by studies demonstrating that irradiation prevents tumor growth while allowing continued cytokine production [59,60], but the status of tumor antigen presentation by such irradiated vaccines has yet to be demonstrated in the clinic.

In addition to the cytokine-gene-transfectant vaccines already discussed, three additional approaches have been studied, which have reached differing stages in their move toward clinical trials. With the goal of overcoming limitations in antigen presentation due to the lack of B7, the ligand for CD28, a number of investigators have shown in preclinical animal models that transfection of tumors with B7 and their subsequent use as vaccines enhance the generation of antitumor responses and subsequent protection against challenge with nontransfected tumor [35,36]. Additional support for the use of this approach in therapy comes from studies by Sue Ostrand-Rosenberg's group [66], which demonstrated in a murine sarcoma model that preexisting nonmodified tumor could be eradicated with vaccination using B7-transfected tumor.

IV. *IN SITU* CYTOKINE GENE TRANSFER TO ENHANCE ANTITUMOR IMMUNITY

As outlined earlier, murine tumors transfected with a variety of cytokine and accessory molecule genes have been shown to induce variable levels of antitumor immunity and in some cases tumor regression when injected into syngeneic mice with preexisting tumors. Although

these results show promise for the approach in developing immunotherapeutic modalities in man, limitations in the ability to harvest, transfect, and reinject a variety of human tumors on a patient-by-patient basis raise questions about its feasibility in a number of human tumors. For this reason, we have developed a strategy of directly inserting the desired cytokine gene into the tumor utilizing vaccinia virus recombinants. As shown in the schematic (Fig. 1), injection of the virus intralesionally or intravesically in the case of bladder cancer (1) would result in the infection of the tumor cells and subsequently (2) the production of cytokine mRNA and (3) the secretion of biologically active protein. Supported by the preclinical tumor transfection studies also described earlier, it is our hypothesis that production of proimmune cytokines locally at the tumor site in this way would enhance the generation of systemic tumor-specific immunity and resultant tumor destruction.

A. Vaccinia Virus Vectors

We have chosen vaccinia virus vectors for our studies for a number of reasons. Members of the *Poxviridae* family of viruses, including vaccinia, are unusual in that replication and transcription of the genome occurs in the cytosol of infected cells, with virally encoded polymerases driving these processes. Thus, recombination of viral DNA into the genome is not of concern with vaccinia, as it is with other vectors, particularly retroviruses. The infectious cycle is divided into three phases. Early phase genes, typically encoding proteins with en-

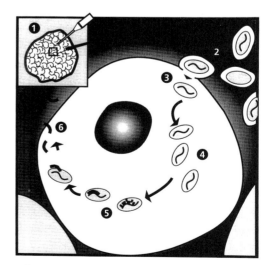

FIGURE 1 Initiating cytokine production around a tumor site. (1) Virus including gene encoding for cytokine is injected into tumor. (2) Virus binds with tumor cell. (3) Virus and gene enter cell. (4) Virus undergoes replication. (5) Cytokines are manufactured. (6) Cytokines are released, initiating a local cytokine-mediated host immune response.

zymatic function, are expressed prior to replication. The expression of a small number of intermediate genes depends on replication of the genome, and intermediate gene expression in turn drives expression of a large set of late genes, typically encoding structural proteins [67]. Generally speaking, "late" vaccinia promoters drive stronger gene expression than "early" vaccinia promoters. Some genes, such as that encoding the 7.5-kDa protein, have both early and late promoters, and are expressed during both phases of infection, with the late-promoter component accounting for approximately 75% of total expression [68].

Vaccinia was best known 20 years ago for its critical role in the eradication of smallpox (caused by the variola poxvirus) [69]. This points to one obvious advantage of vaccinia as a vector, its extensive use in man and characterization in the field. A second key characteristic is its stability. Vaccinia can be carried to remote regions and reconstituted from desiccated material to a highly infectious stock. Over time vaccinia has been recognized as a relatively safe agent with infrequent serious side effects, although it does induce a vigorous immune response and can be lethal for those who are immunocompromised or have eczema [70]. Although variola is highly contagious, transmission being mainly via the respiratory tract, vaccinia is much less so and can be easily confined under standard Biosafety Level 2 practices.

Today, as vaccination of the general population against smallpox has been discontinued, vaccinia is best known as a vector for transient expression of proteins. A clear advantage of vaccinia in this regard is its wide tropism. With variable efficiency vaccinia infects most mammal-derived permanent cell lines and any of the common laboratory animals including mice, rabbits and monkeys. Its large genome (approximately 200 kb) allows for the stable insertion of very large fragments of DNA (25 kb has been reported [71]) into the genome at a single site. This is well above the range of many other vectors. Such recombinants have been used for a wide range of studies including those concerned with folding and oligomerization of proteins [72], signal transduction [73], identifying human immunodeficiency virus coreceptor molecules [74,75], elucidating the antiviral effects of various cytokines [76], and eliciting protective immune responses against pathogens and transformed cells [77,78]. Vaccinia is also capable of rendering some cells very receptive to transfection, and "infection/plasmid transfection" protocols yield high levels of gene expression in a large percentage of cells, bypassing the need for generating a recombinant. In this case, the plasmid need be delivered only to the cytosol of the cell, not the nucleus, provided that the gene of interest can be transcribed in that location. This

has been accomplished by preceding the gene with a vaccinia-specific promoter [79] and also by utilizing the T7 and T3 bacteriophage promoters and infection by T7- or T3-expressing vaccinia recombinants [80,81]. Even if a true recombinant is needed, the procedure can be helpful in confirming the integrity of a construct prior to generation of the recombinant stock, which requires at least several weeks.

Generating a vaccinia recombinant is relatively straightforward. The gene of interest is inserted into a plasmid that minimally contains an origin of replication, an antibiotic resistance gene for cloning purposes, a vaccinia virus promoter to drive expression of the inserted gene, and segments of the vaccinia virus genome, flanking the promoter and inserted gene, to direct site-specific recombination. The $P_{7.5}$ promoter, active during early and late phases of infection, has most often been employed to drive heterologous gene expression. The most popular site of recombination is the viral thymidine kinase (TK) gene, which, as shown in Fig. 2, is disrupted by the recombination event. Recombination is achieved by infection of cells (often the African green monkey–derived CV-1 cells), with wild-type vaccinia stock, followed shortly thereafter by transfection of the infected cells with the recombination plasmid. The resultant vac-

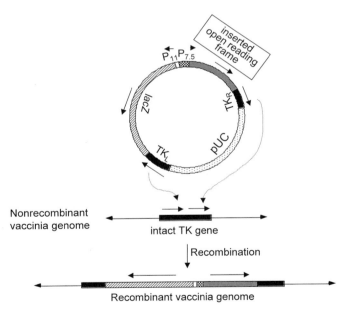

FIGURE 2   Schematic of a typical vaccinia recombination strategy. Shown at top is a recombination plasmid as developed by Chakrabarti et al. [107] featuring the fragmented vaccinia thymidine kinase gene (TK$_L$ and TK$_R$ flanking the *lacZ* and inserted open reading frame driven by the P_{11} (late) and $P_{7.5}$ (early/late) promoters). The pUC-derived segment of the plasmid provides the origin of replication and the β-lactamase gene to confer ampicillin resistance. The result of recombination into the vaccinia genome at the thymidine kinase (TK) gene is depicted at the bottom.

cinia stock will contain recombinant virus at low frequency, on the order of 0.1%, which is then selected and plaque purified in subsequent passages.

Early vaccinia recombination protocols suffered from the consequences of spontaneous inactivation of the TK locus. These TK-mutants cannot be distinguished from true recombinants without assaying for the recombinant gene or protein. The selection process is now straightforward because current vectors are usually designed to allow incorporation of a reporter gene, such as *lacZ*, driven by a second vaccinia promoter. Addition of the substrate X-gal during plaque purification causes β-galactosidase-producing plaques (true recombinants) to turn dark blue. Such plaques are then amplified and carefully titered prior to their use. It is critical that production of the gene of interest be confirmed prior to experimentation because incorporation of a reporter gene into the genome indicates recombination but certainly not integrity of the gene of interest. As mentioned, other loci have been targeted for recombination, and these necessarily require other means of selection, such as neomycin resistance [82]. On occasion a double recombinant is needed. One means of achieving this is by reinserting a TK gene into the vaccinia genome via recombination, as a reporter, and selecting under conditions that require a viable TK gene [83]. Considerable control of heterologous gene expression is possible with the vaccinia recombinant system. Not only can one select among the three types of promoters, expression driven by each can be varied by modification of the promoter sequences from levels that are barely detectable to those that are extraordinarily high [84–86]. Finally, it has been shown recently that, with sufficient care, heterologous DNA, also as large as 25 kb in length, can be directly ligated into the vaccinia genome, obviating the need for a recombination and the associated procedures [87].

Wild-type vaccinia infection is quite cytolytic and, as mentioned earlier, can be lethal in immunocompromised individuals. Much effort has been expended in generating less virulent poxvirus vectors. These include use of viruses such as Modified Vaccinia Ankara (MVA) or fowlpox virus, both of which replicate in avian cells and will infect but not complete replication in most mammalian cells [88–90]. This inhibition of replication can also be achieved with recombinants based upon wild-type vaccinia by treatment of purified virions with psoralen and UV light [91]. An alternative approach to reducing virulence has involved the systematic inactivation of nonessential genes recognized as contributing to virulence [90].

Vaccinia and poxviruses in general are clearly vectors of great utility, and continuing advancements will undoubtedly broaden their applicability. There are some situations for which poxviruses will probably never be suitable. Vaccinia expresses on the order of 100 different proteins, at least some of which stimulate a vigorous immune response, as anyone receiving the smallpox vaccine can attest. Thus, expression of the recombinant gene may be limited in individuals with prior exposure. It must be noted, particularly with respect to the subject of this chapter, that expression is not completely prevented, even following multiple inoculations with the same poxvirus. This could be due to vaccinia-specific antibodies interfering with release of newly assembled virions from the cell rather than with the attachment and entry phase [92].

B. Tumor Transfection by Vaccinia Recombinants

The overall hypothesis behind our studies is that by modulating the immune milieu at the local tumor site and thus recruiting antigen-presenting and effector cell populations, it will be possible to engender a systemic tumor-specific immune response. The result would be to eliminate both localized and disseminated tumor. We have pursued both preclinical and clinical studies to determine the feasibility of the use of recombinant vaccinia as a vector for *in situ* transfection, with the result that highly supportive data have been generated that enhance our enthusiasm for the approach.

Prior to developing recombinant vaccinia for our gene therapy studies, it was necessary to demonstrate that vaccinia virus recombinants are capable of transfecting murine and human tumor cells. A panel of cell lines including the murine melanoma B16, bladder tumors MBT2 and MB49 [93,94], and human melanoma lines produced from our patients [95,96], bladder (T24), and prostate carcinoma (LNCAP, PC3) [97] were examined for their ability to be infected/transfected with vaccinia recombinants. Cell lines were exposed *in vitro* to vaccinia virus recombinants containing the genes for influenza hemagglutinin and nuclear protein antigens termed reporter genes, which allow one to stain for productively infected/transfected cells. After 4.5 hours in culture, the cells were fixed and stained immunohistochemically for the antigens. Figure 3 (see color insert) shows representative murine (MBT2) and human (T24) bladder tumor cell lines expressing the reporter construct following *in vitro* infection. Similar results were obtained using all lines tested (data not shown), demonstrating that vaccinia recombinants have a wide tropism and supporting their use in a variety of human cancers.

To determine if recombinant vaccinia are able to infect/transfect tumor *in vivo*, vaccinia recombinants containing reporter constructs (HA, NP, or the *lacZ* gene) were injected intralesionally into murine B16 melanoma lesions or instilled via urethral catheters into the bladders of C57BL/6 mice bearing the MB49 tumor (see references [93,94,98] and our unpublished results). Eight hours following administration tumors were processed and stained for expression of the reporter genes as a measure of tumor transfection. As noted earlier, given the immunogenicity of vaccinia and the possibility that immunity to the virus would prevent infection/transfection following *in vivo* administration, mice were preimmunized to vaccinia prior to use in these studies, which would model the human state in which patients have received the smallpox vaccine. Figure 4A (see color insert) shows a representative bladder tumor, demonstrating expression of the virally encoded HA reporter genes and showing that the vaccinia recombinant is highly efficient in transducing the tumors following *in vivo* administration. Figure 4B (see color insert) shows expression of vaccinia-encoded *lacZ* activity in a model of B16 melanoma. Thus, systemic immunity to vaccinia, which would be expected to be present in adult patients and following initial vaccinia treatments, does not prevent *in vivo* tumor infection/transfection.

C. Intralesional Vaccinia in Patients with Melanoma

As a prelude to studying the effects of intralesional recombinant vaccinia in human melanoma, we obtained an Investigational New Drug Approval (IND) from the FDA to inject the Wyeth strain of vaccinia (the vaccine used in the United States for smallpox immunization and our nonrecombinant parent) intralesionally in patients with recurrent superficial melanoma [96,99]. Following the demonstration of systemic immunity to vaccinia via an intradermal administration of vaccine to the patients, increasing doses of vaccinia were injected intratumorally. In a representative patient, 10^6 pock forming units (pfu) of Wyeth vaccinia was injected into four sites in a 3-cm superficial melanoma lesion. The lesion was biopsied at 6 hours, and 4 days following administration, the biopsies were processed as frozen sections and stained using the monoclonal antibody TW2-3, which is specific for an early viral protein product of the EL3 gene present at sites of viral replication. Figure 5 (see color insert) shows a high-power field from the biopsy taken 6 hours following intralesional vaccinia containing diffuse numbers of melanoma cells staining intracytoplasmically for the viral antigen. To determine if increasing immunity to vaccinia induced by multiple treatments would block productive infection/transfection, additional biopsies were similarly analyzed over the course of therapy in one patient who received 19 biweekly injections of as high as 10^7 pfu of virus (total

cumulative dose of 13×10^7 pfu). Although the duration of expression was diminished with increasing immunity as measured by antiviral antibody titer, productive infection was seen throughout the treatment course [96,99]. It should be noted that minimal systemic side effects were seen in the trial. These findings demonstrate, as did our murine studies discussed earlier, that vaccinia recombinants are able to infect/transfect tumor *in vivo* following intralesional injection even in the face of systemic immunity to the virus. It is our conclusion from these studies that systemic immunity to the virus acts to protect the patients from toxicity while not preventing local gene expression. Our demonstration of sustained infection in virus-immune individuals strongly supports our approach in demonstrating that infection/transfection using cytokine-gene-encoding vaccinia should result in cytokine production for a prolonged period.

D. Intravesical Vaccinia in Patients with Transitional Cell Carcinoma of the Bladder

Following our demonstration that vaccinia could be safely injected into melanoma lesions, we proceeded to examine its use in human transitional cell carcinoma (TCC) of the bladder. We have previously shown using reporter constructs that intravesical instillation of vaccinia in mice resulted in significant infection/transfection of murine TCC growing in the bladder [94]. To translate these studies into humans, we performed a Phase I study of escalating doses of wild-type vaccinia instilled intravesically in patients with invasive TCC of the bladder prior to cystectomy [100]. After immune competence was assessed as was done in melanoma [96], patients received three escalating doses of vaccinia suspended in 50 mL of sterile saline. Cystectomies were performed on the day following the third dose. Figure 6 (see color insert) shows sections from one such patient demonstrating that vaccinia infected the epithelial cells of the bladder (A) as demonstrated by cell enlargement with nuclear and cytoplasmic vacuolization, with a resultant significant recruitment of inflammatory cells to the submucosa (B) made up of CD4, CD8, and dendritic cells (data not shown). Side effects were limited to slight dysuria, possibly due to the repeated catheterizations only. These studies demonstrate that vaccinia virus is an effective vector for the intravesical introduction of recombinant genes to bladder tumors and that it can be given safely in immune competent patients.

E. Cytokine Gene Delivery Using Vaccinia Recombinants

To determine if vaccinia recombinants could be used to transfect tumors with resultant cytokine production,

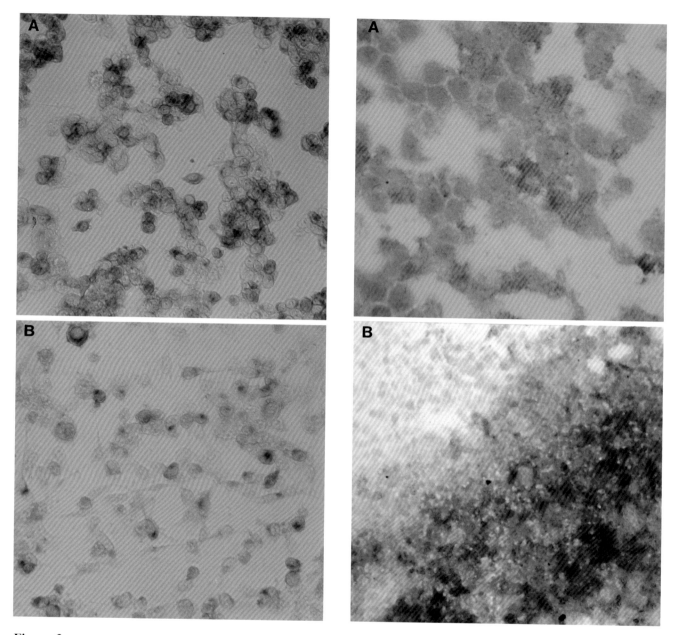

Figure 3 Recombinant vaccinia virus infects/transfects murine and human tumors *in vitro*. The murine bladder carcinoma MBT2 (A) and human TCC T24 (B) were infected *in vitro* with a vaccinia recombinant encoding for the reporter gene HA, and the cells were stained for expression of the HA antigen. Darkly stained cells represent infected/transfected cells.

Figure 4 Recombinant vaccinia virus infects/transfects murine tumors *in vivo*. The murine bladder tumor MB49 growing intravesically expresses the reporter contruct HA after intravesical administration of virus encoding HA (A). Similarly, the murine melanoma B16 expresses the reporter gene *lac-Z* following intratumor inoculation in mice (B).

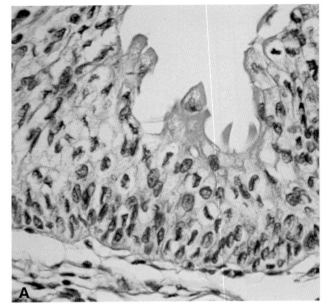

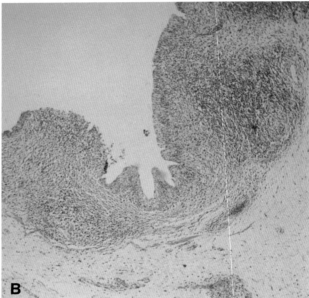

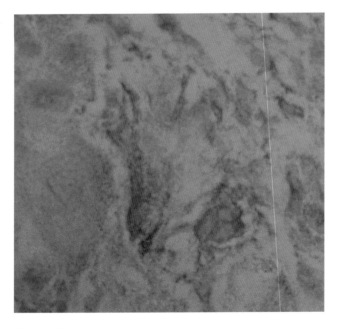

Figure 5 Intralesional injection of vaccinia virus infects human melanoma cells in a treated patient. Patient A.B. was injected intratumorally with vaccinia, and after 6 hours, the tumor was biopsied and frozen sections were processed and stained for the virus–associated protein EL3. Cells staining red were positive for the EL3 antigen and thus demonstrate successful infection/transfection.

Figure 6 Intravesical administration of vaccinia vector in patients with invasive bladder cancer: H&E-stained section of bladder from patient K.B. taken at time of cystectomy 24 hours after the third of three intravesical instillations of vaccinia showing infection of the urothelium (A) and submucosal inflammation (B).

we have established a panel of vaccinia recombinants expressing murine IL-4, IL-5, IFN-γ, and GM-CSF [98,101]. At the time at which these were being produced, a report by Ramshaw *et al.* demonstrated cytokine production by such recombinants and their positive effects on antiviral immunity [102]. In our studies, using methods described by the Moss group (reviewed earlier and in Mackett *et al.* [103]), we inserted the various cytokine genes into a plasmid containing sequences from the thymidine kinase gene behind the early and late P$_{7.5}$ vaccinia promotor (see earlier description of recombinant vaccinia production and Fig. 2). The plasmid was then incorporated into the vaccinia using homologous recombination into the thymidine kinase site, and recombinant virus was isolated, propagated, and tested for activity [98,101,103]. A schematic of the virus-generating protocol is presented in Fig. 2. ELISA and functional cytokine analyses have shown that the vaccinia recombinant-infected tumor cells produce significant levels of cytokine protein (Lee *et al.* [101] and data not shown). Subsequently, *in vivo* studies using vaccinia-specific primers designed in our laboratory that allow the elucidation of encoded cytokine mRNA *in vivo* have added to our earlier findings in demonstrating prolonged cytokine gene expression *in vivo* following intralesional injection (our results submitted for publication), and our preliminary studies demonstrate significant retardation of tumor growth in mice bearing the B16 melanoma treated with recombinant vaccinia expressing GM-CSF (data not shown). Parallel studies by two additional laboratories are consistent with our results demonstrating efficient *in vivo* tumor transfection by vaccinia recombinants in preclinical models [104,105].

V. FUTURE DIRECTIONS

We have demonstrated (1) that vaccinia virus can be effectively used to infect/transfect tumor cells *in vivo* in both preclinical melanoma and bladder systems, (2) that the vaccinia virus vector can be given safely and with continued infectivity in patients despite preexisting or developing immunity to vaccinia, and (3) that vaccinia recombinants expressing the genes for a panel of cytokines effectively induce infected cells to produce high levels of biologically active cytokines. Our ongoing studies are designed to introduce the use of cytokine-encoding vaccinia recombinants in patients with melanoma and bladder cancer. Toward this end, we have produced a human-grade GM-CSF encoding vaccinia recombinant and are studying its safety, pharmacology, and immunologic responses following intralesional injection in patients with superficial melanoma. Initial studies have shown that use of this recombinant is safe

and results in prolonged expression of the encoded GM-CSF molecule with resultant recruitment of immune populations [106]. These studies are continuing and will be followed by the examination of the same parameters in the treatment of bladder cancer patients following intravesical instillation.

VI. CONCLUSIONS

In summary, we have developed an approach to immunologically based gene therapy logically designed from the requirements to generate a productive cellular immune response. As outlined here, numerous strategies have been hypothesized and tested in both preclinical and clinical settings with this goal as an end point.

It is our hypothesis that *in situ* tumor transfection with cytokine genes will provide a logical extension of the vaccine strategies that have been previously studied. By incorporating genes selected based on their known contribution to the generation of systemic immune responses, we anticipate the ability to optimize the generation of an antitumor response. In addition to this logical *in vivo* vaccine design, this methodology will allow the generation of a single reagent in a bottle that will be of use in any tumor type that is accessible to injection. This will preclude the need to have sufficient autologous tumor for harvest and subsequent vaccine production and will overcome the significant limitation of the *in vitro* transfectants for tumor transfection and selection in the lab. As noted earlier, the use of the patient's own tumor as a source of antigens in our system optimizes the generation of a T-cell response and has significant advantages over allogeneic vaccine strategies, which rely on shared antigens restricted by common MHC antigens.

Acknowledgments

Supported by ACS Grants IM-742 and EDT-78842; USPHS Grants CA-42908, CA-55322, CA-69253, and CA-74543; and the Nat Pincus Trust.

References

1. Thomas, L., and Lawrence, H. S., eds. Cellular and Humoral Aspects of the Hypersensitive States, pp. 529–532. New York: Hoeber-Harper, 1959.
2. Burnet, F. M. The concept of immunological surveillance. *Prog. Exp. Tumor. Res.* **13**, 1–27, 1970.
3. Stutman, O. Immunodepression and malignancy. *Adv. Cancer Res.* **22**, 261–422, 1975.
4. Prehn, R. T., and Main, J. M. Immunity to methylcholanthrene-induced sarcomas. *J. Natl. Cancer Inst.* **18**, 769–778, 1957.

5. Bodurtha, A. J., Berkelhammer, J., Kim, Y. H., Laucius, J. F., and Mastrangelo, M. J. A clinical, histologic, and immunologic study of a case of metastatic malignant melanoma undergoing spontaneous remission. *Cancer* **37,** 735–742, 1976.

6. Spontaneous Remission: An Annotated Bibliography, p. 1. Sausalito, CA: Institute of Noetic Sciences, 1993.

7. Bornstein, R. S., Mastrangelo, M. J., Sulit, H., Chee, D., Yarbro, J. W., Prehn, L. M., and Prehn, R. T. Immunotherapy of melanoma with intralesional BCG. *Natl. Cancer Inst. Monogr.* **39,** 213–220, 1973.

8. Laucius, J. F., Bodurtha, A. J., Mastrangelo, M. J., and Creech, R. H. Bacillus Calmette–Guerin in the treatment of neoplastic disease. *J. Reticuloendothelial Soc.* **16,** 347–373, 1974.

9. Lamm, D. L., Thor, D. E., Harris, S. C., Reyna, J. A., Stogdill, V. D., and Radwin, H. M. Bacillus Calmette–Guerin immunotherapy of superficial bladder cancer. *J. Urol.* **124,** 38–42, 1980.

10. Mastrangelo, M. J., Maguire, H. C. J., Lattime, E. C., and Berd, D. Whole cell vaccines. In: V. T. DaVita, S. Hellman, and S. A. Rosenberg, eds., Biological Therapy of Cancer, pp. 648–658. Philadelphia: Lippincott, 1995.

11. Mastrangelo, M. J., Sato, T., Lattime, E. C., Maguire, H. C., Jr., and Berd, D. Cellular vaccine therapies for cancer. In: K. A. Foon, and H. B. Muss, eds., Biological and Hormonal Therapies of Cancer, pp. 35–50. Boston: Kluwer, 1998.

12. Hu, X., Chakraborty, N. G., Sporn, J. R., Kurtzman, S. H., Ergin, M. T., and Mukherji, B. Enhancement of cytolytic T lymphocyte precursor frequency in melanoma patients following immunization with MAGE-1 peptide loaded antigen presenting cell-based vaccine. *Cancer Res.* **56,** 2479–2483, 1996.

13. Rayner, A. A., Grimm, E. A., Lotze, M. T., Chu, E. W., and Rosenberg, S. A. Lymphokine-activated killer (LAK) cells. Analysis of factors relevant to the immunotherapy of human cancer. *Cancer* **55,** 1327–1333, 1985.

14. Grimm, E. A., and Wilson, D. J. The human lymphokine-activated killer cell system. *Cell. Immunol.* **94,** 568–578, 1985.

15. Li, W. Y., Lusheng, S., Kanbour, A., Herberman, R. B., and Whiteside, T. L. Lymphocytes infiltrating human ovarian tumors: Synergy between tumor necrosis factor α and interleukin 2 in the generation of CD8+ effectors from tumor-infiltrating lymphocytes. *Cancer Res.* **49,** 5979–5985, 1989.

16. Schoof, D. D., Jung, S.-E., and Eberlein, T. J. Human tumor-infiltrating lymphocyte (TIL) cytotoxicity facilitated by anti-T-cell receptor antibody. *Int. J. Cancer* **44,** 219–224, 1989.

17. Finke, J. H., Rayman, P., Alexander, J., Edinger, M., Tubbs, R. R., Connelly, R., Pontes, E., and Bukowski, R. Characterization of the cytolytic activity of CD4+ and CD8+ tumor-infiltrating lymphocytes in human renal cell carcinoma. *Cancer Res.* **50,** 2363–2370, 1990.

18. Rosenberg, S. A. Cell Transfer Therapy: Clinical Applications. In: V. T. DaVita, S. Hellman, and S. A. Rosenberg, eds., Biological Therapy of Cancer, pp. 487–506. Philadelphia: Lippincott, 1995.

19. Ciborowski, P., and Finn, O. J. Recombinant epithelial cell mucin (MUC-1) expressed in baculovirus resembles antigenically tumor associated mucin, target for immunotherapy. *Biomed. Pept. Proteins Nucleic Acids* **1,** 193–198, 1995.

20. Bei, R., Kantor, J., Kashmiri, S. V., Abrams, S., and Schlom, J. Enhanced immune responses and anti-tumor activity by baculovirus recombinant carcinoembryonic antigen (CEA) in mice primed with the recombinant vaccinia CEA. *J. Immunother. Emphasis Tumor Immunol.* **16,** 275–282, 1994.

21. Livingston, P. O., Natoli, E. J., Jones Calves, M., Stockert, E., Oettgen, H. F., and Old, L. J. Vaccines containing purified GM2 ganglioside elicit GM2 antibodies in melanoma patients. *Proc. Natl. Acad. Sci.* **84,** 2911–2915, 1987.

22. Maloney, D. G., Liles, T. M., Czerwinski, D. K., Waldichuk, C., Rosenberg, J., Grillo-Lopez, A., and Levy, R. Phase I clinical trial using escalating single-dose infusion of chimeric anti-CD20 monoclonal antibody (IDEC-C2B8) in patients with recurrent B-cell lymphoma. *Blood* **84,** 2457–2466, 1994.

23. Vitetta, E., and Ghetie, V. Immunotoxins in the therapy of cancer: From bench to clinic. *Pharmacol. Ther.* **63,** 209–234, 1994.

24. Pai, L. H., and Pastan, I. Immunotoxins and recombinant toxins for cancer treatment. *Important Adv. Oncol.* 3–19, 1994.

25. Schwartz, R. H. Costimulation of T lymphocytes: The role of CD28, CTLA-4, and B7/BB1 in interleukin-2 production and immunotherapy. *Cell* **71,** 1065–1068, 1992.

26. Disis, M. L., Smith, J. W., Murphy, A. E., Chen, W., and Cheever, M. A. *In vitro* generation of human cytolytic T-cells specific for peptides derived from the HER-2/neu protooncogene protein. *Cancer Res.* **54,** 1071–1076, 1995.

27. Peace, D. J., Smith, J. W., Chen, W., You, S. G., Cosand, W. L., Blake, J., and Cheever, M. A. Lysis of *ras* oncogene-transformed cells by specific cytotoxic T lymphocytes elicited by primary *in vitro* immunization with mutated Ras peptide. *J. Exp. Med.* **179,** 473–479, 1994.

28. Peoples, G. E., Goedegebuure, P. S., Smith, R., Linehan, D. C., Yoshino, I., and Eberlein, T. J. Breast and ovarian cancer-specific cytotoxic T lymphocytes recognize the same HER2/neu-derived peptide. *Proc. Natl. Acad. Sci.* **92,** 432–436, 1995.

29. Van Elsas, A., Nijman, H. W., Van der Minne, C. E., Mourer, J. S., Kast, W. M., Melief, C. J., and Schreir, P. I. Induction and characterization of cytotoxic T-lymphocytes recognizing a mutated p21ras peptide presented by HLA-A*0201. *Int. J. Cancer,* **61,** 389–396, 1995.

30. Storkus, W. J., Zeh, H. J., Maeurer, M. J., Salter, R. D., and Lotze, M. T. Identification of human melanoma peptides recognized by class I restricted tumor infiltrating T lymphocytes. *J. Immunol.* **151,** 3719–3727, 1993.

31. Slingluff, C. L., Hunt, D. F., and Engelhard, V. H. Direct analysis of tumor-associated peptide antigens. *Curr. Opin. Immunol.* **6,** 733–740, 1994.

32. Cox, A. L., Skipper, J., Chen, Y., Henderson, R. A., Darrow, T. L., Shabanowitz, J., Engelhard, V. H., Hunt, D. F., and Slingluff, C. L. Identification of a peptide recognized by five melanoma-specific human cytotoxic T cell lines. *Science* **264,** 716–719, 1994.

33. Janeway, C. A., and Travers, P. Immunobiology: The Immune System in Health and Disease. New York: Garland Publishing, 1995.

34. Linsley, P. S., and Ledbetter, J. A. The role of the CD28 receptor during T cell responses to antigen. *Annu. Rev. Immunol.* **11,** 191–212, 1993.

35. Lenschow, D. J., and Bluestone, J. A. T cell co-stimulation and *in vivo* tolerance. *Curr. Opin. Immunol.* **5,** 747–752, 1993.

36. Gimmi, C. D., Freeman, G. J., Gribben, J. G., Gray, G., and Nadler, L. M. Human T-cell clonal anergy is induced by antigen presentation in the absence of B7 costimulation. *Proc. Natl. Acad. Sci.* **90,** 6586–6590, 1993.

37. Balch, C. M., Riley, L. B., Bae, Y. J., Salmeron, M. A., Platsoucas, C. D., Von Eschenbach, A., and Itoh, K. Patterns of human tumor-infiltrating lymphocytes in 120 human cancers. *Arch. Surg.* **125,** 200–205, 1990.

38. Haas, G. P., Solomon, D., and Rosenberg, S. A. Tumor-infiltrating lymphocytes from nonrenal urological malignancies. *Cancer Immunol. Immunother.* **30,** 342–350, 1990.

39. Yamamura, M., Utemura, K., Deans, R. J., Weinberg, K., Rea, T. H., Bloom, B. R., and Modlin, R. L. Defining protective responses to pathogens: Cytokine profiles in leprosy lesions. *Science* **254**, 277–279, 1991.

40. Scott, P., Pearce, E., Cheever, A. W., Coffman, R. L., and Sher, A. Role of cytokines and CD4+ T-cell subsets in the regulation of parasite immunity and disease. *Immunol. Rev.* **112**, 161–182, 1989.

41. Townsend, S. E., and Allison, J. P. Tumor rejection after direct costimulation of CD8+ T cells by B7-transfected melanoma cells. *Science* **259**, 368–370, 1993.

42. Chen, L., Ashe, S., Brady, W. A., Hellstrom, I., Hellstrom, K. E., Ledbetter, J. A., McGowan, P., and Linsley, P. S. Costimulation of anti-tumor immunity by B7 counterreceptor for the T lymphocyte molecules CD28 and CTLA-4. *Cell* **71**, 1093–1102, 1992.

43. Lattime, E. C., Mastrangelo, M. J., Bagasra, O., Li, W., and Berd, D. Expression of cytokine mRNA in human melanoma tissues. *Cancer Immunol. Immunother.* **41**, 151–156, 1995.

44. Kruger-Krasagakes, S., Krasagakis, K., Garbe, C., Schmitt, E., Huls, C., Blankenstein, T., and Diamantstein, T. Expression of interleukin 10 in human melanoma. *Br. J. Cancer* **70**, 1182–1185, 1994.

45. Chen, Q., Daniel. V., Maher, D. W., and Hersey, P. Production of IL-10 by melanoma cells: Examination of its role in immunosuppression mediated by melanoma. *Int. J. Cancer* **56**, 755–760, 1994.

46. Lattime, E. C., McCue, P. A., Keeley, F. X., Li, W., and Gomella, L. G. Expression of IL10 mRNA in biopsies of superficial and invasive TCC of the human bladder [Abstract]. *Proc. Am. Assoc. Cancer Res.* **36**, 462, 1995.

47. Gastl, G. A., Abrams, J. S., Nanus, D. M., Oosterkamp, R., Silver, J., Liu, F., Chen, M., Albino, A. P., and Bander, N. H. Interleukin-10 production by human carcinoma cell lines and its relationship to interleukin-6 expression. *Int. J. Cancer* **55**, 96–101, 1993.

48. Pisa, P., Halapi, E., Pisa, E. K., Gerdin, E., Hising, C., Bucht, A., Gerdin, B., and Kiessling, R. Selective expression of interleukin 10, interferon γ, and granulocyte-macrophage colony-stimulating factor in ovarian cancer biopsies. *Proc. Natl. Acad. Sci.* **89**, 7708–7712, 1992.

49. McAveney, K. M., Gomella, L. G., and Lattime, E. C. Induction of TH1 and TH2 associated cytokine mRNA in mouse bladder following intravesical growth of the murine bladder tumor MB49 and BCG immunotherapy. Clin. Immunol. *Immunopathol.* **39**, 401–406, 1994.

50. Gorelik, L., Prokhorova, A., and Mokyr, M. B. Low-dose melphalan-induced shift in the production of a Th2-type cytokine to a Th1-type cytokine in mice bearing a large MOPC-315 tumor. *Clin. Immunol. Immunopathol.* **39**, 117–126, 1994.

51. Monken, C. E., Gomella, L. G., Li, W., Fink, E., and Lattime, E. C. IL10 is produced by human transitional cell carcinoma lines immortalized by retroviral transfection with human papilloma virus E6/E7 genes [Abstract]. *Proc. Am. Assoc. Cancer Res.* **37**, 451, 1996.

52. Maguire, H. C., Jr., Ketcha, K. A., Halak, B. K., Holmes, K. L., and Lattime, E. C. Tumor-induced IL10 production *in vivo* suppresses the development of delayed type hypersensitivity (DTH) to tumor associated antigens [Abstract]. *Proc. Am. Assoc. Cancer Res.* **38**, 358, 1997.

53. Ding, L., Linsley, P. S., Huang, L. Y., Germain, R. N., and Shevach, E. M. IL-10 inhibits macrophage costimulatory activity by selectively inhibiting the up-regulation of B7 expression. *J. Immunol.* **151**, 1224–1234, 1993.

54. Fearon, E. R., Pardoll, D. M., Itaya, T., Golumbek, P., Levitsky, H. I., Simons, J. W., Karasuyama, H., Vogelstein, B., and Frost, P. Interleukin-2 production by tumor cells bypasses T helper function in the generation of an antitumor response. *Cell* **60**, 397–403, 1990.

55. Asher, A. L., Mulé, J. J., Kasid, A., Restifo, N. P., Salo, J. C., Reichert, C. M., Jaffe, G., Fendly, B., Kriegler, M., and Rosenberg, S. A. Murine tumor cells transduced with the gene for tumor necrosis factor-α: Evidence for paracrine immune effects of tumor necrosis factor against tumors. *J. Immunol.* **146**, 3227–3234, 1991.

56. Watanabe, Y., Kuribayashi, K., Miyatake, J., Nishihara, K., Nakayama, E., Taniyama, T., and Sakata, T. Exogenous expression of mouse interferon-gamma cDNA in mouse neuroblastoma C1300 cells results in reduced tumorigenicity by augmented antitumor immunity. *Proc. Natl. Acad. Sci.* **86**, 9456–9460, 1989.

57. Tepper, R. I., Pattengale, P. K., and Leder, P. Murine interleukin-4 displays potent anti-tumor activity *in vivo*. *Cell*, **57**, 503–512, 1989.

58. Connor, J., Bannerji, R., Saito, S., Heston, W., Fair, W., and Gilboa, E. Regression of bladder tumors in mice treated with interleukin 2 gene-modified tumor cells. *J. Exp. Med.* **177**, 1127–1134, 1993.

59. Saito, S., Bannerji, R., Gansbacher, B., Rosenthal, F. M., Romanenko, P., Heston, W. D. W., Fair, W. R., and Gilboa, E. Immunotherapy of bladder cancer with cytokine gene-modified tumor vaccines. *Cancer Res.* **54**, 3516–3520, 1994.

60. Dranoff, G., Jaffee, E., Lazenby, A., Golumbek, P., Levitsky, H., Brose, K., Jackson, V., Hamada, H., Pardoll, D. M., and Mulligan, R. C. Vaccination with irradiated tumor cells engineered to secrete murine granulocyte-macrophage colony stimulating factor stimulates potent, specific, and long lasting anti-tumor immunity. *Proc. Natl. Acad. Sci.* **90**, 3539–3543, 1993.

61. Lattime, E. C., McCue, P. A., Ross, R. P., Baltish, M. A., and Gomella, L. T cells bearing >/δ receptor in human transitional cell carcinoma of the bladder [Abstract]. *Proc. Am. Assoc. Cancer Res.* **33**, 334, 1992.

62. Golumbek, P. T., Lazenby, A. J., Levitsky, H. I., Jaffee, L. M., Karasuyama, H., Baker, M., and Pardoll, D. M. Treatment of established renal cancer by tumor cells engineered to secrete interleukin-4. *Science* **254**, 713–716, 1991.

63. Perussia, B., Chan, S. H., D'Andrea, A., Tsuji, K., Santoli, D., Pospisil, M., Young, D., Wolf, S. F., and Trinchieri, G. Natural killer cell stimulatory factor or interleukin-12 has differential effects on the proliferation of TCRαβ+, TCRγδ+ T lymphocytes and NK cells. *J. Immunol.* **149**, 3495–3502, 1992.

64. Tahara, H., Zeh, H. J., Storkus, W. J., Pappo, I., Watkins, S. C., Gubler, U., Wolf, S. F., Robbins, P. D., and Lotze, M. T. Fibroblasts genetically engineered to secrete interleukin 12 can suppress tumor growth and induce antitumor immunity to a murine melanoma *in vivo*. *Cancer Res.* **54**, 182–189, 1994.

65. Lotze, M. T., Rubin, J. T., Carty, S., Edington, H., Ferson, P., Landreneau, R., Pippin, B., Posner, M., Rosenfelder, D., and Watson, C. Gene therapy of cancer: A pilot study of IL-4-gene-modified fibroblasts admixed with autologous tumor to elicit an immune response. *Hum. Gene Ther.* **5**, 41–55, 1994.

66. Basker, S., Glimcher, L., Nabavi, N., Jones, R. T., and Ostrand-Rosenberg, S. Major histocompatibility complex class II+ B7-1+ tumor cells are potent vaccines for stimulating tumor rejection in tumor bearing mice. *J. Exp. Med.* **181**, 619–629, 1995.

67. Baldick, C. J. J., Keck, J. G., and Moss, B. Mutational analysis of the core, spacer, and initiator regions of vaccinia virus intermediate-class promoters. *J. Virol.* **66**, 4710–4719, 1992.

68. Cochran, M. A., Puckett, C., and Moss, B. *In vitro* mutagenesis of the promoter region for a vaccinia virus gene: Evidence for tandem early and late regulatory signals. *J. Virol.* **54,** 30–37, 1985.

69. Fenner, F., Henderson, D. A., and Anita, I. Smallpox and Its Eradication. Geneva, Switzerland: World Health Organization, 1988.

70. Williams, N. R., and Cooper, B. M. Counselling of workers handling vaccinia virus. *Occup. Med.* **43,** 125–127, 1993.

71. Smith, G. L., and Moss, B. Infectious poxvirus vectors have capacity for at least 25,000 base pairs of foreign DNA. *Gene* **25,** 21–28, 1983.

72. Earl, P. L., Moss, B., and Doms, R. W. Folding, interaction with GRP78-BiP, assembly, and transport of the human immunodeficiency virus type 1 envelope protein. *J. Virol.* **65,** 2047–2055, 1991.

73. Scharenberg, A., Lin, S., Cuenod, B., Yamamura, H., and King, F. Reconstitution of interactions between tyrosine kinases and the high affinity IgE receptor which are controlled by receptor clustering. *EMBO J.* **14,** 3385–3394, 1995.

74. Feng, Y., Broder, C. C., Kennedy, P. E., and Berger, M. HIV-1 entry cofactor: Functional cDNA cloning of a seven transmembrane G protein-coupled receptor. *Science,* **272,** 872–877, 1996.

75. Doranz, B. J., Rucker, J., Yi, Y., Smyth, R. J., Samson, M., Peiper, S. C., Parmentier, M., Collman, R. G., and Domzig, W. A dual tropic primary HIV-1 isolate that uses fusin and the beta chemokine receptors CKR-5, CKR-3, and CKR-2b as fusion cofactors. *Cell* **85,** 1149–1158, 1996.

76. Kapuiah, G., Woodhams, C. E., Blanden, R. V., and Ramshaw, I. A. Immunobiology of infection with recombinant vaccinia virus encoding murine IL2. *J. Immunol.* **147,** 4327–4332, 1991.

77. Moss, B. Genetically engineered poxviruses for recombinant gene expression, vaccination, and safety. *Proc. Natl. Acad. Sci.* **93,** 11341–11348, 1996.

78. Restifo, N. P. The new vaccines: Building viruses that elicit antitumor immunity. *Curr. Opin. Immunol.* **8,** 658–663, 1996.

79. Cochran, M. S., Mackett, M., and Moss, B. Eukaryotic transient expression system dependent on transcription factors and regulatory DNA sequences of vaccinia virus. *Proc. Natl. Acad. Sci.* **82,** 19–23, 1985.

80. Fuerst, T. R., Niles, E. G., Studier, F. W., and Moss, B. Eukaryotic transient-expression system based on recombinant vaccinia virus that synthesizes bacteriophage T7 RNA polymerase. *Proc. Natl. Acad. Sci.* **83,** 8122–8126, 1986.

81. Rodriguez, D., Zhou, Y., Durbin, R. K., Jimenez, V., McAllister, W. T., and Esteban, M. Regulated expression of nuclear genes by T3 RNA polymerase and *lac* repressor using recombinant vaccinia virus vectors. *J. Virol.* **64,** 4851–4857, 1990.

82. Perkus, M. E., Limbach, K., and Paoletti, E. Cloning and expression of foreign genes in vaccinia virus using a host range selection system. *J. Virol.* **63,** 3829–3836, 1989.

83. Coupar, B. E. H., Andrew, M. E., and Boyle, D. B. A general method for the construction of recombinant vaccinia viruses expressing multiple foreign genes. *G.* **68,** 1–10, 1988.

84. Davison, A. J., and Moss, B. Structure of vaccinia virus early promoters. *J. Mol. Biol.* **210,** 749–769, 1989.

85. Davison, A. J., and Moss, B. Structure of vaccinia virus late promoters. *J. Mol. Biol.* **210,** 771–784, 1989.

86. Davison, A. J., and Moss, B. New vaccinia virus recombination plasmids incorporating a synthetic late promoter for high level expression of foreign proteins. *Nucleic Acids Res.* **18,** 4285–4286, 1990.

87. Merchlinsky, M., and Moss, B. Introduction of foreign DNA into the vaccinia virus genome by *in vitro* ligation: Recombination-independent selectable cloning vectors. *Virology* **190,** 522–526, 1992.

88. Sutter, G., and Moss, B. Nonreplicating vaccinia vector efficiently expresses recombinant genes. *Proc. Natl. Acad. Sci.* **89,** 10847–10851, 1992.

89. Carroll, M. W., and Moss, B. Host range and cytopathogenicity of the highly attenuated MVA strain of vaccinia virus: Propagation and generation of recombinant viruses in a nonhuman mammalian cell line. *Virology* **238,** 198–211, 1997.

90. Paoletti, E. Application of pox virus vectors to vaccination: An update. *Proc. Natl. Acad. Sci.* **93,** 11349–11353, 1996.

91. Tsung, K., Yim, J. H., Marti, W., Buller, M. L., and Norton, J. A. Gene expression and cytopathic effect of vaccinia virus inactivated by psoralen and long-wave UV light. *J. Virol.* **70,** 165–171, 1996.

92. Vanderplasschen, A., Hillinshead, M., and Smith, G. L. Antibodies against vaccinia virus do not neutralize extracellular enveloped virus but prevent virus release from infected cells and comet formation. *J. Gen. Virol.* **78,** 2041–2048, 1997.

93. Lee, S. S., Eisenlohr, L. C., McCue, P. A., Mastrangelo, M. J., and Lattime, E. C. Intravesical gene therapy: Vaccinia virus recombinants transfect murine bladder tumors and urothelium [Abstract]. *Proc. Am. Assoc. Cancer Res.* **34,** 337, 1993.

94. Lee, S. S., Eisenlohr, L. C., McCue, P. A., Mastrangelo, M. J., and Lattime, E. C. Intravesical gene therapy: *In vivo* gene transfer using vaccinia vectors. *Cancer Res.* **54,** 3325–3328, 1994.

95. Lattime, E. C., Maguire, H. C., Jr., McCue, P. A., Eisenlohr, L. C., Berd, D., Lee, S. S., and Mastrangelo, M. J. Infection of human melanoma cells by intratumoral vaccinia [Abstract]. *J. Invest. Dermatol.* **102,** 568, 1994.

96. Mastrangelo, M. J., Maguire, H. C., Jr., McCue, P. A., Lee, S. S., Alexander, A., Nazarian, L. N., Eisenlohr, L. C., Nathan, F. E., Berd, D., and Lattime, E. C. A pilot study demonstrating the feasibility of using intratumoral vaccinia injections as a vector for gene transfer. *Vaccine Res.* **4,** 55–69, 1995.

97. Gomella, L. G., Mastrangelo, M. J., Eisenlohr, L. C., McCue, P. A., Lee, S. S., and Lattime, E. C. Localized gene therapy for prostate cancer: Strategies for intraprostatic cytokine gene transfection using vaccinia virus vectors [Abstract]. *J. Urol.* **153,** 308A, 1995.

98. Lee, S. S., Eisenlohr, L. C., McCue, P. A., Mastrangelo, M. J., Fink, E., and Lattime, E. C. *In vivo* gene therapy of murine tumors using recombinant vaccinia virus encoding GM-CSF [Abstract]. *Proc. Am. Assoc. Cancer Res.* **36,** 248, 1995.

99. Lattime, E. C., Maguire, H. C. J., McCue, P. A., Eisenlohr, L. C., Berd, D., Lee, S. S., and Mastrangelo, M. J. Gene therapy using vaccinia vectors: Repeated intratumoral injections result in tumor infection in the presence of anti-vaccinia immunity [Abstract]. *Proc. Am. Soc. Clin. Oncol.* **13,** 397, 1994.

100. Gomella, L. G., Mastrangelo, M. J., Eisenlohr, L. C., McCue, P. A., Monken, C. E., Kovatich, A. J., Mulkolland, S. G., Maguire, H. C., Jr., and Lattime, E. C. Phase I study of intravesical vaccinia virus as a vector for gene therapy of bladder cancer [Abstract]. *Proc. Am. Assoc. Cancer Res.* **38,** 9, 1997.

101. Lee, S. S., Eisenlohr, L. C., McCue, P. A., Mastrangelo, M. J., and Lattime, E. C. Vaccinia virus vector mediated cytokine gene transfer for *in vivo* tumor immunotherapy [Abstract]. *Proc. Am. Assoc. Cancer Res.* **35,** 514, 1994.

102. Ramshaw, I., Ruby, J., Ramsay, A., Ada, G., and Karupiah, G. Expression of cytokines by recombinant vaccinia viruses: A model for studying cytokines in virus infections *in vivo. Immunol. Rev.* **127,** 157–182, 1992.

103. Mackett, M., Smith, G. L., and Moss, B. General method for production and selection of infectious vaccinia virus recombinants expressing foreign genes. *J. Virol.* **49,** 857–864, 1984.

104. Elkins, K. L., Ennist, D. L., Winegar, R. K., and Weir, J. P. *In vivo* delivery of interleukin-4 by a recombinant vaccinia prevents tumor development in mice. *Hum. Gene Ther.* **5,** 809–820, 1994.

105. Whitman, E. D., Tsung, K., Paxson, J., and Norton, J. *In vitro* and *in vivo* kinetics of recombinant vaccinia virus cancer-gene therapy. *Surgery,* **116,** 183–188, 1994.

106. Lattime, E. C., Maguire, H. C., Jr., Eisenlohr, L. C., Monken, C. E., McCue, P. A., Kovatich, A. J., and Mastrangelo, M. J. Gene therapy using intratumoral (IT) recombinant GM-CSF encoding vaccinia virus in patients with melanoma [Abstract]. *Proc. Am. Assoc. Cancer Res.* **38,** 11, 1997.

107. Chakrabarti, S., Brechling, K., and Moss, B. Vaccinia virus expression vector: Coexpression of β-galactosidase provides visual screening of recombinant virus plaques. *Mol. Cell Biol.* **5,** 3403–3409, 1985.

CHAPTER 9

Ribozymes in Cancer Gene Therapy

CARMELA BEGER,[1] MARTIN KRÜGER,[1] AND FLOSSIE WONG-STAAL[1,2]
Departments of Medicine[1] and Biology,[2] University of California, San Diego, La Jolla, California 92093

I. INTRODUCTION

Ribozymes are small RNA molecules with endoribonuclease activity that hybridize to complementary sequences of a particular target mRNA transcript through Watson–Crick base pairing. Under appropriate conditions, ribozymes exhibit catalytic sequence-specific cleavage of the target. The cleaved target mRNA is destabilized and subject to intracellular degradation; consequently, the expression of this specific gene and the synthesis of the encoded protein are prevented. Since their discovery in the 1980s [1], the conserved sequences, secondary structure, and biochemistry of different groups of ribozymes have been well characterized, including group I ribozymes, hammerhead ribozymes, hairpin ribozymes, ribonuclease P (RNase P), and hepatitis delta virus ribozymes (for recent reviews see references [2–5]). Their simple structures and ability to cleave RNA molecules in *trans* in a site-specific manner have important therapeutic implications for a variety of diseases in which well-defined key RNA molecules are involved in causing or maintaining a disease state. Theoretically, any RNA involved in a disease state is a potential target for ribozyme cleavage. Practically, the ribozyme approach is limited by certain requirements for the specific recognition sequence of the catalytic center of the ribozyme. Specificity of the recognized target sequence is determined by the binding arms that hybridize to the sequence flanking the cleavage site within the target RNA. The ability to cleave RNA and thereby selectively inhibit the expression of a gene of interest can be used as a tool for manipulation of RNAs *in vitro* and for the inactivation of gene expression and function *in vivo*. The spectrum of potential targets for ribozyme-mediated gene modulation currently ranges from genes involved in malignant diseases to those causing infectious diseases. Most studies have demonstrated cleavage activity of particular ribozymes *in vitro*. Some studies have provided additional data about ribozyme activity in preclinical cellular or animal models. These studies have significantly improved the knowledge about target-specific optimization, intracellular delivery, stability, and intracellular localization of ribozymes as a requirement for successful clinical application. Most investigators have utilized the hammerhead and hairpin ribozymes because their small sizes (35–50 nucleotides) are easily manipulated or synthesized chemically. This article will focus on these two classes of ribozymes, explain their unique features and specific characteristics, review recent advances using these ribozymes in the field of cancer gene therapy, and briefly outline current limitations to the therapeutic application of ribozymes to human diseases.

II. RIBOZYME STRUCTURES AND FUNCTIONS

Hammerhead ribozymes are a group of self-cleaving RNAs that are characterized by a two-dimensional structural motif known as the hammerhead, which enables site-specific cleavage activity. Haseloff and Gerlach originally described the hammerhead structure with three base-paired stems (named as helices 1, 2, 3) flanking the catalytic center and two highly conserved single-stranded regions with defined sequence (see Fig. 1A) [1]. To facilitate hybridization between ribozyme and substrate, helices 1 and 2 of the hammerhead ribozyme sequence have to align to the substrate sequence (target) via complementary Watson–Crick base pairing. Mutagenesis studies defined important nucleotides and functional groups for efficient catalysis. The target site can be any NUH sequence within the substrate [N: any nucleotide; H: adenine (A), cytosine (C), or uridine (U), but not guanine (G)]. The length and composition of the ribozyme-binding arms control a cascade of reactions during hammerhead-mediated cleavage. The length of the flanking sequences (helices 1 and 3) determines not only site specificity, but also the secondary structure of the RNA molecule and the kinetic profile of the ribozyme–substrate interaction [6].

The optimal length and composition of the binding arms can differ between *in vitro* and inside the cell and should therefore be determined individually for each ribozyme/target combination. Usually a length between 9 and 12 nucleotides for the flanking sequences is considered as a good experimental starting point for optimization. In addition, in choosing a suitable target for ribozyme cleavage, the functional importance and secondary structure of the potential target region should also be considered. Although it is not the focus of this review, it should be mentioned that parts of the hammerhead RNA sequence have been successfully substituted by DNA, thereby increasing the stability of the ribozyme. Hendry *et al.* have shown that ribozymes with DNA in helices 1 and 3 showed threefold enhanced cleavage activity compared with all-RNA ribozymes [7]. The site-specific mechanism of a hammerhead-mediated RNA

cleavage has been studied in detail [8] and enables one to successfully discriminate substrate RNAs with a single base mutation [9].

The *hairpin ribozyme* was originally isolated from the negative strand of the satellite RNA of tobacco ringspot virus [4]. It folds into a two-dimensional hairpin structure, consisting of a small catalytic region of four helical domains and five loops. Based on Watson–Crick base pairing, two helices (helix 1 and helix 2) form between the substrate and ribozyme (see Fig. 1B). Helix 2 has to be four nucleotides long, whereas the length of helix 1 can be 4 or more nucleotides, with a functional ribozyme typically having between 6 and 10 nucleotides. The combination of these two helices facilitates binding of the substrate and determines specificity of binding for *trans*-acting hairpin ribozymes. Between the two helices (helix 1 and 2) in the substrate (Fig. 1B, Loop 5) is the cleavage site, consisting of four nucleotides: N*GUC (cleavage occurs at *), where GUC on the 3′ side of the cleavage site is the only sequence required in the substrate for maximal cleavage and N is any nucleotide [10]. Further information about structure and properties, including a practical approach for the design of hairpin ribozymes, can be found in a recent article published by Yu and Burke [11]. Most researchers use disabled hairpin ribozymes as a control to assess the antisense rather than the cleavage-specific effect of the ribozymes. Compared with the hammerhead ribozyme, the requirements for the hairpin target sequence are more restricted, but still within a given substrate RNA of interest numerous potential target sites should be found.

Both groups of ribozymes are usually engineered to cleave the RNA of interest in *trans*. Following substrate cleavage, the two products are released and the ribozyme is free to bind and cleave another substrate. *In vitro,* the ribozymes can be shown to be truly catalytic in that more than one substrate molecule is processed per ribozyme molecule [4]. The "recycling" of ribozymes is considered to be one of the major advances over antisense technology.

Substantial progress and better understanding of ribozyme design and delivery have already led to the first clinical trial of ribozyme safety in humans [12]. During recent years, the site-specific cleavage mechanism of ribozymes became of particular relevance to identify and reduce activity of gene products associated with stimulation of cell growth. The tremendous increase in information about pathways that determine growth signal transduction, regulation of the cell cycle, the mechanism of action of oncogenes and tumor suppressors, and mechanisms of programmed cell death now allow the design of sequence-specific ribozymes. Ribozymes can be designed to target oncogenes, transport proteins, or growth factors to specifically inhibit tumor

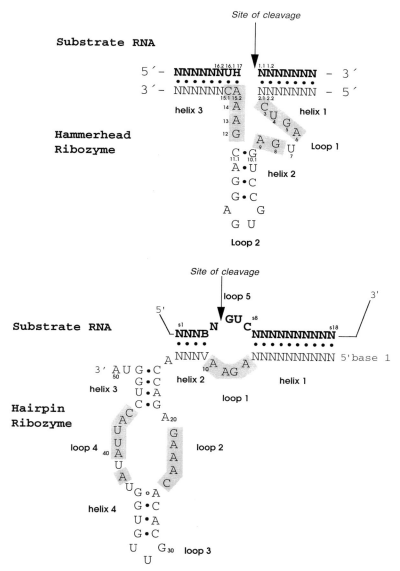

FIGURE 1 Sequence of the hammerhead (A) and the hairpin ribozyme (B) with the corresponding substrate RNA (target). Dots indicate Watson–Crick base pairs. Conserved sequences required for optimal ribozyme activity are shaded. The target site of the hammerhead ribozyme can be any NUH sequence [N: any nucleotide; H: adenine (A), cytosine (C) or uridine (U), but not guanine (G)] with cleavage occuring on the 3′ side of H_{17} as shown by the arrow. Numbering of the hammerhead ribozyme is according to Hertel *et al.* [120]. The hairpin ribozyme requires a GUC triplet on the 3′ side of the cleavage site (marked by an arrow) for optimal activity [10]. The B nucleotide is C, G, or U (not A), and the V nucleotide is G, C, or A (not U). Nucleotides of the hairpin ribozyme are numbered from 1 to 50 [121]. Substrate nucleotides are numbered consecutively.

cell proliferation, drug resistance, or angiogenesis. Genes known to be sufficient for malignant transformation (such as *ras, raf,* and *bcr-abl*) and genes whose expression is important, but not sufficient, for malignant transformation (e.g., *fos*) are both potential targets for ribozymes. Ribozymes that cleave or ligate a particular RNA target sequence can be expressed in tumor cells to prevent or promote expression and translation of RNA molecules comprising the target sequence, thereby leading to a better understanding of the role and importance of the targeted RNA of interest. In comparison to conventional drug therapy, the ribozyme approach offers considerable advantage. Once genes involved in tumorgenicity or tumor pathology are identified, cloned, sequenced, and ideally functionally characterized, this information immediately allows the design of sequence-specific ribozymes aimed to target mRNA molecules, thereby modulating the activity of these genes or their protein products involved in the malignant state of the tumor cell. The following chapters summarize the current experimental approaches and data with ribozymes targeted against a variety of mRNAs involved in malignant cell transformation and proliferation, multidrug resistance, tumor angiogenesis and metastasis as well as malignant complications of viral infections (see Tables 1 and 2).

III. CANCER DISEASE MODELS FOR RIBOZYME APPLICATION

A. Chromosomal Translocations

One possible application of ribozymes for cancer therapy is to target aberrant genes, e.g., fusion transcripts resulting from chromosomal translocations. The best known example is the Philadelphia translocation t(9;22). Several groups have applied ribozymes to this translocation, which can be detected in more than 95%

of patients with chronic myelogenous leukemia [13]. The chimeric bcr-abl hybrid gene [formed by translocation of the proto-oncogene abl from chromosome 9 to the breakpoint cluster region (bcr) on chromosome 22] encodes a fusion protein with a deregulated tyrosine kinase activity [14,15], which leads to cell transformation in hematopoietic cells [16] and causes a leukemia-like phenotype in mice [17–19]. Various different ribozymes have been constructed to cleave within the region of the bcr-abl fusion point. As shown in extracellular cleavage assays, these ribozymes can efficiently cleave chimeric RNAs [20–28]. However, in addition to specific cleavage of the chimeric transcripts, these ribozymes often also nonspecifically cleaved wild-type bcr or abl sequences. Although this unwanted cleavage usually occurred at reduced efficiency, it could limit the application of ribozymes against chimeric fusion genes, in which normal expression of the wild-type genes are important to the cell.

To achieve higher target selectivity, different strategies of designing anti-bcr/abl ribozymes have been developed. Pachuk and colleagues constructed a hammerhead ribozyme cleaving within the abl sequence (19 nt 3' of the fusion point) with a bcr sequence-binding anchor in helix 3 of the ribozyme [22]. As shown by in vitro cleavage assays, these ribozymes can efficiently cleave chimeric RNAs and failed to cleave wild-type

TABLE 1   Malignant Disorders as Potential Targets for Ribozyme Gene Therapy

Target gene	Function of gene product	Ribozyme	Ribozyme-induced change of function	References
bcr/abl	Tyrosine kinase	Hammerhead	Inhibition of cell proliferation and colony formation	20–28
PML/RARα	Unknown	Hammerhead	n.d.	29
AML1/MTG8	Transcription factor	Hammerhead	Inhibition of cell proliferation	32,33
N-ras, Ha-ras, Ki-ras	Signal transduction pathway	Hammerhead	Inhibition of cell proliferation and colony formation, change in morphology, enhanced melanin synthesis, decrease of in vivo tumorigenicity	41–52
c-fos	Transcription regulator	Hammerhead	Change in morphology, reduction in resistance to chemotherapeutic drugs	65,66
mdr-1	Drug-efflux pump	Hammerhead	Reduction in resistance to chemotherapeutic drugs	57–63
CD44	Cell adhesion molecule	Hammerhead	n.d.	78
VLA-6	Adhesion receptor	Hammerhead	Decrease of in vitro invasion and in vivo metastatic ability	79
MMP-9	Matrix metalloproteinase	Hammerhead	Decrease of in vivo metastatic ability	83
CAPL	Unknown	Hammerhead	Decrease of in vivo metastatic ability	85
Pleiotrophin	Growth factor	Hammerhead	Decrease of colony formation, decrease of in vivo tumor growth, tumor angiogenesis, and metastatic ability	88,89
Telomerase	Synthesis of telomeric DNA repeats	Hammerhead	n.d.	93

TABLE 2 Viral Disorders with Malignant Complications as Potential Targets for Ribozyme Gene Therapy

Virus	Malignant complication	Target gene for ribozymes	Ribozyme	Ribozyme-induced change of function	References
Human papilloma virus	Cervical cancer, oral cancer	E6,E7	Hammerhead	Inhibition of cell proliferation and colony formation	95
Epstein–Barr virus	Burkitt lymphoma, nasopharyngeal carcinoma, lymphoproliferative disorders in immunosuppressed patients (AIDS, transplant recipients)	EBNA-1	Hammerhead	Inhibition of cell proliferation	96
Hepatitis B virus	Hepatocellular carcinoma	Pregenomic RNA	Hammerhead, hairpin	Inhibition of viral gene expression	99 100
Hepatitis C virus	Hepatocellular carcinoma	5′ untranslated/core region	Hammerhead, hairpin	Inhibition of viral gene expression	101–103 104

RNAs. A different approach, used by Kearney and co-workers, introduced a single base mismatch in helix 3 [27]. They showed that alteration of the second base 5′ of the cleavage site created a ribozyme with significantly improved specificity for its substrate. Another group tested different possible target sites in the 5′ (*bcr*) and 3′ (*abl*) portions of the fusion point as well as various lengths of helices 1 and 3 within the ribozymes [24]. They found optimal specificity in a ribozyme that cleaves within the *abl* portion of the fusion gene, has a 4-nt helix 1 and a 20-nt helix 3 that overlaps the fusion point, and has part of its binding sequence within the *bcr* portion.

Kronenwett and colleagues designed hammerhead ribozymes with binding to their target sequence occurring either via helix-1- or helix-3-forming antisense arms, such that binding and cleavage occurred on opposite sides of the *bcr-abl* fusion point [26]. They found the best target selectivity for a ribozyme which annealed fast via the *abl* sequences and cleaved within the *bcr* portion of *bcr-abl* RNA. Leopold *et al.* constructed a multiunit ribozyme by fusing three hammerhead ribozymes, with one targeting within the *bcr* sequence, one within the *abl* sequence and one targeting the fusion sequence (cleavage and helix 3 binding sites in *bcr* sequence, helix 1 binding site in *abl* sequence) [20]. This triple-unit ribozyme had enhanced cleavage in comparison to single- and double-unit ribozymes. However, target selectivity could not be achieved using this multi-unit ribozyme.

In addition to the *in vitro* cleavage assays, intracellular applications demonstrate that anti-*bcr/abl*/ribozymes are effective within t(9;22)-positive cells [20,23,25,28]. After transfection into a murine myeloblast cell line (32D) or a human t(9;22)-positive CML cell line established from a patient in blast crisis (K562), the levels of *bcr-abl* RNA were shown to be suppressed [20,28].

Another group used a modified ribozyme with enhanced stability against ribonucleases (RNA in the flanking arms and in a portion of the loop structure was replaced by DNA) to transfect EM-2 cells and demonstrated suppressed *bcr-abl* expression on RNA and protein level as well as inhibition of proliferative capacity and colony formation [23]. Shore and co-workers delivered the ribozymes in K562 cells via retroviral gene transfer and demonstrated elimination of p210 protein-kinase activity in several single cell clones as well as suppression of cell growth in media without recombinant human IL-3 [25].

Another translocation, the t(15;17) is found in acute promyelocytic leukemia. A hammerhead ribozyme has been designed against the fusion RNA [29]. High cleavage specificity could be achieved by this hammerhead ribozyme that cleaves within the gene 3′ to the junction but is able to bind both portions of the fusion genes by containing a noncontiguous helix 3.

Both translocations mentioned here require target selectivity. The translocation (8;21) may be an example of a translocation in which one of the fusion genes is not expressed normally in hematopoietic tissues. Therefore, ribozyme cleavage can occur within this portion of the fusion gene without losing target selectivity. The translocation (8;21) is detected in 7–17% of patients with acute nonlymphocytic leukemia and involves the *AML1* and the *MTG8* (*ETO*) genes [30]. The fusion protein is thought to form a chimeric transcriptional factor [31]. Two different groups have constructed hammerhead ribozymes against the chimeric RNA [32,33]. By *in vitro* cleavage assays they demonstrated cleavage activity for these ribozymes. In addition, activity of the anti-*AML1/MTG8* ribozymes within cells were evaluated in t(8;21)-positive Kasumi-1 [32] or SKNO-1 cells [33]. Both groups demonstrated inhibition of cell growth in ribozyme-transfected cells.

However, cells transfected with antisense oligos or a disabled ribozyme showed almost comparable levels of inhibition. Therefore, further studies will have to evaluate ribozymes with enhanced intracellular activity in comparison to antisense/disabled ribozyme molecules in t(8;21)-positive cells.

B. Malignant Cell Proliferation

The signal transduction pathways regulate cell growth and differentiation. Different genes in the signal transduction cascades are known to be altered in cancer cells [34,35] and have been targeted by ribozymes. The *ras* gene family illustrates some of these studies. The three *ras* genes (N-*ras,* Ha-*ras,* Ki-*ras*) code for proteins that are members of the supergene family of GTP/GDP-binding proteins [36]. Mutations in the *ras* oncogene have been frequently found in a variety of tumors, e.g., in about 90% of pancreatic adenocarcinomas [37], 40–50% of colon adenocarcinomas [38], 30% of lung adenocarcinomas [39], and 45% of melanomas beyond Clark's level II [40]. Several point mutations have been characterized (codons 12, 13, 59, 61) that cause structural changes in the GTP binding site leading to activation of Ras proteins by blocking the capacity of *ras* to induce GTP hydrolysis. Mutant Ras proteins thereby remain activated and stimulate cell growth or differentiation.

Recent studies have investigated the potential of ribozymes in inhibiting such deregulated signal transduction. Hammerhead ribozymes against mutated *ras* oncogene showed selective cleavage of mutated *ras* RNA *in vitro* [41–44] as well as inhibition of *in vivo* function of the oncogene [41–50]. Koizumi and co-workers transfected NIH3T3 cells with ribozymes that specifically target mutated Ha-*ras* oncogene [42,45]. Stable transfectants were then transfected with the activated Ha-*ras* oncogene. An inhibition of foci formation was observed in those clones that expressed active ribozymes. In NIH3T3 cells transformed with cellular DNA from the FEMX-I human melanoma cell line expressing activated Ha-*ras* gene, transfection of the transformants with a hammerhead ribozyme selectively recognizing activated Ha-*ras* RNA resulted in abrogation of the transformed phenotype *in vitro* and suppression of *in vivo* tumorigenicity [51]. Mutant ribozymes lacking catalytic properties and thus functioning as antisense molecules exhibited less dramatic effects. Similar results were obtained when NIH3T3 cells were transfected directly with the activated Ha-*ras* gene [50,52]. In addition, direct injection of ribozyme DNA into tumors induced by transformed NIH3T3 cells in BALB/c mice caused tumor regression [50]. An anti-Ha-*ras* hammerhead ribozyme delivered by retroviral gene transfer into V12Ras-

transformed 3T3 murine fibroblasts as well as rat colon epithelial cells inhibited the expression of V12Ras messenger RNA and V12Ras p21 protein [41]. Reductions in V12Ras expression correlated with the stable expression of the functional ribozyme *in vivo* and decreased tumorigenicity in nude mice. These studies underscore the potential of ribozymes to selectively destroy mutant Ha-*ras* oncogene transcripts, thereby reversing the phenotypic changes of NIH3T3 cells transformation.

Another series of studies were performed using a human bladder carcinoma cell line (EJ) expressing mutant Ha-*ras* gene (G → U transition in codon 12 in exon I). EJ cells stably transfected with a plasmid expressing an anti-Ha-*ras* hammerhead ribozyme revealed suppressed proliferative capacity, colony formation, and cell viability as well as decreased tumor take and enhanced survival after orthotopic inoculation into nude mice [43,49,53]. A highly efficient reversion of the neoplastic phenotype in mutant-Ha-*ras*-expressing tumor cells was observed following adenoviral-mediated delivery of an anti-*ras* hammerhead ribozyme [46]. Tumorigenicity in nude mice after heterotopic injection (s.c.) of transduced cells was completely abrogated during the observation period (30 days). In a human melanoma cell line (FEM) that contains a heterozygous Ha-*ras* gene, mutated in codon 12, transfection with an anti-*ras* ribozyme revealed decreased proliferative capacity and suppressed colony formation, a change in their morphology to a dendritic appearance in monolayer culture associated with enhanced melanin synthesis [47,48]. Scherr and colleagues constructed two different hammerhead ribozymes targeting mutated N-*ras* and demonstrated selective cleavage of mutated N-*ras* RNA whereas wild-type RNA remained unaffected [44].

These different investigations utilizing anti-*ras* ribozymes confirm the oncogenic potential of activated *ras* genes and demonstrate the capability of ribozymes to inhibit gene expression, reverse the transformed phenotype, and suppress *in vivo* tumorigenicity. However, additional studies will have to define the activity of anti-*ras* ribozymes in primary tumor cells and the optimal ribozyme delivery system before anti-*ras* ribozymes can be evaluated as a potential clinical application in the context of current chemotherapeutic protocols. In addition to approaches using anti-*ras* ribozymes, it should be feasible to extend this technology to other genes involved in malignant transformation.

C. Multidrug Resistance

The development of multidrug resistance significantly limits the effectiveness of cytostatic agents and the success of cancer therapy. Chemotherapy-resistant cancer cells may result from alterations at any step in the cell-killing pathway, which includes drug transport, drug metabolism, drug target, cellular repair mechanisms, and toxicity induced apoptosis. One well-documented mechanism is the overexpression of P-glycoprotein (P-GP), usually due to amplification of the multidrug-resistance gene (*mdr-1*). P-GP is localized in the cellular membrane and functions as an ATP-dependent drug-efflux pump [54,55]. This active mechanism results in decreased intracellular levels of structurally unrelated cytotoxic agents [56]. Several groups have designed ribozymes against *mdr-1* and tested their activities *in vitro* [57] or in cell culture [58]. The latter demonstrated decreased levels of *mdr-1* mRNA expression and significant reduction in resistance to daunorubicin when the ribozyme gene was transfected into the human pancreatic carcinoma cell line EPP85-181RDB, which is 1,600-fold more resistant to daunorubicin than the parental cell line (EPP85-181P). This same ribozyme was also applied to an ovarian cancer cell line (A2780AD), which showed 16.6-fold higher resistance to Actinomycin D in comparison with the parental cells [59]. The transfectants restored sensitivity to the chemotherapeutic agent, accompanied by near-complete reversal of cross-resistance to other chemotherapeutic drugs (vincristine, doxorubicin, and etoposide). Northern analysis revealed decreased mRNA for *mdr-1*, c-*fos* and *p53* in the ribozyme-treated cells when compared to drug-resistant control cells. Other studies also demonstrated ribozyme mediated reduction in *mdr-1* expression and drug resistance in a variety of tumor or leukemia cells [60–63].

The nuclear oncogene c-*fos* has been suggested to regulate downstream enzymes associated with DNA synthesis and repair. C-*fos* overexpression is strongly involved in cancer cell resistance to cisplatin chemotherapy [64]. Anti-*fos* ribozymes lead to decreased expression of c-*fos* mRNA, accompanied by reduced mRNA levels of thymidylate (dTMP) synthase, DNA polymerase β, topoisomerase I, and metallothionein IIA. Expression of ribozymes against c-*fos* in cisplatin-resistant A2780DDP and Actinomycin-D-resistant A2780AD cells was successful in restoring their cisplatin-sensitivity [65,66]. Ribozyme expression resulted in decreased expression of c-*fos*, *mdr-1*, c-*jun*, and mutant *p53* mRNAs. Phenotypically, the transformed cells showed altered morphology and restored sensitivity to chemotherapeutic agents composing the MDR phenotype. Interestingly, anti-*fos* ribozymes induced a more rapid reversal of the MDR phenotype in comparison to an anti-*mdr-1*-ribozyme, implicating the central role of c-*fos* in drug resistance [65]. Fos is thought to mediate its various effects through transcriptional activation after interaction with the Jun protein to form the AP-1 complex

[67,68]. The upstream promoter of the *mdr-1* gene contains an AP-1 binding site [69], which is required for full promoter activity in Chinese hamster ovary cells [70] and is suggested to be active in cell lines that overexpress *mdr-1* RNA without gene amplification [71,72]. Therefore, c-*fos* may play an essential role in regulating *mdr-1* gene expression.

Additional factors involved in *mdr-1* regulation may be potential targets for therapeutic application of ribozymes. Recently, overexpression of the transcription factor YB-1 has been shown to be associated with intrinsic expression of *mdr-1* gene and induces the MDR phenotype [73,74]. YB-1 acts through binding to an inverted CCAAT sequence (Y-box sequence), which is present in the human *mdr-1* promoter and is required for basal transcription of the gene [75].

D. Tumor Angiogenesis and Metastasis

Invasion and metastasis still represent the greatest obstacles to successful cancer treatment [76]. Metastasis development involves sequential events, such as detachment of tumor cells from the primary site, invasion of the basement membrane and the underlying extracellular matrix, neovascularization, and tumor cell proliferation. In theory, the process of metastasis can be stopped by targeting molecules at different levels within this cascade. Herein, ribozymes may offer a great potential to target single or multiple steps during metastasis development.

CD44 is a cell adhesion molecule that interacts with matrix molecules and has been shown to be expressed in certain cancer cells. Recently, the involvement of CD44 in human glioma cell invasion has been suggested [77]. Ge and colleagues constructed a hammerhead ribozyme directed against CD44 that cleaved the target RNA *in vitro,* leading to decreased cell surface expression of CD44 when transiently transfected into SNB-19 glioma cells [78].

VLA-6 integrin is the major adhesion receptor for laminin and may play a role in tumor invasion and metastasis. A hammerhead ribozyme directed against the α6 subunit mRNA of VLA-6 was stably transfected into a human fibrosarcoma cell line with high metastatic potential (HT1080) [79]. These cells exhibited reduced expression of α6-containing integrins while the level of other integrins remained unaffected. They were also less adherent to laminin-coated substrates and less invasive into reconstituted basement membrane compared with mock-transfected controls. In nude mice that were i.v. injected with anti-α6 ribozyme transfectants, only 1 out of 35 animals developed lung metastasis, whereas control animals developed multiple metastasis in 22 of 29 animals. These data underline the role of VLA-6 in tumor invasion and metastasis.

Matrix metalloproteinases (MMPs) were found to be overexpressed in tumors and have been linked to tumor progression [80,81]. MMP-9, a member of this family, has a broad range of proteolytic activity against extracellular matrix components [82]. A hammerhead ribozyme directed against MMP-9 mRNA was transfected into a metastatic rat embryo fibroblast line (2.10.10) that was transformed by Ha-*ras* and v-*myc,* and stably expressed high levels of MMP-9 [83]. Ribozyme-transfected cells showed diminished levels of gelatinase activity, a functional assay detecting release and activity of MMP-9, as well as decreased MMP-9 mRNA levels. Furthermore, the metastatic potential of these cells after injection into mice was suppressed. However, ribozyme transfection had no effect on tumorigenicity and tumor growth of the injected cells. These data support a role of MMP-9 expression in metastasis that may be relevant to human cancers.

Finally, the human *CAPL* gene is overexpressed in human melanomas, mammary carcinomas and osteosarcomas, with a trend toward higher expression in more advanced lesions [84]. Stable transfection of a human osteosarcoma cell line expressing high levels of *CAPL* (OHS) with a ribozyme-expressing vector significantly suppressed *CAPL* expression at the RNA and protein levels and changed the morphology of the cells, whereas the growth rates of ribozyme-transfected cells were not significantly different in comparison with controls [85]. Ribozyme-transfected cells showed a marked decrease in skeletal metastasis following intracardial tumor cell injection into nude rats, whereas tumorigenicity of the transfected cells remained unchanged. These effects of specific cleavage of *CAPL* mRNA indicate the involvement of the gene product in key cellular functions associated with the metastatic process.

Besides factors that directly influence the metastatic potential of tumor cells, clinical and experimental evidence suggests that spreading of malignant cells from a localized tumor is directly related to the number of microvessels in the primary tumor [86]. Tumor angiogenesis is thought to be mediated by tumor-cell-derived growth factors. The secreted polypeptide pleiotrophin (PTN) is one of such tumor-derived growth factors [87]. The role of pleiotrophin in tumor growth was recently assessed utilizing anti-PTN ribozymes [88,89]. Two different hammerhead ribozymes targeting PTN mRNA were transfected into PTN-negative (SW13) and -positive cell lines (WM852; 1205LU). Transient and stable coexpression of PTN and the anti-PTN ribozymes inhibited PTN-induced colony formation of PTN-responsive cells (SW13). Colony formation induced by transfections with a closely related growth factor gene

was not affected by active anti-PTN ribozymes. In human melanoma cells expressing high levels of PTN mRNA (WM852; 1205LU), stable transfection with PTN-targeted ribozymes quenched production of PTN and prevented colony formation in soft agar in WM852 cells but not in 1205LU cells. In nude mice, tumor growth was decreased in both cell lines. In addition, anti-PTN ribozymes prevented metastatic spread of the tumors to the lungs in 1205LU cells [89]. These results support a direct link between tumor angiogenesis and metastasis through a secreted growth factor and identify PTN as a candidate cofactor for human melanoma metastasis.

These studies clearly demonstrate that ribozymes can be applied to specifically reduce the expression of key genes involved in the metastatic process, which might ultimately lead to a therapeutic benefit for a variety of metastatic tumors.

E. Telomerase

Telomeres are structures at chromosome ends consisting of highly conserved $(TTAGGG)_n$ repeats that are physiologically lost as a function of cell division. Progressive telomere shortening is linked to the limited proliferative capacity of normal somatic cells (cell senescence; "mitotic clock") [90]. Telomeres are exclusively replicated (elongated) by telomerase, a ribonucleoprotein complex with telomere-specific reverse transcriptase activity [91]. Indefinite proliferation and malignant progression have been found to be associated with high telomerase activity. Telomerase has been found in a majority of malignancies (85–100%), but no activity has been detected in most nonmalignant tissue (except reproductive tissue and hematopoietic cells) [92]. A hammerhead ribozyme directed against the RNA component of human telomerase showed a specific *in vitro* cleavage activity and an inhibitory effect on telomerase activity in cellular extracts from HepG2 or Huh-7, human hepatocellular carcinoma derived lines [93]. The recent discovery of the gene for the human catalytic protein component of telomerase [94] will allow further investigation of the effect of telomerase inhibition on tumor cells, and conceivably to the development of a new class of anticancer drugs.

F. Malignant Complications of Viral Infections

Therapy of human malignancies should ideally target cancer cells while sparing normal cells. This would be most straight forward with infectious agents that are associated with human tumors. Human papilloma virus (HPV) RNA has been found in many cervical and oral cancers and is likely to be an important cofactor in the development of these malignancies. HPV RNA from a HeLa cervical cancer cell line was effectively cleaved *in vitro* by hammerhead ribozymes targeted against the E6 and E7 genes of HPV type 18 (nucleotides 123, 309, and 671 of the viral transcript) [95]. The transfected HeLa cells also exhibited reduced growth rates, increased serum dependency, and reduced colony formation in soft agar. Epstein–Barr virus (EBV) infection is associated with the development of several human malignancies, such as Burkitt's lymphoma. AIDS-associated lymphomas, nasopharyngeal carcinoma, and lymphoproliferative disorders in transplant recipients. Under these conditions, which all share a situation of immunosuppression of the host, the proliferative potential of EBV-infected cells and lifetime exposure through viral persistence are major predisposing factors for the development of EBV-associated neoplasms. Recent studies suggest that expression of the EBV nuclear antigen-1 (EBNA-1) is the major determinant of EBV-associated B-cell neoplasia [96]. EBNA-1 protein is required for the replication and maintenance of the EBV genome. EBV-transformed lymphocytes up-regulate αV integrin expression and were, in contrast to non-transformed B-cells, susceptible to adenovirus infection. Ribozymes targeted against functional EBNA-1 delivered via adenoviral gene transfer down-regulated EBNA-1 RNA and protein expression, thereby inhibiting EBV-induced cell proliferation in these cells [96]. EBV-specific ribozymes might also be useful to further elucidate the mechanism of EBV-induced B-cell activation, transformation, and viral replication by allowing one to target specific steps in these developments.

1. HEPATITIS VIRUSES

Hepatocellular carcinoma is one of the most common malignancies and causes an estimated one million deaths per year worldwide [97]. Chronic infections with hepatitis B (HBV) and hepatitis C virus (HCV) are identified major risk factors for the development of primary hepatocellular carcinoma [98]. Chronic active hepatitis is induced by both infections (HBV, 5–10%; HCV, 50–75%) and in a significant percentage of patients long-term liver inflammation is followed by cirrhosis, leading to the development of liver tumors. It is anticipated that effective treatment regimens against these viral diseases can successfully decrease the incidence of primary liver cancer associated with chronic viral infection worldwide. Both viruses undergo replication through an RNA intermediate and thereby offer the potential for ribozymes to interfere within their replication cycle.

The pregenomic RNA transcribed from HBV genomic DNA has been targeted by three different hammerhead ribozymes derived from a single DNA template [99]. All three target sites within the HBV RNA were cleaved by this construct *in vitro* under physiological conditions. In addition to confirming the feasibility of an *in vitro* cleavage of HBV RNA intermediates, Welch *et al.* demonstrated the intracellular effectiveness of HBV-directed hairpin ribozymes [100]. Several recent reports also documented the use of ribozymes to decrease HCV RNA *in vitro*. The use of ribozymes to target HCV infection seems to be even more attractive, because HCV replicates entirely through an RNA life cycle. The successful use of hammerhead ribozymes [101–103] and hairpin ribozymes [104] targeting highly conserved regions of the 5' end of the 9.4-kb HCV genome underlies the great potential of ribozymes to decrease or eventually eliminate HCV RNA within infected cells. However, due to the lack of tissue culture infection systems for HBV or HCV, it is premature to draw conclusions about the feasibility of clinical application of these ribozymes.

IV. CHALLENGES AND FUTURE DIRECTIONS

The relatively recent discovery of ribozymes and investigation for their therapeutic use in humans are reflected by only a handful of clinical gene therapy protocols planned or approved worldwide. However, given the flexibility in design of ribozymes and their unique potential, an increasing number of clinical protocols will most likely be presented during the next years. As ribozymes are being tested in their first clinical trials [12] to determine their safety and efficacy *in vivo*, researchers are continuing to optimize their intracellular stability, target specificity, and catalytic activities.

The primary structure characteristics of ribozymes can be changed for optimization of ribozyme cleavage activity on a given substrate. In the substrate cleavage region of a hammerhead or hairpin ribozyme, only minor base changes are tolerated to preserve enzyme activity. In contrast, the composition and length of the bases complementary to the ribozyme binding arms can be altered to improve cleavage activity. *In vitro* evolution and selection of ribozymes starting from random sequence nucleotides has led to improved catalytic capabilities of particular ribozymes in comparison to wild-type sequence [105] and significantly improved target specific cleavage activity [106]. Within a given substrate (e.g., oncogene), the accessibility of the particular region must also be taken into account. Secondary structure models of the target sequence obtained by computer programs usually fail to predict *in vivo* accessibility in a dynamic situation. Potential target sites on transcripts of interest can be assayed for local accessibility *in vitro* using analyses such as chemical probing, nuclease mapping or a combination of randomized DNA oligonucleotides and RNase H [107]. However, even these results might not ultimately correlate with *in vivo* efficacy of a ribozyme on a particular substrate cleavage site.

Even with an optimally designed ribozyme against a suitable target sequence, gene delivery and expression is still one of the most difficult steps towards a successful application of ribozymes to human gene therapy. Generally, there are two types of delivery strategies: Exogenous delivery of synthetic or *in vitro* transcribed ribozymes or endogenous expression using a plasmid or viral vector containing a transcriptional unit. Exogenous delivery of ribozymes can be performed using microinjection, transfection, encapsulation in liposomes [108], or electroporation. Ribozymes have been conjugated to polylysine compounds or to lipophilic groups (e.g., cholesterol) to improve stability and uptake [109]. Liposome formulations have been successfully used to increase cellular uptake of ribozymes, intracellular stability, and biological activity [110]. The ribozyme can be chemically modified to enhance cellular uptake or to reduce susceptibility to nuclease degradation, thereby increasing the rate of ribozyme accumulation inside the cell and the likelihood of a biological effect. Because the 2′-hydroxyl group plays an important role in the degradation mechanism by nucleases, a variety of modifications at this site have been exploited to increase ribozyme stability for exogenous delivery: 2′-amino- [111], 2′-*O*-methyl-, and 2′-*O*-allyl- [112] 2′-deoxyribonucleotides together with phosphorothiate linkages [113] and chimeric DNA/RNA molecules [23,114]. In addition sequences have been attached to the 3′ or 5′ end of ribozymes to increase stability against cellular nucleases (e.g., bacteriophage T7 transcriptional terminator) [115]. In addition, chimeric DNA/RNA ribozymes can be made to further increase the resistance to nuclease degradation.

For endogenous expression ribozyme genes are delivered in a suitable vector that facilitates either transient expression (such as adenoviral vectors) or persistent expression following integration into the cellular genome (e.g., adeno-associated viral or retroviral vectors). General concerns persist with the use of vectors derived from viral pathogens such as residual infectivity, toxicity, and rescue of infectivity by recombination. The most extensively utilized viral vectors for gene therapy trials, including the first ribozyme gene therapy trial in humans, have been retroviruses. Retroviral gene delivery (see Chapters 4 and 11) can target a variety of (dividing) cell types, resulting in long-term persistence and continuous expression of ribozymes as a consequence of integration into the host genome. Transient ribozyme expression can be achieved by adenoviral-mediated gene transfer (see Chapter 10). Adenoviral vectors can be produced with high titers. Adenoviruses infect nondividing cells and have been successfully used to target a variety of tissues, showing a high transduction efficiency and high-level expression of the delivered gene. Adenovirus does not integrate into the host genome, thereby reducing the chance of disrupting the cellular genome, but at the same time cannot achieve long-term expression of the encoded gene. Another disadvantage of adenovirus vectors is the immunogenicity of adenoviral proteins, which makes repeated administration of the vector impossible. Adeno-associated virus (AAV) represents a promising system for ribozyme gene delivery (see Chapter 6). AAV is a nonpathogenic replication-defective DNA virus that can stably integrate in the absence of cell division. AAV produces a high copy number within a cell and requires coinfection with a helper virus, such as adenovirus, for productive infection.

To achieve optimal expression of ribozymes from a vector system, an expression cassette is necessary to maximize ribozyme transcription. Self-cleaving ribozymes flanking either site of the therapeutic ribozymes have been used to enhance ribozyme expression [116]. In addition, multiple ribozymes under a single promoter have been constructed to increase intracellular ribozyme expression [117]. The ribozyme coding sequence is inserted into the untranslated regions of genes transcribed by RNA polymerase II (pol II), e.g., SV early promoter, retroviral long terminal repeat, or pol III promoters, such as tRNA [118]. The transduction by retroviral, adenoviral or adeno-associated virus constructs might include signals that localize ribozymes to certain cellular compartments (e.g., utilizing the antigen-binding sites found on most snRNAs). The choice of vector and promoter constructs or transcriptional units for endogenous delivery is also dependent on cell-type-specific differences. Ideally, a delivery vehicle should selectively target the organ or tissue of interest, by exploiting the tissue tropism of different vectors (e.g., adeno–liver) or tissue-specific promoters (albumin promoter–liver).

The complex cellular milieu inside a living cell would influence the efficacy of ribozymes. For example, the optimal hammerhead cleavage activity *in vitro* occurs at a magnesium concentration of around 10 mM, whereas the intracellular concentration of free Mg^{++} is significantly lower (around 500 μM) [20]. On the other hand, the presence of RNA binding proteins may increase the turnover of ribozyme-substrate binding and cleavage [119]. Colocalization of ribozyme and target

RNA in the same cellular compartment is another important parameter of efficient cleavage. Therefore, successful *in vitro* cleavage under optimal conditions might not predict a sufficient biological effect of the particular ribozyme on its substrate RNA inside the living cell.

Significant progress has been made in the development of ribozymes as a potential platform technology for human gene therapy in recent years. Ribozymes might become a valuable tool to target cellular transcripts for destruction, alteration, or repair of genetic information. In the field of cancer gene therapy, ribozymes might be particularly useful to protect cells by either *ex vivo* or *in vivo* gene therapy with transduced, protected, expanded autologous cells or hematopoietic stem cell precursors. However, more information has to be obtained from relevant preclinical models of disease, and innovative strategies are needed to improve ribozyme delivery, expression, colocalization, target specificity, and catalysis properties of ribozymes. Relevant animal models also need to be developed that will allow assessment of the different parameters. Many of these issues are common to the field of gene therapy in general. As the hurdles of gene delivery, expression, and targeting are removed, ribozyme gene therapy can become part of the daily routine in clinics in the prophylaxis and treatment of malignant disorders.

References

1. Haseloff, J., Gerlach, W. L. Simple RNA enzymes with new and highly specific endoribonuclease activities. *Nature* **334,** 585–591, 1988.
2. Marschall, P., Thomson, J. B., Eckstein, F. Inhibition of gene expression with ribozymes. *Cell Mol. Neurobiol.* **14,** 523–538, 1994.
3. Eckstein, F. The hammerhead ribozyme. *Biochem. Soc. Trans.* **24,** 601–604, 1996.
4. Hampel, A., Nesbitt, S., Tritz, R., Altschuler, M. The hairpin ribozyme. Methods: A Companion to *Methods Enzymol* 37–42, 1993.
5. Poeschla, E., Wong-Staal, F. Antiviral and anticancer ribozymes. *Curr. Opin. Oncol.* **6,** 601–606, 1994.
6. Fedor, M. J., Uhlenbeck, O. C. Substrate sequence effects on "hammerhead" RNA catalytic efficiency. *Proc. Natl. Acad. Sci. USA* **87,** 1668–1672, 1990.
7. Hendry, P., McCall, M. J., Santiago, F. S., Jennings, P. A. A ribozyme with DNA in the hybridising arms displays enhanced cleavage ability. *Nucleic Acids Res.* **20,** 5737–5741, 1992.
8. Ruffner, D. E., Stormo, G. D., Uhlenbeck, O. C. Sequence requirements of the hammerhead RNA self-cleavage reaction. *Biochemistry* **29,** 10695–10702, 1990.
9. Hertel, K. J., Herschlag, D., Uhlenbeck, O. C. Specificity of hammerhead ribozyme cleavage. *EMBO J.* **15,** 3751–3757, 1996.
10. Anderson, P., Monforte, J., Tritz, R., Nesbitt, S., Hearst, J., Hampel, A. Mutagenesis of the hairpin ribozyme. *Nucleic Acids Res.* **22,** 1096–1100, 1994.

11. Yu, Q., Burke, J. Design of hairpin ribozymes for *in vitro* and cellular applications, in Turner, P., ed., *Methods in Molecular Biology,* vol. 74. Totowa, NJ, Humana Press, 1997, pp. 161–169.
12. Leavitt, M. C., Yu, M., Wong-Staal, F., Looney, D. J. *Ex vivo* transduction and expansion of CD4+ lymphocytes from HIV+ donors: prelude to a ribozyme gene therapy trial. *Gene Ther.* **3,** 599–606, 1996.
13. Kurzrock, R., Gutterman, J. U., Talpaz, M. The molecular genetics of Philadelphia chromosome-positive leukemias. *N. Engl. J. Med.* **319,** 990–998, 1988.
14. Konopka, J. B., Witte, O. N. Detection of c-abl tyrosine kinase activity *in vitro* permits direct comparison of normal and altered abl gene products. *Mol. Cell Biol.* **5,** 3116–3123, 1985.
15. Lugo, T. G., Pendergast, A. M., Muller, A. J., Witte, O. N. Tyrosine kinase activity and transformation potency of bcr-abl oncogene products. *Science* **247,** 1079–1082, 1990.
16. Daley, G. Q., Baltimore, D. Transformation of an interleukin 3-dependent hematopoietic cell line by the chronic myelogenous leukemia-specific P210bcr/abl protein. *Proc. Natl. Acad. Sci. USA* **85,** 9312–9316, 1988.
17. Daley, G. Q., Van Etten, R. A., Baltimore, D. Induction of chronic myelogenous leukemia in mice by the p210 bcr/abl gene of the Philadelphia chromosome. *Science* **247,** 824–830, 1990.
18. Elefanty, A. G., Hariharan, I. K., Cory, S. Bcr-abl, the hallmark of chronic myeloid leukaemia in man, induces multiple haemopoietic neoplasms in mice. *EMBO J.* **9,** 1069–1078, 1990.
19. Kelliher, M. A., McLaughlin, J., Witte, O. N., Rosenberg, N. Induction of a chronic myelogenous leukemia-like syndrome in mice with v-abl and BCR/ABL. *Proc. Natl. Acad. Sci. USA* **87,** 6649–6653, 1990.
20. Leopold, L. H., Shore, S. K., Newkirk, T. A., Reddy, R. M., Reddy, E. P. Multi-unit ribozyme-mediated cleavage of bcr-abl mRNA in myeloid leukemias. *Blood* **85,** 2162–2170, 1995.
21. Wright, L., Wilson, S. B., Milliken, S., Biggs, J., Kearney, P. Ribozyme-mediated cleavage of the bcr/abl transcript expressed in chronic myeloid leukemia. *Exp. Hematol.* **21,** 1714–1718, 1993.
22. Pachuk, C. J., Yoon, K., Moelling, K., Coney, L. R. Selective cleavage of bcr-abl chimeric RNAs by a ribozyme targeted to non-contiguous sequences. *Nucleic Acids Res.* **22,** 301–307, 1994.
23. Snyder, D. S., Wu, Y., Wang, J. L., *et al.* Ribozyme-mediated inhibition of bcr-abl gene expression in a Philadelphia chromosome-positive cell line. *Blood* **82,** 600–605, 1993.
24. James, H., Mills, K., Gibson, I. Investigating and improving the specificity of ribozymes directed against the bcr-abl translocation. *Leukemia* **10,** 1054–1064, 1996.
25. Shore, S. K., Nabissa, P. M., Reddy, E. P. Ribozyme-mediated cleavage of the BCRABL oncogene transcript: *in vitro* cleavage of RNA and *in vivo* loss of P210 protein-kinase activity. *Oncogene* **8,** 3183–3188, 1993.
26. Kronenwett, R., Haas, R., Sczakiel, G. Kinetic selectivity of complementary nucleic acids: bcr-abl-directed antisense RNA and ribozymes. *J. Mol. Biol.* **259,** 632–644, 1996.
27. Kearney, P., Wright, L. A., Milliken, S., Biggs, J. C. Improved specificity of ribozyme-mediated cleavage of bcr-abl mRNA. *Exp. Hematol.* **23,** 986–989, 1995.
28. Lange, W., Cantin, E. M., Finke, J., Dolken, G. *In vitro* and *in vivo* effects of synthetic ribozymes targeted against BCR/ABL mRNA. *Leukemia* **7,** 1786–1794, 1993.
29. Pace, U., Bockman, J. M., MacKay, B. J., Miller, W. H., Jr., Dmitrovsky, E., Goldberg, A. R. A ribozyme which discriminates *in vitro* between PML/RAR alpha, the t(15;17)-associated fusion RNA of acute promyelocytic leukemia, and PML and RAR alpha, the transcripts from the nonrearranged alleles. *Cancer Res.* **54,** 6365–6369, 1994.

30. Koeffler, H. P. Syndromes of acute nonlymphocytic leukemia. *Ann. Intern. Med.* **107,** 748–758, 1987.

31. Miyoshi, H., Kozu, T., Shimizu, K., *et al.* The t(8;21) translocation in acute myeloid leukemia results in production of an *AML1-MTG8* fusion transcript. *EMBO J.* **12,** 2715–2721, 1993.

32. Matsushita, H., Kobayashi, H., Mori, S., Kizaki, M., Ikeda, Y. Ribozymes cleave the AML1/MTG8 fusion transcript and inhibit proliferation of leukemic cells with t(8;21). *Biochem. Biophys. Res. Commun.* **215,** 431–437, 1995.

33. Kozu, T., Sueoka, E., Okabe, S., Sueoka, N., Komori, A., Fujiki, H. Designing of chimeric DNA/RNA hammerhead ribozymes to be targeted against AML1/MTG8 mRNA. *J. Cancer Res. Clin. Oncol.* **122,** 254–256, 1996.

34. Seemayer, T. A., Cavenee, W. K. Molecular mechanisms of oncogenesis. *Lab. Invest.* **60,** 585–599, 1989.

35. Brunton, V. G., Workman, P. Cell-signaling targets for antitumor drug development. *Cancer Chemother. Pharmacol.* **32,** 1–19, 1993.

36. Kiefer, P. E., Bepler, G., Kubasch, M., Havemann, K. Amplification and expression of protooncogenes in human small cell lung cancer cell lines. *Cancer Res.* **47,** 6236–6242, 1987.

37. Almoguera, C., Shibata, D., Forrester, K., Martin, J., Arnheim, N., Perucho, M. Most human carcinomas of the exocrine pancreas contain mutant c-K-*ras* genes. *Cell* **53,** 549–554, 1988.

38. Forrester, K., Almoguera, C., Han, K., Grizzle, W. E., Perucho, M. Detection of high incidence of K-*ras* oncogenes during human colon tumorigenesis. *Nature* **327,** 298–303, 1987.

39. Rodenhuis, S., Slebos, R. J., Boot, A. J., *et al.* Incidence and possible clinical significance of K-*ras* oncogene activation in adenocarcinoma of the human lung. *Cancer Res.* **48,** 5738–5741, 1988.

40. Ball, N. J., Yohn, J. J., Morelli, J. G., Norris, D. A., Golitz, L. E., Hoeffler, J. P. *Ras* mutations in human melanoma: a marker of malignant progression. *J. Invest. Dermatol.* **102,** 285–290, 1994.

41. Li, M., Lonial, H., Citarella, R., Lindh, D., Colina, L., Kramer, R. Tumor inhibitory activity of anti-*ras* ribozymes delivered by retroviral gene transfer. *Cancer Gene Ther.* **3,** 221–229, 1996.

42. Koizumi, M., Kamiya, H., Ohtsuka, E. Inhibition of c-Ha-*ras* gene expression by hammerhead ribozymes containing a stable C(UUCG)G hairpin loop. *Biol. Pharm. Bull.* **16,** 879–883, 1993.

43. Kashani-Sabet, M., Funato, T., Tone, T., *et al.* Reversal of the malignant phenotype by an anti-*ras* ribozyme. *Antisense Res. Dev.* **2,** 3–15, 1992.

44. Scherr, M., Grez, M., Ganser, A., Engels, J. W. Specific hammerhead ribozyme-mediated cleavage of mutant N-*ras* mRNA *in vitro* and *ex vivo*. Oligoribonucleotides as therapeutic agents. *J. Biol. Chem.* **272,** 14304–14313, 1997.

45. Koizumi, M., Kamiya, H., Ohtsuka, E. Ribozymes designed to inhibit transformation of NIH3T3 cells by the activated c-Ha-*ras* gene. *Gene* **117,** 179–184, 1992.

46. Feng, M., Cabrera, G., Deshane, J., Scanlon, K. J., Curiel, D. T. Neoplastic reversion accomplished by high efficiency adenoviral-mediated delivery of an anti-*ras* ribozyme. *Cancer Res.* **55,** 2024–2028, 1995.

47. Ohta, Y., Kijima, H., Kashani-Sabet, M., Scanlon, K. J. Suppression of the malignant phenotype of melanoma cells by anti-oncogene ribozymes. *J. Invest. Dermatol.* **106,** 275–280, 1996.

48. Ohta, Y., Kijima, H., Ohkawa, T., Kashani-Sabet, M., Scanlon, K. J. Tissue-specific expression of an anti-*ras* ribozyme inhibits proliferation of human malignant melanoma cells. *Nucleic Acids Res.* **24,** 938–942, 1996.

49. Eastham, J. A., Ahlering, T. E. Use of an anti-*ras* ribozyme to alter the malignant phenotype of a human bladder cancer cell line. *J. Urol.* **156,** 1186–1188, 1996.

50. Chang, M. Y., Won, S. J., Liu, H. S. A ribozyme specifically suppresses transformation and tumorigenicity of Ha-*ras*-oncogene-transformed NIH/3T3 cell lines. *J. Cancer Res. Clin. tncol.* **123,** 91–99, 1997.

51. Kashani-Sabet, M., Funato, T., Florenes, V. A., Fodstad, O., Scanlon, K. J. Suppression of the neoplastic phenotypic *in vivo* by an anti-*ras* ribozyme. *Cancer Res.* **54,** 900–902, 1994.

52. Funato, T., Shitara, T., Tone, T., Jiao, L., Kashani-Sabet, M., Scanlon, K. J. Suppression of H-*ras*-mediated transformation in NIH3T3 cells by a *ras* ribozyme. *Biochem. Pharmacol.* **48,** 1471–1475, 1994.

53. Tone, T., Kashani-Sabet, M., Funato, T., *et al.* Supression of EJ cells tumorigenicity. *In Vivo* **7,** 471–476, 1993.

54. Gottesman, M. M., Pastan, I. Biochemistry of multidrug resistance mediated by the multidrug transporter. *Annu. Rev. Biochem.* **62,** 385–427, 1993.

55. Juliano, R. L., Ling, V. A surface glycoprotein modulating drug permeability in Chinese hamster ovary cell mutants. *Biochim. Biophys. Acta* **455,** 152–162, 1976.

56. Kerr, L. D., Holt, J. T., Matrisian, L. M. Growth factors regulate transin gene expression by c-*fos*-dependent and c-*fos*-independent pathways. *Science* **242,** 1424–1427, 1988.

57. Palfner, K., Kneba, M., Hiddemann, W., Bertram, J. Improvement of hammerhead ribozymes cleaving *mdr-1* mRNA. *Biol. Chem. Hoppe Seyler* **376,** 289–295, 1995.

58. Holm, P. S., Scanlon, K. J., Dietel, M. Reversion of multidrug resistance in the P-glycoprotein-positive human pancreatic cell line (EPP85-181RDB) by introduction of a hammerhead ribozyme. *Br. J. Cancer* **70,** 239–243, 1994.

59. Scanlon, K. J., Ishida, H., Kashani-Sabet, M. Ribozyme-mediated reversal of the multidrug-resistant phenotype. *Proc. Natl. Acad. Sci. USA* **91,** 11123–11127, 1994.

60. Kiehntopf, M., Brach, M. A., Licht, T., *et al.* Ribozyme-mediated cleavage of the MDR-1 transcript restores chemosensitivity in previously resistant cancer cells. *EMBO J.* **13,** 4645–4652, 1994.

61. Kobayashi, H., Dorai, T., Holland, J. F., Ohnuma, T. Reversal of drug sensitivity in multidrug-resistant tumor cells by an MDR1 (PGY1) ribozyme. *Cancer Res.* **54,** 1271–1275, 1994.

62. Bertram, J., Palfner, K., Killian, M., *et al.* Reversal of multiple drug resistance *in vitro* by phosphorothioate oligonucleotides and ribozymes. *Anticancer Drugs* **6,** 124–134, 1995.

63. Daly, C., Coyle, S., McBride, S., *et al.* MDR1 ribozyme mediated reversal of the multi-drug resistant phenotype in human lung cell lines. *Cytotechnology* **19,** 199–205, 1996.

64. Kashani-Sabet, M., Lu, Y., Leong, L., Haedicke, K., Scanlon, K. J. Differential oncogene amplification in tumor cells from a patient treated with cisplatin and 5-fluorouracil. *Eur. J. Cancer* **26,** 383–390, 1990.

65. Scanlon, K. J., Jiao, L., Funato, T., *et al.* Ribozyme-mediated cleavage of c-*fos* mRNA reduces gene expression of DNA synthesis enzymes and metallothionein. *Proc. Natl. Acad. Sci. USA* **88,** 10591–10595, 1991.

66. Funato, T., Yoshida, E., Jiao, L., Tone, T., Kashani-Sabet, M., Scanlon, K. J. The utility of an anti-*fos* ribozyme in reversing cisplatin resistance in human carcinomas. *Adv. Enzyme Regul.* **32,** 195–209, 1992.

67. Ransone, L. J., Verma, I. M. Nuclear proto-oncogenes *fos* and *jun. Annu. Rev. Cell Biol.* **6,** 539–557, 1990.

68. Rauscher, F. J. D., Sambucetti, L. C., Curran, T., Distel, R. J., Spiegelman, B. M. Common DNA binding site for Fos protein complexes and transcription factor AP-1. *Cell* **52,** 471–480, 1988.

69. Ueda, K., Pastan, I., Gottesman, M. M. Isolation and sequence of the promoter region of the human multidrug-resistance (P-glycoprotein) gene. *J. Biol. Chem.* **262,** 17432–17436, 1987.

151

70. Teeter, L. D., Eckersberg, T., Tsai, Y., Kuo, M. T. Analysis of the Chinese hamster P-glycoprotein/multidrug resistance gene *pgp1* reveals that the AP-1 site is essential for full promoter activity. *Cell Growth Differ.* **2,** 429–437, 1991.

71. Shen, D. W., Fojo, A., Chin, J. E., *et al.* Human multidrug-resistant cell lines: increased *mdr1* expression can precede gene amplification. *Science* **232,** 643–645, 1986.

72. Gottesman, N. M: How cancer cells evade chemotherapy: sixteenth Richard and Hinda Rosenthal Foundation Award Lecture. *Cancer Res.* **53,** 747–754, 1993.

73. Ohga, T., Koike, K., Ono, M., *et al.* Role of the human Y box-binding protein YB-1 in cellular sensitivity to the DNA-damaging agents cisplatin, mitomycin C, and ultraviolet light. *Cancer Res.* **56,** 4224–4228, 1996.

74. Bargou, R. C., Jürchott, K., Wagener, C., *et al.* Nuclear localization and increased levels of transcription factor YB-1 in primary human breast cancers are associated with intrinsic MDR1 gene expression. *Nat. Med.* **3,** 447–450, 1997.

75. Goldsmith, M. E., Madden, M. J., Morrow, C. S., Cowan, K. H. A Y-box consensus sequence is required for basal expression of the human multidrug resistance (*mdr1*) gene. *J. Biol. Chem.* **268,** 5856–5860, 1993.

76. Stracke, M. L., Liotta, L. A. Molecular mechanisms of tumor cell metastasis, in Mendelsohn, J., Howley, P. M., Israel, M. A., Liotta, L. A. eds. The Molecular Basis of Cancer. Philadelphia, PA, Saunders, 1995, pp 233–247.

77. Merzak, A., Koocheckpour, S., Pilkington, G. J. CD44 mediates human glioma cell adhesion and invasion *in vitro. Cancer Res.* **54,** 3988–3992, 1994.

78. Ge, L., Resnick, N. M., Ernst, L. K., Salvucci, L. A., Asman, D. C., Cooper, D. L. Gene therapeutic approaches to primary and metastatic brain tumors: II. ribozyme-mediated suppression of CD44 expression. *J. Neurooncol.* **26,** 251–257, 1995.

79. Yamamoto, H., Irie, A., Fukushima, Y., *et al.* Abrogation of lung metastasis of human fibrosarcoma cells by ribozyme-mediated suppression of integrin alpha6 subunit expression. *Int. J. Cancer* **65,** 519–524, 1996.

80. Stetler-Stevenson, W. G., Aznavoorian, S., Liotta, L. A. Tumor cell interactions with extracellular matrix during invasion and metastasis. *Annu. Rev. Cell Biol.* **9,** 541–573, 1993.

81. De Clerck, Y. A., Shimada, H., Taylor, S. M., Langley, K. E. Matrix metalloproteinases and their inhibitors in tumor progression. *Ann. N. Y. Acad. Sci.* **732,** 222–232, 1994.

82. Himelstein, B. P., Canete-Soler, R., Bernhard, E. J., Dilks, D. W., Muschel, R. J. Metalloproteinases in tumor progression: the contribution of MMP-9. *Invasion Metastasis* **14,** 246–258, 1994.

83. Hua, J., Muschel, R. J. Inhibition of matrix metalloproteinase 9 expression by a ribozyme blocks metastasis in a rat sarcoma model system. *Cancer Res.* **56,** 5279–5284, 1996.

84. Ebralidze, A., Tulchinsky, E., Grigorian, M., *et al.* Isolation and characterization of a gene specifically expressed in different metastatic cells and whose deduced gene product has a high degree of homology to a Ca^{2+}-binding protein family. *Genes Dev.* **3,** 1086–1093, 1989.

85. Maelandsmo, G. M., Hovig, E., Skrede, M., *et al.* Reversal of the *in vivo* metastatic phenotype of human tumor cells by an anti-*CAPL* (*mts1*) ribozyme. *Cancer Res.* **56,** 5490–5498, 1996.

86. Weidner, N. Intratumor microvessel density as a prognostic factor in cancer. *Am. J. Pathol.* **147,** 9–19, 1995.

87. Li, Y. S., Milner, P. G., Chauhan, A. K., *et al.* Cloning and expression of a developmentally regulated protein that induces mitogenic and neurite outgrowth activity. *Science* **250,** 1690–1694, 1990.

88. Czubayko, F., Riegel, A. T., Wellstein, A. Ribozyme-targeting elucidates a direct role of pleiotrophin in tumor growth. *J. Biol. Chem.* **269,** 21358–21363, 1994.

89. Czubayko, F., Schulte, A. M., Berchem, G. J., Wellstein, A. Melanoma angiogenesis and metastasis modulated by ribozyme targeting of the secreted growth factor pleiotrophin. *Proc. Natl. Acad. Sci. USA* **93,** 14753–14758, 1996.

90. Chiu, C. P., Harley, C. B. Replicative senescence and cell immortality: the role of telomeres and telomerase. *Proc. Soc. Exp. Biol. Med.* **214,** 99–106, 1997.

91. Feng, J., Funk, W. D., Wang, S. S., *et al.* The RNA component of human telomerase. *Science* **269,** 1236–1241, 1995.

92. Kim, N. W., Piatyszek, M. A., Prowse, K. R., *et al.* Specific association of human telomerase activity with immortal cells and cancer. *Science* **266,** 2011–2015, 1994.

93. Kanazawa, Y., Ohkawa, K., Ueda, K., *et al.* Hammerhead ribozyme-mediated inhibition of telomerase activity in extracts of human hepatocellular carcinoma cells. *Biochem. Biophys. Res. Commun.* **225,** 570–576, 1996.

94. Nakamura, T. M., Morin, G. B., Chapman, K. B., *et al.* Telomerase catalytic subunit homologs from fission yeast and human. *Science* **277,** 955–959, 1997.

95. Chen, Z., Kamath, P., Zhang, S., Weil, M. M., Shillitoe, E. J. Effectiveness of three ribozymes for cleavage of an RNA transcript from human papillomavirus type 18. *Cancer Gene Ther.* **2,** 263–271, 1995.

96. Huang, S., Stupack, D., Mathias, P., Wang, Y., Nemerow, G. Growth arrest of Epstein–Barr virus immortalized B lymphocytes by adenovirus-delivered ribozymes. *Proc. Natl. Acad. Sci. USA* **94,** 8156–8161, 1997.

97. Di Bisceglie, A. M. Rustgi, V. K., Hoofnagle, J. H., Dusheiko, G. M., Lotze, M. T. NIH conference. Hepatocellular carcinoma. *Ann. Intern. Med.* **108,** 390–401, 1988.

98. Johnson, P. J. The epidemiology of hepatocellular carcinoma. *Eur. J. Gastroenterol. Hepatol.* **8,** 845–849, 1996.

99. von Weizsäcker, F., Blum, H. E., Wands, J. R. Cleavage of hepatitis B virus RNA by three ribozymes transcribed from a single DNA template. *Biochem. Biophys. Res. Commun.* **189,** 743–748, 1992.

100. Welch, P. J., Tritz, R., Yei, S., Barber, J., Yu, M. Intracellular application of hairpin ribozyme genes against hepatitis B virus. *Gene Ther.* **4,** 736–743, 1997.

101. Sakamoto, N., Wu, C. H., Wu, G. Y. Intracellular cleavage of hepatitis of C virus RNA and inhibition of viral protein translation by hammerhead ribozymes. *J. Clin. Invest* **98,** 2720–2728, 1996.

102. Ohkawa, K., Yuki, N., Kanazawa, Y., *et al.* Cleavage of viral RNA and inhibition of viral translation by hepatitis C virus RNA-specific hammerhead ribozyme *in vitro. J. Hepatol.* **27,** 78–84, 1997.

103. Lieber, A., He, C. Y., Polyak, S. J., Gretch, D. R., Barr, D., Kay, M. A. Elimination of hepatitis C virus RNA in infected human hepatocytes by adenovirus-mediated expression of ribozymes. *J. Virol.* **70,** 8782–8791, 1996.

104. Welch, P. J., Tritz, R., Yei, S., Leavitt, M., Yu, M., Barber, J. A potential therapeutic application of hairpin ribozymes: *in vitro* and *in vivo* studies of gene therapy for hepatitis C virus infection. *Gene Ther.* **3,** 994–1001, 1996.

105. Berzal-Herranz, A., Joseph, S., Burke, J. M. *In vitro* selection of active hairpin ribozymes by sequential RNA-catalyzed cleavage and ligation reactions. *Genes Dev.* **6,** 129–134, 1992.

106. Joseph S., Burke, J. M. Optimization of an anti-HIV hairpin ribozyme by *in vitro* selection. *J. Biol. Chem.* **268,** 24515–24518, 1993.

107. Ho, S. P., Britton, D. H., Stone, B. A., *et al.* Potent antisense oligonucleotides to the human multidrug resistance-1 mRNA are rationally selected by mapping RNA-accessible sites with oligonucleotide libraries. *Nucleic Acids Res.* **24,** 1901–1907, 1996.

108. Sullivan, S. M., Gieseler, R. K., Lenzner, S., *et al.* Inhibition of human immunodeficiency virus-1 proliferation by liposome-encapsulated sense DNA to the 5′ tat splice acceptor site. *Antisense Res. Dev.* **2,** 187–197, 1992.

109. Letsinger, R. L., Zhang, G. R., Sun, D. K., Ikeuchi, T., Sarin, P. S. Cholesteryl-conjugated oligonucleotides: synthesis, properties, and activity as inhibitors of replication of human immunodeficiency virus in cell culture. *Proc. Natl. Acad. Sci. USA* **86,** 6553–6556, 1989.

110. Karik, K., Megyeri, K., Xiao, Q., Barnathan, E. S. Lipofectin-aided cell delivery of ribozyme targeted to human urokinase receptor mRNA. *FEBS Lett.* **352,** 41–44, 1994.

111. Pieken, W. A., Olsen, D. B., Benseler, F., Aurup, H., Eckstein, F. Kinetic characterization of ribonuclease-resistant 2′-modified hammerhead ribozymes. *Science* **253,** 314–317, 1991.

112. Paolella, G., Sproat, B. S., Lamond, A. I. Nuclease resistant ribozymes with high catalytic activity. *EMBO J.* **11,** 1913–1919, 1992.

113. Shibahara, S., Mukai, S., Morisawa, H., Nakashima, H., Kobayashi, S., Yamamoto, N. Inhibition of human immunodeficiency virus (HIV-1) replication by synthetic oligo-RNA derivatives. *Nucleic Acids Res.* **17,** 239–252, 1989.

114. Taylor, N. R., Kaplan, B. E., Swiderski, P., Li, H., Rossi, J. J. Chimeric DNA-RNA hammerhead ribozymes have enhanced *in vitro* catalytic efficiency and increased stability *in vivo*. *Nucleic Acids Res.* **20,** 4559–4565, 1992.

115. Sioud, M., Natvig, J. B., Førre, O. Preformed ribozyme destroys tumour necrosis factor mRNA in human cells. *J. Mol. Biol.* **223,** 831–835, 1992.

116. Ruiz, J., Wu, C. H., Ito, Y., Wu, G. Y. Design and preparation of a multimeric self-cleaving hammerhead ribozyme. *Biotechniques* **22,** 338–345, 1997.

117. Ohkawa, J., Yuyama, N., Takebe, Y., Nishikawa, S., Taira, K. Importance of independence in ribozyme reactions: kinetic behavior of trimmed and of simply connected multiple ribozymes with potential activity against human immunodeficiency virus. *Proc. Natl. Acad. Sci. USA* **90,** 11302–11306, 1993.

118. Yu, M., Ojwang, J., Yamada, O., *et al.* A hairpin ribozyme inhibits expression of diverse strains of human immunodeficiency virus type 1. *Proc. Natl. Acad. Sci. USA* **90,** 6340–6344, 1993.

119. Bertrand, E. L., Rossi, J. J. Facilitation of hammerhead ribozyme catalysis by the nucleocapsid protein of HIV-1 and the heterogeneous nuclear ribonucleoprotein A1. *EMBO J.* **13,** 2904–2912, 1994.

120. Hertel, K. J., Pardi, A., Uhlenbeck, O. C., *et al.* Numbering system for the hammerhead. *Nucleic Acids Res.* **20,** 3252, 1992.

121. Earnshaw, D. J., Gait, M. J. Progress toward the structure and therapeutic use of the hairpin ribozyme. *Antisense Nucleic Acid Drug Dev.* **7,** 403–411, 1997.

PART III

Manipulating Drug Effects through Gene Therapy

CHAPTER 10

In Situ Use of Suicide Genes for Cancer Therapy

SCOTT M. FREEMAN,[1] RAJAGOPAL RAMESH,[2] ANUPAMA MUNSHI,[2]
KATHARINE A. WHARTENBY,[3] JED L. FREEMAN,[4] AND AIZEN J. MARROGI[2]

[1]Department of Pathology, Tulane University Medical School, New Orleans, Louisiana 70122,
[2]Department of Surgery and Gene Therapy Program, Louisiana State University Medical School,
New Orleans, Louisiana 70112, [3]Department of Medicine, Brown University and Rhode Island Hospital,
Providence, Rhode Island 02903, and [4]Department of Oncology, Cancer Care Consultants,
Redding, California 96001

I. INTRODUCTION

The field of cancer therapy has made significant advances in the last three decades, but overall cancer mortality remains virtually unchanged since the 1960s. Further, although approximately 70% of cancer patients achieve a remission, 50% of cancer patients relapse and become nonresponsive to conventional therapeutic regimens. Patients with advanced cancers, particularly those in relapse, are the least responsive to therapy and have a median life expectancy of only one and a half years [1,2]. The progression of disease is often a result of the growth of tumors cells that are resistant to any combination of chemotherapy, surgery, and radiation [3,4]. Thus, effective treatment will depend on the development of new strategies that attack cancers that are resistant to conventional approaches.

Advances in molecular biology have allowed the elucidation of many cellular functions involved in tumorigenesis. These tumor-cell-specific properties are being exploited in a number of novel therapies in which the cancerous cells themselves are targeted in ways that

155

circumvent their resistance to standard agents. Gene therapy is one approach that allows direct treatment of tumor cells by genetically modifying the malignant cells to alter their drug sensitivity phenotype. Although gene therapy was originally envisioned to treat genetic diseases, the potential of genetic manipulation for the treatment of cancer was immediately recognized [5,6]. Investigators had worked toward developing gene therapy to treat genetic diseases for almost a decade beginning in the early 1980s, and animal studies using gene therapy to treat cancer quickly followed [7–9]. Genetic modification of tumor cells was first applied in humans over 8 years ago, when genetically marked tumor infiltrating lymphocytes (TILs) were infused into cancer patients [10]. A bacterial gene that served as a genetic label was inserted into the TIL and served as a stable marker, enabling tracking of the cells after infusion. This gene-marking protocol advanced cell labeling technology because it served as a permanent cell marker, unlike the previously used indium-111, which would only remain in the labeled cells for several days [11]. This trial was not designed to be therapeutic but rather to gain information regarding the distribution of the TIL *in vivo* with the ultimate goal of improving existing adoptive immunotherapy [12–14].

The clinical trial using genetically marked TIL paved the way for gene therapy for cancer, and currently, the majority of clinical gene therapy protocols center on the treatment of cancer [15,16]. The approaches of cancer gene therapy clinical protocols can be categorized generally as follows: genetic marking, cancer vaccination, inhibition of oncogene expression, restoration of tumor suppressor gene function, and the use of suicide genes. The focus of this chapter will be on the use of suicide genes, and in particular the herpes simplex virus thymidine kinase (HSV-TK) gene, to treat cancer [15–20]. A suicide gene is a gene encoding a protein that under appropriate conditions (e.g., exposure to a prodrug) is toxic to the genetically modified cell (i.e., tumor cell).

The treatment of cancer is different from treatment of genetic diseases, because an effective cancer therapy requires complete eradication of all tumor cells, whereas genetic disease therapy may require only a small percentage of cells to be genetically modified for the therapy to be effective. The genetic disease, adenosine deaminase (ADA) deficiency, exemplifies this principle and was the first genetic disease in which a therapeutic goal was applied [21,22]. Patient studies have shown that heterozygous individuals carrying the ADA mutant gene displayed a normal phenotype even if they expressed only 1% of the ADA levels present in homozygous normal individuals [23]. Although an effective cancer therapy requires that all tumor cells be killed, a number of studies using suicide genes have shown that

when less than 100% of the tumor was genetically modified, the entire tumor mass could be eradicated. Thus, this approach could potentially eliminate a cancer even though the technology to genetically modify an entire tumor is not available.

This chapter will focus on the development of suicide gene therapy, beginning with the initial studies showing that genetic modification of tumor cells could impart a new drug sensitivity phenotype to chemotherapy-resistant tumor cells, which would then allow killing of the cells by exposure to a prodrug [17,18,24–29]. Also see the discussion of use of the cytosine deaminase gene for suicide gene therapy in Chapter 12. Initially, chemosensitization was developed for a theoretical prophylactic clinical application because gene therapy technology could not be used to genetically modify all tumor cells in a tumor mass, a prerequisite for a therapeutic application. However, further investigation led to the fundamental discovery, termed the bystander effect, which allowed a therapeutic application of suicide gene therapy [26–29]. The bystander effect demonstrated that a tumor population could be killed if only a fraction of the tumor expressed the HSV-TK gene, and thus was a seminal finding that overcame the limitations of gene therapy technology, i.e., the inability to genetically modify an entire tumor mass [26–29].

The bystander effect allowed gene therapists to overcome the limitation in technology through a biological mechanism. Understanding the mechanism of the bystander effect will allow enhancement of this effect, which should improve the therapeutic potential of suicide gene therapy. Over two dozen clinical gene therapy trials based on the suicide gene model have been initiated in the last 7 years with two phase I trials complete and a phase III trial nearing completion [30,31].

II. NEGATIVE SELECTABLE MARKERS

A negative selectable marker is a gene whose protein product will kill a gene-modified cell under specific conditions. There are three main classes of negative selectable genes used in cancer gene therapy (Table 1). One class is genes that encode a cell surface protein (e.g.,

TABLE 1   Classes of Negative Selectable Markers

1. Cell surface expression of foreign proteins (e.g., HLA-B7).
2. Cytokine proteins that immunopotentiate an antitumor response (e.g., IL-2, GM-CSF).
3. Enzyme conversion of a prodrug to its toxic metabolite (e.g., HSV-TK).

HLA) that is foreign to the host. The expression of this foreign protein on the tumor cell stimulates the host's immune system to recognize and kill the tumor cell [32,33]. A second class is genes (e.g., cytokines) that encode a protein that enhances the host's immune cells ability to recognize the tumor cell as foreign [34,35]. These first two classes are usually considered as immune enhancement genes. The third class of negative selectable markers includes genes whose protein products convert a prodrug into its toxic form. This class is termed suicide genes, and it will be the focus of this chapter.

One of the first approaches to cancer gene therapy was developed by genetically modifying tumor cells with the herpes simplex virus thymidine kinase gene (HSV-TK) and showing that the gene-modified tumor cells were killed by the antiviral drug ganciclovir (GCV) [25,36]. GCV is a derivative of acyclovir, which was developed in the 1970s to treat herpes simplex virus infections [37–39]. These drugs are nucleoside analogs that can be phosphorylated by the viral thymidine kinase gene into a monophosphate form. Normal cellular enzymes can then di- and triphosphorylate the monophosphate form into a toxic drug, which functions as a DNA chain terminator and an inhibitor of the viral DNA polymerase [40–42]. The HSV-TK enzyme is almost 1,000-fold more efficient at monophosphorylating GCV than is the cellular thymidine kinase [40]. Thus, GCV is virtually nontoxic to uninfected cells at therapeutic concentrations of the drug (1–10 μM), although neutropenia can be a clinical manifestation of long-term drug usage [38,41,43]. Because the toxic metabolite of GCV is phosphorylated, its ability to cross cell membranes is impaired and thus its half-life ($t_{1/2}$) within the cell is six times longer (18–24 hours) than that of unmodified GCV [40,41]. The increased cellular $t_{1/2}$ of the phosphorylated GCV is an important feature of the anticancer effects of *HSV-TK*-gene-modified tumors, as discussed later. One advantage to this mechanism of action is that chemotherapy-resistant tumor cells, such as the SKOV-3 ovarian cancer cell line, are sensitive to GCV when they are genetically modified with the HSV-TK gene (Fig. 1) [26,29].

III. CHEMOSENSITIZATION

Initial studies demonstrated that the HSV-TK gene could be transferred into tumor cells and would confer sensitivity for GCV to the gene-modified cells [25]. Further, the drug kinetics in gene-modified cells were similar to those in virally infected cells [18–20,24–29,43]. The HSV-TK-gene-modified tumor cells were killed in culture at a concentration of 0.5 μM GCV, which can be achieved clinically. As shown in Fig. 1, the SKOV-3

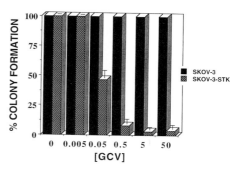

FIGURE 1   SKOV-3, an ovarian tumor cell line resistant to chemotherapy was transduced with the STK retroviral vector, which contained the neomycin resistance gene (*neo*^{R}) and the HSV-TK gene. HSV-TK gene-modified tumor cells were selected in G418 and then plated in culture in varying concentrations of GCV. Ten to fourteen days after GCV exposure, colonies were counted and compared to control cells (SKOV-3). HSV-TK gene-modified tumor cells died at drug concentrations of 0.5 μM and greater. However, even at high concentrations (50 μM), a small percentage of GCV-resistant gene-modified cells were demonstrated.

ovarian tumor cell line transduced with the HSV-TK gene became sensitive to GCV at 0.5 μM concentrations, the trough levels achieved *in vivo*. The maximum killing occurred at 5 μM and higher, which corresponds to the therapeutic peak levels achieved in patients (see Fig. 1). Because the SKOV-3 cell line is resistant to chemotherapeutic agents, the mechanism of drug resistance to cancer chemotherapeutic agents developed by tumor cells does not interfere with drug sensitivity to the antiviral agent GCV in the HSV-TK-gene-modified tumor cells. However, even at the highest GCV concentration, a small percentage of the HSV-TK-gene-modified cells are drug resistance, indicating that in virtually all circumstances drug resistance develops. The significance of drug resistance to GCV will be discussed later.

Further development of this concept was initially limited because gene transfer technology cannot be used to genetically modify an entire tumor mass. Thus, the use of the HSV-TK gene was originally envisioned in the "mosaic theory" for cancer therapy, which was proposed as a prophylactic approach for treating individuals who were genetically predisposed to developing cancer of a particular organ (e.g., bone marrow/leukemia) [25,27,44,45]. In this approach, it was hypothesized that such individuals would undergo genetic modification with a foreign gene of the susceptible organ; the bone marrow, for instance, in patients with myelodysplastic disease could be genetically modified with a suicide gene because these patients are at high risk of developing leukemia [44,45]. If a tumor subsequently developed from a gene-modified cell, then treatment with the appropriate prodrug would eliminate the tumor. Ultimately, mosaicism would be achieved by modifying the organ with several different negative selectable markers, i.e., HSV-TK, cytosine deaminase (CD), and xanthine-guanine phosphoribosyltransferase (XGPRT or *gpt*), such that different cells within the organ expressed a unique drug-sensitive phenotype (e.g., CD, HSV-TK, or XGPRT) [18,44]. In this situation, three populations of gene-modified cells would exist within the organ after genetic modification with three different negative selectable markers (HSV-TK, CD, and gpt). These gene-modified cells (HSV-TK, CD, or XGPRT) would be killed by GCV, 5-fluorocytosine (5-FC), and 6-thioxanthene (6-TX), respectively, if that cell population underwent malignant transformation (Table 2) [18,25,46–49]. Thus, if a cancer developed from one clonally expanded population (e.g., cells expressing HSV-TK), those cells could be selectively eliminated by treatment with the appropriate prodrug (e.g., GCV). The drug therapy would spare nonmalignant stem cells in the organ carrying the two other suicide genes (e.g., CD, XGPRT) that are different from the suicide gene (e.g., HSV-TK) in the transformed cells, thus allowing repopulation of the organ.

Transgenic animal models have been used to demonstrate this prophylactic approach with the HSV-TK gene under the control of the immunoglobulin promoter, which allowed expression of the suicide gene only in lymphoid cells. Lymphomas in these mice were induced by inoculation of the Abelson leukemia virus, and GCV therapy led to tumor remission in most animals [24]. Another study used transgenic mice whose cells contained the SV40 T antigen, which is tumorigenic, and the HSV-TK gene, both of which were under the control of the liver-specific alpha-fetoprotein promoter. GCV therapy delayed the onset of hepatoma formation and prolonged animal survival in the treated group [50].

Although the "mosaic theory" for cancer prevention was a theoretically appealing approach, the technology for a clinical trial has been limiting for a number of reasons. First, *in vivo* gene transfer into cells of an organ is very inefficient, especially when retroviral vectors are used [51]. Therefore, only a small portion of an organ (<1%) could be genetically modified. Recently, adenoviral vectors have been developed that transduce up to 100% of cells *in vivo,* but this vector remains episomal and is lost when the genetically modified cell divides [52]. Thus, these gene-modified cells would not maintain the transgene upon malignant transformation and cell division. Second, *in vivo* gene expression has been relatively short term, with only a small percentage of the transduced cells expressing the recombinant protein after a few months [53]. The inability to maintain long-term expression would severely limit this approach because most tumors develop over a period of years. Third, the killing of HSV-TK-gene-modified tumors may not always be complete in organs prone to develop cancer. In one study, hematopoietic tumors could not be completely killed even though all the cells contained the HSV-TK gene [24]. This suggested that not all genetically modified tumors were killed by GCV and that particular gene-modified cell types (e.g., lymphocytes) may be less sensitive to the toxic effects of GCV. Furthermore, even in the HSV-TK-gene-modified SKOV-3 ovarian cancer cells, in which 100% of the cells contain the HSV-TK gene and >99% were sensitive to GCV, a small percentage of these cells are resistant to GCV (Fig. 1). Although this approach to sensitize chemotherapy-resistant cells to an antiviral agent was developed to overcome drug resistance, GCV-resistant cells were observed even in a population of tumor cells in which every cell was genetically modified with the HSV-TK gene. Thus, as with virtually every drug therapy, antiviral drug-resistant cells develop, posing a potential limitation for this technology. For suicide gene therapy to become clinically feasible as a prophylactic therapy, these three limitations in gene therapy technology must be overcome. Solutions to some of these limitations will be required before negative selectable marker genes can be used in clinical gene therapy trials as a prophylaxis for cancer.

IV. DISCOVERY OF THE "BYSTANDER" EFFECT AND THERAPEUTIC CLINICAL APPLICATION OF HSV-TK SUICIDE GENE THERAPY

Treatment of an existing tumor with suicide gene therapy had a different set of requirements than those for prophylactic therapy, which were overcome in 1991

TABLE 2   Suicide Genes

Gene	Toxic metabolite	Tumoricidal	Bystander
HSV-TK	GCV → GCV-TP 15,17	Yes	Yes
CD	5-FC → 5-FU	Yes	Yes
XGPRT (*gpt*)	6-TX → 6-TX-TP	Yes	?
VZV-TK	araM → araM-MP	Yes	?
DeoD	MeP-dr → 6-MeP	Yes	?
β-glucosidase	Amygdalin → cyanide	?	?
β-lactamase	Vinca-cephalosporin → vinca alkaloid	?	?

because of a seminal discovery by Freeman and colleagues (Table 3) [20,26–29]. The first clinical application of the HSV-TK gene for cancer gene therapy was based on the observation that HSV-TK-gene-modified tumor cells were toxic to nearby unmodified tumor cells when the HSV-TK-modified cells were exposed to GCV [20,26–29]. An important feature of this effect, subsequently termed the bystander effect, was that when as few as 10% of the cells in a tumor contained the HSV-TK-gene-modified cells, the entire population of cells could be killed [26–29]. This discovery allowed the development of a therapeutic approach to the treatment of cancer because it overcame the need to genetically modify the entire tumor (transduction efficiency). Because only a fraction of the tumor mass needed to be genetically modified to cause tumor regression, the state-of-the-art gene transfer technology could be used to transduce a sufficient number of tumor cells in the tumor mass to allow for the bystander effect. In addition, long-term gene expression was not required because tumor cells expressing the HSV-TK suicide gene are killed after only a short exposure to GCV (Table 3). The requirement for short-term GCV exposure is related to the mechanism of cell death. Studies have shown that HSV-TK-gene-modified tumor cells undergo apoptotic or programmed cell death [27,28]. This programmed cell death, which occurs over several days, is initiated when the HSV-TK-gene-modified cells are exposed to GCV for only 6 hours. Thus, only short-term exposure to GCV is required because the triphosphorylated toxic drug metabolite initiates an irreversible protein cascade, leading to programmed cell death within hours of its metabolism.

The discovery of the bystander effect has led to intensive study of this phenomenon to understand its mechanism [54,55]. A number of mechanisms have been proposed to explain the bystander effect, including gap junction intercellular communication (GJIC), apoptotic cell death, activation of an antitumor immune response, release of cytokines, and disruption of tumor vasculature. Studies have shown that the bystander effect is a complex biological process involving several mechanisms, which may involve different mechanisms *in vitro*

TABLE 3   Benefits of Tumor Therapy Using Suicide Genes

1. The toxic metabolite can kill chemotherapy-resistant tumors.
2. Requires short-term gene expression.
3. Requires transduction of only a fraction of the tumor cells.
4. Generation of a bystander effect.

and *in vivo* (see Table 4). *In vitro* studies suggested that the bystander effect resulted from transfer of toxic GCV metabolites (phosphorylated GCV) from HSV-TK-gene-modified cells to nearby unmodified tumor cells because tumor cell proximity was required to observe the effect [28,56–58]. Further studies using tritiated GCV demonstrated that toxic GCV metabolites were transferred from HSV-TK-modified tumor cells to unmodified tumor cells [59]. The two explanations proposed for the intercellular transfer of phosphorylated GCV were (1) phagocytosis by adjacent unmodified tumor cells of the apoptotic vesicles from dying HSV-TK tumor cells, or (2) gap junction communication between adjacent gene-modified and unmodified cells [28,29,60–63].

Apoptotic cell death is morphologically characterized by chromatin condensation, cell shrinkage, and formation of apoptotic vesicles as the cell breaks up [64]. As described earlier, the HSV-TK enzyme phosphorylates GCV into a monophosphate, which does not readily cross cell membranes and thus becomes entrapped in the apoptotic vesicles. Cells within a tumor are in an "equilibrium growth state" in which the cells are continually growing through cell division and dying because of an outpacing of their blood supply. Nearby tumor cells that express cell surface receptors for apoptotic vesicles clear the cell debris from the tumor environment by phagocytosis, which may allow them to take up the apoptotic vesicles containing phosphorylated toxic GCV metabolites [28,60,65–67].

The apoptotic cell death mechanism is consistent with the finding that close proximity of the genetically modified tumor cells to the unmodified cells was necessary for the *in vitro* bystander effect. However, another possible mechanism for the transfer of phosphorylated GCV that would also require cell proximity is through gap junctions [59]. Gap junctions, which are formed by protein called connexins, allow cellular communication by establishing cell-to-cell channels, which can generate ionic gradients [68]. Molecules 1,000 Da in size can pass between cells via gap junctions [69]. A number of studies have shown that gap-junction-deficient cells generated a weak bystander effect and that transfection of the connexin gene into these cells could restore the bystander effect [70–73]. However, as discussed earlier, the importance of transfer of toxic metabolites from gene-modified cells to unmodified cells to produce an *in vivo* bystander effect is only one part of the mechanism. Most studies of intercellular transfer of toxic GCV metabolites relied on *in vitro* models, although some *in vivo* studies demonstrated that gap junctions may play a role in the bystander effect [70,74]. The demonstrated role of gap junctions in these studies did not rule out other possible mechanisms. Additional studies, includ-

Table 4 The Three Phases of the Bystander Effect

Type	Description	In vitro	In vivo
1. Chemosensitization and transfer of drug sensitivity	Gene modification of the tumor with a suicide gene confers a new drug phenotype to the tumor cells and nearby unmodified tumor cells.	+	+
2. Hemorrhagic tumor necrosis (HTN)	Release of proinflammatory cytokines by the dying HSV-TK gene-modified cells leads to HTN and influx of immune cells into the tumor mass.	−	+
3. Immunostimulatory tumor microenvironment	The release of cytokines by tumor-infiltrating lymphocytes within the tumor leads to alteration of the tumor microenvironment with upregulation of immune regulatory molecules (ICAM, B7).	−	+

ing functional studies that block gap-junction communication between tumor cell lines, will be necessary to determine the *in vivo* role of gap junctions in the bystander effect. Thus, the role of toxic ganciclovir metabolites from HSV-TK-gene-modified to nearby unmodified tumor cells is the predominant mechanism of the *in vitro* bystander effect and one of the *in vivo* mechanisms, as will be discussed later (Table 4).

A. *In Vivo* Mechanism of Action of the Bystander Effect

The bystander effect is a complex biological process that allows killing of a tumor mass when only a fraction of the tumor is genetically modified. The *in vivo* mechanism of the bystander effect appears to encompass a broader spectrum of tumoricidal activity than the *in vitro* bystander effect to allow killing of nearby tumor cells by HSV-TK gene-modified tumor cells. This is due to the limited transfer of toxic GCV metabolites to nearby tumor cells and the requirement for a functional immune system in the *in vivo* bystander effect. As described earlier, *in vitro* studies have demonstrated that the bystander effect results from transfer of toxic GCV metabolites from HSV-TK tumor cells to nearby unmodified tumor cells [59]. Although a few studies suggest that transfer of toxic GVC metabolites occurs *in vivo* and contributes to the *in vivo* bystander effect, a number of studies demonstrate that an additional mechanism(s) is operational *in vivo* to account for the diverse findings. Studies analyzing the *in vivo* bystander effect have shown that hemorrhagic tumor necrosis, antitumor immunity, and release of cytokines are important [20,26–29,61,75–84].

B. The *in Vivo* Bystander Effect

Generation of an *in vivo* bystander effect is necessary for an effective treatment because even in a population of 100% HSV-TK-gene-modified tumor cells, resistant cells are present, as shown in Fig. 1. The unique feature of the bystander effect is that a tumor mass consisting of drug-resistant and -sensitive cells will undergo regression when exposed to the drug. Different processes may be active *in vivo* and *in vitro*, because *in vitro* cell death by bystander killing occurred by apoptosis over a 3-day time period, but *in vivo*, a hemorrhagic tumor necrosis (HTN) occurred within 24 hours of treatment (Fig. 2) [27,84]. This demonstrated that a more rapid mechanism of bystander killing was occurring *in vivo* [15,20,26–29,75]. The occurrence of HTN was also reported by other investigators using the HV-TK system, although different experimental approaches led to different hypotheses regarding its mechanism [80]. The first model system studied used inoculation of HSV-TK retroviral vector producer cells intratumorally, and the second used inoculation of HSV-TK-gene-modified tumor cells around or into the tumor. In the former system, because vector producer lines were injected directly into the tumor and HTN occurred with GCV treatment, the investigators hypothesized that the HSV-TK gene was being transferred to endothelial cells lining the tumor's blood vessels from the HSV-TK retroviral producer cells through production of retroviral particles. The HSV-TK-transduced endothelial cells were killed after exposure to GCV, which leads to HTN [80]. In the latter system, which inoculated HSV-TK-gene-modified cells that were not producer cells, HTN was generated in the absence of retroviral particles and could not be attributed to *in vivo* transfer of the HSV-TK gene to endothelial cells within the blood vessels of the tumor [15,20]. Furthermore, in these studies, a centralized hemorrhagic tumor necrosis occurred within 24 hours of inoculation of the HSV-TK-gene-modified cells (Fig. 2) [15,26–29,84,85]. The observation of a centralized HTN was unexpected. If the HTN occurred because of transfer of toxic GCV metabolites to nearby cells, then the HTN should begin at the periphery of the tumor

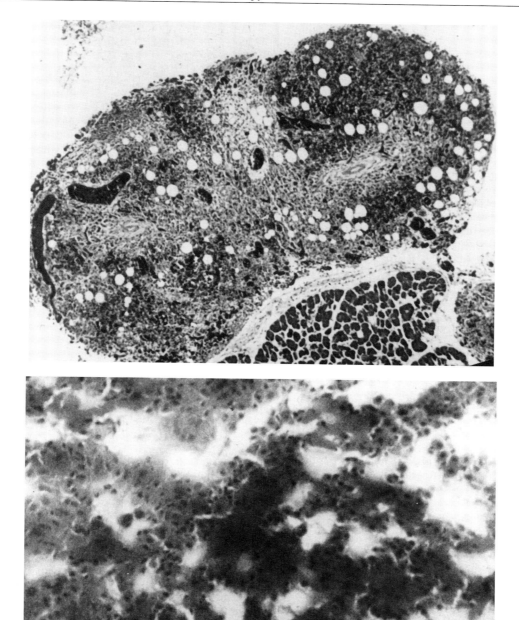

FIGURE 2   Hemorrhagic tumor necrosis. Mice with intraperitoneal tumors were treated with an i.p. injection of HSV-TK gene-modified cells and GCV. Twenty-four hours later the animals were killed and the tumors were removed. (A) A photograph of the tumor histology showing a centralized hemorrhagic necrosis. (B) A photograph of a microscopic section of the tumor showing the hemorrhage in the tumor.

where the HSV-TK-gene-modified tumor cells would contact with the tumor mass. The observation of a relatively rapid and centralized hemorrhagic tumor necrosis after HSV-TK/GCV treatment was unexpected and sug- gested that a mechanism(s) in addition to the transfer of toxic metabolites was responsible for the *in vivo* by- stander effect (Fig. 2). Another striking observation was the demonstration that the bystander effect was diminished or abrogated in athymic nude mice (Table 5) [20,26–29,61,78,79,86]. This indicated that the immune system played an integral role in the bystander effect *in vivo* [76,82,84].

These findings suggested that a complex biological process leads to the *in vivo* bystander effect (Table 4). First, the centralized hemorrhagic necrosis suggests that a soluble factor may be acting on the central tumor vasculature. Second, as previously mentioned, the HSV-TK-gene-modified cells die by apoptosis over a 3-day period whereas HTN occurred within 24 hours of treatment with GCV. Third, the host immune/inflammatory system is required. Thus, the *in vivo* bystander effect apparently relies on three processes to cause tumor regression: chemosensitization of tumor cells to GCV, hemorrhagic tumor necrosis, and antitumor immunity (Table 4). The process of chemosensitization leads to *in vivo* killing of the HSV-TK gene-modified tumor cells and some nearby unmodified tumor cells through transfer of toxic GCV metabolites from the GCV exposed HSV-TK-gene-modified tumor cells to unmodified tumor cells. Proinflammatory cytokines (TNF and IL-1), released from HSV-TK-gene-modified cells undergoing apoptosis, are involved in generating the centralized HTN observed after the inoculation of HSV-TK tumor cells and GCV because of the known effects of these cytokines on the tumor vasculature [84,85,87–89]. Their role was suggested by the presence of TNF, IL-1, and IL-6 mRNA in tumors within 24 hours after treatment of HSV-TK-gene-modified cells with GCV [15,20,83–85,89]. Furthermore, recent *in vivo* functional studies using the soluble TNF receptor and IL-1 receptor antagonist demonstrated that by blocking the functional effects of TNF and/or IL-1, the bystander effect was diminished. Although TNF and IL-1 are important for the *in vivo* bystander effect, they are not sufficient because their blockage did not completely eliminate the bystander effect (Ramesh *et al.,* unpublished data). Each of the first two phases, chemosensitization and hemorrhagic tumor necrosis, leads to killing of *in situ* tumor, the former by a chemical process, GCV toxicity on nearby tumor cells, and the latter through disruption of the tumor vasculature and thus depletion of nutrients

to the tumor mass. The majority of the tumor mass can be killed by these two processes, but residual tumor remains [26,29].

The hemorrhagic tumor necrosis can then lead to a cellular infiltrate within the tumor, especially because the expression of cytokines within the tumor leads to up-regulation of adhesion molecules and the binding of lymphocytes to cells within the tumor [76,77,84]. Studies have shown an increase in T cells and macrophages in the tumor after treatment with HSV-TK-gene-modified cells [15,20,26,29,76,77,83,84,90]. The release of cytokines within the tumor also affects the microenvironment of the tumor by causing the up-regulation of adhesion and immune regulatory molecules [91]. Increased expression of immune regulatory molecules (ICAM-1, B7-1, B7-2, MHC) within the tumor levels have been shown after inoculation of the HSV-TK-gene-modified cells [77]. It has been hypothesized that the tumor-infiltrating cells can become activated to develop an antitumor cytotoxic response within this new environment [15,20,26,29,76,77,83,84,92].

C. Immune Components of the Bystander Effect

Recent studies have shown that T cells isolated from the tumor after inoculation of HSV-TK tumor cells and GCV have a higher proliferative response to syngeneic tumor cells than do T cells isolated from control tumors [76,77,92]. As mentioned previously, the role of the immune system has been further demonstrated in studies in which the "bystander effect" has been abrogated in immunodeficient mice [26,29,61,78,79,81,86]. Taken together, these findings suggest that treatment of tumors with HSV-TK-gene modified cells and GCV leads to an alteration of the tumor microenvironment from one that is immunoinhibitory to one that is stimulatory, and that the stimulatory environment serves to generate an antitumor immune response [20,76,77,83,84]. This antitumor immunity can be generated because GCV is not immunosuppressive, which may be why the bystander effect has not been observed with cancer chemotherapeutic agents. Thus, the bystander effect results from a complex biological process consisting of chemosensitization, hemorrhagic tumor necrosis, and an antitumor immune response (Table 4). Furthermore, the immune response can be divided into two parts, early and late. In the early component, as described earlier, cytokine release leads to alteration of the tumor microenvironment to one that is immunostimulatory. This immunostimulatory microenvironment by itself is usually too weak to completely eradicate tumor growth, but it has

TABLE 5   Role of Immunity

| Tumor mass % HSV-TK | Percent tumor regression | |
	Immune competent	Immune deficient
100	100	25
50	100	0
0	0	0

profound implications. First, it leads to the "late" component of the immune response: the development of an antitimor immune response over several weeks [26,29,47,76,77,79,81,83,84,93,94]. In some reports, this antitumor immunity accounted for an immediate antitumor effect observed in nearby untreated tumors, but this appears to be the exception [74,81]. However, the development of antitumor immunity occurs over several weeks (the late component, delayed hypersensitivity immune response), which allows regrowth of residual tumor cells and outpacing of the antitumor immune response. Thus, mechanisms to enhance or take advantage of the early immune component (i.e., alteration of the tumor microenvironment) to kill residual tumor cells will enhance the bystander effect.

One method to enhance the bystander effect relies on using the early phase of immune activation as an adjuvant to immunotherapy. During this early phase there is an alteration of the tumor microenvironment from one that is immunosuppressive to one that is immunostimulatory, thus generating a microenvironment for immune effector cells to more effectively kill the tumor (Fig. 3). Immunotherapy for tumors largely consists of adoptive and active immunotherapy. In the former, immune effector cells are grown *ex vivo* prior to inocula-tion into the tumor-bearing host. In the latter, tumor-bearing hosts are immunized to their tumor through vaccination. In both cases, the immune effector cells are in the peripheral circulation. One reason that immunotherapy has met with limited success may be because the tumor microenvironment is generally immunosuppressive due to a variety of tumor immunosuppressive factors. Thus, when peripheral blood "activated" immune effector cells generated by adoptive or active immunotherapy traffic into the immunosuppressive tumor microenvironment, they become inactivated. Immunotherapy can be potentiated by vaccinating tumor-bearing mice with tumor cells prior to injection of the HSV-TK cells [29,92]. In this case, the immunized mice have antitumor immune effector cells in their peripheral circulation so that when the mice are treated with HSV-TK therapy, the hemorrhagic tumor necrosis allows the immune effector cells to enter the tumor. The altered immunostimulatory tumor microenvironment provides the immune effector cells an environment to efficiently function and kill residual tumor cells. The combination of suicide gene therapy and tumor vaccine allows synergy of these two therapies. Thus, HSV-TK suicide gene therapy may provide the second event necessary for successful immunotherapy, an immunostimulatory tu-

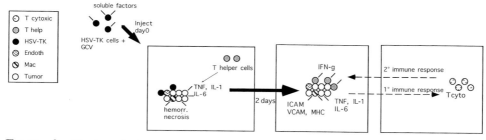

FIGURE 3 The hypothesized mechanism of the *in vivo* bystander effect, which relies on altering the tumor microenvironment from one that is inhibitory to one that is stimulatory. In this schematic representation, initially the HSV-TK (TK) tumor cells are injected i.p. (or intratumorally) and exposed to GCV. The gene-modified tumor cells home to the *in situ* tumor deposit. On day 0, GCV exposure induces the HSV-TK-gene-modified tumor cells to undergo apoptotic cell death and release soluble factors such as cytokines, which activate tumor-infiltrating cells within the tumor to release additional cytokines. This leads to a hemorrhagic tumor necrosis, which allows lymphocytes to infiltrate the tumor. The cytokine release also causes up-regulation of adhesion molecules (e.g., ICAM-1) and immune regulatory cell surface proteins (e.g., B7), causing the lymphocytes to adhere to the tumor and become activated. The tumor microenvironment becomes immunostimulatory, and the process is termed the early phase of the immune component of the bystander effect. This immunostimulatory tumor microenvironment environment leads to antitumor immunity over the next several weeks (1° immune response) which is termed the late phase of the immune component of the bystander effect. In addition, the immunostimulatory tumor microenvironment (early phase) serves as an ideal environment for immune effector cells generated by adoptive or active immunotherapy (2° immune response) to traffic into the tumor and kill it. Thus, the HSV-TK gene therapy can serve as an adjuvant to immunotherapy because it provides a necessary requisite for successfully immunotherapy, an immunostimulatory tumor environment.

mor microenvironment that expresses the immune regulatory molecules (e.g., ICAM, B7) needed for an antitumor immune response to occur when immune effector cells in the peripheral circulation enter the tumor (Fig. 3) [26,29,92].

D. Cytokine Mediators of the Bystander Effect

Another method to enhance the early immune phase of the bystander effect, immunomodulation with cytokines, is also based on the discovery that the bystander effect is dependent on an intact immune system. These early studies have met with varying degrees of success [94]. The combination of HSV-TK therapy and cytokines may lead to an enhanced bystander effect because cytokines are involved in this effect. Initial studies used IL-2-secreting cells, which were administered in conjunction with the injected HSV-TK therapy but failed to demonstrate an enhanced "bystander effect" [95]. It did, however, show an enhanced development of tumor immunity or the late phase of the immune component of the bystander effect. This is consistent with some studies, which have shown that these dying HSV-TK tumor cells are antigenic and in some cases more antigenic than the unmodified irradiated tumor cells [26,29,78]. Although the combination of HSV-TK-gene-modified tumor cells and IL-2-modified tumor cells did not enhance the bystander effect when compared to HSV-TK tumor cells alone, the addition of the IL-2 secreting tumor cells increased the late phase of tumor immunity in the surviving animals [95].

The first study to demonstrate enhancement of the bystander effect using cytokines showed that IFN-α could enhance the effect [96]. This study's findings fit the model previously proposed [76], which showed that inflammatory cytokines (i.e., TNF, IL-1, IL-6) were generated by HSV-TK therapy (Fig. 3). Therefore, interferon appeared to be a more appropriate choice than IL-2 to enhance the early phase of the bystander effect. A second study using IL-2 and GM-CSF in combination with HSV-TK therapy also had some effect on enhancing the bystander effect [97]. This enhancement may again relate to the observation that GM-CSF is released within the tumor after HSV-TK/GCV treatment. In addition to IFN-α, IL-4 has been shown to enhance the bystander effect [98]. The ability to enhance the bystander effect using cytokines must be analyzed in terms of the early and late phases of the immune component. IL-2 enhances the late component (delayed hypersensitivity immunity), whereas IL-4 and IFN-α enhance the early component, the immunostimulatory tumor environment, and immediate immune cell activation [76].

Drugs have been used to alter the bystander effect, apparently through mechanisms other than the immune system. Verapamil and Forskolin have been shown to inhibit the bystander effect [60,99]. Retinoids have been shown to enhance the bystander effect [100]. The enhancement is hypothesized to be secondary to an increase in gap junctions, which are stimulated by retinoids. Other studies have shown that HSV-TK expression in tumor cells leads to increased sensitivity of the gene-modified tumor cells to radiation [101]. Other investigators have focused on increasing HSV-TK expression in tumor cells or genetically engineering the HSV-TK protein to increase its affinity to the GCV substrate to enhance the bystander effect [102–105]. The overall complexity of HSV-TK-gene therapy system necessitates further study to better understand the best agents to enhance this effect [54,55].

V. DELIVERY APPROACHES OF HSV-TK TO THE TUMOR

The first approved HSV-TK clinical trial used HSV-TK-gene-modified tumor cells as the vehicle to genetically modify a tumor mass [26,29]. Murine studies showed that i.p. injection of HSV-TK tumor cells into tumor-bearing mice with an i.p. tumor resulted in prolonged animal survival [26,29]. The study used fluorescein-labeled HSV-TK-gene-modified cells to demonstrate that the i.p. inoculated gene-modified tumor cells would associate with *in situ* tumor deposits, allowing up to 10% of the tumor mass to consist of HSV-TK cells [75]. This study showed that tumor cells could be used as a delivery vehicle to target a tumor (Fig. 4). The mechanism of homing of the inoculated gene-modified tumor cells to tumor deposits is currently being studied, and possible explanations include these: (1) Tumor cells associate more readily with tumor cells than with other nonmalignant cells; (2) tumor masses produce chemotactic factors that attract other tumor cells; and (3) *in situ* i.p. tumor masses are devoid of a "repellent" mesothelial lining, thus allowing adherence of other cells (i.e., tumor cells). More extensive studies will be necessary to determine the mechanism by which injected i.p. tumor cells can associate with an *in situ* tumor [75].

The HSV-TK system has also been used to treat other localized tumor masses [106–108]. A separate approach to genetically alter the tumor with the HSV-TK gene was developed through studies of *in vivo* gene transfer in the brain [109]. Retroviral vectors were used to specifically target tumor cells because only dividing cells are efficiently transduced by these viral vector particles and most normal brain cells do not divide (Fig. 4). HSV-

Comparison of vectors used to deliver HSV-tk as a suicide gene

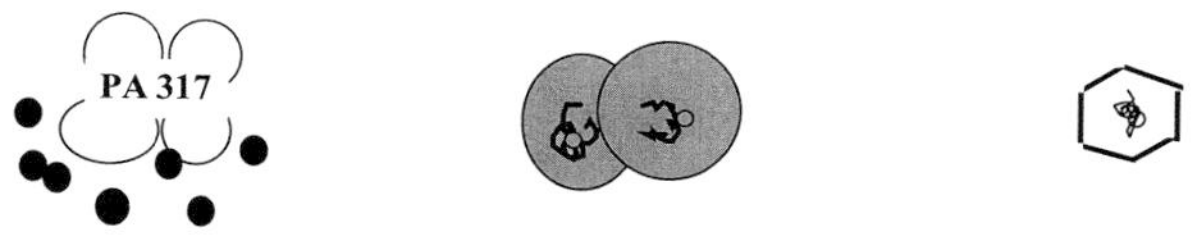

	Vector Producer cells (VPC)	Transduced cells	Adenovirus
	PA 317		
Transduction Efficiency (*in-vivo* or *ex-vivo*)	low (*in-vivo*)	high (*ex-vivo*)	high (*ex-vivo* and *in-vivo*)
Gene Delivery	non-specific	specific	non-specific
Vector Integration	yes	yes	episomal
Cell division requirement	yes	yes	no
Advantage	*in-vivo* transduction	cell targeting	high efficiency gene transfer
Disadvantage	Inactivation by complement	laborious and expensive	host immune response

FIGURE 4 HSV-TK Delivery Modes. Three different methods are used to deliver the HSV-TK gene to a tumor mass:

1. HSV-TK producer cells (VPCs), which generate retroviral particles, have been implanted into a tumor mass to generate *in vivo* transduction of tumor cells within the tumor mass. This method relies on tumor cell division for tumor specificity because retroviral integration is dependent on cell division. The *in vivo* HSV-TK gene transduction leads to a fraction of the tumor mass expressing the HSV-TK gene and thus allows the generation of the bystander effect.
2. HSV-TK gene-modified tumor cells have been used to genetically alter the tumor mass. When injected i.p., these gene-modified tumor cells home to i.p. tumor deposits and thus generate a tumor mass consisting of HSV-TK-positive and -negative tumor cells. Treatment with GCV leads to the bystander effect. Intratumoral injection of the HSV-TK-gene-modified cells also generates the bystander effect.
3. HSV-TK adenoviral vectors have been used to genetically modify the tumor mass. The adenoviral vector has been injected intratumorally into a tumor mass, which leads to transduction of a high portion of the tumor cells, but also nearby normal cells. This vector does not require cell division because it remains episomal, but it is lost to one daughter cell upon cell division.

TK producer cells, which generate retroviral particles encoding the HSV-TK gene, were injected into an established rat glioblastoma and showed a significant antitumor effect after GCV therapy [93,106–114]. This approach has been used with other localized tumors as well [115], although one study showed no evidence of an antitumor effect by the gene-modified cells [116]. Potential shortcomings to this approach include these: (1) retroviral vectors do not efficiently transduce tumor cells *in vivo,* even if they are dividing; (2) the mitotic activity of human tumors is low; and (3) the injection of producer cells into solid tumor masses results in inadequate diffusion of retroviral particles and thus only allows gene transfer to cells near the producer cells. To address some of these difficulties, recent clinical trials to

treat brain tumors using this approach initially surgically debulk the tumor to form a cavity into which the producer cells are infused. The purpose of this procedure is to eliminate residual tumor cells from the debulked tumor cavity and allow better diffusion of retroviral particles to the residual tumor cells [117].

A. Adenoviral Vectors Containing HSV-TK

To overcome the problems of inefficient *in vivo* gene transfer using retroviral vectors, adenoviral vectors have been used to transfer the HSV-TK gene to a tumor mass (Fig. 4) [118,123]. The advantage of these vectors is that they efficiently transduce cells *in vivo*. The disadvantages are these: (1) The adenoviral vector transduces all cells, both dividing and nondividing cells; (2) because the vector remains episomal, the antitumor effect must occur before cells divide because only one daughter cell retains the transgene; (3) adenoviral vectors evoke an immune response to the transduced cells; and (4) adenoviral vectors can be toxic to the host tissue. The first disadvantage relates to all methods used to deliver the HSV-TK gene to tumors (producer cells, adenoviral vectors) using viral vectors. To date, no viral delivery method specifically targets the HSV-TK gene to the tumor, which may be important to prevent killing of normal cells, although the cell-diversity approach selectively targets the gene-modified cells to the tumor, as described earlier (Fig. 4). Investigators have attempted direct injection of HSV-TK adenoviral vectors to treat localized tumors, such as squamous cell carcinoma and melanoma, in an effect to keep the viral vector localized to the tumor [118,120]. Another approach uses inoculation of HSV-TK-containing adenoviral vectors into the pleural or peritoneal cavity of rats with pleural or peritoneal tumors [121–123]. In most experiments, animal survival has been prolonged but long-term cure has been elusive. Further success of both adenoviral-vector-based and other approaches will likely depend on the development of more sophisticated techniques to target the delivery of the HSV-TK gene to an *in situ* tumor. Curiel and colleagues have been successfully attempting to target adenoviral vectors by modifying their viral coat or complexing them to other proteins (e.g., ligands or antibodies) to add a cell targeting capacity to the vector [124–126].

The second disadvantage, that the vector remains episomal, may not be disadvantageous to their use in suicide cancer gene therapy because expression of the transferred HSV-TK gene can occur within several hours of viral cell entry, which would allow for GCV treatment before cell division and subsequent loss of gene expression. The third and fourth disadvantages, immune response and toxicity respectively, are being addressed through genetic modification of the adenoviral vector to decrease its immunogenicity and toxicity. However, some fundamental differences in adenoviral vectors, as compared to the other approaches, may make them less well suited for HSV-TK-gene therapy. One animal study using adenoviral vectors to deliver the HSV-TK gene demonstrated that a better antitumor response was observed in immune-deficient mice than immune-competent mice [72]. Because by bystander effect effect is dependent on an intact immune system, the finding that adenoviral vectors generate a better antitumor effect in immune-deficient animals is contrary to findings in other systems (i.e., cell delivery and retro-

viral vector) and thus eliminates a major advantage of HSV-TK gene therapy, the immune component.

B. Herpes Simplex Vectors

Another approach for delivery of the HSV-TK gene to the tumor uses a genetically modified mutant herpes simplex virus to infect brain tumor cells [127]. The herpes virus has a tropism for neural tissue, and it may be possible to use an avirulent virus to deliver the HSV-TK gene to the brain tumor cells. These studies were based on previous experiments showing that mutant herpes viruses could be oncolytic, independent of the HSV-TK gene [128]. The potential to genetically modify the herpes simplex virus to combine an oncolytic effect and the HSV-TK/GCV suicide effect could have important implications to gene therapy for brain tumors. Recent approaches are combining the oncolytic properties of the defective virus with GCV prodrug sensitivity through increased expression of the HSV-TK viral gene [129]. However, the use of live virus may have toxicity, which will need to be addressed.

Other nonviral approaches to tumor targeting include the use of ligands, liposomes, and antibodies [130–139]. Liposomes are used to encapsulate DNA into a lipid membrane, which can then be used to fuse with a cell membrane to deliver the DNA. By incorporating proteins (e.g., ligands, antibodies) into the liposome, cell targeting can be achieved. In addition, other proteins, ligands and antibodies can be complexed directly to DNA and used to target the DNA to a specific cell [133,134,140,141].

Another targeting approach uses nontumor cells to target the tumor mass. As discussed earlier, tumor cells injected i.p. into mice "home" to i.p. tumor masses. Cells have been previously been shown to target to their organs of origin, as is observed with i.v. injection of bone marrow cells or liver cells. Thus, gene-modified cells may offer an approach for selectively targeting an organ. Another cell based study using gene-modified endothelial cells (GMECs) is being developed by Zwiebel and colleagues, who have demonstrated that i.v. injected gene-modified endothelial cells target to sites of angiogenesis [142–145]. The endothelial cells become incorporated into the newly forming blood vessels of the tumor. By genetically modifying the ECs with either the IL-2 or HSV-TK genes, the GMECs can deliver the gene to the tumor vasculature, and tumor regression in animals has been demonstrated.

Tissue-specific promoters may achieve cell-targeted gene expression without the need to genetically modify and alter the specificity of viral vectors. Promoters are involved in the transcription of the gene into mRNA, and certain promoters are only active in specific tissues. Tissue-specific expression of a suicide gene could thus be achieved by linking the tissue-specific promoter to the transgene in the vector [146–148]. This specificity would allow selective expression of suicide genes to kill tumor cells specifically and spare other normal cells. Expression of a marker gene (β-galactosidase) in mouse glioblastoma cells was shown to depend on the presence of a glial-cell-specific promoter [112]. A regulatory element from the carcinoembryonic antigen (CEA) gene allows tissue specific expression of the HSV-TK gene in lung carcinoma cells and pancreatic cells, and one from the tyrosinase promoter has shown melanoma-specific expression. One limitation to this approach is that few promoters have been shown to be tissue-specific *in vivo*. Further work in developing tissue-specific promoters will have an impact on developing suicide gene therapy.

The ability to target a gene and/or its expression to a specific cell or tumor mass will be critical for further development of suicide gene therapy approaches to treat cancer.

VI. OTHER NEGATIVE SELECTABLE MARKERS

Other genes have also been used as negative selectable markers, including cytosine deaminase (CD), which converts the nontoxic prodrug 5-fluorocytosine (5-FC) to the toxic metabolite 5-fluorouracil (5-FU) (Table 2). Initial studies demonstrated that CD could kill tumor cells when a capsule containing the enzyme was implanted into a tumor and the animals were treated with 5-FC. Diffusion of the enzyme into the tumor cells allowed conversion of 5-FC into 5-FU [149–151]. Genetic modification of tumor cells became possible after the CD gene was cloned [152], and tumor cells genetically modified with the CD gene were shown to be sensitive to 5-FC [46,47,49,79,153]. Although not initially observed with CD, the bystander effect was clearly shown in subsequent studies [46,153]. Some discrepancies between the mechanism of bystander killing using HSV-TK and other negative selective genes may exist. Whereas the toxic metabolite of the HSV-TK/ganciclovir system is a phosphorylated compound, the toxic compound in the CD/5-FC system is not and thus readily crosses cell membranes. It has been proposed that the soluble toxic metabolite (5-FU) causes a more powerful bystander effect because of its ability to readily diffuse to nearby tumor cells and directly kill them. This mechanism has not been confirmed, and further studies will be required to determine the mechanism and efficiency of the bystander killing observed in this system.

The issue to remember, as described earlier, is that the bystander effect has three unique characteristics: chemosensitization, hemorrhagic tumor necrosis, and the requirement of a functional immune system (Table 4). The bystander effect differentiates itself by these three features from other negative selective genes, which rely on the killing of nearby cells only through the diffusion of toxic metabolites. Some suicide genes may produce a more potent toxic metabolite from a prodrug but without the other characteristic features would not be classified as generating a bystander effect and thus may have a weaker antitumor effect. Therefore, each suicide gene must be evaluated for its ability to generate the bystander effect, and to date only HSV-TK and CD have shown these characteristic features.

The requirement for a functional immune system has been demonstrated for both the HSV-TK/GCV and CD systems, because both therapies were less effective in immunodeficient and T-cell-depleted mice [26–29,79,153]. Some *in vivo* studies have demonstrated that an antitumor immune response develops to CD-gene-modified tumors after 5-FC infusion [79,153]. The dying CD-gene-modified tumor cells served as a tumor vaccine [79]. These results are similar to those described above in which dying HSV-TK-gene-modified cells served as a tumor vaccine [26–29,77,78,83,84,146]. The mechanism of cell death from 5-FC has not been studied, but one possible explanation is apoptosis, which would result in the generation of soluble factors. Comparison studies between HSV-TK and CD have been performed to determine which suicide gene therapy approach is more effective. Although initial reports indicated that CD generated a more potent bystander effect based on a smaller fraction of the tumor mass that needed to be genetically modified, the studies were performed in nude mice and thus eliminated a key feature of the bystander effect [154–156]. Recent studies have demonstrated that a more potent effect is observed with cells expressing both the HSV-TK and CD suicide genes [157]. Further studies using a therapeutic model in immune-competent animals need to be performed before this issue can be resolved.

Other genes that can serve as negative selectable markers are shown in Table 2 [17,18]. However, they have not been as well characterized as the HSV-TK or CD systems and the existence of the bystander effect has not been extensively studied. One of the first negative selectable genes used to genetically modify tumor cells was the varicella-zoster virus thymidine kinase (VZV-TK) gene [48,158]. The prodrug for the VZV-TK enzyme, 6-methoxypurine arabinonucleoside (araM), becomes monophosphorylated into araM monophosphate. In addition, several *Escherichia coli* genes have been used. The *gpt* gene, which encodes the enzyme xanthine-guanine-phosphoribosyl transferase (XGPRT), can convert 6-thioxanthene into its monophosphate form, which can then be triphosphorylated into its toxic metabolite [18,159]. Although the *gpt* gene may be less effective than the HSV-TK gene as a suicide gene, it has the advantage that it can also be used for positive selection by exposing the gene-modified cells to mycophenolic acid [160,161]. The *E. coli* purine nucleoside phosphorylase and nitroreductase genes, which convert 6-methylpurine-2′-deoxyribonucleoside (MeP-dr) and the weak alkylating agent 5-azauridine-1-yl-2,4-dinitrobenzadine (CB1954) to 6-methylpurine and a strong alkylating agent, respectively, are toxic to the cell, with resultant bystander killing [17–19,162–165]. Other genes that could also be used as suicide genes are β-lactamase and β-glucosidase, which can enzymatically alter vinca-cephalosporin to vinca alkaloid and amygdalin to cyanide, [17–19]. Furthermore, genes that do not metabolize a prodrug can also generate the bystander effect. Gene transfer of the wild-type p53 gene into tumor cells with a mutated p53 gene has shown a bystander effect [166,167,168].

A. Clinical Trials

Over 200 clinical gene therapy trials have been approved, more than 50% of which are related to cancer [30,31]. Suicide genes are employed in a significant number of the cancer trials because of the several advantages described earlier (Table 6): First, suicide genes and their prodrug are toxic to chemotherapy-resistant tumors; second, only short-term gene expression is required; and third, only a fraction of the tumor cells within the tumor mass (>10%) need to express the suicide gene to kill the entire tumor, i.e., the bystander effect (Tables 3 and 4). The bystander effect allows for cytoreduction of the tumor mass through the killing of chemosensitized gene-modified tumor cells and some nearby unmodified tumor cells by a prodrug (i.e., GCV).

The first clinical gene therapy trial using the HSV-TK gene was approved in 1991 to treat ovarian cancer (Table 6) [26,29]. Ovarian cancer is well suited for this therapeutic approach for a number of reasons: First, the disease is confined to the peritoneal cavity in over 80% of patients; second, the CA125 tumor marker allows for the detection of relapse at an early stage, when the residual tumor burden is small; and third, the peritoneal cavity is easily accessible. As described earlier, these studies are based on the premise that the gene-modified cells will home to *in situ* ovarian tumor deposits after i.p. infusion so that at least 10% of the *in situ* tumor will be comprised of gene-modified cells. This approach may have the further advantage of serving as a tumor vaccine because GCV is nonimmunosuppressive, the dying tumor mass can stimulate an antitumor immune response, and the allogeneic HSV-TK-gene-modified ovarian tumor cells may express common ovarian tumor antigens [169,170]. The patient population for the phase I clinical trial consisted of patients with stage III disease in relapse after receiving at least platinum and taxol therapy, the main drugs that have shown efficacy in treating ovarian cancer. An HSV-TK-gene-modified ovarian tumor cell line termed PA-1 was infused i.p. into patients with recurrent ovarian cancer. In the dose-escalating phase I clinical trial, patients received up to 1×10^{10} PA-1STK cells infused i.p. followed by GCV treatment. The results showed that this treatment had minimal to moderate toxicity, mostly grade I and II fever and abdominal pain. Of the 18 patients treated 4 responded, as evidenced by either decreased CA125 or tumor, with 1 partial and 3 complete responses. Further, average patient survival was similar to other salvage therapies for this patient group [92].

Subsequent trials to treat ovarian cancer that are currently in progress use other delivery methods, including HSV-TK producer cells and adenoviral vectors to deliver the HSV-TK gene. Similarly, brain tumors, particularly glioblastoma, have been an early focus of clinical trials using suicide gene therapy [71]. Prognosis for patients with glioblastoma is very poor, with a mean survival of less than 9 months. Brain tumors readily lend themselves to suicide gene therapy because they are localized, often with only one tumor mass. Because of their location, these tumors are often not treatable by surgery or radiation. In this trial, producer cells generating retroviral particles containing HSV-TK cells were infused into the tumor beds of patients with glioblastoma. Because retroviral vectors efficiently transduce dividing cells, most cells susceptible to retroviral transduction in the brain will be tumor cells. There has been some toxicity related to this study, which probably resulted from inoculating the fluid volume into a small enclosed vital organ. Antitumor responses have been suggested in several patients treated, and one patient in this trial has also been characterized as being in complete remission for over one year. This trial has been expanded into a phase III trial by Genetic Therapy, Inc., with results expected in summer 1998. Other clinical gene therapy trials using intratumoral injection of producer cells to treat localized tumors such as brain tumors and head and neck tumors are underway (Table 6).

As described earlier, a number of methods are being developed to improve the clinical efficacy of suicide gene therapy. Adenoviral vectors containing the HSV-TK gene are being used to also treat ovarian cancer, brain tumors, and pleural mesothelioma [30,31,72,73].

TABLE 6 HSV-TK Clinical Protocols

RAC review date	Primary investigators	Delivery system	Disease
7/91	S. Freeman Tulane U.	HSV-TK transduced ovarian cancer cells	Ovarian cancer
6/92	E. Oldfield NIH	HSV-TK vector producer cells (VPCs)	Brain tumors
3/93	J. Van Gilder M. Berger M. Pradow U. of Iowa	HSV-TK VPCs	Brain tumors
6/93	C. Raffel Child. Hosp., LA	HSV-TK VPCs	Brain tumors
9/93	L. Kun *et al.* St. Jude Hosp.	HSV-TK VPCs	Brain tumors
12/93*	E. Oldfield Z. Ram NIH	HSV-TK VPCs	Leptomeningeal cancer
9/94	S. Eck J. Alavi U. of Penn.	HSV-TK/adenovirus	CNS malignancy
9/94	S. Albelda U. of Penn.	HSV-TK/adenovirus	Mesothelioma
12/94	R. Grossman S. Woo Baylor U.	HSV-TK/adenovirus	CNS malignancy
2/95	M. Fetell Columbia U.	HSV-TK VPCs	Glioma
3/95	C. Link D. Moorman Iowa Methodist	HSV-TK VPCs	Ovarian cancer
12/95	R. Alvarez D. Curiel U. of Alabama	HSV-TK/adenovirus	Ovarian cancer

Note. The first dozen HSV-TK gene therapy clinical trials approved by the NIH Recombinant DNA Committee. More than a dozen additional clinical trials have been approved in the past several years.

Pleural mesothelioma provides another good model system in that it is localized to the pleural cavity, it is universally fatal, with a poor median survival, and the pleural cavity is easy to access. Injection of adenoviral vectors into the pleural cavity of mice with mesotheliomas resulted in the transduction of only the mesothelioma and pleural mesothelium, showing that the mesothelial cell lining in the pleural cavity protected the underlying parenchymal tissue from adenoviral transduction [121–123]. The advantage of this clinical gene therapy approach is similar to that in the ovarian cancer trial, in that the *in situ* tumor is easily accessible in the pleural cavity. However, results from this trial have not been encouraging. Even with relatively high doses of adenovirus (1×10^{10} pfu), there was often undetectable gene expression. This was apparently the result of the large amount of fibrosis that is characteristic of the disease. A follow-up trial in which the patient's tumor is debulked prior to intrapleural injection of the adenoviral vector is being initiated in 1998. Adenoviruses are also being used to deliver the HSV-TK gene to other tumors [72,74].

Suicide gene therapy is still in its infancy, yet there have been some very encouraging results over the past few years. Clinical trials using HSV-TK suicide gene therapy have been shown to have minimal toxic side effects. Several phase I trials have been completed, with published reports expected in spring 1998. Preliminary results indicate that the toxicity is mild to moderate, with some responses reported in phase I trials. A phase III trial will be completed in 1998. These encouraging reports indicate that suicide gene therapy will have future impact on cancer therapy. However, continued studies leading to improvements in tissue specific expression of suicide genes and immune enhancement of the bystander effect will be necessary to maximize the potential of this technology. Because tumor immunization has been shown to potentiate the antitumor effects of HSV-TK tumor cells, a follow-up study by Freeman and colleagues is being initiated to combine tumor vaccination with HSV-TK suicide gene therapy [15,20,26,29,30,76,92]. The premise is that HSV-TK gene therapy alters the tumor microenvironment to one that is immunostimulatory and allows immune effector cells to efficiently kill tumor cells in this altered microenvironment. Thus, identifying tumor specific antigens as targets for immunization and developing effective tumor vaccines will be an important component of future suicide gene therapy approaches [169,170].

B. Other Uses of the HSV-TK System

The HSV-TK/GCV system is also being used to enhance existing cancer therapies by modifying normal cells (i.e., leukocytes) for their eventual destruction after performing an *in vivo* function. One approach focuses on the antileukemic effects of leukocytes from the bone marrow donor of a patient undergoing allogeneic bone marrow transplantation (ABMT) for cancer. These cells have been shown to prolong patient remission through their graft-versus-leukemia (GVL) effect. A negative side effect for patients undergoing ABMT has been graft-versus-host disease (GVHD) also associated with these cells. Because the GVL occurs prior to GVHD, a role for gene therapy exists by infusing HSV-TK-gene-modified donor leukocytes into ABMT patients and then eliminating them after GVL but before GVHD toxicity occurs. This approach is also being developed to treat multiple myeloma and other cancers such as chronic myelogenous leukemia [175–177].

A second related approach involves using donor leukocytes to treat Epstein–Barr virus (EBV)–associated lymphomas in ABMT patients. EBV-associated lymphoma results from EBV-infected cells in the donor bone marrow, which become transformed when placed in the immunosuppressive environment of the transplant patient [178,179]. In one study, the HSV-TK gene is being used to modify donor leukocytes of patients undergoing ABMT for a GVL effect or for immune therapy for patients with EBV-associated lymphoma. Once the genetically modified donor leukocytes have killed the tumor, they can be eliminated by GCV therapy [180,181]. Eight patients have been treated to date, and the gene-modified lymphocytes have been shown to remain for up to 12 months. An antitumor response was observed in five of the patients [182].

VII. CONCLUSIONS

Because conventional cancer treatments are not able to cure approximately 50% of cancers in patients in relapse, novel therapies must be developed. A number of different cancer gene therapy protocols are currently underway, including those using suicide genes. Gene therapy provides one approach by which tumors that are resistant to conventional agents may be treated. Inserting a suicide gene into tumor cells will lead to tumor cell death upon exposure to a prodrug, even in chemotherapy-resistant tumor cells. One major limitation to using gene therapy for cancer is that modification of an entire tumor is not possible and untreated cells can grow back into tumors. However, the HSV-TK and CD systems have a critical advantage, because the gene-modified cells are toxic to nearby unmodified cells when exposed to the prodrug. This has been termed the bystander effect. Although the mechanism of this bystander effect is complex, the *in vitro* mechanism relies on transfer of toxic metabolites from gene-modified to unmodified cells, whereas the *in vivo* mechanism appears to be more complex and related to chemosensitization, hemorrhagic tumor necrosis, and activation of an immune response.

Clinical trials are underway using the HSV-TK gene in localized tumors including ovarian cancer, brain tumors, and pleural mesotheliomas. Reports of several phase I trials and a phase III trial evaluating delivery using tumor cells, retroviral producer cells, and adenoviral vectors should be published in 1998. To date, it appears that the HSV-TK suicide gene therapy clinical approach is relatively nontoxic, and preliminary results show that some patients have achieved partial and complete clinical remission. Future studies will involve better systems to more efficiently and specifically modify the tumor with the suicide gene and focus on methods to enhance the immune aspect of the bystander effect.

References

1. Boring, C. C., Squires, T. S., and Tong, T. Cancer statistics 1992. *Cancer: A Cancer Journal for Clinicians*, 1992. **42**, 19.
2. Feldman, A. R., Kessler, L., Myers, M. H., and Naughton, M. D. The prevalence of cancer. Estimates based on the Connecticut Tumor Registry. *N. Engl. J. Med.*, 1986. **315**, 1394–1397.
3. Goldstein, L. J., Pastan, I., and Gottesman, M. M. Multidrug resistance in human cancer. *Drit. Rev. Oncol. Hematol.*, 1992. **12**, 243–253.
4. Goldstein, L. J., Gottesman, M. M., and Pastan, I. Expression of the MDR1 gene in human cancers. *Cancer Treat Res.*, 1991. **57**, 101–119.

5. Freeman, S. M., Whartenby, K. A., and Abraham, G. N. Gene therapy: applications to diseases associated with aging. *Generations,* 1992. **16,** 45–48.

6. Freeman, S. M., Whartenby, K. A., Abboud, C. N., and Abraham, G. N. Clinical trials in cancer gene therapy. *Adv. Drug Delivery,* 1993. **12,** 169–183.

7. Anderson, W. F. Prospects for human gene therapy. *Science,* 1984. **226,** 401–409.

8. Friedman, T. Progress toward human gene therapy. *Science,* 1989. **244,** 1275–1281.

9. Whartenby, K. A., Muenchau, D. D., Zwiebel, J. A., Abraham, G. N., and Freeman, S. M. A prospective for gene therapy in the treatment of disease. *J. Univ. Rochester Med. Center,* 1991. **3,** 15–19.

10. Clinical protocol: the N2-TIL human gene transfer clinical protocol. *Hum. Gene Ther.,* 1990. **1,** 73.

11. Fisher, B., Packard, B. S., Read, E. J., *et al.* Tumor localization of adoptively transferred indium-111 labeled tumor infiltrating lymphocytes in patients with metastatic melanoma. *J. Clin. Oncol.,* 1989. **7,** 250–261.

12. Clinical protocol: TNF/TIL human gene therapy clinical protocol. *Hum. Gene Ther.,* 1990. **1,** 443.

13. Kasid, A., Aebersold, P., Cornetta, K., Culver, K., Freeman, S., Director, E., Lotze, M., Blaese, R. M., Anderson, W. F., and Rosenberg, S. A. Human gene transfer: retroviral-mediated gene transfer in man. *Proc. Natl. Acad. Sci. USA,* 1990. **87,** 473–477.

14. Rosenberg, S. A., Aebersold, P., Cornetta, K., Kasid, A., Morgan, R., Moen, R., Karson, E., Lotze, M., Yang, J., Topalian, S., Merino, M., Culver, K., Miller, A. D., Blaese, R. M., and Anderson, W. F. Gene transfer into humans—immunotherapy of patients with advanced melanoma, using tumor-infiltrating lymphocytes modified by retroviral gene transduction. *N. Engl. J. Med.,* 1990. **323,** 570–578.

15. Whartenby, K. A., Abboud, C. N., Marrogi, A. J., Ramesh, R., and Freeman, S. M. The biology of cancer gene therapy. *Lab. Invest.,* 1995. **72,** 131–141.

16. Freeman, S. M., and Zwiebel, J. A. Gene therapy of cancer. *Cancer Invest.,* 1993. **11,** 676–678.

17. Tiberghien, P. Use of suicide genes in gene therapy. *J. Leukoc. Biol.,* 1994. **56,** 203–209.

18. Moolten, F. L. Drug sensitivity ("suicide") genes for selective cancer chemotherapy. *Cancer Gene Ther.,* 1994. **4,** 279–287.

19. Deonarain, M. P., and Epenetos, A. A. Targeting enzymes for cancer therapy: old enzymes in new roles. *Br. J. Cancer,* 1994. **5,** 786–794.

20. Freeman, S. M., Whartenby, K. A., Freeman, J. L., Abboud, C. N., and Marrogi, A. J. *In situ* use of suicide genes for cancer therapy. *Semin. Oncol.,* 1996. **23**(1), 31–45.

21. Clinical protocol: the ADA human gene therapy clinical protocol. *Hum. Gene Ther.,* 1990. **1,** 327–362.

22. Bordignon, C., Mavilio, F., Ferrari, G., Servida, P., Ugazio, A. G., Notarangelo, L. D., Gilboa, E., Rossini, S., O'Reilly, R. J., Smith, C. A., *et al.* Transfer of the ADA gene into bone marrow cells and peripheral blood lymphocytes for the treatment of patients affected by ADA-deficient SCID. *Hum. Gene Ther.,* 1993. **4**(4), 513–520.

23. Hirschborn, R. Adenosine deaminase deficiency. *Immunodefic. Rev.,* 1990. **2,** 175–198.

24. Moolten, F. L., Wells, J. M., Heyman, R. A., and Evans, R. M. Lymphoma regression induced by ganciclovir in mice bearing a herpesthymidine kinase transgene. *Hum. Gene Ther.,* 1990. **1,** 125–134.

25. Moolten, F. L. Tumor chemosensitivity conferred by inserted herpes thymidine kinase genes: paradigm for a prospective cancer control strategy. *Cancer Res.,* 1986. **46,** 5276–5281.

26. Freeman, S. M., McCune, C., Robinson, W., Abboud, C. N., Angel, C., Abraham, G. N., and Marrogi, A. J. Treatment of ovarian cancer using a gene-modified vaccine. *Hum. Gene Ther.,* 1995. **6,** 927–939.

27. Freeman, S. M., Whartenby, K. A., Koeplin, D. S., Moolten, F. L., Abboud, C. N., and Abraham, G. N. Tumor regression when a fraction of the tumor mass contains the HSV-TK gene. *J. Cell Biochem.,* 1992. **16F,** 47.

28. Freeman, S. M., Abboud, C. N., Whartenby, K. A., and Abraham, G. N. The bystander effect: tumor regression when a fraction of the tumor mass is genetically modified. *Cancer Res.,* 1993. **53,** 5274–5284.

29. Freeman, S. M., McCune, C., Angel, C., Abraham, G. N., and Abboud, C. N. Treatment of ovarian cancer using HSV-TK gene-modified vaccine-regulatory issues. *Hum. Gene Ther.,* 1992. **3,** 342–349.

30. Marcel, T., and Grausz, J. D. The TMC Worldwide Gene Therapy Enrollment Report (June 1996). *Hum. Gene Ther.,* 1996. **7**(16), 2025–2046.

31. Roth, J. A., and Cristiano, R. J. Gene therapy for cancer: what have we done and where are we going? *J. Natl. Cancer Inst.,* 1997. **89,** 21–29.

32. Nabel, G. J., Nabel, E. G., Yang, Z. Y., Fox, B. A., Plautz, G. E., Gao, X., *et al.* Direct gene transfer with DNA-liposome complexes in melanoma: expression, biologic activity, and lack of toxicity in humans. *Proc. Natl. Acad. Sci. USA* 1993. **90,** 11307–11311.

33. Nabel, G. J. Immunotherapy for cancer by direct gene transfer into tumors. *Hum. Gene Ther.,* 1994. **5,** 57–77.

34. Clinical protocol: a pilot study of immunization with interleukin-2 secreting allogeneic HLA-A2 matched renal cell carcinoma cells in patients with advanced renal cell carcinoma. *Hum. Gene Ther.,* 1992. **3,** 691.

35. Clinical protocol: phase I study of non-replicating autologous tumor cell injections using cells prepared with or without granulocyte-macrophage colony stimulating factor gene transduction in patients with metastatic renal cell carcinoma. *Hum. Gene Ther.,* 1994. **5,** 112.

36. Moolten, F. L., and Wells, J. M. Curability of tumors bearing herpes thymidine kinase genes transferred by retroviral vectors. *J. Natl. Cancer Inst.,* 1990. **82,** 297–300.

37. Oliver, S., and Bubley, G. Inhibition of HSV-transformed murine cells by nucleoside analogs, 2′-NDG and 2′-nor-cGMP: mechanisms of inhibition and reversal by exogenous nucleosides. *Virology,* 1985. **145,** 84–93.

38. Field, A. K., Davies, M. E., DeWitt, C., Perry, H. D., Liou, R., Germershausen, J., Karkas, J. D., Ashton, W. T., Johnston, D. B. R., and Tolman, R. L. 9-{[2-hydroxy-1-(hydroxy-methyl) ethoxy methyl} guanine: a selective inhibitor of herpes group virus replication. *Proc. Natl. Acad. Sci. USA,* 1983. **80,** 4139–4143.

39. Nishiyama, Y., and Rapp, F. Anticellular effects of 9-(2-hydroxy-ethoxymethyl) guanine against herpes simplex virus-transformed cells. *J. Genet. Virol.,* 1979. **45,** 227–230.

40. Elion, G. B., Furman, P. A., Fyfe, J. A., deMiranda, P., Beauchamp, L., and Schaeffer, H. J. Selectivity of action of an antiherpetic agent, 9-(2-hydroxyethoxymethyl) guanine. *Proc. Natl. Acad. Sci. USA,* 1977. **74,** 5716–5720.

41. Elion, G. B. The chemotherapeutic exploitation of virus-specified enzymes. *Adv. Enzyme Regul.,* 1980. **18,** 53–60.

42. Davidson, R. L., Kaufman, E. R., Crumpacker, C. S., and Schnipper, L. E. Inhibition of herpes simplex virus transformed and nontransformed cells by acycloguanosine: mechanisms of uptake and toxicity. *Virology,* 1981. **113,** 9–19.

43. Shepp, D. H., Dandiliker, P., DeMiranda, P., Burnette, T. C., Cederberg, D. M., Kirk, L. E., and Meyers, J. D. Activity of 9-[2-hydroxy-1-(hydroxymethyl) ethoxymethyl] guanine in the treatment of cytomegalovirus pneumonia. *Ann. Intern. Med.,* 1985. **103,** 368–373.

44. Moolten, F. L. Mosaicism induced by gene insertion as a means of improving chemotherapeutic selectivity. *Immunology,* 1990. **10,** 203–233.

45. Sanz, G. F., Sanz, M. A., Vallespi, T., Canizo, M. C., Torrabadella, M., Garcia, S., Irriguible, D., and San Miguel, J. F. Two regression models and a scoring system for predicting survival and planning treatment in myelodysplastic syndromes: a multivariate analysis of prognostic factors in 370 patients. *Blood,* 1989. **74,** 395–408.

46. Mullen, C. A., Kilstrup, M., Blaese, R. M. Transfer of the bacterial gene for cytosine deaminase to mammalian cells confers lethal sensitivity to 5-fluorocytosine: a negative selection system. *Proc. Natl. Acad. Sci.,* 1992. **89,** 33–37.

47. Mullen, C. A., Coale, M. M., Lowe, R., and Blaese, R. M. Tumors expressing the cytosine deaminase suicide gene can be eliminated in vivo with 5-fluorocytosine and induce protective immunity to wild type tumor. *Cancer Res.,* 1994. **54,** 1503–1506.

48. Huber, B. E., Richards, C. A., and Krenitsky, T. A. Retroviral-mediated gene therapy for the treatment of hepatocellular carcinoma: an innovative approach for cancer therapy. *Proc. Natl. Acad. Sci. USA,* 1991. **88,** 8039–8043.

49. Huber, B. E., Austin, E. A., Good, S. S., Knick, V. C., Tibbels, S., and Richards, C. A. In vivo antitumor activity of 5-fluorocytosine on human colorectal carcinoma cells genetically modified to express cytosine deaminase. *Cancer Res.,* 1993. **53**(19), 4619–4626.

50. Macri, P., and Gordon, J. W. Delayed morbidity of albumin/SV40 T-antigen transgenic mice after insertion of an alphafetoprotein/herpes virus thymidine kinase transgene and treatment with ganciclovir. *Hum. Gene Ther.,* 1994. **5,** 175–182.

51. Price, J., Turner, D., and Cepko, C. Lineage analysis in the vertebrate nervous system by retrovirus mediated gene transfer. *Proc. Natl. Acad. Sci. USA,* 1987. **84,** 156–160.

52. Kozacsky, K. F., and Wilson, J. Gene therapy: adenovirus vectors. *Curr. Opin. Genet. Dev.,* 1993. **3,** 499–503.

53. Miller, A. D. Progress toward human gene therapy. *Blood,* 1990. **76,** 271–278.

54. Kolberg, R. The bystander effect in gene therapy: great, but how does it work. *J. NIH Res.,* 1994. **6,** 62–64.

55. Seachrist, L. Successful gene therapy has researchers looking for the bystander effect. *J. Natl. Cancer Inst.,* 1994. **86,** 82–83.

56. Kuriyama, S., Nakatani, T., Masui, K., Sakamoto, T., Tominaga, K., Yoshikawa, M., Fukui, H., Ikenaka, K., and Tsujii, T. Bystander effect caused by suicide gene expression indicates the feasibility of gene therapy for hepatocellular carcinoma. *Hepatology,* 1995. **22**(6), 1838–1846.

57. Vrionis, F. D., Wu, J. K., Qi, P., Cano, W., and Cherington, V. A more potent bystander cytocidal effect elicited by tumor cells expressing the herpes simplex virus-thymidine kinase gene than by fibroblast virus-producer cells in vitro. *J. Neurosurg.,* 1995. **83**(4), 698–704.

58. Wu, J. K., Cano, W. G., Meylaerts, S. A., Qi, P., Vrionis, F., and Cherington, V. Bystander tumoricidal effect in the treatment of experimental brain tumors. *Neurosurgery,* 1994. **35**(6), 1094–1102; discussion 1102–1103.

59. Bi, W. L., Parysek, L. M., Warnick, R., and Stambrook, P. J. In vitro evidence that metabolic cooperation is responsible for the Bystander Effect observed with HSV tk retroviral gene therapy. *Hum. Gene Ther.,* 1993. **4,** 725–731.

60. Samejima, Y., and Meruelo, D. "Bystander killing" induces apoptosis and is inhibited by forskolin. *Gene Ther.,* 1995. **2,** 50–58.

61. Colombo, B. M., Benedetti, S., Ottolenghi, S., Mora, M., Pollo, B., Poli, G., and Finocchiaro, G. The "bystander effect": association of U87 cell death with ganciclovir-mediated apoptosis of nearby cells and lack of effect in athymic mice. *Hum. Gene Ther.,* 1995. **6,** 763–772.

62. Hamel, W., Magnelli, L., Chiarugi, V. P., and Israel, M. A. Herpes simplex virus thymidine kinase/ganciclovir-mediated apoptotic death of bystander cells. *Cancer Res.,* 1996. **56,** 2697–2702.

63. Wygoda, M. R., Wilson, M. R., Davis, M. A., Trosko, J. E., Rehemtulla, A., and Lawrence, T. S. Protection of herpes simplex virus thymidine kinase-transduced cells from ganciclovir-mediated cytotoxicity by bystander cells: the Good Samaritan effect. *Cancer Res.,* 1997. **57**(9), 1699–1703.

64. Farber, E. Programmed cell death: necrosis versus apoptosis. *Mod. Pathol.,* 1994. **5,** 605–609.

65. Sambrano, G. R., and Steinberg, D. Recognition of oxidatively damaged and apoptotic cells by an oxidized low density lipoprotein receptor on mouse peritoneal macrophages: role of membrane phosphotidylserine. *Proc. Natl. Acad. Sci. USA,* 1995. **92,** 1396–1400.

66. Hall, S. E., Savill, J. S., Henson, P. M., and Haslett, C. Apoptotic neutrophils are phagocytosed by fibroblasts with participation of the fibroblast vitronectin receptor and involvement of a mannose/fucose-specific lectin. *J. Immunol.,* 1994. **153,** 3218–3227.

67. Savill, J., Fadok, V., and Henson, P. Phagocyte recognition of cells undergoing apoptosis. *Immunol. Today,* 1993. **14,** 131–136.

68. Doble, B. W., and Kardani, E. Basic fibroblast growth factor stimulates connexin-43 expression and intercellular communication of cardiac fibroblasts. *Mol. Cell. Biochem.,* 1995. **143,** 81–87.

69. Yeager, M., and Nicholson, B. J. Structure of gap junction intercellular channels. *Curr. Opin. Struct. Biol.,* 1996. **6,** 183–192.

70. Vrionis, F. D., Wu, J. K., Qi, P., Waltzman, M., Cherington, V., and Spray, D. C. The bystander effect exerted by tumor cells expressing the herpes simplex virus thymidine kinase (HSVtk) gene is dependent on connexin expression and cell communication via gap junctions. *Gene Ther.,* 1997. **4**(6), 577–585.

71. Elshami, A. A., Kucharczuk, J. C., Zhang, H. B., Smythe, W. R., Hwang, H. C., Litzky, L. A., Kaiser, L. R., and Albelda, S. M. Treatment of pleural mesothelioma in an immunocompetent rat model utilizing adenoviral transfer of the herpes simplex virus thymidine kinase gene. *Hum. Gene Ther.,* 1996. **7**(2), 141–148.

72. Ishii-Morita, H., Agbaria, R., Mullen, C. A., Hirano, H., Koeplin, D. A., Ram, Z., Oldfield, E. H., Johns, D. G., and Blaese, R. M. Mechanism of "bystander effect" killing in the herpes simplex thymidine kinase gene therapy model of cancer treatment. *Gene Ther.,* 1997. **4**(3), 244–251.

73. Mesnil, M., Piccoli, C., Tiraby, G., Willecke, K., and Yamasaki, H. Bystander killing of cancer cells by herpes simplex virus thymidine kinase gene is mediated by connexins. *Proc. Natl. Acad. Sci.,* 1996. **93**(5), 1831–1835.

74. Dilber, M. S., Abedi, M. R., Christensson, B., Bjorkstrand, B., Kidder, G. M., Naus, C. C., Gahrton, G., and Smith, C. I. Gap junctions promote the bystander effect of herpes simplex virus thymidine kinase in vivo. *Cancer Res.,* 1997. **56,** 1523–1528.

75. Freeman, S. M., Ramesh, R., Marrogi, A. J., Jensen, A., and Aboud, C. N. In vivo studies on the mechanism of the "Bystander Effect." *Cancer Gene Ther.,* 1994. **4,** 326.

76. Freeman, S. M., Ramesh, R., and Marrogi, A. J. Immune system in suicide-gene therapy. *Lancet,* 1997. **349**(9044), 2–3.

77. Ramesh, R., Munshi, A., Abboud, C. N., Marrogi, A. J., and Freeman, S. M. Expression of costimulatory molecules: B7 and ICAM up-regulation after treatment with a suicide gene. *Cancer Gene Ther.,* 1996. **3**(6), 373–384.

78. Vile, R. G., Nelson, J. A., Castleden, S., Chong, H., and Hart, I. R. Systemic gene therapy of murine melanoma using tissue specific expression of the HSVtk gene involves an immune component. *Cancer Res.,* 1994. **54**(23), 6228–6234.

79. Consalvo, M., Mullen, C. A., Modesti, A., Piero, M., Allione, A., Cavallo, F., Giovarelli, M., and Forni, G. 5-fluorocytosine-induced eradication of murine adenocarcinomas engineered to express the cytosine deaminase suicide gene requires host immune competence and leaves an efficient memory. *J. Immunol.,* 1995. **154,** 5302–5312.

80. Ram, Z., Walbridge, S., Shawker, T., Culver, K. W., Blaese, R. M., and Oldfield, E. H. The effect of thymidine kinase transduction and ganciclovir therapy on tumor vasculature and growth of 9L gliomas in rats. *J. Neurosurg.,* 1994. **81,** 256–260.

81. Bi, W., Kim, Y. G., Feliciano, E. S., Pavelic, L., Wilson, K. M., Pavelic, Z. P., and Stambrook, P. J. An HSVtk-mediated local and distant antitumor bystander effect in tumors of head and neck origin in athymic mice. *Cancer Gene Ther.,* 1997. **4**(4), 246–252.

82. Kruse, C. A., Roper, M. D., Kleinschmidt-DeMasters, B. K., Banuelos, S. J., Smiley, W. R., Robbins, J. M., and Burrows, F. J. Purified herpes simplex thymidine kinase Retrovector particles. I. *In vitro* characterization, in situ transduction efficiency, and histopathological analyses of gene therapy-treated brain tumors. *Cancer Gene Ther.,* 1997. **4**(2), 118–128.

83. Vile, R. G., Castleden, S., Marshall, J., Camplejohn, R., Upton, C., and Chong, H. Generation of an anti-tumour immune response in a non-immunogenic tumour: HSVtk killing *in vivo* stimulates a monoculear cell infiltrate and a th1-like profile of intratumoural cytokine expression. *Intl. J. Cancer,* 1997. **71,** 267–274.

84. Ramesh, R., Marrogi, A. J., Munshi, A., Abboud, C. N., and Freeman, S. M. *In vivo* analysis of the "bystander effect": a cytokine cascade. *Exp. Hematol.,* 1996. **24**(7), 829–838.

85. Freeman, S. M., Ramesh, R., Shastri, M., Munshi, A., Jensen, A. K., and Marrogi, A. J. The role of cytokines in mediating the bystander effect using HSV-TK xenogeneic cells. *Cancer Lett.,* 1995. **92**(2), 167–174.

86. Gagandeep, S., Brew, R., Green, B., Christmas, S. E., Klatzmann, D., Poston, G. J., and Kinsella, A. R. Prodrug-activated gene therapy: involvement of an immunological component in the "bystander effect." *Cancer Gene Ther.,* 1996. **3**(2), 83–88.

87. Schilling, P. J. Novel tumor necrosis factor toxic effects. *Cancer,* 1992. **69,** 256–260.

88. Robertson, P. O., Ross, H. J., and Figlin, R. A. Tumor necrosis factor induces hemorrhagic necrosis of a sarcoma. *Ann. Intern. Med.,* 1989. **111,** 682–684.

89. Tracey, K. J., Beutler, B., Lowry, S. F., *et al.* Shock and tissue injury induced by recombinant human cachectin. *Science,* 1986. **234,** 470–474.

90. Caruso, M., Panis, Y., Gagandeep, S., Houssin, D., Salzmann, J. L., and Klatzmann, D. Regression of established macroscopic liver metastases after *in situ* transduction of a suicide gene. *Proc. Natl. Acad. Sci.,* 1993. **90,** 7024–7028.

91. Chelen, C. J., Fang, Y., Freeman, G. J., Secrist, H., Marshall, J. D., Hwang, P. T., Frankel, L. R., DeKruyff, R. H., and Umetsu, D. T. Human alveolar macrophages present antigen ineffectively

due to defective expression of B7 costimulatory cell surface molecules. *J. Clin. Invest.,* 1995. **95,** 1415–1421.

92. Yamamoto, S., Suzuki, S., Hoshino, A., Akimoto, M., and Shimada, T. Herpes simplex virus thymidine kinase/ganciclovir-mediated killing of tumor cell induces tumor specific cytotoxic T cells in mice. *Cancer Gene Ther.,* 1997. **4**(2), 91–96.

93. Barba, D., Hardin, J., Sadelain, M., and Gage, F. H. Development of anti-tumor immunity following thymidine kinase-mediated killing of experimental brain tumors. *Proc. Natl. Acad. Sci. USA,* 1994. **91,** 4348–4352.

94. Vile, R. G., Diaz, R. M., Castleden, S., and Chong, H. Targeted gene therapy for cancer: herpes simplex virus thymidine kinase gene-mediated cell killing leads to anti-tumour immunity that can be augmented by co-expression of cytokines in the tumour cells. *Biochem. Soc. Trans.,* 1997. **25**(2), 717–722.

95. Chen, S. H., Li Chen, X. H., Wang, Y., Kosai, K. I., Finegold, M. J., Rich, S. S., and Woo, S. L. C. Combination gene therapy for liver metastasis of colon carcioma *in vivo. Proc. Natl. Acad. Sci. USA,* 1995. **92**(March), 2577–2581.

96. Santodonato, L., Ferrantini, M., Gabriele, L., Proietti, E., Venditti, M., Musiani, A., Modesti, A., Modica, A., Lupton, S. K., and Belardelli, F. Cure of mice with established metastatic friend leukemia cell tumors by a combined therapy with tumor cells expressing both interferon-alpha 1 and herpes simplex thymidine kinase followed by ganciclovir. *Hum. Gen. Ther.,* 1996. **7,** 1–10.

97. Chen, S. H., Kosai, K., Xu, B., Pham-Nguyen, K., Contant, C., Finegold, M. J., and Woo, S. L. Combination suicide and cytokine gene therapy for hepatic metastases of colon carcinoma: sustained antitumor immunity prolongs animal survival. *Cancer Res.,* 1996. **56**(16), 3758–3562.

98. Benedetti, S., Dimeco, F., Pollo, B., Cirenei, N., Colombo, B. M., Bruzzone, M. G., Cattaneo, E., Vescovi, A., Didonato, S., Colombo, M. P., and Finocchiaro, G. Limited efficacy of the HSV-TK/GCV system for gene therapy of malignant glioma and perspectives for the combined transduction of the Interleukin-4 gene. *Hum. Gene Ther.,* 1997. **8**(11), 1345–1353.

99. Marini, F. Cr., Pan, B. F., Nelson, J. A., and Lapeyre, J. N. The drug verapamil inhibits bystander killing but not cell suicide in thymidine kinase-ganciclovir prodrug-activated gene therapy. *Cancer Gene Ther.,* 1996. **3**(6), 405–412.

100. Park, J. Y., Elshami, A. A., Amin, K., Rizk, N., Kaiser, L. R., and Albelda, S. M. Retinoids augment the bystander effect *in vitro* and *in vivo* in herpes simplex virus thymidine kinase/ganciclovir-mediated gene therapy. *Gene Ther.,* 1997. **4**(9), 909–917.

101. Kim, S. H., Kim, J. H., Kolozsvary, A., Brown, S. L., and Freytag, S. O. Preferential radiosensitization of 9L glioma cells transduced with HSV-TK gene. *J. Neurooncol.,* 1997. **33**(3), 189–194.

102. Elshami, A. A., Cook, J. W., Amin, K. M., Choi, H., Park, J. Y., Coonrod, L., Sun, J., Molnar-Kimber, K., Wilson, J. M., Kaiser, L. R., and Albelda, S. M. The effect of promoter strength in adenoviral vectors containing herpes simplex virus thymidine kinase on cancer gene therapy *in vitro* and *in vivo. Cancer Gene Ther.,* 1997. **4**(4), 213–221.

103. Black, M. E., Rechtin, T. M., and Drake, R. R. Effect on substrate binding of an alteration at the conserved aspartic acid-162 in herpes simplex virus type 1 thymidine kinase. *J. Gen. Virol.,* 1996. **77**(Pt 7), 1521–1527.

104. Black, M. E., and Loeb, L. A. Random sequence mutagenesis for the generation of active enzymes. *Methods Mol. Biol.,* 1996. **57,** 335–349.

105. Black, M. E., Newcomb, T. G., Wilson, H. M., and Loeb, L. A. Creation of drug-specific herpes simplex virus type 1 thymidine

kinase mutants for gene therapy. *Proc. Natl. Acad. Sci. USA,* 1996. **93**(8), 3525–3529.

106. Culver, K. W., Ishii, H., Blaese, R. M., Ram, Z., Wallbridge, S., and Oldfield, E. H. *In vivo* gene transfer with retroviral vector producer cells for treatment of experimental brain tumors. *Science,* 1992. **256**, 1550–1552.

107. Ezzedine, Z. D., Maruza, R. L., *et al.* Selecting killing of glioma cells in culture and *in vivo* by retrovirus transfer of the herpes simplex virus thymidine kinase gene. *New Biol.,* 1991. **3**, 608–614.

108. Takamiya, Y., Short, M. P., Ezzeddine, Z. D., Moolten, F. L., Breakefield, X. O., and Martuza, R. L. Gene therapy of malignant brain tumors: a rat glioma line bearing the herpes simplex virus type 1-thymidine kinase gene and wild type retrovirus kills other tumor cells. *J. Neurosci. Res.,* 1992. **33**, 493–503.

109. Short, M. P., Choi, B. C., Lee, J. K., *et al.* Gene delivery to glioma cells in rat brain by grafting of a retrovirus packaging cell line. *J. Neurosci.,* 1990. **27**, 427–433.

110. Barba, D., Hardin, J., Ray, J., and Gage, F. H. Thymidine kinase-mediated killing of rat brain tumors. *J. Neurosurg.,* 1993. **79**, 729–735.

111. Ram, Z., Culver, K. W., Walbridge, B., and Oldfield, E. H. *In situ* retroviral-mediated gene transfer for the treatment of brain tumors in rats. *Cancer Res.,* 1993. **53**, 83–88.

112. Miyao, Y., Shimizu, K., Moriuchi, S., Yamada, M., Nakahira, K., Nakajima, K., Nakao, J., Kuriyama, S., Tsujii, T., Mikoshiba, K., *et al.* Selective expression of foreign genes in glioma cells: use of the mouse myelin basic protein gene promoter to direct toxic gene expression. *J. Neurosci. Res.,* 1993. **36**(4), 472–479.

113. Kim, S. H., Kim, S. H., Brown, S. L., and Freytag, S. O. Selective enhancement by an antiviral agent of the radiation-induced cell killing of human glioma cells transduced with HSV-tk gene. *Cancer Res.,* 1994. **54**, 6053–6056.

114. Ram, Z., Culver, K. W., Walbridge, S., Frank, J. A., Blaese, R. M., and Oldfield, E. H. Toxicity studies of retroviral-mediated gene transfer for the treatment of brain tumors. *J. Neurosurg.,* 1993. **79**, 400–407.

115. Yosha, K., Kawami, H., Yamaguchi, Y., Kuniyasu, H., Nishiyama, M., Hirai, T., Yanagihara, K., Tahara, E., and Toge, T. Retrovirally transmitted gene therapy for gastric carcinoma using herpes simplex virus thymidine kinase gene. *Cancer,* 1995. **75**, 1467–1471.

116. Tapscott, S. J., Miller, A. D., Olson, J. M., Berger, M. S., Groudine, M., and Spence, A. M. Gene therapy of rat 9L gliosarcoma tumors by transduction with selectable genes does not require drug selection. *Proc. Natl. Acad. Sci. USA,* 1994. **91**(17), 8185–8189.

117. Clinical protocol: gene therapy for the treatment of malignant brain tumors with *in vivo* tumor transduction with the herpes simplex thymidine kinase gene/ganciclovir system. *Hum. Gene Ther.,* 1994. **5**, 343.

118. Chen, S.-H., Shine, H. D., Goodman, J. C., *et al.* Gene therapy for brain tumors: regression of experimental gliomas by adenovirus-mediated gene transfer *in vivo. Proc. Natl. Acad. Sci. USA,* 1994. **91**, 3054–3057.

119. O'Malley, B. W., Chen, S.-H., Schwartz, M. R., and Wood, S. L. C. Adenovirus-mediated gene therapy for human head and neck squamous cell cancer in a nude mouse model. *Cancer Res.,* 1995. **55**, 1080–1085.

120. Bonnekoh, D., Greenhalgh, D. A., Bundman, D. S., Eckhardt, J. N., Longley, M. A., Chen, S.-H., Woo, S. L. C., and Roop, D. R. Inhibition of melanoma growth by adenoviral-mediated HSV thymidine kinase gene transfer *in vivo. J. Invest. Dermatol.,* 1995. **104**, 313–317.

121. Smythe, W. R., Hwang, H. C., Amin, K. M., *et al.* Use of recombinant adenovirus to transfer the herpes simplex virus thymidine kinase (HSVtk) gene to thoracic neoplasms: an effective *in vitro* drug sensitization system. *Cancer Res.* 1994. **54**, 3055–3059.

122. Smythe, W. R., Kaiser, L. R., Amin, K. M., *et al.* Successful adenovirus mediated gene transfer in an *in vivo* model of human malignant mesothelioma. *Ann. Thorac. Surg.,* 1994. **57**, 1395–1401.

123. Smythe, W. R., Hwang, H. C., Elshami, A. A., *et al.* Treatment of experimental human mesothelioma using adenovirus transfer of the herpes simplex thymidine kinase gene. *Ann. Surg.,* 1995. **222**(1), 78–86.

124. Douglas, J. T., and Curiel, D. T. Strategies to accomplish targeted gene delivery to muscle cells employing tropism-modified adenoviral vectors. *Neuromusc. Disord.,* 1997. **7**(5), 284–298.

125. Ebbinghaus, S. W., Vigneswaran, N., Miller, C. R., Chee-Awai, R. A., Mayfield, C. A., Curiel, D. T., and Miller, D. M. Efficient delivery of triplex forming oligonucleotides to tumor cells by adenovirus-polylysine complexes. *Gene Ther.,* 1996. **3**(4), 287–297.

126. Hong, S. S., Karayan, L., Tournier, J., Curiel, D. T., and Boulanger, P. A. Adenovirus type 5 fiber knob binds to MHC class I alpha2 domain at the surface of human epithelial and B lymphoblastoid cells. *EMBO J.,* 1997. **16**(9), 2294–2306.

127. Markert, J. M., Malick, A., Goen, D. M., and Martuza, R. L. Reduction and elimination of encephalitis in an experimental glioma therapy model with attenuated herpes simplex mutants that retain susceptibility to acyclovir. *Neurosurgery,* 1993. **32**, 597–603.

128. Martuza, R. L., Malick, A., Markert, J. M., Ruffner, K. L., and Coen, D. M. Experimental therapy of human glioma by means of a genetically engineered virus mutant. *Science,* 1991. **252**, 854–856.

129. Miyatake, S., Martuza, R. L., and Rabkin, S. D. Defective herpes simplex virus vectors expressing thymidine kinase for the treatment of malignant glioma. *Cancer Gene Ther.,* 1997. **4**(4), 222–228.

130. Primus, F. J., Finch, M. D., Wetzel, S. A., Masci, A. M., Schlom, J., and Kashmiri, S. V. Monoclonal antibody gene transfer. Implications for tumor-specific cell-mediated cytotoxicity. *Ann. N. Y. Acad. Sci.,* 1994. **716**, 154–165; discussion 165–166.

131. Batra, R. K., Wang-Johanning, F., Wagner, E., Garver, R. I., Jr., and Curiel, D. T. Receptor-mediated gene delivery employing lectin-binding specificity. *Gene Ther.,* 1994. **1**(4), 255–260.

132. Ledley, F. D. Nonviral gene therapy: the promise of genes as pharmaceutical products. *Hum. Gene Ther.,* 1995. **6**(9), 1129–1144.

133. Cooper, M. J. Noninfectious gene transfer and expression systems for cancer gene therapy. [Review, 176 refs.]. *Semin. Oncol.,* 1996. **23**(1), 172–187.

134. Kasahara, N., Dozy, A. M., and Kan, Y. W. Tissue-specific targeting of retroviral vectors through ligand-receptor interactions [see comments]. *Science,* 1994. **266**(5189), 1373–1376.

135. Chonn, A., and Cullis, P. R. Recent advances in liposomal drug-delivery systems. [Review, 120 refs.]. *Curr. Opin. Biotechnol.,* 1995. **6**(6), 698–708.

136. Gao, X., and Huang, L. Cationic liposome-mediated gene transfer. [Review, 92 refs.]. *Gene Ther.,* 1995. **2**(10), 710–722.

137. Rihova, B. Targeting of drugs to cell surface receptors. *Crit. Rev. iotechnol.,* 1997. **17**(2), 149–169.

138. Aoki, K., Yoshida, T., Matsumoto, N., Ide, H., Hosokawa, K., Sugimura, T., and Terada, M. Gene therapy for peritoneal dissemination of pancreatic cancer by liposome-mediated transfer

of herpes simplex virus thymidine kinase gene. *Hum. Gene Ther.*, 1997. **8**(9), 1105–1113.

139. Grim, J., Deshane, J., Feng, M., Lieber, A., Kay, M., and Curiel, D. T. erbB-2 knockout employing an intracellular single-chain antibody (sFv) accomplishes specific toxicity i erbB-2-expressing lung cancer cells. *Am. J. Respir. Cell Mol. Biol.*, 1996. **15**, 348–354.

140. Cristiano, R. J., and Roth, J. A. Molecular conjugates: a targeted gene delivery vector for molecular medicine. *J. Mol. Med.*, 1995. **73**(10), 479–486.

141. Poncet, P., Panczak, A., Goupy, C., Gustafsson, K., Blanpied, C., Chavanel, G., Hirsch, R., and Hirsch, F. Antifection: an antibody-mediated method to introduce genes into lymphoid cells *in vitro* and *in vivo*. *Gene Ther.*, 1996. **3**, 731–738.

142. Ojeifo, J. O., Forough, R., Paik, S., Maciag, T., and Zwiebel, J. A. Angiogenesis-directed implantation of genetically modified endothelial cells in mice. *Cancer Res.*, 1995. **55**(11), 2240–2244.

143. Zwiebel, J. A., Freeman, S. M., Kantoff, P. W., Cornetta, K., Ryan, U. S., and Anderson, W. F. High-level recombinant gene expression in rabbit endothelial cells transduced by retroviral vectors. *Science*, 1989. **243**(4888), 220–222.

144. Zwiebel, J. A., Freeman, S. M., Cornetta, K., Forough, R., Maciag, T., and Anderson, W. F. Recombinant gene expression in human umbilical vein endothelial cells transduced by retroviral vectors. *Biochem. Biophys. Res. Comm.*, 1990. **170**(1), 209–213.

145. Zwiebel, J. A., Freeman, S. M., Newman, K., Dichek, D., Ryan, U. S., and Anderson, W. F. Drug delivery by genetically engineered cell implants. [Review, 33 refs.]. *Ann. N. Y. Acad. Sci.*, 1991. **618**, 394–404.

146. Vile, R. G., and Hart, I. R. Use of tissue-specific expression of the herpes simplex virus thymidine kinase gene to inhibit growth of established murine melanomas following direct intratumoral injection of DNA. *Cancer Res.*, 1993. **53**, 3860–3864.

147. Manome, Y., Abe, M., Hagen, M. F., Fine, H. A., and Kufe, D. W. Enhancer sequences of the DF3 gene regulate expression of the herpes simplex virus thymidine kinase gene and confer sensitivity of human breast cancer cells to ganciclovir. *Cancer Res.*, 1994. **54**, 5408–5413.

148. Osaki, T., Tanio, Y., Tachibana, I., Hosoe, S., Kumagai, T., Kawase, I., Oikawa, S., and Kishimoto, T. Gene therapy for carcinoembryonic antigen-producing human lung cancer cells by cell type-specific expression of herpes simplex virus thymidine kinase gene. *Cancer Res.*, 1994. **20**, 5258–5261.

149. Nishiyama, T., Kawamura, Y., Kawamoto, K., Matsumura, H., Yamamoto, N., Ito, T., Ohyama, A., Katsuragi, T., and Sakai, T. Antineoplastic effects in rats of 5-fluorocytosine in combination with cytosine deaminase capsules. *Cancer Res.*, 1985. **45**, 1753–1761.

150. Sakai, T., Katsuragi, T., Tonomura, K., Nishiyama, T., and Kawamura, Y. Implantable encapsulated cytosine deaminase havine 5-fluorocytosine deaminating activity. *J. Biotechnol.*, 1985. **2**, 13–21.

151. Katsuragi, T., Sakai, T., and Tonomura, K. Implantable enzyme capsules for cancer chemotherapy from bakers yeast cytosine deaminase immobilized on epoxy-acrylic resin and urethane prepolymer. *Appl. Biochem. Biotechnol.*, 1987. **16**, 61–69.

152. Austin, E. A., and Huber, B. E. A first step in the development of gene therapy for colorectal carcinoma: cloning, sequencing, and expression of *Escherichia coli* cytosine deaminase. *Mol. Pharmacol.*, 1992. **43**, 380–387.

153. Huber, B. E., Austin, E. A., Richards, C. A., Davis, S. T., and Good, S. S. Metabolism of 5-fluorocytosine to 5-fluorouracil in human colorectal tumor cells transduced with the cytosine deaminase gene: significant antitumor effects when only a small per-

centage of tumor cells express cytosine deaminase. *Proc. Natl. Acad. Sci. USA*, 1994. **91**, 8302–8306.

154. Trinh, Q. T., Austin, E. A., Murray, D. M., Knick, V. C., and Huber, B. E. Enzyme/prodrug gene therapy: comparison of cytosine deaminase/5-fluorocytosine versus thymidine kinase/ganciclovir enzyme/prodrug systems in a human colorectal carcinoma cell line. *Cancer Res.*, 1995. **55**(21), 4808–4812.

155. Rogers, R. P., Ge, J. Q., Holley-Guthrie, E., Hoganson, D. K., Comstock, K. E., Olsen, J. C., and Kenney, S. Killing Epstein-Barr virus-positive B lymphocytes by gene therapy: comparing the efficacy of cytosine deaminase and herpes simplex virus thymidine kinase. *Hum. Gene Ther.*, 1996. **7**(18), 2235–2245.

156. Hoganson, D. K., Batra, R. K., Olsen, J. C., and Boucher, R. C. Comparison of the effects of three different toxin genes and their levels of expression on cell growth and bystander effect in lung adenocarcinoma. *Cancer Res.*, 1996. **56**(6), 1315–1323.

157. Rogulski, K. R., Kim, J. H., Kim, S. H., and Freytag, S. O. Glioma cells transduced with an *Escherichia coli* CD/HSV-1 TK fusion gene exhibit enhanced metabolic suicide and radiosensitivity. *Hum. Gene Ther.*, 1997. **8**(1), 73–85.

158. Grignet-Debrus, C., and Calberg-Bacq, C. M. Potential of Varicella zoster virus thymidine kinase as a suicide gene in breast cancer cells. *Gene Ther.*, 1997. **4**(6), 560–569.

159. Besnard, C., Monthioux, E., and Jami, J. Selection against expression of the *Escherichia coli* gene gpt in hprt+ mouse teratocarcinoma and hybrid cells. *Mol. Cell Biol.*, 1987. **7**, 4139–4141.

160. Mroz, P. J., and Moolten, F. L. Retrovirally transduced *Escherichia coli* gpt genes combine selectability with chemosensitivity capable of mediating tumor eradication. *Hum. Gene Ther.*, 1993. **4**, 589–595.

161. Mulligan, R., and Gerg, P. Selection for animal cells that express the *Escherichia coli* gene coding for xanthine-guanine phosphoribosyltransferase. *Proc. Natl. Acad. Sci. USA*, 1981. **78**, 2072–2076.

162. Green, N. K., Youngs, D. J., Neoptolemos, J. P., Friedlos, F., Knox, R. J., Springer, C. J., Anlezark, G. M., Michael, N. P., Melton, R. G., Ford, M. J., Young, L. S., Kerr, D. J., and Searle, P. F. Senzitization of colorectal and pancreatic cancer cell lines to the prodrug 5-(aziridin-1-yl)-2,4-dinitrobenzamide (CB1954) by retroviral transduction and expression of the *E. coli* nitroreductase gene. *Cancer Gene Ther.*, 1997. **4**(4), 229–238.

163. Hughes, B. W., Wells, A. H., Bebok, Z., Gadi, V. K., Garver, R. I., Jr., Parker, W. B., and Sorscher, E. J. Bystander killing of melanoma cells using the human tyrosinase promoter to express the *Escherichia coli* purine nucleoside phosphorylase gene. *Cancer Res.*, 1995. **55**(15), 3339–3345.

164. Drabek, D., Guy, J., Craig, R., and Grosveld, F. The expression of bacterial nitroreductase in transgenic mice results in specific cell killing by the prodrug CB1954 [see comments]. *Gene Ther.*, 1997. **4**(2), 93–100.

165. Bridgewater, J. A., Knox, R. J., Pitts, J. D., Collins, M. K., and Springer, C. J. The bystander effect of the nitroreductase/CB1954 enzyme/prodrug system is due to a cell-permeable metabolite. *Hum. Gene Ther.*, 1997. **8**(6), 709–717.

166. Munshi, A., Ramesh, R., Marrogi, A. J., and Freeman, S. M. Evaluation of adenovirus p53 mediated "bystander effect" *in vivo*. *Cancer Gene Ther.*, 1997. **4**, 513.

167. Fujiwava, T., Cai, D. W., Georges, R. N., Mukopadhyay, T., Grimm, E. A., and Roth, J. A. Therapeutic effect of a retroviral wild-type p53 expression vector in an orthotopic lung cancer model. *J. Natl. Cancer Inst.*, 1994. **86**, 1458–1462.

168. Lesoon-Wood, L. A., Kim, W. H., Klwinman, H. K., Weintraub, B. D., and Mixson, A. J. Systemic gene therapy with p53 reduces

growth and metastases of a malignant human breast cancer in nude mice. *Hum. Gene Ther.,* 1995. **6,** 395–405.

169. Ioannides, C. G., Fisk, B., Jerome, K. R., Irimura, T., Wharton, J. T., and Finn, O. J. Cytotoxic T cells from ovarian malignant tumors can recognize polymorphic epithelial mucin core peptides. *J. Immunol.,* 1993. **151,** 3693–3703.

170. Kuiper, M., Peakman, M., and Farzaneh, F. Ovarian tumour antigens as potential targets for immune gene therapy. *Gene Ther.,* 1995. **2,** 7–15.

171. Clinical protocol: gene therapy for the treatment of brain tumors using intratumoral transduction with the thymidine kinase gene and intravenous ganciclovir. *Hum. Gene Ther.,* 1993. **4,** 39.

172. Clinical protocol: HSV-TK gene therapy. *Cancer Gene Ther.,* 1994. **1,** 291–292.

173. Alvarez, R. D., and Curiel, D. T. A phase I study of recombinant adenovirus vector-mediated intraperitoneal delivery of herpes simplex virus thymidine kinase gene and intravenous ganciclovir for previously treated ovarian cancer and extraovarian cancer patients. *Hum. Gene Ther.,* 1997. **8**(5), 597–613.

174. Eck, S. L., Alavi, J. B., Alavi, A., Davis, A., Hackney, D., Judy, K., Mollman, J., Phillips, P. C., Wheeldon, E. B., and Wilson, J. M. Treatment of advanced CNS malignancies with the recombinant adenovirus H5.010RSVTK: a phase I trial. *Hum. Gene Ther.,* 1996. **7**(12), 1465–1482.

175. Tiberghien, P., Cahn, J. Y., Brion, A., Deconinck, E., Racadot, E., Herve, P., Milpied, N., Lioure, B., Gluckman, E., Bordigoni, P., Jacob, W., Chiang, Y., Marcus, S., Reynolds, C., and Longo, D. Use of donor T-lymphocytes expressing herpes-simplex thymidine kinase in allogeneic bone marrow transplantation: a phase I-II study. *Hum. Gene Ther.,* 1997. **8**(5), 615–624.

176. Porter, D. L., Roth, M. S., McGarigle, C., Ferrara, J. L. M., and Antin, J. H. Induction of graft-versus-host disease as immuno-therapy for relapsed chronic myeloid leukemia. *N Engl. J. Med.,* 1994. **330,** 100–106.

177. Tiberghien, P., Reynolds, C. W., Keller, J., Spence, S., Deschaseaux, M., Certoux, J. M., Contassot, E., Murphy, W. J., Lyons, R., Chiang, Y., Herve, P., Longo, D. L., and Ruscetti, F. W. Ganciclovir treatment of herpes simplex thymidine kinase-transduced primary T lymphocytes: an approach for specific *in vivo* T-cell depletion after bone marrow transplantation? *Blood,* 1994. **84,** 1333–1341.

178. Clinical protocol: transfer of the HSV-tk gene into donor peripheral blood lymphocytes for *in vivo* modulation of donor anti-tumor immunity after allogeneic bone marrow transplantation. *Hum. Gene Ther.,* 1995. **6,** 813–819.

179. Riddell, S. R., Watanabe, K. S., Goodrich, J. M., Li, C. R., Agha, M. D., and Greenberg, P. D. Restoration of viral immunity in immunodeficient humans by the adoptive transfer of T cell clones. *Science,* 1992. **257,** 238–241.

180. Bonini, C., Verzeletti, S., Servida, P., Rossini, S., Traversari, C., Ferrari, G., Novili, N., Mavilio, F., and Bordignon, C. Transfer of the HSV-tk gene into donor peripheral blood lymphocytes for *in vivo* immunodulation of donor anti-tumor immunity after allo-BMT. *Blood,* 1994. **84,** 110a.

181. Clinical protocol: administration of neomycin resistance gene marked EBV specific cytotoxic T lymphocytes to recipients of mismatched-related or phenotypically similar unrelated donor marrow grafts. *Hum. Gene Ther.,* 1994. **83,** 1988.

182. Bonini, C., Ferrari, G., Verzeletti, S., Servida, P., Zappone, E., Ruggieri, L., Ponzoni, M., Rossini, S., Mavilio, F., Traversari, C., and Bordignon, C. HSV-TK gene transfer into donor lymphocytes for control of allogeneic graft-versus-leukemia [see comments]. *Science,* 1997. **276**(5319), 1719–1724.

Transfer of Drug-Resistance Genes into Hematopoietic Progenitors

OMER N. KOÇ, BRIAN M. DAVIS, JANE S. REESE, SARAH E. FRIEBERT, AND
STANTON L. GERSON

Division of Hematology/Oncology, Case Western Reserve University/University Hospitals Ireland Cancer Center,
Cleveland, Ohio 44106

I. INTRODUCTION

A better understanding of the molecular mechanisms of drug-induced myelosuppression and improvements in eukaryotic gene transfer technology have stimulated interest in gene transfer strategies to enhance drug-resistance mechanisms in hematopoietic cells. This chapter focuses on the use of gene transfer to introduce drug-resistance genes with known resistance phenotypes into hematopoietic progenitors to increase marrow tolerance to chemotherapy.

Two important considerations underlie this technol-

177

ogy. First, if an adequate number of clonogenic hemato-poietic cells can be transduced with a drug-resistance gene that transmits a high degree of resistance to a chemotherapeutic agent, drug-induced myelosuppression and other marrow toxicities such as myelodysplasia and leukemogenesis could be eliminated. Second, drug-resistance genes may serve as dominant selectable markers to allow *in vivo* selection of transduced clones of hematopoietic cells expressing both a drug-resistance gene and a therapeutic gene. In this chapter, we review the rationale for drug-resistance gene therapy, the known drug-resistance mechanisms under consideration for hematopoietic gene therapy, and the *in vitro* animal and clinical trial studies that support further development of this exciting technology.

II. RATIONALE FOR DRUG-RESISTANCE GENE THERAPY

A. Myelosuppression Is the Dose-Limiting Toxicity of Antineoplastic Agents

Myelosuppression and the consequent risks of infection and bleeding represent the predominant dose-limiting toxicity of many chemotherapeutic agents. The mechanism responsible for this hypersensitivity may vary among agents, but, in general, this sensitivity is perhaps not surprising because most chemotherapeutic drugs are identified based on their efficacy against highly proliferative tumor models. Human marrow is extremely proliferative, producing 4×10^{11} cells every day while maintaining a pool of stem cells sufficient for the lifetime of the organism. Thus, it is not surprising to observe sensitivity to cell-cycle-specific chemotherapeutic agents and to those that affect actively proliferating cells. Sensitivity to agents that are not cell cycle specific depends upon the drug's mechanism of action and the cell's ability to detoxify the drug; the latter varies with the degree of maturation within the hematopoietic hierarchy.

Commonly used cytotoxic drugs such as cyclophosphamide, paclitaxel, anthracyclines, and epipodophyllotoxins are predominantly cytotoxic to committed progenitors in the marrow and cause a relatively transient pause in hematopoiesis. In contrast, early hematopoietic progenitors are spared from the cytotoxic effects of these agents because of their relatively higher levels of aldehyde dehydrogenase [1] and P-glycoprotein [2,3]. In addition, early progenitors are quiescent and less susceptible to cell-cycle-specific agents. They also express low topoisomerase I and II, and thus are less sensitive to topoisomerase poisons. On the other hand, nitrosoureas and busulfan are equally cytotoxic to late

and early progenitors, in part due to formation of DNA interstrand crosslinks. These crosslinks are very poorly repaired, regardless of the time given to repair, resulting in prolonged cytopenias or even aplasia with repeated exposures [4–6].

B. Bone Marrow as the Target for Drug-Resistance Gene Therapy

As was noted by Williams *et al.* in 1984 [7], the challenge of gene therapy is to provide a biologically meaningful outcome in a specific target tissue. In this case, the cells that make up the myeloid and lymphoid cell population derived from the bone marrow become the targets of interest. A major limitation for effective gene therapy *in vivo* is the inefficiency of gene transfer into a sufficient number of cells to create a phenotypic change at the level of a tissue rather than an individual cell. One approach has been to use bone marrow ablation followed by transplantation of transduced cells. Unless a population of early progenitors can be genetically altered, the population of transduced cells will decline to extinction over time. Therefore, the emphasis of hematopoietic gene therapy has been to target the stem cells. Biological evidence of therapeutic gene transfer will be facilitated if a survival advantage is conferred to transduced cells to maintain or increase the proportion of genetically modified cells over time.

The polyclonal nature of hematopoiesis is both a blessing and a challenge. Drug-resistance gene transfer into multiple clones is an attractive strategy because different clones may contribute differentially to hematopoietic lineages over time and be of different immaturity. Recent clinical studies with cord blood hematopoietic progenitor gene transfer indicate that gene transduction of an oligoclonal population may limit the spectrum of cells transduced, even though there may be a survival advantage of the transduced cells over a number of years [8]. With drug-resistance gene transfer, drug treatment will not only allow selective survival of hematopoietic progenitors, it should also result in enrichment of the transduced cells, perhaps allowing reconstituted marrow to be replaced with genetically altered cells over time.

Currently the most consistent stable transduction of hematopoietic cells is accomplished by retroviral vectors. Because the murine leukemia virus class of retrovirus requires breakdown of the nuclear matrix for transport into the nucleus, as described in Chapter 4, quiescent cells cannot be transduced. For this reason, the ease of gene transfer into hematopoietic progenitors varies inversely with the degree of immaturity of the cell, because the most primitive stem cells are quiescent and therefore unlikely to proliferate during retroviral exposure. It is not clear whether hematopoietic stem cells can truly self-renew or whether they are derived successively from a pool of progenitor cells that transiently proliferate and then quiesce. Whether it will be possible to affect long-term gene transfer into hematopoietic cells by targeting cells forced to undergo cell division during the gene transfer process is thus unknown.

On the other hand, early progenitors as well as repopulating stem cells can maintain hematopoiesis for prolonged periods of time. This suggests that if a clone of progenitor cells is rendered drug resistant, it may contribute to hematopoiesis for an extended period, particularly under selection pressure during drug therapy. In current studies using hematopoietic progenitor cells as targets, there remains concern that it may be difficult if not impossible to target the truly repopulating stem cell, particularly for primates and humans. Recently described HIV-based lentiviral vectors have been shown to stably transduce nondividing cells, making it potentially feasible to transduce quiescent stem cells. However, these are just beginning to be used to transduce hematopoietic progenitors, and their advantage over conventional murine-leukemia-based virus has not been established. These classes of viruses used for gene transfer into hematopoietic progenitors may not only have potential therapeutic benefit, they may also help resolve a number of fundamental issues remaining in hematopoietic stem cell biology.

Despite the potential limited ability to transduce the stem cell, drug-resistance gene transduction of even a few early progenitors can lead to repopulation of the marrow with cells carrying the gene of interest. Data from a number of groups describing these results in murine models, discussed later, establish that if the correct early progenitor cell can be targeted, enrichment will be possible.

Successful clinical applications of drug-resistance gene transfer would provide both significant decreases in chemotherapy-induced cytopenia and reduction in long-term cumulative marrow toxicity. There is well-established evidence that successful therapy with alkylating agents [9–11], epipodophyllotoxins [12,13], and anthracyclines [14] is associated with a significant risk of myelodysplasia and secondary leukemia in long-term cancer survivors. This suggests that hematopoietic progenitors are vulnerable not only to the myelotoxicity of chemotherapeutic agents but also to their carcinogenic effects. Overexpression of drug-resistance proteins may decrease the mutagenic and carcinogenic effects of antineoplastic agents. For example, overexpression of the DNA repair gene, MGMT, has been demonstrated in transgenic systems to reduce mutation frequency in the thymus of mice significantly [15] and thereby reduce the incidence of nitrosourea-induced T-cell leukemias/lymphomas [16].

III. DRUG-RESISTANCE MECHANISMS

In this section, we describe the drug-resistance genes of current interest for gene transfer and their mechanisms of action, followed by a brief description of experimental models that analyze gene transfer into hematopoietic progenitors (Table 1).

IV. P-GLYCOPROTEIN

The multiple drug-resistance gene-1 (MDR-1) encodes the 170-kDa P-glycoprotein (P-gp), an ATP-dependent transmembrane efflux pump for a diverse group of lipophilic compounds [17,18]. Although two MDR genes exist in humans, only the 3.8-kb MDR-1 is responsible for multidrug resistance [19,20], whereas the main function of MDR-2 (also called MDR-3) appears to mediate biliary phosphatidylcholine secretion. MDR-1 provides resistance to a variety of drugs that are typically cationic, weakly basic, and hydrophobic but are structurally unrelated. These include the anthracyclines, the vinca alkaloids, the epipodophyllotoxins, actinomycin D, and paclitaxel. A variety of other compounds are competitive inhibitors of MDR-1; examples are verapamil, a calcium channel blocker, and cyclosporin A, an immunosuppressant [21].

A. Mechanism of Drug Resistance

The most widely accepted mechanism of the MDR-1 transporter entails that drugs are detected and removed as they enter the plasma membrane, and that the transport occurs through a single channel formed by the transmembrane domains [22]. P-gp consists of two intracellular ATP-binding domains and two transmembrane domains that span the membrane multiple times, forming the channel used for translocation of cytotoxic agents (Fig. 1). The evidence to support direct removal of drugs from the plasma membrane comes from kinetic data of drug efflux [23], the hydrophobicity of target drugs [24], and the demonstration that doxorubicin is removed directly from the plasma membrane by MDR-1 [22]. Photoaffinity labeling [19], mutational analysis [25], and inhibitor studies [21] have all provided evidence of single-channel transport. Drugs that are transported by MDR-1 stimulate ATPase activity, and

TABLE 1   Drug-Resistance Mechanisms

Gene	Mechanism of action	Drug resistance
MDR-1 (P-glycoprotein)	ATP-dependent transmembrane efflux pump that removes a variety of hydrophobic drugs from the plasma membrane.	Anthracyclines, vinca alkaloids, epipodophyllotoxins, actinomycin D, paclitaxel, and colchicine
DHFR (Dihydrofolate reductase)	NADPH-dependent reduction of dihydrofolate to tetrahydrofolate. Antifolates are folic acid analogs that bind to the active site of DHFR, inactivate the enzyme, and inhibit the generation of tetrahydrofolate.	Antifolates such as methotrexate (MTX) and trimetrexate (TMTX)
MGMT (Alkyltransferase)	Repair of DNA alkylations at the O^6 position of guanine by covalent transfer of the adduct to alkyltransferase active site.	Monofunctional methylating agents dacarbazine, temozolomide, streptozotocin, and procarbazine; bifunctional chloroethylating agents BCNU, ACNU, and CCNU.
ALDH (Aldehyde dehydrogenase)	Oxidizes the cyclophosphamide derivative aldophosphamide, forming the relatively nontoxic carboxyphosphamide.	Cyclophosphamide, 4-hydroxycyclophosphamide
GST (Glutathione-S-transferase)	(1) Conjugates the thiol group in glutathione with drugs containing an electrophilic center and (2) directly binds to nonsubstrate ligands.	Alkylating agents such as melphalan, as well as cisplatin, adriamycin, etoposide, and xenobiotics and metabolites containing carbonyl, peroxide, and epoxide groups.
SOD (superoxide dismutase)	Conversion of superoxide anion to hydrogen peroxide (subsequently converted to water by catalase) and oxygen.	Anthracyclines, mitomycin C, and paraquat. Ionizing irradiation.
Cytosine deaminase	Deaminates cytosine nucleoside analog to form inactive uracil derivatives.	Cytosine arabinoside (Ara-C)

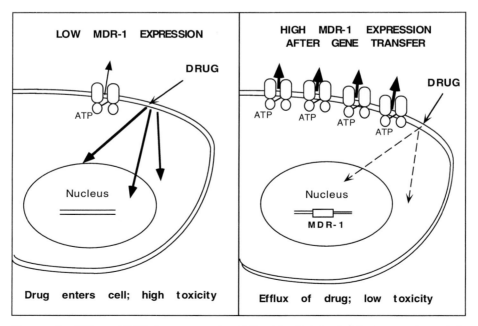

FIGURE 1   Effects of MDR-1 overexpression. Cells with a high level of P-glycoprotein on their surface are capable of exporting a number of cytotoxic drugs as they enter into the cell membrane or from the cytosol more efficiently than those with a low level of P-glycoprotein. This mechanism prevents accumulation of drugs to cytotoxic concentrations within the cell.

it is believed that P-gp utilizes ATP hydrolysis to translocate the drugs directly across the plasma membrane. Other proposed mechanisms of P-gp generation of multidrug resistance include (1) perturbation of the electric membrane potential or the intracellular pH, which alters retention or translocation of the chemotherapeutic agents, and (2) efflux of drugs in association with its role as a Cl⁻ channel or channel regulator.

MDR-1 is normally expressed on the apical or lumenal surface of secretory cells in the adrenal glands, liver, colon, and kidney [26,27]. This pattern of expression is consistent with its proposed role as a drug-transport protein. Endothelial cells of the blood-brain barrier [28], the testes [29], and the placenta [30] also exhibit MDR-1 expression. Most relevant for gene therapy, MDR-1 is expressed in hematopoietic cells [2]. The relatively highest level of MDR-1 expression occurs in the most immature subsets of CD34+ precursors [31] and accounts for the low levels of accumulated rhodamine, an observation used in the selection of early hematopoietic progenitors [3]. Thus, transfer of MDR-1 into hematopoietic cells would be expected to increase resistance in differentiated cells normally sensitive to the treatment with MDR-1 substrate drugs. Myelosuppression

associated with these drugs would be reduced, but the impact on the early progenitor/stem cell pool would be expected to be less.

B. Experimental Models of MDR-1-Gene Transfer

Gene transfer of MDR-1 into murine and human hematopoietic progenitors has been documented in many separate studies, all of which demonstrate increased drug resistance (see Tables 2 and 3). Increased expression of P-gp in MDR-1-transduced cells has been documented using P-gp-specific antibodies and rhodamine efflux assays. A landmark MDR-1 gene transfer study demonstrated *in vivo* selection of MDR-1-transduced cells in mice treated with paclitaxel [32]. Long-term expression of human MDR-1 in repopulating stem cells was demonstrated in mice by serially transplanting MDR-1-transduced bone marrow cells in six successive cohorts of mice with paclitaxel selection [33]. Each successive transplant produced recipient mice with greater resistance to paclitaxel than donor mice. MDR-1 expression was observed 17 months after the initial

TABLE 2   Retroviral Transduction of Murine Hematopoietic Cells with Cytotoxic Drug Resistance Genes

Gene	Vectors	Outcome	References
hMDR1	HaMSV LNL6	*In vivo* selection of the transduced cells. Multilineage engraftment and survival in secondary transplant recipients. Serial transplantation of 6 cohorts of mice with gradual reduction in myelosuppression.	32–34,40
mDHFR (L22R)	MoMLV	*In vivo* selection of the transduced cells. Reduced myelosuppression and increased survival with MTX in secondary transplant recipients.	57–59
hDHFR (L22Y)	HaMSV	CFU TMTX resistance *in vitro;* protection against TMTX-induced myelosuppression *in vivo.*	63
hDHFR (F31S)	MoMLV (DC)	Reduced myelosuppression and increased survival with MTX in recipient mice.	60,61
ada	MoMLV	Increased survival with O⁶BG/BCNU in recipient mice.	96
MGMT	MPSV MoMLV N2/Zip MESV	*In vivo* enrichment for transduced cells. Protection from BCNU myelosuppression and lymphopenia. Increased survival following BCNU.	80–83,85
ΔMGMT (G156A)	MFG	Increased survival with O⁶BG/BCNU in recipient mice 10- to 30-fold enrichment after repeated doses.	100
MDR1-IRES-DHFR (L22Y)	HaMSV	2.9-fold resistance to paclitaxel and 140-fold resistance to TMTX.	137
MGMT-IRES-MDR1	HaMSV	Simultaneous vincristine and ACNU resistance.	138

Note. hMDR-1, human MDR-1 gene; HaMSV, Harvey murine sarcoma virus; mDHFR, murine dihydrofolate reductase gene; hDHFR, human dihydrofolate reductase gene; L22R, leucine to arginine mutation at codon 22; F31S, phenylalanine to serine mutation at codon 31; L22Y, leucine to tyrosine mutation at codon 22; MTX, methotrexate; TMTX, trimetrexate; MoMLV, Moloney murine leukemia virus; DC, double copy; *ada,* bacterial gene encoding alkyltransferase; MGMT, methylguanine methyltransferase; ΔMGMT, mutant methylguanine methyltransferase; G156A, glycine to alanine mutation at codon 156; MPSV, myeloproliferative sarcoma virus; MESV, murine embryonic stem cell virus; IRES, internal ribosomal entry site.

TABLE 3   Retroviral Transduction of Human Hematopoietic Cells with Cytotoxic Drug Resistance Genes and Clinical Protocols

Gene	Vector	Drug resistance	References
hMDR-1	HaMSV	Paclitaxel, doxorubicin, colchicine resistant human PB, BM, and CB CFU	37–39
hDHFR (F31S)	MoMLV	MTX-resistant human PB CFU	62
MGMT	MPSV	BCNU-resistant human PB and BM CFU	83
ΔMGMT (G156A)	MFG	O^6-BG/BCNU resistant human PB CFU	99

Gene	Vector	PI (disease)	Target	Transduction condition	In vivo detection of transduced cells	References
hMDR-1	HaMSV	O'Shaughessy (breast cancer)	PB/BM CD34+	Supernatant only	Pending	151
hMDR-1	MoMLV	Deisseroth (ovarian cancer)	BM CD34+	Supernatant only Over stroma	0 of 10 patients 5 of 8 patients	141, 149
hMDR-1	HaMSV	Hesdorffer (breast-ovarian-brain)	BM CD34+	Supernatant only	2 of 5 patients	142, 150
ΔMGMT	MFG	Gerson–Koç (solid tumors)	PB CD34+	Over stroma	Pending	152

Note. hMDR-1; human MDR-1 gene; HaMSV, Harvey murine sarcoma virus; PB, peripheral blood; BM, bone marrow; CB, cord blood; CD34+, enriched for CD34 expression; hDHFR, human dihyrofolate reductase gene; F31S, phenylalanine to serine mutation at codon 31; CFU, Colony forming unit; MoMLV, Moloney murine leukemia virus; MTX, methotrexate; MGMT, methylguanine methyltransferase; ΔMGMT, mutant methylguanine methyltransferase; G156A, glycine to alanine mutation at codon 156; MPSV, myeloproliferative sarcoma virus.

transduction, implying that continuous expression can be achieved during *in vivo* selection. Both *in vitro* and *in vivo* selection for transduced MDR-1 cells have been demonstrated, even though a truncated MDR-1 mRNA resulting from cryptic splice sites within the MDR-1 cDNA has been detected, possibly resulting in decreased expression of P-gp [34].

Although most studies utilize retroviral vectors, others have evaluated different gene delivery systems of MDR-1 into hematopoietic precursors. Adeno-associated viral (AAV) vectors encoding MDR-1 [35], which demonstrate site-specific integration, have been shown to transduce hematopoietic cells. Liposomal-mediated delivery of the retroviral vector pHaMDR-1/A has also been described, with P-gp over-expression detectable one month after transplant into mice [36].

In addition to the use of bone-marrow-derived cells as targets for gene transfer, MDR transduction has been achieved in murine and human hematopoietic cells mobilized from peripheral blood [37], human cord blood [38], and fetal liver cells [39]. Purified murine early progenitors (Lin⁻ MHCII⁻ Sca1⁺ cells) have been transduced with MDR-1 and transplanted into sublethally irradiated severe combined immunodeficient (SCID) mice. MDR-1⁺ cells were observed 4 to 6 months after transplantation, and 6 weeks after secondary transplantation, implying that the Lin⁻ MHCII⁻ Sca1⁺ cells possess stem cell capacity [40]. A higher rate of MDR-1

gene transfer efficiency was observed in mouse peripheral blood stem cells mobilized by G-CSF and stem cell factor (SCF) than in bone-marrow-derived hematopoietic cells [41]. Furthermore, *ex vivo* enrichment for high P-gp expression after transduction has been achieved using fluorescence activated cell sorting (FACS) of transduced midgestational fetal liver cells [39]. After transplantation of the sorted cells, MDR-1⁺ progenitors were detected for up to 12 months.

Drug resistance of MDR-1-transduced human MO-7e (CD34+, c-kit⁺) cells was evaluated in immunocompromised mice. Paclitaxel treatment of mice transplanted with MDR-1-transduced MO-7e cells resulted in no reduction in MO-7e cellularity whereas mice transplanted with untransduced cells exhibited a 93% decrease in human cells [42]. Recently, a model of simultaneous chemoprotection of bone marrow and chemosensitization of tumor cells was described by Hanania *et al.* [43]. In these experiments, nude mice were transplanted with MDR-1-transduced hematopoietic cells and treated with paclitaxel. The mice tolerated high doses of paclitaxel, leading to a significant decrease in tumor growth.

A protein related to MDR-1 called the multidrug resistance associated protein (MRP), confers resistance to a similar group of chemotherapeutic agents as MDR-1, but is not sensitive to P-gp inhibitors. NIH 3T3 cells transduced with MRP cDNA demonstrated increased resistance to doxorubicin, vincristine, and etoposide

[44]. Transplantation of MRP-transduced hematopoietic stem cells along with MDR-1 inhibitor treatment has also been suggested as a method to sensitize multidrug-resistant tumors while protecting the marrow.

V. DIHYDROFOLATE REDUCTASE

Dihydrofolate reductase (DHFR) is a 22-kDa enzyme that catalyzes the nicotinamide adenine dinucleotide phosphate (NADPH) dependent reduction of dihydrofolate to tetrahydrofrate, an essential carrier of one-carbon units in the biosynthesis of thymidylate, purine nucleotides, and methyl compounds (Fig. 2) [45]. Antifolates such as methotrexate (MTX) and trimetrexate (TMTX) are folic acid analogs that bind tightly to the active site of DHFR, inactivate the enzyme, and therefore inhibit the generation of reduced folate [46]. MTX but not TMTX is converted to polyglutamates after entry into the cell and inhibits enzymes in the purine nucleotide pathway other than DHFR, such as thymidylate synthase and transformylases.

The use of MTX in treating malignancies is limited by the emergence of resistant tumors and myelosuppression [47]. MTX resistance can originate from amplification of the wild-type DHFR gene [48,49] or from mutations that either reduce MTX transport into the cell [50] or lessen the affinity of DHFR for MTX [51]. Quiescent hematopoietic stem cells are thought to be resistant to MTX because of the low requirement for reduced folate and the lack of DNA synthesis. Normal hematopoietic progenitors can also escape from the cytotoxicity of

antifolates by importing thymidine and hypoxanthine from serum. For this reason, elimination of the untransduced cells is less efficient than with other cytotoxic agents. Recently, a number of thymidine transport inhibitors have been shown to block the scavenger pathway and potentiate antifolate cytotoxicity in unmodified cells [52]. These agents have been used in hematopoietic gene transfer studies as mentioned later.

In addition to wild-type DHFR, several naturally occurring and site-directed mutations of DHFR that alter its affinity for antifolates have been characterized. A DHFR molecule with arginine substituted for leucine at position 22 (L22R) was isolated from a MTX-resistant mutant mouse cell line and was shown to have 1/270 the binding affinity for MTX with only a threefold decrease in dihydrofolate catalysis [53,54]. Structure–function analysis of the naturally occurring mutants has led to an exploration for mutants that possess the lowest antifolate affinity while retaining high catalytic efficiency. The human DHFR mutants F31S and L22Y have been recently characterized and appear most promising. Of note, the human L22Y DHFR retains 13% of the catalytic activity compared to wild-type DHFR and yet has 3300-fold less binding affinity to MTX and TMTX.

A. Experimental Models of Mutant DHFR Gene Transfer

Retroviral-mediated gene transfer of the murine L22R DHFR cDNA into primary human, canine, and mouse hematopoietic cells revealed a transduction effi-

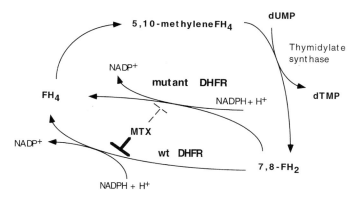

FIGURE 2   The mechanism of action of antifolate drugs. Dihydrofolate reductase (DHFR) catalyzes the reduction of dihydrofolate (7,8-FH2) to tetrahydrofolate (FH4) using reduced nicotinamide adenine dinucleotide phosphate (NADPH+H) as an electron donor. Antifolate drugs such as methotrexate (MTX) exert their cytotoxicity by tightly binding to the active site of DHFR and thereby inactivating the enzyma. Loss of FH4 results in inhibition of nucleotide production and DNA synthesis, which results in cell death. Mutants of DHFR have decreased affinity to MTX and therefore are not inhibited by this drug.

ciency of 3 to 16% as measured by increased MTX resistance of colony forming unit–granulocyte macrophage (CFU-GM) *in vitro* [55–57]. Murine bone marrow transplantation with L22R DHFR–transduced marrow cells, followed by treatment with escalating doses of MTX, resulted in identification of 10 to 20% transduced cells from the marrow and spleen 8 weeks after transplant [58]. MTX treatment was lethal to mice transplanted with cells transduced with a neomycin resistance gene, whereas MTX-treated mice previously transplanted with L22R DHFR–transduced cells had significantly higher peripheral blood counts and marrow and spleen cellularity [58]. CFU-GM harvested from DHFR+ mice had increased MTX resistance [57,59] and serial transplantation of mutant DHFR-transduced cells into secondary and tertiary recipient mice increased resistance to MTX and improved survival (Table 2).

These data give a strong indication that mutant DHFR is a suitable *in vivo* selectable gene able to provide MTX resistance to long-term repopulating murine stem cells. Similar results were obtained with human F31S DHFR both *in vitro* [60] and in mouse transplant experiments [61]. In addition, CFU-GM derived from human peripheral blood progenitors transduced with the human F31S DHFR were significantly more resistant to MTX immediately after infection compared to mock infected controls. *Ex vivo* expansion and MTX selection of transduced peripheral blood progenitors resulted in PCR identification of the provirus in 80% of the CFU-GM tested [62].

Murine hematopoietic progenitors transduced with a human L22Y DHFR retroviral vector showed high-level TMTX resistance *in vitro*. Transplantation of these cells into mice was associated with protection from TMTX-induced neutropenia and reticulocytopenia [63]. Recently, Sorrentino and co-workers constructed a bicistronic murine stem cell virus-based vector containing both the L22Y DHFR and green fluorescence protein cDNAs. Transduced cells displayed a high degree of fluorescence *in vitro* and *in vivo* following transplantation. Furthermore, *in vitro* selection of transduced cells by FACS resulted in a high degree of TMTX resistance *in vitro*. This strategy should allow transplantation of only L22Y DHFR-transduced, antifolate resistant progenitors and improve hematopoietic protection against antifolate challenge [64].

VI. METHYLGUANINE METHYLTRANSFERASE

The human methylguanine–DNA–methyltransferase (MGMT) gene encodes the 207 amino acid DNA repair protein, O^6-alkylguanine DNA alkyltransferase

(AGT), which removes alkyl lesions from the O^6 position of guanine and to a much lesser degree from the O^4 position of thymine. Expression of MGMT provides resistance to monofunctional methylating agents such as dacarbazine, temozolomide, streptozotocin, and procarbazine [65,66] as well as bifunctional chloroethylating agents such as 1,3-bis(2-chloroethyl)-1-nitrosourea (BCNU), 1-(4-amino-2-methyl-5-pyrimidinyl)methyl-3-(2-chloro)-3 nitrosourea (ACNU) and 3-cyclohexyl-1-chloroethyl-nitrosourea (CCNU). The primary toxicity of nitrosoureas, tetrazines, and triazines occurs by alkylation of the O^6 position of guanine [67]. Although there are several known mechanisms of resistance, including glutathione-*S*-transferase [68] and polyamines [69], direct DNA repair of the adduct by AGT is the predominant repair mechanism associated with drug resistance [70–73].

A. Mechanism of Drug Resistance Due to AGT

The mechanism of action of the alkyltransferase is unique among DNA repair enzymes. The protein serves as the acceptor of DNA alkylations at the O^6 position of guanine, the site of one of the most cytotoxic lesions formed by both chloroethylating and methylating agents [74,75]. Repair proceeds by covalent transfer of the adduct to the active site of the protein, an irreversible "suicide" process that inactivates the transfer activity of the protein. O^6-Chloroethylguanine lesions undergo rapid intramolecular rearrangement to the more stable O^6-N^1-ethanoguanine. The adduct is repairable by AGT, forming a covalent protein–DNA crosslink using the cysteine residue in the active site [67,76]. If unrepaired, O^6-N^1-ethanoguanine will form a highly toxic interstrand DNA crosslink with the complementary cytosine nucleotide residue [77]. Cytotoxicity by methylating agents is due to recognition of the O^6-methylguanine:cytosine or O^6-methylguanine: thymine base mispair (formed after one round of replication) by the mismatch repair complex and induction of aberrant repair processes. This process, termed abortive mismatch repair, leads to multiple DNA strand breaks [78].

There is considerable variation in AGT expression in mammalian tissues. In humans, liver contains the highest AGT activity and hematopoietic CD34+ cells possess the lowest activity, explaining the observation of myelosuppression after nitrosourea treatment [79]. The low activity of AGT in human hematopoietic progenitors compared with high-level expression of other drug-resistance genes (such as MDR-1 and ALDH-1) suggests that targeting these cells for MGMT-gene transfer may result in more dramatic protection from myelosuppression after chemotherapy than that observed with other drug-resistance genes.

B. Experimental Models of MGMT-Gene Transfer

Retroviral gene transfer of wild-type (wt) MGMT cDNA has been shown to successfully confer nitrosourea resistance *in vitro* to primary murine hematopoietic progenitors [80–82] and human committed myeloid progenitors derived from CD34+ cells [83] (see Tables 2 and 3). Following lethal irradiation, transplantation of wtMGMT-transduced murine hematopoietic progenitors resulted in increased resistance to BCNU *in vivo* and a 10- to 40-fold increase in AGT expression in hematopoietic tissues [80]. Furthermore, repetitive *in vivo* BCNU administration increased the proportion of cells containing the wtMGMT provirus in mice transplanted with wtMGMT-transduced cells [84]. BCNU resistance was conferred to both myeloid [80–82] and lymphoid [85] lineages in wtMGMT-transplanted mice. In contrast to mock-infected controls, mice infused with wtMGMT⁺ cells had increased leukocyte counts, platelet counts, and hematocrit, as well as a normal distribution of T-cell subsets. In fact, the controls were pancytopenic, and, consistent with the supposition that BCNU acts as a stem cell toxin, had evidence of profound delayed myeloid and lymphoid suppression [85]. More than 6 months after treatment, the wtMGMT⁺ mice appeared to have maintained hematopoiesis with only a mild decrease in cellularity. Thus, transfer of wtMGMT generates enhanced BCNU resistance and a reduction in myelosuppression.

The prospects for successful translation of this observation to the clinical setting must take several factors into account: the relative degree of drug resistance observed, the opportunity for selection of drug-resistant cells *in vivo,* and the identification of a potent inhibitor of AGT, O^6-benzylguanine, now in clinical trials to overcome AGT-mediated tumor resistance to BCNU and related compounds [86]. In the course of initial gene transfer studies, it became apparent that wtMGMT transduction increases BCNU resistance only twofold above endogenous resistance. A likely explanation for this finding is that hematopoietic progenitors express MGMT, albeit at low levels [79], thereby limiting the utility of overexpression of wtMGMT. This limits the relative degree of protection for transduced cells exposed to BCNU; it also curtails the potential to use MGMT as a selection gene to enrich for genetically-altered progenitors carrying a second therapeutic gene. With just two- to fourfold relative resistance, selection

pressure for transduced early progenitor cells is modest and drug selection takes many cycles of treatment.

C. O^6-Benzylguanine and O^6-Benzylguanine-Resistant MGMT

O^6-benzylguanine (BG), a pseudosubstrate that inactivates endogenous AGT, has been shown *in vitro* [74,87,88] and in multiple xenograft applications [89,90] to improve the antitumor effect of BCNU and other chloroethylating agents, and methylating agents such as temozolomide. BG forms an *S*-benzylcysteine moiety at the cysteine acceptor site [91], resulting in irreversible inactivation of AGT. This observation led to clinical phase I trials with this compound in combination with BCNU at our institution, Case Western Reserve University [86], and at the University of Chicago. The end point of these trials is biochemical modulation (depletion) of AGT in tumors.

In our study, inactivation of tumor AGT is monitored by pre- and posttreatment BG. BG inactivation of tumor AGT has been found at doses as low as 10 mg/m², but complete inactivation requires much higher doses, approaching 120 mg/m² [92]. In the course of the preclinical toxicity profile analysis of BG, increased toxicity to human hematopoietic progenitors was noted *in vitro* with BG and BCNU [79] and myelosuppression was observed with the combination of BG and BCNU in both murine [93] and dog systems [94].

Because BG is progressing through clinical trials and appears to offer the potential for a much greater increase in cytotoxic effect against hematopoietic progenitors, the use of wtMGMT in gene transfer experiments could be questioned. Instead, the observation that the bacterial AGT gene *ada* was resistant to BG [88] prompted the possibility of using a BG-resistant form of AGT to increase the differential survival of transduced cells compared to unmodified cells following BG plus BCNU. *ada* is resistant to BG inactivation because of differences in the amino acid sequence near the active site cysteine [95]; as expected, retroviral gene transfer of *ada* resulted in improved resistance of murine hematopoietic progenitors to the BG and BCNU combination both *in vitro* and *in vivo* [96]. However, its effectiveness was limited by the relatively low efficacy for DNA repair noted with bacterial AGT in mammalian cells, possibly due to the low proportion of the protein retained in the nucleus. In addition, the bacterial protein may be immunogenic in humans.

Recently, several mutant human AGT proteins (including G156A, P140A, V139F, V139L, and G160R) have been characterized as resistant to BG, presumably because the active site is sterically hindered from accepting the bulky hydrophobic benzyl group of BG [97,98]. All retain the ability to remove methyl and chloroethyl groups from the O^6 position of guanine in DNA and thus retain drug-resistance activity *in vitro* (Fig. 3).

We have studied the mutant G156A human MGMT (termed ΔMGMT), which Pegg and co-workers identified as having 240-fold increased resistance to BG inactivation [98]. ΔMGMT has been retrovirally transduced into human CD34+ cells [99]. After BG pretreatment, ΔMGMT-transduced cells had a survival advantage over wild-type MGMT of twofold at the IC_{50} and 10-fold at the IC_{90}, and a sixfold advantage over mock, *lacZ*, or untransduced cells at the IC_{50} (Fig. 4). These results demonstrated that survival of ΔMGMT-transduced cells after BG and BCNU is markedly superior to that of wtMGMT-transduced cells treated with BCNU alone. Furthermore, 95% of mice transplanted with ΔMGMT-transduced primary hematopoietic progenitors survived 50 days after 2 doses of BG and BCNU compared to only 23% of mock transduced controls [100]. This represents a greater differential survival than we observed with BCNU alone after wtMGMT transfer [80], using much lower doses of BCNU than required by Williams and co-workers in survival experiments using wtMGMT and BCNU alone [81,82]. After BG and BCNU administration, *in vivo* selection for ΔMGMT-transduced bone marrow cells was improved over the degree of selection obtained after transplant of wtMGMT-transduced cells with BCNU treatment alone. In these studies, one cycle of BG and BCNU increased the proportion of cells expressing ΔAGT from 30% to 60% (Fig. 5) [100]. These results further support that BCNU is cytotoxic to un-transduced quiescent cells and allows enrichment of ΔMGMT-transduced cells.

When extrapolated from *in vitro* studies, *in vivo* results suggest that there is approximately a 10-fold survival advantage to ΔMGMT-transduced hematopoietic progenitors compared to a two- to threefold advantage to wtMGMT-transduced cells. After three cycles of treatment, this predicts transduced hematopoietic progenitor cell enrichment of 1,000-fold for ΔMGMT compared to 8- to 27-fold with wtMGMT. In fact, infusion of ΔMGMT-transduced progenitors into nonmyeloablated mice and subsequent BG and BCNU administration generated 10- to 30-fold enrichment with each cycle [101]. Based on these differences, ΔMGMT appears to offer a clear advantage both for drug resistance and for use as a selectable gene in dual-gene transfer. Our recent data also indicate that ΔMGMT-transduced hematopoietic progenitors have a clear survival advantage over tumor cells following treatment with BG plus BCNU, providing a therapeutic benefit over that of drug resistance alone [102]. In fact, rather than simply making the marrow cells more resistant, the use of BG in combination with ΔMGMT gene transfer results in hematopoietic progenitors uniquely resistant to BCNU, thereby increasing the therapeutic window of this drug combination.

VII. GLUTATHIONE-*S*-TRANSFERASE

The glutathione-*S*-transferase (GST) family of proteins detoxify a wide range of xenobiotics and metabo-

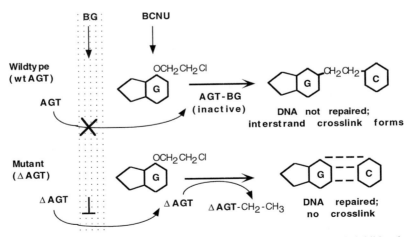

FIGURE 3   Protection from O^6-Benzylguanine (BG) and BCNU. BG inhibits the repair protein O^6-alkylguanine-DNA-alkyltransferase (wtAGT) and renders the cell susceptible to BCNU-induced DNA interstrand crosslinks, which are cytotoxic. Mutants of AGT have been developed (ΔAGT) that are resistant to inhibition by BG yet efficiently repair the chloroethyl adducts (OCH2CH2Cl) induced by BCNU prior to formation of crosslinks.

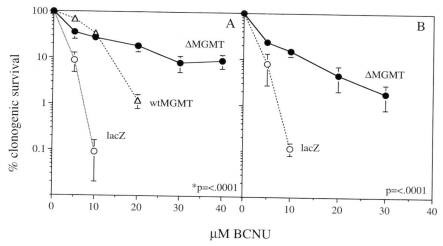

FIGURE 4 Clonogenic survival of ΔMGMT-transduced CD34+ cells following treatment with BG plus BCNU. CD34+ cells were transduced by coculture with MFGΔMGMT retroviral producer cells (A) or MFGΔMGMT retroviral supernatant in the presence of allogeneic human marrow stroma (B). Cells were treated with 10 μM BG and 0–40 μM BCNU and plated in methylcellulose in triplicate, and colonies were enumerated in 7 to 10 days. Data points represent the mean ±SEM of five experiments (A) and three experiments (B) from separate donors. $P < 0.001$ for the comparision ΔMGMT vs *lacZ*. (From Reese *et al.* [99] with permission.)

lites containing carbonyl, peroxide, and epoxide groups resulting from oxidative stress, including oxidized DNA and lipids. GST proteins can be soluble or membrane bound, the soluble forms being the best candidates for chemo-protection. Three cytosolic isoforms of GST consist of the alpha, mu, and pi protein families, which have basic, neutral, and acidic subunits, respectively [103]. Additional cytosolic isoforms include the sigma and theta families. Of all GST isoforms, GST alpha Y2c is believed to confer the greatest level of resistance [104]. These proteins confer resistance to many drugs, including alkylating agents such as melphalan, as well as cisplatin, Adriamycin, and etoposide.

There are two mechanisms by which GST proteins protect against chemotherapeutic agents. GST conjugates the thiol group in the reduced tripeptide gluta-

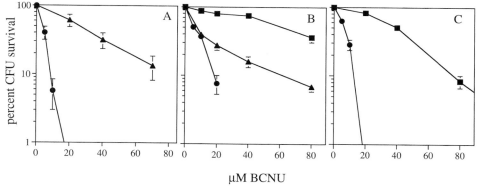

FIGURE 5 ΔMGMT transduction of murine bone marrow cells. BG and BCNU resistance in bone-marrow-derived CFU (A) after MFGΔMGMT transduction and before transplantation, (B) from transplanted mice sacrificed 13 weeks posttransplant (*lacZ*, $n = 5$; unselected ΔMGMT, $n = 3$; selected ΔMGMT, $n = 5$) and (C) from transplanted mice killed 23 weeks posttransplant (selected ΔMGMT, $n = 4$; untransduced $n = 2$). Bone marrow cells were treated with 20 μM BG for 1 hour at 37°C followed by 0–80 μM BCNU for 2 h at 37°C, then plated in triplicate in methylcellulose plates and CFU growth was enumerated. Curves represent survival of *lacZ* transduced or untransduced CFU (●), unselected ΔMGMT transduced CFU (▲) and selected ΔMGMT transduced CFU (■). Bars, SE. (From Davis *et al.* [100], with permission from National Academy of Sciences, U.S.A.)

thione (whose reduction is catalyzed by gammaglu-tamylcysteine synthetase) with drugs containing an electrophilic center. GST proteins also directly bind to nonsubstrate ligands and can detoxify oxidative damage that results from drug exposure. The GST proteins are most active as homodimers, but they are also believed to function as monomers. GST alpha is especially adept at detoxification of alkylating agents.

Different isoforms of GST exhibit tissue-specific expression. GST alpha has the greatest level of expression in the liver, and its degree of expression varies among individuals. GST mu is expressed in the liver and in thymocytes but curiously is not present in 50% of the population [105]. GST pi is expressed in a variety of organs, including the lung, kidney, GI tract, and erythrocytes, and is often overexpressed in human tumors [103]. The expression of many of these genes is induced in many cells, but not hematopoietic cells, by drugs that are detoxified by GST. For this reason, gene transfer of GST into hematopoietic cells, especially GST alpha, could increase resistance to alkylating agents and potentially to a wide range of chemotherapeutic agents.

Increased resistance to various alkylating agents has been observed after retroviral transduction of the rat GST-alpha Y2c in both K562 cell line and murine hematopoietic progenitors. Transduced K562 cells displayed threefold increased resistance to chlorambucil, fivefold resistance to mechlorethamine, and twofold resistance to melphalan. Transduced murine progenitors showed a less pronounced increase in resistance to mechlorethamine and chlorambucil. NIH3T3 cells transduced with either the GST-alpha Y2c [106] or GST-pi gene [107] were protected from xenobiotics and alkylating agents. Recently, human CD34+ progenitors were transduced with GSTπ and protected from Adriamycin and cyclophosphamide [108].

VIII. ALDEHYDE DEHYDROGENASE

Of the multiple isozymes of aldehyde dehydrogenase (ALDH), class 1 and 3 ALDH are cytosolic and have been shown to confer resistance to the oxazaphosphorines such as cyclophosphamide and ifosfamide. The cDNA of ALDH-1 is 1.6 kb in length and codes for a 340 amino acid protein [109]. The cDNA of ALDH-3, or tumor-associated ALDH, is also 1.6 kb, but codes for a 453 amino acid protein [110]. Incubation of resistant cells with ALDH inhibitors, including disulfiram, cyanamide [111], and diethylaminobenzaldehyde [112], reestablishes sensitivity to oxazaphosphorine drugs, proving that resistance can be at least partially attributed to ALDH.

Hydroxylation of cyclophosphamide catalyzed by cytochrome P$_{450}$ and hydrolysis of mafosfamide results in the formation of 4-hydroxycyclophosphamide (4-HC). 4-HC exists in tautomeric equilibrium with its open chain form, aldophosphamide, and cellular uptake of the drug occurs in this form. Toxicity occurs when aldophosphamide undergoes a β-elimination reaction, generating the alkylating agents acrolein and phosphoramide mustard, the latter also acting as a DNA cross-linking agent. ALDH oxidizes aldophosphamide, forming the relatively nontoxic carboxyphosphamide.

Human hematopoietic progenitors (CD34+) express high levels of ALDH-1, whereas differentiation is associated with a decrease in expression [1]. Purging of bone marrow with 4-HC selectively eliminates leukemic cells and maturing hematopoietic cells, including growth-factor-responsive CFU-GM and burst forming unit–erythroid (BFU-E) cells; neither more quiescent long-term culture initiating cells (LTCICs) nor hematopoietic repopulating cells that have higher expression of ALDH are affected. ALDH-1 gene transfer provides a 4- to 10-fold increased resistance to cyclophosphamide in human hematopoietic progenitor cells [113]. Studies evaluating survival advantage *in vivo* in animal models are currently ongoing and, if validated, would provide the basis for using ALDH-1 as a drug-resistance gene in clinical trials. However, the ALDH-1 cDNA appears to have sequences that reduce the stability of the transcript, resulting in undetectable protein expression [114].

IX. SUPEROXIDE DISMUTASE

An oxygen molecule requires four electrons to be completely reduced to two water molecules by the successive production of reactive oxygen species (ROS) such as superoxide anions, hydrogen peroxide, and free hydroxyl radicals via the Haber–Weiss or Fenton-type reactions [115,116]. The free hydroxyl radical reacts with many biological molecules including DNA, proteins, and lipids, resulting in cellular damage. Because these oxidative processes are ubiquitous in intermediary metabolism, cells have developed sophisticated detoxification mechanisms to reduce the level of free radical formation.

Superoxide dismutase (SOD) is an enzyme that protects cells from oxidative damage by catalyzing the conversion of superoxide anion to hydrogen peroxide (subsequently converted to water by catalase) and oxygen in a two-step reaction that alternatively oxidizes and reduces the active site metal [117,118].

$$M^{3+} + O_2^- \rightarrow M^{2+} + O_2$$
$$M^{2+} + O_2^- + 2H^+ \rightarrow M^{3+} + H_2O_2$$

There are two known types of eukaryotic SODs with differences in metal cation dependence and cellular localization. Manganese SOD (MnSOD) is localized to the mitochondrial matrix [119,120], and copper/zinc SOD (CuZnSOD) to the cytoplasm and extracellular fluid [121]. Each form is constitutively expressed from different genes, and expression is enhanced under oxidative conditions [122]. SOD forms dimers or tetramers that keep the superoxide anion substrate in direct contact with the metal cation [123–125].

Overexpression of MnSOD in various cell lines and primary hematopoietic cells results in increased resistance to ionizing radiation [126]. This resistance is significantly reduced in hypoxic conditions [127], implying that MnSOD contributes to irradiation resistance primarily in the presence of oxygen radicals. SOD is also believed to contribute to resistance to the anthracyclines doxorubicin and daunorubicin, mitomycin C, and paraquat. Based upon these results, it may be possible to increase the resistance of human hematopoietic cells to both anthracyclines and ionizing irradiation after gene transfer of SOD. However, SOD gene transfer may be limited due to the observations that (1) human myeloid progenitors are sensitive to ROS due to low expression of catalase and accumulation of hydrogen peroxide [128], (2) CuZnSOD may act as an inhibitor of erythroid progenitor cycling [129], and (3) overexpression of CuZnSOD results in an increase in hydrogen peroxide production and apoptosis in the thymus of transgenic mice [130].

Interestingly, transfer of MnSOD into tumors may generate increased sensitivity to ionizing radiation. Nude mice were infused with a tumor cell line derived from a murine spontaneous fibrosarcoma, Fsa-II, and irradiated under hypoxic conditions. Tumors transduced with MnSOD before infusion demonstrated reduced tumorigenicity [127]. Transfer of MnSOD also has been shown to inhibit tumor growth in human oral squamous carcinoma cells in nude mice [131]. This suggests that transfer of MnSOD directly into tumors may provide therapeutic benefit during radiation therapy. However, overexpression of other proteins involved in oxygen free radical metabolism may protect cells from oxidative stress and could be used as a method to protect hematopoietic cells from chemotherapeutic agents or radiation.

X. CYTOSINE DEAMINASE

Cytosine deaminase (CD) catalyzes the deamination of cytidine or deoxycytidine to uridine or deoxyuridine, respectively (see previous chapters for a detailed description of its function). It can also deaminate cytosine nucleoside analogues such as cytosine arabinoside (Ara-C), a potent antileukemic drug, forming inactive uracil derivatives. Leukemia cells expressing high levels of cytosine deaminase were shown to be resistant to Ara-C *in vitro* [132]. The human cytosine deaminase cDNA has been cloned and encodes a 146 amino acid protein [133].

Overexpression of this gene in murine hematopoietic progenitors resulted in 90 to 100% clonogenic survival *in vitro* at concentrations of Ara-C that allowed survival of 0 to 2.5% of untransduced progenitors [134]. Thus, selective expression of CD in normal hematopoietic cells would protect them from high-dose Ara-C and could have value either for drug-resistance gene therapy in diseases such as lymphoma or as a selectable marker gene. CD gene therapy has also been used to sensitize tumor cells to therapy with 5-fluorocytine [5-FC] because it rapidly converts 5-FC to 5-fluorouracil, a much more potent chemotherapeutic agent. Therefore, individuals receiving CD-transduced hematopoietic cells should not be exposed to 5-FC unless it is desirable to eliminate the transduced cells.

XI. COMBINED EXPRESSION OF DRUG-RESISTANCE GENES

In order to generate broad-range drug resistance to a variety of chemotherapeutic agents, retroviral vectors that express two drug-resistance genes are being developed. Because eukaryotic cells will only translate the first gene encountered on a mRNA transcript, researchers had to design approaches to generate expression of two proteins from one vector. Four methods have been utilized to express multiple genes from one vector.

The first method entails the use of a strong promoter such as those derived from the viruses SV40 or CMV to drive transcription of a second gene within the vector (the viral LTR acts as the promoter for the first gene in the vector). The orientation of the internal promoter can be either the same as or opposite from the LTR. Unfortunately, in either orientation, a potential problem involves suppression of transcriptional activity of the internal promoter because of high LTR transcriptional activity, a phenomenon termed promoter interference. Kane *et al.* [135] have developed double-gene vectors in which the MDR-1 is under the control of the Harvey sarcoma virus LTR, and a second gene is transcribed from another viral or cellular promoter. Constructs containing MDR-1 and either GST or glutathione peroxidase (GP) were used to transduce NIH3T3 cells. After selection in colchicine, a three- and fivefold increase in activity was observed for GST and GP, respectively.

A second method utilizes the endogenous splice sites encoded within the retrovirus to generate two differently expressed transcripts, one full-length and one spliced. Problems with this method include variability in splice site recognition in primary cells and therefore variability in the ratio of full-length to spliced transcripts produced in transduced cells.

A third approach is to express a gene that encodes a fusion protein that retains the activity of the two original proteins. Of course, the activity of the fusion protein may be reduced compared to the original proteins and the fusion protein may be recognized as foreign, resulting in an immunologic response against transduced cells. An MGMT–human AP endonuclease fusion protein has been generated that provides enhanced nitrosourea protection both by increasing direct removal of the O^6-alkylguanine lesion from DNA via AGT activity and by enhancing base excision repair of other lesions that may contribute to cytotoxicity [136]. This represents a promising approach to broadening the efficacy of drug-resistance approaches and awaits confirmation *in vivo* in murine studies.

The fourth and most promising approach utilizes the picornavirus-derived internal ribosome entry site (IRES). The IRES is a sequence that folds into a secondary structure recognized by ribosomes, allowing the initiation of 7-methylguanine cap-independent translation of the second gene in one transcript. An advantage of this approach is that one promoter (the viral LTR) controls the transcription of one mRNA transcript containing both genes. Selection for cells expressing one of the genes should generate cells with higher levels of proviral mRNA, allowing expression of the second gene. In bicistronic constructs consisting of the sequence LTR–gene A–IRES–gene B, gene A is translated from the LTR and gene B is translated from the IRES. In all instances, the B protein expressed from the IRES has been shown to be translated less efficiently than the A protein translated from the 7-methylguanine mRNA cap [107]. This may influence the choice of gene sequence in the construct, depending on whether it is more important to have high expression of gene B or a high degree of selection pressure from gene A.

A bicistronic vector containing LTR–MDR-1–IRES–L22Y DHFR was transduced into CEM cells, resulting in significant resistance to paclitaxel, vinblastine, TMTX, and MTX [137]. These results demonstrate that selection for one gene can result in elevated levels of a second drug-resistance gene at levels sufficient to produce broad-range drug resistance. Another recently described IRES-based vector expresses both the MDR-1 gene and the wtMGMT gene [138]. Transduced cell lines selected in colchicine for MDR-1 expression demonstrated increased nitrosourea resistance based on

MGMT expression, with increased AGT when analyzed by Western blotting. *In vivo* studies with murine or human hematopoietic progenitors have yet to be reported but may yield a vector with a broad range of drug resistance.

XII. CLINICAL APPLICATIONS

A. Optimization of Transduction Protocols

Although murine experiments have proven the feasibility of generating drug-resistant bone marrow with *ex vivo* retroviral gene transfer, the clinical application of this strategy in humans has a number of obstacles. Preliminary human studies indicate that transduction of early human hematopoietic progenitors with retroviral vectors is significantly less efficient compared to their murine counterparts. Clinical trials to date show that after reinfusion, transduced cells are detected in peripheral blood and/or bone marrow at low frequency and in many instances only transiently [139–143]. One of two factors may be at work here: either gene transfer occurs at low frequency *in vitro* using the experimental conditions of early clinical trials; or transduced cells are present in low proportion immediately following infusion because of the remarkable dilutional effect of endogenous cells [141]. Both of these issues need to be addressed in the next generation of clinical trials both by improving the efficiency of gene transfer *in vitro* and by using drug-selection genes of high potency, which will allow *in vivo* selection after infusion.

Of interest are recent reports by Kohn and coworkers that retroviral transduction of the adenosine deaminase (ADA) gene into cord blood CD34+ cells of two children with severe combined immunodeficiency, followed by transplantation without prior myeloablation, resulted in persistent transduced lymphocytes 4 years later [144]. A confounding factor, though, is that these transduced cells clearly have a survival advantage over their ADA-defective counterparts. Also noteworthy is work by Barranger and co-workers, who used a centrifugation transduction technique to transduce human CD34+ cells with the MFG–glucocerebrosidase gene vector. The cells were then infused into adult recipients with Gaucher's Disease who were not given conditioning therapy. This group reported that up to 4% of recipient CD34+ cells had evidence of the provirus and produced glucocerebrosidase [145]. These results suggest that myeloablation will not be necessary in all instances for successful gene therapy, although the general applicability of the techniques used by these two groups may be limited.

Recent efforts have focused on methods to optimize conditions that allow transduction of early hematopoietic progenitors. Infection protocols with cytokines, stromal support or fibronectin for 48- to 72-hour incubation periods result in more consistent and longer-term detection of transgene-positive cells *in vivo,* compared to protocols not using stromal support or shorter (6 hour) incubations [141,146,147]. The use of the fibronectin fragment CH-296 improves gene transfer efficiency into human CD34$^+$ cells and earlier progenitors [147]. Likewise, early acting hematopoietic cytokines such as Flt3-ligand (FL), in conjunction with IL-3, IL-6 and/or SCF during the transduction period, improve the proliferative capacity of primitive human hematopoietic cells and increase gene transfer into early progenitors.

In each instance, the key to improved gene transfer appears to be the ability to force early hematopoietic progenitors into the cell cycle, where they are much more susceptible to retroviral integration. At the same time, they need to be maintained in their primitive state without differentiating into more lineage-restricted progenitors. Although direct analysis of such early repopulating cells is best done in the setting of clinical trials, surrogate markers of these cells such as LTCICs and immunocompromised mouse (SCID or NOD-SCID) repopulating cells (SRCs) can also guide clinical trial design. The presence of bone marrow stroma was shown to increase gene transfer efficiency and the survival of long-lived human hematopoietic progenitors in NOD-SCID mice [148].

B. Clinical Trials

A number of clinical trials are underway using retroviral-mediated gene transfer of MDR-1 into hematopoietic progenitor cells (Table 3). In these trials, bone marrow or peripheral blood CD34+ cells are transduced with or without cytokines and stroma coculture for short periods of time and then infused into the recipient. Most current trials call for administration of intensive chemotherapy prior to infusion to increase the probability of engraftment of transduced cells, and the cells are infused as part of an autologous bone marrow and/or peripheral blood transplantation. Following engraftment, patients are given P-gp substrate drugs such as paclitaxel [149–151]. Preliminary results indicate that MDR-transduced cells can be detected in low frequency (0.01–1%) for up to 7 months after reinfusion [141–143]. Despite multiple cycles of therapy with paclitaxel and doxorubicin, no enrichment in the frequency of MDR-transduced cells has been reported [143].

In contrast to MDR substrate antineoplastic drugs, and as noted earlier, nitrosoureas are more toxic to early hematopoietic progenitors and therefore may exert a higher degree of selection pressure to cells transduced with MGMT. In a study in pediatric patients with CNS tumors, an aliquot of MGMT-transduced CD34+ cells collected prior to chemotherapy will be infused after each cycle of nitrosourea chemotherapy. The nitrosourea will be given every 4 weeks in a dose-intensive schedule, rather than the 6-week conventional dosing schedule (D. Williams, Indiana University). Marrow and blood samples will monitor the presence of gene-transduced blood and marrow cells over time in this pediatric population.

We have proposed a complementary trial in adult patients with solid tumors. Our intent is to use the MFG vector carrying the G156AMGMT gene and to infuse transduced CD34+ cells after the first cycle of BG plus BCNU given at standard Phase II doses without prior myeloablation [152]. Because our own preclinical data support the use of a conventional dosing schedule without dose-intensive therapy, we expect to treat patients at the standard 6-week interval. Patients will be monitored for evidence of transduced cells, which we predict will increase with each cycle of therapy. In patients receiving 3 cycles of therapy, the proportion of cells with proviral elements may exceed 5% and less myelosuppression may be observed. Because tumors will be sensitized to BCNU by BG, the potential exists for improving the therapeutic value of the BG plus BCNU combination by protecting the marrow from the cumulative toxicity expected from the two drugs.

McIvor and Verfaillie have generated a novel retroviral vector containing both L22YDHFR and *bcr-abl* antisense [153]. Transduction of this vector into CD34+ cells of patients in chronic phase chronic myelogenous leukemia (CML) will make the transduced cells resistant to MTX. If CML cells are transduced, the expression of *bcr-abl* antisense will kill the cells. Subsequent therapy with MTX is designed to kill untransduced cells, particularly residual CML cells.

In each of these studies, the objectives are safety of gene transduction and infusion, patient tolerance to the chemotherapy regimen, and ability to detect transduced cells after each cycle of chemotherapy. Because an important end point of these studies is the longevity of the genetically altered cells, data will also be forthcoming regarding the ability to transduce hematopoietic progenitors capable of long-term (>6 months) reconstitution in humans.

C. Safety of Drug-Resistance Gene Therapy

Although there have been no apparent toxicities associated with retroviral gene transfer in any clinical trials as yet, a number of safety concerns exist. Most important is the risk of insertional mutagenesis and carcinogenesis, known to occur with replication-competent virus. Laboratory documentation of such toxicity has been limited to immunocompromised primates exposed to helper virus containing large amounts of retroviral vector [154]. The risk of insertional mutagenesis and carcinogenesis has not been defined in humans but is thought not to exceed the overall risks associated with chemotherapy administration. Careful screening of clinically used retroviral vector preparations to exclude the presence of replication-competent retrovirus is mandated by the FDA [155]. Likewise, the FDA suggests that patients who are exposed to retroviral vectors be monitored for prolonged periods of time for late toxicity. These late toxicities may become more apparent with improvement in the efficiency of gene transfer techniques, resulting in a greater number of transduced cells.

Another concern is the inadvertent transduction of tumor cells with a drug-resistance phenotype during the transduction of CD34+ isolated cells. This can be avoided with careful patient selection and improved techniques of progenitor cell separation prior to gene transduction. Currently, these selection techniques result in 2–4 log selection of CD34+ cells relative to myeloma and breast cancer cells. However, because even transduced tumor cells would only be resistant to the drugs in the class of agents protected by the resistance gene, chemotherapeutic drugs in other classes would still be effective and could still be administered. Incorporation of safety (suicide) genes into vectors [156] might provide a way to eliminate transduced tumor cells but has the clear disadvantage of killing transduced hematopoietic cells as well.

XIII. APPLICATION OF DRUG-RESISTANCE GENES FOR SELECTION

One end point of current drug-resistance gene therapy protocols is establishment of a drug-resistant marrow, which will allow dose intensification for cancer patients. However, another clearly important role for this technique exists for genetic disorders of the hematopoietic system. The success of corrective gene therapy of disorders such as hemoglobinopathies relies on the expression of the transgene in a significant portion of mature blood cells. Cotransduction of a drug-resistance gene with a corrective gene would allow a transduced clone to survive and become enriched in the presence of selection pressure. If this selection pressure could be administered over time and not require myeloablation with its attendant risks and toxicity, this dominant selec-

tion capability could become an important clinical tool in the treatment of genetic disorders, possibly representing the most important long-term utility of this approach.

Children with a variety of single-gene defects and acquired disorders represent potential beneficiaries of gene therapy. The requirements for candidate genes include those in which the morbidity and mortality of the associated disease outweigh the potential toxicities of myelosuppressive therapy or, preferably, those for which preparative regimens are not required. Children with genetic defects present a special case in which successful correction of a devastating phenotype may be possible with only a small level of efficient transduction. Although much gene transfer research has been hampered by inefficiency of transgene expression and inability to sustain long-term correction, early hematopoietic transduction of a few progenitors may alleviate much clinical symptomatology when done early and in the small child. In fact, severe combined immunodeficiency caused by deficiency of the purine metabolic enzyme adenosine deaminase was the first genetic disease successfully corrected using retroviral-vector-mediated gene transfer, as discussed earlier [144].

Because initial fusions of the *ada* gene with MDR-1 and *lacZ* demonstrated selectability and bifunctionality, several other candidate second genes are under investigation to improve transduction efficiency and selectivity. Pediatric diseases that show promise for bicistronic vector correction include lysosomal storage diseases such as Gaucher's (glucocerebrosidase deficiency) and the mucopolysaccharidoses; AIDS, especially in the context of modification of viral drug resistance through intracellular immunization with suicide genes under the control of an HIV promoter; other immunodeficiency syndromes such as chronic granulomatous disease (CGD); and hematologic disorders such as thalassemia and Factor IX deficiency (Table 4).

For each of these conditions, genetic constructs need to be generated and tested for expression *in vitro* in the context of bicistronic vectors or other dual-gene-vector systems outlined earlier and in Chapter 4. Animal models of efficacy will allow determination of whether drug selection increases the proportion of transduced cells into the therapeutically acceptable range, a value that is likely to differ with each gene. Ultimately, limited clinical trials will be necessary to define whether transduction with the dual-gene vector followed by drug selection will result in both expression of the corrective gene and clinical improvement in the congenital defect.

In summary, drug-resistance gene therapy has developed into one of the most promising modalities of gene therapy, particularly due to the ability to enrich for genetically altered, drug-resistant cells that may reduce

TABLE 4 Candidate Hematopoietic Diseases for Correction with Bicistronic Retroviral Vectors as Dominant Selectable Genes

Single gene disease of interest	Aberrant/target enzyme or gene defect	cDNA size, kb	Chromosomes involved/inheritance	Retroviral backbone	Clinical trial?	References
Severe combined immunodeficiency	Adenosine deaminase			Moloney MLV/LASN	Yes	144,157–158
Chronic granulomatous disease	NADPH-oxidase 4 subunits:			Murine stem cell virus (MSCV); MFG		
	gp91phox	1.7	X		Yes	159–165
	p22phox	0.6	Autosomal recessive		Yes	
	p47phox	1.2	Autosomal recessive		Yes	
	p67phox	1.6	Autosomal recessive		No	
Hemophilia A	Factor VIII	6.9	X	Various	No	166–168
Hemophilia B	Factor IX	1.5	X	Various	No	169–170
Fanconi anemia	FACC (Fanconi anemia C complementing)	2.0	Autosomal recessive	Moloney MLV	No	171
β-Thalassemia	β-globin, chromosome 11		Autosomal recessive	Moloney MLV	No	172–173
Sickle cell disease	β-globin (valine for glutamic acid substitution, codon 6)		Autosomal recessive	Moloney MLV	No	174
Pyruvate kinase deficiency	Pyruvate kinase	2.3	Autosomal recessive	pMNSM-LPK	No	175
AIDS	Human immunodeficiency virus (HIV), RevM10 gene		N/A		Yes	176
Lysosomal storage diseases/ mucopolysaccharidoses (MPSs)	(1) Gaucher's Disease: glucocerebrosidase	2.2 or 1.6	Autosomal recessive	Moloney MLV, Harvey murine sarcoma virus (HaMSV)	Yes	177–182
	(2) Hurler syndrome: α-L-iduronidase (IDUA)—MPS I			Moloney MLV	No	183
	(3) Sly syndrome: β-glucuronidase—MPS VII		X	Moloney MLV	No	184
	(4) Hunter syndrome: lysosomal iduronate-2-sulfatase (IDS)—MPS II	1.4	X	Moloney MLV	Yes	185–187

myelosuppression, prevent cumulative toxicity and enable therapeutic gene transfer and enrichment.

Acknowledgments

This work was supported in part by Public Health Service Grants P30CA43703, RO1CA63193, RO1ES06288, RO1CA73062, MO1RR00080-35, and UO1CA75525, and by Grant ACS-PRTA-35 from the American Cancer Society.

References

1. Kastan, M. B., Schlaffer, E., Russo, J., Colvin, O., Civin, C., and Hilton, J. (1990). Direct demonstration of elevated aldehyde dehydrogenase in human hematopoietic progenitor cells. *Blood* **75**, 1947–1950.
2. Chaudhary, P., and Roninson, I. (1991). Expression and activity of P-glycoprotein, a multidrug efflux pump, in human hematopoietic stem cells. *Cell* **66**, 85–94.
3. Smeets, M., Raymakers, R., Viervinden, G., Pennings, A., van de Locht, L., Wessels, H., Boezeman, J., and de Witte, T. (1997). A low but functionally significant MDR1 expression protects primitive hematopoietic progenitor cells from anthracycline toxicity. *Br. J. Haematol.* **96**, 346–355.
4. Maze, R., Mortiz, T., and Williams, D. (1994). Increased survival and multilineage hematopoietic protection from delayed and severe myelosuppressive effects of a nitrosourea with recombinant interleukin-11. *Cancer Res.* **54**, 4947–4951.
5. Naben, S., Hemman, S., Montegomery, M., Ferrera, J., and Mauch, P. (1993). Hematopoietic stem cell deficit of transplanted bone marrow previously exposed to cytotoxic agents. *Exp. Hematol.* **21**, 156–162.
6. Weiss, R., and Issell, B. (1982). The nitrosoureas: carmustine (BCNU) and lomustine (CCNU). *Cancer Treat. Rev.* **9**, 313–330.
7. Williams, D. A., Lemishka, I. R., Nathan, D. G., and Mulligan, R. C. (1984). Introduction of new genetic material into pluripotent haematopoietic stem cells of the mouse. *Nature* **310**, 476–480.
8. Kohn, D., Weinberg, K., Shigeoka, A., Carbonaro, D., Brooks, J., Smogorzewska, E., Barsky, L., Annett, G., and Nolta, J. (1997). PEG-ADA reduction in recipients of ADA gene-transduced autologous umbilical cord blood CD34+ cells. *Blood* **90**, 404a.
9. Pedersen-Bjergaard, J., Ersboll, J., Sorensen, H., Keiding, N., Larsen, S., Philip, P., Larsen, M., Schultz, H., and Nissen, N. (1985). Risk of acute leukemia and preleukemia in patients treated with cyclophosphamide for non-Hodgkin's lymphomas. *Ann. Intern. Med.* **103**, 195–200.
10. Curtis, R., Boice, J., Stovall, M., Bernstein, L., Greenberg, R., Flannery, J., Schwartz, A., Weyer, P., Moloney, W., and Hoover, N. (1992). Risk of leukemia after chemotherapy and radiation treatment for breast cancer. *N. Engl. J. Med.* **326**, 1745–1751.
11. Devereux, S., Selassie, T. G., Hudson, G. V., Hudson, B. V., and Linch, D. C. (1990). Leukemia complicating treatment for Hodgkin's disease: the experience of the British National Lymphoma Investigation. *Br. Med. J.* **301**, 1077–1080.
12. Pedersen-Bjergaard, J., Daugaard, J., Hansen, S., Philip, P., Larsen, S., and Rorth, M. (1991). Increased risk of myelodysplasia and leukaemia following etoposide, cisplatin and bleomycin for germ cell tumours. *Lancet* **338**, 359–363.
13. van Leeuwen, F., Stiggelbout, A., van den Belt-Dusebout, A., Noyon, R., Eliel, M., van Kerkhoff, E., Delemarre, F., and Som-

ers, R. (1993). Second cancer risk following testicular cancer: a follow-up study of 1,909 patients. *J. Clin. Oncol.* **11**, 415–424.
14. Pedersen-Bjergaard, J., Sigsgaard, T., Nielsen, D., Gjedde, S., Philip, P., Hansen, M., Larsen, S., Rorth, M., Mouridsen, H., and Dombernowsky, P. (1992). Acute monocytic or myelomonocytic leukemia with balanced chromosome translocations to band 11q23 after therapy with 4-epi-doxorubicin and cisplatin or cyclophosphamide for breast cancer. *J. Clin. Oncol.* **10**, 1444–1451.
15. Allay, E., McGuire, E., Koç, O., Marko, D., Pincus, E., and Gerson, S. (1994). Transgenic human O^6-alkylguanine-DNA alkyltransferase decreases incidence and increases latency of MNU-induced thymic lymphomas in Ttg-1 transgenic mice. *Proc. Am. Assoc. Cancer Res.* **35**, 115.
16. Dumenco, L. L., Allay, E., Norton, K., and Gerson, S. L. (1993). The prevention of thymic lymphomas in transgenic mice by human O^6-alkylguanine-DNA alkyltransferase. *Science* **259**, 219–222.
17. Gottesman, M. (1993). How cancer cells evade chemotherapy: sixteenth Richard and Hinda Rosenthal Foundation award lecture. *Cancer Res.* **53**, 747–754.
18. Gottesman, M., and Pastan, I. (1993). Biochemistry of multidrug resistance mediated by the multidrug transporter. *Annu. Rev. Biochem.* **62**, 385–427.
19. Roninson, I., Chin, J., Choi, K., Gros, P., Housman, D., Fojo, A., Shen, D., Gottesman, M., and Pastan, I. (1986). Isolation of human MDR DNA sequences amplified in multidrug-resistant KB carcinoma cells. *Proc. Natl. Acad. Sci. USA* **83**, 4538–4542.
20. Shen, D.-W., Fojo, A., Chin, J., Roninson, I., Richert, N., Pastan, I., and Gottesman, M. (1986). Human multidrug-resistant cell lines: increased MDR1 expression can precede gene amplification. *Science* **232**, 643–645.
21. Cano-Gauci, D., and Riordan, J. (1987). Action of calcium antagonists on multidrug resistant cells. *Biochem. Pharmacol.* **36**, 2115–2123.
22. Raviv, Y., Pollard, H., Bruggemann, E., Pastan, I., and Gottesman, M. (1990). Photosensitized labeling of a functional multidrug transporter in living drug-resistant tumor cells. *J. Biol. Chem.* **7**, 3975–3980.
23. Inaba, M., Kobayashi, H., Sakurai, Y., and Johnson, R. (1979). Active efflux of daunorubicin and Adriamycin in sensitive and resistant sublines of P388 leukemia. *Cancer Res.* **39**, 2200–2203.
24. Zamora, J., Pearce, H., and Beck, W. (1988). Physical-chemical properties shared by compounds that modulate multidrug resistance in human leukemic cells. *Mol. Pharmacol.* **33**, 454–462.
25. Pastan, I., Willingham, M., and Gottesman, M. (1991). Molecular manipulations of the multidrug transporter: a new role for transgenic mice. *FASEB J.* **5**, 2523–2528.
26. Thiebaut, F., Tsuruo, T., Hamada, H., Gottesman, M., Pastan, I., and Willingham, M. (1987). Cellular localization of the multidrug resistance gene product P-glycoprotein in normal human tissues. *Proc. Natl. Acad. Sci. USA* **84**, 7735–7738.
27. Fojo, A., Ueda, K., Slamon, D., Poplack, D., Gottesman, M., and Pastan, I. (1987). Expression of a multidrug-resistance gene in human tumors and tissues. *Proc. Natl. Acad. Sci. USA* **84**, 265–269.
28. Thiebaut, F., Tsuruo, T., Hamada, H., Gottesman, M., Pastan, I., and Willingham, M. (1989). Immunohistochemical localization in normal tissues of different epitopes in the multidrug transport protein P170: evidence for localization in brain capillaries and crossreactivity of one antibody with a muscle protein. *J. Histochem. Cytochem.* **37**, 159–164.
29. Cordon-Cardo, C., O'Brien, J., Casals, D., Rittman-Grauer, L., Biedler, J., Melamed, M., and Bertino, J. (1989). Multidrug-resistance gene (P-glycoprotein) is expressed by endothelial cells

at blood-brain barrier sites. *Proc. Natl. Acad. Sci. USA* **86,** 695–698.

30. Sugawara, I., Kataoka, I., Morishita, Y., Hamada, H., Tsuruo, T., Itoyama, S., and Mori, S. (1988). Tissue distribution of P-glycoprotein encoded by a multidrug-resistant gene as revealed by a monoclonal antibody, MRK16. *Cancer Res.* **48,** 1926–1929.

31. Drach, D., Zhao, S., Mahadevia, R., Gattringer, C., Huber, H., and Andreeff, M. (1992). Subpopulations of normal peripheral blood and bone marrow cells express a functional multidrug resistant phenotype. *Blood* **80,** 2729–2734.

32. Sorrentino, B., Brandt, S., Bodine, D., Gottesman, M., Pastan, I., Cline, A., and Nienhuis, A. (1992). Selection of drug-resistant bone marrow cells *in vivo* after retroviral transfer of human MDR1. *Science* **257,** 99–103.

33. Hanania, E. G., and Deisseroth, A. B. (1994). Serial transplantation shows that early hematopoietic precursor cells are transduced by MDR-1 retroviral vector in a mouse gene therapy model. *Cancer Gene Ther.* **1,** 21–25.

34. Sorrentino, B. P., McDonagh, K. T., Woods, D., and Orlic, D. (1995). Expression of retroviral vectors containing the human multidrug resistance 1 cDNA in hematopoietic cells of transplanted mice. *Blood* **86,** 491–501.

35. Baudard, M., Flotte, T., Aran, J., Thierry, A., Pastan, I., Pang, M., Kearns, W., and Gottesman, M. (1996). Expression of the human multidrug resistance and glucocerebrosidase cDNAs from adeno-associated vectors: efficient promotor activity of AAV sequences and *in vivo* delivery systems via liposomes. *Hum. Gene Ther.* **7,** 1309–1322.

36. Aksentijevich, I., Pastan, I., Lunardi-Iskandar, Y., Gallo, R., Gottesman, M., and Thierry, A. (1996). *In vitro* and *in vivo* liposome mediated gene transfer leads to human MDR1 expression in mouse bone marrow progenitor cells. *Hum. Gene Ther.* **7,** 1111–1122.

37. Ward, M., Richardson, C., Pioli, P., Smith, L., Podda, S., Goff, S., Hesdorffer, C., and Bank, A. (1994). Transfer and expression of the human multiple drug resistance gene in human CD34+ cells. *Blood* **84,** 1408.

38. Bertolini, F., Battaglia, M., Corsini, C., Lazzari, L., Soligo, D., Zibera, C., and Thalmeier, K. (1996). Engineered stromal layers and continuous flow culture enhance multidrug resistance gene transfer in hematopoietic progenitors. *Cancer Res.* **56,** 2566–2572.

39. Richardson, C., and Bank, A. (1995). Preselection of transduced murine hematopoietic stem cell populations leads to increased long-term stability and expression of the human multiple drug resistance gene. *Blood* **86,** 2579–2589.

40. Licht, T., Aksentijevich, I., Gottesman, M. M., and Pastan, I. (1995). Efficient expression of functional human MDR1 gene in murine bone marrow after retroviral transduction of purified hematopoietic stem cells. *Blood* **86,** 111–121.

41. Bodine, D. M., Seidel, N. E., Gale, M. S., Nienhuis, A. W., and Orlic, D. (1994). Efficient retrovirus transduction of mouse pluripotent hematopoietic stem cells mobilized into the peripheral blood by treatment with granulocyte colony stimulating factor and stem cell factor. *Blood* **84,** 1482–1491.

42. Schwartzenberger, P., Spence, S., Lohrey, N., Kmiecik, T., Longo, D., Murphy, W., Ruscetti, F., and Keller, J. (1996). Gene transfer of multidrug resistance into a factor-dependent human hematopoietic cell line: *in vivo* model for genetically transferred chemoprotection. *Blood* **88,** 2723–2731.

43. Hanania, E., and Deisseroth, A. (1997). Simultaneous genetic chemoprotection of normal marrow cells and genetic chemosensitization of breast cancer cells in a mouse cancer gene therapy model. *Clin. Cancer Res.* **3,** 281–286.

44. D'Hondt, V., Caruso, M., and Bank, A. (1997). Retrovirus mediated gene transfer of the multidrug resistance-associated protein (MRP) cDNA protects cells from chemotherapeutic agents. *Hum. Gene Ther.* **8,** 1745–1751.

45. Blakley, R. L. (1984). Folates and pterins, in Blakley, R. L., Benkovic, S. J. (eds.): Chemistry and biochemistry of folates. New York, NY, Wiley. 191

46. Werkheiser, W. (1961). Specific binding of 4-amino folic acid analogues by folic acid reductase. *J. Biol. Chem.* **236,** 888.

47. Bertino, J., Donohue, D., Simmons, B., Gabrio, B., Silber, R., and Huennekens, F. (1963). The "induction" of dihydrofolic reductase activity in leukocytes and erythrocytes of patients treated with amethopterin. *J. Clin. Invest.* **42,** 466.

48. Alt, F., Kellems, R., Bertino, J., and Schimke, R. (1978). Selective multiplication of dihydrofolate reductase genes in methotrexate-resistant variants of cultured murine cells. *J. Biol. Chem.* **253,** 1357–1370.

49. Raunio, R., and Hakala, M. (1967). Comparison of folate reductases of sarcoma 180 cells, sensitive and resistant to amethopterin. *Mol. Pharmacol.* **3,** 279–283.

50. Sirotnak, F., Kurita, S., and Hutchison, D. (1968). On the nature of a transport alteration determining resistance to amethopterin in the L1210 leukemia. *Cancer Res.* **28,** 75–80.

51. Flintoff, W., Davidson, S., and Siminovitch, L. (1976). Isolation and partial characterization of three methotrexate-resistant phenotypes from chinese hamster ovary cells. *Somatic Cell Genet.* **2,** 245–261.

52. Allay, J., Spencer, H., Wilkinson, S., Belt, J., Blakley, R., and Sorrentino, B. (1997). Sensitization of hematopoietic stem and progenitor cells to trimetrexate using nucleoside transport inhibitors. *Blood* **90,** 3546–3554.

53. Haber, D. A., Beverley, S. M., Kiely, M. L., and Schimke, R. T. (1981). Properties of an altered dihydrofolate reductase encoded by amplified genes in cultured mouse fibroblasts. *J. Biol. Chem.* **256,** 9501–9510.

54. Simonsen, C. C., and Levinson, A. D. (1983). Isolation and expression of an altered mouse dihydrofolate reductase cDNA. *Proc. Natl. Acad. Sci. USA* **80,** 2495–2499.

55. Hock, R., and Miller, A. (1986). Retrovirus-mediated transfer and expression of drug resistance genes in human hematopoietic progenitor cells. *Nature* **320,** 275–277.

56. Kwok, W. W., Schuening, F., Stead, R. B., and Miller, A. (1986). Retroviral transfer of genes into canine hemopoietic progenitor cells in culture: a model for human gene therapy. *Proc. Natl. Acad. Sci. USA* **83,** 4552–4555.

57. Zhao, S., Li, M., Banerjee, D., Schweitzer, B., Mineishi, S., Gilboa, E., and Bertino, J. (1994). Long-term protection of recipient mice from lethal doses of methotrexate by marrow infected with a double-copy vector retrovirus containing a mutant dihydrofolate reductase. *Cancer Gene. Ther.* **1,** 27–33.

58. Williams, D., Hsieh, K., DeSilva, A., and Mulligan, R. (1987). Protection of bone marrow transplant recipients from lethal doses of methotrexate by the generation of methotrexate-resistant bone marrow. *J. Exp. Med.* **166,** 210–218.

59. Corey, C., DeSilva, A., Holland, C., and Williams, D. (1990). Serial transplantation of methotrexate-resistant bone marrow: protection of murine recipients from drug toxicity by progeny of transduced stem cells. *Blood* **75,** 337–343.

60. Banerjee, D., Schweitzer, B., and Volkenandt, M. (1994). Transfection with cDNA encoding a ser[31] or ser[34] mutant human dihydrofolate reductase into chinese hamster ovary and mouse marrow progenitor cells confers methotrexate resistance. *Gene* **139,** 269–274.

61. Li, M.-X., Banerjee, D., Zhao, S., Schweitzer, B. I., Mineishi, S., Gilboa, E., and Bertino, J. R. (1994). Development of a retroviral construct containing a human mutated dihydrofolate reductase cDNA for hematopoietic stem cell transduction. *Blood* **83,** 3403–3408.

62. Flasshove, M., Banerjee, D., Mineishi, S., Li, M. X., Bertino, J. R., and Moore, M. A. S. (1995). *Ex vivo* expansion and selection of human CD34+ peripheral blood progenitor cells after introduction of a mutated dihydrofolate reductase cDNA via retroviral gene transfer. *Blood* **85,** 566–574.

63. Spencer, H., Sleep, S., Rehg, J., Blakley, R., and Sorrentino, B. (1996). A gene transfer strategy for making bone marrow cells resistant to trimetrexate. *Blood* **87,** 2579–2587.

64. Persons, D., Allay, J., Allay, E., Smeyne, R., Ashmun, R., Sorrentino, B., and Nienhuis, A. (1997). Retroviral-mediated transfer of the green fluorescent protein gene into murine hematopoietic cells facilitates scoring and selection of transduced progenitors *in vitro* and identification of genetically modified cells *in vivo*. *Blood* **90,** 1777–1786.

65. Foster, B. J., Newell, D. R., Carmichael, J., Harris, A. L., Gumbrell, L. A., Jones, M., Goodard, P. M., and Calvert, A. H. (1993). Preclinical phase I and pharmacokinetic studies with the dimethyl phenyltriazene CB10-277. *Br. J. Cancer* **67,** 362–368.

66. Stevens, M. F. G., and Newlands, E. S. (1993). From triazines and triazenes to temozolomide. *Eur. J. Cancer* **29A,** 1045–1047.

67. Gonzaga, P. E., and Brent, T. P. (1989). Affinity purification and characterization of human O^6-alkylguanine-DNA alkyltransferase complexed with BCNU-treated, synthetic oligonucleotide. *Nucleic Acids Res.* **17,** 6581–6590.

68. Hansson, J., Edgren, M., Ehrsson, H., Ringborg, U., and Nilsson, B. (1988). Effect of D,L,-buthionine-*S, R*-sulfoximine on cytotoxicity and DNA cross-linking induced by bifunctional DNA-reactive cytostatic drugs in human melanoma cells. *Cancer Res.* **48,** 19–26.

69. Seidenfeld, J., and Komar, K. A. (1985). Chemosensitization of cultured human carcinoma cells to 1,3-bis(2-chloroethyl)-1-nitrosourea by difluoromethylornithine-induced polyamine depletion. *Cancer Res.* **45,** 2132–2138.

70. Imperatori, L., Damia, G., and Taverna, P. (1994). 3T3 NIH murine fibroblasts and B78 murine melanoma cells expressing the *Escherichia coli* N^3-methyladenine-DNA glycosylase I do not become resistant to alkylating agents. *Carcinogenesis* **15,** 533–537.

71. Matijasevic, Z., Boosalkis, M., and Mackay, W. (1993). Protection against chloroethylnitrosourea cytotoxicity by eukaryotic 3-methyladenine DNA glycosylase. *Proc. Natl. Acad. Sci. USA* **90,** 11855–11859.

72. Lukash, L. L., Boldt, J., Pegg, A. E., Dolan, M. E., Maher, V. M., and McCormick, J. J. (1991). Effect of O^6-alkylguanine-DNA alkyltransferase on the frequency and spectrum of mutations induced by *N*-methyl-*N*'-nitro-*N*-nitrosoguanidine in the HPRT gene of diploid human fibroblasts. *Mutat. Res.* **250,** 397–409.

73. Pegg, A. (1984). Methylation of the O^6-position of guanine in DNA is the most likely initiating event in carcinogenesis by methylating agents. *Cancer Invest.* **2,** 223–231.

74. Gerson, S. L., and Willson, J. K. (1995). O^6-alkylguanine-DNA alkyltransferase. A target for the modulation of drug resistance. *Hematol. Oncol. Clin. North Am.* **9,** 431–450.

75. Pegg, A. E. (1990). Mammalian O^6-alkylguanine-DNA alkyltransferase: regulation and importance in response to alkylating carcinogenic and therapeutic agents. *Cancer Res.* **50,** 6119–6129.

76. Brent, T., Lestrud, S., and Smith, D. (1987). Formation of DNA interstrand cross-links by the novel chloroethylating agent 2-chloroethyl(methylsulfonyl)-methanesulfonate: suppression of

77. Tong, W. P., Kirk, M. C., and Ludlum, D. B. (1982). Formation of the cross link 1-(N^3-deoxycytidyl), 2-(N^1-deoxy-guanosinyl)-ethane in DNA treated with N,N^1-bis(2-chloroethyl)-*N*-nitrosourea. *Cancer Res.* **42,** 3102–3105.

78. Karran, P., Macpherson, P., Ceccotti, S., Dogliotti, E., Griffin, S., and Bignam, M. (1993). O^6-Methylguanine residues elicit DNA repair synthesis by human cell extracts. *J. Biol. Chem.* **268,** 15878–15886.

79. Gerson, S., Phillips, W., Kastan, M., Dumenco, L., and Donovan, C. (1996). Human CD34 hematopoietic progenitors have low, cytokine-unresponsive O^6-alkylguanine-DNA alkyltransferase and are sensitive to O^6-benzylguanine plus BCNU. *Blood* **88,** 1649–1655.

80. Allay, J., Dumenco, L., Koç, O., Liu, L., and Gerson, S. (1995). Retroviral transduction and expression of the human alkyltransferase cDNA provides nitrosourea resistance to hematopoietic cells. *Blood* **85,** 3342–3351.

81. Moritz, T., Mackay, W., Glassner, B. J., Williams, D. A., and Samson, L. (1995). Retrovirus-mediated expression of a DNA repair protein in bone marrow protects hematopoietic cells from nitrosourea-induced toxicity *in vitro* and *in vivo*. *Cancer Res.* **55,** 2608–2614.

82. Maze, R., Carney, J., Kelley, M., Glassner, B., Williams, D., and Samson, L. (1996). Increasing DNA repair methyltransferase levels via bone marrow stem cell transduction rescues mice from the toxic effects of 1,3-bis(2-chloroethyl)-1-nitrosourea, a chemotherapeutic alkylating agent. *Proc. Natl. Acad. Sci. USA* **93,** 206–210.

83. Allay, J., Koç, O., Davis, B., and Gerson, S. (1996). Retroviral-mediated gene transduction of the human alkyltransferase cDNA confers nitrosourea resistance to human hematopoietic progenitors. *Clin. Cancer Res.* **2,** 1353–1359.

84. Allay, J., Davis, B., and Gerson, S. (1997). Human alkyltransferase-transduced murine myeloid progenitors are enriched *in vivo* by BCNU treatment of transplanted mice. *Exp. Hematol.* **25,** 1069–1076.

85. Maze, R., Kapur, R., Kelley, M., Hansen, W., Oh, S., and Williams, D. (1997). Reversal of 1,3-bis(2-chloroethyl)-1-nitrosourea-induced severe immunodeficiency by transduction of murine long-lived hematopoietic progenitor cells using O^6-methylguanine DNA methyltransferase complementary cDNA. *J. Immunol.* **158,** 1006–1013.

86. Spiro, T., Willson, J., Haaga, J., Hoppel, C., Liu, L., Majka, S., and Gerson, S. (1996). O^6-benzylguanine and BCNU: establishing the biochemical modulatory dose in tumor tissue for O^6-alkylguanine DNA alkyltransferase directed DNA repair. *Proc. Am. Soc. Clin. Oncol.* **15,** 177.

87. Gerson, S. L., Berger, N. A., Arce, C., Petzold, S. J., and Willson, J. K. (1992). Modulation of nitrosourea resistance in human colon cancer by O^6-methylguanine. *Biochem. Pharmacol.* **43,** 1101–1107.

88. Dolan, M. E., Mitchell, R. B., Mummert, C., Moschel, R. C., and Pegg, A. E. (1991). Effect of O^6-benzylguanine analogues on sensitivity of human tumor cells to the cytotoxic effects of alkylating agents. *Cancer Res.* **51,** 3367–3372.

89. Gerson, S., Zborowska, E., Norton, K., Gordon, N., and Willson, J. (1993). Synergistic efficacy of O^6-benzylguanine and 1,3-bis(2-chloroethyl)-1-nitrosourea (BCNU) in a human colon cancer xenograft completely resistant to BCNU alone. *Biochem. Pharmacol.* **45,** 483–491.

90. Dolan, M. E., Pegg, A. E., Moschel, R. C., and Grindey, G. B. (1993). Effect of O^6-benzylguanine on the sensitivity of human

colon tumor xenografts to 1,3-bis(2-chloroethyl)-1-nitrosourea (BCNU). *Biochem. Pharmacol.* **46**, 285–290.

91. Pegg, A., Boosalis, M., and Samson, L. (1993). Mechanism of inactivation of human O^6-alkylguanine-DNA alkyltransferase by O^6-benzylguanine. *Biochemistry* **32**, 11998–20006.

92. Spiro, T., Gerson, S., Hoppel, C., Liu, L., Schupp, J., Majka, S., Haaga, J., and Willson, J. (1998). O^6-benzylguanine totally depletes alkylguanine DNA alkyltransferase in tumor tissue: a phase I pharmacokinetic/pharmacodynamic study. *Proc. Am. Soc. Clin. Onc.* **17**, 212a.

93. Chinnasamy, N., Rafferty, J., Hickson, I., Ashby, J., Tinwell, H., Margison, G., Dexter, M., and Fairbairn, L. (1997). O^6-benzylguanine potentiates *in vivo* toxicity and clastogenicity of temozolomide and BCNU in mouse bone marrow. *Blood* **89**, 1566–1573.

94. Page, J., Giles, H. D., Phillips, W., Gerson, S. L., Smith, A. C., Tomaszewski, J. E. (1994). Preclinical toxicology study of O^6-benzylguanine (NSC-637037) and BCNU (carmustine, NSC-409962) in male and female Beagle dogs. *Proc. Am. Assoc. Cancer Res.* **35**, 328.

95. Moore, M. H., Gulbis, J. M., Dodson, E. J., Demple, B., and Moody, P. C. (1994). Crystal structure of a suicidal DNA repair protein: the Ada O^6-methylguanine-DNA methyltransferase from *E. coli*. *EMBO J.* **13**, 1495–1501.

96. Harris, L. C., Marathi, U. K., Edwards, C. C., Houghton, H. P., Srivastava, D. K., Vanin, E. F., Sorentino, B. J., and Brent, T. P. (1995). Retroviral transfer of a bacterial alkyltransferase gene into murine bone marrow protects against chloroethylnitrosourea cytotoxicity. *Clin. Cancer Res.* **1**, 1359–1365.

97. Crone, T., and Pegg, A. (1993). A single amino acid change in human O^6-alkylguanine-DNA alkyltransferase decreasing sensitivity to inactivation by O^6-benzylguanine. *Cancer Res.* **53**, 4750–4753.

98. Crone, T., Goodtzova, K., Edara, S., and Pegg, A. (1994). Mutations in human O^6-alkylguanine-DNA alkyltransferase imparting resistance to O^6-benzylguanine. *Cancer Res.* **54**, 6221–6227.

99. Reese, J., Koç, O., Lee, K., Liu, L., Allay, J., Phillips, W., and Gerson, S. (1996). Retroviral transduction of a mutant methylguanine DNA methyltransferase gene into human CD34 cells confers resistance to O^6-benzylguanine plus 1,3-bis(2-chloroethyl)- 1-nitrosourea. *Proc. Natl. Acad. Sci. USA* **93**, 14088–14093.

100. Davis, B., Reese, J., Koç, O., Lee, K., Schupp, J., and Gerson, S. (1997). Selection for G156A O^6-methylguanine DNA methyltransferase gene-transduced hematopoietic progenitors and protection from lethality in mice treated with O^6-benzylguanine and 1,3-bis(2-chloroethyl)-1-nitrosourea. *Cancer Res.* **57**, 5093–5099.

101. Davis, B., Koç, O., and Gerson, S. (1997). Detection of long term hematopoiesis by G156A MGMT transduced progenitors in non myeloablated mice after enrichment with O^6-benzylguanine and BCNU. *Blood* **90**, 554a.

102. Koç, O., Davis, B., Reese, J., Liu, L., and Gerson, S. (1997). ΔMGMT transduced bone marrow infusion increases tolerance to O^6-benzylguanine and BCNU and allows intensive therapy of BCNU resistant xenografts in mice. *Blood* **90**, 243a.

103. Sugimoto, M. (1995). Glutathione S-transferases (GSTs). *Jpn. J. Clin. Med.* **53**, 1253–1259.

104. Schecter, R., Alaoui-Jamali, M., Woo, A., Fahl, W., and Batist, G. (1993). Expression of rat glutathione S-transferase Yc complementary DNA in rat mammary carcinoma cells: impact upon alkylator-induced toxicity. *Cancer Res.* **53**, 4900.

105. Hayes, J., and Pulford, D. (1995). The glutathione S-transferase supergene family: regulation of GST and the contribution of the isoenzymes to cancer chemoprotection and drug resistance. *Crit. Rev. Biochem. Mol. Biol.* **30**, 445–600.

106. Greenbaum, M., Letourneau, S., Assar, H., Schecter, R., Batist, G., and Cournoyer, D. (1994). Retrovirus-mediated gene transfer of rat glutathione S-transferase Yc confers alkylating drug resistance in NIH 3T3 mouse fibroblasts. *Cancer Res.* **54**, 4442–4447.

107. Doroshow, J., Metz, M., Matsumoto, L., Winters, K., and Muramatsu, M. (1995). Transduction of NIH 3T3 cells with retrovirus carrying both human MDR1 and glutathione S-transferase pi produces broad range multidrug resistance. *Cancer Res.* **55**, 4073–4078.

108. Kuga, T., Sakamaki, S., Matsunaga, T., Hirayama, Y., Kuroda, H., Takahashi, Y., Kusakabe, T., Kato, I., and Niitsu, Y. (1997). Fibronectin fragment-facilitated retroviral transfer of the glutathione-S-transferase pi gene into CD34+ cells to protect them against alkylating agents. *Hum. Gene Ther.* **8**, 1901–1910.

109. Hsu, C., Tani, K., Fujiyoshi, T., Kurachi, K., and Yoshida, A. (1985). Cloning of cDNAs for human aldehyde dehydrogenases 1 and 2. *Proc. Natl. Acad. Sci. USA* **82**, 3771–3775.

110. Hsu, L. C., Chang, W. C., Shibuya, A., and Yoshida, A. (1992). Human stomach aldehyde dehydrogenase cDNA and genomic cloning, primary structure, and expression in *Escherichia coli*. *J. Biol. Chem.* **267**, 3030–3037.

111. Sladek, N. E., and Landkamer, G. J. (1985). Restoration of sensitivity to oxazaphosphorines by inhibitors of aldehyde dehydrogenase activity in cultured oxazaphosphorine resistant L1210 and crosslinking agent resistant P388 cell lines. *Cancer Res.* **45**, 1549–1555.

112. Bunting, K. D., Lindahl, R., and Townsend, A. J. (1994). Oxazaphosphorine specific resistance in human MCF7 breast carcinoma cell lines expressing transfected rat class 3 aldehyde dehydrogenase. *J. Biol. Chem.* **269**, 23197–23203.

113. Magni, M., Shammah, S., and Schiro, R., Mellado, W., Dalla-Favera, R., Gianni, A. (1996). Induction of cyclophosphamide resistance by aldehyde dehydrogenase gene transfer. *Blood* **87**, 1097–1103.

114. Bunting, K., Webb, M., Giorgianni, F., Galipeau, J., Blakley, R., Townsend, A., and Sorrentino, B. (1997). Coding region-specific destabilization of mRNA transcripts attenuates expression from retroviral vectors containing class 1 aldehyde dehydrogenase cDNAs. *Hum. Gene Ther.* **8**, 1531–1543.

115. Fridovich, I. (1978). The biology of oxygen radicals. *Science* **201**, 875–880.

116. Khan, A. U., and Kasha, M. (1994). Singlet molecular oxygen in the Haber-Weiss reaction. *Proc. Natl. Acad. Sci. USA* **91**, 12365–12367.

117. McAdam, M., Levelle, F., Fox, R., and Fielden, E. (1977). A pulse-radiolysis study of the manganese-containing superoxide dismutase from *Bacillus stearothermophilus*. *Biochem. J.* **165**, 81–87.

118. Lavelle, F., McAdam, M., Fielden, E., and Roberts, P. (1977). A pulse-radiolysis study of the catalytic mechanism of the iron-containing superoxide dismutase from *Photobacterium leiognathi*. *Biochem. J.* **161**, 3–11.

119. Weisiger, R., and Fridovich, I. (1973). Mitochondrial superoxide dismutase. Site of synthesis and intramitochondrial localization. *J. Biol. Chem.* **248**, 4793–4796.

120. Weisiger, R., and Fridovich, I. (1973). Superoxide dismutase. Organelle specificity. *J. Biol. Chem.* **248**, 3582–3592.

121. Marklund, S., Holme, E., and Hellner, L. (1982). Superoxide dismutase in extracellular fluids. *Clin. Chim. Acta* **126**, 41–51.

122. Jones, P., Kucera, G., Gordon, H., and Boss, J. (1995). Cloning and characterization of the murine manganous superoxide dismutase-encoding gene. *Gene* **153**, 155–161.

123. Borgstahl, G., Parge, H., Hickey, M., Beyer, W. J., Hallewell, R., and Tainer, J. (1992). The structure of human mitochondrial manganese superoxide dismutase reveals a novel tetrameric interface of two 4-helix bundles. *Cell* **71**, 107–118.

124. Cooper, J., McIntyre, K., Badasso, M., Wood, S., Zhang, Y., Garbe, T., and Young, D. (1995). X-ray structure analysis of the iron-dependent superoxide dismutase from *Mycobacterium tuberculosis* at 2.0 angstroms resolution reveals novel dimer-dimer interactions. *J. Mol. Biol.* **246**, 531–44.

125. Wagner, U., Pattridge, K., Ludwig, M., Stallings, W., Werber, M., Oefner, C., Frolow, F., and Sussman, J. (1993). Comparison of the crystal structures of genetically engineered human manganese superoxide dismutase and manganese superoxide dismutase from *Thermus thermophilus:* differences in dimer-dimer interaction. *Protein Sci.* **2**, 814–825.

126. Suresh, A., Tung, F., and Zucali, J. (1994). Retroviral mediated gene transfer of human manganese superoxide dismutase into bone marrow stem cells. *Exp. Hematol.* **22**, 90.

127. Urano, M., Kuroda, M., Reynolds, R., Oberley, T. D., and St. Clair, D. K. (1995). Expression of manganese superoxide dismutase reduces tumor control radiation dose: gene-radiotherapy. *Cancer Res.* **55**, 2490–2493.

128. Meagher, R., Salvado, A., and Wright, D. (1988). An analysis of the multilineage production of human hematopoietic progenitors in long term marrow culture: evidence that reactive oxygen intermediates derived from mature phagocytic cells have a role in limiting cell self-renewal. *Blood* **72**, 273–281.

129. Pluthero, F., and Axelrad, A. (1991). Superoxide dismutase as an inhibitor of erythroid progenitor cell cycling. *Ann. N. Y. Acad. Sci.* **628**, 222–232.

130. Peled, K., M, Lotem, J., Okon, E., Sachs, L., and Groner, Y. (1995). Thymic abnormalities and enhanced apoptosis of thymocytes and bone marrow cells in transgenic mice overexpressing Cu/Zn superoxide dismutase: implications for Down syndrome. *EMBO J.* **14**, 4985–4993.

131. Liu, R., Oberly, T., and Oberly, L. (1997). Transfection and expression of MnSOD cDNA decreases tumor malignancy of human oral squamous carcinoma SCC-25 cells. *Hum. Gene Ther.* **8**, 585–595.

132. Onetto, N., Momparler, R., Momparler, L., and Gyger, M. (1987). *In vitro* biochemical tests to evaluate the response to therapy of acute leukemia with cytosine arabinoside or 5-aza-2-deoxycytidine. *Semin. Oncol.* **14**, 231–237.

133. Laliberte, J., and Momparler, R. (1994). Human cytosine deaminase: purification of enzyme, cloning, and expression of its cDNA. *Cancer Res.* **54**, 5401–5407.

134. Momparler, R., Eliopoulos, N., Bovenzi, V., Letourneau, S., Greenbaum, M., and Cournoyer, D. (1996). Resistance to cytosine arabinoside by retrovirally mediated gene transfer of human cytosine deaminase into murine fibroblast and hematopoietic cells. *Cancer Gene Ther.* **3**, 331–338.

135. Kane, S., and Gottesman, M. (1989). Multidrug resistance in the laboratory and clinic. *Cancer Cells* **1**, 33.

136. Kelley, M., Hansen, W., Xu, Y., Yacoub, A., Williams, D., and Deutsch, W. (1997). Combining O^6-methylguanine DNA methyltransferase (MGMT) and AP endonuclease (APE) DNA repair activities in chimeric proteins: use in retroviral gene therapy and chemotherapeutic dose intensification. *Proc. Am. Assoc. Cancer Res.* **38**, 383.

137. Galipeau, J., Beniam, E., Spencer, H., Blakley, R., and Sorrentino, B. (1997). A bicistronic retroviral vector for protecting hematopoietic cells against antifolates and P-glycoprotein effluxed drugs. *Hum. Gene Ther.* **8**, 1773–1783.

138. Suzuki, M., Sugimoto, Y., Tsukahara, S., Okochi, E., Gottesman, M., and Tsuruo, T. (1997). Retroviral coexpression of two different types of drug resistance genes to protect normal cells from combination therapy. *Clin. Cancer Res.* **3**, 947–954.

139. Dunbar, C. E., Cottler-Fox, M., O'Shaughnessy, J. A., Doren, S., Carter, C., Berenson, R., Brown, S., Moen, R. C., Greenblatt, J., Stewart, F. M., Leitman, S. F., Wilson, W. H., Cowan, K., Young, N. S., and Nienhius, A. W. (1995). Retrovirally marked CD34-enriched peripheral blood and bone marrow cells contribute to long-term engraftment after autologous transplantation. *Blood* **85**, 3048–3057.

140. Brenner, M., Rill, D., Holladay, M., Heslop, H., Moen, R., Buschle, M., Krance, R., Santana, V., Anderson, W., and Ihle, J. (1993). Gene marking to determine whether autologous marrow infusion restores long-term haemopoiesis in cancer patients. *Lancet* **342**, 1134–1137.

141. Hanania, E. G., Giles, R. E., Kavanagh, J., Ellerson, D., Zu, Z., Want, T., Su, Y., Kudelka, A., Rahman, Z., Holmes, G., Hortobagyi, G., Claxton, D., Bachier, C., Thall, P., Cheng, S., Hester, J., Ostove, J. M., Bird, R. E., Chang, A., Korbling, M., Seong, D., Cote, R., Holzmayer, T., Mechetner, E., Heimfeld, S., Berenson, R., Burtness, B., Edwards, C., Bset, R., Andreeff, M., Champlin, R., and Deisseroth, A. B. (1996). Results of MDR-1 vector modification trial indicate that granulocyte/macrophage colony-forming unit cells do not contribute to posttransplant hematopoietic recovery following intensive systemic therapy. *Proc. Natl Acad. Sci. USA* **93**, 15346–15351.

142. Hesdorffer, C., Ayello, J., Kaubisch, A., Vahdat, L., Balmaceda, C., Garrett, T., Fetell, M., Reiss, R., Bank, A., and Antman, K. (1998). Phase I trial of retroviral-mediated transfer of the human MDR1 gene as marrow chemoprotection in patients undergoing high-dose chemotherapy in autologous stem-cell transplantation. *J. Clin. Oncol.* **16**, 165–172.

143. Moscow, J., Zujewski, J., Huang, H., Sorrentino, B., Chiang, Y., Wilson, W., Cullen, E., McAtee, N., Gottesman, M., Pastan, I., Dunbar, C., Neinhuis, A., and Cowan, K. (1997). Hematopoietic reconstitution with CD34 selected cells transduced with a retroviral vector containing the MDR1 gene in patients with metastatic breast cancer. *Proc. Am. Assoc. Cancer Res.* **38**, 343.

144. Kohn, D., Weinberg, K., Nolta, J., Heiss, L., Lenarsky, C., Crooks, G., Hanley, M., Annett, G., Brooks, J., El-Khoureiy, A., Lawrence, K., Wells, S., Moen, R., Bastian, J., Williams-Herman, D., Elder, M., Wara, D., Bowen, T., Hershfield, M., Mullen, C., Blaese, R., and Parkman, R. (1995). Engraftment of gene-modified umbilical cord blood cells in neonates with adenosine deaminase deficiency. *Nature Med.* **1**, 1017–1023.

145. Barranger, J., Rice, E., Sansieri, C., Bahnson, A., Mohney, T., Swaney, W., Takiyama, N., Dunigan, J., Beeler, M., Lucot, S., Schierer-Fochler, S., and Ball, E. (1997). Transfer of the glucocerebrosidase gene to CD34 cells and their autologous transplantation in patients with Gaucher disease. *Blood* **90**, 405a.

146. Emmons, R., Doren, S., Zujewski, J., Cottler-Fox, C., Carter, C., Hines, K., O'Shaughnessy, J., Leitman, S., Greenblatt, J., Cowan, K., and Dunbar, C. (1997). Retroviral gene transduction of adult peripheral blood or marrow derived CD34+ cells for 6 hours without growth factors or on autologous stroma does not improve marking efficiency assessed *in vivo*. *Blood* **89**, 4040–4046.

147. Hanenberg, H., Hashino, K., Konishi, H., Hock, R., Kato, I., and Williams, D. (1997). Optimization of fibronectin-assisted retroviral gene transfer into human CD34+ hematopoietic cells. *Hum. Gene Ther.* **8**, 2193–2206.

148. Nolta, J. A., Smogorzewska, E. M., and Kohn, D. B. (1995). Analysis of optimal conditions for retroviral-mediated transduction of primitive human hematopoietic cells. *Blood* **86**, 101–110.

149. Deisseroth, A., Kavanagh, J., and Champlin, R. (1994). Use of safety-modified retroviruses to introduce chemotherapy resistance sequences into normal hematopoietic cells for chemoprotection during the therapy of ovarian cancer: a pilot trial. *Hum. Gene Ther.* **5,** 1507–1522.

150. Hesdorffer, C., Antman, K., and Bank, A. (1994). Human MDR gene transfer in patients with advanced cancer. *Hum. Gene Ther.* **5,** 1151–1160.

151. O'Shaughessy, J., Cowan, K., and Nienhuis, A. (1994). Retroviral mediated transfer of the human multidrug resistance gene (MDR-1) into hematopoietic stem cells during autologous transplantation after intensive chemotherapy for metastatic breast cancer. *Hum. Gene Ther.* **5,** 891–911.

152. Gerson, S., and Koç, O. (1997). Mutant MGMT gene transfer into human hematopoietic progenitors to protect hematopoiesis during O^6-benzylguanine and BCNU therapy of advanced solid tumors, Case Western Reserve University *Clinical Protocol* 2Y97.

153. Zhao, R. C., McIvor, R. S., Griffin, J. D., and Verfaillie, C. M. (1997). Gene therapy for chronic myelogenous leukemia (CML): a retroviral vector that renders hematopoietic progenitors methotrexate-resistant and CML progenitors functionally normal and nontumorigenic *in vivo. Blood* **90,** 4687–4698.

154. Donahue, R., Kessler, S., Bodine, D., McDonagh, K., Dunbar, C., Goodman, S., Agricola, R., Byrne, E., Raffeld, M., Moen, R., Bacher, J., Zsebo, K., and Neinhuis, A. (1992). Helper virus induced T cell lymphoma in non human primates after retroviral mediated gene transfer. *J. Exp. Med.* **176,** 1125–1135.

155. Wilson, C., Ng, T., and Miller, A. (1997). Evaluation of recommendations for replication competent retrovirus testing associated with use of retroviral vectors. *Hum. Gene Ther.* **8,** 869–874.

156. Sugimoto, Y., Hryeyna, C., Aksentijevich, I., Pastan, I., and Gottesman, M. (1995). Coexpression of a multidrug-resistance gene (MDR1) and herpes simplex virus thymidine kinase gene as part of a bicistronic messenger RNA in a retrovirus vector allows selective killing of MDR1-transduced cells. *Clin. Cancer Res.* **1,** 447–457.

157. Blaese, R. (1993). Development of gene therapy for immunodeficiency: adenosine deaminase deficiency. *Pediatr. Res.* **33,** 49–53.

158. Mullen, C., Snitzer, K., Culver, K., Morgan, R., Anderson, W., and Blaese, R. (1996). Molecular analysis of T lymphocyte-directed gene therapy for adenosine deaminase deficiency: long-term expression *in vivo* of genes introduced with a retroviral vector. *Hum. Gene Ther.* **7,** 1123–1129.

159. Bjorgvinsdottir, H., Ding, C., Pech, N., Gifford, M., Li, L., and Dinauer, M. (1997). Retroviral-mediated gene transfer of gp91phox into bone marrow cells rescues defect in host defense against *Aspergillus fumigatus* in murine X-linked chronic granulomatous disease. *Blood* **89,** 41–48.

160. Ding, C., Kume, A., Bjorgvinsdottir, H., Hawley, R., Pech, N., and Dinauer, M. (1996). High-level reconstitution of respiratory burst activity in a human X-linked chronic granulomatous disease (X-CGD) cell line and correction of murine X-CGD bone marrow cells by retroviral-mediated gene transfer of human gp91phox. *Blood* **88,** 1834–1840.

161. Li, F., Linton, G., Sekhsaria, S., Whiting-Theobald, N., Katkin, J., Gallin, J., and Malech, H. (1994). CD34+ peripheral blood progenitors as a target for genetic correction of the two flavocytochrome b558 defective forms of chronic granulomatous disease. *Blood* **84,** 53–58.

162. Mardiney, M., Jackson, S., Spratt, S., Li, F., Holland, S., and Malech, H. (1997). Enhanced host defense after gene transfer in the murine p47phox-deficient model of chronic granulomatous disease. *Blood* **89,** 2268–2275.

163. Sekhsaria, S., Gallin, J., Linton, G., Mallory, R., Mulligan, R., and Malech, H. (1993). Peripheral blood progenitors as a target for genetic correction of p47phox-deficient chronic granulomatous disease. *Proc. Natl. Acad. Sci. USA* **90,** 7446–7450.

164. Sokolic, R., Sekhsaria, S., Sugimoto, Y., Whiting-Theobald, N., Linton, G., Li, F., Gottesman, M., and Malech, H. (1996). A bicistronic retrovirus vector containing a picornavirus internal ribosome entry site allows for correction of X-linked CGD by selection for MDR1 expression. *Blood* **87,** 42–50.

165. Weil, W., Linton, G., Whiting-Theobald, N., Vowells, S., Rafferty, S., Li, F., and Malech, H. (1997). Genetic correction of p67phox deficient chronic granulomatous disease using peripheral blood progenitor cells as a target for retrovirus mediated gene therapy. *Blood* **89,** 1754–1761.

166. Chuah, M., Vandendriessche, T., and Morgan, R. (1995). Development and analysis of retroviral vectors expressing human factor VIII as a potential gene therapy for hemophilia A. *Hum. Gene Ther.* **6,** 1363–1377.

167. Hoeben, R., Fallauz, F., Van Tilburg, N., Cramer, S., Van Ormondt, H., Briet, E., and Van Der Eb, A. (1993). Toward gene therapy for hemophilia A: long-term persistence of factor VIII-secreting fibroblasts after transplantation into immunodeficient mice. *Hum. Gene Ther.* **4,** 179–186.

168. Dwarki, V., Belloni, P., Nijjar, T., Smith, J., Couto, L., Rabier, M., Clift, S., Berns, A., and Cohen, L. (1995). Gene therapy for hemophilia A: production of therapeutic levels of human factor VIII *in vivo* in mice. *Proc. Natl. Acad. Sci. USA* **92,** 1023–1027.

169. Wang, L., Zoppe, M., Hackeng, T., Griffin, J., Lee, K., and Verma, I. (1997). A factor IX-deficient mouse model for hemophilia B gene therapy. *Proc. Natl. Acad. Sci. USA* **94,** 11563–11566.

170. Hao, Q., Malik, P., Salazar, R., Tang, H., Gordon, E., and Kohn, D. (1995). Expression of biologically active human factor IX in human hematopoietic cells after retroviral vector-mediated gene transduction. *Hum. Gene Ther.* **6,** 873–880.

171. Walsh, C., Grompe, M., Vanin, E., Buchwald, M., Young, N., Nienhuis, A., and Liu, J. (1994). A functionally active retrovirus vector for gene therapy in Fanconi anemia group C. *Blood* **84,** 453–459.

172. Ren, S., Wong, B., Li, J., Luo, X., Wong, P., and Atweh, G. (1996). Production of genetically stable high-titer retroviral vectors that carry a human β-globin gene under the control of the β-globin Locus Control Region. *Blood* **87,** 2518–2524.

173. Villeval, J., Rouyer-Fessard, P., Blumenfeld, N., Henri, A., Vainchenker, W., and Beuzard, Y. (1994). Retrovirus-mediated transfer of the erythropoietin gene in hematopoietic cells improves the erythrocyte phenotype in murine β-thalassemia. *Blood* **84,** 928–933.

174. Takekoshi, K., Oh, Y., Westerman, K., London, I., and Leboulch, P. (1995). Retroviral transfer of a human β-globin/α-globin hybrid gene linked to β locus control region hypersensitive site 2 aimed at the gene therapy of sickle cell disease. *Proc. Natl. Acad. Sci. USA* **92,** 3014–3018.

175. Tani, K., Yoshikubo, T., Ikebuchi, K., Takahashi, K., Tsuchiya, T., Takahashi, S., Shimane, M., Ogura, H., Tojo, A., Ozawa, K., Takahara, Y., Nakauchi, H., Markowitz, D., Bank, A., and Asano, S. (1994). Retrovirus-mediated gene transfer of human pyruvate kinase (PK) cDNA into murine hematopoietic cells: implications for gene therapy of human PK deficiency. *Blood* **83,** 2305–2310.

176. Bauer, G., Valdez, P., Kearns, K., Bahner, I., Wen, S., Zaia, J., and Kohn, D. (1997). Inhibition of human immunodeficiency

virus-1 (HIV-1) replication after transduction of granulocyte colony-stimulating factor-mobilized CD34+ cells from HIV-1-infected donors using retroviral vectors containing anti-HIV-1 genes. *Blood* **89**, 2259–2267.

177. Dunbar, C., Kohn, D., Karlsson, S., Barton, N., Brady, R., Cottler-Fox, M., and Crooks, G. (1996). Retroviral mediated transfer of the cDNA for human glucocerebrosidase into hematopoietic stem cells of patients with Gaucher Disease. A phase I study. *Hum. Gene Ther.* **7**, 231–253.

178. Aran, J., Licht, T., Gottesman, M., and Pastan, I. (1996). Complete restoration of glucocerebrosidase deficiency in Gaucher fibroblasts using a bicistronic MDR retrovirus and a new selection strategy. *Hum. Gene Ther.* **7**, 2165–2175.

179. Bahnson, A., Nimgaonkar, M., Fei, Y., Boggs, S., Robbins, P., Ohashi, T., Dunigan, J., Li, J., Ball, E., and Barranger, J. (1994). Transduction of CD34+ enriched cord blood and Gaucher bone marrow cells by a retroviral vector carrying the glucocerebrosidase gene. *Gene Ther.* **1**, 176–184.

180. Medin, J., Migita, M., Pawliuk, R., Jacobson, S., Amiri, M., Kluepfel-Stahl, S., Brady, R., Humphries, R., and Karlsson, S. (1996). A bicistronic therapeutic retroviral vector enables sorting of transduced CD34+ cells and corrects the enzyme deficiency in cells from Gaucher patients. *Blood* **87**, 1754–1762.

181. Migita, M., Medin, J., Pawliuk, R., Jacobson, S., Nagle, J., Anderson, S., Amiri, M., Humphries, R., and Karlsson, S. (1995). Selection of transduced CD34+ progenitors and enzymatic correction of cells from Gaucher patients, with bicistronic vectors. *Proc. Natl. Acad. Sci. USA* **92**, 12075–12079.

182. Nimgaonkar, M., Bahnson, A., Boggs, S., Ball, E., and Barranger, J. (1994). Transduction of mobilized peripheral blood CD34+ cells with the glucocerebrosidase. *Gene Ther.* **1**, 201–207.

183. Fairbairn, L., Lashford, L., Spooncer, E., McDermott, R., Lebens, G., Arrand, J. E., Arrand, J. R., Bellantuono, I., Holt, R., Hattons, C., Cooper, A., Besley, G., Wraity, J., Anson, D., Hopwood, J., and Dexter, T. (1996). Long-term *in vitro* correction of α-L-iduronidase deficiency (Hurler syndrome) in human bone marrow. *Proc. Natl. Acad. Sci. USA* **93**, 2025–2030.

184. Taylor, R., and Wolfe, J. (1997). Decreased lysosomal storage in the adult MPS VIII mouse brain in the vicinity of grafts of retroviral vector-corrected fibroblasts secreting high levels of β-glucuronidase. *Nature Med.* **3**, 771–774.

185. Braun, S., Aronovich, E., Anderson, R., Crotty, P., McIvor, R., and Whitley, C. (1993). Metabolic correction and cross-correction of mucopolysaccharidosis type II (Hunter syndrome) by retroviral-mediated gene transfer and expression of human iduronate-2-sulfatase. *Proc. Natl. Acad. Sci. USA* **90**, 11830–11834.

186. Braun, S., Pan, D., Aronovich, E., Jonsson, J., McIvor, R., and Whitley, C. (1996). Preclinical studies of lymphocyte gene therapy for mild Hunter syndrome (mucopolysaccharidosis type II). *Hum. Gene Ther.* **7**, 283–290.

187. Whitley, C., McIvor, R., Aronovich, E., Berry, S., Blazar, B., Burger, S., Kersey, J., King, R., Faras, A., Latchaw, R., McCullough, J., Pan, D., Ramsay, N., and Stroncek, D. (1996). Retroviral-mediated transfer of the iduronate-2-sulfatase gene into lymphocytes for treatment of mild Hunter syndrome (mucopolysaccharidosis type II). *Hum. Gene Ther.* **7**, 537–549.

CHAPTER 12

Cytosine Deaminase as a Suicide Gene in Cancer Gene Therapy

CRAIG A. MULLEN

Departments of Experimental Pediatrics and Immunology, University of Texas, M. D. Anderson Cancer Center, Houston, Texas 77030

I. INTRODUCTION

This article describes the cytosine deaminase (CD) metabolic suicide gene system and possible applications to cancer therapy. The CD gene from bacteria can be transferred to and expressed in mammalian tumor cells. CD-expressing cells can deaminate the relatively nontoxic prodrug 5-fluorocytosine (5-FC) to the highly toxic drug 5-fluorouracil (5-FU). As a result CD-expressing tumor cells can be killed by 5-FC whereas normal cells not expressing the gene are not. This review discusses the metabolism of 5-FC and 5-FU, the effect of CD gene transfer to tumors *in vitro* and *in vivo*, and the pharmacological and immunologic effects of the process on tumor cells not expressing the suicide gene. Potential application of this suicide gene system to human cancer is also discussed. Also see the description of suicide gene transfer with the herpes simplex thymidine kinase gene in Chapter 10.

II. METABOLIC SUICIDE GENES

Most effective traditional anticancer drugs produce substantial toxicity to normal tissues. An important goal in cancer drug development has been the identification of agents that have selective antitumor toxicity. However, this goal has been difficult to achieve because tumor cells and cycling normal cells are in many respects metabolically similar. The search for selective antimicrobial and antiviral drug toxicity has been more rewarding because these organisms have many novel metabolic reactions that do not occur in mammalian cells. The identification of microbial genes underlying these novel metabolic processes and the development of gene transfer technology for mammalian cells have created prospects for creating novel chemosensitivities in tumor cells that would not be shared by normal tissues.

Metabolic suicide genes encode enzymes that are not normally present in mammalian cells and that can acti-

201

vate otherwise nontoxic prodrugs. A tumor cell expressing a metabolic suicide gene would in essence commit suicide by producing a lethal intracellular toxin. The best-known example is the herpes simplex thymidine kinase (TK) gene, which activates the antiherpes prodrugs acyclovir and ganciclovir. Observations in the TK system have incited searches for other novel microbial enzymes that are capable of activating prodrugs [1]. Cytosine deaminase (CD) is one such enzyme, and its potential as a metabolic suicide gene in cancer gene therapy has been explored over the past several years.

III. CYTOSINE DEAMINASE IN MICROBES

Cytosine deaminase is an enzyme found in many bacteria and fungi. Its normal function in times of nutritional stress is to provide uracil by deamination of cytosine [2]. Mammalian cells do not perform this reaction. Over three decades ago this metabolic difference between microbes and mammalian cells led to the development of 5-fluorocytosine (5-FC) as an antibiotic for use in humans. 5-FC is a substrate for CD, and CD-expressing organisms deaminate 5-FC to 5-fluorouracil (5-FU), a potent toxin. Although it is not widely used, 5-FC is still clinically employed as an adjunct to amphotericin in treatment of fungal infections [3]. The CD gene has been cloned from several organisms, and the *Escherichia coli* gene is approximately 1.3 kb [4,5], small enough to fit into most gene transfer vectors.

IV. 5-FLUOROCYTOSINE AND 5-FLUOROURACIL IN HUMANS

A brief consideration of 5-FC and 5-FU in humans will help later consideration of CD/5-FC as a suicide gene/prodrug system in cancer gene therapy. These drugs are concisely but thoroughly reviewed elsewhere[3,6], but certain features will be noted here. Both are fluorinated pyrimidines and are small uncharged molecules. As such they distribute to the total volume of body water. Both penetrate the cerebrospinal fluid. They enter mammalian cells by diffusion, whereas in fungi 5-FC relies on cytosine permease for effective entry. With typical oral 5-FC dosing, serum levels of >500 μM can be achieved. 5-FC is largely eliminated by the kidneys and has a half-life of 3 to 6 hours. 5-FU is typically administered intravenously, and when it is given by continuous infusion, serum concentrations of approximately 1 μM are achieved. 5-FU is degraded rapidly in the liver by the enzyme dihydropyrimidine

dehydrogenase, and the half-life of 5-FU is less than 20 minutes.

As mentioned earlier, 5-FC is sometimes used as an adjunct to amphotericin [3]. It is not used alone as an antifungal because of the frequent and rapid emergence of resistant fungi. Resistance in fungi can be due to loss of cytosine permease, to decreased expression of CD, or to decreased expression of uridine monophosphate (UMP) pyrophosphorylase, which converts 5-FU to 5-FUMP. Human toxicity with 5-FC can occur and is primarily manifest by myelotoxicity, especially in patients with limited bone marrow function. Activation of 5-FC to 5-FU may occur in the gut and is due to CD expressing microbes that reside in the gastrointestinal tract [7].

5-FU is a commonly used anticancer drug [6]. It has a limited clinical spectrum, with efficacy seen primarily in adenocarcinomas of the colon, liver, and breast. Not all cancers are sensitive to 5-FU, and even within the malignancies mentioned earlier, only half or fewer patients will have a substantial response to single-agent 5-FU. 5-FU also produces gut and marrow toxicity. The metabolism of 5-FU is complex, and its nucleotide metabolites can affect both RNA and DNA metabolism (Fig. 1). 5-FU can be converted to 5-UMP by several routes, and following further metabolism to floxuridine triphosphate (FUTP) it can be incorporated into RNA, resulting in dysfunctional RNA. Its antineoplastic activity may depend more on interference with DNA metabolism. 5-FU can be converted by thymidine phosphorylase and normal cellular thymidine kinases to fluorodeoxyuridine monophosphate (FdUMP). Alternatively, FUDP can be converted by ribonucleotide reductase to FdUDP with further dephosphorylation to FdUMP. A critical reaction involves thymidylate synthase, which normally functions to produce thymidine triphosphate (TTP), which is essential for DNA synthesis. dUMP is a natural substrate for thymidylate synthase and is converted to TMP. However, FdUMP inhibits

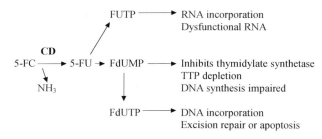

FIGURE 1   Metabolism of 5-FC in tumor cells expressing the cytosine deaminase gene. Key products are identified in this figure, but other intermediate products have been omitted for the sake of simplicity. CD, cytosine deaminase; 5-FC, 5-fluorocytosine; 5-FU, 5-fluorouracil; FUTP, floxuridine triphosphate; FdUMP, fluorodeoxyuridine monophosphate; FdUTP, fluorodeoxyuridine triphosphate; TTP, thymidine triphosphate.

thymidylate synthase and together with tetrahydrofolate forms a stable inactive complex. TTP depletion results and DNA synthesis is impeded. FdUMP can also be converted to FdUTP, which can be incorporated into DNA, leading to DNA damage. Tumor cells can become resistant to 5-FU through multiple mechanisms. Upregulation of thymidylate synthase, insufficiency of tetrahydrofolate, and loss of enzymes that convert 5-FU to its nucleotide forms have been described.

V. CYTOSINE DEAMINASE AS A METABOLIC SUICIDE GENE IN MAMMALIAN CELLS

Given that mammalian cells do not efficiently metabolize 5-FC to 5-FU, it appears that cytosine deaminase could function as a metabolic suicide gene. In principle three conditions would need to be fulfilled for this to occur. First, expression of CD in tumor cells would need to be high enough to efficiently deaminate 5-FC. Second, intracellular 5-FC concentrations would need to be high enough to generate toxic levels of 5-FU. Third, the tumor cell should not exhibit intrinsic or acquired resistance to 5-FU. Although all three have been shown to be achievable in experimental systems, it must be borne in mind that failure to achieve any of the three would block CD from being an effective anticancer suicide gene *in vivo*.

The CD gene from *E. coli*, after minor modification, can be expressed in mammalian cells using any of a variety of vectors [4,5]. Many studies have been performed using CD in retroviral vectors; in such systems a single copy of the CD gene is stably inserted into the target tumor cell *in vitro* and is usually constitutively expressed using a viral promoter. Using such systems CD expression can be detected by several methods, including the measurement of tritiated uracil generated from tritiated cytosine. Constitutive expression of CD does not alter the viability or growth rate *in vitro* of most cell lines tested. However, CD-expressing cell lines do exhibit concentration-dependent inhibition when grown in medium containing 5-FC, whereas unmodified cells are not similarly impaired (Fig. 2A) [4,8–10]. CD-expressing cells may exhibit anywhere from 100- to 10,000-fold greater sensitivity to 5-FC. However, variability of sensitivity can be observed and can be due to several factors. First, if individual clones of cells transduced with a CD-containing vector are analyzed, one can identify a range of enzyme expression. Such variability of gene expression is common with retrovirus vectors. For any given concentration of 5-FC, sensitivity to this prodrug is largely explained by the level of CD activity: high-expressing cells are efficiently killed; clones with

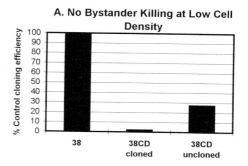

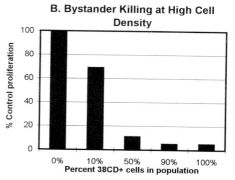

FIGURE 2   5-FC inhibition of 38 adenocarcinoma tumor cells engineered to express the CD gene. (A) Clonogenic assay of tumor cells at low cell density: 1,000 tumor cells were placed in a well of a 6-well plate and grown in medium containing 0.125 mg/mL 5-FC; 1 week later the cultures were fixed and stained, and the colonies were counted. "38" is the unmodified tumor cell line, "38CD" was a heterogeneous population of cells selected in G418 transduced with a retroviral vector containing a neomycin resistance gene and the CD gene, and "38CD cloned" was a clone from the 38CD population that expressed the CD gene very efficiently. The uncloned 38CD population contained a substantial (~50%) subpopulation that did not efficiently express CD. Percent control cloning efficiency is calculated relative to the efficiency of unmodified 38 cells growing in 5-FC. (B) Proliferation of cells at high cell density: 10,000 cells were plated in a well of a 96-well plate and grown in 0.125 mg/mL 5-FC, and proliferation was assessed by tritiated thymidine incorporation about 4 days later. The cells were unmodified 38 and a cloned 38CD line, and the percentage of CD+ cells in the wells is indicated in the legend. Twelve replicates of each condition were plated.

lower levels of enzyme activity are not. Second, there appear to be tumor- or tissue-specific differences in sensitivity to 5-FC, even when the variable of CD activity is held constant. For example, in our laboratory most sarcomas are only modestly sensitive to the CD suicide gene system, whereas a colon adenocarcinoma line is quite sensitive [10]. Many reports of the CD system have shown considerable selective toxicity in colon or liver tumor cell lines [11]. This may reflect the underlying sensitivity of different tumors or tissues to 5-FU, and/ or the capacity of the particular tumor line to acquire relative 5-FU resistance.

The CD suicide gene system can also function *in vivo* (Fig. 3A). As noted earlier, relatively high concentrations of 5-FC can be achieved without serious toxicity to otherwise normal hosts, and the reported achievable serum levels are equal to or greater than the concentrations of 5-FC necessary to eliminate CD-expressing tumor cells *in vitro*. When tumor cells engineered *in vitro* to express CD are injected into syngeneic mice, normal tumor growth is usually observed. However, treatment with systemic 5-FC, by bolus intraperitoneal or intravenous injection, or by continuous infusion, prevents the growth of CD-expressing tumors. The effectiveness of tumor inhibition *in vivo* seems to correlate with *in vitro* sensitivity.

The effectiveness of 5-FC-mediated growth inhibition of CD-expressing tumors can be limited by tumor burden [10]. If CD-expressing tumor cells (stably transduced *in vitro* with a retroviral vector) are injected and allowed to grow for 1 to 3 weeks before initiation of 5-FC treatment, a progressive increase in tumor incidence is seen as the time to initiation of prodrug treatment increases. However, in animals failing to exhibit complete tumor regression there may still be observed growth arrest of the tumor during the period of prodrug administration, with a subsequent acceleration of the growth rate after cessation of treatment. Several explanations for the difficulty in achieving equally effective regression of large, established CD-expressing tumor are possible. First, this observation is common even with traditional cancer chemotherapy drugs and may be related to decreased intratumoral drug levels seen in portions of tumors. Because 5-FC-mediated cytotoxicity is concentration dependent, failure to effectively deliver prodrug to all portions of the tumor may account for treatment failure. Second, during growth *in vivo* the tumor cell population greatly expands, allowing the cell to develop variants that either fail to efficiently express CD or to develop acquired resistance to 5-FU. Tumors reisolated from mice even in the absence of 5-FC treatment have been shown to have decreased levels of CD expression [10].

VI. EFFECTS ON NONTRANSDUCED TUMOR CELLS

The premise of suicide gene cancer therapy is that tumor cells *in situ* can be transduced with genetic vectors and then eliminated by systemic administration of prodrug. A significant problem with this approach is the inefficiency of gene transfer. It is well known from decades of cancer therapy research that tumors can regrow if microscopic residual disease remains after therapy. Thus, one would expect that if even a small fraction of a tumor failed to be killed by a suicide gene/prodrug system, the likely result would be regrowth of the tumor after initial regression. Given that all known methods of *in vivo* gene transfer are inefficient, the expectation is that local regrowth of tumor would occur because tumor cells not transduced with the suicide gene should not activate the prodrug.

However, both *in vitro* and *in vivo* killing of tumor cells not expressing the suicide gene can be observed when they are in the microenvironment of a suicide gene transduced tumor cell. This effect has been extensively investigated in the herpes thymidine kinase system, and the killing of nontransduced cells has been called the bystander effect [12–14]. Bystander killing of nontransduced cells has also been observed in the CD system [15].

In vitro tests of 5-FC/CD-mediated killing can show both the presence and the absence of bystander killing, depending on the cell density (Fig. 2). When cultures

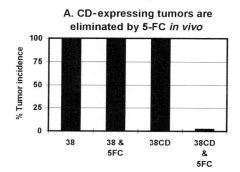

FIGURE 3   Tumor growth *in vivo*. (A) Growth of unmodified 38 or 38CD cells in normal mice: 100,000 cells were injected subcutaneously into syngeneic C57BL/6 mice. Some mice received 37.5 mg of 5-FC intraperitoneally twice daily for 10 days. Animals (*n* = 10 per group) were monitored for 6 weeks for tumor growth. (B) Growth of unmodified 38 tumors in naïve mice or in syngeneic C57BL/6 mice that had been pretreated 6 weeks earlier with 38CD cells and 5-FC (as described earlier): 100,000 unmodified 38 cells were injected subcutaneously, and the animals were monitored for up to 6 weeks for tumor growth (*n* = 21 in the prior treatment group, *n* = 15 in the naïve group).

consisting of both CD-expressing and nontransduced cells are established at low density (e.g., 1,000 cells in a 3-cm well) and clonogenic assays are performed to determine the sensitivity of individual CD+ cells to 5-FC killing, no bystander killing is seen. Rather, untransduced cells give rise to colonies and CD-expressing cells in the same culture are eliminated. However, if mixed cultures are established at high concentration (e.g., 10,000 cells in a well of a 96-well plate), significant bystander killing of nontransduced cells is observed. Indeed, some have reported that under some circumstances, using cells that are apparently fairly sensitive to 5-FU, growth arrest can be seen when as little as 2% of the cells *in vitro* express CD [16]. The explanation for the bystander killing in the CD system includes both release of CD enzyme and release of 5-FU. As CD+ cells die, they may release CD; the deamination reaction requires no cellular cofactors and is not energy dependent, so it can occur wherever there is enzyme and substrate. Therefore, 5-FU can be generated in the extracellular space and diffuse into nontransduced cells. Similarly, preformed 5-FU from CD expressing cells may diffuse out of transduced cells and affect neighboring cells. The failure to observe bystander killing at low cell concentrations probably represents dilution of extracellular 5-FU to nontoxic concentrations.

Similar bystander killing can be seen *in vivo* [17]. If a mouse is simultaneously injected with a CD-expressing tumor in one flank and a nontransduced tumor in the contralateral flank, subsequent treatment with 5-FC prevents outgrowth of the CD+ tumor but not the unmodified tumor. However, if CD+ and unmodified tumor cells are mixed together and injected in the same site, bystander killing can be observed. The efficiency of the bystander effect varies from tumor to tumor, but some have reported that highly sensitive tumors composed of 1 to 10% CD+ cells can be inhibited. However, when using sarcoma cells that are relatively resistant to CD/5-FC *in vitro*, little bystander killing is seen *in vivo*.

VII. COMPARISON OF CD WITH THE TK SUICIDE GENE SYSTEM

Several reports have compared the CD and TK suicide gene systems. Several studies have concluded that in certain tumor models CD appears more effective [18,19]. Our experience with a variety of tumor lines indicates that in some lines CD is more efficient, in others TK is. There are some theoretical reasons CD may be more effective, primarily on the basis of bystander killing. As noted earlier, the CD-enzyme-mediated deamination of 5-FC requires no cellular cofactors or energy sources and can occur in the extra-

cellular space, whereas TK requires a phosphate donor to activate ganciclovir to ganciclovir monophosphate, and the reaction must take place in a cell. In addition, 5-FU is an uncharged molecule that can diffuse through cells, whereas ganciclovir monophosphate and the di- and triphosphates are charged and pass to neighboring cells through gap junctions and not by simple diffusion. However, many factors determine the sensitivity of a cell to 5-FU-mediated killing (as noted earlier), and it is unlikely that one can correctly conclude that one suicide gene system is inherently better or worse than another.

VIII. IMMUNOLOGIC EFFECTS

The phenomenon of bystander killing of tumor cells not expressing the CD or other suicide genes can be largely explained pharmacologically. However, it has been observed that the efficiency of bystander killing of suicide-gene-expressing and normal tumor cells is reduced in immunodeficient mice [13,20]. Late local regrowth of tumor after cessation of prodrug treatment is seen more commonly in nude mice than in normal mice. Thus, although suicide gene/prodrug chemoablation acts locally by generation of toxins, it is possible that there may be immunologic sequelae to the process. In the CD system there is evidence for induction of systemic antitumor immunity against unmodified tumor cells after 5-FC-mediated elimination of CD-expressing tumors [10,21]. In several models mice pretreated with CD-transduced cells and 5-FC have exhibited tumor-specific, T-cell-dependent resistance to rechallenge 3 to 6 weeks later with unmodified tumor cells (Fig. 3B). Earlier it was noted that animals simultaneously challenged with suicide-gene-modified and normal tumor cells at distant locations do not show inhibition of unmodified tumor growth. However, in some models of micrometastatic disease there is some evidence that progression of distant micrometastases may be retarded by the simultaneous local treatment of a larger subcutaneous tumor. This discrepancy can be explained by the kinetics of T-cell immune responses, which typically require 7 to 14 days for full induction, and by the repeated observation in multitudes of tumor immunology models in which a small tumor burden is more sensitive to immunologic control than is a large established tumor.

Induction of antitumor immunity may be facilitated by induction of tumor necrosis by prodrug [21]. *In vitro* suicide-gene-modified cells die primarily due to DNA damage, and it is likely that triggering of apoptosis occurs. Apoptosis occurs naturally in the body and is usually a isolated event that does not recruit a host inflammatory response. However, *in vivo* suicide-gene-modified tumors exhibit necrosis after treatment with prodrug. Studies in the CD system have shown a heavy host-cell infiltrate composed largely of polymorphonuclear cells and both CD4+ and CD8+ T cells, and analyses of other models using the TK system have produced similar findings. It is possible that this induction of necrosis changes the microenvironment and that as a result antigens are presented more effectively to the immune system.

IX. POTENTIAL APPLICATIONS IN CLINICAL ONCOLOGY

Many of the studies of the CD and other suicide gene systems have used tumor cells stably transduced *in vitro* with retroviral vectors. However, retroviral vectors are in general very inefficient for transducing cells *in vivo* and are unsuitable for cancer gene therapy. Although some bystander effect killing can be seen, effective application of suicide gene systems to tumors in humans will certainly require efficient intratumoral generation of suicide gene enzyme products.

One approach to this problem is monoclonal-antibody-directed delivery of enzyme to the tumor [22,23]. Here the enzyme is chemically coupled to or genetically engineered as a fusion protein with a monoclonal antibody with specificity for tumor. In theory there will be accumulation of antibody and enzyme in tumor foci. Because CD can function extracellularly, the extracellular conversion of 5-FC to 5-FU in the tumor microenvironment should occur and intratumoral 5-FU should exert an antitumor effect. Effects have been reported both *in vitro* and *in vivo*. The principal limitation of the approach is the distribution of antibody *in vivo*. The majority of studies with monoclonal-antibody-directed drug or isotope delivery have demonstrated that although significant tumor targeting is possible, there is still considerable accumulation of antibody in other organs such as liver, spleen, bone marrow, and lungs. Delivery of prodrug converting enzyme to these locations might produce 5-FU in these organs, producing organ damage, and might possibly create systemically significant 5-FU concentrations.

Adenoviral vectors are being explored as efficient means of *in vivo* gene transduction in a variety of gene therapy schemes. Their principal advantages are twofold: First, they can be produced at very high titer (>10^9 pfu/mL), and second, both replicating and nonreplicating cells can be transduced and produce gene products from the vector. As such they represent good tools for delivery of suicide genes to a localized tumor mass. The principal disadvantage of adenoviral vectors is that they are expressed only transiently, unlike retro-

viral vectors, which stably integrate into the target cell genome. This is not important for suicide gene schemes because the transduced cells are to be eliminated by prodrug shortly after gene delivery. A number of groups have cloned the cytosine deaminase gene into adenoviral vectors and have demonstrated both delivery of gene and antitumor activity *in vivo* [24,25]. Other methods of suicide gene transfer have also been explored, including nonviral liposomal vectors [26,27] and even anaerobic bacteria [28]. Whether these approaches will prove efficacious in established clinical tumors remains to be proven.

Direct injection of adenovirus vectors (or nonviral vectors) into a localized tumor may be helpful in certain situations in which localized disease is the primary problem (e.g., brain tumors or unresectable hepatic or thoracic tumors) [29,30]. However, metastatic tumor remains the greatest threat to life for the majority of cancer patients. No genetic vector currently available is capable of efficient, tissue-targeted, systemic gene delivery. This remains a considerable technical challenge for the gene therapy field, but it is conceivable that advances will be made. Should systemic delivery of genes be feasible, tumor-specific expression of suicide genes would be necessary for selective toxicity. One approach to achieve this is the use of tissue-specific or tumor-specific promoters of the suicide gene. A number of investigators have demonstrated preferential expression of suicide genes in tumors using this approach [31–34].

X. CONCLUSION

The narrow therapeutic index of current cancer chemotherapy drugs provides barriers to cure of many cancers. The development of metabolic suicide gene systems such as cytosine deaminase and herpes thymidine kinase raises the possibility of creating novel chemosensitivities in tumors without inducing serious host toxicity. The successful application of suicide genes to cancer therapy will require relatively efficient methods of tumor-specific gene delivery and expression. This remains a formidable technical problem that deserves considerable scientific attention.

References

1. Mullen, C. A. (1994). The use of suicide vectors for the gene therapy of cancer. *Pharmacol. Ther.* **63,** 199–207.
2. Danielsen, S., Kilstrup, M., Barilla, K., Jochimsen, B., and Neuhard, J. (1992). Characterization of the *Escherichia coli* codBA operon encoding cytosine permease and cytosine deaminase. *Mol. Microbiol.* **6,** 1335–1344.

3. Bennett, J. E. (1996). Antimicrobial agents: antifungal agents. In: *Goodman and Gilman's The Pharmacological Basis of Therapeutics*, pp. 1175–1190. Eds. Hardman, J. G., Limbird, L. E., Molinoff, P. B., Ruddon, R. W., and Gilman, A. G. McGraw-Hill, New York.

4. Mullen, C. A., Kilstrup, M., and Blaese, R. M. (1992). Transfer of the bacterial gene for cytosine deaminase to mammalian cells confers lethal sensitivity to 5-fluorocytosine: a negative selection system. *Proc. Natl. Acad. Sci. U.S.A.* **89**, 33–37.

5. Austin, E. A., and Huber, B. E. (1993). A first step in the development of gene therapy for colorectal carcinoma: cloning, sequencing, and expression of *Escherichia coli* cytosine deaminase. *Mol. Pharmacol.* **43**, 380–387.

6. Chabner, B. A., Allegra, C. J., Curt, G. A., and Calabrisi, P. (1996). Antineoplastic agents. In: *Goodman and Gilman's The Pharmacological Basis of Therapeutics*, pp. 1233–1287. Eds. Hardman, J. G., Limbird, L. E., Molinoff, P. B., Ruddon, R. W., and Gilman, A. G. McGraw-Hill, New York.

7. Harris, B. E., Manning, B. W., Federle, T. W., and Diasio, R. B. (1986). Conversion of 5-fluorocytosine to 5-fluorouracil by human intestinal microflora. *Antimicrob. Agents Chemother.* **29**, 44–48.

8. Kuriyama, S., Masui, K., Sakamoto, T., Nakatani, T., Tominaga, K., Fukui, H., Ikenaka, K., Mullen, C. A., and Tsuji, T. (1995). Bacterial cytosine deaminase suicide gene transduction renders hepatocellular carcinoma sensitive to the prodrug 5-fluorocytosine. *Int. Hepatol. Commun.* **4**, 72–79.

9. Khil, M. S., Kim, J. H., Mullen, C. A., Kim, S. H., and Freytag, S. O. (1996). Radiosensitization by 5-fluorocytosine of human colorectal carcinoma cells in culture transduced with cytosine deaminase gene. *Clin. Cancer Res.* **2**, 53–57.

10. Mullen, C. A., Coale, M. M., Lowe, R., and Blaese, R. M. (1994). Tumors expressing the cytosine deaminase suicide gene can be eliminated *in vivo* with 5-fluorocytosine and induce protective immunity to wild type tumor. *Cancer Res.* **54**, 1503–1506.

11. Huber, B. E., Austin, E. A., Good, S. S., Knick, V. C., Tibbels, S., and Richards, C. A. (1993). *In vivo* antitumor activity of 5-fluorocytosine on human colorectal carcinoma cells genetically modified to express cytosine deaminase. *Cancer Res.* **53**, 4619–4626.

12. Culver, K. W., Ram, Z., Wallbridge, S., Ishii, H., Oldfield, E. H., and Blaese, R. M. (1992). *In vivo* gene transfer with retroviral vector-producer cells for treatment of experimental brain tumors. *Science* **256**, 1550–1552.

13. Ishiimorita, H., Agbaria, R., Mullen, C. A., Hirano, H., Koeplin, D. A., Ram, Z., Oldfield, E. H., Johns, D. G., and Blaese, R. M. (1997). Mechanism of bystander effect killing in the herpes simplex thymidine kinase gene therapy model of cancer treatment. *Gene Ther.* **4**, 244–251.

14. Mullen, C. A. (1994). Herpes thymidine kinase suicide gene transfer: a review of preclinical models and proposals to treat brain tumors. In: *Cytokine-Induced Tumor Immunogenicity*, pp. 455–465. Academic Press. London; San Diego.

15. Rowley, S., Lindauer, M., Gebert, J. F., Haberkorn, U., Oberdorfer, F., Moebius, U., Herfarth, C., and Schackert, H. K. (1996). Cytosine deaminase gene as a potential tool for the genetic therapy of colorectal cancer. *J. Surg. Oncol.* **61**, 42–48.

16. Hirschowitz, E. A., Ohwada, A., Pascal, W. R., Russi, T. J., and Crystal, R. G. (1995). *In vivo* adenovirus-mediated gene transfer of the *Escherichia coli* cytosine deaminase gene to human colon carcinoma-derived tumors induces chemosensitivity to 5-fluorocytosine. *Hum. Gene Ther.* **6**, 1055–1063.

17. Huber, B. E., Austin, E. A., Richards, C. A., Davis, S. T., and Good, S. S. (1994). Metabolism of 5-fluorocytosine to 5-fluorouracil in human colorectal tumor cells transduced with the cytosine deaminase gene: significant antitumor effects when only a small percentage of tumor cells express cytosine deaminase. *Proc. Natl. Acad. Sci. U.S.A.* **91**, 8302–8306.

18. Hoganson, D. K., Batra, R. K., Olsen, J. C., and Boucher, R. C. (1996). Comparison of the effects of three different toxin genes and their levels of expression on cell growth and bystander effect in lung adenocarcinoma. *Cancer Res.* **56**, 1315–1323.

19. Trinh, Q. T., Austin, E. A., Murray, D. M., Knick, V. C., and Huber, B. E. (1995). Enzyme/prodrug gene therapy: comparison of cytosine deaminase/5-fluorocytosine versus thymidine kinase/ganciclovir enzyme/prodrug systems in a human colorectal carcinoma cell line. *Cancer Res.* **55**, 4808–4812.

20. Gagandeep, S., Brew, R., Green, B., Christmas, S. E., Klatzmann, D., Poston, G. J., and Kinsella, A. R. (1996). Prodrug-activated gene therapy: involvement of an immunological component in the "bystander effect". *Cancer Gene Ther.* **3**, 83–88.

21. Consalva, M., Mullen, C. A., Modesti, A., Musiani, P., Allione, A., Cavallo, F., Giovarelli, M., and Forni, G. (1995). 5-Fluorocytosine-induced eradication of murine adenocarcinomas engineered to express the cytosine deaminase suicide gene requires host immune competence and leaves an efficient memory. *J. Immunol.* **154**, 5302–5312.

22. Wallace, P. M., MacMaster, J. F., Smith, V. F., Kerr, D. E., Senter, P. D., and Cosand, W. L. (1994). Intratumoral generation of 5-fluorouracil mediated by an antibody-cytosine deaminase conjugate in combination with 5-fluorocytosine. *Cancer Res.* **54**, 2719–2723.

23. Wallace, P. M., and Senter, P. D. (1994). Selective activation of anticancer prodrugs by monoclonal antibody-enzyme conjugates. *Methods Find. Exp. Clin. Pharmacol.* **16**, 505–512.

24. Li, Z., Shanmugam, N., Katayose, D., Huber, B., Srivastava, S., Cowan, K., and Seth, P. (1997). Enzyme/prodrug gene therapy approach for breast cancer using a recombinant adenovirus expressing *Escherichia coli* cytosine deaminase. *Cancer Gene Ther.* **4**, 113–117.

25. Evoy, D., Hirschowitz, E. A., Naama, H. A., Li, X. K., Crystal, R. G., Daly, J. M., and Lieberman, M. D. (1997). *In vivo* adenoviral-mediated gene transfer in the treatment of pancreatic cancer. *J. Surg. Res.* **69**, 226–231.

26. Kreuzer, J., Denger, S., Reifers, F., Beisel, C., Haack, K., Gebert, J., and Kubler, W. (1996). Adenovirus-assisted lipofection: efficient *in vitro* gene transfer of luciferase and cytosine deaminase to human smooth muscle cells. *Atherosclerosis* **124**, 49–60.

27. Szala, S., Missol, E., Sochanik, A., and Strozyk, M. (1996). The use of cationic liposomes DC-CHOL/DOPE and DDAB/DOPE for direct transfer of *Escherichia coli* cytosine deaminase gene into growing melanoma tumors. *Gene Ther.* **3**, 1026–1031.

28. Fox, M. E., Lemmon, M. J., Mauchline, M. L., Davis, T. O., Giaccia, A. J., Minton, N. P., and Brown, J. M. (1996). Anaerobic bacteria as a delivery system for cancer gene therapy: *in vitro* activation of 5-fluorocytosine by genetically engineered clostridia. *Gene Ther.* **3**, 173–178.

29. Ohwada, A., Hirschowitz, E. A., and Crystal, R. G. (1996). Regional delivery of an adenovirus vector containing the *Escherichia coli* cytosine deaminase gene to provide local activation of 5-fluorocytosine to suppress the growth of colon carcinoma metastatic to liver. *Hum. Gene Ther.* **7**, 1567–1576.

30. Dong, Y., Wen, P., Manome, Y., Parr, M., Hirschowitz, A., Chen, L., Hirschowitz, E. A., Crystal, R., Weichselbaum, R., Kufe, D. W., and Fine, H. A. (1996). *In vivo* replication-deficient adenovirus vector-mediated transduction of the cytosine deaminase

gene sensitizes glioma cells to 5-fluorocytosine. *Hum. Gene Ther.* **7,** 713–720.

31. Kanai, F., Lan, K. H., Shiratori, Y., Tanaka, T., Ohashi, M., Okudaira, T., Yoshida, Y., Wakimoto, H., Hamada, H., Nakabayashi, H., Tamaoki, T., and Omata, M. (1997). *In vivo* gene therapy for alpha-fetoprotein-producing hepatocellular carcinoma by adenovirus-mediated transfer of cytosine deaminase gene. *Cancer Res.* **57,** 461–465.

32. Lan, K. H., Kanai, F., Shiratori, Y., Okabe, S., Yoshida, Y., Wakimoto, H., Hamada, H., Tanaka, T., Ohashi, M., and Omata, M. (1996). Tumor-specific gene expression in carcinoembryonic antigen-producing gastric cancer cells using adenovirus vectors. *Gastroenterology* **111,** 1241–1251.

33. Judde, J. G., Spangler, G., Magrath, I., and Bhatia, K. (1996). Use of Epstein-Barr virus nuclear antigen-1 in targeted therapy of EBV-associated neoplasia. *Hum. Gene Ther.* **7,** 647–653.

34. Richards, C. A., Austin, E. A., and Huber, B. E. (1995). Transcriptional regulatory sequences of carcinoembryonic antigen: identification and use with cytosine deaminase for tumor-specific gene therapy. *Hum. Gene Ther.* **6,** 881–893.

CHAPTER 13

Preemptive and Therapeutic Uses of Suicide Genes for Cancer and Leukemia

FREDERICK L. MOOLTEN[1,2] AND PAULA J. MROZ[1]

[1]Edith Nourse Rogers Memorial Veterans Hospital, Bedford, Massachusetts 01730, and [2]Boston University School of Medicine, Boston, Massachusetts 02118

I. INTRODUCTION

The emergence of cancer gene therapy as a new discipline bears testimony to a need unmet by conventional therapies, selectivity. Cytokine gene therapy, suppressor genes, and antisense/ribozymes each aim at targeting cancer cells selectively. Implicit in these approaches is the presumption that there will be something about neoplastic cells that distinguishes them sufficiently from vital normal cells to permit therapeutic modalities to suppress or kill them without subjecting their normal counterparts to intolerable host toxicity.

The presumption is probably true for some cancers but false for others, perhaps for a majority.

Suicide genes constitute an alternative approach. Rather than manipulating, positively or negatively, existing cellular functions, they introduce new functions that sensitize cells to drugs at concentrations that would otherwise be innocuous. Most suicide genes encode enzymes that catalyze the conversion of prodrugs into cytotoxic antimetabolites. The best known among these genes, the herpes thymidine kinase (HSV-TK) gene, sensitizes cells to the guanosine analog ganciclovir (GCV) as a consequence of HSV-TK-catalyzed phosphorylation of GCV to intermediates that lethally inhibit DNA synthesis. Since the initial reports introducing the suicide gene concept [1,2], many animal studies have demonstrated that systemically administered GCV can eradicate transplanted tumors bearing transduced HSV-TK genes (reviewed in Moolten [3] and Tiberghien [4]) (Fig. 1). A serendipitous property of the HSV-TK/GCV combination is a "bystander effect," a phenomenon that manifests itself as an ability of GCV to kill not only HSV-TK-transduced cells but also untransduced cells in their proximity. The mechanism probably involves the transfer of activated GCV metabolites [1,5–8], at least *in vitro,* although stimulation of host immunity and damage to tumor blood vessels may also play a role *in vivo* (see Chapter 10).

Numerous other suicide gene/prodrug systems have since been described [9–17]. Some of the better characterized combinations are listed in Table 1. Of interest is the fact that the p450-2B1 and nitroreductase genes generate products that are not antimetabolites but alkylating agents, and therefore potentially more effective than antimetabolites in proliferatively quiescent cells. The product of the Fas/FKBP fusion gene is an apoptosis inducer. Both of its components are human proteins; similarly, p450-2B1, although of rat origin, has human counterparts. Both Fas/FKBP and p450 genes, therefore, pose less risk of inducing host immune reactions against transduced cells than other suicide genes that generate proteins of bacterial or viral origin.

II. THERAPEUTIC USES OF SUICIDE GENES

In theory suicide genes can be used both *therapeutically* in cancer patients and also *preemptively* in individuals not yet afflicted with a cancer, as described later. Clinical trials to date, however, have been limited to their use in patients with established malignancies. These trials have principally involved tumors that are limited to identified locations in the body but are nevertheless highly lethal even in the absence of distant metastases. They include brain tumors [18,19], ovarian cancer that has extended to the peritoneal cavity [20–22], and mesotheliomas [23]. Most of the trials utilize the HSV-TK/GCV combination, and the majority of these employ a modified virus as a vehicle ("vector") for introducing the HSV-TK gene into tumor cells after intratumoral injection or other instillation techniques that restrict the gene to the known location of the tumor. HSV-TK transduction is then followed by systemic GCV administration. A comprehensive listing of trials as of the end of 1996 has been compiled [24].

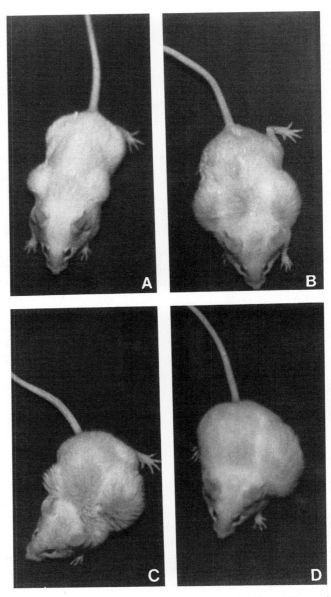

FIGURE 1   Differential effects of GCV on HSV-TK-positive and -negative tumors in the same mouse. HSV-TK-positive sarcoma cells were injected subcutaneously into the right flank and HSV-TK negative cells into the left flank. (A) At 13 days, small tumors were observed at each site. (B) At day 16, the tumors were growing progressively; an 8-day course of GCV administered intraperitoneally was begun. (C) By day 23, the gene-positive tumor had shrunk while the negative tumor had enlarged. (D) At day 37, the positive tumor had regressed completely, while the negative tumor continued to grow. Reproduced from Moolten [1].

Ongoing clinical trials have utilized one of two different vector systems to transduce the tumor cells [25]. In each case, the vector consists of a virus capable of infecting human cells that has been genetically engineered to eliminate genes responsible for viral replication and

TABLE 1 Suicide Gene/Prodrug Combinations

Gene [ref.]	Prodrug	Active product
HSV-TK [3]	GCV	GCV mono- and diphosphates
Cytosine deaminase [9]	5-Fluorocytosine	5-Fluorouracil
gpt [10]	6-Thioxanthine	6-Thioxanthine ribonucleotide
P450-2B1 [11,12]	Cyclophosphamide	Phosphoramide mustard
Purine nucleoside phosphorylase [13]	6-Methylpurine deoxyribonucleoside	6-Methylpurine
Deoxycytidine kinase [14]	Ara-C	Ara-C monophosphate
Nitroreductase [15]	CB1954	5-Azaridin-1-yl-4-hydroxylamino-2-nitrobenzamide
Fas-FKBP [16,17]	AP1510[a]	Multimerized Fas

[a] A proprietary synthetic dimerizer compound; a description can be found at the Internet site of Ariad Pharmaceuticals: http://www.ariad.com/reagent.html. Strictly speaking, AP1510 is not a prodrug because it is not activated by the product of the suicide gene, but rather activates that product by cross-linking it to form the multimers needed for the Fas protein to trigger apoptotic pathways.

cellular pathology, substituting in their place the suicide gene to be used for therapy. The first entails the use of vectors derived from murine retroviruses. Retroviral vectors mediate transduction that is relatively stable, at least in the short term, as a consequence of the integration of vector sequences into the DNA of the host genome, but to date it has been difficult to produce cell-free suspensions containing these vectors at titers sufficient to yield more than minimal transduction levels *in vivo*. Because of this limitation, most protocols have not attempted to introduce the vectors themselves into tumors, but rather producer cells, murine fibroblasts that generate and release the HSV-TK vectors at their *in vivo* injection site to yield a continuous supply until the cells are rejected by the host or killed by the administration of GCV.

The second system entails the use of vectors derived from human adenoviruses. Because adenoviral vectors do not integrate into genomic DNA, they mediate only transient transduction, but they possess the advantage of high titers that obviate the need for producer cells. Neither adenoviral nor retroviral vector systems, however, are currently capable of transducing suicide genes into more than a minority of tumor cells *in vivo*. A major component of the rationale underlying current trials is the expectation that bystander effects might permit GCV to eradicate untransduced cells by virtue of their proximity to transduced cells.

Most of the clinical trials are in early stages. To date, reported results include encouraging signs of tumor regression in some individuals [26], but not yet evidence of remissions durable enough to qualify as cures. A clear limitation is the difficulty of delivering suicide genes to all areas of a large tumor, even if the tumor has not metastasized. This problem is not fully solved by bystander effects because these effects tend to be powerful only at short ranges.

Another obvious limitation of a localized injection approach stems from the fact that the lethality of most cancer results from metastatic rather than localized disease. Metastatic disease will require systemic approaches that expose normal as well as neoplastic cells to the therapeutic modality. Attempts to address this problem include the linkage of suicide genes to promoters that might be highly active in tumor cells but have little or no activity in vital normal tissues. These include a tyrosinase promoter for melanomas [27,28], an alpha-fetoprotein promoter for hepatomas [29,30], Epstein-Barr Virus (EBV)-encoded transcriptional regulatory elements for EBV-related lymphomas and other EBV-associated malignancies [31,32], an osteocalcin promoter for osteosarcomas [33], and an ErbB2 promoter for breast carcinomas (Harris *et al.* [34], based on evidence that an occasional breast cancer may exhibit ErbB2 promoter hyperactivity). Promising initial evidence for therapeutic specificity has been reported in murine systems involving transplanted tumors [27,28,30,35], including a reduction in lung metastases of a melanoma after intravenous administration of a retroviral HSV-TK vector followed by GCV therapy [28]. It remains to be determined how much specificity might be achievable with these promoters in human cancers that arise endogenously and whether these genes can be delivered in bulk to metastatic deposits in sufficient quantity and uniformity to ensure tumor eradication by prodrug therapy. In addition, these cancers are exceptional; most cancers have yet to exhibit evidence of promoter activities unshared by vital normal stem cells.

In aggregate, this evidence illustrates potential therapeutic benefits that might ensue from the use of suicide genes, but it also compels the realization that the use of these genes as a cancer treatment modality resembles other cancer gene therapy modalities in that it requires the presumption of some selective property of the cancers that would permit the genes to be acquired or expressed selectively. As with other therapies, cancers that lack such a property will remain beyond the reach of this approach.

III. PREEMPTIVE USES OF SUICIDE GENES IN CANCER

Our recent work in murine systems has focused on exploring the feasibility of a different application of suicide genes, their preemptive use before a cancer develops, with particular emphasis on individuals at excessive risk for cancer. The goal of preemption is to achieve selectivity without requiring neoplastic cells to possess the one property whose frequent absence has confounded other approaches to cancer therapy, genetic or otherwise, a targetable difference from normal cells. To obviate the need for targetability, it is designed to exploit the clonal (i.e., single-cell) origin of human cancers [36–38] by introducing suicide genes not into an established cancer, but into a tissue from which cancers may arise. Because it is clonal, any cancer that subsequently arises within that tissue from a transduced cell should uniformly carry the suicide gene in all its cells as a clonal property, including metastases. Within a transduced clone of cancer cells, suicide gene expression might be lost in an occasional cell through mutations that delete or inactivate the gene, but in theory such cells might be susceptible to bystander killing by their proximity to gene-positive cells. The several studies that report the curability of tumors that arise from transplanted clonal populations of HSV-TK positive tumor cells [1,2,6,39], even when the tumors are known to harbor gene-negative mutants [1,39,40], are consistent with this expectation.

In a nonvital tissue such as breast or prostate epithelium, preemption can aim at transducing a chosen suicide gene into as many cells as possible to maximize the probability that a subsequent cancer will arise from a transduced cell. Cells outside the transduced tissue would remain unsensitized, and measures to promote selectivity within the transduced tissue itself would be unnecessary, because loss of nonneoplastic breast or prostate epithelial cells during cancer therapy would not be life-threatening. For preemptive sensitization of a vital tissue such as bone marrow or gastrointestinal epithelium, selectivity must be achieved differently, by introducing one or more suicide genes in mosaic rather than homogeneous fashion [1,41]. Mosaicism creates

selectivity by ensuring that whatever cell later spawns a cancer will share its clonal sensitivity pattern with only a fraction of the normal cells (Fig. 2).

As a first step in testing the preemption paradigm, we have asked whether suicide gene transduction into cells that are not yet malignant might permit effective therapy of cancers that later arose from them [42]. TM4 is a line of preneoplastic murine mammary epithelial cells that can be propagated in tissue culture for subsequent in vivo transplantation [43,44]. A retroviral vector, STK [2], was used to transduce the HSV-TK gene into these cells in vitro. The cells were then injected subcutaneously into syngeneic BALB/c mice, where they formed small, nongrowing nodules from which cancers later arose in 40 percent of the mice. When the mice were treated with GCV, 7/20 responded with complete and durable tumor regressions, and the remainder exhibited a significant retardation of tumor growth (Table 2). Control tumors (transduced and untreated, or untransduced and GCV treated) invariably exhibited progressive growth. In comparison with controls, the HSV-TK gene by itself exerted no adverse effects on cancer incidence or growth rates, indicating that its presence was not a liability for the preemptively transduced preneoplastic cells and that its observable therapeutic effects operated through GCV.

The results represent the first experimental validation of the principle of preemption, demonstrating that a process applied to premalignant cells could alter the response to therapy of a future cancer. They also illustrate a number of obstacles that stand between this principle and its effective human implementation. The majority of mice were not cured. In tumors that were not eradicated, HSV-TK enzyme activity was low, consistent with an in vivo down-regulation of gene expression that occurred during the brief (weeks to months) interval of experimental observation. Durable regressions were limited to small tumors; larger ones responded only with growth delays. Finally, the study was feasible only because the epithelial cells at risk for cancer could be cultivated and transduced in vitro and later reintroduced into host mice, thus obviating the need to reach mammary epithelial cells in situ. These limitations define issues that must ultimately be addressed to convert the principle of preemption into a modality that can be applied to individuals at risk for breast cancer or other malignancies. Paramount among them are the need to achieve high efficiency integration of suicide genes into the genomic DNA of tissues in vivo and the need to improve the long-term stability of suicide functions in cells harboring the integrated genes beyond what is currently achievable with retroviral transduc-

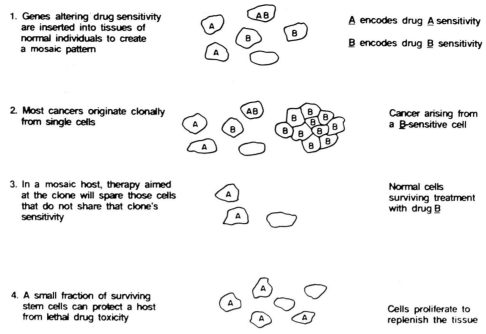

1. Genes altering drug sensitivity are inserted into tissues of normal individuals to create a mosaic pattern

A encodes drug A sensitivity

B encodes drug B sensitivity

2. Most cancers originate clonally from single cells

Cancer arising from a B-sensitive cell

3. In a mosaic host, therapy aimed at the clone will spare those cells that do not share that clone's sensitivity

Normal cells surviving treatment with drug B

4. A small fraction of surviving stem cells can protect a host from lethal drug toxicity

Cells proliferate to replenish the tissue

FIGURE 2 Preemptive introduction of suicide genes in *mosaic* fashion to ensure that later cancers are sensitized to eradication by a prodrug while their tissue of origin is protected by the presence of cells that do not share their sensitivity [41]. In the diagram, genes A and B are suicide genes that sensitize cells to prodrugs A and B respectively. The clonal origin of a cancer arising from one of the B-sensitized cells renders it uniformly sensitive to the prodrug B. Normal cells with the same sensitivity are also killed, but the remaining normal cells in the mosaic survive to repopulate the tissue. It should be noted that the principle of mosaicism illustrated in the diagram with suicide genes can also be implemented with resistance genes [41]. Thus, for example, if genes A and B encode resistance to drugs A and B, administration of drug A will selectively kill the cells that lack gene A, and drug B will kill cells that lack gene B. Unlike sensitivity mosaicism, however, mosaicism created by resistance genes cannot utilize drug doses that fully exploit the difference between sensitive and resistant cells without risking excessive drug toxicity to other tissues that have not acquired the genes. A potential advantage of resistance mosaicism is stability. Thus, a cell that carries a suicide gene can rid itself of its selective sensitivity through loss or inactivation of the gene via a one-step mutation, whereas a cell that is selectively sensitive because it lacks a gene for high-level drug resistance will in many cases require multiple independent mutations to achieve that level of resistance.

tion. Stable chemosensitivity is threatened not only by changes in gene regulation that cells experience consequent to exposure to an *in vivo* environment, but also by mutations that permanently delete or inactivate transduced genes. An additional concern, the possibility of long-term ill effects of transduction by retroviral vectors, including oncogenesis, has been ameliorated by theoretical calculations [45] and by the absence of vector-related cancers over the course of multiple gene therapy studies in animals and humans [46,47].

The efficiency problem is one that has long vexed much of the gene therapy field and may yield only to the eventual development of redesigned, and perhaps synthetic, vectors. Given the constraints of available technology, we have recently focused on the second

issue, long-term stability, limiting our current efforts to cells that can be manipulated *ex vivo*. Cells in this category include lymphocytes or hematopoietic stem cells that might be transfused into recipients after *ex vivo* manipulation, or fibroblastic or epithelial cells that might be cultured as a source of tissue replacement or for the introduction of therapeutic genes to correct a deficiency (e.g., Factor IX in some hemophiliacs, growth hormone in deficient individuals, insulin in diabetics). Introduction of a suicide gene into such cells is an attractive prospect as a "fail-safe" maneuver to permit their subsequent ablation if they later exhibit malignant or other aberrant behavior [45,48]. Critical to the prospective use of genes in this fashion is the requirement that all, or almost all, sensitized cells and their progeny

TABLE 2   GCV Therapy of Tumors Arising from
Preoplastic Mammary Epithelial Cells [42]

Preoplastic cells injected	GCV therapy of subsequent tumors[a]	Mice with durable tumor regressions/total[b]	Median survival (days)
HSV-TK transduced[c]	0	0/25	46
	+	7/20	152
Untransduced	0	0/8	70
	+	0/11	72
Transduced with control vector[c]	0	0/8	38
	+	0/7	53

[a] Tumor-bearing mice received 150 mg/kg GCV twice daily for 5 days by intraperitoneal injection.

[b] Seven GCV-treated mice in the HSV-TK group exhibited durable regressions, defined as complete tumor regressions without recurrence over a 300-day observation interval; tumor regressions were not observed in any of the other mice.

[c] The STK vector used to transduce the HSV-TK gene was constructed by inserting this gene into vector LNL6 [2]; the latter was used as an HSV-TK negative control.

retain chemosensitivity over intervals that may range from months to years and encompass many cell generations.

IV. CREATION OF STABLE SUICIDE FUNCTIONS BY COMBINING SUICIDE GENE TRANSDUCTION WITH ENDOGENOUS GENE LOSS

A. Loss of Purine or Thymidine Salvage Pathways Creates Chemosensitivity

Our effort to maximize the stability of suicide functions exploited the observation that when endogenous cellular functions are lost through mutation, the frequency with which they are regained is typically much lower than the frequency with which the functions of exogenously transduced genes are lost, that is, most "loss of function" mutations are highly stable. Two well-characterized loss of function mutations are those involving the genes for hypoxanthine/guanine phosphoribosyltransferase (HPRT) and cellular thymidine kinase (not to be confused with the HSV-TK gene, which encodes a different enzyme). The salient feature of HPRT and cellular TK is that mutational loss of either creates chemosensitivity, HPRT deficiency sensitizing cells to inhibitors of purine synthesis and thymidine kinase deficiency creating sensitivity to inhibitors of thymidylate synthesis. A regimen that inhibits both of these biosynthetic pathways is hypoxanthine/aminopterin/thymidine (HAT, see Sybalski and Sybalska [49]). HAT is well

tolerated by normal cells because unlike HPRT-deficient cells, they can utilize the hypoxanthine, and unlike TK-deficient cells, they can utilize the thymidine to circumvent the respective blocks in purine and thymidylate synthesis imposed by the antifolate drug aminopterin. HPRT-negative cells can be selected by virtue of their resistance to 6-thioguanine (6TG, see Sybalski and Sybalska[49]), and TK-negative cells can be selected for their resistance to iododeoxyuridine or bromodeoxyuridine [50].

The advantage of exploiting the stability that might characterize HAT-sensitizing mutations is offset by a significant potential limitation. The normal role of the HPRT and TK enzymes is to incorporate hypoxanthine and thymidine respectively into "salvage pathways" that reclaim these compounds for nucleic acid synthesis. Although loss of these pathways is not lethal, it appears to create subtle growth disadvantages in certain cell populations, eventually resulting, for example, in the loss of detectable HPRT-negative cells in cell populations of hematopoietic origin in women who begin life with both HPRT-positive and -negative cells [51]. The growth disadvantage appears not to be universal, because some other tissues (fibroblasts, hair follicles) retain their mosaic character in these women. Nevertheless, in the disadvantaged tissues, HAT sensitivity will ultimately prove unstable, not because cells lose sensitivity but because the cells themselves will fail to persist in the absence of a substitute means of accomplishing salvage pathway functions.

To determine whether stable suicide functions could be created, a two-pronged approach was utilized that combined HPRT or TK deficiency with addition of a new gene that replaced the lost salvage pathway functions and also mediated a suicide function of its own. HPRT-negative cells were obtained by 6TG selection and then exposed to a retroviral vector [10] that transduced the *Escherichia coli gpt* gene, which sensitizes cells to 6-thioxanthine (6TX). Like HPRT, the enzyme encoded by the *gpt* suicide gene, xanthine/guanine phosphoribosyltransferase (XGPRT), is capable of catalyzing the incorporation of hypoxanthine or guanine into nucleotide synthesis salvage pathways; its suicide function derives from its additional ability to use 6TX as a substrate. The *gpt* gene thus served a dual purpose: It added to the already stable chemosensitivity of HPRT-deficient cells by introducing an additional suicide function that must be lost by mutation for cells to lose all chemosensitivity. At the same time, it preserved the salvage pathway competency that HPRT-deficient cells would otherwise lack.

Based on the same rationale, the stability of the suicide function was also examined in cells that were deficient in cellular TK and had acquired the HSV-TK gene.

Before transduction, each of five HPRT-deficient clones tested exhibited a growth rate that was slightly to moderately slower than that of their wild-type parents (Table 3). Transduction of the *gpt* gene into HPRT-deficient cells yielded a new population that substituted 6TX sensitivity for their previously acquired HAT-sensitivity (Table 4). Unlike their predecessors, these *gpt*-transduced clones exhibited a broad range of doubling times, some growing slowly and others growing as fast or faster than parental cells. The observed variations may reflect position effects or other attributes of the integrated *gpt* vector.

B. Stability of Suicide Functions in HPRT-/*gpt*+ Cells

When four HPRT-deficient, *gpt*-transduced subclones of murine K3T3 fibrosarcoma cells were exposed to 6TX, surviving colonies ranged in number from 1 to 14 per 1.2×10^3 cells, representing mutant frequencies (corrected for plating efficiency) of 1.3×10^{-3} to 1.9×10^{-2}. The loss of 6TX sensitivity was accompanied in each case by reacquisition of HAT sensitivity, consistent with loss of *gpt*-mediated salvage functions (Table 4).

To determine the frequency with which both suicide functions were lost, expanded populations of three 6TX resistant subclones, each grown in 17 T-75 flasks containing 10^6 cells/flask, were tested for the presence of

TABLE 3 Growth Rates of Wild-Type Cell Lines and Subclones

Cells[a]	Salvage pathway competency	Doubling time (hours)[b]
K3T3 H+G-	+	15.6
K3T3 H-G-	-	16.2, 18.6
K3T3 H-G+	+	15.3, 15.8, 17.1, 19.2, 20.6, 22.6, 24.0
CLS1 H+G-	+	19.5
CLS1 H-G-	-	21.7
CLS1 H-G+	+	13.7, 15.7, 16.0, 24.4, 29.1, 29.4, 41.0, 43.1
LY18 H+G-	+	20.1
LY18 H-G-	-	21.5, 27.3
LY18 H-G+	+	18.5, 20.0, 21.6, 22.1, 22.6, 23.8, 25.1, 27.5

[a] The cell lines tested were of fibroblastic (K3T3), epithelial (CLS1), and pre-B-lymphocytic (LY18) origin.
[b] Doubling times represent the mean calculated from one to four replicates of duplicate cultures for each line and subclone tested.
[c] H, hprt; G, *gpt*.

HAT-resistant mutants. Of the 54 flasks, 53 yielded no HAT-resistant colonies and the remaining flask yielded a single colony. The result corresponds to a corrected mutant frequency of 3.0×10^{-8}. The subclones originated from an HPRT-negative, *gpt*-transduced clone that yielded 6TX-resistant mutants at a frequency of 5.4×10^{-3}. When this figure is multiplied by the frequency of HAT-resistant mutants, the resulting value,

$$5.4 \times 10^{-3} \times 3.0 \times 10^{-8} = 1.6 \times 10^{-10}$$

constitutes an estimate of the predicted frequency of the combined loss of both suicide functions. The same calculations applied to the clones with the greatest and poorest *gpt* stability yielded a frequency range of 3.9×10^{-11} to 5.7×10^{-10}. The rarity of HAT-resistant colonies among HPRT-negative K3T3 cells appeared to be matched by an HPRT-negative subclone of CLS1 cells, which yielded no surviving colonies among 20 flasks totaling 2×10^7 cells exposed to HAT.

C. Stability of Suicide Functions in TK⁻/HSV-TK⁺ Cells

NIH3T3 fibroblasts that lack cellular TK but had been transduced with the HSV-TK gene were exposed to 8.8 μM GCV to select for mutants that had lost the HSV-TK suicide function. GCV-resistant clones were obtained from replicate cultures in numbers that corresponded to mutant frequencies of 1.5×10^{-4} to 3.4×10^{-3}. Subsequent reexposure to GCV confirmed their resistant status. Analysis of two GCV-resistant clones confirmed that they were now HAT sensitive, as expected, and revealed TK enzyme levels that were only 1.5 and 1.8 percent of the levels of the GCV-sensitive cells from which they were derived, a decline consistent with their loss of GCV sensitivity and reacquisition of HAT sensitivity. When the two clones were subsequently exposed to HAT, HAT-resistant mutants were obtained at a frequency of 2×10^{-7} and 1.2×10^{-6}. The combined frequencies, representing the frequency with which a GCV-chemosensitive population would be expected to revert to a GCV-insensitive, HAT-insensitive wild-type phenotype, thus ranged from 3.0×10^{-11} to 4.1×10^{-9}. This implies a stability similar to the stability of suicide functions observed in HPRT-negative, *gpt*-transduced K3T3 cells.

In vivo studies are in progress to determine whether the high stabilities achieved by this strategy will preclude the emergence of cells that have lost all chemosensitivity from tumors bearing HSV-TK or *gpt* suicide genes and treated with GCV or 6TX respectively.

TABLE 4 Frequency of Acquisition and Subsequent Loss of Phenotypes
Associated with Chemosensitivity

Transition process	Phenotype	Isolation frequency	Sensitivity profile		
			HAT	6TX	6TG
None (wild type)	H^+G^-		R	R	S
6TG selection	H^-G^-	5×10^{-6}–1.3×10^{-5}	S	R	R
gpt transduction	H^-G^+	~0.5	R	S	S
Natural mutation	H^-G^-	1.3×10^{-3}–1.9×10^{-2}	S	R	R
Natural mutation	HAT^R	3.0×10^{-8}	R	R	NT

Note. The parental K3T3, CLS1, and LY18 cells used in the study were subjected to 6TG selection to yield clones that lacked HPRT enzyme activity and were sensitive to HAT, thus manifesting a suicide function consistent with their loss of hypoxanthine salvage capacity. Their 6TX ID_{50} exceeded 200 μM for K3T3 and LY18, and 250 μM for CLS1. After *gpt* transduction, the ID_{50} of the tested clones ranged from 0.5 to 10 μM for K3T3, from 1 to 3 μM for CLS1 cells, and from 2 to 10 μM for LY18. Data on transitions from H^+G^- to H^-G^- to H^-G^+ were obtained for all three cell lines; further transitions to H^-G^- and then HAT^R were measured only for K3T3. In addition, an H^-G^- subclone of CLS1 cells that had never been subjected to *gpt* transduction was tested and yielded no HAT-resistant colonies from 2×10^7 cells.

H, hprt; G, *gpt*; S, sensitive (all cells destroyed except for resistant mutants); R, resistant (cells grow at normal or near normal rates in the presence of the drug. NT, not tested; ID_{50}, 6TX concentration reducing cell numbers to 50 percent of the numbers in untreated control cultures during a 6 day assay interval.

V. PREEMPTIVE USES OF SUICIDE GENES TO CONTROL GRAFT-VERSUS-HOST DISEASE IN LEUKEMIA

The relevance of suicide genes to neoplastic disease extends beyond their direct presence in neoplastic cells. A promising area currently under active investigation involves the use of the HSV-TK gene to impart GCV sensitivity not to malignant cells but rather to cells used to treat the malignancy, that is, donor T lymphocytes administered in conjunction with allogeneic bone marrow in patients with leukemia and related diseases. Allogeneic bone marrow transplantation (allo-BMT) is currently associated with long-term remissions in a substantial number of patients with acute leukemia, chronic myelogenous leukemia, multiple myeloma, and myelodysplasia at a frequency that may exceed 50 percent in favorable circumstances; many of these remissions are thought to represent cures [52,53]. Most of the reduction in leukemic cell numbers is accomplished by the intensive chemoradiotherapy that precedes the allo-BMT, but donor T cells play a critical role in eradicating residual cells. This achievement comes at a cost, the frequent occurrence of graft-versus-host disease (GVHD) severe enough to result, directly or indirectly, in substantial treatment-related mortality. In addition, the severity of GVHD reflects in part the degree of antigenic disparity between donor and recipient, and thus limits the availability of suitable donors; HLA mis-matching poses the greatest threat of lethal GVHD, and despite HLA matching, unrelated donors represent a greater hazard than HLA-matched sibling donors. Because immunosuppressive drugs have often failed to control GVHD adequately and impose hazards of their own, T-cell-depleted marrow has been employed in an effort to avoid this complication. Unfortunately, the absence of T cells has been associated with poor leukemia control, reduced marrow engraftment, and a serious immunodeficiency that renders patients vulnerable to a variety of infections. Among the infectious sequelae are severe cytomegalovirus infections and potentially lethal EBV-induced lymphoproliferative disease [53].

One approach to preserving T-cell function involves allo-BMT with T-cell-depleted marrow followed by infusions of donor peripheral blood leukocytes, a rich source of T cells. In some cases, the infusions have been delayed until specifically necessitated by leukemia relapse or viral sequelae [53]. Delayed infusion of T cells appears to reduce the threat of GVHD but does not eliminate it. An additional advantage of utilizing separate marrow and T-cell infusions, however, is the opportunity to manipulate the T cells. In particular, this opportunity has been exploited to transduce the HSV-TK gene into donor T cells to sensitize them to GCV and thereby permit their subsequent ablation for severe GVHD [54–57]. The administration of HSV-TK-transduced T cells thus extends the benefits of T cells to all patients, later eliminating the cells only in those patients in whom they induce life-threatening pathology. This rationale is the basis for ongoing clinical trials in Milan [54,57] and more recently in this country (at the University of Arkansas [56]), as well as additional protocols that have been approved or are under review, all involving patients receiving allo-BMT for leukemia or myeloma [55]. In most cases, donor peripheral blood leukocytes are isolated, stimulated to proliferate, transduced with retroviral vectors bearing the HSV-TK gene plus a selectable marker, subjected to selection, and infused into patients after further growth to achieve adequate cell numbers.

Results reported to date from the first clinical trial (from Milan) are preliminary but encouraging [57]. Transduced cells retained their ability to exert antileukemic effects in most cases, including complete remissions in 3 of 8 patients. Two patients developed acute GVHD; in each case administration of GCV quickly eliminated the transduced cells from the circulation and induced nearly complete resolution of the clinical and biochemical signs of GVHD. The T cells thus appear to have responded as expected. In the Arkansas study, significant graft-versus-myeloma effects have been noted in some but not all patients treated with transduced cells for relapses after T-cell-depleted allo-BMT (G. Tricot, personal communication).

VI. FUTURE PROSPECTS FOR PREEMPTIVE USE OF SUICIDE GENES

Until *in vivo* transduction efficiency improves, the fail-safe use of suicide genes is likely to remain a phenomenon that can only be applied to cells that are manipulated *in vitro* and later reintroduced into human hosts. Potential applications include their use as a precaution against either malignant behavior of the reintroduced cells or immune pathology that they might induce [1,2,4,48]. Additionally, suicide genes added to cells transplanted to supply a missing function constitute a potential mechanism to control hyperactivity of the transplanted cells, for example, hyperinsulinemia resulting from excessive growth or function of cells transduced with insulin genes [58].

If suicide genes of nonhuman origin are to be used preemptively, their success will require that their presence not provoke host immune reactions that result in the elimination of the transduced cells. Such reactions have been observed in some [48,59] but not other [42,60] studies involving cells transduced with the HSV-TK gene. The development of improved methods for inducing immune tolerance, the use of genes transcribed from inducible promoters that remain inactive until an appropriate stimulus is applied, or the creation of suicide

genes that are expressed at the level of nucleic acid rather than protein (e.g., as catalytic RNA [61,62]) are possible approaches to this problem. If reliable methods for controlling immune rejection are developed to the point that they permit the use of xenografted tissues in humans, the introduction of suicide genes as transgenes into animals used as a source of the xenografts constitutes a further fail-safe use of suicide genes, one designed to protect against undesired effects of the grafted cells.

An intriguing application of the HSV-TK gene that is currently nearing clinical trial involves the ability of HSV-TK to phosphorylate the pyrimidine analog, 5-iodo-2′-fluoro-2′-deoxy-1-β-D-arabinofuranosyluracil (FIAU). Tjuvajev *et al.* have shown that when ^{131}I-labeled FIAU is administered to mice bearing tumors carrying transduced HSV-TK genes, the location and extent of HSV-TK expression can be precisely delineated by *in vivo* imaging with a gamma camera and single-photon emission tomography (SPECT) [63]. Extending this concept, they have also demonstrated that when the HSV-TK vector also transduces a separate gene *(lacZ)*, the imaging analysis correlates not only with HSV-TK expression but also locates and quantifies expression of the linked gene [64]. This use of HSV-TK as a marker harbors the potential for it to serve a dual purpose: measuring the function of whatever therapeutic gene it might be linked to in a gene therapy subject, and additionally serving to protect that subject against unwanted behavior by the transduced cells.

A final prospect relates to the possibility, discussed earlier, that efficient incorporation of one or more suicide genes into one or more tissues might eventually permit cancers that arise later to be treated effectively based on their clonal origin from a sensitized cell. The previous discussion emphasized the prospect that clonality might ensure the presence of a suicide gene even in metastatic or disseminated cancers, that is, the late stages of a cancer/host relationship. It is also possible, however, that early, preclinical stages might be targetable as well. Recent evidence indicates that DNA derived from cancer cells is sometimes detectable in blood or secretions by PCR analysis. Thus, mutant K-*ras* [65–68] genes have been detected in both plasma [65,66] and feces [67,68] of patients with colorectal [66,67] and pancreatic [65,68] carcinomas, mutant p53 genes have been demonstrated in the urine of patients with bladder cancer [69], and specific microsatellite DNA alterations have been detected in the plasma [70] and sputum [71] of lung cancer patients and in serum from patients with head and neck cancer [72]. Some of the detected alterations represented changes that were also present in premalignant lesions that accompanied the cancer or, in one case, that were found in the absence of a cancer

[68]. In theory, suicide genes harbored by the cells of cancers that arose in preemptively transduced tissues would also be detectable, and analysis of flanking genomic sequences could be used to determine whether they represented the monoclonal pattern of a neoplasm or the polyclonal pattern of nonneoplastic tissue. If the detection sensitivity of this type of DNA analysis increases to the point at which incipient clonal proliferations are detectable in individuals who harbor suicide genes in various vulnerable tissues (breast, lung, bone marrow, etc.), that detection would permit early action, for example, a search for the neoplasm, biopsy, and surgery or radiotherapy as indicated. If the neoplasm is found, prodrug administration could be added to surgery or radiotherapy in an adjuvant role. If the neoplasm is small enough to elude attempts to locate it, administration of a prodrug could be used to ablate it before it surfaces clinically, in essence exploiting preemption as a form of cancer prevention.

If *in vivo* transduction efficiency in nonvital tissues such as breast or prostate eventually improves to the point at which a suicide gene can be transduced into almost all the epithelial cells of these tissues, prevention should also be feasible at an even earlier stage if desired. Thus, individuals at high risk for breast or prostate cancer might, at some stage in their life, choose to receive a prodrug as a form of molecular "epitheliectomy" in preference to surgical bilateral mastectomy or prostatectomy.

References

1. Moolten, F. L. (1986). Tumor chemosensitivity conferred by inserted herpes thymidine kinase genes: paradigm for a prospective cancer control strategy. *Cancer Res.* **46,** 5276–5281.
2. Moolten, F. L., and Wells, J. M. (1990). Curability of tumors bearing herpes thymidine kinase genes transferred by retroviral vectors. *J. Natl. Cancer Inst.* **82,** 297–300.
3. Moolten, F. L. (1994). Drug sensitivity ("suicide") genes for selective cancer chemotherapy. *Cancer Gene Ther.* **1,** 279–287.
4. Tiberghien, P. (1994). Use of suicide genes in gene therapy. *J. Leukocyte Biol.* **56,** 203–209.
5. Bi, W. L., Parysek, L. M., Warnick, R., and Stambrook, P. J. (1993). *In vitro* evidence that metabolic cooperation is responsible for the bystander effect observed with HSV tk retroviral gene therapy. *Human Gene Ther.* **4,** 725–731.
6. Freeman, S. M., Abboud, C. N., Whartenby, K. A., Packman, C. H., Koeplin, D. S., Moolten, F. L., and Abraham, G. N. (1993). The "bystander effect": tumor regression when a fraction of the tumor mass is genetically modified. *Cancer Res.* **53,** 5274–5283.
7. Hooper, M. L., and Subak-Sharpe, J. H. (1981). Metabolic cooperation between cells. *Int. Rev. Cytol.* **69,** 45–104.
8. Culver, K. W., Ram, Z., Wallbridge, S., Ishii, H., Oldfield, E. H., and Blaese, R. M. (1992). *In vivo* gene transfer with retroviral vector producer cells for treatment of experimental brain tumors. *Science* **256,** 1550–1552.

9. Mullen, C. A., Kilstrup, M., and Blaese, R. M. (1992). Transfer of the bacterial gene for cytosine deaminase to mammalian cells confers lethal sensitivity to 5-fluorocytosine: a negative selection system. *Proc. Natl. Acad. Sci. U.S.A.* **89,** 33–37.
10. Mroz, P. J., and Moolten, F. L. (1993). Retrovirally transduced *Escherichia coli gpt* genes combine selectability with chemosensitivity capable of mediating tumor eradication *Hum. Gene Ther.* **4,** 589–595.
11. Wei, M. X., Tamiya, T., Chase, M., Boviatsis, E. J., Chang, T. K. H., Kowall, N. W., Hochberg, F. H., Waxman, D. J., Breakefield, X. O., and Chiocca, E. A. (1994). Experimental tumor therapy in mice using the cyclophosphamide-activating cytochrome P450 2B1 gene. *Hum. Gene Ther.* **5,** 969–978.
12. Chen, L., Waxman, D. J., Chen, D., and Kufe, D. W. (1996). Sensitization of human breast cancer cells to cyclophosphamide and ifosfamide by transfer of a liver cytochrome P450 gene. *Cancer Res.* **56,** 1331–1340.
13. Sorscher, E. J., Peng, S., Bebok, Z., Allan, P. W., Bennett, L. L., and Parker, W. B. (1994). Tumor cell bystander killing in colonic carcinoma utilizing the *Escherichia coli* DeoD gene to generate toxic purines. *Gene Ther.* **1,** 233–238.
14. Manome, Y., Wen, P. Y., Dong, Y., Tanaka, T., Mitchell, B. S., Kufe, D. W., and Fine, H. A. (1996). Viral vector transduction of the human deoxycytidine kinase cDNA sensitizes glioma cells to the cytotoxic effects of cytosine arabinoside *in vitro* and *in vivo. Nature Med.* **2,** 567–573.
15. Bridgewater, J. A., Knox, R. J., Pitts, J. D., Collins, M. K., and Springer, C. J. (1997). The bystander effect of the nitroreductase/CB1954 enzyme/prodrug system is due to a cell-permeable metabolite. *Hum. Gene Ther.* **8,** 709–717.
16. Spencer, D. M., Belshaw, P., Chen, L., Ho, S. N., Randazzo, F., Crabtree, G. R., and Schreiber, S. L. (1996). Functional analysis of Fas signaling *in vivo* using synthetic inducers of dimerization. *Curr. Biol.* **6,** 839–847.
17. Freiberg, R. A., Spencer, D. M., Choate, K. A., Peng, P. D., Schreiber, S. L., Crabtree, G. R., and Khavari, P. A. (1996). Specific triggering of the Fas signal transduction pathway in normal human keratinocytes. *J. Biol. Chem.* **271,** 31666–31669.
18. Oldfield, E. H. (1993). Gene therapy for the treatment of brain tumors using intra-tumoral transduction with the thymidine kinase gene and intravenous ganciclovir. *Hum. Gene Ther.* **4,** 39–69.
19. Culver, K. W., van Gilder, J. (1994). Gene therapy for the treatment of malignant brain tumors with in vivo tumor transduction with the herpes simplex thymidine kinase gene/ganciclovir system. *Human Gene Ther.* **5,** 343–379.
20. Freeman, S. M., McCune, C., Angel, C., Abraham, G. N., and Abboud, C. N. (1992). Treatment of ovarian cancer using HSV-TK gene modified vaccine–regulatory issues. *Hum. Gene Ther.* **3,** 342–349.
21. Link, C. J., and Moorman, D. (1996). A phase I trial of *in vivo* gene therapy with the herpes simplex thymidine kinase/ganciclovir system for the treatment of refractory or recurrent ovarian cancer. *Hum. Gene Ther.* **7,** 1161–1179.
22. Alvarez, R. D., and Curiel, D. T. (1997). A phase I study of recombinant adenovirus vector-mediated intraperitoneal delivery of herpes simplex virus thymidine kinase (HSV-TK) gene and intravenous ganciclovir for previously treated ovarian and extraovarian cancer patients. *Hum. Gene Ther.* **8,** 597–613.
23. Treat, J., Kaiser, L. R., Sterman, D. H., Litzky, L., Davis, A., Wilson, J. M., and Albelda, S. M. (1996). Treatment of advanced mesothelioma with the recombinant adenovirus H5.010RSVTK: a phase I trial (BB-IND 6274). *Hum. Gene Ther.* **7,** 2047–2057.
24. Marcel, T., and Grausz, J. D., (1997). The TMC worldwide gene therapy enrollment report, end 1996. *Hum. Gene Ther.* **8,** 775–800.

25. Jolly, D. (1994). Viral vector systems for gene therapy. *Cancer Gene Ther.* **1,** 51–64.

26. Freeman, S. M., Whartenby, K. A., Freeman, J. L., Abboud, C. N., and Marrogi, A. J. (1996). *In situ* use of suicide genes for cancer therapy. *Semin. Oncol.* **23,** 31–45.

27. Vile, R. G., and Hart, I. R. (1993). Use of tissue-specific expression of the herpes simplex virus thymidine kinase gene to inhibit growth of established murine melanomas following direct intratumoral injection of DNA. *Cancer Res.* **53,** 3860–3864.

28. Vile, R. G., Nelson, J. A., Castleden, S., Chong, H., and Hart, I. R. (1994). Systemic gene therapy of murine melanoma using tissue specific expression of the HSVtk gene involves an immune component. *Cancer Res.* **54,** 6228–6234.

29. Huber, B. E., Richards, C. A., and Krenitsky, T. A. (1991). Retroviral-mediated gene therapy for the treatment of hepatocellular carcinoma: an innovative approach for cancer therapy. *Proc. Natl. Acad. Sci. U.S.A.* **88,** 8039–8043.

30. Macri, P., and Gordon, J. W. (1994). Delayed morbidity and mortality of albumin/SV40 T-antigen transgenic mice after insertion of an alphafetoprotein/herpes virus thymidine kinase transgene and treatment with ganciclovir. *Hum. Gene Ther.* **5,** 175–182.

31. Judde, J-G., Spangler, G., MacGrath, I., and Bhatia, K. (1996). Use of Epstein-Barr virus nuclear antigen-1 in targeted therapy of EBV-associated neoplasia. *Hum. Gene Ther.* **7,** 647–653.

32. Franken, M., Estabrooks, A., Cavacini, L., Sherburne, B., Wang, F., and Scadden, D. T. (1996). Epstein-Barr virus-driven gene therapy for EBV-related lymphomas. *Nature Med.* **2,** 1379–1382.

33. Ko, S-C., Cheon, J., Kao, C., Gotoh, A., Shirakawa,T., Sikes, R. A., Karsenty, G., and Chung, L. W. K. (1996). Osteocalcin promoter-based toxic gene therapy for the treatment of osteosarcoma in experimental models. *Cancer Res.* **56,** 4614–4619.

34. Harris, J. D., Gutierrez, A. A., Hurst, H. C., Sikora, K., and Lemoine, N. R. (1994). Gene therapy for cancer using tumour-specific prodrug activation. *Gene Ther.* **1,** 170–175.

35. Kaneko, S., Hallenbeck, P., Kotani, T., Nakabayashi, H., Mc-Garrity, G., Tamaoki, T., Anderson, W. F., and Chiang, Y. L. (1995). Adenovirus-mediated gene therapy of hepatocellular carcinoma using cancer-specific gene expression. *Cancer Res.* **55,** 5283–5287.

36. Fialkow, P. J. (1976). Clonal origin of human tumors. *Biochim. Biophys. Acta* **458,** 283–321.

37. Fearon, E. R., Hamilton, S. R., and Vogelstein, B. (1987). Clonal analysis of human colorectal tumors. *Science* **238,** 193–197.

38. Fujii, H., Marsh, C., Cairns, P., Sidransky, D., and Gabrielson, E. (1996). Genetic divergence in the clonal evolution of breast cancer. *Cancer Res.* **56,** 1493–1497.

39. Moolten, F. L., Wells, J. M., Heyman, R. A., and Evans, R. M. (1990). Lymphoma regression induced by ganciclovir in mice bearing a herpes thymidine kinase transgene. *Hum. Gene Ther.* **1,** 125–134.

40. Moolten, F., Wells, J. M., and Mroz, P. J. (1992). Multiple transduction as a means of preserving ganciclovir chemosensitivity in sarcoma cells carrying retrovirally transduced herpes thymidine kinase genes. *Cancer Lett.* **64,** 257–263.

41. Moolten, F. L. (1990). Mosaicism induced by gene insertion as a means of improving chemotherapeutic selectivity. *Crit. Rev. Immunol.* **10,** 203–233.

42. Moolten, F. L., Vonderhaar, B. K., and Mroz, P. J. (1996). Transduction of the herpes thymidine kinase gene into premalignant murine mammary epithelial cells renders subsequent breast cancers responsive to ganciclovir therapy. *Hum. Gene Ther.* **7,** 1197–1204.

43. Jerry, D. J., Ozbun, M. A., Kittrell, F. S., Lane, D. P., Medina, D., Butel, J. S. (1993). Mutations in p53 are frequent in the preneo-plastic stage of mouse mammary tumor development. *Cancer Res.* **53,** 3374–3381.

44. Kittrell, F. S., Oborn, C. J., and Medina, D. (1992). Development of mammary preneoplasias *in vivo* from mouse mammary epithelial cell lines *in vitro. Cancer Res.* **52,** 1924–1932.

45. Moolten, F. L., and Cupples, L. A. (1992). A model for predicting the risk of cancer consequent to retroviral gene therapy. *Hum. Gene Ther.* **3,** 479–486.

46. Cornetta, K., Morgan, R. A., and Anderson, W. F. (1991). Safety issues related to retroviral-mediated gene transfer in humans. *Hum. Gene Ther.* **2,** 5–14.

47. Cornetta, K., Morgan, R. A., Gillio, A., Sturm, S., Baltrucki, L., O'Reilly, R., and Anderson, W. F. (1991). No retroviremia or pathology in long-term follow-up of monkeys exposed to a murine amphotropic retrovirus. *Hum. Gene Ther.* **2,** 215–220.

48. Riddell, S. R., Elliott, M., Lewinsohn, D. A., Gilbert, M. J., Wilson, L., Manley, S. A., Lupton, S. D., Overell, R. W., Reynolds, T. C., Corey, L., and Greenberg, P. D. (1996). T-cell mediated rejection of gene-modified HIV-specific cytotoxic T lymphocytes in HIV-infected patients. *Nature Med.* **2,** 216–223.

49. Sybalski, W., and Sybalska, E. H. (1961). A new chemotherapeutic principle for the treatment of drug resistant neoplasms. *Cancer Chemother. Rep.* **11,** 87–89.

50. Littlefield, J. W. (1966). The use of drug-resistant markers to study the hybridization of mouse fibroblasts. *Exp. Cell Res.* **41,** 190–196.

51. Rossiter, B. J. F., and Caskey, C. T. (1995). Hypoxanthine-guanine phosphoribosyltransferase deficiency: Lesch-Nyhan syndrome and gout. *In* Scriver, C. R., Beaudet, A. L., Sly, W. S., Valle, D., eds., *The Metabolic and Molecular Bases of Inherited Disease,* 7th ed., McGraw-Hill, New York, pp. 1679–1706.

52. Beutler, E., Lichtman, M. K., Coller, B. S., and Kipps, T. J., eds. (1995). *Williams Hematology,* 5th ed., McGraw-Hill, New York.

53. Shlomchik, W. D., and Emerson, S. G. (1996). The Immunobiology of T cell therapies for leukemias. *Acta Haematol.* **96,** 189–213.

54. Bordignon, C., and Bonini, C. (1995). Transfer of the HSV-TK gene into donor peripheral blood lymphocytes for *in vivo* modulation of donor anti-tumor immunity after allogeneic bone marrow transplantation. *Hum. Gene Ther.* **6,** 813–819.

55. Tiberghien, P. (1997). Use of donor T-lymphocytes expressing herpes-simplex thymidine kinase in allogeneic bone marrow transplantation: a phase I-II study. *Hum. Gene Ther.* **8,** 615–624.

56. Munshi, N. C., Govindarajan, R., Drake, R., Ding, L. M., Iyer, R., Saylors, R., Kornbluth, J., Marcus, S., Chiang, Y., Ennist, D., Kwak, L., Reynolds, C., Tricot, G., and Barlogie, B. (1997). Thymidine kinase (TK) gene-transduced human lymphocytes can be highly purified, remain fully functional, and are killed efficiently with ganciclovir. *Blood* **89,** 1334–1340.

57. Bonini, C., Ferrari, G., Verzeletti, S., Servida, P., Zappone, E., Ruggieri, L., Ponzoni, M., Rossini, S., Mavilio, F., Traversari, C., and Bordignon, C. (1997). HSV-TK gene transfer into donor lymphocytes for control of allogeneic graft-versus-host leukemia. *Science* **276,** 1719–1724.

58. Yoshimoto, K., Murakami, R., Moritani, M., Ohta, M., Iwahana, H., Nakauchi, H., and Itakura, M. (1996). Loss of ganciclovir sensitivity by exclusion of thymidine kinase gene from transplanted proinsulin-producing fibroblasts as a gene therapy model for diabetes. *Gene Ther.* **3,** 230–234.

59. Tapscott, S. J., Miller, A. D., Olson, J. M., Berger, M. S., Groudine, M., and Spence, A. M. (1994). Gene therapy of rat 9L gliosarcoma tumors by transduction with selectable genes does not require drug selection. *Proc. Natl. Acad. Sci. U.S.A.* **91,** 8185–8189.

60. Pavlovic, J., Nawrath, M., Tu, R., Heinicke, T., and Moelling, K. (1996). Anti-tumor immunity is involved in the thymidine kinase-

mediated killing of tumors induced by activated Ki-*ras* (G12V). *Gene Ther.* **3,** 635–643.

61. Prudent, J. R., Uno, T., and Schultz, P. G. (1994). Expanding the scope of RNA catalysis. *Science* **264,** 1924–1927.

62. Wilson, C., and Szostak, J. W. (1995). *In vitro* evolution of a self-alkylating ribozyme. *Nature* **374,** 777–782.

63. Tjuvajev, J. G., Finn, R., Watanabe, K., Joshi, R., Oku, T., Kennedy, J., Beattie, B., Koutcher, J., Larson, S., Blasberg, R. G. (1996). Noninvasive imaging of herpes virus thymidine kinase gene transfer and expression: a potential method for monitoring clinical gene therapy. *Cancer Res.* **56,** 4087–4095.

64. Tjuvajev, J., Safer, M., Sadelain, M., Avril, N., Oku, T., Joshi, R., Finn, R., Larson, S., and Blasberg, R. (1997). Noninvasive imaging of the HSV1-tk marker gene for monitoring the expression of other target genes *in vivo*. *J. Neurooncol.* **35** (Suppl. 1), S45.

65. Sorenson, G. D., Pribish, D. M., Valone, F. H., Memoli, V. A., and Yao, S. L. (1993). Mutated K-*ras* sequences in plasma from patients with pancreatic carcinoma. *Proc. Am. Assoc. Cancer Res.* **34,** A174.

66. Lefort, L., Anker, P., Vasioukhin, V., Lyautey, J., Lederrey, C., and Stroun, M. (1995). Point mutations of the K-*ras* gene present in the DNA of colorectal tumors are found in the blood plasma DNA of the patients. *Proc. Am. Assoc. Cancer Res.* **36,** A3319.

67. Sidransky, D., Tokino, T., Hamilton, S. R., Kinzler, K. W., Levin, B., Frost, P., and Vogelstein, B. (1992). Identification of *ras* oncogene mutations in the stool of patients with curable colorectal tumors. *Science* **256,** 102–105.

68. Caldas, C., Hahn, S., Hruban, R. H., Yeo, C., and Kern, S. (1994). Detection of K-*ras* mutations (mut) in the stool of patients (pts) with pancreatic adenocarcinoma (PCa). *Proc. Am. Soc. Clin. Oncol.* **13,** A294.

69. Sidransky, D., Von Eschenbach, A., Tsai, Y. C., Jones, P., Summerhayes, I., Marshall, F., Meera, P., Green, P., Hamilton, S. R., Frost, P., and Vogelstein, B. (1991). Identification of p53 gene mutations in bladder cancers and urine samples. *Science* **252,** 706–709.

70. Chen, X. Q., Stroun, M., Magnenat, J-L., Nicod, L. P., Kurt, A-M, Lyautey, J., Lederrey, C., and Anker, P. (1996). Microsatellite alterations in plasma DNA of small cell lung cancer patients. *Nature Med.* **2,** 1033–1035.

71. Miozzo, M., Sozzi, G., Musso, K., Pilotti, S., Incarbone, M., and Pastorino, U. (1996). Microsatellite alterations in bronchial and sputum specimens of lung cancer patients. *Cancer Res.* **56,** 2285–2288.

72. Nawroz, H., Koch, W., Anker, P., Stroun, M., and Sidransky D. (1996). Microsatellite alterations in serum DNA of head and neck cancer patients. *Nature Med.* **2,** 1035–1037.

PART IV

Targeting Oncogenes and Growth Factors for Gene Therapy

CHAPTER 14

Antisense Strategies in the Treatment of Leukemias

BRUNO CALABRETTA,[1] TOMASZ SKORSKI,[1] GERALD ZON,[2] MARIUSZ Z. RATAJCZAK,[3] AND ALAN M. GEWIRTZ[3,4]

[1]Department of Microbiology and Jefferson Cancer Institute, Thomas Jefferson University, Philadelphia, Pennsylvania 19107; [2]Lynx Therapeutics Inc., Hayward, California 34545; and the Departments of [3]Pathology and [4]Internal Medicine, University of Pennsylvania, Philadelphia, Pennsylvania 19104

I. INTRODUCTION

Gene expression may be disrupted by a variety of methods. It is convenient to group the available technologies according to whether they target the gene itself (e.g., homologous recombination) or the gene's transcriptional product, a messenger RNA. Among methods that are RNA directed, the most widely used are catalytic RNA molecules, or ribozymes, and "antisense" oligodeoxynucleotides (ODNs) [1]. ODNs are short-nucleotide DNA sequences synthesized as exact complements of the nucleotide sequence of the targeted mRNA. In theory, stable hybridization only occurs between exact complementary sequences such as the antisense DNA molecule with its mRNA target. Once the RNA–DNA duplex forms, translation of the message is prevented and destruction of the molecule by RNase H is promoted. The major appeal of the antisense approach is its simplicity.

Since the earliest attempts by Zamecnick and Stephenson to inhibit Rous sarcoma virus replication and cell transformation by a specific oligodeoxynucleotide [2], these compounds have been proposed as potential therapeutic agents in a spectrum of pathological processes ranging from viral infections to neoplastic disorders. Genes to target using the antisense approach to treat cancer were identified through the pioneering observations of the Weinberg, Barbacid, and Wigler laboratories on activating point mutations of the *ras* transforming genes isolated from epithelial neoplasia [3], and the subsequent elucidation of two other common modalities of oncogene activation in cancer cells, amplification (e.g., erb-B2 amplification in breast and ovarian cancers) and translocation [e.g., juxtaposition of the *bcr* and *c-abl* genes in chronic myelogenous leukemia (CML)].

The potential for highly specific gene targeting contrasts with the mechanism(s) of action of conventional anticancer chemotherapeutic agents, which block enzymatic pathways or randomly interact with nucleic acids irrespective of the cell phenotype. Anticancer chemotherapeutic agents exploit differences in biochemical or metabolic processes (e.g., growth rate) between normal and cancer cells for the preferential killing of neoplastic cells over normal cells. In contrast, antisense oligodeoxynucleotides exploit the presence of genetically defined characteristics that distinguish neoplastic cells and are responsible for their growth advantage over normal cells. In recent years, the antisense strategy for cancer therapy has progressed from *in vitro* culture studies, to investigations in animal models, and now to clinical studies. We described here the current state of progress toward mRNA-directed antisense-based cancer therapy primarily from the viewpoint of initial proof-of-concepts studies in animal models of human leukemias and phase I clinical investigations.

II. TARGETING *bcr-abl* mRNA IN A SCID MOUSE MODEL OF PHILADELPHIA[1] LEUKEMIA

The ideal strategy for the treatment of leukemia would be selective elimination of leukemic cells and restoration of normal hematopoiesis. An example of such rational antisense drug design is the targeting of BRC/ABL transcripts found in leukemic patients carrying the Philadelphia chromosome translocation [4–6]. The pathogenic role of the *bcr-abl* genes in CML has been strongly suggested by the appearance of CML-like syndromes in mice bearing *bcr-abl* constructs [7–9]. Synthetic oligodeoxynucleotides complementary to the junction of BCR/ABL transcripts produced from the splicing of either the second or the third exon of the *bcr* gene to the second exon of *c-abl* were shown to suppress Philadelphia[1] leukemic cell proliferation *in vitro* and to spare the growth of normal marrow progenitors [10]. A prerequisite for the *in vivo* utilization of antisense oligonucleotides as anticancer drugs is the de-

velopment of animal models of human malignancies that mimic the natural course of the disease in patients. Unlike other types of human neoplasia, leukemic cells obtained directly from marrows of patients and transplanted into immunodeficient SCID mice show a pattern of leukemic spread reminiscent of that observed during the natural course of the disease [11]. SCID mice injected with Philadelphia[1] BV173 and systemically treated with nuclease-resistant 26-mer b2/a2 antisense phosphorothioate oligodeoxynucleotides ([S]ODNs) at 1 mg/day for 9 consecutive days showed a marked decrease in three different measures of leukemia burden: percentage of CALLA-positive cells, number of clonogenic leukemic cells, and amounts of BCR/ABL transcripts in mouse tissues [12]. SCID mice injected with 10^6 BV173 cells and then treated intravenously with BCR/ABL antisense [S]ODNs had only molecular evidence of minimal residual disease at 42 and 56 days after injection of leukemic cells [12]. By contrast, untreated mice and mice treated with sense or 6-base mismatched [S]ODNs had macroscopic, microscopic, and molecular evidence of active leukemia [12]. These differences among the groups of mice were reflected in their mortality rates; all 9 untreated and 10 BCR/ABL sense-oligodeoxynucleotide-treated (5 each treated starting day +7 and day +21) mice died with diffuse leukemia, as confirmed by necropsy, 8–13 weeks after injection of BV173 leukemia cells (median survival time, 9.7 ± 0.9 weeks). In marked contrast, the 10 BCR/ABL antisense-treated mice (5 each treated starting day +7 and day +21) died of leukemia 18 to 23 weeks after injection of leukemic cells (median survival time 19.4 ± 1.4 weeks, $P < 0.001$ as compared to control groups) (Table 1). Similar studies conducted in SCID mice carrying Philadelphia[1] cells directly taken from a patient with CML in blast crisis confirmed the ability of BCR/ABL antisense [S]ODNs to suppress temporarily the spreading of leukemia (Table 1).

Although the disease process was temporarily suppressed and survival was prolonged in leukemic mice treated with BCR/ABL antisense oligodeoxynucleotides, mice invariably succumbed to their disease. This outcome may have reflected the inability to maintain a sufficient tissue concentration of ODNs after systemic injection to induce a sustained suppression of oncogenic expression leading to leukemic cell death and/or an intrinsic biological limitation of targeting a single oncogene.

In this regard, it is now becoming clear that the ability of the BCR/ABL oncoprotein to transform hematopoietic cells rests in the activation of downstream effectors that transmit the oncogenic signal from the cytoplasm to the nucleus [13–15]. One such effector is the proto-oncogene c-Myc, which is required for BCR/ABL or

TABLE 1   Survival of Leukemic SCID Mice Treated with Antisense Oligonucleotide-Based Therapies

Treatment	Leukemia	Survival
Control oligonucleotides (ODNs)	BV173[a]	7–9 weeks
bcr/abl antisense ODNs	BV173	16–23 weeks
Control ODNs	CML-BC[b]	17–20 weeks
bcr/abl antisense ODNs	CML-BC	24–48 weeks
bcr/abl + c-myc antisense ODNs	BV 173	26–38 weeks
bcr/abl + c-myc antisense ODNs	CML-BC	32–50 weeks
bcr/abl antisense ODNs and cyclophosphamide	CML-BC	38–60 weeks (50%) >60 weeks (50%)

[a] BV173 is a Philadelphia[1] cell line derived from a patient with lymphoid CML-blast crisis.
[b] CML-BC are primary cells obtained from a patient with myeloid CML-blast crisis.

V-ABL transformation of hematopoietic cells [16]. Thus, use of a combination of ODNs targeting two different components of the signal transduction pathway required for proliferation of CML cells might yield more profound cell killing *in vitro* and *in vivo*. In fact, results of analytical assays (immunofluorescence detection of leukemic cells in tissues colony formation from mouse tissue cell suspensions and detection of BCR/ABL transcripts in mouse tissues), showed that the leukemic cell load in SCID mice systemically treated with the combination of BCR/ABL and c-Myc ODNs was reduced as compared with that of mice treated with the individual ODNs [17,18].

The more potent effects of the ODN combination therapy were reflected in the different survival of the antisense-ODNs-treated mice. Median survival of the control mice was 15.5 ± 2.1 weeks; median survival of the mice treated with BCR/ABL or c-Myc antisense oligodeoxynucleotides was 23.0 ± 1.1 weeks and 23.3 ± 1.4 weeks, respectively; mice treated with an equal dose of both antisense [S]ODNs survived significantly longer (32.5 ± 6.9 weeks; $P < 0.001$, compared to mice receiving a single ODN) (Table 1).

The reasons underlying the more potent antileukemia effects of the ODN combination therapy remain unclear. One possibility is that ODNs targeting a second oncogene involved in the disease process arrest the growth of cells that escape the proliferation inhibitory effect associated with individual gene targeting. It is also possible that down-regulation of gene expression by a single antisense ODN at the relatively low concentration reached *in vivo* [19] might be insufficient to inhibit cell proliferation, whereas "partial" inhibition of two cooperating oncogenes might induce a longer-lasting suppression of cell proliferation. The more potent antileukemia effects obtained by targeting two cooperating oncogenes is reminiscent of the effects induced by com-

bination chemotherapy strategies in which different drugs are designed to target tumor cells as they progress through distinct stages of the cell cycle; however, the potential advantage of antiocogene therapy rests in its selectivity for disease-inducing agents and, in turn, its sparing of normal cells. The therapeutic potential of the combination therapy approach may be further improved by optimal combinations of agents that inhibit gene expression at different stages of the disease process once the underlying pathogenic mechanisms are better known.

III. COMBINATIONS OF ANTISENSE ODNs WITH OTHER ANTILEUKEMIC AGENTS

Another strategy that might enhance the therapeutic potential of oncogene-targeted ODNs involves the use of these compounds in conjunction with conventional antineoplastic drugs. It is becoming clear that the therapeutic effects of most of these drugs depend not only on their interference with distinct aspects of tumor cell metabolism, but also on their ability to modulate the levels of proteins involved in regulating the process of apoptosis. This raises the possibility of adjusting drug concentrations to limit the effects on metabolic processes (which are also toxic for normal cells) while preserving the apoptosis-inducing function.

CML cells are effectively killed *in vitro* when exposed to suboptimal concentrations of mafosfamide, and BCR/ABL antisense ODNs [20]. Such improvement over the effects of either mafosfamide or antisense ODNs alone correlated with an enhanced susceptibility of leukemic cells to undergo apoptosis when exposed to increasing concentrations of mafosfamide followed by treatment with a constant amount of BCR/ABL ODNs [21]. At

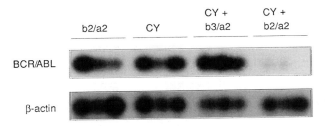

FIGURE 1 Detection of Philadelphia[1] leukemia cells in the bone marrow of severe combined immunodeficient mice by means of BCR/ABL transcript amplification. BCR/ABL and β-actin gene transcripts were separately detected by means of coupled reverse transcription and the polymerase chain reaction, using total RNA from 10^6 bone marrow cells for each sample, which was isolated 20 weeks after leukemia cell injection. The amplification products were resolved by electrophoresis and blotted to Nytran-Plus membranes. The membranes were then hybridized with [γ-^{32}P]-end-labeled c-*abl* or β-actin oligomers complementary to regions of the amplified 257-base-pair (bp) *bcr-abl* and 210-bp β-actin segments, respectively. The hybridized membranes were exposed to X-ray films for 24 (BCR/ABL) and 2 (β-actin) hours, respectively. The results shown are representative of two different experiments using three mice per group. CY, cyclophosphamide.

the highest concentration of mafosfamide (100 μg/mL), almost all cells underwent apoptosis; however, in cultures exposed to lower concentrations of the drug, only subsequent treatment with b2/a2 antisense ODNs increased the proportion of apoptotic cells. Treatment with mafosfamide increased levels of apoptosis-associated p53 protein and of BAX, a member of the BCL-2 family, which is positively regulated by p53 [22,23]. In cultures treated with b2/a2 antisense ODNs, the most marked effect was the suppression of BCL-2 expression, whereas p53 and BAX levels were essentially unchanged. Overall, cultures treated with mafosfamide and b2/a2 antisense ODNs exhibited increased expression of positive effectors (p53 and BAX) and decreased expression of a negative regulator (BCL-2) in the pathway leading to apoptotic cell death [21].

In leukemic SCID mice treated with a single injection of cyclophosphamide (25 mg/kg) followed by multiple injections of antisense ODNs (1 mg/day for 12 consecutive days), the disease process was slowed as indicated by the inability to detect BCR/ABL transcripts in peripheral blood (Fig. 1) and by histopathologic analysis (Fig. 2) to a much greater extent than that induced by treatment with either agent alone. Mice treated with BCR/ABL antisense oligonucleotides or with a single injection of cyclophosphamide had increased survival compared to control mice (median survival of 22 ± 2.5 weeks vs. 14 ± 1 week) (Table 1). However, mice treated with BCR/ABL antisense ODNs plus cyclophosphamide showed the most favorable therapeutic response; 50% were healthy more than 60 weeks after the injection of leukemic cells, whereas the remaining 50% died of

leukemia after 36 to 48 weeks (median survival 39 ± weeks; $P < 0.001$ compared with survival of mice treated with cyclophosphamide or BCR/ABL antisense ODNs alone) (Table 1). Consistent with the increased susceptibility of leukemic cells to apoptosis after *in vitro* treatment with mafosfamide and BCR/ABL antisense ODNs, the improved therapeutic response of mice to treatment with cyclophosphamide and BCR/ABL ODNs suggests that there is also an enhanced propensity for apoptosis among leukemic cells in vivo.

IV. TARGETING c-*myb* mRNA IN A SCID MOUSE MODEL OF HUMAN LEUKEMIA

MYB, the encoded product of the proto-oncogene c-*myb* is preferentially expressed in hematopoietic cells, in which it functions as a DNA binding specific transcription factor [24]. Down-regulation of c-Myb expression correlates with hematopoietic cell differentiation, and its overexpression has been shown to inhibit mouse erythroleukemia cell differentiation induced by erythropoietin or dimethyl sulfoxide [25,26]. An unmodified 18-mer antisense (but not sense) oligodeoxynucleotide targeted to codons 2–7 of c-*myb* was used to demonstrate that c-Myb gene function is required for proliferation of normal and malignant hematopoietic cells [27,28]. Subsequently, it was shown that this c-Myb 18-mer antisense (but not sense) oligomer strongly inhibited or completely abolished clonogenic growth in 78% (18 of 23) of primary acute myelogenous leukemia (AML) cases examined and in 80% (4 of 5) of primary CML cases in blast crisis. The c-Myb antisense had more limited effects on normal hematopoietic progenitor cells [29], suggesting that leukemia cells were more dependent than their normal counterparts on c-Myb function for proliferation and/or cell maintenance.

Inhibition of Myb function based on its differential expression in leukemic vs. normal cells and/or the differential requirements of Myb function in these cells represents a fundamentally different use of the antisense approach from that of specific inhibition of BCR/ABL expression based on its presence in Philadelphia-positive leukemic but not in Philadelphia-negative normal cells. On the other hand, the targeting of c-*myb* is applicable not only to leukemias, but also to other malignancies with activated levels of Myb expression [30,31].

To extend these *in vitro* antisense findings for c-*myb* obtained with a unmodified, nuclease-sensitive oligomer, to an *in vivo* setting, Gewirtz and co-workers first confirmed the *in vitro* antiproliferative activity expected of a nuclease-resistant 24-mer antisense phosphorothioate oligodeoxynucleotide to codons 2–9 and then sought to develop a suitable animal model of human CML. The

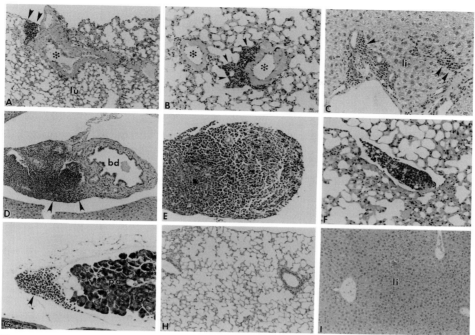

FIGURE 2   Gallery of light micrographs of tissue sections from leukemic mice treated with cyclophosphamide (A–C), b2/a2 antisense phosphorothioate oligodeoxynucleotides ([S]ODNs) (D–G), or a combination of cyclophosphamide and b2/a2 antisense [S]ODNs (H–I). Note the presence of leukemic infiltrates (arrowheads) in the areas around blood vessels (asterisks in A and B) of pulmonary parenchyma (lu). Note also the presence of leukemic infiltrates in the portal area of the liver (li in C) and around the bile duct (bd in D). A large hepatic leukemic infiltrate is shown in E. In some instances, leukemic infiltrates are seen in the expanded lymphatic of the lung (F) and the peripancreatic region (pa in G). In contrast, lung (H) and liver (I) tissues from animals receiving the combination treatment show no residual leukemia (original magnification: A, C, D, G, H, and I = ×120; B, E, and F = ×240).

K562 cell line was used in those studies because it is derived from a CML patient in blastic transformation and, more important, because these cells grow well in mice and carry the human tumor-specific marker, BCR/ABL. In this work, residual immune cells in SCID mice were diminished by treatment with cyclophosphamide (150 mg/kg; ~3 mg per mouse) on each of 2 successive days prior to tail-vein injection of 10^7 K562 human leukemia cells. Onset of disease was monitored by PCR detection of BCR/ABL transcripts in periodically sampled blood obtained from the retroorbital plexus. BCR/ABL-expressing Philadelphia-positive cells were detected after 5–6 weeks. Shortly thereafter the mice became weak and sickly in apperance, and rapidly manifested hydrocephalus and abdominal enlargement, with mean survival of 6 ± 3 days; none surviving longer than 12 days. Experiments were therefore carried out in which an osmotic pump was implanted s.c. in mice after leukemia was detected by PCR such that mice received a constant infusion of 100 mg/day of a 24-mer phosphorothioate oligomer consisting of either the antisense or control sequences. In the 2-week treatment study, mice receiving the antisense compound lived up to 42 days, whereas the control oligomer-treated and untreated mice had a mean survival of ~7 days, with the longest survival being only 11–17 days [32].

A. Pharmacokinetics and Biodistribution of Phosphorothioate Oligodeoxynucleotides in Animals

The *in vivo* efficacy (and toxicity) of any antisense agent is controlled not only by its ability to interact with a biologically relevant target, but also by its pharmacokinetics and biodistribution, which refer respectively to the time course of appearance in plasma and urine and the precentage of dose found in an organ at any given time. Early investigations by Iversen [33] used uniformly ^{35}S-labeled [S]ODNs to explore the pharmacokinetics of a 27-mer oligomer in rats following a relatively low-dose, single i.v. injection. Biphasic plasma elimination was characterized by $t_{1/2}a$ of 15–25 min for distribution out of the plasma compartment, and $t_{1/2}b$ of 20–40 hours for elimination from the body. Repeated daily injections of the 27-mer oligomer provided steady-state concentrations in 6–9 days, confirming the relatively long half-life measured for single injections. In a subsequent systematic extension of this work, Iversen [34] compared the pharmacokinetics of a 20-mer [S]ODN in mice (4 mg/kg s.c.), rats (3–300 mg/kg i.v.), monkeys (8.3 mg/kg/day intraatrial infusion for 6–15 days) and patients undergoing phase I treatment for AML (0.05– 0.25 mg/kg/h for 10 days). Steady-state plasma concentrations of 1.5–5.6 mM were reported for monkeys after 4–9 days. In patients, it was reported that 36% of the recovered material in urine and 90% of the material recovered in plasma was intact, whereas 53% of the material in urine was degraded to mononucleotide.

In parallel with the aforementioned studies, several research groups conducted detailed pharmacokinetic measurements with [S]ODNs in various animal models. Findings by Isis Pharmaceuticals, which have been reviewed by Crooke [35], indicated that serum protein binding provides a repository for [S]ODNs and prevents rapid renal excretion. Plasma-clearance parameters concur with those mentioned earlier, and broad distribution to all peripheral tissues was seen, with relatively high percentage-of-dose accumulation in liver, kidney, bone marrow skeletal muscle, and skin, but no significant penetration of the blood–brain barrier. Rates of incorporation and clearance from tissues are organ/dependent.

Pharmacokinetic and biodistribution findings generated with compounds tested by Hybridon have been reviewed by Agrawal and colleagues [36], and include results for rats, monkeys, and humans. The latter studies involve 2-hour i.v. infusions of a ^{35}S-labeled 25-mer [S]ODN to HIV-infected patients at a dose of 0.1 mg/ kg [37]. Plasma clearance was characterized by $t_{1/2}$a of ~12 min and $t_{1/2}$b of ~27 h, which agrees with the previously mentioned results in animal models. PAGE analysis of plasma indicated intact material, whereas radioactivity in the urine was associated with low-molecular-weight material. In a more recent publication related to this clinical study [38], it was reported that degradation of the [S]ODN found in plasma had occurred from the 3' end to give a series of shorter metabolites missing either 1, 2, or 3 nucleotide units.

Relatively early pharmacokinetic and biodistribution measurements in a SCID mouse model of human leukemia included distribution analysis of BCR/ABL antisense [S]ODNs in mouse tissues by DNA extraction and hybridization with a ^{32}P-labeled oligomer complementary to the injected [S]ODN [12]. This analysis revealed intact (i.e., hybridizable) antisense oligodeoxynucleotides throughout the body with accumulation in the liver at 24 hours and 72 hours after the last oligodeoxynucleotide injection. Moreover, intact oligodeoxynucleotides were detected in the kidney and liver up to 14 days after the last injection. Accumulation of the BCR/ABL phosphorothioate oligodeoxynucleotide in various organs was also assessed by measuring the amount of ^{35}S-labeled material in weighted organ samples; tissue concentrations correlated with the relative levels of intact oligodeoxynucleotides detected in the same tissues and ranged from 3 to 26 mM. Evidently, phosphorothioate oligodeoxynucleotides undergo relatively slow degradation in mouse tissues, and the 9-day treatment schedule in SCID mice appeared to reach tissue concentrations in every tissue, except brain, that would be sufficient to inhibit the growth of primary leukemic cells while sparing that of normal cells [10,17]. Accessibility of the brain to efficacious amounts of administered [S]ODNs may, however, be possible if there is disease-induced disruption of the blood–brain barrier. This possibility was first evidenced in an investigation of human leukemia–SCID mouse chimera treated by continuous s.c. infusion of an [S]ODN via a microosmotic pump [32], and most recently in a study of intracranial U-87 tumors in mice given daily i.p. injections of an [S]ODN [39]. Regarding infusion pumps, efficacy of an [S]ODN in a mouse model of Burkitt's lymphoma has also been reported [40] using a s.c. pump/implant analogous to that employed by Ratajczak *et al.* [32].

A very recent publication by Cotter and co-workers [41] has provided pharmacokinetic, biodistribution, and metabolic data for a ^{35}S-labeled [S]ODN after administration by i.v. injections or continuous s.c. infusion with a microosmotic pump. Interestingly, these data reportedly show that s.c. infusion resulted in less excretion and metabolism of the administered dose, although for both routes there was accumulation of radioactivity in liver and kidneys. Extensive biodistribution of [S]ODNs to liver and kidney, largely independent of sequence, length, and route of administration, has been confirmed recently by Rifai *et al.* [42], who reported that >85% of a ^{125}I-labeled 14-mer [S]ODN was processed by the liver and kidneys, which was 25-fold more than the amount in the gastrointestinal tract and 40- and 50-fold more than that in the heart, lungs, skin, muscle, and bone. Based on light and electron microscope autoradiographs, the administered material in the liver was said to be localized predominantly in sinusoidal cells, with uptake primarily by Kupffer's cells. Other investigations [43] using a ^{35}S-labeled 18-mer [S]ODN have confirmed that there is high accumulation in the kidney and have used PAGE to demonstrate intact material in this organ tissue, but not in urine. Light-microscopic autoradiography showed that the administered material was predominantly in the early proximal tubule. Rappaport *et al.* [44], who used a ^{32}P-labeled 20-mer [S]ODN to study in detail transport in kidney, suggested that the kidney was an excellent target for organ-directed antisense therapy but was also a site of [S]ODN toxicity (discussed later). In this regard, it remains to be seen whether formulation of [S]ODN can be used to maintain or enhance biodistribution to target organs, as well as lessen undesirable side effects. The first of such studies exemplified by formulations with cationic lipids have been reported for normal mice [45,46] and mice harboring LOX acites tumors [45].

In concluding this section, it is worth noting that various types of modified ODNs other than fully phosphorothioated [S]ODNs have now been investigated [47–49] with regard to the impact of alternative chemical modifications on pharmacokinetic and biodistribution parameters, as well as efficacy and toxicity.

B. Safety Studies in Animals and Humans

The only substantial body of data available to date on either tolerance or toxicity of antisense agents is for phosphorothioate oligodeoxynucleotide analogs. Overall, these data indicate that phosphorothioates are relatively well tolerated and nontoxic in rodents, rabbits, monkeys, and humans when administered either as single i.v. injections or as continuous i.v. infusion for up to 10 days. For example, in the case of a 20-mer phosphorothioate antisense oligodeoxynucleotide to a p53 mRNA, a Rhesus monkey given either a single i.v. dose of ~15 mg/kg or continuous i.v. infusion of ~2 g over 10 days (~2000 mg/day), showed no ill effects in behavior, food intake, excretion patterns, hematocrit, blood count, blood electrolytes, blood pressure, heart rate, and cardiac output (P. Iversen, personal communication). Other studies with Rhesus monkeys using a phosphorothioate complementary to c-myb mRNA, found no significant side effects in repeated cycles of treatment, namely, ~1 g given by continuous i.v. infusion for 7 days (~140 mg/day) repeated twice more at 21-day intervals (~3 g total over 3 months) (P. Iversen, personal communication).

Unlike some oligopeptides and protein drug products, no detectable antigenicity is associated with repeated administration of phosphorothioate oligodeoxynucleotides, perhaps due in part to their structural similarity to natural DNA, which is a very poor antigen. Nevertheless, certain sequences of [S]ODNs have been found to have immunostimulatory properties [50] that include amplification of antibody production [51].

Dose-dependent side effects observed in mice following repeated i.v. administration of [S]ODNs include splenomegaly, thrombocytopenia, elevation of the levels of liver enzymes, monocellular infiltrate in liver and kidney, and renal tubule necrosis [53]. Crooke [35] has concluded that, based on evaluation of a significant

number of [S]ODNs in animals, the most prominent toxicities in rodents appear to be related to induction of cytokine release. Local inflammation with T- and B-cell involvement at the injection site of an [S]ODN has been found after intradermal injections in rats [53], whereas in monkeys complement activation, abnormalities in clotting, and hypotension are found. A decrease in platelet levels that was found in HIV-infected patients after 10 days of [S]ODN administered at 3.2 mg/kg/day has recently led to discontinuation of a phase II study [54]. Despite the aforementioned side effects, potentially therapeutic dosages of various [S]ODNs have been found to be well tolerated by patients in various clinical settings. More important, there have been preliminary reports of efficacy in clinical trials of [S]ODNs involving CMV-related retinitis [54] and non-Hodgkin's lymphoma [55].

V. ONCOGENE-TARGETED ANTISENSE OLIGODEOXYNUCLEOTIDES: POTENTIAL CLINICAL APPLICATIONS IN HEMATOLOGIC MALIGNANCIES

Antisense oligodeoxynucleotides can in principle be added to the armamentarium of antileukemic drugs and utilized for *ex vivo* or systemic treatment of leukemia.

A. *Ex Vivo* Treatment

Among the various experimental applications of autologous cell therapy, *ex vivo* bone marrow purging has had a definite place in the treatment of several neoplasms, including acute and chronic leukemias [55]. In previous studies, the marrow has been treated in an effort to destroy leukemic cells with immunologic reagents [56] and chemotherapeutic drugs [57,58], and then reinfused into patients treated with ablative chemotherapy. Theoretically, antisense oligodeoxynucleotides targeted against an oncogene that confers a growth advantage to leukemic cells should prove therapeutically useful and, most important, more selective than conventional chemotherapeutic agents in killing leukemic cells while sparing normal progenitor cells. However, several issues must be addressed before devising effective protocols for *ex vivo* use of antisense oligodeoxynucleotide. One issue relates to the half-life of the mRNA target and, as a consequence, to the time of incubation of marrow cells in the presence of oligodeoxynucleotides. For example, the half-life of Myb protein

(10–30 min) is considerably shorter of that of p210$^{bcr/abl}$ protein (18–24 h), suggesting that a 24- to 48-hour incubation of marrow cells might be adequate if the target is c-*myb* but not *bcr-abl* mRNA. A second issue relates to the potential benefit of enriching hematopoietic progenitor cells before the *ex vivo* treatment to compensate for the relatively low proportion of clonogenic cells. The selection of such enriched progenitor cell populations (e.g., CD34+ cells) could also offset the likely differential uptake of oligodeoxynucleotides among marrow cells, which might result in ineffective targeting of leukemic cells. A protocol for *ex vivo* treatment of marrow cells enriched in CD34+ cells is now being implemented at the University of Pennsylvania under the direction of Alan Gewirtz.

Oncogene-targeted oligodeoxynucleotides might be utilized in combination with conventional purging agents under conditions that favor the killing of malignant cells and the sparing of a high number of normal progenitor cells. To this end, the bone-marrow-purging drug, mafosfamide, was utilized at low doses in combination with BCR/ABL antisense oligodexoynucleotides to eradicate Philadelphia[1] cells from a mixture of normal and leukemic cells. The full eradication of leukemic cells and the sparing of a significant number of normal progenitors was demonstrated by *in vitro* clonogenic assays and reconstitution experiments in immunodeficient mice [20].

B. Systemic Treatment

The SCID mouse models of human leukemia in the studies noted earlier offer the opportunity to evaluate the effectiveness of antisense oligodeoxynucleotides as a function of predetermined leukemic cell burden. As expected, BCR/ABL antisense oligodeoxynucleotides prolonged the survival of mice bearing Philadelphia[1] leukemic cells more effectively when 10^6 (compared to 10^7) BV173 cells were injected. Also, the antileukemic effects were more pronounced when the systemic treatment was initiated 1 week (compared to 3 weeks) after leukemic cell implantation. However, the leukemic process was only temporarily suppressed in SCID mice, perhaps reflecting an inadequate uptake of oligodeoxynucleotides by the leukemic cells or an inefficient treatment schedule. Repeated oligodeoxynucleotide injections may prolong survival or even cure leukemic mice if the tumor burden is greatly diminished. Alternatively, a "cocktail" of oligodeoxynucleotides or a combination of these reagents with low doses of conventional antitumor chemotherapeutic agents might enhance the therapeutic effect. In this regard, preliminary evidence suggests that, in combination, BCR/ABL and c-Myc antisense [S]ODNs exert a synergistic antiproliferative effect, *in vitro* and *in vivo,* using BV173 and CML-BC primary cells [20,21], at concentrations at which individual [S]ODNs were only partially effective or completely ineffective.

VI. USE OF ANTISENSE OLIGONUCLEOTIDES IN A CLINICAL SETTING

Early clinical experience with oligonucleotides has been reported by several groups. In the context of CML, we have reported in preliminary communication results of our clinical trials to evaluate the effectiveness of phosphorothioate-modified ODN antisense to the c-*myb* gene as marrow-purging agents for chronic phase (CP) or accelerated phase (AP) chronic myelogenous leukemia (CML) patients, and a phase I intravenous infusion study for blast crisis (BC) patients and patients with other refractory leukemias [59]. ODN purging was carried out for 24 hours on CD34+ marrow cells. Patients received busulfan and cytoxan, followed by reinfusion of previously cryopreserved P-ODN-purged MNC. In the pilot marrow-purging study 7 CP and 1 AP CML patients were treated, and 7/8 engrafted. In 4/6 evaluable CP patients, metaphases were 85–100% normal 3 months after engraftment, suggesting that a significant purge had taken place in the marrow graft. Five CP patients demonstrated marked, sustained, hematologic improvement with essential normalization of their blood counts. Follow-up ranged from 6 months to ~2 years. In an attempt to further increase purging efficiency we incubated patient MNC for 72 hours in the P-ODN. Although PCR and LTCIC studies suggested a very efficient purge had occurred, engraftment in 5 patients was poor. In the phase I systemic infusion study of 18 refractory leukemia patients (2 patients were treated at 2 different dose levels; 13 had AP or BC CML), Myb antisense ODN was delivered by continuous infusion at dose levels ranging between 0.3 and 2.0 mg/kg/day for 7 days). No recurrent dose-related toxicity has been noted, though idiosyncratic toxicities, not clearly drug related, were observed (1 transient renal insufficiency; 1 pericarditis). One BC patient survived ~14 months with transient restoration of CP disease. These studies show that ODN may be administered safely to leukemic patients. Whether patients treated on either study derived clinical benefit is uncertain, but the results of these studies suggest to us that ODN may eventually demonstrate therapeutic utility in the treatment of human leukemias.

Some clinical experience with BCR/ABL-targeted antisense oligonucleotides in CML has also been reported. de Fabritis and colleagues treated a patient with chronic myeloid leukemia in accelerated phase with autologous bone marrow transplantation [60]. Before reinfusion, cells were purged in vitro with a 26-mer phosphorothioate antisense oligodeoxynucleotide specific for the BCR exom 2/Abl exom 2 junction. This treatment resulted in a 24 and 41% reduction of CFU-GM and CD34+ cells, respectively. The patient was successfully engrafted with the purged marrow cells after 17 and 25 days for platelets and neutrophils, respectively. Using fluorescence *in situ* hybridization in interphase nuclei, some Ph-negative cells were found after the autograft. The patient was reported to be in a complete hematologic remission at 9 months posttreatment. A more recent study [61] in 8 patients with advanced CML confirmed the lack of toxicity of BCR/ABL ODNs used as purging agents and showed some therapeutic effects as indicated by karyotypic response and direction of a second chronic phase.

In the context of acute leukemia, Bayever *et al.* and Bishop *et al.* have reported using an antisense p53 oligonucleotide for bone marrow purging and for systemic therapy of treatment refractory leukemias [62,63]. In the most recent study, a 20-nucleotide phosphorothioate oligonucleotide complementary to p53 mRNA [OL(1)p53] was employed in a phase I dose-escalating trial conducted to determine the toxicity of the oligodeoxynucleotide following systemic administration to patients with hematologic malignancies. Sixteen patients with either refractory acute myelogenous leukemia ($n = 6$) or advanced myelodysplastic syndrome ($n = 10$) were treated. Patients were given OL(1)p53 at doses of 0.05 to 0.25 mg/kg/h for 10 days by continuous intravenous infusion. No specific toxicity directly related to the administration of the compound was observed. One patient developed transient nonoliguric renal failure; one patient died of anthracycline-induced cardiac failure. Approximately 36% of the administered dose of OL(1)p53 was recovered intact in the urine. Plasma concentrations and areas under the plasma concentration curves were linearly correlated with dose. Leukemic cell growth *in vitro* was inhibited as compared with pretreatment samples. There were no clinical complete responses. The authors concluded that this particular phosphorothioate oligonucleotide can be administered systematically without complications and speculated that it might prove useful in combination with currently available chemotherapy agents for the treatment of malignancies. Whether any of the effects reported was due to an antisense mechanism is uncertain because this was not specifically looked for.

Within the past year Webb and colleagues treated 9 patients with refractory non-Hogkin's lymphoma with a BCL-2 targeted phosphorothiate oligodeoxy-

nucleotide [54] based on preclinical studies, which suggested that the oligonucleotide could cause disease regression in an animal model [64]. The oligonucleotide employed was targeted to a portion of the BCL-2 mRNA open reading frame of the BCL-2 mRNA and was delivered as a daily subcutaneous infusion for 2 weeks. Toxicity was scored by the common toxicity criteria, and tumor response was assessed by computed tomography scan. This study was also more rigorously controlled in terms of mechanism because an antisense effect was specifically assessed by quantification of BCL-2 expression and BCL-2 protein levels in treated patients as assessed by flow cytometry. During the course of the study, the daily dose of BCL-2 antisense was increased incrementally from 4.6 mg/m² to 73.6 mg/m². Save for local inflammation at the site of infusion, no treatment-related toxic effects occurred. In two patients on the study, computed tomography scans showed reductions in tumor size that were stated to be "minor" and "major" respectively. In two patients, the number of circulating lymphoma cells also found to be decreased during treatment. Other indicators of response, such as decrease in LDH were also reported. BCL-2 protein levels were measured by flow cytometry in five patients and were found to be decreased in two. The authors concluded that in these patients the antisense therapy led to an improvement in symptoms and was accompanied by some objective biochemical and radiological evidence of tumor response. Based on more favorable responses then might have been predicted that authors also speculated that BCL-2 antisense therapy might also prime cells for an improved chemotherapeutic response. In common with the experience of others, then, the antisense compounds appear to be relatively nontoxic, and there is at least the suggestion that in some patients useful responses may be observed. However, these are likely due to a combination of aptameric and antisense effects.

VII. CONCLUSION

At the present time, the antisense approach links biology and medicine in a realistic hope of "rational therapy" of human malignancies. The rationale for using antisense oligonucleotides as oncogene-targeted therapeutic agents in human leukemias requires the demonstration that specific gene abnormalities have a role in maintaining the leukemic phenotype and that synthetic oligonucleotides preferentially or selectively affect the survival of leukemic cells. The results of preclinical investigations to assess the therapeutic potential of antisense oligodeoxynucleotides as *ex vivo* and *in vivo* antileukemia agents have been encouraging. However, the application of such compounds as therapeutic agents is still in its infancy. Only time will tell whether antisense compounds are effective as "magic bullets" [65,66] or adjuvants of conventional chemotherapy, or whether they have a more limited role in gene therapy.

Acknowledgments

Supported in part by grants from the National Institutes of Health to Drs. Calabretta and Gewirtz. Dr. Gewirtz is the recipient of a Translational Research Grant from the Leukemia Society of America.

References

1. Van der Kral, A. R., Mol. J. N. M., Sturtje, A. R., *et al.* (1988). Modulation of eukaryotic gene expression by complementary RNA or DNA sequences. *Biotechniques* **6**, 958–970.

2. Zamecnik, P., and Stephenson, M. (1978). Inhibition of Rous sarcoma virus replication and cell transformation by a specific oligodeoxynucleotide. *Proc. Natl. Acad. Sci. USA* **75**, 280–284.

3. Varmus, H. (1987). *Cellular and Viral Oncogenes. The Molecular Basis of Blood Diseases*. Philadelphia: W. B. Saunders.

4. Rowley, J. D. (1982). Identification of athe constant chromosome regions involved in human hematologic malignant diseases. *Science* **216**, 749–751.

5. Shtivelman, E., Lifshitz, B., Gale, R. B., Roe, B. A., and Canaani, E. (1986). Alternative splicing of RNAs transcribed from the human *abl* gene and from *bcr-abl* fused gene. *Cell* **47**, 277–284.

6. Fainstein, E., Marcell, C., Rosner, A., Canaani, E., Gale, R. P., Dreazen, O., Smith, S. D., and Croce, C. M. (1987). A new fused transcript in Philadelphia chromosome positive acute lymphocytic leukemia. *Nature* **330**, 386–389.

7. Daley, G. R., Van Etten, R. A., and Baltimore, D. (1990). Induction of chronic myelogenous leukemia in mice by the p120[bcr/abl] gene of the Philadelphia chromosome. *Science* **247**, 824–830.

8. Heisterkamp, N., Jenster, G., ten Hoeve, J., Zovich, D., Pattengale, P. K., and Groffin, J. (1990). Acute leukemia in BCR/ABL transgenic mice. *Nature* **344**, 251–253.

9. Elefanty, A. G., Hariharan, I. K., and Cory, S. (1990). *ber-abl,* the hallmark of chronic myeloid leukemia in man, induces multiple hematopoietic neoplasms in mice. *EMBO J.* **9**, 1069–1078.

10. Szczylik, C., Skorski, T., Nicolaides, N. C., Manzella, L., Malaguarnera, L., Venturelli, D., Gewirtz, A. M., and Calabretta, B. (1991). Selective inhibition of leukemia cell proliferation by BCR/ABL antisense oligodeoxynucleotides. *Science* **253**, 562–565.

11. Kamel-Reid, S., Letarte, M., Sirard, C., Doedens, M., Grunberger, T., Fulop, G., Freedman, M. H., Phillips, R. A., and Dick, J. A. (1989). A model of human acute lymphoblastic leukemia in immunodeficient SCID mice. *Science* **246**, 1597–1601.

12. Skorski, T., Nieborowska-Skorska, M., Nicolaides, N. C., Szczylik, C., Iversen, P., Iozzo, R. V., Zon, G., and Calabretta, B. (1994). Suppression of Philadelphia[1] leukemia cell growth in mice by BCR/ABL antisense oligodeoxynucleotides. *Proc. Natl. Acad. Sci. USA* **91**, 4504–4508.

13. Sawyers, C. J., McLaughlin, J., and Witte, O. N. (1995). Genetic requirement for *ras* in the transformation of fibroblasts and hematopoietic cells by the *bcr-abl* oncogene. *J. Exp. Med.* **181**, 307–313.

14. Goga, A., McLaughlin, J., Afar, D. E. H., Saffran, D. C., and Witte, O. N. (1995). Alternative signals to ras for hematopoietic transformation by the bcr-abl oncogene. *Cell* **82**, 981–988.

15. Skorski, T., Kanakaraj, P., Nieborowska-Skorska, M., Ratajczak, M. Z., Wen, S.-C., Zon, G., Gewirtz, A. M., Perussia, B., and Calabretta, B. (1995). Phosphatidylinositol-3 kinase activity is regulated by *bcr-abl* and is required for the growth of Philadelphia chromosome-positive cells. *Blood* **86**, 726–736.

16. Sawyers, C. L., Callahan, W., and Witte, O. N. (1992). Dominant negative Myc blocks transformation by *abl* oncogenes. *Cell* **70**, 901–910.

17. Skorski, T., Nieborowska-Skorska, M., Campbell, K., Iozzo, R. V., Zon, G., Darzynkiewicz, Z., and Calabretta, B. (1995). Leukemia treatment in SCID mice by antisense oligodeoxynucleotides targeting cooperating oncogenes. *J. Exp. Med.* **182**, 1645–1653.

18. Skorski, T., Nieborowska-Skorska, M., Wlodarski, P., Zon, G., Iozzo, R. V., and Calabretta, B. (1996). Antisense oligodeoxynucleotide combination therapy of primary chronic myelogenous leukemia blast crisis in SCID mice. *Blood* **88**, 1005–1012.

19. Agrawal, S., Temsamani, J., and Tang, J. Y. (1991). Pharmacokinetics, biodistribution, and stability of oligodeoxynucleotide phosphorothioates in mice. *Proc. Natl. Acad. Sci. USA* **88**, 7595–7599.

20. Skorski, T., Nieborowska-Skorska, M., Barletta, C., Malaguarnera, L., Szczylik, C., Chen, S.-T., Lange, B., and Calabretta, B. (1993). Highly efficient elimination of Philadelphia[1] leukemia cells by exposure to BCR/ABL antisense oligodeoxynucleotides combined with mafosfamide. *J. Clin. Invest.* **92**, 194–202.

21. Skorski, T., Nieborowska-Skorska, M., Wlodarski, P., Perrotti, D., Hoser, G., Kawick, J., Majewski, M., Christensen, L., Iozzo, R. V., and Calabretta, B. (1997). Treatment of Philadelphia[1] leukemia in SCID mice by combination of cyclophosphamide and BCR/ABL antisense oligodeoxynucleotide. *J. Natl. Cancer Inst.* **89**, 124–133.

22. Oltvai, Z. N., Milliman, C. L., and Korsmeyer, S. J. (1993). Bcl-2 heterodimerizes in *vivo* with a conserved homology, Bax, that accelerates programmed cell death. *Cell* **74**, 609–619.

23. Miyashita, T., and Reed, J. C. (1995). Tumor suppressor p53 is a direct transcriptional activator of the human *bax* gene. *Cell* **80**, 293–299.

24. Calabretta, B., and Nicolaides, N. C. (1992). c-*myb* and growth control. *Crit. Rev. Eukaryot. Gene Expr.* **2**, 225–235.

25. Todokoro, K., Watson, R. J., Higo, H., Amanuma, H., Kuramuchi, S., Yanagisawa, H., and Ikawa, Y. T. (1988). Downregulation of c-*myb* gene expression is a requisite for erythropoietin-induced erythroid differentiation. *Proc. Natl. Acad. Sci. USA* **85**, 8900–8904.

26. Clarke, M. F., Kukowska-Latallo, J. F., Westin, E., Smith, M., and Prochownick, E. V. (1988). Constitute expression of a c-*myb* cDNA blocks Friend murine erythroleukemia cell differentiation. *Mol. Cell. Biol.* **8**, 884–892.

27. Gewirtz, A. M., and Calabretta, B. (1988). A c-Myb antisense oligodeoxynucleotide inhibits normal human hematopoiesis in vitro. *Science* **242**, 1303–1306.

28. Anfossi, G., Gewirtz, A. M., and Calabretta, B. (1989). An oligomer complementary to c-*myb* encoded mRNA inhibits proliferation of human myeloid leukemia cell lines. *Proc. Natl. Acad. Sci. USA* **86**, 3779–3383.

29. Calabretta, B., Sims, R. R., Valtieri, M., Caracciolo, D., Szczylik, C., Venturelli, D., Beran, M., and Gewirtz, A. M. (1991). Normal and leukemic hematopoietic cells manifest differential sensitivity to inhibitory effects of c-Myb antisense oligodeoxynucleotides: an *in vitro* study with relevance to bone marrow purging. *Proc. Natl. Acad. Sci. USA* **88**, 2351–2355.

30. Hijiya, N., Zhang, J., Ratajczak, M. Z., Kant, J. A., de Riel, K., Herly, M., Zon, G., and Gewirtz, A. M. (1994). Biologic and therapeutic significance of Myb expression in human melanoma. *Proc. Natl. Acad. Sci. USA* **91,** 4499–4503.

31. Del Bufalo, D., Cucco, C., Leonetti, C., D'Agnano, I., Amedeo, C., Geiser, T., Calabretta, B., and Zupi, G. (1996). Effect of cisplatin and a c-Myb antisense phosphorothioate oligonucleotide combination on a human colon cancer cell line *in vitro* and *in vivo. Br. J. Cancer* **74,** 387–393.

32. Ratajczak, M. Z., Kant, J. A., Luger, S. M., Hijiya, N., Zhang, J., Zon, G., and Gewirtz, A. M. (1992). *In vivo* treatment of human leukemia in a SCID mouse model with c-Myb antisense oligodeoxynucleotides. *Proc. Natl. Acad. Sci. USA* **89,** 11823–11837.

33. Inversen, P. (1991). *In vivo* studies with phosphorothioate oligonucleotides: pharmacokinetics prologue. *Anticancer Drug Des.* **6,** 531–538.

34. Iversen, P., Copple, B. L., and Tewery, H. K. (1996). Pharmacology and toxicology of phosphorothioate oligonucleotides in the mouse, rat, monkey, and man. *Toxicol. Lett.* **82/83,** 425–530, 1995.

35. Crooke, S. T. (1996). Progress in antisense therapeutics. *Med. Res. Rev.* **10,** 319–344.

36. Agrawal, S., Temsamani, J., Galbraith, W., and Tang, J. (1995). Pharmacokinetics of antisense oligonucleotides. *Clin. Pharmacol. Kinet.* **28,** 7–16.

37. Zhang, R., Yan, J., Shahinian, H., Amin, G., Zhihong, L., Liu, T., Saag, M., Jiang, Z., Temsamani, J., Martin, R., Schechter, P., Agrawal, S., and Diasio, R. (1995). Pharmacokinetics of an anti-human immuno-deficiency virus antisense oligodeoxynucleotide phosphorothioate (GEM 91) in HIV-infected subjects. *Clin. Pharmacol. Ther.* **58,** 44–53.

38. Temsamani, J., Roskey, A., Choix, C., and Agrawal, S. (1996). *In vivo* metabolic profile of a phosphorothioate oligodeoxynucleotide. *Antisense Nucleic Acid Drug Dev.* **7,** 159–165.

39. Glazer, R. I. (1997). Protein kinase C as a target for cancer therapy. *Antisense Nucleic Acid Drug Dev.* **7,** 235–238.

40. Huang, Y., Snyder, R., Kligshteyn, M., and Wickstrom, E. (1995). Prevention of tumor formation in a mouse model of Burkitt's lymphoma by 6 weeks of treatment with anti-c-*myc* DNA phosphorothioate. *Mol. Med.* **1,** 647–658.

41. Raynaud, F., Orr, R., Goddard, P., Lacey, H., Lancashire, H., Judson, I., Beck, T., Bryan, B., and Cotter, F. (1997). Pharmacokinetics of G3139, a phosphorothioate oligodeoxynucleotide antisense to *bcl-2,* after intravenous administration or continuous subcutaneous infusion to mice. *J. Pharm. Exp. Therapeut.* **281,** 420–427.

42. Rifai, A., Brysch, W., Fadden, K., Clark, J., and Schlingensiepen, K.-H. (1996). Clearance kinetics, biodistribution, and organ saturability of phosphorothioate oligodeoxynucleotides in mice. *Am. J. Pathol.* **149,** 717–725.

43. Oberbauer, R., Schreiner, G., and Meyer, T. (1995). Renal uptake of an 18-mer phosphorothioate oligonucleotide. *Kidney Int.* **48,** 1226–1232.

44. Rappaport, J., Hanss, B., Kopp, J., Copeland, T., Bruggeman, L., Coffman, T., and Klotman, P. (1995). Transport of phosphorothioate oligonucleotides in kidney: implications for molecular therapy. *Kidney Int.* **47,** 1462–1469.

45. Saijo, Y., Perlaky, L., Wang, H., and Busch, H. (1994). Pharmacokinetics, tissue distribution, and stability of antisense oligodeoxynucleotide phosphorothioate ISIS 3466 in mice. *Oncol. Res.* **6,** 243–249.

46. Bennett, C. F., Zuckerman, J., Kornbrust, D., Sasmor, H., Leeds, J., and Crooke, S. (1996). Pharmacokinetics in mice of a [³H]-labeled phosphorothioate oligonucleotide formulated in the presence and absence of a cationic lipid. *Tibtech* **41,** 121–130.

47. Crooke, S., Graham, M., Zuckerman, J., Brooks, D., Conklin, B., Cummins, L., Grieg, M., Guinosso, C., Kornbrust, D., Manoharan, M., Sasmor, H., Schleich, T., Tivel, K., and Griffey, R. (1996). Pharmacokinetic properties of several novel oligonucleotide analogs in mice. *J. Pharm. Exp. Therapeut.* **277,** 923–937.

48. Skorski, T., Perrotti, D., Nieborowska-Skorska, M., Gryaznov, S., and Calabretta, B. (1997). Antileukemia effect of c-*myc* N3'P5' phosphoramidate antisense oligonucleotides *in vivo. Proc. Natl. Acad. Sci. USA* **94,** 3966–3971.

49. Agrawal, S., Jiang, Z., Zhao, Q., Shaw, D., Cai, Q., Roskey, A., Channavajjala, L., Saxinger, C., and Zhang, R. (1997). Mixed-backbone oligonucleotides as second generation antisense oligonucleotides: *in vitro* and *in vivo* studies. *Proc. Natl. Acad. Sci. USA* **94,** 2620–2625.

50. Krieg, A., Ae-Kyung, Y., Matson, S., Waldschmidt, T., Bishop, G., Teasdale, R., Koretzky, G., and Klinman, D. (1995). CpG motifs in bacterial DNA trigger direct B-cell activation. *Nature* **374,** 546–548.

51. Branda, R., Moore, A., Lafayette, A., Matthews, L., Hong, R., Zon, G., Brown, T., and McCormack, J. (1996). Amplification of antibody production by phosphorothioate oligodeoxynucleotides. *J. Lab. Clin. Med.* **128,** 329–338.

52. Henry, S., Grillone, L., Orr, J., Bruner, R., and Kornbrust, D. (1997). Comparison of the toxicity profiles of ISIS 1082 and ISIS 2105, phosphorothioate oligonucleotides, following subacute intradermal administration in Sprague-Dawley rats. *Elsevier Tibtech* **116,** 77–88.

53. Henry, S., Giclas, P., Leeds, J., Pangburn, M., Auletta, C., Levin, A., and Kornbrust, D. (1997). Activation of the alternative pathway of complement by a phosphorothioate oligonucleotide: potential mechanism of action. *J. Pharm. Exp. Therapeut.* **281,** 810–816.

54. Webb, A., Cunningham, D., Cotter, F., Clarke, P. A., DiStefano, F., Ross, P., Corbo, M., and Dziewenowska, Z. (1997). BCL-2 antisense therapy in patients with non-Hodgkin's lymphoma. *Lancet* **349,** 1137–1141.

55. Santos, G. W. (1990). Bone marrow transplantation in hematologic malignancies: current status. *Cancer* **65,** 786–791.

56. Ball, E. D. (1988). *In vitro* purging of bone marrow for autologous marrow transplantation in acute myelogenous leukemia using myeloid specific monoclonal antibodies. *Bone Marrow Transplant.* **3,** 387–393.

57. Yeager, A. M., Kaizer, H., Santos, G. W., Saral, R., Colvin, O. M., Stuart, R. K., Braine, H. G., Burke, P. J., Ambinder, R. F., Burns, W. H., Fuller, D. J., Davis, J. M., Karp, J. E., Stratford, M. W., Rowley, S. D., Senebremer, L. L., Vogelsong, G. B., and Wingard, J. R. (1986). Autologous bone marrow transplantation in patients with acute nonlymphocytic leukemic using *ex vivo* marrow treatment with 4-hydroperoxycyclophos-phamide. *N. Engl. J. Med.* **315,** 141–145.

58. Gorin, M. C., Douay, L., Laporte, J. P., Lopez, M., Mary, J. Y., Majman, A., Salmon, O., Aegerter, P., Stachowak, J., David, R., Pene, F., Kantor, G., Deloux, J., Duhamel, E., van den Akker, J., Gerota, J., Parlier, Y., and Duhamel, G. (1986). Autologous bone marrow transplantation using marrow incubated with Asta Z 7557 in adult acute leukemia. *Blood* **67,** 1367–1373.

59. Gewirtz, A. M., Luger, S., Sokol, D., Gowdin, B., Stadtmauer, E., Reccio, A., nd Ratajczak, M. Z. (1996). Oligodeoxynucleotide therapeutics for human myelogenous leukemia: interim results. *Blood* **88** Suppl. 1 (10), 270a.

60. de Fabritiis, P., Amadori, S., Petti, M. C., Mancini, M., Montefusco, E., Picardi, A., Geiser, T., Campbell, K., Calabretta, B., and Mandelli, F. (1995). *In vitro* purging with BCR/ABL antisense oligodeoxynucleotides does not prevent haematologic reconstitution after autologous bone marrow transplantation. *Leukemia* **9**(4): 662–664.

61. de Fabritiis, P., Petti, M. C., Montefusco, E., De Propris, M. S., Salie, R., Mancini, M., Lisci, A., Bonetto, F., Geiser, T., Celebretta, B., and Mandelli, F. (1998). BCR-ABL antisense oligodeoxynucleotide in vitro purging and autologous bone marrow transplantation for patients with chronic myelogenous leukemia in advanced phase. *Blood* **91**, 3156–3161.

62. Bayever, E., Iversen, P. L., Bishop, M. R., Sharp, J. G., Tewary, H. K., Arneson, M. A., Pirruccello, S. J., Ruddon, R. W., Kessinger, A., Zon, G., *et al.* (1993). Systemic administration of a phosphorothioate oligonucleotide with a sequence complementary to p53 for acute myelogenous leukemia and myelodysplastic syndrome: initial results of a phase I trial. *Antisense Res. Dev.* **3**(4), 383–390.

63. Bishop, M. R., Iversen, P. L., Bayever, E., Sharp, J. G., Greiner, T. C., Copple, B. L., Ruddon, R., Zon, G., Spinolo, J., Arneson, M., Armitage, J. O., and Kessinger, A. (1996). Phase I trial of an antisense oligonucleotide OL(1)p53 in hematologic malignancies. *J. Clin. Oncol.* **14**(4), 1320–1326.

64. Cotter, F. E., Johnson, P., Hall, P., Pocock, C., al Mahdi, N., Cowell, J. K., and Morgan, G. (1994). Antisense oligonucleotides suppress B-cell lymphoma growth in a SCID-hu mouse model. *Oncogene* **9**(10), 3049–3055.

65. Stein, C. A., and Cheng, Y.-C. (1993). Antisense oligonucleotides as therapeutic agents. Is the bullet really magical? *Science* **261**, 1004–1012.

66. Ma, D. D. F., and Doan, T. L. (1994). Antisense oligonucleotide therapies: are they the "magic bullet"? *Ann. Intern. Med.* **120**, 161–166.

CHAPTER 15

Selectively Replicating Viruses as Therapeutic Agents Against Cancer

DAVID KIRN

Director of Clinical Research, ONYX Pharmaceuticals, Richmond, California 94806; Assistant Professor (Adjunct), University of California, San Francisco, California 94143

I. INTRODUCTION

The vast majority of human cancers are incurable once metastatic. Chemotherapy and radiotherapy can induce tumor growth inhibition or regression in some cases, but solid tumor progression and resistance to these standard therapeutic modalities inevitably develops. Further dose escalation to overcome resistance is generally limited by the resultant toxicity to normal tissues. Therefore, new agents with larger therapeutic indices between cancer cells and normal cells are needed. Such agents would have improved efficacy and/or reduced toxicity when compared to currently available modalities.

Viruses have been used as gene delivery vectors to cause cancer cell inhibition or killing. These viruses have been constructed to deliver tumor suppressor genes, antisense constructs for specific oncogenes, immunostimulatory genes, or drug-activating enzymes. One of the major difficulties with this approach, however, is the daunting goal of delivering genes to every cancer cell in the body. Although local effects on noninfected tumor cells (i.e., bystander effects) might be possible with drug-activating enzyme systems [1] or immunostimulatory gene therapy, most gene therapy approaches will have effects that are limited almost exclusively to cells that are initially infected. In contrast, a replicating viral therapeutic agent can overcome this limitation. Virus replication in a small fraction of the tumor cells leads to amplification and spread of the antitumoral effect [2]. Cell killing can be due to viral replication and cell lysis exclusively, or this could be augmented by including additional immunostimulatory or toxin-producing genes.

The idea that viruses might be used as selective anticancer agents dates back almost a century. Patients with various malignancies were noted to have transient remissions following viral illnesses or rabies vaccination [3]. Although direct viral lysis of cancer cells was not proven in these patients, these case reports set off a flurry of experiments aimed at identifying oncolytic viruses. By 1970, an estimated 38 different viruses with *in vivo* antitumoral activity in animals or man had been identified [4–7].

Why then, given the favorable attributes of a selectively replicating oncolytic virus, was this approach not pursued more aggressively over the last two to three decades? First of all, the vast majority of patients in the earlier clinical trials had end-stage cancer and life expectancies of less than 3 months. Even the most efficacious chemotherapies currently available would look dismal in this patient population. In addition, patients with multiple tumor types were treated with widely varying virus dosages and routes of administration, thus making useful and definitive conclusions impossible. In addition to these methodological problems, technical limitations with virus manufacturing and purification led to suboptimal dosing. Poorly purified crude cell lysates were used, which limited the amount of virus that could be administered. Although some of the viruses resulted in little or no toxicity [6,7], other agents (e.g., West Nile virus) were not selective enough and severe toxicity to normal tissues occurred [8]. Finally, the mechanisms leading to efficacy, tumor resistance, and toxicity were left unexplored in most trials.

Over the last several decades there has been an explosion of knowledge in molecular biology. Many viruses have been characterized genetically, and the specific gene products responsible for necessary viral functions have been identified. For example, the genes necessary for control of the cell cycle, pathogenesis, and avoidance of cellular immunity have been identified for a number of viruses. Genetic engineering of viruses is now possible, allowing deletion or augmentation of specific viral functions. Development of virus purification and concentration techniques has facilitated large-scale production of high-titer purified virus in many cases. The in-crease in our knowledge of immunology will allow identification of the mechanisms limiting virus efficacy [9]. The virus and host mechanisms modulating the immune response can be modified to decrease antiviral immunity and consequently to enhance efficacy.

II. HUMAN EXPERIENCE WITH ONCOLYTIC VIRUSES

Between 1950 and 1975, a large number of oncolytic viruses were studied in human cancer patients, including adenovirus, Bunyamwara, coxsackie, dengue, feline panleukemia, Ilheus, mumps, Newcastle Disease virus (NDV), vaccinia, and West Nile. One of the most illustrative studies was performed at the U.S. National Cancer Institute in 1956 [10]. Wild-type human adenoviruses of ten different serotypes were used. Thirty patients with locally advanced cervical carcinoma were treated with direct intratumoral injection, with intra-arterial injection (of the artery perfusing the tumor), or by both routes. The patients were treated with a serotype to which they had no neutralizing antibodies, when possible, and corticosteroids were coadministered in roughly half of the cases. Sixty-five percent of patients had a "marked to moderate" (not strictly defined) local tumor response, with liquefaction and ulceration of the injected tumor mass. None of the seven control patients treated with either virus-free tissue culture fluid or heat-inactivated virus responded. Responses were only seen with replication-competent virus. Neutralizing antibodies developed within 7 days, but viral replication and efficacy persisted for up to several weeks thereafter. Viable infectious adenovirus was cultured out of the tumor in two thirds of the cases and was present for up to 17 days postinoculation. Three patients on steroids had a viral syndrome lasting 2–7 days; this viral syndrome resolved spontaneously. No toxicity was noted systemically or to adjacent normal tissues. However, the total dose of virus administered was probably many orders of magnitude less than is possible with currently available adenovirus production and purification techniques. In addition, all treatments were administered locally. Therefore, treatment with much higher doses of adenovirus, either locally or especially intravenously, are likely to result in substantially greater toxicity. Nevertheless, this study illustrates the promise of oncolytic viruses. Tumor lysis and shrinkage can be achieved, and persistent localized virus replication was possible, even without immunosuppression. In similar studies with other viruses, however, toxicity was seen if the natural pathogenicity of the virus was not attenuated.

III. SELECTION OF A CANDIDATE VIRUS FOR CLINICAL USE

When considering a virus candidate for development as an oncolytic therapeutic agent, any potential efficacy, safety, and manufacturing issues need to be considered. First and foremost, the virus must replicate in and lyse human tumor cells. The magnitude of oncolysis can be altered, but the parent virus should ideally be intrinsically lytic. Second, the parent virus should preferably cause only mild well-characterized human disease. Therefore, any reversion to wild-type or contamination with wild-type virus will not pose serious risks to the patient or the public. If effective treatment for the virus is not available, then a "suicide" drug-activating gene (e.g., herpes simplex virus thymidine kinase) can be added to the genome for an added level of safety. Knowledge of the genes modulating replication (e.g., adenovirus early region-1 genes) or pathogenesis (e.g., HSV-1 neurovirulence gene g34.5) allows precise modification of the virus to improve selectivity and safety. A genetically stable, nonintegrating virus has obvious safety advantages as well. Finally, high-titer virus production amenable to large-scale manufacturing for clinical use is critical.

IV. POTENTIAL MECHANISMS FOR TUMOR-SELECTIVE REPLICATION

The life cycle of a DNA virus includes the following sequential steps: cell surface binding, entry, uncoating, early gene expression (often leading to cell cycle progression into the S phase), DNA replication, virus packaging, and release from the cell. Restriction of viral gene expression or replication can be accomplished by the modification of several of these steps.

A. Selective Cell Entry

The tissue tropism of certain viruses can be altered by modifying surface proteins responsible for binding to target cells. Both antibodies and receptor ligands have been fused to viral coat proteins for targeting purposes. Viral envelope protein–single-chain antibody fragments have been used to target retroviruses to antigen-bearing cells [11]. Another group fused erythropoietin to a retroviral particle, resulting in selective uptake into erythroid precursor cells expressing the erythropoietin receptor [12]. The same group subsequently achieved specific infection of HER2/*neu* overexpressing cells by fusing the heregulin molecule to retro-

viral coat proteins [13]. Therefore, the tissue specificity of a replication-competent virus might be determined through selective cell entry. Whether this approach is possible with DNA viruses remains to be seen. Cellular uptake has been improved, albeit nonspecifically, through adenoviral coat protein modification resulting in binding to heparin sulfate [14].

B. Selective Transcription of Genes Necessary for Replication

Selectivity can be achieved by increasing the expression of genes necessary for replication in tumor cells or by decreasing expression of such genes in normal cells. Tumor-specific viral gene expression might be accomplished through the use of tissue-specific promoters [15,16]. Such a promoter would be activated preferentially in the tissue of tumor origin. Tumors arising from nonvital organs might be amenable to such an approach (e.g., prostate cancer or melanoma). Tissue-specific gene expression has been documented using adenovirus and retroviral vectors. The E1B promoter was replaced with albumin or immunoglobulin promoters, resulting in liver- or myeloma-specific E1B mRNA transcription, respectively [16,17]. The E1A enhancer can markedly increase transcription from the albumin promoter, particularly in the presence of the E1A protein [16]. Two groups have applied this approach to replicating adenoviruses. The prostate-specific promoter and the alpha-fetoprotein promoter have each been exchanged with the E1A promoter to achieve enhanced replication in cells from the prostate (cancer and normal cells) and in hepatocellular carcinomas, respectively. How selective these promoters are for cancer cells versus normal cells remains to be proven *in vivo* (e.g., regenerating liver expresses alpha-fetoprotein). Therefore, this approach might be used to express genes necessary for replication only in specific tissues.

C. Deletion of Genes Necessary for Replication in Normal Cells but Not in Tumor Cells

Another approach to abrogating replication in normal cells, but not in tumor cells, is to delete/mutate genes necessary for replication in normal cells that are dispensible in tumor cells. For example, HSV-1 ribonucleotide reductase and thymidine kinase are essential for viral DNA synthesis in quiescent cells, whereas in rapidly proliferating cells these enzymes are not necessary [18,19]. In a clinical situation such as primary brain cancer, the tumor cells are proliferating and therefore are susceptible to replication and lysis by such deletion mutants. In contrast, the surrounding neurons are quiescent and resistant to viral replication. However, this approach results in a proliferating-cell-specific, and not a cancer-specific, viral therapeutic agent. Another example of the genetic deletion approach is seen with ONYX-015 (dl1520), an attenuated adenovirus that lacks expression of the E1B-55-kDa protein. The binding of this protein to p53 is necessary for efficient replication in normal cells [20]. However, E1B-55-kDa is dispensable in tumor cells lacking functional p53 [21]. p53 function is lost in over 50 percent of all human cancers. Similarly, the deletion of specific RB-binding sites in the E1A protein may lead to selective replication in cancer cells with abnormalities in the RB pathway and in proliferating normal cells (D. Kirn *et al.,* ONYX Pharmaceuticals, in press). Such attenuation of viruses may potentially decrease replication in tumor cells as well, but to a lesser degree than in normal cells, if the deleted gene product serves multiple functions [2,18]. Replication competence can be better maintained by selectively mutating the gene rather than deleting it completely. For example, second generation adenovirus E1B-55-kDa gene mutants have been constructed that still lack p53-binding but retain the viral mRNA transport function (which is lost with the total E1B-55-kDa gene deletion); these viruses replicate more efficiently in p53-deficient cells than do mutants completely lacking the 55-kDa gene (T. Hermiston, personal communication).

V. MECHANISMS LEADING TO EFFICACY *IN VIVO*

Oncolytic viruses can lead to tumor destruction through several distinct mechanisms (Table 1). Direct cell lysis secondary to virus replication and shedding has been demonstrated *in vitro* and *in vivo* with numerous viruses [2,5,22]. The exact mechanism by which lytic viruses induce cell membrane rupture is unclear, but the result is a single large burst of virus released upon cell lysis.

A second mechanism involves the production of a toxic protein during the replication cycle in tumor cells. For example, the autonomous parvoviruses (e.g., MVMP, H-1) encode NS-1, a nonstructural protein that can induce apoptosis of infected cells [23]. Although this protein is produced during virus replication in tumor cells, NS-1 can kill the cell independent of replication. Adenovirus encodes a "death protein" that causes cell death approximately 48 hours after infection [24].

Nonspecific immunologic mechanisms can also play a role in tumor destruction following tumor-specific viral replication. Infection of a tumor cell with a replicating virus can increase the cell's sensitivity to killing by cytokines such as TNF-α and interferons. Newcastle disease virus and adenovirus, for example, increase tumor cell killing by TNF-α *in vitro* [25,26]. *In vivo* studies with HSV-1 mutants in immunocompetent rats also suggest that nonspecific immunologic mechanisms may mediate tumor destruction. Tumor inhibition is associated with minimal evidence of viral replication [27], in contrast to the active replication seen in athymic mouse–tumor xenograft models.

Finally, the expression of viral antigens in a tumor cell and their presence on the cell surface bound to MHC class I complexes can lead to tumor cell destruction by antigen-specific cytotoxic T lymphocytes. An intriguing possibility is that tumor-selective viral replication may induce or augment the development of tumor-specific immunity [28,29]. Tumor infection will induce infiltration by antigen-presenting cells and lymphocytes while inducing local cytokine production (e.g., TNF-α) [25].

TABLE 1   Potential Mechanisms of Antitumoral Efficacy with
Oncolytic Viruses

Mechanism	Examples
1. Direct cell lysis during viral shedding.	Herpes simplex type 1 Adenovirus
2. Direct cytotoxicity due to viral protein production.	Nonstructural proteins of autonomous parvovirus
3. Increased sensitivity to antitumoral cytokines (e.g., TNF).	Newcastle Disease virus Adenovirus
4. Augmentation of antitumoral immunity. Immune cell infiltration Tumor cell lysis, antigen release Immunostimulatory cytokine production Antitumoral cytokine production (e.g., TNF)	Adenovirus Unknown Adenovirus Newcastle Disease virus

Note. TNF, tumor necrosis factor-alpha

Tumor specific T lymphocytes may be activated in such a milieu.

VI. FACTORS POTENTIALLY INFLUENCING CLINICAL EFFICACY (Table 2)

A. Antiviral Immunity

The antiviral immune response will be a major determinant of the efficacy and toxicity of oncolytic viruses (Table 3). The timing and magnitude of viral antigen-specific humoral and cell-mediated immunity, as well as induction of nonspecific elements (e.g., cytokines, phagocytes), will vary from virus to virus. Virus characteristics such as cytopathogenicity, kinetics of replication, and tissue distribution will greatly influence immune response characteristics.

Primary infection with a cytopathic virus activates both the humoral and cellular arms of the immune system. Neutralizing antibodies are generated to specific antigenic determinants within 7 to 14 days. Binding of antibody to virus is able to prevent infection of target cells [9]. Therefore, the development of neutralizing antibodies may eventually limit virus dissemination to distinct tumor deposits and decrease the efficacy of repeat intravenous treatment. However, the efficacy of intratumoral injections with a viral agent will not necessarily be limited by neutralizing antibodies due to their poor penetration into solid tumor masses (D. Kirn, unpublished data). The induction of virus-specific cytotoxic T lymphocytes could potentially result in the kill-ing of infected cells prior to completion of the viral replication cycle [30]. Through evolution, however, viruses have developed mechanisms to avoid immunologic recognition and clearance [26,31,32]. Modulation of these viral gene functions alters immune response characteristics. The net effect of cell-mediated immunity on antitumoral efficacy is therefore unclear and will require further study. Several additional strategies exist to deal with these responses should they be problematic clinically, including immunomodulation.

Repeat infection following neutralizing antibody development has been accomplished. Influenza viruses, for example, modulate their antigenic coat proteins and thereby cause recurrent infections in the host [9]. Using a similar approach, investigators have shown that switching adenovirus serotypes allows for repeat gene delivery to the airways of rats preimmunized with another serotype. In addition to modifying antigenic viral coat proteins, the host immune response can also be altered to minimize neutralizing antibody development. Interferon-gamma and interleukin-12, for example, have been shown to prevent mucosal antibody development and to allow repeat gene delivery when coadministered with an adenoviral vector [33]. Finally, the accessibility of replicating virus to antibody binding should be minimal following direct intratumoral injection. Therapeutic antibody studies have shown that antibodies do not effectively penetrate the core of a solid tumor; extravasation is primarily in the tumor periphery.

The development of virus-specific cell-mediated immunity can also be reduced through both virus- and host-mediated mechanisms. Many viruses encode for proteins that down-regulate viral protein expression on the surface of infected cells. The adenovirus E3-gp-19-kDa protein, for example, binds the major histocompatibility complex (MHC) class I proteins and suppresses their expression on the cell surface. Virus-specific cytotoxic T-lymphocyte recognition is thereby minimized [34]. The HSV-1 protein ICP47 has a similar function [9]. In addition to viral mechanisms, the host cellular immune response can be suppressed through agents such as CTLA4Ig that target T lymphocytes.

Finally, antiviral cytokine effects might also be diminished through up-regulation of protective viral genes. Tumor necrosis factor (TNF) has been shown to inhibit the replication of many viruses *in vitro*. Infected cell lysis is also mediated by TNF. Many viruses contain genes that encode for cytokine-inhibitory proteins. For example, the adenovirus E3 proteins 14.7 kDa and 10.4 kDa/14.5 kDa act to protect virus-infected cells from TNF-mediated cytolysis [26]. Once again, however, the net effect of protecting infected tumor cells from

TABLE 2   Factors Influencing Clinical Efficacy of Oncolytic Viruses

Virus
  Physical characteristics affecting intratumoral penetration, spread (e.g., size).
  Time from infection to viral replication, cell lysis.
  Number of virions produced per infected cell.
  Antigenicity (induction of humoral and cellular immunity).

Tumor
  Intratumoral virus distribution: dependent on distribution of vasculature, leakiness of vessels to virus.
  Physical barriers to virus spread: intermixed normal fibroblasts, fibrosis.

Host
  Immune response
    cell-mediated immunity
    humoral immunity
    timing of immune response
  Reticuloendothelial cell uptake of virus from circulation.

TABLE 3   Viral Therapy and Immune Responses

Potentially beneficial immune responses	Potentially detrimental immune responses
Induction of tumor-specific CTL. Induction of susceptibility to TNF-α or IFN-α Induction of virus-specific CTL.	Antibody binding, prevention of virus entry into cell. Reticuloendothelial cell uptake of virus, prevention of distribution of virus via circulation to multiple distant tumor sites. CTL-mediated lysis of infected tumor cells before virus replication or spread.
Approaches to increase effect Select "antigenic" tumor types in nonimmunosuppressed cancer patients. Delete *ICP47* (HSV-1) and the gene encoding E3-19-kDa (Ad) (gene products inhibit viral antigen presentation). Addition of cytokine genes (such as that encoding IL-2) to replicating virus genome. Delete genes encoding E3-14.7-kDa/14.5-kDa (TNF-α signal transduction inhibitors). Immunomodulation to increase cell-mediated immunity (such as IL-12, IFN-γ).	**Approaches to minimize effect** Direct intratumoral administration (antibodies do not penetrate solid tumors efficiently). Serotype switching, selection to avoid serotypes to which the patient has pre-existing or treatment-induced antibodies. Immunosuppression (e.g., corticosteroids, cyclosporine), immunomodulation (e.g., IL-12, IFN-gamma). Formulation change (e.g., liposome encapsulation). Reticuloendothelial cell blockade (such as intravenous IgG). Increase expression of the E3-19-kDa gene (Ad), and of *ICP47* (HSV-1) (decrease viral antigen presentation).

Note. Ad, adenovirus; CTLs, cytotoxic T lymphocytes; HSV-1, herpes simplex virus 1; IFN, interferon; IL, interleukin; TNF, tumor necrosis factor.

cytokine-mediated killing is unclear and further *in vivo* studies will be needed to determine the best strategy.

B. Physical Barriers to Virus Distribution in Tumors

The antitumoral efficacy of monoclonal antibodies and other macromolecules has been limited by physiological barriers within tumors [35]. Three main barriers have been identified: (1) variable distribution of permeable blood vessels, (2) elevated interstitial pressure, and (3) large transport distances in the interstitium.

The periphery of a solid tumor mass tends to be well vascularized with permeable blood vessels [36]. Because the pore sizes of leaky tumor vessels are estimated to be roughly 400–600 nm [37], most viruses should readily be deposited in the tumor periphery following intravenous, injection or dissemination (e.g., adenovirus is roughly 80–100 nm) [21]. However, the central cores of solid tumor masses tend to be poorly vascularized and necrotic. The high interstitial pressure should also resist inward diffusion of viruses present in the periphery of the tumor. This problem can be overcome by direct intratumoral injections or surgical tumor debulking prior to therapy.

Infection of a tumor mass will therefore result in a dynamic interplay between (1) viral replication and spread; (2) tumor infection, lysis, and proliferation; and (3) host antiviral immune responses. The clinical efficacy of oncolytic virus treatment will depend on the interplay between these variables.

VII. SPECIFIC ONCOLYTIC VIRUSES IN DEVELOPMENT (Table 4)

A. Herpes Simplex Virus Type 1 for Cancer Therapy

Herpes simplex virus type 1 (HSV-1) is a double-stranded DNA virus that remains latent in neurons. HSV-1 causes clinical disease ranging from mucocutaneous ulcerations ("cold sores") to fatal encephalitis. Given the "neurotropism" of HSV-1, attenuated virus variants have been studied as potential anticancer agents for tumors of the central nervous system (CNS). In fact, these viruses efficiently infect many epithelial tissues.

1. FIRST GENERATION VIRUSES

The initial HSV-1 mutant studied for tumor-selective replication contained a 360-bp deletion in the thymidine kinase (tk) gene UL23 [18]. This virus (dlsptk) is unable to efficiently replicate in quiescent cells such as neurons because of this mutation. In contrast, rapidly proliferating cells, such as some tumor cells, have endogenous levels of nucleotides that are high enough to allow viral replication in spite of the tk deletion. Therefore, dlsptk was studied as a therapeutic agent for CNS malignancies. *In vitro* data shows that tk-attenuated HSV-1 mutants lyse proliferating cells effectively [18].

The initial studies with tk-deleted mutants were in immunodeficient, athymic mouse models. The efficacy of a replicating virus in an immunocompetent host was

TABLE 4 Oncolytic Viruses in Development

Agent	Virus strain	Disease[a]	Cell phenotype allowing selective replication	Genetic alterations
dlsptk (first generation)	HSV-1	Encephalitis	Proliferating cells	*Thymidine kinase* gene deletion
G207 (third generation)	HSV-1	Encephalitis	Proliferating cells	Ribonucleotide reductase disruption (*lacZ* insertion into *ICP6* gene) Neuropathogenesis gene mutation (*γ34.5* gene)
73-T	Newcastle Disease virus (chicken paramyxovirus)	Conjunctivitis Laryngitis	Unknown	Unknown (serial passage on tumor cells)
H-1, MVM	Autnomous parvoviruses	Unknown	Transformed cells ↑ proliferation ↓ differentiation *ras*, p53 mutation	None
ONYX-015	Ad2/5 chimera	Upper respiratory infection Conjunctivitis	Cells lacking p53 function (e.g., deletion, mutation, HPV infection)	E1B-55-kDa gene deletion

[a] Disease typically caused by wild-type parent strain of virus. HSV, herpes simplex virus.

still unknown. Subsequently, another group used a similar vector in the immunocompetent rat glioma model [27]. As in previous studies, tumor necrosis and increased animal survival was noted. However, replication was mimimal by immunohistochemical staining. Therefore, replication-independent mechanisms such as cytokine-mediated killing may have been active in this model. The specific mechanism(s) leading to efficacy, however, have not been elucidated to date.

2. SECOND GENERATION HSV-1 CONSTRUCTS

Despite the encouraging activity with dlsptk outlined earlier, investigators remained concerned about the following factors: (1) persistent, albeit reduced, encephalitis risk [38], and (2) the virus's resistance to acyclovir and ganciclovir [18]. Second generation HSV-1 mutants have been developed that have further reductions in neurovirulence and increased sensitivity to ganciclovir [19].

The gamma 34.5 gene is necessary for neurovirulence in some animal models [39]. This gene is not essential for *in vitro* viral replication, but the deletion mutant R3616 was avirulent following intracerebral injection in mice at doses up to approximately 10^6 pfu. Minimal, if any, viral replication occurred with R3616. Therefore, the gamma 34.5 gene was a candidate for deletion from HSV-1 oncolytic viruses. The exact function of this gene product *in vitro* is unclear.

The second cause for concern was the resistance of dlsptk to the standard antiviral ganciclovir, due to the lack of thymidine kinase expression [19]. In contrast, mutants attenuated in the ribonucleotide reductase gene have ehanced sensitivity to ganciclovir. Mutants lacking functional ribonucleotide reductase had previously been identified in which the *lacZ* gene had been inserted into the *ICP6* gene [19]. These mutants still replicated in Vero cells, although virus production was reduced four- to fivefold versus wild-type HSV-1.

A group at Georgetown University has published data on an HSV-1 mutant that incorporates a number of the genetic changes outlined earlier into a single virus, G207 [19]. The gamma 34.5 gene deletions and *lacZ* gene insertion into the *ICP6* ribonucleotide reductase gene result in further reductions in neurovirulence, a useful histochemical marker, and ganciclovir hypersensitivity (Fig. 1). The resultant G207 virus has a peak viral production burst size *in vitro* that is 1.5–2.0 logs less than the parent virus (3616), whereas the safe intracerebral dose appears to be at least 4 logs higher. Intracerebral toxicity studies in owl monkeys confirm the lethal effects of 10^3 pfu HSV-1 wild-type (strain F), whereas 10^7 pfu G207 was not associated with any clinical toxicity in 3 cases over the 14 weeks postinoculation. Subsequent studies have followed monkeys for up to one year following intracerebral virus injection and have revealed no toxicity. Nude mouse, human tumor xenograft studies using direct injection of virus into glioblastoma tumors report results similar (although perhaps somewhat reduced) to those seen with dlsptk. In summary, G207 is an example of an oncolytic virus with several attractive features, including multiple genetic changes (making reversion to wild-type less likely), safety in primates, temperature sensitivity, and gangclovir hypersensitivity. The replication competence of this virus, however, has been significantly reduced as a result of these safety features; whether or not this results in reduced efficacy will require further study *in vivo*.

B. Newcastle Disease Virus

Newcastle Disease virus (NDV) is a chicken paramyxovirus associated with minimal disease in humans (e.g., conjunctivitis, laryingitis). *In vitro* cytolysis with NDV has been reported in numerous human tumor cell lines including both carcinomas and sarcomas; in contrast, nine human fibroblast lines were resistant to NDV [40]. In a nude mouse, human neuroblastoma tumor model, 17 of 18 subcutaneous tumors underwent complete regression following injection with 10^7 pfu of NDV [41]; intratumoral virus replication was documented. Definitive data on antitumoral efficacy in immunocompetent tumor models is not available to date.

Experience with NDV in human cancer patients has involved both viral oncolysate (tumor cell–virus suspensions) and free virus injection. Direct intratumoral injection of strain 73-T of NDV into locally advanced cervical carcinomas resulted in partial responses in 8 of 33 patients [42]. NDV has also been used as component of viral tumor cell oncolysates in melanoma and colon cancer patients [43]. No well-controlled, randomized trials have been conducted, however, and no conclusions about the clinical utility of NDV alone or as a component of an oncolysate can be made.

Several questions remain regarding the use of NDV as an oncolytic virus. First of all, the selectivity of the virus for replication in tumor cells versus normal cells is unclear. Toxicity studies have been limited to rodents *in vivo;* it is unclear how permissive rodents are for NDV replication as compared to humans. *In vitro* studies showing a lack of cytopathic effect with NDV on normal cells have been limited to rodent cells or human fibroblasts; normal human epithelial cells have not been systemically studied. Definitive primate toxicity studies have not been reported. In addition, the mechanism by which 73-T causes "selective" tumor cell lysis is unclear. Tumor cell killing may involve virus replication and

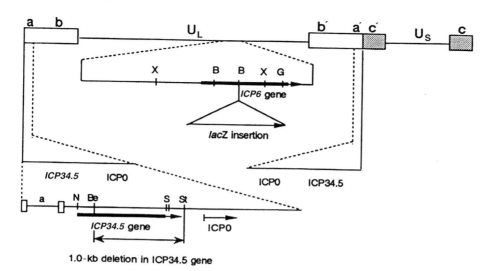

FIGURE 1   Structure of G207 DNA. The boxes (top line) represent inverted repeat sequences flanking the long (U_L) and short (U_S) unique sequences of herpes simplex virus type 1 (HSV-1) DNA (thin lines). The *ICP6* gene is located as indicated in the expanded domain of U_L (thick line indicates transcribed region). The *Escherichia coli lacZ* coding sequence was inserted into the indicated *Bam*HI site. G207 contains 1.0-kb deletions in both copies of *ICP34.5*, in the expanded domain of the long repeat region. B, *Bam*HI; Be, *Bst*EII; G, *Bgl*II; N, *Nco*I; S, *Sca*I; St, *Stu*I; X, *Xho*I.

direct cell lysis and/or induction of TNF secretion and increased sensitivity of tumor cells to TNF-mediated killing [42,44]. Further clinical development will be augmented by answers to these questions. Groups at several academic centers continue work in this field.

C. Autonomous Parvoviruses

Parvoviruses are nonenveloped, single-stranded DNA viruses with relatively small genomes (roughly 5,000 bases). Due to this relative simplicity these viruses are dependent on either (1) helper viruses (as with adeno-associated virus, AAV) or on, (2) specific cellular functions for replication. The latter parvoviruses are classified as "autonomous" and include murine (e.g., minute virus of mice, MVM) and human (e.g., H1) viruses. The ability of autonomous parvoviruses to replicate and cause toxicity in a given cell type is dependent on factors associated with proliferation and differentiation. The cytopathic effect of autonomous parvoviruses in a given cell type is markedly increased after cell transformation in many cases. The cytotoxicity does not appear to be dependent on viral replication, *per se,* but rather on products of toxic nonstructural proteins.

The exact mechanism by which transformed cells become sensitive to parvoviral cytopathic effects is unknown. Numerous human and rodent cell types that are resistant to parvovirus toxicity can be made sensitive through transformation with SV40, HPV-16, or H-*ras*.

Transformation leads to increased viral DNA replication, gene expression, and cytolysis following infection [45]. A human chronic myelogenous leukemia cell line subclone that developed resistance to parvoviral cytolysis had reverted to wild-type p53 expression [46]. Subsequent transfection of a rat fibroblasts line with a dominant-negative p53 gene construct and mutant H-*ras* resulted in enhanced sensitivity to cytolysis; transformation by H-*ras* and c-*myc* did not induce sensitivity [46]. Therefore, transformation through multiple genetic mechanisms has led to enhanced sensitivity to cytolysis, but transformation does not uniformly lead to enhanced sensitivity.

The mechanism by which parvovirus infection leads to cell death in sensitive cells is also poorly understood. Production of infectious virus does not appear to be necessary or sufficient for cytolysis. Many sensitive cell types do not support active viral replication [47], and a cell line resistant to parvoviral cytopathic effects has been shown to support virus production [46]. Subsequent studies have shown that induction of NS-1 (nonstructural) protein expression alone in transformed fibroblasts leads to apoptosis whereas equivalent expression in normal fibroblasts is nontoxic [48]. NS-1 expression leads to the accumulation of proliferating cells in the G_2 phase of the cell cycle, and transformed cells subsequently undergo apoptosis. Clones that have acquired resistance to NS-1 are no longer arrested in G_2 [49]. Therefore, although the cytopathic effect is generally associated with viral DNA replication, expression of nonstructural gene products and enhanced sensitivity to their effects may be the actual mechanism by which toxicity is achieved [23,50]. Further work is needed to determine whether these viruses can replicate and spread well enough within a solid tumor *in vivo* to cause significant tumor necrosis.

D. Vaccinia

Vaccinia is a member of the poxvirus family of enveloped, double-stranded DNA viruses. Although the intrinsic antitumoral efficacy of replication-competent vaccinia has not been studied in depth, vaccinia virus vectors are being utilized to deliver immunostimulatory cytokines in early phase clinical trials to patients with melanoma [51]. A replication-competent vector may act as an immune adjuvant in this gene therapy approach. However, the intrinsic replication and tumor cell lysis of other, more aggressive poxviruses may lead to efficacy that is independent of immune stimulation.

E. Adenovirus

Adenoviruses are nonintegrating, double-stranded DNA viruses that infect a broad range of human cell types. Infections with human adenovirus are mild and widespread, with 70–80% of adults having evidence of antibodies directed against adenoviruses serotypes 2 and/or 5, which usually arose during childhood infections. Wild-type adenovirus exposure can lead to self-limited symptoms such as upper respiratory tract infection, bronchitis, gastroenteritis, conjunctivitis, and cystitis. Other favorable attributes of adenovirus as an oncolytic agent include its well-described genome, its ability to infect both dividing and nondividing cells and its amenability to high-titer manufacturing and purification.

VIII. DEVELOPMENT OF ONYX-015

p53 is mutated in roughly 50 percent of all human cancers, including non-small-cell lung (60%), colon (50%), breast (40%), head and neck (60%), and ovarian (60%) cancers in the advanced stages. Loss of p53 function is associated with resistance to chemotherapy and/or decreased survival in numerous tumor types, including breast, colon, bladder, ovarian, and non-small-cell lung cancers. Therefore, effective therapies for tumors that lack functional p53 are clearly needed.

p53 mediates cell cycle arrest and/or apoptosis in response to DNA damage (e.g., due to chemotherapy

or radiation) or foreign DNA synthesis (e.g., during virus replication). Consequently, DNA tumor viruses such as adenovirus, SV40, and human papilloma virus encode for proteins that inactivate p53 and thereby allow efficient viral replication. For example, the adenovirus E1B-region 55-kDa protein binds and inactivates p53 in complex with the E4orf6 protein (Fig. 2, see color insert). Because p53 function must be blocked to allow efficient virus replication, Dr. Frank McCormick hypothesized that an adenovirus lacking E1B-55-kDa gene expression might be severely limited in its ability to replicate in normal cells; however, cancer cells that lack p53 function should support virus replication and resultant cell destruction. ONYX-015 (ONYX Pharmaceuticals, Richmond, CA) is an attenuated adenovirus type 2/5 chimera (dl1520) with two mutations in the early region E1B-55-kDa gene; this virus was created in the laboratory of Dr. Arnie Berk [20]. The cytopathic effects of wild-type adenovirus and ONYX-015 were studied on a pair of cell lines that are identical except for p53 function: the RKO human colon cancer cell line with normal p53 function (the parent line), and an RKO subclone transfected with dominant-negative p53 (courtesy of Dr. Michael Kastan) (Fig. 3, see color insert) [2]. As predicted, ONYX-015 induced cytopathic effects identical to wild-type adenovirus in the subclone lacking functional p53, whereas cytopathic effects with ONYX-015 were reduced by approximately two orders of magnitude in the parental tumor line harboring normal p53 (Fig. 2). Subsequently, a tumor cell line that was resistant to ONYX-015 due to normal p53 function (U2OS) became sensitive to ONYX-015 following transfection and expression of the E1B-55-kDa gene. Therefore, ONYX-015 is able to replicate selectively in p53-deficient cancer cells due to a deletion in the E1B-55-kDa gene.

Subsequent experiments demonstrated that primary (nonimmortalized) human endothelial cells, fibroblasts, small airway cells, and mammary epithelial cells highly resistant to ONYX-015 replication and cytolysis, in contrast to effects seen with wild-type adenovirus. Replication-dependent cytopathic effects were demonstrated in human tumor cell lines of many different histologies following infection with ONYX-015. Tumor cells that lack p53 function through different mechanisms (p53 gene mutation and/or deletion, or p53 degradation by human papilloma virus E6 protein) were shown to be destroyed by ONYX-015. In addition, several carcinoma lines with a normal p53 gene sequence, including two chemotherapy-resistant ovarian cancer subclones, were efficiently lysed. ONYX-015 had significant *in vivo* antitumoral activity against subcutaneous human tumor xenografts in nude mice following intratumoral or intravenous injection; the *in vivo* efficacy against each tumor type correlated with the *in vitro* sensitivity of the cell line to ONYX-015. Efficacy against intraperitoneal carcinoma was documented following intraperitoneal virus administration (C. Heise and I. Ganley, in press). Due to the lack of efficient replication in rodent cells, however, immunocompetent (syngeneic) tumor models have not been useful for studying replication-dependent effects. Therefore, the role of the antiviral and antitumoral immune responses may only be determined in cancer patients until a novel model is developed.

To study potential interactions between ONYX-015 and chemotherapy *in vivo,* the following experiments were carried out. Cisplatin and 5-fluorouracil (5-FU), two chemotherapeutic agents commonly used to treat head and neck cancer patients, were selected for use in combination with ONYX-015 in the nude mouse, HLaC (head and neck) human tumor xenograft model [21]. Tumors were treated with ONYX-015 as described earlier, followed by treatment with cisplatin or 5-FU by intraperitoneal injection on days 8–12. Four groups of mice were treated with one of the following regimens: ONYX-015 plus chemotherapy (cisplatin or 5-FU), chemotherapy alone (cisplatin or 5-FU), ONYX-015 alone, or vehicles alone. All treatment groups received identical injections of the active agent or vehicle control into both the tumor and peritoneum. Unlike cisplatin or 5-FU alone, treatment with ONYX-015 alone significantly increased survival times versus placebo ($P = 0.01$). The combination of cisplatin or 5-FU with ONYX-015 was more effective than chemotherapy or virus treatment alone. Median survival was significantly increased with the combination therapy compared with chemotherapy alone: 38 days vs. 24 days for the 5-FU treated groups ($P = 0.005$), 44 vs. 27 days for the cisplatin-treated groups ($P = 0.003$). Similar results have been reported with other tumor types *in vivo.*

IX. CLINICAL DEVELOPMENT OF ONYX-015

A. Approach

ONYX-015 is a novel agent with a novel mechanism of action. We predicted that both toxicity and efficacy would depend on the intrinsic ability of a given tumor to replicate the virus, on the location of the tumor to be treated (e.g., intracranial vs. peripheral), and on the route of administration of the virus. In addition, data on viral replication, antiviral immune responses, and their relationship to antitumoral efficacy were critical in the early stages of development. We therefore elected to treat patients with recurrent head and neck carcinomas initially.

B. Phase I Trial: Head and Neck Cancer

The rationale for targeting this population is this: These tumors are frequently amenable to direct injection and biopsy in the outpatient clinic setting. p53 abnormalities are very common; gene mutations or deletions are present in up to 70% of recurrent tumors [52,53], and other p53-inactivating mechanisms such as *mdm-2* overexpression and HPV E6 expression appear to be present in another 15–20% of these tumors. Finally, most patients suffer severe morbidity, and even mortality, from the local/regional progression of these tumors. Up to two thirds of these patients die due to local complications. Therefore, a local therapy might lead to significant palliation and even survival prolongation.

Patients enrolled onto the phase I trial had recurrent squamous cell carcinoma of the head and neck that was not surgically curable and had failed either prior radiation or chemotherapy [54]. p53 gene sequence and immunohistochemical staining were determined on all tumors but were not used as entry criteria. Other baseline tests included lymphocyte subsets (CD3, 4, 8), delayed-type hypersensitivity skin testing (including mumps and candida), and neutralizing antibodies to ONYX-015. This was a standard phase I dose escalation trial in which at least three patients are treated per dose level prior to escalation to the next cohort; intrapatient dose escalation was not allowed. Six patient cohorts received single intratumoral injections of ONYX-015 every 4 weeks (until progression) at levels of 10^7 to 10^{11} pfu per dose. Two additional cohorts received five consecutive daily doses of 10^9 or 10^{10} pfu per day (total dose 5×10^9 or 5×10^{10}) every 4 weeks. Following treatment, patients were observed for toxicity and for target (injected) tumor response. Additional biological end points included changes in neutralizing antibodies, the presence of virus in the blood (PCR on days 3, 8), viral replication within the injected tumor (in tumor biopsies on days 8 and 22), and associated immune cell infiltration.

No significant toxicity was seen in any of the 32 patients treated. Eleven patients received repeat treatments (2–7 total). Grade 3 tumor site pain was noted on a single occasion in one patient. Otherwise, tumor injection site pain was either nonexistent or mild. Flulike symptoms were noted in approximately one third of patients on the single-dose regimen and two thirds of patients on the daily × 5 regimen. Symptoms included low-grade fevers (less than 38.5 degrees), grade 1–2 myalgias, and grade 1 nausea. Symptoms typically started within 12 to 24 hours of injection and lasted for 1–5 days. Following ONYX-015-induced tumor necrosis, nonbleeding ulcerations developed over several injected tumors. However, no significant local complications occurred.

Neutralizing antibodies were positive in approximately 70% of the cases prior to treatment. Following treatment, all patients had positive antibody titers, and all patients had an increase in antibody titer. Replication was identified infrequently on day 8 tumor biopsies in patients on the single-injection protocol, whereas day 8 biopsies were almost uniformly positive in tumors from patients on the multidose regimen (Fig. 4, see color insert). Day 22 biopsies were negative for viral replication.

Three of the 23 patients on the single-dose regimen had formal partial responses (PRs) of the injected tumor and 9 had tumor stabilization (8–16 weeks). In addition, 3 patients with stable disease had >50% necrosis of the injected tumor. In contrast, 3 of 9 patients on the multidose regimen had PRs and an additional 3 had stabilization with significant necrosis; only 2 patients had progressive disease. One patient received seven treatments over 7 months while maintaining a partial remission. These results are consistent with experiments comparing these two regimens in nude mouse, human tumor xenograft models (D. Kirn, in press). Responding patients included some with positive baseline-neutralizing antibodies and tumors with a normal p53 gene sequence. However, definitive correlations between these variables and the degree of tumor response cannot be made until larger phase II trials are completed.

Based on these results, two phase II trials in head and neck cancer patients were initiated. In a study using ONYX-015 treatment alone, approximately 30 patients refractory to chemotherapy or radiotherapy following recurrence are being treated with ONYX-015 alone; final data are pending. In a second phase II trial, patients are treated simultaneously over 5 days with ONYX-015 intratumorally and cisplatin (day 1 bolus) and continuous infusion 5-FU (days 1–5) intravenously. These patients are all chemotherapy-naive in the setting of recurrent disease. This study is to be completed and analyzed by June 1998.

Additional local or regional tumor targets include ovarian cancer (phase I intraperitoneal injection trial underway), pancreatic cancer (phase I intratumoral injection trial underway), colorectal liver metastases (hepatic arterial infusion trial to be initiated in 1998), superficial recurrent bladder cancer (intravesical administration), and malignant astrocytomas (including glioblastoma multiforme) (Table 5).

TABLE 5 Potential Tumor Targets for Replication-Competent Viruses Following Different Routes of Administration

Route of administration	Target cancer examples[a]
Local	
Direct intratumoral injection	Head and neck, GBM
Postoperative surgical field perfusion	Head and neck, GBM
Regional	
Body cavity instillation	Peritoneum: ovarian
	Pleural cavity
	Metastatic lung, breast
	Mesothelioma
	Bladder: superficial disease
Regional intra-arterial instillation	Liver-directed
	Metastatic colon
	Hepatocellular
Systemic	
Intravenous	Systemic metastases

[a] Tumor types potentially amenable to viral therapy delivered by the specified route. Local and regional indications include tumors for which a relatively localized antitumoral effect could be beneficial to a patient (i.e., local complications cause significant morbidity and mortality). GBM, glioblastoma multiforme (primary brain cancer).

X. SUMMARY

Selectively replicating viruses may offer a new approach to cancer treatment. If they are successful in clinical trials, these agents will constitute a new category in the antitumoral armamentarium. Many viruses are currently being studied, and an adenovirus (ONYX-015) entered clinical trials in 1996; herpesvirus agents are scheduled to enter clinical trials in 1998. Critical issues need to be addressed if the utility of these agents is to be optimized. For each virus, the effect of antiviral immunity on antitumoral efficacy must be better understood. For all viruses, physical barriers to spread within tumors (e.g., fibrosis, pressure gradients) must be overcome. Although proof-of-concept experiments with chemotherapy and ONYX-015 have been encouraging, further studies are required to determine optimal treatment regimen sequencing. Combination studies with radiation therapy are also underway with ONYX-015. Finally, these agents will require modification (e.g., coat modification) to be effective against systemic metastases following intravenous administration.

Second generation virus constructs will be developed based on clinical and preclinical data. Enhanced replication and virulence against tumor cells will be a major goal. The necessary degree of selectivity for tumor cells versus normal cells will depend on the route of administration; normal tissue will be exposed to much higher

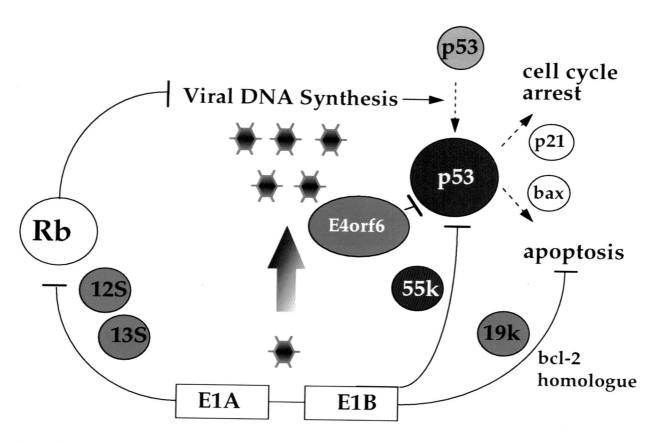

Figure 2 Adenovirus early gene products control the cell cycle (E1A), inhibit p53 (E1B-55-kDa, E4orf6), and block apoptosis (E1B-19-kDa). Following binding and internalization of adenovirus particles, E1A gene products bind and inactivate pRB (and other pocket proteins), leading to viral DNA synthesis. E1A expression leads to up-regulation of p53. The adenoviral proteins from E1B-55-kDa and E4orf6 subsequently bind and inactivate p53. In addition, the E1B-19-kDa gene product (a *bcl-2* homolog) blocks both p53-dependent and p53-independent apoptosis, allowing completion of the replicative cycle prior to cell destruction.

wild type Adenovirus ONYX-015 (dl1520)

m.o.i.   0   0.01   0.1   1   0   0.01   0.1   1

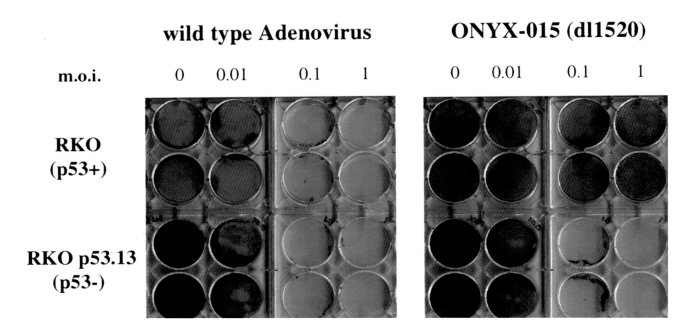

RKO (p53+)

RKO p53.13 (p53-)

Figure 3   Effects of ONYX-015 (dl1520) and wt Ad (wild-type adenovirus) on RKO cells (p53+) and RKO.p53.13 cells (p53–). Cells were infected at the MOI (multiplicity of infection) shown and stained for viability 8 days later. RKO cells (p53+) have functional p53 and are resistant to ONYX-015. RKO.p53.13 cells are p53-deficient and are lysed by ONYX-015. wt Ad replicates in both lines nonselectively.

Figure 4   *In situ* hybridization and electron microscopy demonstrate replicating ONYX-015 (dl1520) virus particles in human tumor cells (day 8 biopsy following 10^9 pfu on days 1–5). Squamous cell carcinoma cells stained with hematoxylin and eosin show adenovirus-induced cytopathic effects at low (A) and high (C) magnification. *In situ* hybridization for adenoviral DNA demonstrates replicating virus in blue-staining cells (B and D). Electron microscopy of squamous tumor cells demonstrates replicating ONYX-015 in nucleus (E). Arrows indicate dense aggregates of virus particles in crystalline arrays. Higher magnification of boxed area shows icosahedral-shaped adenovirus particles (F). Panels E and F courtesy of Dr. Carla Heise, Onyx Pharmaceuticals.

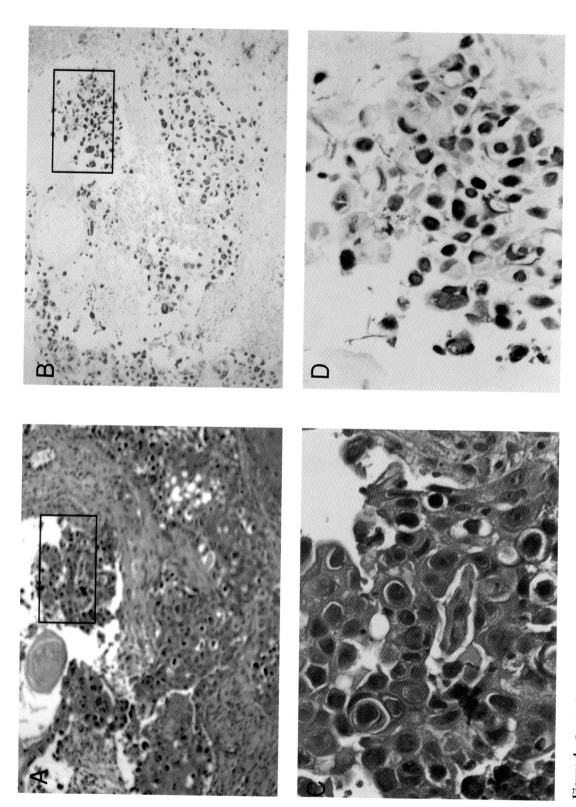

Figure 4 - *Continued.*

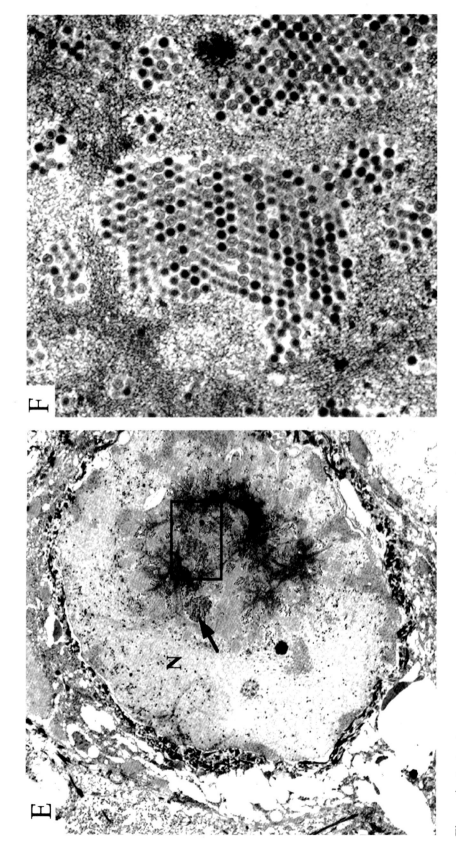

Figure 4 - *Continued.*

doses of virus following intravenous injection than following intratumoral injection, for example. Finally, replicating viruses have been constructed that carry genes encoding prodrug-activating enzymes (e.g., cytosine deaminase or thymidine kinase) or immunomodulatory cytokines, for example. This approach will allow the beneficial attributes of gene therapy agents to be combined with the advantages of selectively replicating vectors.

References

1. Trinh, Q. T., et al. Enzyme/prodrug gene therapy: comparison of cytosine deaminase/5-fluorocytosine versus thymidine kinase/ganciclovir enzyme/prodrug systems in a human colorectal carcinoma cell line. Cancer Res., 1995. 55(21), 4808–4812.

2. Bischoff, J. R., et al. An adenovirus mutant that replicates selectively in p53-deficient human tumor cells. Science, 1996. 274(5286), 373–376.

3. Dock, G. Am. J. Med. Sci. 1904. 127, 563.

4. Southam, C. M., and Moore, A. E. Clinical studies of viruses as antineoplastic agents, with particular reference to Egypt 101 virus. Cancer, 1952. 5, 1025–1034.

5. Webb, H. E., and Smith, C. E. Viruses in the treatment of cancer. Lancet, 1970. 1(658), 1206–1208.

6. Smith, R., et al. Studies on the use of viruses in the treatment of carcinoma of the cervix. Cancer, 1956. 9, 1211–1218.

7. Asada, T. Treatment of human cancer with mumps virus. Cancer, 1974. 34(6), 1907–1928.

8. Southam, C. M., and Moore, A. E. Clinical studies of viruses as antineoplastic agents, with particular reference to Egypt 101 virus. Cancer, 1952. 5, 1025–1034.

9. Zinkernagel, R. M. Immunology taught by viruses. Science, 1996. 271(5246), 173–178.

10. Smith, R., et al. Studies on the use of viruses in the treatment of carcinoma of the cervix. Cancer, 1956. 9, 1211–1218.

11. Russell, S. J., Hawkins, R. E., and Winter, G. Retroviral vectors displaying functional antibody fragments. Nucleic Acids Res., 1993. 21(5), 1081–1085.

12. Kasahara, N., Dozy, A. M., and Kan, Y. W. Tissue-specific targeting of retroviral vectors through ligand-receptor interactions [see comments]. Science, 1994. 266(5189), 1373–1376.

13. Han, X., Kasahara, N., and Kan, Y. W. Ligand-directed retroviral targeting of human breast cancer cells. Proc. Natl. Acad. Sci. U.S.A., 1995. 92(21), 9747–9751.

14. Wickham, T. J., Evans, T., Shears, L., Roelvink, P., Li, Y., Lee, G., Brough, D., Lizonova, A., and Kovesdi, I. Increased in vitro and in vivo gene transfer by adenovirus vectors containing chimeric fiber proteins. J. Virol., 1997. 71(11), 8221–8229.

15. Vile, R. G., et al. Tissue-specific gene expression from Mo-MLV retroviral vectors with hybrid LTRs containing the murine tyrosinase enhancer/promoter. Virology, 1995. 214(1), 307–313.

16. Babiss, L. E., Friedman, J. M., and Darnell, J. E., Jr. Cellular promoters incorporated into the adenovirus genome: effects of viral regulatory elements on transcription rates and cell specificity of albumin and beta-globin promoters. Mol. Cell Biol., 1986. 6(11), 3798–3806.

17. Friedman, J. M., et al. Cellular promoters incorporated into the adenovirus genome: cell specificity of albumin and immunoglobulin expression. Mol. Cell Biol., 1986. 6(11), 3791–3797.

18. Martuza, R. L., et al. Experimental therapy of human glioma by means of a genetically engineered virus mutant. Science, 1991. 252(5007), 854–856.

19. Mineta, T., et al. Attenuated multi-mutated herpes simplex virus-1 for the treatment of malignant gliomas. Nat. Med., 1995. 1(9), 938–943.

20. Barker, D. D., and Berk, A. J. Adenovirus proteins from both E1B reading frames are required for transformation of rodent cells by viral infection and DNA transfection. Virology, 1987. 156, 107–121.

21. Heise, C., et al. ONYX-015, an E1B gene-attenuated adenovirus, causes tumor-specific cytolysis and antitumoral efficacy that can be augmented by standard chemotherapeutic agents [see comments]. Nat. Med., 1997. 3(6), 639–645.

22. Southam, C. M. Present status of oncolytic virus studies. N. Y. Acad. Sci., 1960, 656–673.

23. Mousset, S., et al. The cytotoxicity of the autonomous parvovirus minute virus of mice nonstructural proteins in FR3T3 rat cells depends on oncogene expression. J. Virol., 1994. 68(10), 6446–6453.

24. Tollefson, A. E., et al. The E3-11.6-kDa adenovirus death protein (ADP) is required for efficient cell death: characterization of cells infected with adp mutants. Virology, 1996. 220(1), 152–162.

25. Lorence, R. M., Rood, P. A., and Kelley, K. W. Newcastle Disease virus as an antineoplastic agent: induction of tumor necrosis factor-alpha and augmentation of its cytotoxicity. J. Natl. Cancer Inst., 1988. 80(16), 1305–1312.

26. Gooding, L. R. Regulation of TNF-mediated cell death and inflammation by human adenoviruses. Infect. Agents Dis., 1994. 3(2–3), 106–115.

27. Jia, W. W., et al. Selective destruction of gliomas in immunocompetent rats by thymidine kinase-defective herpes simplex virus type 1 [see comments]. J. Natl. Cancer Inst., 1994. 86(16), 1209–1215.

28. Webb, H. E., and Smith, C. E. Viruses in the treatment of cancer. Lancet, 1970. 1(658), 1206–1208.

29. Cassel, W. A., and Garrett, R. E. Tumor immunity after viral oncolysis. J. Bacteriol., 1966. 92(3), 792.

30. Yang, Y., et al. Cellular immunity to viral antigens limits E1-deleted adenoviruses for gene therapy. Proc. Natl. Acad. Sci. U.S.A., 1994. 91(10), 4407–4411.

31. Day, D. B., Zachariades, N. A., and Gooding, L. R. Cytolysis of adenovirus-infected murine fibroblasts by IFN-gamma-primed macrophages is TNF- and contact-dependent. Cell Immunol., 1994. 157(1), 223–238.

32. Hermiston, T. W., et al. Deletion mutation analysis of the adenovirus type 2 E3-gp19K protein: identification of sequences within the endoplasmic reticulum lumenal domain that are required for class I antigen binding and protection from adenovirus-specific cytotoxic T lymphocytes. J. Virol., 1993. 67(9), 5289–5298.

33. Yang, Y., Trinchieri, G., and Wilson, J. M. Recombinant IL-12 prevents formation of blocking IgA antibodies to recombinant adenovirus and allows repeated gene therapy to mouse lung [see comments]. Nat. Med., 1995. 1(9), 890–893.

34. Ginsberg, H. S., and Prince, G. A. The molecular basis of adenovirus pathogenesis. Infect. Agents Dis., 1994. 3(1), 1–8.

35. Jain, R. K. Barriers to drug delivery in solid tumors. Sci. Am., 1994. 271(1), 58–65.

36. Dvorak, H. F., et al. Identification and characterization of the blood vessels of solid tumors that are leaky to circulating macromolecules. Am. J. Pathol., 1988. 133(1), 95–109.

37. Yuan, F., et al. Vascular permeability in a human tumor xenograft: molecular size dependence and cutoff size. Cancer Res., 1995. 55(17), 3752–3756.

38. Markert, J. M., *et al.* Reduction and elimination of encephalitis in an experimental glioma therapy model with attenuated herpes simplex mutants that retain susceptibility to acyclovir. *Neurosurgery,* 1993. **32**(4), 597–603.

39. Chou, J., *et al.* Mapping of herpes simplex virus-1 neurovirulence to gamma 34.5, a gene nonessential for growth in culture. *Science,* 1990. **250**(4985), 1262–1266.

40. Reichard, K. W., *et al.* Newcastle Disease virus selectively kills human tumor cells. *J. Surg. Res.,* 1992. **52**(5), 448–453.

41. Lorence, R. M., *et al.* Complete regression of human neuroblastoma xenografts in athymic mice after local Newcastle Disease virus therapy [see comments]. *J. Natl. Cancer Inst.,* 1994. **86**(16), 1228–1233.

42. Cassel, W. A., and Garrett, R. E. Newcastle Disease virus as an antineoplastic agent. *Cancer,* 1965. **18**(7), 863–868.

43. Cassel, W. A., Murray, D. R., and Phillips, H. S. A phase II study on the postsurgical management of stage II malignant melanoma with a Newcastle Disease virus oncolysate. *Cancer,* 1983. **52**(5), 856–860.

44. Lorence, R. M., Rood, P. A., and Kelley, K. W. Newcastle Disease virus as an antineoplastic agent: induction of tumor necrosis factor-alpha and augmentation of its cytotoxicity. *J. Ntl. Cancer Inst.,* 1988. **80**(16), 1305–1312.

45. Rommelaere, J., and Cornelis, J. J. Antineoplastic activity of parvoviruses. *J. Virol. Methods,* 1991. **33**(3), 233–251.

46. Telerman, A., *et al.* A model for tumor suppression using H-1 parvovirus. *Proc. Natl. Acad. Sci. U.S.A.,* 1993. **90**(18), 8702–8706.

47. Dupressoir, T., *et al.* Inhibition by parvovirus H-1 of the formation of tumors in nude mice and colonies *in vitro* by transformed human mammary epithelial cells. *Cancer Res.,* 1989. **49**(12), 3203–3208.

48. Dupont, F., *et al.* Use of an autonomous parvovirus vector for selective transfer of a foreign gene into transformed human cells of different tissue origins and its expression therein. *J. Virol.,* 1994. **68**(3), 1397–1406.

49. Op De Beeck, A., *et al.* The nonstructural proteins of the autonomous parvovirus minute virus of mice interfere with the cell cycle, inducing accumulation in G2. *Cell Growth Differ.,* 1995. **6**(7), 781–787.

50. Van Pachterbeke, C., *et al.* Parvovirus H-1 inhibits growth of short-term tumor-derived but not normal mammary tissue cultures. *Int. J. Cancer,* 1993. **55**(4), 672–677.

51. Lattime, E. C., *et al. In situ* cytokine gene transfection using vaccinia virus vectors. *Semin. Oncol.,* 1996. **23**(1), 88–100.

52. Boyle, J. O., *et al.* The incidence of p53 mutations increases with progression of head and neck cancer. *Cancer Res.,* 1993. **53**(19), 4477–4480.

53. Brennan, J. A., *et al.* Association between cigarette smoking and mutation of the p53 gene in squamous-cell carcinoma of the head and neck. *N. Engl. J. Med.,* 1995. **332**(11), 712–717.

54. Ganly, I., Kirn, D., Rodriguez, G., *et al.* Phase I trial of intratumoral injection with an E1B-deleted adenovirus. ONYX-015, in patients with recurrent head and neck cancer. *Proc. Am. Soc. Clin. Oncol.,* 1997.

PART V

Molecular Vaccine Strategies for Cancer

ras Oncogene Products as Tumor-Specific Antigens for Activation of T-Lymphocyte-Mediated Immunity

SCOTT I. ABRAMS, J. ANDREW BRISTOL, AND JEFFREY SCHLOM
Laboratory of Tumor Immunology and Biology, National Cancer Institute, National Institutes of Health, Bethesda, Maryland 20892

I. INTRODUCTION

Central to the investigative design of cancer immunotherapy is a fundamental understanding of the interplay between the host immune system and its ca-pacity to functionally distinguish normal from neoplastic cells. The exquisite specificity of this interaction is largely governed by cellular immune recognition of tumor-associated antigens (TAAs) or tumor-specific antigens (TSAs) expressed and presented by the antigen

(Ag)-bearing target cell in the context of self–major histocompatibility complex (MHC) class I or II molecules. Point mutations in the *ras* p21 proto-oncogenes (K-*ras,* H-*ras,* N-*ras*) have been found in a wide range and high proportion of human and rodent tumors, suggesting a strong link and role in the pathogenesis of the neoplastic process. The resulting oncoproteins are distinct from the normal proto-*ras* forms at very specific sites, typically at codons 12, 13, 59, or 61, as they contain single amino acid substitutions at these positions. Importantly, from an immunologic perspective, these "neodeterminants" may now not only be defined as TSAs, they also may bear unique epitopes for T-cell (CD4+ and/or CD8+) recognition and their subsequent exploitation for the induction of cell-mediated and antitumor immunity. Analysis and identification of *ras* oncogene products as potential T-cell epitopes could be determined biologically by the rational design and application (*in vivo* and/ or *in vitro*) of short synthetic peptides that precisely reflect those sequences surrounding the sites of mutations. Several laboratories have established preclinical models (in both murine and human systems) to explore the underlying hypothesis that *ras* oncogenes encode for TSAs that express mutant-*ras*-specific T-cell peptide epitopes, resulting in the production of Ag-specific cellular immune (CD4+ and/or CD8+) effector mechanisms. Taken collectively, these studies support that hypothesis and provide the rationale for the development of peptide-based immunotherapies directed against human cancers harboring *ras* oncogene products. Moreover, in studies in which weak T-cell responses are induced, an important objective of future experiments is to modify and enhance the immunogenicity of intrinsically weak peptide epitopes for the generation and amplification of more potent cellular immune reactions.

II. PRINCIPLES OF CANCER IMMUNOTHERAPY

Understanding the mechanisms by which tumor cells escape immune recognition in an otherwise immune-competent host and the development of effective therapies for metastatic disease have both been long-standing challenges of biomedical research. The inability of the immune system to effectively battle malignancy may be compromised further by the nature of current therapeutic interventions, such as chemotherapy, which may facilitate or accelerate the extent of host immune suppression. Thus, considerable interest has been directed toward the development of additional or alternative modalities for metastatic therapy, particularly for solid cancers, which may possess enhanced tumor-specific targeting capabilities with less biological or immunologic-

associated systemic toxicity. One such example is immunotherapy, which may be divided into two major components: active, specific immunotherapy (ASI) and passive or adoptive cellular immunotherapy [1–5], although each is still in relatively early stages of clinical development.

The purpose of immunotherapy strategies is not necessarily to modify the endogenous antigenicity of the tumor, but rather to molecularly define these potential TAAs or TSAs so that they may be developed and employed as purified or recombinant immunogens (i.e., proteins or peptides in processed forms or their sequences encoded by plasmid DNA or live viral vectors) in the generation and expansion of the relevant tumor-specific CD4+ and/or CD8+ T-cell effector mechanisms. This philosophy is consistent with the notion that tumor cells are indeed "antigenic," but weakly if at all "immunogenic," as they appear to lack sufficient immunogenicity of their own for the initiation of a productive host immune response. ASI is analogous to vaccination, which is based on the induction and/or enhancement of intrinsic host immune responses by immunization with antigenic determinants derived from, or endogenous to, the tumor. This may be accomplished by the presentation and formulation of such purified or recombinant molecules in an immunogenic chemical and/or biological adjuvant setting and administered, perhaps, distal from the sites of tumor burden, where host immune system interactions may be most functionally compromised. An added advantage of ASI would be the potential induction of systemic, long-term immunologic memory for recall responses to recurrent disease. Passive immunotherapy, in contrast, involves the *ex vivo* activation and expansion of large quantities of autologous tumor-specific effector cells, which are then returned to the host. This may be accomplished by *in vitro* stimulation (IVS) of such CD4+ and/or CD8+ precursors using minimally defined peptide epitopes so that the resulting T-cell response may exhibit high avidity for recognition of tumor cells expressing extremely low densities of endogenously produced antigens (Ags).

Because both types of immunotherapy strategies have distinct characteristics, they may be combined to elicit a more comprehensive response. This may be particularly relevant in the context of metastasis, whereby ASI alone would unlikely generate *in vivo,* at least quantitatively, adequate numbers of tumor-reactive lymphocytes capable of mediating effective eradication of advanced, disseminated disease. Perhaps the rationale and intent of ASI under these circumstances would be to produce and provide a source of Ag-specific lymphocytes, which could subsequently be employed for adoptive immunotherapy applications. Although it remains largely unsettled whether activation and expansion of one or the other or, perhaps, both CD4+ and CD8+ T-cell subsets would be more beneficial in an antitumor setting, it is generally well accepted that optimal development and regulation of the cellular immune response typically require cellular cooperation between these two subpopulations [6–9]. Moreover, because both CD4+ and CD8+ T-cell subsets express distinct requirements for Ag recognition and possess nonoverlapping effector functions, concurrent activation of both T-cell responses might enhance the diversity and breadth of potential antitumor pathways. In terms of Ag recognition, the T-cell receptors (TCRs) of CD8+ or CD4+ T cells recognize antigenic peptides displayed on the extracellular surface of host Ag-presenting cells (APCs; monocytes/macrophages, dendritic cells or B cells) or tumor target cells in the context of self–MHC class I or II molecules, respectively [10–13]. In terms of functional properties, CD4+ T cells have generally been proposed to play an integral role in immunoregulation through the production of cytokines [6,7], whereas CD8+ T cells have been described as cytotoxic T lymphocytes (CTLs), which mediate the destruction of Ag-bearing targets [14,15]. More recently, however, multiple functional subtypes of both CD4+ and CD8+ T cells, which reflect differences in cytotoxic potential/mechanisms and cytokine phenotype patterns, have been described [6,16–21]. Accordingly, the development of animal models provides a preclinical *in vivo* environment to address not only general safety and toxicity issues but also certain biological and mechanistic principles that may have broader conceptual implications for understanding fundamental requirements important for effective induction of cell-mediated and antitumor immunity in human neoplasia. For a further discussion of immune mechanisms, see Chapter 3.

III. TUMOR-SPECIFIC VERSUS TUMOR-ASSOCIATED ANTIGENIC DETERMINANTS

As stated earlier, it is generally thought that tumor cells may be antigenic, but lack sufficient immunogenicity of their own for the initiation and induction of a productive immune response. For immunotherapy to be effective, it must satisfy the following major premises: (1) the host immune system is competent and contains the relevant precursor cells and functional APC populations, despite the presence of metastatic tumor burden; and (2) tumor cells are "antigenic" (i.e., they express or overexpress Ags that selectively or specifically distinguish them from normal host tissues for effective immune recognition and attack). Indeed, in the context of human cancers, a host of tumor Ags and peptides de-

rived from those Ags, which generally can be classified as TAAs or TSAs, have been defined and characterized [1–5,22].

The overall rationale for the identification of tumor-specific or tumor-associated peptides in cancer immunotherapy reflects a fundamental concept: The peptide epitope constitutes an essential component for T-cell recognition of Ag, which is requisite for the subsequent induction and development of cell-mediated and antitumor immunity. Moreover, it is important to stress that the binding interaction between a given MHC class I or II molecule is a genetically restricted event [11,23–25]. Certain amino acid residues within the peptide are critical for MHC binding; others are important for TCR recognition [26,27]. In both murine and human systems, such MHC "consensus anchor motifs" have been defined for a variety of MHC molecules. In contrast to class I/peptide interactions, class II/peptide interactions involve anchor sites that appear to be more degenerate in specificity. Additionally, MHC class I and II molecules present peptides of different sizes, which generally are 8–10 and 13–18 amino acid residues in length, respectively [11,23–25]. Although peptide binding to MHC molecules is a prerequisite for TCR recognition, not all MHC-reactive peptides are antigenic, which may reflect fundamental mechanisms controlling peripheral tolerance or clonal deletion.

TAAs represent the overexpression of normal self-proteins, such as oncofetal, differentiation, or nuclear proteins, and include, for example, carcinoembryonic Ag (CEA) [28,29], a variety of melanoma-associated proteins (e.g., melanoma antigen [MAGE], melanoma antigen recognized by T cells [MART-1], gp100, tyrosinase) [4,5,22] and wild-type p53 [30,31], respectively. Although TAAs may not be considered foreign moieties, because of their overexpression in many human cancers, peptides derived from those proteins may become selective targets for immune recognition. However, the capacity to initiate cellular immune responses against TAAs may be largely determined by the extent of tolerance mechanisms imposed intra- or extrathymically on the selection, development, and maturation of "self-reactive" lymphocytes. For cancer immunotherapy strategies to be effective under conditions of peripheral lymphoid anergy (or nonresponsiveness), for example, they must be able to overcome or "break" such "tolerance" mechanisms, without inducing undesirable autoimmune reactions. In contrast to TAAs, TSAs represent the expression of altered self-proteins as a consequence of viral transformation or genetic or somatic mutations in DNA encoded by oncogenes, such as point-mutated *ras* [32–35]. These aberrant proteins may contain unique antigenic determinants surrounding the sites of mutations for immune recognition and subsequent activation and expansion of Ag-specific lymphocyte (CD4+ and/or CD8+ T-cell) precursors, thereby making oncogene products potentially attractive tumor-specific target molecules for cancer immunotherapy. Because oncogene products represent foreign proteins, it is also likely that the appropriate lymphoid precursors exist within the host immune repertoire and that the resulting immune response would be restricted toward recognition of the aberrant target cell population and not normal cells. An important purpose of this chapter is to focus on an understanding of the nature of the biological interactions between, and consequences resulting from, the host immune system and products of activated *ras* genes.

IV. *ras* ONCOGENES AND TUMORIGENESIS

For years, oncogenes have been implicated in the pathways of cellular transformation and the pathogenesis of neoplasia [35,36]. The *ras* p21 oncogenes illustrate one such example; point mutations in these genes are commonly found in a wide diversity and high frequency of human malignancies (further reviewed in references [32–34,37]). The *ras* p21 proto-oncogenes normally encode a family of evolutionary-conserved, 21-kDa intracellular guanosine triphosphate (GTP) binding proteins (189 amino acids in length) important for cellular signal transduction, which ultimately regulate cellular differentiation, proliferation, and function in a wide variety of cell types [38,39]. In mammalian cells, the *ras* proto-oncogenes consist of three highly homologous members, K-*ras,* H-*ras,* and N-*ras,* and each gene has the capacity to become oncogenic as a consequence of somatic mutation. *ras* p21 was originally discovered as an oncogene product; that is, the activated form of *ras* was identified as a transforming gene of an acute transforming murine sarcoma virus. Subsequently, genomic DNA from human tumors and tumor cell lines was demonstrated to contain activated *ras* genes. Cloning and sequencing of the *ras* oncogenes from acute transforming retroviruses, human tumor DNA, and the proto-oncogenes from normal cells revealed that the activated forms of the *ras* genes contained mutations restricted to specific sites, principally at codons 12, 13, 59, or 61. In human tumors, the vast majority of *ras* mutations are found at codon 12; whereas in rodent tumors, such as those induced by chemical carcinogens, the majority of *ras* mutations are found at codon 61.

Under normal physiological conditions, the basic function of *ras* p21 is to bind one molecule of GTP or guanosine diphosphate (GDP) per molecule of *ras* p21 protein. With regard to normal signal transduction and

cellular function, *ras* p21 acts as a binary switch and is in the active or ON state when bound to GTP and in the inactive or OFF state when bound to GDP. The presence of mutations at those codons results in proteins containing single amino acid substitutions at those corresponding positions. In a functional sense, such *ras* proteins are aberrant as they become constitutively GTP bound due to a loss both in intrinsic GTPase activity and responsiveness to GAPs (GTPase-activating proteins), which help to facilitate the hydrolysis or dephosphorylation of GTP to GDP. Hence, mutated *ras* proteins in the activated form appear to be in an ON or signal transmitting mode and consequently can mediate transformation of fibroblasts in culture.

A number of animal tumor models illustrate the central role that *ras* mutations can play in the pathogenesis and development of neoplasia. The experimental skin models for the development of epidermal tumors, papillomas, and squamous cell carcinoma in rodents and rabbits employ an initiation and promotion system for inducing tumor growth. The initiation step is provided by a chemical carcinogen (e.g., N-methylnitrosourea [MNU], 1-methyl-3-nitro-1-nitrosoguanidine [MNNG], 3-methylcholanthrene [MCA], 7,12-dimethylbenz(a) anthracene [DMBA]), which appears to induce specific mutations in the DNA. This is followed by treatment with phorbol esters, such as 12-O-tetradecanoylphorbol-13-acetate [TPA], to induce proliferation of cells harboring mutant *ras* genes and results in clonal expansion of the mutated cell populations. Using this system to induce mutations in H-*ras,* for example, analysis of tumor DNA demonstrated that MNNG is specific for G to A transitions at codon 12 and DMBA is specific for transversion mutations from A to T at codon 61. Similarly, UV radiation has been demonstrated to induce amplification of H-*ras* and A to G transitions at codon 61 in N-*ras* in mice. The presence of the mutation at different stages of tumor development indicates that the mutations are early events in tumorigenesis. The specificity of carcinogens for types and location of the mutations in the *ras* genes suggests that mutations may be an initiating event. Furthermore, transgenic mice containing mutated H-*ras* genes develop papillomas and squamous cell carcinomas without chemical or UV treatment as a required initiating step.

The role of mutated *ras* genes in the initiation and progression of a range of human tumors has been the focus of numerous investigations. The most widely used assay system for detection of *ras* mutations originally was the NIH-3T3 transfection assay; this assay was based on the ability of *ras* genes to transform the mouse NIH-3T3 cells. Although it was able to detect point mutations, it was not suitable for screening large numbers of tumor samples. The development of rapid assay systems for the identification of mutated *ras* genes has facilitated the analysis of large numbers of human tumor samples. These methods include (1) selective hybridization of tumor DNA with synthetic oligodeoxynucleotide probes specific for mutations in codons 12, 13, and 61 of the three *ras* genes [40,41]; (2) RNase mismatch cleavage assays [42]; and (3) use of polymerase chain reaction (PCR) to amplify segments of the *ras* genes [43,44].

In human cancers, the majority of point mutations in the *ras* genes are observed at codon 12. Mutations of codon 12 can result in the amino acid replacement of the normal Gly residue at position 12 with either an Asp, Val, Cys, Ser, Arg, or Ala residue. Carcinomas, especially adenocarcinomas, and hematologic malignancies of the myelomonocytic lineage are more frequently associated with *ras* mutations than are tumors of neuroectodermal origin or differentiated lymphoid malignancies, whereas carcinomas of the breast and cervix rarely contain *ras* mutations. Furthermore, there does appear to be a relationship between the expression of a given *ras* gene (H-, K-, or N-*ras*) and certain tumor types. For example, the predominant *ras* oncogene expressed in pancreatic, lung, and colon adenocarcinomas is K-*ras*. An extensive survey of the literature [37] has revealed that a large percentage of carcinomas of the colon/rectum ($\geq$38%), pancreas ($\geq$81%), lung ($\geq$19%), endometrium ($\geq$16%), and thyroid ($\geq$22%) contain *ras* position 12 mutations. Furthermore, *ras* position 12 mutations occur in a proportion of several other tumor types, including melanoma, hepatocellular, bile duct, basal cell, and squamous cell carcinomas and some pediatric malignancies (e.g., juvenile chronic myelogenous leukemia). Overall, for a diversity of such prevalent tumor types, more mutations are observed with the substitution of Gly to Asp ($\sim$40%), Val ($\sim$28%), or Cys ($\sim$16%), than Ala ($\sim$6%), Arg ($\sim$5%), or Ser ($\sim$5%) [37]. In the United States alone, for example, over 140,000 new cases of cancers are reported each year in patients whose tumors harbor *ras* codon 12 mutations. Based on these estimates, the total number of cancer patients in the United States currently available for treatment whose tumors contain position 12 *ras* mutations is over 800,000.

V. *ras* ONCOGENES AS TARGETS FOR IMMUNE EFFECTOR MECHANISMS AND IMMUNOTHERAPY

The rationale for the study of *ras* oncogenes as targets for immune effector mechanisms and immunotherapy thus reflects a number of reasons, some of which are highlighted here (and in Table 1): (1) First and foremost, *ras* oncoproteins result from somatic mutations, and from an immunologic perspective, these represent the ultimate type of TSAs. The production and presentation of such neo-epitopes underlies the basis of the central investigative hypothesis, which is to exploit the innate properties and abilities of the immune system to distinguish "self" (proto-oncogene) from "nonself" (oncogene) forms. Thus, in contrast to TAAs, issues of tolerance imposed on the negative selection and development of the Ag-specific immune precursor cells, and autoimmune reactions directed against self-tissues, become much less of a factor. Also, it is important to emphasize that although *ras* p21 proteins, whether normal or mutant, are found intracellularly (in association with the plasma membrane), peptide fragments produced by exogenous or endogenous pathways of Ag processing may associate with *de novo* synthesized MHC molecules, which can then be transported to and presented on the external surface of the APC/tumor cell for selective immune recognition. Although biochemical predictions can be made and even tested to evaluate and confirm expression of peptide epitopes (in association with MHC molecules), the most informative observations will result from functional studies that measure the specificity and intensity of the immune response. (2) *ras* mutations are thought to occur early in the neoplastic process and, indeed, have been found in premalignant lesions. Most notably, *ras* mutations have been identified in and associated with the early development, pathogenesis, and progression of colorectal and pancreatic adenocarcinomas [45,46]. Immunotherapy, therefore, is likely to be most effective when the disease burden is minimal and the immune system is intact and most functionally competent. (3) *ras* mutations are restricted and limited to only a few "hot spots" (codons) in the genes, thereby making the development of purified or recombinant reagents for cancer immunotherapy feasible and practical; and (4) *ras* mutations

TABLE 1   Rationale for the Study of *ras* Oncogenes as Targets for Immunotherapy

---

ras Mutations

1. Result from somatic mutations, which encode for ultimate type of TSAs for immune system recognition.

2. Found in a broad spectrum and high frequency of human cancers.

3. Expressed early in the neoplastic process.

4. Linked to the malignant phenotype, making loss of epitope expression unlikely.

5. Restricted to only a few hot spots in the genes.

are linked to the maintenance and support of the malignant phenotype, thus making loss of epitope expression unlikely. Because the presence of the *ras* mutation is intimately linked to tumorigenic phenotype, it is conceivable that the production of the resulting oncoprotein and expression of the corresponding peptide epitope(s) would also be maintained. Although mutated *ras* epitopes may be processed and presented by APCs/tumor cells, perhaps a more revealing and provocative question is whether a sufficient density will be expressed by the Ag-bearing target cell beyond a critical biological threshold to initiate and mediate productive host cellular and antitumor immune responses.

Because the *ras* proto-oncogenes are highly conserved in evolution between mice and humans, and because activated *ras* genes result from somatic mutations, the mouse becomes a potentially useful preclinical model to study the role and modulation of the host immune response against tumors harboring *ras* oncogenes. A number of laboratories, including ours, have developed animal models to explore the underlying hypothesis that *ras* oncogenes encode for TSAs, which stimulate the production of Ag-specific cellular and antitumor immune responses. In the context of that hypothesis, several fundamental immunologic issues emerge (Table 2). The first two questions reflect and call into focus the most fundamental innate properties of the immune system, as they raise issues of induction of immune response, immune specificity and immune recognition/discrimination of self versus nonself peptide/ MHC complexes. The third and fourth questions reflect an understanding of the phenotypic and functional

---

TABLE 2   Immunologic Questions that Form the Basis for the Study of Host Immune Responses Directed Against *ras* Oncogene Products

---

1. Can T-cell responses be mounted against mutated *ras* gene products or, more specifically, mutated *ras* peptides surrounding the sites of mutations?

2. Can these T-cell responses discriminate recognition of mutant from wild-type *ras* forms?

3. Can both CD4+ and CD8+ T-cell responses be induced and targeted against the same point-mutated *ras* determinants?

4. What is the nature of these anti-*ras* immune effector mechanisms?

5. Can these peptide-induced T-cell responses recognize processed forms of the mutant *ras* proteins presented by host APC populations and/or tumor cells?

6. Can immunogenicity or specific immunogenic parameters be modified to further enhance the sensitivity and/or potency of cellular and antitumor immune reactions?

---

depth of the induced cellular immune response (i.e., the composition of the relevant immunologic cell types), and raise important issues of multiple T-cell (subset/ subtype) epitopes and delineation and characterization of the immune mechanisms requisite for cell-mediated and antitumor activities. The fifth question introduces an important transition and reflects an understanding of the biological relevance of the interaction between the peptide-induced TCR with MHC/peptide ligand expressed by the Ag-bearing target cell (i.e., third-party APC or tumor cell population). This is extremely important in the context of peptide-based immunotherapy, as the peptide-induced T-cell response must see and appropriately respond to the Ag-bearing target cell naturally processing and presenting the epitope(s), either endogenously (in the case of tumor cells) or exogenously (in the case of specialized APC). The sixth and final question addresses the nature of some future directions, and reflects an understanding of diverse immunological principles, including those outlined in the first five, and their application towards the synthesis of comprehensive approaches to improve and maximize immunogenicity.

VI. MURINE CD4+ T-CELL RESPONSES TO *ras* ONCOGENE PRODUCTS

Peace *et al.* [47] originally reported on the efficient *in vivo* priming of CD4+ T cells from mice (C57BL/ 6; H-2^b) challenged with mutant *ras* 12-mer or 13-mer peptides, spanning positions 5–16 or 5–17, respectively, and containing the Gly to Arg substitution at position 12. Furthermore, 13-mer *ras* peptides containing the Gly to Val or Gly to Cys substitution at position 12 were shown to be immunogenic in C57BL/6 or BALB/c (H-2^d) backgrounds. Lymphocyte reactivity, as measured by cellular proliferation, occurred in response to the immunizing peptide, but not to the normal *ras* peptide, and could be blocked by anti-Ia^b monoclonal antibody (MAb) in the C57BL/6 model. These observations revealed that the mutated *ras* peptides, in certain genetic backgrounds, were immunogenic and primed for MHC class-II-restricted, Ag-specific CD4+ T-cell activity. Peace *et al.* [47] later showed that murine CD4+ T-cell clones specific for the mutant *ras* Arg12 peptide proliferated in response to the corresponding recombinant mutant protein, supporting the possibility that T cells induced *in vivo* by peptide immunization could recognize native, exogenously processed oncoprotein. In a reciprocal fashion, they found that anti-*ras*-specific CD4+ T cells could be derived from mice administered a point-mutated H-*ras* protein (Gln to Leu at codon 61) [48], suggesting that in certain circumstances, abnormal *ras* p21 proteins may be sufficiently immunogenic to generate a cellular immune response.

Our studies were initiated and conducted in BALB/c mice (H-2^d) using purified peptides reflecting the *ras* codon 12 mutation, Gly to Val, as a model target cell determinant [49,50]. The rationale for the selection of this murine/*ras* oncogene combination was based on the development of a preclinical model to mimic a human cancer mutational hot spot and on the prediction that tumor-specific *ras* Val12 peptides can be defined that activate both CD4+ and CD8+ T-cell responses (see Table 3 for peptide sequences identified as epitopes and employed in these studies). First, we showed that splenic T cells derived from BALB/c mice challenged with the *ras*5–17(Val12) peptide *in vivo* displayed a specific proliferation response to the immunizing peptide. A T-cell line directed against this Val12 mutation was established that expressed strong Ag-specific proliferation and displayed a classical $\alpha\beta$-TCR+, CD3+, CD4+, CD8− phenotype [49]. MHC class II restriction analysis was determined using peptide-pulsed L-cell transfectants as APCs, each expressing one of the three major Ia^d class II alleles, AαAβ, EαEβ, or EαAβ. Only L cells expressing the EαAβ^d chain supported proliferation, and this response was strongly peptide specific. These data demonstrated that anti-*ras* CD4+ activity, as measured by lymphoblastic transformation, was MHC class II restricted and, in this setting, mapped to the EαAβ^d heterodimer. Cytokine release assays revealed the production of TNF, granulocyte-macrophage colony-stimulating factor (GM-CSF), IFN-γ, and IL-2, but not IL-4, suggesting that the anti-*ras*5–17(Val12) CD4+ T-cell line was characteristic of the Th1 subtype [6,51]. Consistent with Th1-type CD4+ T cells [51], Ag-

specific cytotoxicity was demonstrable against Ia^d-bearing A20 tumor cells incubated with exogenously bound *ras*5–17(Val12) peptide. Peptide-specific CD4+-mediated cytotoxicity was MHC class II restricted, as revealed by the absence of lysis against MHC class II$^-$ targets, inhibition of A20 lysis with anti-Ia^d MAbs, and induction of lysis against L-cell targets transfected with EαAβ^d [49].

The relative success of peptide-based immunotherapy depends on the capacity of these peptide-induced T cells to recognize and respond productively to autologous tumor expressing endogenous mutant *ras* Ag. Because we were unable to identify a naturally occurring murine tumor cell line that expressed both the appropriate *ras* mutation and the MHC class II Ia^d molecule to test this hypothesis, we developed an alternative approach by the introduction of the full-length K-*ras* gene encoding the Gly to Val mutation into A20 tumor cells via retrovirus transduction [49]. Furthermore, the construction of such cell lines allows one to test this hypothesis specifically, because any molecular differences and potential functional responses observed could be compared to and correlated with control targets that lack that gene. The sensitivity of these transduced A20 tumor cells to CD4+-mediated lysis was thus examined in the absence of exogenous *ras*5–17(Val12) peptide as a demonstration for TCR recognition of endogenously produced mutant *ras* Ag. Compared to the control targets, specific lysis was found against A20 cells transduced with the *ras* oncogene. With control A20 targets, lytic activity was expressed only in the presence of added *ras*5–17(Val12) peptide. Independent isolation of a second anti-*ras*5–17(Val12) CD4+ T-cell line revealed a very similar cytolytic and MHC class-II-restricted profile

TABLE 3   Mutant *ras* Peptide Sequences Reflecting the Substitution of Gly to Val at Position 12 Identified as Murine T-Cell Epitopes[a]

ras Peptides	Amino acid sequence	Immune response
	3 4 5 6 7 8 9 10 11 12 13 14 15 16 17	
*ras*3–17(Gly12)[b]	Glu-Tyr-Lys-Leu-Val-Val-Val-Gly-Ala-<u>Gly</u>-Gly-Val-Gly-Lys-Ser	Wild-type sequence
*ras*5–17(Val12)	Lys-Leu-Val-Val-Val-Gly-Ala-<u>Val</u>-Gly-Val-Gly-Lys-Ser	CD4+
*ras*4–12(Val12)	Tyr-Lys-Leu-Val-Val-Val-Gly-Ala-<u>Val</u>	CD8+
*ras*3–12(Val12)	Glu-Tyr-Lys-Leu-Val-Val-Val-Gly-Ala-<u>Val</u>	CD8+
*ras*4–12(Leu12)	Tyr-Lys-Leu-Val-Val-Val-Gly-Ala-<u>Leu</u>	CD8+
*ras*4–16(Val12)	Tyr-Lys-Leu-Val-Val-Val-Gly-Ala-<u>Val</u>-Gly-Val-Gly-Lys	CD4+ and CD8+

[a] Based on ability to generate or mediate anti-*ras* Val12 Ag-specific cellular immune responses using T-cell lines established from appropriately immunized BALB/c (H-2^d) mice.

[b] Normal sequence of *ras* p21 from positions 3–17 of the protein, which encompasses the various peptides analyzed in these studies.

(Fig. 1A). Importantly, as with the first line, this second CD4+ line also recognized and lysed A20 targets expressing the *ras* oncogene in the absence of exogenous peptide (Fig. 1B) [49]. Because such functional properties were consistent with the Th1 subset, we have demonstrated a previously unrecognized arm of the cellular immune response against *ras* oncogene products. Thus, the induction of a CD4+ Th1 response may be important for direct interaction with and lysis of the Ag-bearing target cell. Moreover, Th1-mediated antitumor effects may result indirectly from interactions with MHC class II+ APCs, leading to the secretion of cytokines that in turn alter tumor cell viability (e.g., TNF-α/β) or cause recruitment and further activation (e.g., interleukins, IFN-γ, GM-CSF) of other cytotoxic effector cells, such as CD8+ T cells, macrophages, or natural killer (NK) cells [2,52–54]. It is also important to emphasize that although lytic activity against these *ras*-transduced A20 cells was demonstrable, it was suboptimal as compared with the control A20 target population incubated in the presence of exogenous peptide for maximal response (Fig. 1B), suggesting that the level of endogenous Ag was limiting and/or that these anti-*ras*Val12 CD4+ T cells expressed weak TCR affinity for recognition of the intracellularly derived peptide/MHC class II cell surface complexes.

VII. MURINE CD8+ T-CELL RESPONSES TO *ras* ONCOGENE PRODUCTS

In terms of anti-*ras* CD8+ CTL immunity in murine models, several investigators have analyzed the Gln to Leu or Lys substitution at codon 61 of the H- or N-*ras* oncogene. Skipper and Stauss [55] examined the immunogenicity of both point-mutated and normal N-*ras* in an H-2^b model using vaccinia virus recombinants. In mice immunized with a vaccinia virus recombinant encoding point-mutated *ras* (Gln to Lys at codon 61), a specific CTL response was demonstrated against targets expressing the point-mutated, but not the normal *ras* gene. Interestingly, in mice administered a vaccinia virus recombinant expressing the normal *ras* proto-oncogene, they found evidence for the induction of a CTL response against target cells overexpressing normal *ras* (i.e., via transfection). Although the biological relevance of this response remains unclear, these CTLs did not lyse non-transfected targets, suggesting that the levels of endogenous normal *ras* expression were insufficient for TCR recognition. Peace *et al.* [56] examined similarly the immunogenicity of point-mutated *ras* p21 at codon 61 (Gln to Leu) in an H-2^b model, albeit by IVS of naive splenic T cells with a complex mixture of overlapping peptides. CTL elicited by this approach were shown to

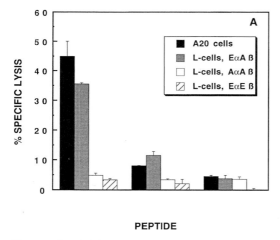
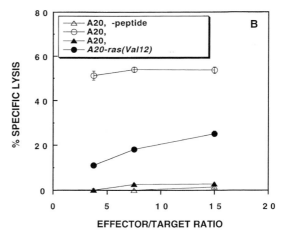

FIGURE 1   (A) CD4+ T-cell-mediated lysis is MHC class II restricted, which maps to the EαAβ allele. MHC class II restriction analysis as determined by ^{51}Cr-release cytotoxic assays using A20 cells (positive control) or L-cell transfectants expressing the different Ia^d molecules as targets, in the absence and presence of exogenous peptide (2 μg/mL; effector/target ratio, 5/1). (B) CD4+ T-cell recognition of A20 tumor cells expressing the point-mutated *Ras* oncogene. Recognition and lysis of A20 tumor cells transduced with the *ras*Val12 oncogene [i.e., A20-*ras*(Val12) gene] in the absence of exogenous peptide. Controls included lysis against vector-transduced A20 cells in the absence and presence of mutant or wild-type *ras* peptide (2 μg/mL). In both panels, data are expressed as means ± SEM of triplicate cultures. (Data are from Abrams *et al.* [49].)

specifically lyse target cells transfected with the point-mutated H-*ras* oncogene. With respect to codon 12, Fenton *et al.* [57] described the induction of an anti-*ras* CTL response *in vivo* using recombinant mutant protein, which was specific for the substitution of Gly to Arg of the H-*ras* oncogene. Immunization of BALB/c mice with recombinant H-*ras* protein containing the Arg12 substitution also led to the rejection of syngeneic tumor cells transfected with the corresponding oncogene. This work demonstrated the capacity of a mutated *ras* immunogen to mediate antitumor activity *in vivo*.

As an approach toward the identification of putative CD8+ CTL epitopes [50], we have analyzed the mutant *ras* protein surrounding position 12 for potential MHC class I binding peptide sequences based on predicted anchor motifs for murine H-2^d [23,58]. Emphasis was placed on position 12, as this represented a potential focal point for the coexistence of both CD4+ and CD8+ T-cell epitope(s) surrounding that same *ras* point mutation (i.e., Gly to Val at codon 12). Activation of both anti-*ras* immune mechanisms (i.e., Th1 and CD8+) may be important to promote more comprehensive, specific, and potent cellular and antitumor responses. As a result of this analysis, two such candidate peptide sequences were identified (Table 3), *ras*4–12(Val12), a 9-mer, and *ras*3–12(Val12), a 10-mer, as they possessed a primary amino acid MHC anchor motif for the H-2K^d class I allele. The motif for H-2K^d illustrates Tyr at positions 1 or 2 and a hydrophobic residue, such as Leu, Ile, or Val, at the C terminus of an 8- to 10-mer peptide. An added attraction to inducing a CTL response to these particular peptides is that they may represent *de novo* antigenic peptides that the immune system has not previously encountered. The corresponding sequences from wild-type *ras* lack a putative C-terminus anchor and so are unlikely to ever have been presented as an H-2K^d-restricted epitope. Consequently, the anti-*ras* CD8+ CTL response would result from TCR recognition of a MHC/peptide complex previously unrecognized, providing, in part, a novel understanding of the basis of immunogenicity to mutant, but not normal *ras*.

We first demonstrated that both mutant *ras*4–12(Val12) and mutant *ras*3–12(Val12) peptides, but not the counterpart wild-type sequences, specifically inhibited in a dose-dependent fashion a positive control H-2K^d-restricted CTL response, thus displaying functional interaction with the H-2K^d molecule. However, relative to both positive and negative control (model) H-2^d class I allele peptides, both mutant *ras* 9-mer and 10-mer peptides exhibited comparable and weak binding to H-2K^d. Because both mutant *ras* peptides appeared to behave similarly (at least by functional binding to the H-2K^d molecule), we opted to select the *ras*4–12(Val12) peptide as the putative minimal or optimal size sequence

to illustrate potential CD8+ T-cell immunogenicity. To that end, BALB/c mice were injected with the peptide in adjuvant and CTL activity was assessed following the immunization schedule. Although a weak primary CTL response was observed, an anti-*ras* CTL line was established from this bulk culture by continuous IVS with antigenic peptide and IL-2. The resulting CTL line displayed a classical $\alpha\beta$-TCR+, CD3+, CD8+, CD4− phenotype and mediated specific lysis of syngeneic tumor targets incubated with the immunizing mutant, but not the wild-type *ras* 9-mer sequence. Cytotoxicity was restricted by H-2K^d, as MAbs directed against H-2K^d, but not H-2D^d or H-2L^d (the two other major class I alleles in H-2^d mice), molecules strongly inhibited CTL activity (Fig. 2A). In addition to *ras*4–12(Val12), cytotoxicity was induced by incubation with the closely related mutant *ras* 10-mer peptide, *ras*3–12(Val12) (Fig. 2B), which paralleled its capacity to bind to H-2K^d. However, binding to H-2K^d alone was insufficient for stimulating anti-*ras* CTL activity. For example, NP$_{147-155}$ peptide, a potent H-2K^d binding immunodominant peptide of influenza virus, failed to activate the anti-*ras* CTL response (Fig. 2B). Furthermore, point-mutated *ras* peptides that contained the 9-mer core sequence, were longer (i.e., residues 3–15) or of the appropriate length, but completely lacked a relevant anchor(s) (i.e., residues 5–14 or 8–16) were unable to sensitize targets for lysis (Fig. 2B). Thus, these data emphasized the critical importance of both amino acid sequence and peptide length for anti-*ras* CD8+ CTL-mediated lysis.

As with the CD4+ response, the efficiency of an antitumor CD8+ response and the extent of its biological relevance *in vivo* require TCR recognition of tumor-derived endogenous Ag in the context of MHC class I. CTL recognition of endogenously derived Ag was determined against the *ras*-transduced A20 target cells, as it expressed both the appropriate *ras* codon 12 mutation and the MHC class I restriction element. Such Ag-specific CTL lysed these *ras*-transduced A20 cells in the absence of peptide (Fig. 3), suggesting productive TCR recognition of an endogenously produced target epitope. In contrast, no lysis was detectable against control (vector-transduced) A20 cells in the absence of peptide, supporting the nature of target specificity of this anti-*ras* CTL/A20 interaction. Lysis against control A20 targets was inducible, however, in the presence of the mutated, but not the wild-type, *ras* 9-mer peptide (Fig. 3), revealing that sensitivity to lysis likely reflected the availability of the appropriate peptide/MHC class I complexes for TCR recognition. Furthermore, as with the anti-*ras* CD4+ T-cell response, it is important to stress here that although cytotoxicity against these *ras*-transduced A20 cells was demonstrable, it was suboptimal. This was determined

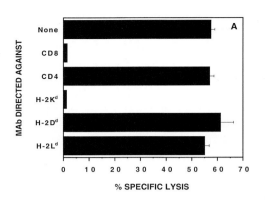
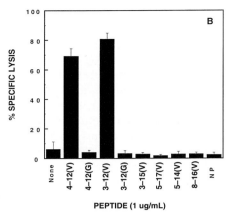

FIGURE 2   (A) CD8+ CTL-mediated lysis is MHC class I restricted, which maps to the H-2K^d allele. Role of CD8 and MHC class I H-2K^d molecules in cytotoxicity as determined by ^{51}Cr release using peptide-coated (1 μg/mL) P815 cells as targets (effector/target ratio, 5/1), in the absence and presence of the different MAbs as shown. (B) Importance of peptide size and primary sequence for anti-*ras* CTL-mediated lysis. The CD8+ CTL line was tested for lytic specificity using a panel of normal (Gly12) and mutant (Val12) *ras* peptides (including 9-, 10-, and 13-mer sequences each at 1 μg/mL). Also, NP$_{147-155}$ was included as a negative control peptide. Data represent percent specific lysis against ^{51}Cr-labeled P815 targets in the absence and presence of peptide. In both panels, data are expressed as means ± SEM of triplicate cultures. (Data are from Abrams *et al.* [50].)

by comparing the lytic sensitivity of control A20 cells to *ras*-transduced A20 cells incubated in the presence of a saturable dose of exogenous relevant peptide for maximal response (Fig. 3). The fact that lytic activity under these conditions could be further boosted supported the hypothesis that the density of endogenous Ag was limiting and/or that these CTLs expressed weak TCR affinity for recognition of the endogenously derived MHC class I/peptide ligand cell surface complexes.

VIII. ENHANCING PEPTIDE EPITOPE IMMUNOGENICITY AND POTENCY

Because of their potentially relevant role as CD4+ and/or CD8+ tumor-specific peptide epitopes, but low affinity for MHC binding and weak *in vivo* immunogenicity, the ability to generate and mediate effective cellular and antitumor immune reactions will necessitate, in part, targeted modifications in cancer immunotherapy designs that aim to amplify their immunogenic potency. From a cancer biology perspective, these findings may also have insights into understanding potential mechanisms of tumor escape, which may occur at the level of the APC/tumor cell or TCR, reflecting both quantitative and qualitative aspects of the mechanism of T cell recognition of

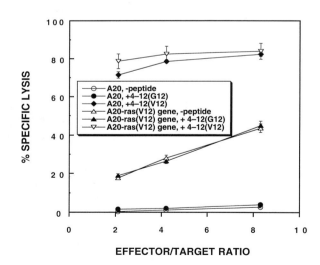

FIGURE 3   CD8+ T-cell recognition of A20 tumor cells expressing the point-mutated *ras* oncogene. Cytotoxicity against A20 tumor cells transduced with the *ras*(Val12) oncogene [i.e., A20-*ras*(V12) gene] tested in the absence and presence of exogenous mutant [*ras*4–12(Val12)] or wild-type [*ras*4–12(Gly12)] peptide (1 μg/mL). Controls included lysis against vector-transduced A20 cells, also in the absence and presence of mutant or wild-type *ras* peptide. Data are expressed as means ± SEM of triplicate cultures. (Data are from Abrams *et al.* [50].)

Ag. For example, quantitative factors may reflect low *epitope density* expressed by the target cell population, which may be due to (1) the intracellular generation and/or extracellular expression of suboptimal MHC class I or II/peptide ligand complexes below a critical biological threshold for efficient TCR (CD4+ or CD8+) activation, an outcome that may be largely influenced by low peptide/MHC binding affinity and the formation of unstable (short-lived) immunogenic complexes, which rapidly disassemble from the cell surface or fail to compete effectively with a multitude of stronger binding exogenously or endogenously derived peptides; or (2) the presentation of the relevant epitope predominantly in the context of an inappropriate MHC molecule (e.g., class II alleles/restriction patterns, which tend to display greater degeneracy in binding interactions with a given peptide). Qualitative factors may reflect properties of the T-cell response, such as weak TCR affinity/avidity for recognition the intracellularly derived peptide/MHC class I or II cell surface complexes.

ASI or adoptive cellular immunotherapy thus becomes an alternative strategy to circumvent, perhaps, some of these inherent biological limitations. The overall goal of these studies, therefore, is to further alter or manipulate immunogenicity *in vivo* (via ASI) and/or *in vitro* (via cell expansion for adoptive transfer) in order to achieve a maximum biological response, consisting of abundant T-cell clonal expansion (i.e., "quantitative response") and high TCR affinity, sensitivity, and potency (i.e., "qualitative response") for recognition of limiting amounts of endogenously or exogenously derived tumor Ag. Furthermore, the interaction between higher-affinity TCRs with MHC/peptide ligand complexes may be of sufficient strength to reduce or obviate the need for Ag-independent interactions [59,60], such as those involved in cell–cell adhesion, which may be important to overcome another potential step of the tumor escape process. The discussion that follows is a laboratory overview of preclinical studies designed to improve the immunogenic potency of minimally defined mutant *ras* CD4+ and/or CD8+ T-cell peptide epitopes for the induction, expansion, and sensitivity of the relevant cellular immune response(s). Thus, these studies focus first on an understanding of the nature of TCR–MHC/peptide interaction, which is fundamental to the mechanism of Ag recognition and eventual induction of effective cell-mediated immunity.

A. Immunogenicity of the *ras*5–17(Val12) Peptide Epitope in Linear Versus MAP Form

Our earlier studies had established that the *ras*5–17(Val12) linear peptide epitope was immunogenic, in the context of an adjuvant setting, for the induction of CD4+ T-cell responses. Using the *ras*5–17(Val12) peptide as a model antigenic determinant, we examined whether immunogenicity could be enhanced by construction of the peptide epitope in a branched peptide form, known to MAP (multiple antigenic peptide) [61]. Furthermore, due to the sophisticated and complex nature of this immunogen structure, we examined whether adjuvant was necessary for the induction of the cellular immune response. The utility of MAPs for the induction of humoral and cellular immune responses in animal models has been demonstrated for a variety of peptide epitopes involved in infectious disease [62,63]. The MAP structure typically consists of two, four, or eight branches of the peptide epitope on a lysine core backbone [64]. In comparative immunization studies, MAPs have been shown to elicit stronger Ab responses than linear peptides [65]. However, it remains unclear whether MAPs are superior to linear peptides for the induction of cellular immune responses. Overall, we demonstrated, in a comparative analysis, that the linear peptide form was as efficient as three different MAP structures (i.e., 2, 4, and 8 branches) in the generation and potency of specific CD4+ T-cell responses. Furthermore, we found that adjuvant was required for the induction of these cellular immune responses, whether the immunogen was delivered in linear or branched forms. This work represented the first reported investigation of a cellular immune response using MAP immunogens incorporating a tumor-specific peptide epitope and demonstrated, in a comparative fashion, that although MAP forms induced epitope-specific T-cell responses *in vivo,* the magnitude of those responses were no greater than those for the linear peptide itself. This does not preclude the possibility that with appropriate modifications in MAP structure or immunization parameters (e.g, prime with MAP and boost with linear peptide), they may prove to be more effective.

B. Development of Mutant *ras* CD8+ CTL Peptide Epitope Variants

Although the *ras*4–12(Val12) peptide was immunogenic in BALB/c mice, it displayed weak binding to the MHC class I H-2K^d molecule *in vitro* and generated a weak primary CTL response. Thus, the relative degree of weak immunogenicity *in vivo* may have correlated with the weak binding characteristics of the peptide to MHC molecules and/or the availability of a limited precursor CTL population. To discern between these possibilities, we have initiated studies to focus first on the nature of the MHC/peptide interaction. The overall rationale of these studies is to evaluate in a model system the ability to develop a more stable antigenic peptide/ MHC complex with enhanced immunogenicity for the production of the relevant antitumor CD8+ T-cell populations. Positive correlations between affinity of peptide binding to MHC class I molecules and *in vivo* immunogenicity for the generation of CTL responses have been described previously in murine models of infectious disease [27,66].

As a model to test this hypothesis, we have sought to define potential variants of the *ras*4–12(Val12) peptide epitope, which may enhance both binding to H-2K^d and potency of the CTL response without altering immune specificity and TCR recognition of the naturally expressed point-mutated determinant [95]. Because the motif for H-2K^d may accommodate Tyr at positions 1 or 2 and a hydrophobic residue, such as Leu, Ile, or Val, at the C terminus of an 8 to 10-mer peptide sequence, the approach taken was to either replace or reposition an existing anchor to promote more favorable binding to MHC class I H-2K^d. Because other hydrophobic amino acids, such as Leu or Ile, have been reported to serve as preferred, dominant C-terminus anchor residues for binding to H-2K^d with higher affinities [23,25,58], the first strategy was to synthesize *ras*4–12 peptides containing those residues in place of Val12. The ability to manipulate the Val12 residue is consistent with the hypothesis that the mutation at that site led to the creation of a C-terminus anchor or MHC contact site, rather than a TCR contact site. In fact, this was confirmed using single, sequential Ala-substituted peptides, in which residues 6–10 of the *ras*4–12(Val12) peptide were found to be critical for TCR recognition.

Using the *ras*4–12(Leu12) sequence as a model peptide variant (Table 3), in comparative studies with the *ras*4–12(Val12) peptide, we have thus far demonstrated: (1) enhanced functional binding to H-2K^d, as determined by more efficient inhibition (or reduction in lysis) of a control H-2K^d restricted, antiflu (NP$_{147–155}$-peptide-specific) CTL response (Fig. 4A); and (2) enhanced activation of the established anti-*ras*4–12(Val12)-derived CTL line, as determined by their improved sensitivity to limiting peptide dose and half-maximal responses for stimulation of cytotoxicity (Fig. 4B) and proliferation (Fig. 4C) activities. In addition to *in vitro* immunogenicity, injection of BALB/c mice with *ras*4–12(Leu12) peptide (in adjuvant) led to the production of an Ag-specific CD8+ CTL line, which lysed tumor cells expressing Ag, either exogenously (i.e., pulsed with the mutant or variant peptide, but not wild-type peptide) or endogenously (i.e., A20 targets transduced with the *ras* Val12 oncogene; Fig. 4D). Moreover, molecular and functional analyses of the TCRs of both anti-*ras* Val12- and Leu12-derived

CD8+ CTL lines revealed predominant usage of the same Vα/Vβ chains (i.e., Vα1/Vβ9) and recognition of the identical amino acid contact sites in the *ras*4–12(Val12) peptide epitope, respectively, indicating that the peptide variant essentially elicited *in vivo* the biologically relevant CTL populations.

To gain a more thorough understanding of the mechanism of action of mutant *ras* peptide epitope variants *in vivo*, however, it is important to further evaluate and compare the immunogenicity of the unmodified mutant sequence to the modified sequence at both quantitative and qualitative cellular levels. Quantitative assessment of immunogenicity may involve precursor frequency analysis of immune lymphocytes at limiting dilution for recognition of the relevant Ag (i.e., unmodified mutant *ras* peptide). Qualitative assessment of immunogenicity may involve comparing immune lymphocytes from the different experimental groups for the intensity of functional response following stimulation *in vitro* with titrating doses of the relevant Ag. Taken collectively these studies may provide important conceptual insights into the hypothesis that targeted modifications at primary and, perhaps, secondary MHC amino acid anchor residues can be introduced into oncogene-derived tumor-specific peptide epitopes, such as point-mutated *ras*, to enhance both MHC binding activity and immunogenicity for the induction, expansion ("quantity"), and potency ("quality") of the antigenically relevant T-cell response without cross-reaction against self (or proto-*ras* forms). Moreover, the fact that this concept can be applied directly to oncogene-derived peptide at sites of mutations that create MHC anchor positions, producing surrogate peptide epitopes with enhanced immunogenic potency, may help broaden our understanding about the development and augmentation of not only CD8+, but also CD4+ T-cell-mediated immunity against malignancies expressing neo-epitopes encoded by a variety of mutated cellular proto-oncogenes and/or tumor suppressor genes.

C. Induction and Amplification of Ag-Specific CD4+ and CD8+ T-Cell Responses

The finding that overlapping CD4+ [i.e., *ras*5–17(Val12)] and CD8+ [i.e., *ras*4–12(Val12)] T-cell epitopes exist that reflect the same *ras* mutation may have important conceptual and mechanistic implications. Indeed, overlapping CD4+ and CD8+ T-cell epitopes have been described in experimental models of influenza [67–69], HIV [70], and p53 [71]. The biological significance of overlapping T-cell epitopes is as yet unclear, but may impact on the coordination, efficiency,

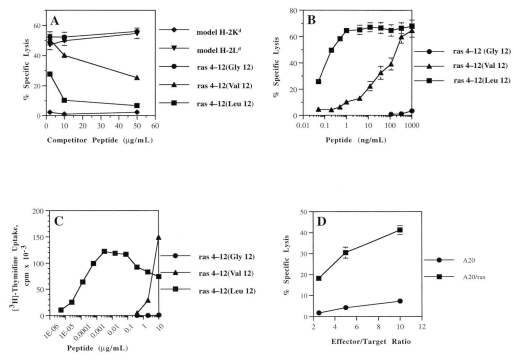

FIGURE 4 Correlation between MHC class I binding activity and CD8+ T-cell immunogenicity. (A) *ras* peptide variant 4–12(Leu12) displays enhanced binding to H-2K^d as compared with the unmodified mutant *ras*4–12(Val12) sequence. The different *ras* peptides were tested for their ability to bind to H-2K^d using a functional competition bioassay, which measured specific inhibition (or reduction in lysis) of a control H-2K^d-restricted CD8+ CTL line reactive with an immunodominant 9-mer peptide epitope of influenza virus (i.e., NP$_{147–155}$). (See Abrams *et al.* [50] for details of assay design). 9-mer peptide sequences containing known H-2K^d and H-2L^d consensus anchor motifs in a polyalanine backbone (i.e., model peptides) were used as positive and negative controls, respectively. Effector/target ratio, 3/1; NP$_{147–155}$ peptide, 0.3 ng/mL; time of assay, 4 hours. No lysis was detectable in the absence of NP peptide or in the presence of any competitor peptide. (B) *ras* peptide variant 4–12(Leu12) enhances cytotoxic response of anti-*ras* 4–12(Val12)-derived CD8+ CTL line. Cytolytic activity of anti-*ras*4–12(Val12)-derived CD8+ CTL was tested against P815 target cells incubated with the indicated concentrations of unmodified or modified (variant) mutant *ras* 9-mer peptides, as determined by ^{51}Cr release (effector/target ratio, 5/1). (C) *ras* peptide variant 4–12(Leu12) enhances proliferative response of anti-*ras*4–12(Val12)-derived CD8+ CTL line. Proliferation of anti-*ras*4–12(Val12)-derived CD8+ CTL tested in response to stimulation with irradiated, syngeneic splenocytes as APCs plus the indicated concentrations of the *ras* 9-mer peptides, as determined by [^{3}H]thymidine uptake. (D) *ras* peptide variant 4–12(Leu12) generates a CD8+ CTL line that lyses A20 tumor cells expressing the *ras*(Val12) oncogene. Recognition and lysis of A20 tumor cells transduced with the *ras*(Val12) oncogene (i.e., A20/*ras*) or vector only as control, in the absence of exogenous peptide. In all four panels, data are expressed as means ± SEM of triplicate cultures. (See reference 95).

and potency of the resultant cellular and antipathogen immune response. The combination of both Ag-specific CD4+ and CD8+ T-cell responses may be important at both the induction and effector phases of the immune response. At the induction phase, lymphokines produced by the Ag-primed CD4+ T-cell response, such as IL-2, may further amplify the clonal expansion of the Ag-primed CD8+ T-cell population. Moreover, at the effector phase, the activation of both subsets may enhance the diversity or repertoire of antipathogen im-

mune mechanisms. To evaluate the role of Ag-specific CD4+ T-cell help in the generation and potency of the resulting CD8+ T-cell response, immunizations may employ (1) a combination of two separate mutant *ras* peptides, one reflecting a CD4+ T-cell epitope [i.e., *ras*5–17(Val12)] and one reflecting a CD8+ T-cell epitope [i.e., *ras*4–12(Val12) or peptide variant], and (2) a single mutant *ras* peptide immunogen containing both CD4+ and CD8+ T-cell epitopes in a nested configuration.

In regard to the second approach, a 13-mer mutant *ras* peptide has been identified that encompasses both epitopes [i.e., *ras*4–16(Val12); Table 3] and, consistent with this hypothesis, this peptide has been found to be immunogenic (in adjuvant) for the induction of both CD4+ and CD8+ T-cell responses at the population level (Fig. 5; J. A. Bristol, unpublished observations). T lymphocytes isolated *ex vivo* from immunized BALB/c mice responded specifically to the immunizing peptide in a dose-dependent fashion (Fig. 5A), as measured by proliferation. In contrast, minimal activity was observed against the wild-type *ras* peptide (Gly12), demonstrating specificity for the induction and TCR recognition of the mutant *ras* sequence. Ag-specific proliferation was completely inhibited by MAbs directed against the CD4 molecule, confirming the phenotype of the functional T-cell subset as CD4+. In addition to the induction of a CD4+ T-cell response, immunization of these mice with this "nested" peptide sequence led to the generation of a CD8+ T-cell response (Fig. 5B). To that end, in parallel cultures, a portion of these same T lymphocytes were cultured *in vitro* under conditions containing the mutant *ras* 9-mer CD8+ CTL peptide sequence, and cytotoxicity against peptide-pulsed tumor targets was subsequently determined. Specific lytic activity was observed against targets incubated with the mutant *ras*4–12(Val12) peptide at multiple concentrations, but not with the wild-type *ras* 9-mer sequence. Moreover, Ag-specific cytotoxicity was strongly blocked by MAbs directed against

the CD8, but not the CD4 molecule, confirming the phenotype of the functional T-cell subset as CD8+. As with the studies on the mutant *ras* peptide epitope variants, to gain a more thorough understanding of the role of Ag-specific CD4+ T-cell help in the generation and amplification of the CD8+ T-cell response, it is important to further evaluate the relative degree of improved immunogenicity at both quantitative and qualitative cellular levels.

IX. PRINCIPLES OF ADOPTIVE CELLULAR IMMUNOTHERAPY USING EPITOPE-SPECIFIC T CELLS

The principles for developing adoptive immunotherapy for human disease may be guided by insights derived from experimental animal models, such as viral infection [72] and cancer [2,3,5,73] that investigate the cellular requirements for an effective host response. Because of the advanced nature of metastatic disease, an important goal of adoptive cellular transfer is to generate *ex vivo* such Ag-sensitized and specific T cells (lines or clones) as efficiently as possible so that they may be returned to the patient as promptly as possible. To that end, the *in vitro* methodologies developed and utilized for the isolation and appropriate propagation of these tumor-reactive T cells are of paramount importance. The use of nominal T-cell peptide epitopes as Ag for *ex vivo* stimulation offers the advantage of specifically targeting

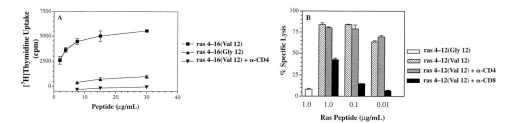

FIGURE 5   Induction of both anti-*ras*Val12-specific CD4+ and CD8+ T-cell responses using a nested peptide epitope. (A) BALB/c mice were primed and boosted twice *in vivo* with the mutant *ras*4–16(Val12) peptide (100 μg/inoculation; admixed with adjuvant). Splenic T cells were isolated and restimulated *in vivo* with either nothing (medium control) or the *ras* peptides corresponding to the wild-type (Gly12) or mutant (Val12) forms at multiple concentrations for measurement of proliferative response. Additionally, MAbs directed against the murine CD4 molecule were included to confirm the phenotype of the functional T-cell subset. (B) A T-cell line was established from immune lymphocytes described in (A), and tested for cytolytic activity against ^{51}Cr-labeled P815 targets (effector/target ratio, 10/1) incubated *in vitro* with the mutant *ras*4–12(Val12) CTL peptide epitope at multiple concentrations ± MAbs directed against the murine CD4 or CD8 molecule to confirm the phenotype of the functional T-cell subset. Prior to testing, however, T-cell cultures were propagated by weekly IVS cycles with relevant Ag to amplify the frequency of CD8+ precursors (i.e., initial stimulations with the immunizing peptide, which contained the minimal epitope, and subsequent stimulations with the nominal 9-mer sequence). In both panels, data are expressed as means ± SEM of triplicate cultures. (Data are from Bristol *et al.,* unpublished observations).

the biologically relevant clones within bulk cultures for selective T-cell activation and clonal expansion. This is in contrast with other types of immunogens employed for the production of immune lymphocytes for adoptive immunotherapy, such as irradiated tumor cells [74,75] or anti-CD3 MAbs [76–78], as these represent broader-based polyclonal methods of stimulation, which also have a greater potential to generate and amplify contaminating nonspecific and/or nonbiologically active cell populations.

Preclinical studies in animal models involving the adoptive transfer of epitope-specific T cells allow a focused and controlled investigation of a variety of fundamental immunologic principles, including (1) reconstitution and repopulation of the Ag-specific immune cells in lymphoid compartments, (2) persistence of immunologic memory, (3) maintenance of functional activity and Ag specificity, and (4) trafficking and targeting to tumor deposits. Initial studies may focus on experimental parameters to optimize the recovery of functional immune cells from lymphoid tissues and the phenotypic tracking of their migration and localization patterns *in vivo,* such as by specific anti-TCR (Vα/Vβ) measurements. Overall, these concepts address several important issues related to an understanding of the functional stability and Ag specificity of the immune cells, the nature and roles of different T-cell subsets (analyzed independently and in combination) in the particular adoptive transfer model, the ability to sustain and potentiate cellular immune reactions *in vivo* by periodic antigenic challenge, and whether the expression of specific lymphoid phenotypic markers correlate with their functional properties *in vivo,* such as homing efficiency to target sites. Furthermore, these studies may provide important insights into an understanding of the combined actions of both active and passive immunotherapies on the regression of established tumor.

In preliminary experiments with point-mutated *ras,* we have begun to explore some of these basic issues. We have initiated studies to examine both the CD4+ and CD8+ T-cell "recall" response to the *ras*(5–17(Val12) and *ras*4–12(Val12) peptides, respectively, as model systems to explore principles of adoptive transfer of epitope-specific immune lymphocytes (in tumor-free hosts). We have demonstrated that the adoptive transfer of *ras*5–17(Val12) peptide-specific immune lymphocytes into naive (sublethally irradiated) syngeneic recipients persisted *in vivo* ≥ 8 weeks posttransfer, as determined by the specificity and potency of the proliferative response recovered from splenic T cells. Moreover, we found that Ag stimulation via peptide boosting 2–3 days posttransfer markedly enhanced the recovery of the CD4+ proliferative response. Similarly, in the CD8+ T-cell model, we have demonstrated that the adoptive

transfer of a *ras*4–12(Val12) peptide-specific CD8+ T-cell clone into naive (sublethally irradiated) syngeneic recipients persisted *in vivo* ≥ 2 weeks posttransfer, as determined by specificity and potency of the CTL response recovered from T cells isolated from both spleen and lung. Ag stimulation via peptide boosting 2 days posttransfer was necessary for the recovery of the CTL response.

Finally, although studies in animal tumor models have revealed that the efficacy of T-cell-based therapy is quantitative, with larger numbers of T cells causing a higher degree of tumor regression, the functional potency of the transferred cells is perhaps just as important as or even more than the number of cells transferred. This may be accomplished by IVS of such CD4+ and/or CD8+ precursors using (sub)dominant peptide epitopes (or variants), coupled with modulation of the Ag dose [79,80] so that the resulting T-cell clones may posses high avidity for recognition of APCs or tumor cells expressing extremely low densities of exogenously or endogenously derived Ag. This approach is consistent with the hypothesis that TCR affinity for MHC/peptide expressed by the APC or target cell is determined, in part, by Ag density [80] and the number of cell surface TCRs that must be engaged to trigger the appropriate signal transduction pathways leading to functional clonal activation and proliferation [81]. For example, T-cell clones with high-affinity TCRs become activated at a low Ag dose, as only few numbers of TCRs must be engaged, and conversely for T-cell clones with low-affinity TCRs. Thus, under limiting Ag conditions, T-cell clones with high-affinity TCRs will selectively outgrow those of low affinity. Although under high-Ag conditions T-cell clones with low-affinity TCRs may begin to proliferate, due to the magnitude of Ag exposure and strength of the induced signal, T-cell clones with high-affinity TCRs, alternatively, may become anergized or even die via the activation of apoptotic pathways [80]. Accordingly, in the context of peptide epitope-based IVS methodology, such Ag-primed T lymphocytes could be specifically amplified and manipulated *ex vivo* to achieve not only a desired *quantity,* but also a desired *quality* of immune effector cells that may mediate a potentially more biologically meaningful and therapeutic antitumor response upon return to the autologous tumor-bearing host. In a further discussion of adoptive immunotherapy see Chapters 22 and 24.

X. FUTURE DIRECTIONS

Due to their prevalence in cancer and functional link to the development and maintenance of the malignant phenotype, *ras* oncogenes present themselves as attractive and specific targets for the induction of tumor-specific, cellular immune responses. Preclinical studies in mice strongly support the hypothesis that although *ras* p21 may not be expressed on the cell surface as integral membrane proteins, point-mutated *ras* peptides may be presented in the context of the appropriate MHC class I and II molecules for Ag-specific immune recognition, cytokine production, and cytotoxicity by CD8+ CTLs and CD4+ Th1-type cells, respectively. Moreover, the fact that peptide-induced CD4+ and CD8+ T cells can be directed toward target cells expressing activated *ras* molecules provides the rationale for the development of immunotherapies against human cancers that express products of *ras* oncogenes.

It is important to emphasize that although mutated *ras* epitopes may be processed and presented by APC/tumor cells, perhaps the more profound hypothesis is whether a sufficient density will be expressed by the Ag-bearing target cell beyond a critical biological threshold to initiate and mediate productive host antitumor immune responses *in vivo*. The fundamental basis of effective cancer immunotherapy and the ability to overcome a potential diversity of tumor escape mechanisms will likely require a synergism between both quantitative and qualitative elements of the T-cell-mediated immune response. Preclinical studies in animal models, such as those outlined here, allow the analysis and dissection of such immunologic principles and mechanisms, which may help guide the investigation and understanding of host immune responses directed against human neoplasia, in general, and human tumors harboring *ras* oncogenes, specifically. The identification of human CD4+ and/or CD8+ T-cell peptide epitopes reflecting specific *ras* point mutations at codons 12, 13, or 61 by a variety of investigators [28,82–92] may have important and direct implications for the development of oncogene-specific vaccines or gene therapy approaches in cancer immunotherapy. Clinical studies [93,94] are thus warranted to fully explore the potential immunogenicity, specificity, and efficacy of immunotherapies directed against such human malignancies that harbor *ras* oncogenes. Appropriate modifications in immunogen design and immunization parameters (i.e., ASI), in concert with the *ex vivo* isolation and expansion of epitope-specific CD4+ and/or CD8+ T lymphocytes for adoptive transfer may be paramount toward the generation of tumor-specific T-cell clones, which display sufficiently high sensitivity for Ag recognition and the potential to mediate desirable clinical responses.

Note Added in Proof

Subsequent to submission of this book chapter, a manuscript by Bristol et al. (initially noted on pages 262 and 263 of the page proofs as "submitted for publication") was published in the *Journal of Immunology* (see reference 95).

References

1. Bystryn, J.-C. Tumor vaccines. *Cancer Metastasis Rev.* **9,** 81–91, 1990.
2. Greenberg, P. D. Adoptive T cell therapy of tumors: mechanisms operative in the recognition and elimination of tumor cells. *Adv. Immunol.* **49,** 281–355, 1991.
3. Melief, C. J. M., and Kast, W. M. T-cell immunotherapy of tumors by adoptive transfer of cytotoxic T lymphocytes and by vaccination with minimal essential epitopes. *Immunol. Rev.* **146,** 167–177, 1995.
4. Rosenberg, S. A. The development of new cancer therapies based on the molecular identification of cancer regression antigens. *Cancer J.* **1,** 90–100, 1995.
5. Salgaller, M. L. Monitoring of cancer patients undergoing active or passive immunotherapy. *J. Immunother.* **20,** 1–14, 1997.
6. Abbas, A. K., Murphy, K. M., and Sher, A. Functional diversity of helper T lymphocytes. *Nature* **383,** 787–793, 1996.
7. Swain, S. L., Croft, M., Dubey, C., Hayes, L., Rogers, P., Zhang, X., and Bradley, L. M. From naive to memory T cells. *Immunol. Rev.* **150,** 143–167, 1996.
8. Abrams, S. I., Hodge, J. W., McLaughlin, J. P., Steinberg, S. M., Kantor, J. A., and Schlom, J. Adoptive immunotherapy as an *in vivo* model to explore antitumor mechanisms induced by a recombinant anticancer vaccine. *J. Immunother.* **20,** 48–59, 1997.
9. Stuhler, G., and Schlossman, S. F. Antigen organization regulates cluster formation and induction of cytotoxic T lymphocytes by helper T cell subsets. *Proc. Natl. Acad. Sci. USA* **94,** 622–627, 1997.
10. Brodsky, F. M., and Guagliardi, L. E. The cell biology of antigen processing and presentation. *Annu. Rev. Immunol.* **9,** 707–744, 1991.
11. Rothbard, J. B., and Gefter, M. L. Interactions between immunogenic peptides and MHC proteins. *Annu. Rev. Immunol.* **9,** 527–565, 1991.
12. Jorgensen, J. L., Reay, P. A., Ehrich, E. W., and Davis, M. M. Molecular components of T-cell recognition. *Annu. Rev. Immunol.* **10,** 835–873, 1992.
13. Germain, R. N., and Margulies, D. H. The biochemistry and cell biology of antigen processing and presentation. *Annu. Rev. Immunol.* **11,** 403–450, 1993.
14. Henkart, P. A. Lymphocyte-mediated cytotoxicity: two pathways and multiple effector molecules. *Immunity* **1,** 343–346, 1994.
15. Kagi, D., Vignaux, F., Ledermann, B., Burki, K., Depraetere, V., Nagata, S., Hengartner, H., and Golstein, P. Fas and perforin pathways as major mechanisms of T cell-mediated cytotoxicity. *Science* **265,** 528–530, 1994.
16. Salgame, P., Abrams, J. S., Clayberger, C., Goldstein, H., Convit, J., Modlin, R. L., and Bloom, B. R. Differing lymphokine profiles of functional subsets of human CD4 and CD8 T cell clones. *Science* **254,** 279–282, 1991.
17. Croft, M., Carter, L., Swain, S. L., and Dutton, R. W. Generation of polarized antigen-specific CD8 effector populations: reciprocal action of interleukin (IL)-4 and IL-12 in promoting type 2 versus type 1 cytokine profiles. *J. Exp. Med.* **180,** 1715–1728, 1994.
18. Romagnani, S. Lymphokine production by human T cells in disease states. *Annu. Rev. Immunol.* **12,** 227–257, 1994.

19. Sad, S., Marcotte, R., and Mosmann, T. R. Cytokine-induced differentiation of precursor mouse CD8+ T cells into cytotoxic CD8+ cells secreting Th1 or Th2 cytokines. *Immunity* **2,** 271–279, 1995.
20. Dutton, R. W. The regulation of the development of CD8 effector T cells. *J. Immunol.* **157,** 4287–4292, 1996.
21. Fowler, D. H., Breglio, J., Nagel, G., Eckhaus, M. A., and Gress, R. E. Allospecific CD8+ Tc1 and Tc2 populations in graft-versus-leukemia effect and graft-versus-host disease. *J. Immunol.* **157,** 4811–4821, 1996.
22. Boon, T., and van der Bruggen, P. Human tumor antigens recognized by T lymphocytes. *J. Exp. Med.* **183,** 725–729, 1996.
23. Falk, K., Rotzschke, O., Stevanovic, S., Jung, G., and Rammensee, H.-G. Allele-specific motifs revealed by sequencing of self-peptides eluted from MHC molecules. *Nature* **351,** 290–296, 1991.
24. Rudensky, A. Y., Preston-Hurlburt, P., Hong, S.-C., Barlow, A., and Janeway, Jr, C. A. Sequence analysis of peptides bound to MHC class II molecules. *Nature* **353,** 622–627, 1991.
25. Rammensee, H.-G., Friede, T., and Stevanovic, S. MHC ligands and peptide motifs: first listing. *Immunogenetics* **41,** 178–228, 1995.
26. Bjorkman, P. J., Saper, M. A., Samraoui, B., Bennett, W. S., Strominger, J. L., and Wiley, D. C. The foreign antigen binding site and T cell recognition regions of class I histocompatibility antigens. *Nature* **329,** 512–518, 1987.
27. Boehncke, W.-H., Takeshita, T., Pendleton, C. D., Sadegh-Nasseri, S., Racioppi, L., Houghten, R. A., Berzofsky, J. A., and Germain, R. N. The importance of dominant negative effects of amino acid side chain substitution in peptide-MHC molecule interactions and T cell recognition. *J. Immunol.* **150,** 331–341, 1993.
28. Tsang, K. Y., Nieroda, C. A., DeFilippi, R., Chung, Y. K., Yamaue, H., Greiner, J. W., and Schlom, J. Induction of human cytotoxic T cell lines directed against point-mutated p21 *ras*-derived synthetic peptides. *Vaccine Res.* **3,** 183–193, 1994.
29. Hodge, J. W. Carcinoembryonic antigen as a target for cancer vaccines. *Cancer Immunol. Immunother.* **43,** 127–134, 1996.
30. Houbiers, J. G. A., Nijman, H. W., van der Burg, S. H., Drijfhout, J. W., Kenemans, P., van de Velde, C. J. H., Brand, A., Momburg, F., Kast, W. M., and Melief, C. J. M. *In vitro* induction of human cytotoxic T lymphocyte responses against peptides of mutant and wild-type p53. *Eur. J. Immunol.* **23,** 2072–2077, 1993.
31. Nijman, H. W., Van der Burg, S. H., Vierboom, M. P., Houbiers, J. G., Kast, W. M., and Melief, C. J. p53, a potential target for tumor-directed T cells. *Immunol. Lett.* **40,** 171–178, 1994.
32. Bos, J. L. The *ras* gene family and human carcinogenesis. *Mut. Res.* **195,** 255–271, 1988.
33. Bos, J. L. *ras* oncogenes in human cancer: a review. *Cancer Res.* **49,** 4682–4689, 1989.
34. Kiaris, H., and Spandidos, D. A. Mutations of *ras* genes in human tumors (review). *Int. J. Oncol.* **7,** 413–421, 1995.
35. Spandidos, D. A., Sourvinos, G., and Koffa, M. *ras* genes, p53 and HPV as prognostic indicators in human cancer (review). *Oncol. Rep.* **4,** 211–218, 1997.
36. Bishop, J. M. Molecular themes in oncogenesis. *Cell* **64,** 235–248, 1991.
37. Abrams, S. I., Hand, P. H., Tsang, K. Y., and Schlom. J. Mutant *ras* epitopes as targets for cancer vaccines. *Semin. Oncol.* **23,** 118–134, 1996.
38. Grand, R. J. A., and Owen, D. The biochemistry of *ras* p21. *Biochemistry* **279,** 609–631, 1991.
39. Satoh, T., and Kaziro, Y. *Ras* in signal transduction. *Semin. Cancer Biol.* **3,** 169–177, 1992.
40. Bos, J. L., Verlaan-de Vries, M., Jansen, A. M., Veeneman, G. H., van Boom, J. H., and van der Eb, A. J. Three different mutations in codon 61 of the human N-*ras* gene detected by synthetic oligonucleotide hybridization. *Nucleic Acids Res.* **12,** 9155–9163, 1984.
41. Verlaan-de Vries, M., Bogaard, M. E., van den Elst, H., van Boom, J. H., van der Eb, A. J., and Bos, J. L. A dot-blot screening procedure for mutated *ras* oncogenes using synthetic oligodeoxynucleotides. *Gene* **50,** 313–320, 1986.
42. Winter, E., Yamamoto, F., Almoguera, C., and Perucho, M. A method to detect and characterize point mutations in transcribed genes: amplification and overexpression of the mutant c-Ki-*ras* allele in human tumor cells. *Proc. Natl. Acad. Sci. USA* **82,** 7575–7579, 1985.
43. McMahon, G., Davis, E., and Wogan, G. N. Characterization of c-Ki-*ras* oncogene alleles by direct sequencing of enzymatically amplified DNA from carcinogen-induced tumors. *Proc. Natl. Acad. Sci. U.S.A.* **84,** 4974–4978, 1987.
44. Collins, S. J., Direct sequencing of amplified genomic fragments documents N-*ras* point mutations in myeloid leukemia. *Oncogene Res.* **3,** 117–123, 1988.
45. Bos, J. L., Fearon, E. R., Hamilton, S. R., Verlaan-de Vries, M., van Boom, J. H., van der Eb, A. J., and Vogelstein, B. Prevalence of *ras* gene mutations in human colorectal cancers. *Nature* **327,** 293–297, 1987.
46. Moskaluk, C. A., Hruban, R. H., and Kern, S. E. p16 and K-*ras* gene mutations in the intraductal precursors of human pancreatic adenocarcinoma. *Cancer Res.* **57,** 2140–2143, 1997.
47. Peace, D. J., Chen, W., Nelson, H., and Cheever, M. A. T cell recognition of transforming proteins encoded by mutated *ras* proto-oncogenes. *J. Immunol.* **146,** 2059–2065, 1991.
48. Peace, D. J., Smith, J. W., Disis, M. L., Chen, W., and Cheever, M. A. Induction of T cells specific for the mutated segment of oncogenic p21*ras* protein by immunization *in vivo* with the oncogenic protein. *J. Immunother.* **14,** 110–114, 1993.
49. Abrams, S. I., Dobrzanski, M. J., Wells, D. T., Stanziale, S. F., Zaremba, S., Masuelli, L., Kantor, J. A., and Schlom, J. Peptide-specific activation of cytolytic CD4+ T lymphocytes against tumor cells bearing mutated epitopes of K-*ras* p21. *Eur. J. Immunol.* **25,** 2588–2597, 1995.
50. Abrams, S. I., Stanziale, S. F., Lunin, S. D., Zaremba, S., and Schlom, J. Identification of overlapping epitopes in mutant *ras* oncogene peptides that activate CD4+ and CD8+ T cell responses. *Eur. J. Immunol.* **26,** 435–443, 1996.
51. Chang, J. C., Zhang, L., Edgerton, T. L., and Kaplan, A. M. Heterogeneity in direct cytotoxic function of L3T4 T cells. Th1 clones express higher cytotoxic activity to antigen-presenting cells than Th2 clones. *J. Immunol.* **145,** 409–416, 1990.
52. Baskar, S., Azarenko, V., Garcia Marshall, E., Hughes, E., and Ostrand-Rosenberg, S. MHC class II-transfected tumor cells induce long-term tumor-specific immunity in autologous mice. *Cell. Immunol.* **155,** 123–133, 1994.
53. Nagarkatti, M., Clary, S. R., and Nagarkatti, P. S. Characterization of tumor-infiltrating CD4+ T cells as Th1 cells based on lymphokine secretion and functional properties. *J. Immunol.* **144,** 4898–4905, 1990.
54. Kahn, M., Sugawara, H., McGowan, P., Okuno, K., Nagoya, S., Hellstrom, K. E., Hellstrom, I., and Greenberg, P. CD4+ T cell clones specific for the human p97 melanoma-associated antigen can eradicate pulmonary metastases from a murine tumor expressing the p97 antigen. *J. Immunol.* **146,** 3235–3241, 1991.
55. Skipper, J., and Stauss, H. J. Identification of two cytotoxic T lymphocyte-recognized epitopes in the Ras protein. *J. Exp. Med.* **177,** 1493–1498, 1993.

56. Peace, D. J., Smith, J. W., Chen, W., You, S.-G., Cosand, W. L., Blake, J., and Cheever, M. A. Lysis of *ras* oncogene-transformed cells by specific cytotoxic T lymphocytes elicited by primary *in vitro* immunization with mutated ras peptide. *J. Exp. Med.* **179,** 473–479, 1994.

57. Fenton, R. G., Taub, D. D., Kwak, L. W., Smith, M. R., and Longo, D. L. Cytotoxic T-cell response and *in vivo* protection against tumor cells harboring activated *ras* proto-oncogenes. *J. Natl. Cancer Inst.* **85,** 1294–1302, 1993.

58. Romero, P., Corradin, G., Luescher, I. F., and Maryanski, J. L. H-2K^d restricted antigenic peptides share a simple binding motif. *J. Exp. Med.* **174,** 603–612, 1991.

59. Alexander, M. A., Damico, C. A., Wieties, K. M., Hansen, T. H., and Connolly, J. M. Correlation between CD8 dependency and determinant density using a peptide-induced, L^d-restricted cytotoxic T lymphocytes. *J. Exp. Med.* **173,** 849–858, 1991.

60. Theobald, M., Biggs, J., Dittmer, D., Levine, A. J., and Sherman, L. A. Targeting p53 as a general tumor antigen. *Proc. Natl. Acad. Sci. USA* **92,** 11993–11997, 1995.

61. Schott, M. E., Wells, D. T., Schlom, J., and Abrams, S. I. Comparison of linear and branched peptide forms (MAPs) in the induction of T helper responses to point-mutated *ras* immunogens. *Cell. Immunol.* **174,** 199–209, 1996.

62. Zavala, F., and Chai, S., Protective anti-sporozoite antibodies induced by chemically defined synthetic vaccine. *Immunol. Lett.* **25,** 271–274, 1990.

63. Wang, C. Y., Looney, D. J., Li, M. L., Walfield, A. M., Ye, J., Hosein, B., Tam, J. P., and Wong-Staal, F. Long-term high-titer neutralizing activity induced by octameric synthetic HIV-1 antigen. *Science* **254,** 285–288, 1991.

64. Tam, J. P. Synthetic peptide vaccine design: synthesis and properties of a high-density multiple antigenic peptide system. *Proc. Natl. Acad. Sci. U.S.A.* **85,** 5409–5413, 1988.

65. Francis, M. J., Hastings, G. Z., Brown, F., McDermed, J., Lu, Y.-A., and Tam, J. P. Immunological evaluation of the multiple antigen peptide (MAP) using the major immunogenic site of foot-and-mouth disease virus. *Immunology* **73,** 249–254, 1991.

66. Tourdot, S., Oukka, M., Manuguerra, J. C., Magafa, W., Vergnon, I. Riche, N., Bruley-Rosset, M., Cordopatis, P., and Kosmatopoulos, K. Chimeric peptides: a new approach to enhancing the immunogenicity of peptides with low MHC class I affinity. *J. Immunol.* **159,** 2391–2398, 1997.

67. Wysocka, M., Eisenlohr, L. C., Otvos, L., Jr., Horowitz, D., Yewdell, J. W., Bennink, J. R., and Hackett, C. J. Identification of overlapping class I and class II H-2^d-restricted T cell determinants of influenza virus N1 neuraminidase that require infectious virus for presentation. *Virology* **201,** 86–94, 1994.

68. Perkins, D. L., Lai, M.-Z., Smith, J. A., and Gefter, M. L. Identical peptides recognized by MHC class I- and II-restricted T cells. *J. Exp. Med.* **170,** 279–289, 1989.

69. Carreno, B. M., Turner, R. V., Biddison, W. E., and Coligan, J. E. Overlapping epitopes that are recognized by CD8+ HLA class I-restricted and CD4+ class II-restricted cytotoxic T lymphocytes are contained within an influenza nucleoprotein peptide. *J. Immunol.* **148,** 894–899, 1992.

70. Takeshita, T., Takahashi, H., Kozlowski, S., Ahlers, J. D., Pendleton, C. D., Moore, R. L., Nakagawa, Y., Yokomuro, K., Fox, B. S., Margulies, D. H., and Berzofsky, J. A. Molecular analysis of the same HIV peptide functionally binding to both a class I and class II MHC molecule. *J. Immunol.* **154,** 1973–1986, 1995.

71. Noguchi, Y., Chen, Y.-T., and Old, L. J. A mouse mutant p53 product recognized by CD4+ and CD8+ cells. *Proc. Natl. Acad. Sci. USA* **91,** 3171–3175, 1994.

73. Sussman, J. J., Shu, S., Sondak, V. K., and Chang, A. E. Activation of T lymphocytes for the adoptive immunotherapy of cancer. *Ann. Surg. Oncol.* **1,** 296–306, 1994.

74. Shu, S., Chou, T., and Rosenberg, S. A. *In vitro* differentiation of T-cells capable of mediating the regression of established syngenic tumors in mice. *Cancer Res.* **47,** 1354–1360, 1987.

75. Ward, B. A., Shu, S., Chou, T., Perry-Lalley, D., and Chang, A. E. Cellular basis of immunologic interactions in adoptive T cell therapy of established metastases from syngeneic murine sarcoma. *J. Immunol.* **141,** 1047–1053, 1988.

76. Crossland, K. D., Lee, V. K., Chen, W., Riddell, S. R., Greenberg, P. D., and Cheever, M. A. T cells from tumor-immune mice nonspecific expanded in vitro with anti-CD3 plus IL-2 retain specific function in vitro and can eradicate disseminated leukemia in vivo. *J. Immunol.* **146,** 4414–4420, 1991.

77. Yoshizawa, H., Sakai, K., Chang, A. E., and Shu, S. Y. Activation by anti-CD3 of tumor-draining lymph node cells for specific adoptive immunotherapy. *Cell. Immunol.* **134,** 473–479, 1991.

78. Peng, L., Shu, S., and Krauss, J. C. Treatment of subcutaneous tumor with adoptively transferred T cells. *Cell. Immunol.* **178,** 24–32, 1997.

79. Alexander-Miller, M. A., Leggatt, G. R., and Berzofsky, J. A. Selective expansion of high- or low-avidity cytotoxic T lymphocytes and efficacy for adoptive transfer. *Proc. Natl. Acad. Sci. U.S.A.* **93,** 4102–4107, 1996.

80. Alexander-Miller, M. A., Leggatt, G. R., Sarin, A., and Berzofsky, J. A. Role of antigen, CD8, and cytotoxic T lymphocyte (CTL) avidity in high dose antigen induction of apoptosis of effector CTL. *J. Exp. Med.* **184,** 485–492, 1996.

81. Kwan-Lim, G. E., Ong, T., Aosai, F., Stauss, H., and Zamoyska, R. Is CD8 dependence a true reflection of TCR affinity for antigen? *Int. Immunol.* **5,** 1219–1228, 1993.

82. Jung, S., and Schluesener, H. J. Human T Lymphocytes recognize a peptide of single point-mutated, oncogenic Ras proteins. *J. Exp. Med.* **173,** 273–276, 1991.

83. Gedde-Dahl, III, T., Eriksen, J. A., Thorsby, E., and Gaudernack, G. T-cell responses against products of oncogenes: generation and characterization of human T-cell clones specific for p21 *ras*-derived synthetic peptides. *Hum. Immunol.* **33,** 266–274, 1992.

84. Fossum, B., Gedde-Dahl, III, T., Hansen, T., Eriksen, J. A., Thorsby, E., and Gaudernack, G. Overlapping epitopes encompassing a point mutation (12 Gly → Arg) in p21 *ras* can be recognized by HLA-DR, -DP and -pDQ restricted T cells. *Eur. J. Immunol.* **23,** 2687–2691, 1993.

85. Fossum, B., Breivik, J., Meling, G. I., Gedde-Dahl, III, T., Hansen, T., Knutsen, I., Rognum, T. O., Thorsby, E., and Gaudernack, G. A K-*ras* 13Gly → Asp mutation is recognized by HLA-DQ7 restricted T cells in a patient with colorectal cancer. Modifying effect of DQ7 on established cancers harbouring this mutation? *Int. J. Cancer* **58,** 506–511, 1994.

86. Fossum, B., Gedde-Dahl, III, T., Breivik, J., Eriksen, J. A., Spurkland, A., Thorsby, E., and Gaudernack, G. p21-*ras*-peptide-specific T-cell responses in a patient with colorectal cancer. CD4+ and CD8+ T cells recognize a peptide corresponding to a common mutation (13Gly → Asp). *Int. J. Cancer* **56,** 40–45, 1994.

87. Gedde-Dahl, III, T., Nilsen, E., Thorsby, E., and Gaudernack, G. Growth inhibition of an colonic adenocarcinoma cell line (HT29) by T cells specific for mutant p21 *ras. Cancer Immunol. Immunother.* **38,** 127–134, 1994.

88. Gjertsen, M. K., Bakka, A., Breivik, J., Saeterdal, I., Solheim, B. G., Soreide, O., Thorsby, E., and Gaudernack, G. Vaccination with mutant ras peptides and induction of T-cell responsiveness in pancreatic carcinoma patients carrying the corresponding *RAS* mutation. *Lancet* **346,** 1399–1400, 1995.

89. Gjertsen, M. K., Bakka, A., Breivik, J., Saeterdal, I., Gedde-Dahl, I. T., Stokke, K. T., Solheim, B. G., Egge, T. S., Soreide, O., Thorsby, E., and Gaudernack, G. *Ex vivo ras* peptide vaccination in patients with advanced pancreatic cancer: results of a phase I/II study. *Int. J. Cancer* **65,** 450–453, 1996.

90. Gjertsen, M. K., Saeterdal, I., Thorsby, E., and Gaudernack, G. Characterization of immune responses in pancreatic carcinoma patients after mutant p21 *ras* peptide vaccination. *Br. J. Cancer* **74,** 1828–1833, 1996.

91. Fossum, B., Olsen, A. C., Thorsby, E., and Gaudernack, G. CD8+ T cells from a patient with colon carcinoma, specific for a mutant p21-*ras*-derived peptide (Gly13 → Asp), are cytotoxic towards a carcinoma cell line harbouring the same mutation. *Cancer Immunol. Immunother.* **40,** 165–172, 1995.

92. Van Elsas, A., Nijman, H. W., Van der Minne, C. E., Mourer, J. S., Kast, W. M., Melief, C. J. M., and Schrier, P. I. Induction and characterization of cytotoxic T-lymphocytes recognizing a mutated p21 *ras* peptide presented by HLA-A*0201. *Int. J. Cancer* **61,** 389–396, 1995.

93. Gjertsen, M. K., Bakka, A., Breivik, J., Saeterdal, I., Gedde-Dahl III, T., Stokke, K. T., Sohlheim, B. G., Egge, T. S., Soreide, O., Thorsby, E., and Gaudernack, G. *Ex vivo* ras peptide vaccination in patients with advanced pancreatic cancer: results of a phase I/II study. *Int. J. Cancer* **65,** 450–453, 1996.

94. Abrams, S. I., Khleif, S. N., Bergmann-Leitner, E. S., Kantor, J. A., Chung, Y., Hamilton, J. M., and Schlom, J. Generation of stable CD4+ and CD8+ T cell lines from patients immunized with *ras* oncogene-derived peptides reflecting codon 12 mutations. *Cell. Immunol.* **182,** 137–151, 1997.

95. Bristol, J. A., Schlom J., and Abrams, S. I. Development of a murine Ras CD8+ CTL peptide epitope variant that possesses enhanced MHC class I binding and immunogenic properties. *J. Immunol.* **160,** 2433–2441, 1998.

Polynucleotide-Mediated Immunization Therapy of Cancer

STEPHEN ANDREW WHITE, ROBERT MARTIN CONRY, THERESA V. STRONG, DAVID TERRY CURIEL, AND ALBERT FRANCES LoBUGLIO

Gene Therapy Program, Comprehensive Cancer Center, University of Alabama at Birmingham, Birmingham, Alabama 35294

I. INTRODUCTION

In 1988, investigators at the Surgery Branch of the National Cancer Institute reported objective tumor regression in a significant proportion of patients with metastatic melanoma following the adoptive transfer of tumor-derived, culture-expanded T lymphocytes coadministered with interleukin-2 [1]. This regression of established metastatic tumors from an immune-mediated strategy lent credibility to the development of immunotherapeutic approaches for the treatment of cancer. To date, most active tumor immunotherapy trials in humans have utilized autologous or allogeneic tumor cells or tumor cell extracts coadministered with various immune adjuvants [2]. Recently, multiple factors including an emerging understanding of the initiation of the immune response (see Chapter 3) [3], the recognition and molecular characterization of specific immunogenic tumor antigens [4,5], the increasing use of recombinant DNA technology for immune applications [6], and the development of sensitive *in vitro* assays of antigen-specific T-cell function have produced a shift in interest toward the development of defined antigen vaccines targeting specific tumor-associated antigens or immunogenic peptides derived from these antigens. These defined antigen vaccine strategies usually involve administration of intact tumor antigens, peptides (T-cell epitopes) derived from tumor antigens, recombinant viruses encoding tu-

271

mor antigens, or plasmid DNA constructs encoding tumor antigens (polynucleotide vaccination) [2]. This chapter will review the biological processes and postulated immune mechanisms underlying polynucleotide vaccination (PNV) and will discuss current preclinical and clinical applications of this form of active immunotherapy in the treatment of malignancy.

II. DIRECT GENE TRANSFER WITH POLYNUCLEOTIDE REAGENTS

In 1990, Wolff et al. described in vivo expression of foreign genes by mouse myofiber cells following intramuscular injection of plasmid DNA and mRNA expression vectors [7]. Subsequent studies have demonstrated myocyte expression of reporter genes, microbial proteins, and tumor-associated antigens in a broad range of animal species following intramuscular injection of plasmid DNA [8–13]. This form of gene transfer and expression appears to utilize myocyte caveolae and T tubules for nucleic acid transfection [11,14]. Transgene expression occurs without integration into host chromosomal DNA [11,15,16]. The uptake and expression of naked DNA and mRNA molecules is not restricted to striated muscle but can occur in cardiac muscle [17] or dermis [18]. In addition, particle bombardment-mediated (gene gun) transfer of plasmid DNA-coated gold beads into skin or mucosa can generate expression of encoded proteins [19,20]. Finally, some degree of hepatocyte expression of plasmid-encoded reporter genes has been observed following direct needle injection of high doses of plasmid DNA into murine and feline liver [21].

The level of gene expression generated by administration of naked mRNA or DNA molecules is influenced by multiple factors including the amount of polynucleotide inoculated, the nature of gene expression regulatory elements utilized, physical characteristics of the nucleic acid construct, the technique of polynucleotide administration, the coadministration of myocyte lytic agents, the age of the inoculated animal, and the size of the inoculated muscle.

In mice, maximum levels of plasmid-encoded protein synthesis are generally achieved by intramuscular injection of 50 to 100 μg doses of plasmid DNA [22]. Higher doses of polynucleotide reagents may be required in larger species for optimization of gene expression.

Alteration of promoters and other regulatory elements within a plasmid DNA construct modulates the level of gene expression generated. For example, replacing weaker promoters (e.g., the Rous sarcoma virus promoter/enhancer) with very strong promoter sequences (e.g., the cytomegalovirus intermediate early promoter/enhancer) leads to up-regulation of expression of plasmid-encoded genes [23]. Similarly, manipulation of 3' untranslated regions and transcription terminators in DNA expression constructs influences the level of gene expression [24]. However, up-regulation of expression of toxic or immunogenic proteins encoded by plasmid DNA may curtail the duration of gene expression by direct or immune-mediated destruction of plasmid-transfected cells [25,26]. In addition, coadministration of multiple plasmids incorporating highly efficient promoters within the same muscle appears to diminish the expression of each construct, presumably reflecting competition for cellular transcriptional machinery [24]. Messenger RNA constructs utilized for direct gene transfer appear to require incorporation of a modified guanacil cap for translational competence, and levels of transgene expression may be modulated by inclusion of 5' and 3' untranslated regions from heterologous mRNA transcripts [27].

Physical characteristics of nucleic acid molecules influence the efficiency of transgene expression following intramuscular or intradermal delivery. Circular DNA constructs produce significantly higher levels of expression than linearized plasmid DNA containing identical regulatory and coding elements [24]. In general, plasmid DNA constructs induce a greater level and duration of protein synthesis than do mRNA transcripts [7], presumptively secondary to the relative instability of mRNA [28]. However, transgene expression from mRNA constructs may be augmented and prolonged by the utilization of recombinant self-replicative transcripts. [29]. These transcripts are generated by the incorporation of heterologous coding sequences into self-replicative genomic RNA derived from the togaviruses Sindbis virus [29] and Semliki Forest virus [30].

The efficiency of expression of nucleic acid molecules is a function of the technique of polynucleotide administration. For plasmid constructs, needle injection of muscle or dermis may require up to 100 times the dose of DNA to produce levels of gene expression equivalent to those generated by particle bombardment-mediated epidermal delivery [31]. However, the duration of gene expression following epidermal gene gun delivery is limited by a loss of plasmid-transfected cells with cutaneous desquamation. Injection of plasmid DNA into rat myocardium has been reported to generate a 40-fold higher level of gene expression than that observed with inoculation of an equivalent dose of plasmid into skeletal muscle [17].

Various other factors influence the level of gene expression generated by administration of nucleic acid molecules. Myocyte lytic agents (e.g., bupivacaine, cardiotoxin), when administered several days prior to plasmid constructs, lead to enhanced transfection of re-generating myofiber cells and thus increase the level of plasmid-encoded protein synthesis [32–34]. The level of transgene expression following administration of naked polynucleotide constructs appears to be inversely pro-portional to the age of the inoculated animal [31] and the size of the inoculated muscle [28].

III. POLYNUCLEOTIDE VACCINATION

In 1992, Tang *et al.* described the development of humoral immune responses against human growth hor-mone (GH) in mice following intramuscular inoculation of a GH-expressing plasmid DNA construct [35]. There-after, Ulmer's group utilized plasmid DNA encoding influenza virus nucleoprotein (NP) to elicit NP-specific antibody and cytolytic T-cell responses with resultant immunoprotection against intranasal challenge with in-fluenza A virus [36]. Since these initial reports, numer-ous investigators have reported induction of immune responses against a variety of microbial proteins and tumor-associated antigens with direct injection of plas-mid DNA reagents. This phenomenon, now termed polynucleotide vaccination, has been demonstrated in a broad range of animal species including mice, guinea pigs [37], rabbits [38], dogs [12], chickens [39], cattle [40], and nonhuman primates [41,42]. Though initially described with intramuscular inoculation of plasmid DNA, this process can occur (with varying degrees of efficiency) following intravenous, intradermal, or subcu-taneous injection or particle bombardment-mediated polynucleotide transfer to skin or mucosa [39].

A few investigators have demonstrated elicitation of antigen-specific humoral immune responses following administration of naked mRNA polynucleotides. Qui *et al.* utilized gene gun-mediated transfer of a transcript encoding α-1 antitrypsin into mouse epidermis to gener-ate antigen-specific antibodies [43]. Zhou *et al.* reported the induction of NP-specific humoral immunity in mice following intramuscular administration of a self-replicative mRNA construct encoding influenza virus nucleoprotein [30]. Our group noted priming of a carci-noembryonic antigen (CEA)-specific antibody response in mice following multiple intramuscular injections of an mRNA construct encoding CEA [44]. However, the immune response elicited by mRNA expression vectors is generally poor relative to the immune response achieved with plasmid DNA expression vectors, likely secondary to the limited magnitude and duration of transgene expression with mRNA constructs [29].

IV. DETERMINANTS OF THE MAGNITUDE OF IMMUNE RESPONSE TO PNV

The magnitude of the immune response induced by plasmid DNA vaccine constructs is affected by multiple factors, including the immunogenicity of the plasmid-encoded antigen, the level of genetic expression of the encoded immunogen, the route of administration, the age of the immunized animal, the use of myocyte lytic agents, and the concurrent inoculation of plasmid-encoded immunomodulatory molecules.

Increased immunogen expression, generated by mod-ulation of plasmid dose, gene-expression regulatory elements, and other factors, generally augments the induction of immune responses. However, enhanced production of toxic or immunogenic antigens may atten-uate immune responses by direct or immune-mediated destruction of plasmid-transfected cells and consequent curtailment of gene expression. For example, Xiang and Ertl noted abbreviated gene expression and no augmen-tation of immune responses following substitution of the early promoter of CMV for the weaker simian virus 40 (SV40) promoter in a plasmid expression vector en-coding the rabies virus glycoprotein [45].

The route of administration of a polynucleotide con-struct may influence the magnitude of the resultant im-mune response. Some authors have reported that epi-dermal delivery of plasmid DNA molecules induces immune responses quantitatively superior to those gen-erated with intramuscular inoculation of an equivalent dose of plasmid [31]. In a series of murine studies of a plasmid reagent encoding the influenza virus hemagglu-tinin glycoprotein, Robinson's group noted that 250–2,500 times less DNA was required for induction of measurable immune responses by gene gun epidermal delivery than for generation of immunity by saline intra-muscular injection [39]. This observation may reflect differences in the efficiency of plasmid transfection and expression between these modes of vaccine administra-tion. A second potential explanation for this effect is the relatively high density of antigen-presenting cells (APCs) in the skin and the consequent possibility of direct APC transfection with cutaneous administration. The route of administration of a polynucleotide vaccine may also influence the character of the immune response generated, as discussed later.

The magnitude of the immune response noted with plasmid DNA administration may be a function of the age of the immunized animal. Johnston and Barry noted an inverse relationship between antigen-specific anti-body production and animal age in mice inoculated with a luciferase-expressing plasmid construct [31]. This effect was only partially explained by variations in the level of plasmid expression with age.

Intramuscular injection of myolytics (e.g., bupivacaine, cardiotoxin) prior to polynucleotide vaccine reagents appears to augment resultant immune responses [46]. This effect probably results from a combination of mechanisms. First, the level of expression of the plasmid-encoded antigen is enhanced as described previously [32–34]. Second, myolytic administration may induce differentiation of myogenic stem cells (i.e., satellite cells) into myoblasts, which overexpress MHC class I molecules relative to mature myocytes [47]. Third, intramuscular introduction of these agents may produce local inflammatory responses with recruitment and subsequent direct plasmid transfection of bone marrow-derived APCs.

Immune responses induced by polynucleotide vaccine constructs may be augmented by concurrent use of a variety of plasmid-encoded immunomodulatory molecules including granulocyte-macrophage colony-stimulating factor (GM-CSF), various cytokines, costimulatory molecules (B7-1 and B7-2), and endoplasmic-reticulum-targeting sequences. Coadministration of plasmid-encoded GM-CSF with plasmid constructs encoding immunogens appears to consistently enhance humoral immune responses but has a variable effect on T-cell responses. Ertl and Xiang reported augmentation of B-cell and T-helper-cell activity against the rabies virus glycoprotein following co-inoculation of a plasmid encoding GM-CSF with a plasmid expressing the rabies virus antigen [45]. Kim *et al.* noted quantitative enhancement of antigen-specific humoral responses and T-helper-cell proliferation but no augmentation of CTL responses in mice coimmunized with polynucleotide constructs encoding HIV-1 *gag-pol* and GM-CSF [48]. Wands' group has described augmentation of both T-helper and cytolytic T-cell responses against plasmid-encoded hepatitis C proteins by co-inoculation of a GM-CSF-encoding polynucleotide construct [49]. Our group observed enhanced CEA-specific antibody responses and lymphoblastic transformation in mice when plasmid-encoded GM-CSF was administered by gene gun epidermal delivery 3 days prior to gene gun administration of a plasmid DNA construct encoding CEA [50].

Several animal studies have demonstrated modulation of the magnitude and character of immune responses generated by immunogen-encoding plasmid reagents by coadministration of polynucleotide constructs expressing cytokines. Kim *et al.* reported augmentation of cytolytic T-cell responses against multiple plasmid-encoded HIV-1 antigens following coadministration of a polynucleotide construct encoding IL-12 [48]. A Japanese group described similar findings with co-inoculation of separate plasmid reagents expressing HIV-1 *env* and IL-12 [51]. In studies of a polynucleotide vaccine encoding a hepatitis C virus antigen, Wands' group noted enhanced cytolytic T-cell responses following coadministration of plasmid-encoded IL-2 but suppressed cytolytic T-cell responses following coadministration of plasmid-encoded IL-4 [49]. Conversely, Xiang *et al.* reported reduction of immune responses to polynucleotide-encoded rabies virus glycoprotein with co-delivery of plasmid-encoded gamma interferon [45].

Plasmid-encoded costimulatory molecules, including B7-1 (CD80) and B7-2 (CD86), appear to augment antigen-specific immune responses when co-delivered with immunogen-encoding polynucleotide constructs. Our group has reported that co-delivery of B7-1 cDNA within a dual-expression plasmid encoding CEA produced anti-CEA antibody responses and antitumor effects superior to those generated by the same amount of plasmid DNA encoding CEA alone or separate plasmids encoding CEA and B7-1 [50]. Weiner's group noted augmentation of anti-*env* cytolytic T-cell responses following coadministration of separate plasmid constructs encoding B7-2 and HIV-1 *env*. [52] Co-inoculation of separate plasmids encoding B7-1 and HIV-1 *env* minimally increased anti-*env* cytolytic T-cell immunity in this study. Barber's group at the University of Toronto noted that co-delivery of B7-2 cDNA but not B7-1 cDNA within a dual-expression plasmid encoding a mutant nonimmunogenic influenza nucleoprotein augmented antigen-specific cytolytic T-cell responses [53].

Several investigators have described the use of plasmid constructs specifically designed to promote MHC class I-restricted antigen presentation and thereby enhance cytolytic T-lymphocyte induction. Carbone's group noted augmentation of cytolytic T-cell responses following fusion of a plasmid-encoded mutant p53 epitope to cDNA encoding the adenovirus E3 leader sequence, a signal sequence targeting nascent polypeptide to the endoplasmic reticulum [54]. In another study, utilization of a plasmid-expression vector incorporating immunogen cDNA fused in frame with ubiquitin cDNA resulted in elicitation of enhanced cytolytic T-cell responses but abrogated antibody responses [55].

V. DETERMINANTS OF THE CHARACTER OF IMMUNE RESPONSES TO PNV

The character of the immune response (i.e., T helper 1 vs. T helper 2) elicited by polynucleotide vaccine constructs is influenced by multiple factors including the presence of certain immunostimulatory nucleotide sequences in noncoding regions of the plasmid vector, the route of vaccine administration, and the coadministration of plasmid-encoded immunomodulatory molecules. The T-helper-1 (Th1) immune response, crucial to antitumor immunity, is characterized by gamma interferon and IL-2 lymphokine release, complement-dependent antibody (IgG2a, IgG2b) production, enhanced cytolytic T-cell immunity, recruitment and activation of macrophages, and increased phagocytic responses. The T-helper-2 (Th2) immune response is characterized by IL-4, IL-5, IL-6, and IL-10 lymphokine release and complement-independent antibody (IgG1, IgE) production, as well as recruitment and activation of mast cells and eosinophils. For a more complete discussion of Th1 versus Th2 immunity see Chapter 3.

Recent studies have identified specific bacterial DNA sequences in noncoding regions of plasmid constructs that enhance immunogenicity of polynucleotide vaccines [56–58]. These immunostimulatory motifs, consisting of unmethylated CpG dinucleotides in the six-residue nucleotide sequence 5′purine-purine-CpG-pyrimidine-pyrimidine3′, appear to promote generation of Th1 proinflammatory responses by induction of gamma interferon and IL-12. Klinman *et al.* demonstrated reduction of the immunogenicity of a plasmid construct following methylation of CpG motifs [59]. Several investigators have described augmentation of antigen-specific immune responses to very low (nonimmunogenic) doses of polynucleotide vaccines by coadministration of CpG-rich oligonucleotides or exogenous CpG-containing bacterial DNA [59,60]. However, codelivery of CpG-containing oligonucleotides with immunogenic doses of plasmid vaccines does not augment immune responses, and administration of optimally immunogenic doses (50 μg) of polynucleotide vaccines to mice appears to produce immune effects equivalent to those elicited by coadministration of CpG-rich moieties and low doses of vaccine. Plasmid DNA constructs in current preclinical and clinical use typically contain 15 to 20 CpG motifs, and insertion of additional immunostimulatory sequences does not appear to confer enhanced immunogenicity.

Numerous investigators have reported differences in the character of immune responses elicited by various routes of plasmid DNA delivery. Intramuscular injection of polynucleotide constructs appears to preferentially induce Th1 responses [61]. Particle bombardment-mediated plasmid transfer into the skin is frequently associated with generation of Th2 immunity. For example, in our laboratory, administration of a CEA-encoding plasmid by intramuscular inoculation elicited IgG1, IgG2a, and IgG2b humoral responses and immunoprotection against challenge with syngeneic, CEA-expressing colon carcinoma cells, whereas gene gun-mediated epidermal delivery of this polynucleotide reagent induced an IgG1 isotype antibody response and did not confer immunoprotection against tumor challenge [60]. Another group has reported preferential induction of Th1 humoral immune responses against influenza hemagglutinin with saline injection of a plasmid construct into skin or muscle but elicitation of Th2 antibody responses following particle bombardment-mediated delivery of the same plasmid into skin or muscle [62]. These differences were found to be independent of PNV reagent dose and suggest that the mode of plasmid delivery may be as important as the vaccine target tissue in determining the character of the immune response generated by a polynucleotide vaccination. Johnston and Barry have suggested that observed differences in immune response patterns may sometimes be dose dependent rather than a function of route of administration [31]. In murine studies, they noted elicitation of IgG1 humoral responses (Th2) following intramuscular or gene gun delivery of minute doses of plasmid constructs expressing human α-1 antitrypsin and human growth hormone. Intramuscular or gene gun administration of 50 μg doses of these plasmid reagents generated IgG2a humoral responses (Th1). These observations suggest that the frequently observed preferential induction of Th1 responses following intramuscular inoculation of polynucleotide vaccines may be secondary to antigen-independent adjuvant effects of high doses of bacterial DNA.

Finally, the character of the immune response elicited by polynucleotide vaccine constructs is affected by codelivery of plasmid-encoded immunomodulatory molecules. For example, Chow *et al.* reported the preferential elicitation of Th1 immune responses following coadministration of an IL-2-expressing plasmid with a DNA construct encoding a hepatitis B virus antigen [63]. Other investigators have noted augmentation of Th1 responses to plasmid immunization with co-delivery of plasmid-encoded IL-12 [48].

VI. IMMUNE MECHANISMS IN POLYNUCLEOTIDE VACCINATION

The specific roles of myocytes and bone marrow-derived antigen presenting cells in antigen synthesis and MHC-restricted antigen presentation following intramuscular plasmid DNA immunization are not fully understood. Three hypotheses regarding this process have been proposed and are listed in Table 1.

The synthesis and subsequent MHC-restricted presentation of antigen by myocytes is supported by several lines of evidence. First, immunohistochemical analysis

TABLE 1   Mechanism of Polynucleotide Vaccination

1. Antigen synthesis and presentation by myocytes.
2. Direct plasmid transfection of bone marrow-derived APCs with subsequent antigen expression and presentation.
3. Antigen synthesis by myocytes with subsequent transfer to bone marrow-derived APCs for presentation.

of muscle tissue inoculated with plasmid has conclusively demonstrated synthesis of plasmid-encoded antigen within myofiber cells [14]. Second, *in vitro* studies have revealed up-regulation of MHC class I expression and induction of MHC class II expression in cultured muscle cell lines following exposure to gamma interferon [47,64]. In addition, myoblasts treated with gamma interferon have been shown to present antigen to previously primed T cells with resultant cytotoxicity and/or T-cell proliferation [65].

Several *in vivo* observations appear to support direct polynucleotide vaccine transfection of bone marrow-derived APCs with subsequent APC antigen synthesis and presentation. First, the efficacy of polynucleotide vaccination following cutaneous or intravenous plasmid administration suggests that myocytes are not required for immune induction. Second, numerous investigators have described an immunogenic effect of GM-CSF when administered with polynucleotide vaccines, presumptively secondary to dendritic cell stimulation [45,48,50]. Third, Torres *et al.* recently reported preservation of antigen-specific humoral and cellular immune responses despite ablation of inoculated muscle tissue within 1–10 minutes of plasmid DNA injection, suggesting early escape of plasmid molecules or plasmid-transduced cells from muscle and subsequent antigen synthesis and presentation at distant sites [66]. Fourth, several investigators have demonstrated transfection of bone marrow-derived APCs by plasmid DNA reagents following intramuscular PNV inoculation. For example, Caseres *et al.* described polymerase chain reaction (PCR)-based detection of plasmid sequences in dendritic cells from draining lymph nodes after intramuscular injection of a plasmid construct [67]. Finally, murine chimeric bone marrow transplant experiments have demonstrated restriction of cytolytic T-cell responses following intramuscular polynucleotide immunization to the MHC class I haplotypes found on bone marrow-derived APCs [68–70].

The third potential mechanism of immune induction following intramuscular polynucleotide vaccination, the transfer of intracellular antigens from plasmid-transduced myocytes to bone marrow-derived APCs, has been hypothesized to occur via the secretion of antigen-containing vesicles from myofiber cells, the release of apoptotic blebs from myofiber cells, contact-mediated transfer, and CTL-induced myocyte lysis [25]. This model of plasmid DNA-induced T-cell immunity is supported by recent observations from Ulmer's group [68,71]. In this series of studies, transplantation of myoblasts stably transfected with the influenza NP gene into parent → F1 bone marrow chimeric mice was sufficient to induce NP-specific CTL responses, and these responses were restricted to the MHC class I type of bone marrow.

In summary, most current data suggests that bone marrow-derived APCs are the predominant cells involved in antigen presentation following PNV inoculation. The relative importance of myocytes and APCs in antigen synthesis following intramuscular polynucleotide vaccination is unclear. Further delineation of the mechanisms governing elicitation of immune responses by inoculation of plasmid DNA constructs will facilitate the rational design of strategies to enhance the efficacy of this form of immunotherapy.

VII. SAFETY CONSIDERATIONS IN POLYNUCLEOTIDE VACCINATION

Theoretical adverse events associated with the utilization of PNV reagents in the immunotherapy of cancer include the provocation of destructive immune responses against plasmid-transfected cells, the induction of immune responses against host tissues expressing antigens homologous to plasmid-encoded immunogens, the elicitation of anti-DNA autoantibodies with consequent development or exacerbation of autoimmune disease, and the integration of plasmid sequences into host chromosomal DNA with the attendant risks of insertional mutagenesis or activation of endogenous oncogene sequences. However, extensive *in vivo* evaluation of this form of active immunotherapy has failed to demonstrate significant toxicity. Numerous animal studies employing intramuscular inoculation of high-dose plasmid DNA constructs have revealed no induction of anti-myocyte autoantibodies or myositis despite the elicitation of potent antigen-specific humoral and cellular immune responses. Similarly, intradermal and epidermal delivery of immunogenic PNV reagents has not been associated with the development of dermatitis.

Immune destruction of host tissues expressing antigens homologous to plasmid-encoded proteins, a particular concern with the use of PNV constructs encoding tumor-associated self-antigens that are expressed at low levels in normal tissues (e.g., CEA), has not been reported in preclinical investigations. Our group has conducted preclinical trials of polynucleotide immunization to human CEA in macaques using intramuscular injection and particle bombardment of the epidermis by gene gun. No evidence of immune destruction of host tissues was observed despite the elicitation of potent antibody and cellular immune responses to human CEA in animals possessing a CEA-like molecule sharing 90 percent homology with this immunogen [72].

A recent study reported production of small increases in serum IgG anti-DNA autoantibodies in normal mice following repeated polynucleotide vaccinations without the development of clinically apparent autoimmune disease [73]. In this series of investigations, multiple immunizations of lupus-prone mice with plasmid DNA constructs neither accelerated the onset nor exacerbated the severity of systemic autoimmune disease.

As noted previously, PCR-based studies of myocytes transfected with PNV reagents have suggested an absence of plasmid integration into host chromosomal DNA [16]. This finding is likely secondary to multiple factors, including a lack of homology between host and plasmid nucleotide sequences, the absence of a functional eukaryotic origin of replication on plasmid DNA molecules, and the general absence of chromosomal replication and cell division in cells transfected with PNV reagents [74]. In this regard, polynucleotide vaccines are likely to be safer than recombinant defined antigen vaccines employing viral vectors. In addition, the utilization of plasmid constructs for immunization does not confer the risk of recombinational events leading to the derivation of replication-competent viral species. Collectively, these observations suggest that PNV represents an active immunotherapy strategy with a favorable toxicity profile.

VIII. PRECLINICAL AND CLINICAL EXPERIENCE WITH PNV IN INFECTIOUS DISEASES

Since the initial description of the elicitation of antigen-specific immunity following administration of a polynucleotide construct by Tang *et al.*, numerous investigators have utilized immunogen-encoding plasmid DNA molecules for the induction of immune responses against microbial antigens in animals [75]. These studies have demonstrated generation of antigen-specific humoral and cellular immune responses and immunoprotection against microbe challenge in a broad range of animal species following immunization with plasmid expression vectors encoding a variety of parasitic, bacterial, and viral antigens (reviewed in Donnelly *et al.* [74] and Chattergoon *et al.* [75]).

Based on these preclinical findings, multiple early phase human clinical trials utilizing polynucleotide constructs encoding microbial antigens are underway in the United States and Europe. For example, a group in Sweden is currently immunizing asymptomatic HIV-seropositive individuals with plasmid vaccines encoding either *rev, tat,* or *nef.* Therapeutic and prophylactic trials of a polynucleotide construct targeting HIV gp 160 are underway at the University of Pennsylvania [75]. Investigators at the University of Washington are currently immunizing normal volunteers and genital herpes patients with a plasmid encoding the HSV antigen gd [75]. A prophylactic trial of a PNV reagent encoding influenza core antigen is being conducted at Johns Hopkins University [76]. A plasmid encoding hepatits B surface antigen (HBsAg) is being studied in humans at the University of Cincinnati [76]. Finally, the U.S. Navy is conducting a trial of a plasmid vaccine encoding a malarial antigen in healthy volunteers [76].

IX. PRECLINICAL EXPERIENCE WITH PNV REAGENTS ENCODING CEA

Our group has conducted a series of preclinical studies of a plasmid DNA construct encoding the full-length cDNA for human carcinoembryonic antigen under transcriptional regulatory control of the CMV early promoter/enhancer, termed pCEA. Intramuscular inoculation of mice with this reagent generated CEA-specific antibody and cellular immune responses with resultant immunoprotection against challenge with syngeneic, CEA-expressing colon carcinoma cells [77,78].

The immune responses elicited in mice by pCEA were affected by the dose, schedule, and route of plasmid administration, and by the concurrent use of plasmid-encoded immunomodulatory molecules. Although 4 μg intramuscular doses of pCEA were sufficient to generate measurable CEA-specific antibody and lymphoproliferative responses and to provide immunoprotection against tumor cell challenge [60], optimal immune responses were induced by intramuscular inoculation of 50 to 100 μg of the polynucleotide construct [77,78]. The magnitude of antigen-specific humoral responses generated by administration of low doses (1 μg) of pCEA was significantly augmented by co-delivery of irrelevant plasmid DNA, as depicted in Fig. 1 [60].

The schedule of administration of pCEA affected the time to onset of CEA-specific immunity. Mice immunized with four weekly administrations of pCEA reliably generated antigen-specific humoral and lymphoproliferative responses and protection against tumor challenge within 6 weeks of initiation of vaccination, whereas administration of pCEA every 3 weeks elicited comparable immune responses and antitumor effects

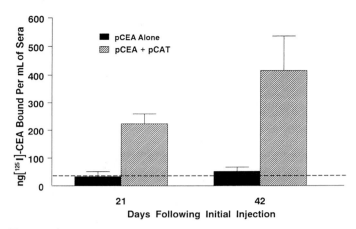

FIGURE 1   Augmentation of the anti-CEA antibody response to CEA polynucleotide immunization by co-delivery of irrelevant plasmid DNA. Groups of seven mice received low doses (1 μg) of plasmid DNA encoding CEA (pCEA) by i.m. injection alone or mixed with 50 μg doses of irrelevant plasmid DNA encoding chloramphenicol acetyltransferase (pCAT) for two injections 3 weeks apart. Sera for anti-CEA antibody response were obtained 21 and 42 days following the initial injection. Results are reported as ng of [^{125}I]-CEA bound per mL of sera, with the dashed line at 35 indicating the threshold for a positive result.

within 8 to 9 weeks [77,78]. The relatively long period (6–9 weeks) required for the induction of antitumor immunity by pCEA appears to limit the utility of this reagent in the treatment of established murine MC38 colon carcinoma tumors, which typically display tumor outgrowth within 2 to 3 weeks of implantation.

Administration of the CEA-expressing polynucleotide by intramuscular or intradermal needle injection induced CEA-specific antibody and lymphoproliferative responses as well as protection against syngeneic tumor cell challenge [60]. Gene gun epidermal delivery of comparable doses of pCEA elicited anti-CEA humoral and lymphoproliferative responses but failed to generate immunoprotection against tumor cell challenge. In addition, the CEA-specific humoral immune response induced by gene gun administration was exclusively of the IgG1 isotype, whereas intramuscular injection produced IgG1, IgG2a, and IgG2b antibodies characteristic of a Th1 response, which may be necessary for efficacious antitumor immunity. Intraperitoneal and subcutaneous administration of pCEA produced no measurable CEA-specific humoral immune responses (Table 2).

Subsequent studies have suggested that CD4+ T cells are the primary antitumor effector cells induced by pCEA administration [60]. Immunohistochemical analysis of tumor transplantation sites in mice immunized with pCEA demonstrated extensive infiltration of tumor by predominantly CD4+ T cells within 48 hours of challenge with syngeneic, CEA-expressing colon carcinoma

cells. Similar analysis of tumor transplantation sites in naïve mice following tumor challenge revealed no T-cell infiltration. The mediation of pCEA antitumor effects by CD4+ T cells was confirmed using the Winn assay. Nylon wool-enriched immune splenic T cells harvested from pCEA-immunized mice prevented tumor growth when coadministered to naïve mice with syngeneic, CEA-expressing colon cancer cells at a ratio of two effector cells to one tumor cell. A CD8-depleted population of immune splenic T cells, selected by magnetic activated cell sorting, provided protection against tumor growth equivalent to that of unfractionated T cells. However, CD4 depletion of immune splenic T cells obtained from immunized mice resulted in diminution of immunoprotection against tumor challenge of naïve mice.

The specificity of the antitumor immune response elicited by pCEA was demonstrated in a series of investigations in which vaccinated mice were challenged with either CEA-transfected syngeneic colon carcinoma MC38 cells or a mixture of CEA-transfected and untransfected MC38 cells [60]. Animals immunized with pCEA reliably survived challenge with a homogenous population of CEA-transfected cells but displayed tumor outgrowth comparable to that of naïve mice when challenged with a mixture of MC38 cells with and without CEA expression.

Our preclinical investigations of plasmid-encoded carcinoembryonic antigen included a series of nonhuman primate studies of a dual-expression plasmid con-

TABLE 2   Summary of our Experience with pCEA Immunization by a Variety of Routes in a Variety of Species

| | Species | | | | | |
| | Mouse | | | Rabbit | Dog | |
Route of administration	Antibody response	Lymphoproliferation	Tumor protection	Antibody response	Antibody response	Lymphoproliferation
Intramuscular (needle injection)	+	+	+	+	+	+
Intramuscular (Biojector)	N/D	N/D	N/D	+	+	+
Intradermal (needle injection)	+	+	+	+	+	N/D
Intradermal (gene gun)	+	+	−	N/D	N/D	
Intraperitoneal	−	N/D	N/D	N/D	N/D	N/D
Subcutaneous	−	N/D	N/D	N/D	N/D	N/D

Note. +, Evidence of immune responses or protection against tumor challenge; −, absence of immune responses or protection against tumor challenge; N/D, not done.

struct containing independent expression cassettes for CEA and HBsAg [72]. cDNA encoding HBsAg was included in this reagent to provide an internal positive control for polynucleotide immunization efficacy. In these studies, three pigtail macaques received 500 μg doses of plasmid (pCEA/HBsAg) into each quadriceps muscle (1 mg total dose) at weeks 0, 3, 9, 16, and 25. A second group of three pigtail macaques received 3 μg doses of pCEA/HBsAg by particle bombardment of inguinal skin at weeks 0, 10, 19, and 27. No local inflammation or other toxicities were observed in immunized animals. CEA-specific humoral immune responses were demonstrable in all monkeys prior to completion of vaccinations, as illustrated in Fig. 2. CEA-specific lymphoproliferative responses were noted in two of three macaques following intramuscular vaccination and in two of three macaques immunized by particle bombardment. Delayed-type hypersensitivity (DTH) responses to CEA were observed in two of three monkeys following intramuscular plasmid immunization, consistent with the induction of a Th1 immune response. These animals developed large (>3 cm) regions of induration and erythema coincident with sites of intradermal CEA protein inoculation, and the presence of DTH responses in these areas was confirmed by histopathologic examination. No DTH responses to intradermal CEA were noted in the three macaques immunized by particle bombardment or in five naïve macaques.

X. OTHER PRECLINICAL EXPERIENCE WITH PNV IN MALIGNANCY

Several other investigators have reported the utilization of polynucleotide constructs encoding tumor-associated antigens for the induction of measurable immune responses and antitumor effects in animal studies. Syrengelas *et al.* immunized mice with a plasmid DNA construct encoding the idiotype (Id) of a murine B-cell lymphoma, 38C13, with elicitation of humoral responses against the tumor idiotype and resultant immunoprotection against challenge with 38C13 tumor cells [79]. In this study, use of a plasmid DNA molecule encoding an Id/GM-CSF fusion protein (pId-GM) produced earlier and more frequent idiotype-specific humoral responses

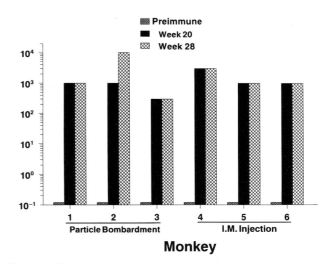

FIGURE 2   Anti-CEA antibody response elicited by CEA polynucleotide immunization in nonhuman primates. A group of three macaques received 3 μg doses of plasmid DNA encoding CEA (pCEA/HBsAg) by particle bombardment of inguinal skin at weeks 0, 10, 19, and 22. A second group of three macaques received 1 mg doses of pCEA/HBsAg by i.m. injection at weeks 0, 3, 9, 16, and 25. Sera for anti-CEA antibody responses were obtained before and 20 and 28 weeks following the initial vaccination. Results represent the limiting dilution of sera in an ELISA assay yielding positive results. Values of dilution > 0 represent a positive result.

than did immunization with the DNA construct expressing Id alone. A subsequent study by this group described generation of immunoprotection against 38C13 tumor cell challenge following administration of a polynucleotide construct encoding immunoglobulin variable region single-chain Fv (scFv) fused to an immunostimulatory peptide sequence derived from IL-1β [80]. Immunizations with plasmid DNA constructs encoding scFv alone or an scFv-GM-CSF fusion protein did not confer protection against tumor challenge.

Concetti *et al.* noted the induction of antibodies against p185, the erbB2/neu oncoprotein, in mice following intramuscular or intradermal inoculation of a plasmid DNA construct encoding rat neu [81]. Anti-p185 IgG obtained from immunized animals in this study produced growth inhibition and antibody-dependent lysis of p185^{erbB2}-positive SKBR3 human cancer cells *in vitro*. In addition, immunohistochemical studies revealed binding of anti-p185 polyclonals derived from the sera of immunized animals to murine renal tubular cells (which physiologically express the *neu* protooncogene), SKBR3 cells, and human breast carcinoma specimens. Binding of anti-p185 murine antibodies to human ErbB2 was confirmed by Western blots of cell lysates obtained from three cell lines expressing human ErbB2 (SKBR3, T28.1, and T19.7 cells).

Bueler and Mulligan described the generation of antigen-specific cellular immune responses in mice following intramuscular injection of plasmid DNA encoding the tumor-associated antigens MAGE-1 and MAGE-3 [82]. Utilization of dual-expression polynucleotide constructs encoding tumor-associated antigens and GM-CSF or B7.1 enhanced antitumor immunity and provided immunoprotection against challenge with MAGE-transfected B16 melanoma cells. Another group has noted a reduction in tumor formation in a murine melanoma model following immunization with a gp100-encoding plasmid DNA molecule [83]. Our group has demonstrated elicitation of immune responses to MART-1 (melanoma antigen recognized by T cells-1) following intramuscular injection of plasmid DNA encoding MART-1 in mice (unpublished observations).

Carbone's group has reported the elicitation of epitope-specific cytolytic T-cell responses in mice following gene gun administration of a plasmid DNA construct encoding a single T-cell epitope (T1272 peptide) from mutant p53 [54]. In this study, splenocytes harvested from immunized animals and restimulated in culture with the mutant p53 epitope were capable of peptide-specific lysis of T1272 pulsed target cells. Immunization with a polynucleotide construct encoding complete mutant p53 generated comparable epitope-specific lytic activity. Cytolytic T-cell responses were significantly enhanced by utilization of a plasmid containing

the T1272 epitope sequence fused with the adenovirus E3 leader sequence, which facilitates transport of nascent polypeptide into the endoplasmic reticulum for peptide-driven assembly of class I MHC molecules. Animals immunized with this construct encoding a T-cell epitope fused to an endoplasmic reticulum targeting sequence survived intravenous challenge with P815 mastocytoma cells transfected with and expressing a minigene encoding T1272.

Rosenberg's group immunized mice with plasmid DNA encoding β-galactosidase (β-gal) with the elicitation of β-gal-specific humoral and cellular immune responses and resultant immunoprotection against challenge with CT26 tumor cells transfected with the gene for β-gal [84]. This immunization alone had a minimal effect on established CT26-β-gal pulmonary metastases. However, the adoptive transfer of splenocytes harvested from immunized animals and restimulated *in vitro* with a T-cell epitope derived from β-gal induced regression of established CT26-β-gal lung nodules. In addition, coadministration of various recombinant cytokines, including human rIL-2, human rIL-7, mouse rIL-6, and mouse rIL-2, with the β-gal-expressing plasmid construct resulted in regression of established tumor implants.

Schirmbeck *et al.* reported generation of antigen-specific cytolytic T-cell responses and resultant CTL-mediated rejection of syngeneic tumor grafts in H-2^d mice following intramuscular administration of a plasmid DNA construct encoding the SV40 large tumor antigen (T-Ag) [85]. Cellular immune responses and rejection of T-Ag-positive tumor grafts were not observed in H-2^d mice following immunization with exogenous T-Ag protein or SV40 infection.

XI. CLINICAL EXPERIENCE WITH PNV IN MALIGNANCY

Based on the induction of antitumor immunity in preclinical studies of plasmid reagents encoding tumor-associated antigens, several early phase clinical trials of PNV in malignancy are underway. Our group is conducting a human dose-escalation trial of a dual-expression plasmid encoding CEA and the small and middle (S2.S) proteins of HBsAg. In this study, patients with metastatic, CEA-expressing colorectal adenocarcinoma are receiving deltoid inoculations of plasmid according to the following schedule: group 1 (three patients) receives 0.1 mg of plasmid on day 1; group 2 (three patients) receives 0.3 mg of plasmid on day 1; group 3 (three patients) receives 1 mg of plasmid on day 1; group 4 (three patients) receives 0.3 mg of plasmid on days 1, 22, and 43; and group 5 (three patients) receives 1 mg of plasmid on days 1, 22, and 43. All patients receive a 10 μg recombinant HBsAg protein boost on day 64. At present, 12 patients (groups 1–4) have completed this study. Priming of the anti-HBsAg humoral immune response by plasmid inoculation was noted in one patient at the lowest dose level. HBsAg-specific antibody responses were demonstrable in one of three patients at each subsequent dose level within 3 to 9 weeks of initiation of vaccination, prior to the HBsAg protein boost. Analysis of humoral and cellular immune responses against CEA and HBsAg in patients at the highest plasmid dose level (group 5) is ongoing. Polynucleotide immunizations in this trial have produced no local inflammation, serological or clinical evidence of autoimmune disease, or unanticipated acute toxicities.

We are planning to conduct a dose-escalation trial of a PNV construct encoding MART-1 in patients with resected melanoma at significant risk for relapse in the near future. Investigators at the University of Pennsylvania are currently conducting a clinical trial of a plasmid reagent encoding idiotypic T-cell receptor in patients with cutaneous T-cell lymphoma [75]. Other phase I trials of polynucleotide vaccination are anticipated in B-cell lymphoma and metastatic carcinomas.

XII. CONCLUSION

The elicitation of humoral and cellular immune responses against tumor-associated antigens has been demonstrated in numerous preclinical models. Recent observations have defined potential strategies for the preferential induction of antitumor Th1 and cytolytic T-cell immunity. These strategies include the co-inoculation of immunogen-encoding constructs with plasmid-encoded cytokines or costimulatory molecules, the generation of fusion cDNAs that specifically promote MHC class I-restricted antigen presentation, and the optimization of plasmid delivery variables including dose, site, and technique of administration. Utilization of these strategies in murine tumor models has augmented antitumor immunity. However, the efficacy and safety of polynucleotide vaccination in humans remains unproven. The first generation of clinical trials of PNV is underway, and results from these studies should provide valuable insight into the potential clinical utility and limitations of this form of immunotherapy.

Acknowledgments

We would like to thank Sharon Garrison for assistance with manuscript preparation and Susan Moore, Karen Allen, Daunte Barlow, and Joyce Pike for expert technical advice and assistance.

References

1. Rosenberg, S. A., Packard, B. S., Aebersold, P. M., et al. Use of tumor infiltrating lymphocytes and interleukin-2 in the immunotherapy of patients with metastatic melanoma. Preliminary report. N. Engl. J. Med. 316, 1676–1680, 1988.
2. Restifo, N., Sznol, M. Cancer Vaccines in Cancer: Principles and Practice of Oncology (ed. 5). Philadelphia, Lippincott-Raven Publishers, 1997, pp 3023–3043.
3. Huang, A. Y. C., Golumbek, P., Ahmadzadeh, M., et al. Role of bone marrow-derived cells in presenting MHC class I-restricted tumor antigens. Science 264, 961–965, 1994.
4. Schreiber, H., Ward, P. L., Rowley, D. A., Stauss, H. J. Unique tumor-specific antigens. Ann. Rev. Immunol. 6, 465–483, 1988.
5. Rosenberg, S. A. The development of new cancer therapies based on the molecular identification of cancer regression antigens. Cancer J. 1, 90–100, 1995.
6. Irvine, K., Restifo, N. P. The next wave of recombinant and synthetic anti-cancer vaccines. Semin. Cancer Biol. 6, 337–347, 1995.
7. Wolff, J. A., Malone, R. W., Williams, P., et al. Direct gene transfer into mouse muscle in vivo. Science 247, 1465–1468, 1990.
8. Wolff, J., Williams, P., Ascadi, G., et al. Conditions affecting direct gene transfer into rodent muscle in vivo. Biotechniques 11, 474–484, 1991.
9. Jiao, S., Williams, P., Berg, R., et al. Direct gene transfer into non-human primate myofibers in vivo. Hum. Gene Ther. 3, 21–33, 1992.
10. Cox, G. J. M., Zamb, T. J., Babiuk, L. A. Bovine herpesvirus 1: immune responses in mice and cattle injected with plasmid DNA. J. Virol. 67, 5664–5667, 1993.
11. Danko, I., Wolff, J. A. Direct gene transfer into muscle. Vaccine 12, 1499–1502, 1994.
12. Conry, R. M., Curiel, D. T., Smith, B., et al. CEA polynucleotide immunization elicits humoral and cellular immune responses in a canine model without toxicity. Presented at the Fourth International Conference on Gene Therapy of Cancer, San Diego, CA, November 9–11, 1995 (abstr.).
13. Hansen, E., Fernandes, K., Goldspink, G., et al. Strong expression of foreign genes following direct injection into fish muscle. Fed. Europ. Biochem. Soc. 290, 73–76, 1991.
14. Wolff, J. A., Dowty, M. E., Jiao, S., Repetto, G., et al. Expression of naked plasmids by cultured myotubes and entry of plasmids into T-tubules and caveolae of mammalian skeletal muscle. J. Cell Sci. 103, 1249–1259, 1993.
15. Wolff, J., Ludtke, J., Ascadi, G., et al. Long term persistence of plasmid DNA and foreign gene expression in mouse muscle. Hum. Mol. Genet. 1, 363–369, 1992.
16. Nichols, W. W., Ledwith, B. J., Manam, S. V., Troilo, P. J. Potential DNA vaccine integration into host cell genome. Ann. N. Y. Acad. Sci. 772, 30–39, 1995.
17. Kitsis, R. N., Buttrick, P. M., McNally, E. M., et al. Hormonal modulation of a gene injected into rat heart in vivo. Proc. Natl. Acad. Sci. U.S.A. 88, 4138–4142, 1991.
18. Raz, E., Carson, D. A., Parker, S. E., et al. Intradermal gene immunzation: the possible role of DNA uptake in the induction of cellular immunity to viruses. Proc. Natl. Acad. Sci. U.S.A. 91, 9519–9523, 1994.
19. Keller, E. T., Burkholder, J. K., Shi, F., Pugh, T. D., et al. In vivo particle-mediated cytokine gene transfer into canine oral mucosa and epidermis. Cancer Gene Ther. 3, 186–191, 1996.
20. Yang, N.-S., Burkholder, J., Roberts, B., et al. In vivo and in vitro gene transfer to mammalian somatic cells by particle bombardment. Proc. Natl. Acad. Sci. U.S.A. 87, 9568–9572, 1990.

21. Malone, R. W., Hickman, M. A., Lehmann-Bruinsma, K., et al. Dexamethasone enhancement of gene expression after direct hepatic DNA injection. *J. Biol. Chem.* **269,** 29903–29907, 1994.

22. Manthorpe, M., Cornefert-Jensen, F., Hartikka, J., et al. Gene therapy by intramuscular injection of plasmid DNA: studies on firefly luciferase gene expression in mice. *Hum. Gene Ther.* **4,** 419–431, 1993.

23. Norman, J. A., Hobart, P., Manthorpe, M., et al. Development of improved vectors for DNA-based immunization and other gene therapy applications. *Vaccine* **15,** 801–803, 1997.

24. Kass-Eisler, A., Li, K., Leinwand, L. A. Prospects for gene therapy with direct injection of polynucleotides. *Ann. N. Y. Acad. Sci.* **772,** 232–240, 1995.

25. Ertl, H. C. J., Xiang, Z. Q. Genetic immunization. *Viral Immunol.* **9,** 1–9, 1996.

26. Xiang, Z. Q., Spitalnik, S. L., Cheng, J., et al. Immune responses to nucleic acid vaccines to rabies virus. *Virology* **209,** 569–579, 1995.

27. Malone, R. W., Felanger, P. L., Verma, I. M. Cationic liposome-mediated RNA transfection. *Proc. Natl. Acad. Sci. U.S.A.* **86,** 6077–6081, 1989.

28. Davis, H. L., Michel, M.-L., Whalen, R. G. Use of plasmid DNA for direct gene transfer and immunization. *Ann. N. Y. Acad. Sci.* **772,** 21–29, 1995.

29. Johanning, F. W., Conry, R. M., LoBuglio, A. F., et al. A Sindbis virus mRNA polynucleotide vector achieves prolonged and high level heterologous gene expression in vivo. *Nucleic Acids Res.* **23,** 1495–1501, 1995.

30. Zhou, X., Berglund, P., Rhodes, G., et al. Self-replicating Semliki Forest virus RNA as recombinant vaccine. *Vaccine* **12,** 1510–1514, 1994.

31. Barry, M. A., Johnston, S. A. Biological features of genetic immunization. *Vaccine* **15,** 788–791, 1997.

32. Wells, D. J. Improved gene transfer by direct plasmid injection associated with regeneration in mouse skeletal muscle. *Fed. Eur. Biochem. Soc.* **332,** 179–182, 1993.

33. Vitadello, M., Schiaffino, M. V., Picard, A., et al. Gene transfer in regenerating muscle. *Hum. Gene Ther.* **5,** 11–18, 1994.

34. Danko, I., Fritz, J. D., Jiao, S., et al. Pharmacological enhancement of in vivo foreign gene expression in muscle. *Gene Ther.* **1,** 114–121, 1994.

35. Tang, D. C., DeVit, M., Johnston, S. A. Genetic immunization is a simple method for eliciting an immune response. *Nature* **356,** 152–154, 1992.

36. Ulmer, J. B., Donnelly, J. J., Parker, S. E., et al. Heterologous protection against influenza by injection of DNA encoding a viral protein. *Science* **259,** 1745–1749, 1993.

37. McClements, W. L., Armstrong, M. E., Keys, R. D., Liu, M. A. The prophylactic effect of immunization with DNA encoding herpes simplex virus glycoproteins on HSV-induced disease in guinea pigs. *Vaccine* **15,** 857–860, 1997.

38. Yang, K., Mustafa, F., Valsamakis, A., et al. Early studies on DNA-based immunizations for measles virus. *Vaccine* **15,** 888–892, 1997.

39. Fynan, E. F., Webster, R. G., Fuller, D. H., et al. DNA vaccines: protective immunizations by parenteral, mucosal, and gene-gun inoculations. *Proc. Natl. Acad. Sci. U.S.A.* **90,** 11478–11482, 1993.

40. Lewis, P. J., Cox, G. J. M., et al. Polynucleotide vaccines in animals: enhancing and modulating responses. *Vaccine* **15,** 861–864, 1997.

41. Prince, A. M., Whalen, R., Brotman, B. Successful nucleic acid based immunization of newborn chimpanzees against hepatitis B virus. *Vaccine* **15,** 916–919, 1997.

42. Fuller, D. H., Corb, M. M., Barnett, S., et al. Enhancement of immunodeficiency virus-specific immune responses in DNA-immunized rhesus macaques. *Vaccine* **15,** 924–926, 1997.

43. Qui, P., Ziegelhoffer, P., Sun, J., Yang, N. S. Gene gun delivery of mRNA in situ results in efficient transgene expression and genetic immunization. *Gene Ther.* **3,** 262–268, 1996.

44. Conry, R. M., LoBuglio, A. F., Wright, M., et al. Characterization of a messenger RNA polynucleotide vaccine vector. *Cancer Res.* **55,** 1397–1400, 1994.

45. Xiang, Z. Q., Ertl, H. C. J. Manipulation of the immune response to a plasmid-encoded viral antigen by coinoculation with plasmids expressing cytokines. *Immunity* **2,** 129–135, 1995.

46. Davis, H. L., Michel, M.-L., Whalen, R. G. DNA-based immunization induces continuous secretion of hepatitis B surface antigen and high-levels of circulating antibody. *Hum. Mol. Genet.* **2,** 1847–1851, 1993.

47. Hohlfeld, R., Engel, A. G. The immunobiology of muscle. *Immunol. Today* **15,** 269–274, 1994.

48. Kim, J. J., Ayyavoo, V., Bagarazzi, M. L., et al. In vivo engineering of a cellular immune response by coadministration of IL-12 expression with a DNA immunogen. *J. Immunol.* **158,** 816–826, 1997.

49. Geissler, M., Gesien, A., Tokushige, K., Wands, J. R. Enhancement of cellular and humoral immune responses to hepatitis C virus core protein using DNA-based vaccines augmented with cytokine-expressing plasmids. *J. Immunol.* **158,** 1231–1237, 1997.

50. Conry, R. M., Widera, G., LoBuglio, A. F., et al. Selected strategies to augment polynucleotide immunization. *Gene Ther.* **3,** 67–74, 1996.

51. Tsuji, T., Hamajima, K., Fukushima, J. Enhancement of cell-mediated immunity against HIV-1 induced coinoculation of plasmid-encoded HIV-1 antigen with plasmid expressing IL-12. *J. Immunol.* **158,** 4008–1013, 1997.

52. Kim, J. J., Bagarazzi, M. L., Trivedi, N., Hu, Y., et al. Engineering of in vivo immune responses to DNA immunization via codelivery of costimulatory molecule genes. *Nat. Biotech.* **15,** 641–646, 1997.

53. Iwasaki, A., Stiernholm, B. J. N., Chan, A. K., et al. Enhanced CTL responses mediated by plasmid DNA immunogens encoding costimulatory molecules and cytokines. *J. Immunol.* **158,** 4951–4601, 1997.

54. Ciernik, I. F., Berzofsky, J. A., Carbone, D. P. Induction of cytotoxic T lymphocytes and antitumor immunity with DNA vaccines expressing single T cell epitopes. *J. Immunol.* **156,** 2369–2375, 1996.

55. Rodriguez, F., Zhang, J., Whitton, J. L. DNA immunization: ubiquitination of a viral protein enhances cytotoxic T-lymphocyte induction and antiviral protection but abrogates antibody induction. *J. Virol.* **71,** 8497–8503, 1997.

56. Sato, Y., Roman, M., Tighe, H., et al. Immunostimulatory DNA sequences necessary for effective intradermal gene immunization. *Science* **273,** 352–354, 1996.

57. Klinman, D. M., Yi, A., Beaucage, S. L., et al. CpG motifs expressed by bacterial DNA rapidly induce lymphocytes to secrete IL-6, IL-12, and IFN-γ. *Proc. Natl. Acad. Sci. U.S.A.* **93,** 2879–2883, 1996.

58. Halpern, M. D., Kurlander, R. J., Pisetsky, D. S. Bacterial DNA induces murine interferon-gamma production by stimulation of IL-12 and tumor necrosis factor-alpha. *Cell Immunol.* **167,** 72–78, 1996.

59. Klinman, D. M., Yamshchikov, G., Ishigatsubo, Y. Contribution of CpG motifs to the immunogenicity of DNA vaccines. *J. Immunol.* **158,** 3635–3639, 1997.

60. Conry, R. M., LoBuglio, A. F., Curiel, D. T. Polynucleotide-mediated immunization therapy of cancer. *Semin Oncol.* **23,** 135–147, 1996.

61. Pertmer, T. M., Roberts, T. R., Haynes, J. R. Influenza virus nucleoprotein-specific immunoglobulin G subclass and cytokine

responses elicited by DNA vaccination are dependent on the route of vector DNA delivery. *J. Virol* **70**, 6119–6125, 1996.

62. Feltquate, D. M., Heaney, S., Webster, R. G., Robinson, H. L. Different T helper cell types and antibody isotypes generated by saline and gene gun DNA immunization. *J. Immunol.* **158**, 2278–2284, 1997.

63. Chow, Y. H., Huang, W. L., Chi, W. K., Chu, Y. D., Tao, M. H. Improvement of hepatitis B virus DNA vaccines by plasmids coexpressing hepatitis B surface antigen and interleukin-2. *J. Virol.* **71**, 169–178, 1997.

64. Goebels, N., Michaelis, D., Wekerle, H., *et al.* Human myoblasts as antigen-presenting cells. *J. Immunol.* **149**, 661–667, 1992.

65. Fabry, Z., Sandor, M., Gajewski, T. F., *et al.* Differential activation of Th1 and Th2 CD4+ cells by murine brain microvessel endothelial cells and smooth muscle/pericytes. *J. Immunol.* **151**, 38–47, 1993.

66. Torres, C. A., Iwasaki, A., Barber, B. H., Robinson, H. L. Differential dependence on target site tissue for gene gun and intramuscular DNA immunizations. *J. Immunol.* **158**, 4529–4532, 1997.

67. Casares, S., Inaba, K., Brumeanu, T.-D., *et al.* Antigen presentation by dendritic cells after immunization with DNA encoding a major histocompatibility complex class II-restricted viral epitope. *J. Exp. Med.* **186**, 1481–1486, 1997.

68. Fu, T. M., Ulmer, J. B., Caulfield, M. J., *et al.* Priming of cytotoxic T lymphocytes by DNA vaccines: requirement for professional antigen presenting cells and evidence for antigen transfer from myocytes. *Mol. Med.* **3**, 362–371, 1997.

69. Corr, M., Lee, D. J., Tighe, H. Gene vaccination with naked plasmid DNA: mechanism of CTL priming. *J. Exp. Med.* **184**, 1555–1560, 1996.

70. Iwasaki, A., Torres, A. T., Ohashi, P. S., *et al.* The dominant role of bone marrow-derived cells in CTL induction following plasmid DNA immunization at different sites. *J. Immunol.* **159**, 11–14, 1997.

71. Ulmer, J. B., Deck, R. R., DeWitt, C. M., *et al.* Expression of a viral protein by muscle cells *in vivo* induces protective cell-mediated immunity. *Vaccine* **15**, 839–841, 1997.

72. Conry, R., Khazaeli, M. B., Curiel, D., LoBuglio, A. Carcinoembryonic antigen polynucleotide immunization by gene gun versus i.m. injection in nonhuman primates. *Proc. Am. Assoc. Cancer Res.* **38**, 399, 1997 (abstr. #2679).

73. Mor, G., Singla, M., Steinberg, A. D., *et al.* Do DNA vaccines induce autoimmune disease? *Hum. Gene Ther.* **8**, 293–300, 1997.

74. Donnelly, J. J., Ulmer, J. B., Shiver, J. W., Liu, M. A. DNA vaccines. *Ann. Rev. Immunol.* **15**, 617–648, 1997.

75. Chattergoon, M., Boyer, J., Weiner, D. Genetic immunization: a new era in vaccines and immune therapeutics. *FASEB J.* **11**, 753–763, 1997.

76. Taubes, G. Salvation in a snippet of DNA? *Science* **278**, 1711–1714, 1997.

77. Conry, R. M., LoBuglio, A. F., Loechel, F., *et al.* A carcinoembryonic antigen polynucleotide vaccine has *in vivo* antitumor activity. *Gene Ther.* **2**, 59–65, 1995.

78. Conry, R. M., Lobuglio, A. F., Loechel, F., *et al.* A carcinoembryonic antigen polynucleotide vaccine for human clinical use. *Cancer Gene Ther.* **2**, 33–38, 1995.

79. Syrengelas, A. D., Chen, T. T., Levy, R. DNA immunization induces protective immunity against B-cell lymphoma. *Nat. Med.* **2**, 1038–1041, 1996.

80. Hakim, I., Levy, S., Levy, R. A nine-amino acid peptide from IL-1beta augments antitumor immune responses induced by protein and DNA vaccines. *J. Immunol.* **157**, 5503–5511, 1996.

81. Concetti, A., Amici, A., Petrelli, C., *et al.* Autoantibody to p185[erbB2/neu] oncoprotein by vaccination with xenogenic DNA. *Cancer Immunol. Immunother.* **43**, 307–315, 1996.

82. Bueler, H., Mulligan, R. C. Induction of antigen-specific tumor immunity by genetic and cellular vaccines against MAGE: enhanced tumor protection by coexpression of granulocyte-macrophage colony-stimulating factor and B7-1. *Mol. Med.* **2**, 545–555, 1996.

83. Molling, K. Naked DNA for vaccine or therapy. *J. Mol. Med.* **75**, 242–246, 1997.

84. Irvine, K. R., Rao, J. B., Rosenberg, S. A., Restifo, N. P. Cytokine enhancement of DNA immunization leads to effective treatment of established pulmonary metastases. *J. Immunol.* **156**, 238–245, 1996.

85. Schirmbeck, R., Bohm, W., Reimann, J. DNA vaccination primes MHC class-I-restricted simian virus 40 large tumor antigen-specific CTL in H-2^d mice that reject syngeneic tumors. *J. Immunol.* **157**, 3550–3558, 1996.

DNA and Dendritic Cell-Based Genetic Immunization Against Cancer

LISA H. BUTTERFIELD, ANTONI RIBAS, AND JAMES S. ECONOMOU
Division of Surgical Oncology, UCLA Medical Center, 54-140 CHS, Los Angeles, California 90095

I. INTRODUCTION

Genetic immunization is a new and potentially powerful technique in cancer gene therapy that utilizes DNA, in plasmid or virus form, to stimulate a tumor-antigen-specific immune response. With the identification and cloning of several human tumor rejection antigens and a better understanding of the molecular requirements for induction of T-cell immune responses, genetic immunization has the potential to become a novel treatment for cancer.

II. BACKGROUND

The recent identification of genes encoding cancer regression antigens has allowed the development of new strategies for the immunotherapy of human cancer (see Table 1). In malignant melanoma, these antigens were identified by tumor-infiltrating lymphocytes (TILs) capable of recognizing shared major histocompatibility complex (MHC) class-I-restricted epitopes present on tumors from different patients with the same HLA type [1,2]. The genes for the common melanocyte/melanoma lineage antigens MART-1/Melan-A (MART-1), gp100, tyrosinase, and tyrosinase-related-protein (TRP-1) were cloned by cDNA library transfection and TIL screening [3–7]. The MAGE family of tumor antigens, with at least 14 genes, has been cloned and analyzed [8,9]. Their expression occurs in a wide variety of tumor types [10–12], whereas in normal tissues, MAGE expression appears to be restricted to testes and perhaps early stages of wound healing [13]. Many MAGE epitopes have been identified and used to induce cytotoxic T lymphocytes

285

TABLE 1   Types of Tumor Antigens

Normal differentiation antigens	MART-1/Melan-A	gp100	tyrosinase	TRP-1
Tumor-specific antigens	MAGE family (1–14)	BAGE	GAGE	
Overexpressed proteins	Her2/neu	CEA	PSA	p53
Mutated oncogenes	p21 *ras*	p53		

(CTLs) from normal donors and melanoma patients [14–18]. For common and well-characterized class I alleles (e.g. *A1, A2, A3, A24, A31*) the immunodominant peptides for many melanoma tumor antigens have been defined by a variety of techniques, including screening of synthetic overlapping peptides and high-performance liquid chromatography (HPLC) fractionation of eluted peptides. Some of these immunogenic tumor antigen epitopes are currently in clinical testing [19,20].

Many of the described melanoma-associated antigens are normal nonmutated differentiation antigens present in melanomas and normal melanocytes, which are present in the skin and retina. Any immune response generated to these antigens on tumor cells should also induce a response to the same antigens if they are present on normal tissues. Thus, the almost 29% incidence of vitiligo (patchy loss of pigmentation) in responding patients undergoing IL-2 immunotherapy is now understood to be the result of a cell-mediated immune response directed toward these normally expressed antigens on cutaneous melanocytes [21]. In different immunotherapy strategies tested in melanoma thus far, vitiligo appears to be the only manifestation of autoimmunity, and no vision impairment or other toxicities in responding patients have been observed. There has been one report of the involvement of melanoma-antigen-reactive CTLs in a melanocyte autoimmunity disorder, Vogt–Koyanagi–Harada disease [22].

The *HLA-A2.1* genotype is the most common allele among Caucasians (approximately 50% are *HLA-A2.1*) and it is also well represented in other populations [23]. HLA-A2.1 has been crystallized and its binding groove well characterized [24]. The sequences of naturally processed and presented peptides from HLA-A2.1 have been thoroughly analyzed. Many peptides have been sequenced from immunoprecipitated A2 peptide complexes from JY lymphoblastoid cells (*HLA-A2.1* homozygous) [25]. Similarly, *A2.1* peptide complexes have been purified and sequenced from lymphoblastoid cells transfected with *A2.1*.

These studies began to define the sequences of HLA A2.1–binding peptides. Synthetic peptides have been used in binding competition studies [26–28], and different HLA-A2 subtypes have been analyzed [29,30]. This extensive body of work has defined important peptide

motifs. A2.1 binds peptides from 8 to 10 amino acids in length, but primarily 9 mers. The amino acids L, I, and M are considered important anchor residues in peptide position 2, and V, L, and I are anchors in position 9 (or 10 depending on peptide length) [25,28,31,32].

A number of class I epitopes are derived from the protein sequence of tumor antigens that have been shown to bind to class I molecules due to the presence of either one or two of the optimal anchor residues and other important amino acids in their sequence. These epitopes derive from intracellular proteins that are naturally processed and presented by the intracellular machinery: digested by the proteosome to free candidate epitopes, transported to the ER where they bind to nascent MHC molecules, and displayed on the cell surface (see Figure 1). However, many epitopes that are displayed on the surface of a tumor cell may not be capable of stimulating T cells without some change in the tumor environment. Some examples of epitopes that are naturally processed and presented and have been demonstrated to be able to generate antigen-specific immune responses are in Table 2 [33].

The melanoma story has significantly revised our thinking about the human T-cell repertoire and immunologic tolerance. It is clear that there are T-cell receptors capable of recognizing these self-peptides and that these T cells have not been deleted from the immune system [34]. For MART-1 and gp100 immunodominant peptides, the HLA-A2 binding affinities are categorized

Peptide Epitope Processing

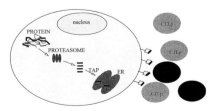

FIGURE 1   Peptide epitope processing. A cytosolic protein is digested by the proteosome into peptide fragments that are transported into the endoplasmic reticulum by TAP molecules. There, the peptides interact with newly synthesized class I molecules and are transported to the cell surface. The epitopes on the cell surface are available to interact with T-cell receptors of CTL precursors.

TABLE 2 Immunogenic Epitopes for Melanoma Antigens

Antigen	HLA type	Peptides
MART-1	A2	AAGIGILTV
		ILTVILGVL
		GIGILTVL
gp100	A2	YLEPGPVTA
		KTWGQYWQV
		LLDGTATLRL
Tyrosinase	A2	MLLAVLYCL
		YMNGTMSQV
	B44	SEIWRDIDF
MAGE-1	A1	EADPTGHSY
	Cw1601	SAYGEPRKL
MAGE-3	A1	EVDPIGHLY
	A2	FLYGPRALV
	B44	MEVDPIGHLY
TRP1(gp75)	A31	MSLQRQFLR

as intermediate due to the absence of optimal residues at either the second or ninth anchor positions [35,36]. No high-affinity peptides from these two tumor antigens appear to serve as targets recognized by TILs. One hypothesis that has been forwarded is that high-affinity peptides, expressed at high levels on the cell surface, may generate tolerance. Lower binding or "subdominant" determinants may be capable of stimulating peptide-responsive T cells not deleted from the T-cell repertoire [37]. In contrast, tyrosinase contains immunogenic epitopes restricted by HLA-A2 that contain both anchor residues. MAGE-3 also contains an immunogenic two-anchor epitope for A2. It is also possible that a strong-binding, stable peptide is required in an immunogenic epitope, and that tolerance to such peptides could be broken if high enough antigen density could be attained in a vaccine. In the case of lower-affinity peptides, some have been shown to make up in stability or slow off-kinetics what they lack in initial binding capacity [38].

These results have spawned investigations of self-antigens expressed by other human cancers that might serve as suitable targets for immunotherapy. A number of these proteins, generally overexpressed by human cancer cells, have been identified: CEA [39], PSA [40], Her2/neu [41], mutated Ras [42] and p53 (wt and mutant) [43,44]. Both peptide-based and DNA-based genetic immunization clinical trials are in progress testing a number of these putative antigens.

CEA: Carcinoembryonic antigen (CEA) is expressed in the majority of colorectal, gastric, and pancreatic cancers, and also in some breast cancers and non-small-cell lung cancers (NSCLCs) [45]. It is also expressed in

some normal colon epithelium. Many different strategies have been investigated for therapy against CEA+ tumors, including generation of anti-idiotype antibodies, vaccination with vaccinia and avipox virus containing the CEA cDNA, vaccination with plasmid-CEA cDNA and recombinant CEA protein, and use of the immunodominant HLA-A2–restricted CEA peptide. In a clinical trial in which patients received vaccinations with recombinant vaccinia–CEA, an increase in CEA-reactive T cells after treatment was found [39]. In this trial, in vitro restimulation of T cells from patients could generate CEA peptide (CAP-1)/HLA-A2–specific killing of CEA+/A2+ tumor cells. These results indicate that CEA is a very promising target for a variety of potential genetic immunization strategies.

PSA: Prostate specific antigen (PSA) is overexpressed in many prostate tumors. Use of PSA as a tumor antigen has been reported in a murine model [40] in which vaccination of mice with a human PSA-transfected tumor lead to PSA-specific killing by the T cells generated in vivo. Recently, in humans, normal donor blood could be used in vitro to generate PSA-peptide-specific killing of peptide-pulsed HLA-A2+/ PSA+ tumors, but not peptide pulsed, non-A2 tumors [46]. Although the findings are not as strong to date as those for CEA, PSA continues to be a potential prostate cancer tumor antigen target for therapy.

Her2/neu: Her2/neu is overexpressed by a subset of both breast and ovarian tumors. In vitro, CTLs that cross-react with common HLA-A2–restricted peptides have been generated from both breast and ovarian cancer patients. Some of these peptides have been HPLC fractionated, purified, and partially sequenced [41,47]. This has lead to the identification of several potential peptide targets restricted by HLA-A2 for Her2/neu. In vivo, a study in which rats were immunized with either rat Neu peptides or the whole Neu protein demonstrated that tolerance to rat Her2/neu could be broken to this self-protein with peptide-based immunization, but not whole protein [48]. This result also supports the notion that for some proteins, tolerance to an immunodominant peptide can mask an immune reaction to a subdominant epitope unless that subdominant epitope is separated from the whole protein before intracellular processing. This was shown for the model antigen hen egg lysozyme [49].

Ras: The mutated ras oncogene has been detected in a wide variety of tumors. The fact that there are three common mutations in the Ras protein that create the constitutively active form of p21 Ras makes it a particularly attractive target for genetic immunization. This should preclude any concerns over autoimmunity and tolerance because the immune reaction would be directed toward a tumor-specific mutated epitope. To address the feasibility of mutated *ras* as a target, CTLs were initially raised in mice against the human Ras-12 mutation [42]. Since then, *in vitro,* Ras-61 mutant-specific human CTLs have been raised that are HLA-A2 restricted, but none of the CTLs have exhibited mutant Ras+ tumor cell killing [50]. An additional problematic finding has been that the precursor frequency of CTLs against mutant Ras is extremely low in the donors analyzed thus far. Recently, a CD4+ T-cell clone that exhibited a proliferative response and specific cytokine release in response to mutated *RAS* epitopes or the mutant protein was isolated from a gastric cancer patient [51]. For a complete discussion of the *ras* system, see Chapter 17.

p53: The most commonly mutated (and often over expressed) gene in all cancer appears to be the tumor suppressor gene p53. Unlike *ras,* p53 has a wide range of mutational hot spots, located over 11 exons. This makes peptide-epitope-based strategies more difficult to conceive of. To begin the study of using p53 as a tumor target, peptides from the wild-type p53 sequence were screened for binding to class I molecules [52]. Several epitopes were found that were sufficiently immunogenic to generate peptide-specific CTLs *in vitro.* More recently, peptides (also from wild-type p53 sequences) that bind HLA-A2 were used to generate CTLs from a normal donor that could lyse an allogeneic mutant p53+ tumor line in an HLA-A2–restricted fashion [53], which indicates further that tolerance can be broken to the normal self-protein p53. It is somewhat troubling that the CTLs generated from normal p53 epitopes can kill mutant p53 tumor cells. Because expression of normal p53 is essential to the normal growth control function of most cells in the body, the issue of autoimmunity against normally expressed p53 is raised. Perhaps the short half-life of normal p53 will protect nonmalignant cells from CTL killing. In addition, it has been proposed that wt p53 epitopes would have to be used due to the wide potential mutation spectrum present across a large patient population. Some recent studies [44,54] utilize *HLA-A2* transgenic mice to generate high-affinity CTLs against human p53 epitopes, trying to circumvent tolerance [55].

III. RECENT ADVANCES: METHODS OF GENETIC IMMUNIZATION

A powerful means of inducing both cellular and humoral immunity is the expression of antigen-encoding DNA sequences in host antigen-presenting cells (APCs). Two methods, DNA immunization and dendritic cell transduction, appear to be effective in inducing antitumor immunity (see Figure 2).

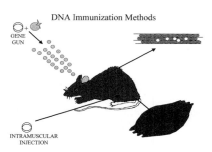

FIGURE 2 DNA immunization methods. Plasmid DNA can be coated onto the surface of gold beads and pulsed into the skin layers by helium gas. Once in the skin cells, the DNA dissociates from the gold. Alternatively, plasmid DNA can be injected into muscle cells, where it is taken up by both muscle cells and local APCs.

Antigen-encoding DNA plasmids administered intramuscularly (i.m.) or intradermally (i.d.) can induce cellular and humoral immune responses to pathogenic viruses, parasites, bacteria, and tumors [56–58]. The primary advantages of using DNA as the immunogen are purity, ease of production, stability of a DNA vaccine, and the long-term immunity that can be generated, as well as the preferential generation of CD8+ T cells and Th1-based responses [59]. In an early study using commercially available plasmids containing the human CEA gene [60], vaccinated mice were able to generate anti-CEA antibodies as well as T cells that proliferated and secreted IL-2 and IL-4 in response to CEA. In a follow-up study [61], improvements were made to the plasmid constructs, including the addition of the B7-1 costimulatory molecule to increase antigen presentation by the transfected muscle cells (which had to be on the same DNA construct as the CEA antigen gene to obtain optimal results). A second improvement was injecting a granulocyte-macrophage colony-stimulating factor (GM-CSF)-containing plasmid 3 days before treatment with the CEA plasmid to increase potential APC numbers in the injected muscle environment, which would better present the CEA molecule to the immune system. These changes led to increased T-cell responses and slowed the growth of a CEA+ tumor in the mice. These studies demonstrate not only the feasibility of plasmid immunization, but also the importance of stimulating APCs with either antigen plus B7-1 costimulation in a single transfected cell or drawing professional APCs to the vaccination site to optimize antigen presentation.

The plasmids employed in these studies have been found to contain immune-system-stimulating sequences in the plasmid backbone. The hypomethylated state of CpG motifs (in the context of AACpGTT in particular [62]) in plasmids grown in bacteria can yield a nonspecific inflammatory response to the plasmid itself. This can be important for the overall immune response to the specific antigen encoded by the plasmid, by increasing the number and activation state of the cells drawn to the vaccination site. To make plasmid backbones that can be used *in vivo* for human genetic immunization therapies, many avenues have been investigated to maximize both safety and efficacy [63]. These include replacement of the ampicillin resistance gene for bacterial propagation with the kanamycin resistance gene (to protect against a possible penicillin allergic response). To increase expression of the antigen gene, promoters, enhancers, intron sequences for optimal processing, termination, and polyadenylation signals continue to be thoroughly investigated and optimized.

The mechanism of immune response induction by genetic immunization, particularly MHC class-I-restricted CTLs, is currently being defined. Gene expression can be detected for many months in myocytes, but it is unlikely that these transfected cells can serve as the primary APC. Because the epidermis is a tissue that is known to contain professional APCs and is easily accessible, this tissue has also been investigated as a route for genetic immunization.

Recently, extensive use has been made of the helium-driven gene gun, primarily for epidermal delivery of plasmid DNA coated onto gold beads to an organism, as well as for *in vitro* transfection of cells. For *in vivo* use, this method is based on the observation that the skin is a rich source of the APCs known as Langerhans cells (LCs). Therefore, direct transfection of LCs using the gene gun could yield APCs that would be more efficient than muscle cells at generating an immune response to the tumor antigen of interest. In one recent study, the gene gun was used to deliver either mutant p53 or HIV gp120 epitopes to murine ear epidermis [64]. This strategy utilized the adenovirus E3 gene leader sequence to target the antigen epitope to the endoplasmic reticulum for potentially more efficient presentation by class I molecules. These investigators were able to observe epitope-specific T-cell responses to the gene-gun-delivered epitopes. Use of surrogate tumor antigens such as β-galactosidase (β-gal) in gene-gun-based experiments has also generated clearly defined antigen-specific responses. Irvine *et al.* [65] showed that immunization with a β-gal plasmid yielded both cytotoxic and antibody responses against the antigen and slowed tumor growth of a β-gal expressing tumor line. This response was improved with the addition of recombinant cytokines.

A number of methods to increase the response to DNA vaccines have been reported, including injection of DNA into regenerating skeletal muscle following injection of agents such as snake venom toxins, bupivacaine, or edematous muscle (after an injection of a hypertonic 25% sucrose solution) [66]. These measures

may improve expression greater than 80-fold. Irvine *et al.* [65] have shown that intraperitoneal treatment with recombinant IL-2, IL-6, IL-7, and IL-12 enhanced anti-tumoral responses after particle-mediated gene delivery to the epidermis. Geissler [67] has further shown that i.m. coimmunization with IL-2, IL-4, and GM-CSF DNA expression constructs enhances the response elicited by a plasmid encoding a viral epitope. Insertion of certain cytokine genes into the same expression cassette as the antigen-encoding DNA plasmid might attract APCs into the site of DNA delivery. As mentioned, this was shown by Conry *et al.* [61] to be optimal strategy to enhance the immune response to CEA using an i.m. injection model.

Several recent studies have addressed the mechanism of the immune response generated by these two plasmid DNA-based methods of genetic immunization. Many of these reports implicate bone-marrow-derived professional APCs in the induction of protective immunity, whether the DNA immunization was via intramuscular injection or gene gun to the epidermis [68–72]. Although these mechanistic studies indicate that the most important cells are professional APCs, other studies using transfection of myocytes with both antigens and the costimulatory molecule B7-1 and/or B7-2 [73–75] support the notion that direct antigen presentation by non-professional APCs does occur and can be made more efficient by giving a "second signal" of costimulation. In experimental systems, this second signal is most often B7-1, but B7-2 and ICAM-1 are also being investigated.

In a system in which mice were vaccinated via the gene gun with an ovalbumin (OVA) plasmid, the treated mice generated OVA-specific CTLs and showed increased survival against a stably transfected B16/OVA tumor line versus the parental B16 [76]. The mechanism of this gene-gun-mediated immunity was investigated with direct gene transfer of the green fluorescent protein (GFP) reporter gene. Gene transfer of the GFP into these skin APCs with the gene gun was shown. Therefore, as expected, LCs were directly transfected with the model tumor antigen in a way that results in systemic immunity to an antigen-bearing tumor challenge.

Dendritic cells (DCs) are among the most potent APCs described [77]. DCs are at least 100x more potent in presenting antigen than the monocyte/macrophage [77]. They are bone-marrow-derived and characterized by dendritic morphology (elongated, stellate processes), high mobility, and the ability to present antigen to naive T cells and stimulate primary T-cell responses in an MHC-restricted fashion [78,79]. DCs express high levels of MHC class I and II molecules; costimulatory molecules B7-1, B7-2, CD40 receptor, and CD1a; and adhesion molecules ICAM-1 (CD54), LFA-3 (CD58), and CD11a, b, and c [77]. DCs do not express classical T,

B, NK, or monocyte/macrophage markers. DCs are present in low numbers in the circulation but can migrate from tissues to lymphoid organs and interact with antigen-reactive T cells [80,81].

DCs acquire antigen in at least three major ways: first, through constitutive macropinocytosis, in which many soluble antigens can be taken up in the fluid phase; second, via receptor-mediated endocytosis with the mannose receptor (in which the yeast cell wall constituent zymosan has been shown to be taken up) and $Fc\gamma II$ (for antigen-antibody-complex uptake); and third, by utilizing phagocytosis, for example, to take up either bacteria or latex beads [82,83].

Once an antigen is taken up, DCs process intracellular proteins within the proteosome complex into 8- to 11-amino-acid peptides and transport these to the endoplasmic reticulum by transporter molecules TAP-1 and TAP-2. There they associate with synthesized MHC class I molecules, and this complex is expressed on the cell surface [84]. The class I pathway has recently been shown to be accessible by the nonclassical endocytic pathway via macropinocytosis in addition to the classical cytoplasmic endogenous antigen processing pathway [85]. Processing for class II expression is also very efficient [86–88].

It is now possible to readily prepare large numbers of DCs from murine and human progenitors *in vitro*. This has made possible the explosion in DC research as well as the potential to treat patients with therapeutic doses of autologous DCs. DCs may be differentiated from loosely adherent peripheral blood mononuclear leukocytes in 7-day cultures in the presence of GM-CSF and IL-4 [89]. (In our hands, we usually obtain 10^6 human DCs from every 50 mL of blood). CD34 progenitor cells, cultured in stem cell factor (SCF), GM-CSF, and tumor necrosis factor (TNF)-α, likewise will yield enriched DC populations in humans [90]. In mice, DCs can be differentiated from bone marrow precursors in a 7-day culture with GM-CSF and IL-4 [91].

Another method of obtaining DCs is harvesting them *in vivo* [92]. In humans, this can be achieved using cell separation machines that are able to collect circulating DCs (usually less than 0.5% of blood cells). Although yields are still limited, some of these machines have become efficient enough to generate DCs for clinical testing [93]. This approach may become more feasible if the circulating number of DCs can be increased. In mice, numbers of DCs can be dramatically increased with *in vivo* administration of Flt-3 ligand [94,95].

There are four major methods for utilizing DC in genetic immunization strategies; first, pulsing the DC with tumor cell extracts containing a complex mixture of known and unknown tumor antigens and proteins or pulsing with a purified, recombinant protein for a known

tumor antigen. This allows the DC to endocytose, process, and present any number of epitopes in an MHC-restricted fashion without requiring a specific MHC type for the patient. Second, the DC can be pulsed with mild-acid-eluted tumor peptides containing a mixture of known and unknown tumor epitopes or with a high concentration of a purified, synthetic peptide epitope. This requires eluting peptides from the same patient's tumor or from an HLA-matched tumor. Employing the synthetic peptide approach requires knowing the tumor antigen and the immunogenic epitope for the HLA type of each patient. Because a number of shared tumor antigen epitopes have been identified for common HLA types, use of synthetic peptides is a promising strategy. Third, DCs can be pulsed or transfected with mRNA or cDNA via lipids or viruses, or without carriers. The mRNA or cDNA can be synthesized from a specific tumor antigen or from patient tumor, and this leads to multiple epitopes being presented. Finally, DCs can also be fused with tumor directly, allowing the DC antigen-presentation machinery to present known and unknown tumor antigens from the tumor. Each of these strategies has been demonstrated to be feasible experimentally, and some are currently in clinical testing. Examples of each strategy are described in Figure 3.

To investigate the use of unfractionated tumor extracts as an immunogen, Nair and colleagues compared DCs and macrophages as the APC after pulsing each with poorly immunogenic murine tumor lysates [96]. They

DENDRITIC CELL-BASED STRATEGIES

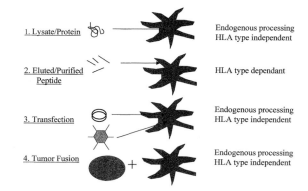

FIGURE 3 Dendritic cell-based strategies. Four methods of utilizing DCs to present antigenic epitopes are shown. (1) Tumor cell lysates or purified proteins are added to DCs, which take up, process, and present the peptide epitopes; (2) mild-acid-eluted peptides or purified, synthetic peptides are pulsed onto the surface of the DC for direct presentation; (3) plasmids or viruses can be used to transfect DCs with entire tumor antigen genes, leading to protein synthesis, processing, and presentation by the DC, and (4) DCs can be fused to tumor cells to allow the cellular proteins of the tumor cell to be processed and presented by the DC. Only the use of eluted or synthetic peptides depends upon a specific HLA type for correct presentation by the DC.

were able to demonstrate the presence of tumor-specific CTLs after as little as one immunization. In addition, they observed increased survival of tumor-challenged mice. This demonstrates how effective a complex, unpurified mix of normal and tumor antigens can be when used in conjunction with DCs. Whole, purified protein antigens used in the context of DC-based immunization include the model tumor antigen β-gal [43,97]. These studies used murine DC loaded *in vitro* with soluble β-gal protein as a vaccine against stably transfected β-gal-expressing tumors. CTLs specific to β-gal antigen were generated, and tumor immunity resulted.

DCs can be pulsed with peptide antigens, bypassing the need for processing. DCs pulsed with the peptide of genuine or surrogate tumor antigens can induce protective immunity, even to established tumors. For example, Mayordomo, Celluzi, and co-workers [98,99] investigated three different tumor antigen systems, in one of which they pulsed murine DCs with an OVA peptide that protected mice against a lethal challenge of OVA-transfected tumor cells. MUT-1 and an E7 HPV viral peptide were also used successfully, and the effects were dependant upon CD8+ cells. Using mutant p53 peptide-pulsed DC as a vaccine, Gabrilovich and co-workers were able to protect mice from a lethal challenge with a mutant-p53-transfected tumor line [100]. This effect was better than that seen with mice treated with systemic recombinant IL-12, but weaker than the effect of administering both mutant p53-peptide-pulsed DCs and systemic IL-12. This provides another example of the value of vaccination with a tumor antigen or its epitope at the same time as cytokine treatment.

This method of genetic immunization has been extended to human trials. There has been a phase I human trial using PSMA peptide-pulsed autologous DC in treatment of HLA A2–positive prostate cancer patients [101]. The vaccine was not toxic, and there was some decrease in serum PSA levels after vaccination. In another trial, HLA A1 melanoma patients carrying MAGE-1-positive tumors have been treated with MAGE-1/A1-peptide-pulsed autologous GM-CSF-treated APCs [20,102]. These patients showed an increased CTL precursor frequency for MAGE-1-reactive T cells, as well as MAGE-1-reactive and autologous melanoma-reactive CTLs at both the vaccination site and distant metastases. These encouraging data indicate that peptide-pulsed DCs have detectable physiological effects in advanced cancer patients, generating cytotoxic T cells with the desired specificity at the desired site. Although these trials are just beginning, it is important to know that these strategies are headed in the right direction.

In the setting in which the tumor antigens or peptides are unknown, several groups have described the antitu-

moral effect of DCs pulsed with peptides obtained by mild acid treatment of immunogenic tumor cells. Zitvogel and co-workers [103] have shown that peptides eluted from three weakly immunogenic tumor lines, the MCA205 sarcoma, the CL8.1 melanoma, and the TS/A mammary carcinoma, induced transient tumor stasis when pulsed onto cultured syngeneic DCs. Only immunization with eluted peptides derived from the more immunogenic C3 tumor elicited true tumor regressions. Nair *et al.* [96,104] used a similar protocol to elute peptides from the OVA-transfected E.G7-OVA cell line, the EL-4 parental thymoma cell line, and the F10.9 clone of the B16 melanoma. The eluted peptides were pulsed onto RMA-S cells (a TAP peptide-transporter-defective line) or onto APCs purified from adherent splenocytes. They observed a significant reduction in lung metastases when immunizing with peptides eluted from the OVA-transfected cell line and the F10.9 clone, but not when using peptides derived from the EL-4 thymoma. In our experience, peptide-pulsed DCs could effectively immunize mice only against immunogenic tumors, whereas no protection was observed when using peptides from the nonimmunogenic NFSA tumor (Ribas *et al.,* manuscript submitted). Genetically engineering the peptide-pulsed DC to produce IL-2 was not able to elicit greater protection in the immunogenic tumors, nor was it able to alter the nonimmunogenicity of peptides eluted from NFSA tumors.

One of the potential drawbacks of this approach compared to the use of defined peptide epitopes is a reduced efficiency due to the low concentration of the effective tumor antigens in the eluted peptide mixture. However, it has been shown in the OVA system that peptides eluted from an OVA-transfected cell line are as effective as purified, synthetic OVA peptide to generate OVA-specific class-I-restricted CTLs [96,104]. Another potential drawback is the possibility of loading the most powerful APCs known with autologous peptides to which an autoimmune response could be generated. However, we and others [96,103,104] have not observed any adverse side effects in treated mice that might suggest an autoimmune response.

There have been recent advances in transfection of DC with whole tumor-antigen-coding sequences in DNA or virus form. Recent interesting and unexpected work has shown that even RNA can be used directly to transfect DCs with model tumor antigens [105]. Using an OVA model, this group was able to show that *in vitro*–transcribed specific OVA mRNA, OVA peptide, tumor-derived mRNA or tumor-derived total RNA pulsed onto murine DCs were each able to induce antigen-specific CTLs *in vitro*. In this model, *in vitro*–transcribed specific mRNA was a better immunogen than the OVA peptide. Several questions remain unanswered, such as the level of transgene expression, the efficiency of transfection, and the mechanism by which foreign RNA can be internalized into DCs and translated into protein in the host cell.

Lipofection has also been used to transfect DCs with antigen genes [106]. Although direct demonstration of gene expression can be hard to come by with this less efficient gene transfer method, perhaps by using "nature's adjuvant," the DC, a small amount of antigen transgene expression may be sufficient to stimulate T cells. Indeed, Alijagic and co-workers were able to demonstrate antigen-specific T-cell clustering around only tumor-antigen-transfected DCs, not CAT-transfected DCs.

Several viruses have been used to transduce DCs very efficiently. Human DCs have been transduced with retroviruses carrying the β-gal reporter gene [107]. With three cycles of viral infection, 35–67% of 7-day-cultured DC were positive for β-gal, and that expression lasted for at least 20 days as shown by PCR. The genomic PCR demonstrates that the gene was incorporated into the DCs' genomic DNA. A retrovirus encoding the melanoma tumor antigen MART-1 has been used to transduce CD34+ hematopoietic progenitors [108]. Transduced cells were then differentiated *in vitro* into DCs. These transduced DCs stimulated cytokine release from TILs of known MART-1 specificity and also could be used to generate MART-1-specific CTLs. This study demonstrates that tumor-antigen-transduced DCs can be used to generate CTLs, an important first step toward a clinical trial. More recently, recombinant vaccinia viruses encoding β-gal [109] or MART-1 [110] have been used to transduce murine or human DCs, respectively, which were able to stimulate antigen-specific CTLs *in vitro*.

We investigated many potential methods of DC transfection [111]. Although we could not detect transgene expression with any physical method of gene transfer utilized (calcium phosphate, electroporation, lipofection), we found that adenovirus (AdV) was a particularly easy, reliable and, most important, highly efficient vehicle for transgene expression in human and murine DC. Adenoviral vectors can transduce up to 100% of the DCs in a population, and when transduced with IL-2 or IL-7AdV vectors, DCs synthesize up to nanogram amounts of cytokine per milliliter of culture medium per 24 hours. Others have also found adenovirus to be an important vector for DC [112]. In an important study, Zhai *et al.* created adenoviruses encoding the melanoma tumor antigens MART-1 and gp100. These vectors could effectively transduce human DCs, which then stimulated cytokine release from TILs specific for the tumor antigen.

We [113] and others [108,114] have demonstrated that both murine and human DCs genetically engi-

neered to express an entire tumor antigen polypeptide could induce immunity by processing and presenting relevant immunodominant peptides. Transduction of the OVA gene into murine DC using replication-defective adenoviral vectors protects against challenge with an OVA-expressing tumor cell line. This antitumoral protection has been shown to be superior to the one elicited by direct injection of the adenoviral vector alone [114]. Although there was a initial fear that use of adenovirus vectors might induce an overwhelming adenovirus antigen response that could mask the desired tumor antigen response, this recent work has also shown that this is not an insurmountable problem.

We have developed a murine model of genetic immunotherapy using an adenovirus that encodes the human melanoma antigen MART-1 (AdVMART1) (see Figure 4). In this model, vaccination of mice with adenovirus-transduced DC gave antitumor protection superior to that of direct injection of the adenovirus alone or i.m. injection of naked DNA.

Our observations with this AdVMART1 DC-based model can be summarized as follows: (1) MART-1-specific immunity can be generated, (2) only one or two immunizing injections are needed, (3) spleens from immunized mice contain class-I-restricted CTL specific for MART-1 targets, (4) growth of both microscopically and macroscopically established tumor is arrested by MART-1-engineered DC treatment, and (5) intraperitoneal DC administration is somewhat inferior to intravenous or subcutaneous. We are currently using this animal model to test strategies that employ cytokine transduction and alternate methods of DC gene transfer.

DC-AdVMART1 IMMUNIZATION

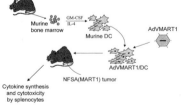

FIGURE 4   Murine model using DCs transduced with AdVMART1. Mice are vaccinated with murine bone-marrow-derived DCs transduced with the MART-1-expressing adenovirus AdVMART1. Some animals are challenged with NFSA tumor cells stably transfected with MART-1 [NFSA(MART1)] or control tumor cells and followed for tumor growth. Alternatively, splenocytes from the vaccinated mice are harvested and restimulated *in vitro* with NFSA(MART1). The splenocytes from AdVMART1/DC-vaccinated animals synthesize Th1 cytokines IL-2 and IFN-γ compared to little or no synthesis from animals vaccinated with nothing, DC only, or DC transduced with an irrelevant adenovirus (RR5). In addition, the splenocytes from AdVMART-1/DC-vaccinated animals specifically lyse MART-1-expressing targets, whereas control animal splenocytes do not.

We have performed a series of *in vitro* studies to test the utility of our AdVMART1 vector in a human system (see Figure 5). It was first important to confirm that transduction with the adenovirus resulted in MART-1 mRNA transcription in MART-1-negative cells. This has been shown by RT-PCR in human dendritic cells. Increasing the multiplicity of infection (MOI) correlates with increasing amounts of MART-1 mRNA. A time-course analysis demonstrates that this mRNA synthesis continues for at least 8 days *in vitro*. To show that the MART-1 protein is made in transduced cells, a monoclonal antiMART-1 antibody has been used to stain AdVMART1-transduced DCs. This antibody reacts with MART-1-expressing melanoma cells and AdVMART1-transduced DCs, but not untransduced DCs or DCs transduced with an irrelevant vector. To confirm that the immunodominant HLA-A2.1 MART-1$_{27-35}$ peptide is correctly processed and presented, the AdVMART1 was used to transduce MART-1-/HLA-A2+ cells, which successfully sensitized these cells to lysis by MART-1$_{27-35}$-specific CTLs. To demonstrate that this AdVMART1 could be used to efficiently generate CTLs, normal donor DCs were used as APCs after transduction with the AdVMART1. These cells were incubated with autologous CD8+ responder T cells. After as little as one week, anti-MART-1-specific killing has been observed against MART-1+ melanoma cells, as well as HLA-A2+/MART-1- cells transduced with the same virus but not against MART-1+/A2- or MART-1-/A2+ melanomas. An important finding of this method of CTL generation has been that anti-adenovirus antigen responses do not overwhelm the desired anti-MART-1 tumor antigen response (Butterfield, *et al.*, manuscript submitted).

DCs can be fused with tumor cells using techniques similar to those used to create hybridomas. Gong *et al.*, [115] have reported that murine DCs fused with a mu-

rine adenocarcinoma cell line expressing the MUC-1 tumor-associated antigen were able to induce MUC-1-specific CTLs *in vivo* and reject established lung metastases. In this system, tumor cells would supply tumor-associated antigens that would then be optimally presented to the host immune system by the DC. Benefits of this approach are that tumor-associated antigens do not need to be previously identified, that the antigenic peptides would be physiologically processed and presented by the DC, and that once the fused population is cloned, the cells used for vaccination would be immortal. A major problem is that this technique requires a fusion for each patient. The time required to select and grow the fused cells is likely to be months, requiring extensive *in vitro* manipulation to differentiate the fused cells from the contaminating tumor cells, which makes this approach difficult to translate into a human trial. A more practical approach might be to establish allogeneic DC/tumor fused cell lines using DC from common HLA subtypes and tumors that express common shared tumor antigens [116].

Naked DNA intramuscular immunization and DNA/gene gun approaches continue to be of great interest. If these methods could be refined with the best cytokine adjuvant and improved plasmid backbones, they would be a very attractive strategy for genetic immunization because of their safety, ease of use, and inexpensive and stable materials, and, as with all whole gene/whole protein methods, they are not restricted to HLA type. However, to date, the impressive immune-stimulating capacity of DCs, in particular with regard to naive T cells, makes them the most promising vehicle for genetic immunization. *Ex vivo* differentiation and expansion of these cells has the added benefit of overcoming the inhibitory state of the DC that has been observed in the tumor-bearing host [117]. Even in prostate cancer patients who have received prior radiation therapy, DCs can be expanded and differentiated *in vitro* that have the important immunostimulatory functions of those obtained from healthy donors [118]. The field of genetic immunization has already demonstrated impressive tumor immunotherapeutic results with not only model tumor antigens such as β-galactosidase, ovalbumin, and sperm whale myoglobin, but also known human tumor antigens. These methods can not only generate immunity to these known human tumor rejection antigens in mice, in which there is only partial homology in protein sequence, but also in humans, both patients and normal donors, *in vitro* and *in vivo*.

An emerging issue to be addressed with any of these tumor-antigen-driven immunotherapies is the emergence of antigen-loss variants. Subpopulations of antigen-negative tumor cells have been demonstrated to be resistant to tumor antigen epitope CTL induction methods [19,119]. One potential way to overcome this

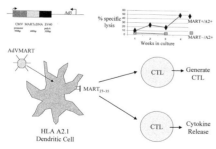

FIGURE 5 AdVMART-1 transduction of human DCs *in vitro*. The AdVMART1 adenovirus (with a map of the MART-1 expression cassette shown above it) has been used to transduce HLA A2.1 DCs. These transduced DCs have been used to generate CTLs *in vitro* that can lyse MART-1+, HLA-matched melanoma cells, but not MART-1−, HLA-matched melanoma cells. In addition, these AdVMART1 transduced cells can elicit cytokine release from MART$_{27-35}$-peptide-specific CTL lines.

is to immunize with multiple peptide antigens or to use the entire tumor antigen (as a naked DNA injection or as transfected DC) as an immunogen to allow multiple epitopes from the same antigen to be presented. In some cases, lack of CTL activity is due to defects in the TCR zeta chain signal transduction that can be overcome with adjuvant cytokine therapy such as IL-2. Treatment with IFN-γ can cause sufficient up-regulation of class I antigen presentation in target cells to cause CTL-killing susceptibility. Again, these situations support the use of a combination treatment including cytokines as part of the antitumor arsenal.

IV. TRANSLATION TO CLINIC

Several issues need to be addressed in order to take these new strategies to the clinic. The cytokines needed to culture DCs for clinical trials, for example, must be prepared under GMP (good manufacturer's practice) conditions. There are limited sources for cytokines prepared this way. With regard to synthetic peptides and proteins, these must also be of sufficient purity and safety for patient use, and the criteria for direct infusion are more strict than the criteria for *ex vivo* culture, with extensive rinsing. When considering the use of patient tumor cells, there must be sufficient live tumor to elute peptides, obtain tumor lysate, or fuse with autologous DCs.

For *ex vivo* expansion of DCs to be reinfused, the amount of blood needed for monocyte-lineage IL-4/GM-CSF DCs can be daunting despite the increases in DC yields that can be obtained. For example, to infuse a patient with 10^7 DCs in three weekly doses, one must obtain at least 2×10^8 peripheral blood mononuclear cells for each dose, which can necessitate 100 mL of blood from a healthy donor each of 3 weeks, and possibly more from a cancer patient. Leukopheresis is sometimes a requirement.

To use a virus for direct administration or to transduce DCs, then wash off the excess, even with an E1-deleted, replication-deficient adenovirus, extensive and expensive pharmacology and toxicity testing must be performed. Perhaps, if the results of the early trials with these reagents continue to indicate a total lack of toxicity and side effects, some of the safety concerns will disappear. This could reduce the time and cost of performing these trials and allow faster progress in the genetic immunization field.

V. FUTURE DIRECTIONS

There are a variety of improvements to look forward to in the near future. In addition to creating better DNA backbones for plasmid-based strategies, there is considerable heterogeneity in studies with cytokines used in different tumor systems. There is much work to be done to either prove the universality in cytokine use or to show which particular cytokines might be best for particular tumor types.

There is also room for improvement in DC generation methodology. There is currently a great deal of controversy over the use of "immature" DCs, which are still proficient at antigen uptake, versus "mature" DCs, which have greater T-cell stimulatory activity and increased display of MHC molecules. Although there are valid arguments on both sides, it should be kept in mind that there will be little control over the cytokine environment of the reinfused DC encountered in the patient.

There is much activity in the area of virus improvement. "Third generation" adenoviral vectors are being developed that express fewer and fewer viral genes. These will be used to maintain the transduction efficiency and high-transgene-expression levels of viruses while eliminating the virus-associated antigenicity. Other methods of transgene expression in professional APCs are also actively studied.

The studies in animal models, the preclinical laboratory data, and the results of pioneer clinical trials suggest that tumor-antigen-based genetic immunization holds promise as a novel immunotherapy approach for cancer.

References

1. Anichini, A., Maccalli, C., Mortarini, R., Salvi, S., Mazzocchi, A., Squarcina, P., Herlyn, M., and Parmiani, G. (1993). Melanoma cells and normal melanocytes share antigens recognized by HLA-A2-restricted cytotoxic T cell clones from melanoma patients. *J. Exp. Med.* **177,** 989–998.
2. Kawakami, Y., Zakut, R., Topalian, S. L., Stotter, H., and Rosenberg, S. A. (1992). Shared human melanoma antigens. Recognition by tumor-infiltrating lymphocytes in HLA-A2.1-transfected melanomas. *J. Immunol.* **148,** 638–643.
3. Bakker, A. B., Schreurs, M. W., de Boer, A. J., Kawakami, Y., Rosenberg, S. A., Adema, G. J., and Figdor, C. G. (1994). Melanocyte lineage-specific antigen gp100 is recognized by melanoma-derived tumor-infiltrating lymphocytes. *J. Exp. Med.* **179,** 1005–1009.
4. Brichard, V., Van Pel, A., Wolfel, T., Wolfel, C., De Plaen, E., Lethe, B., Coulie, P., and Boon, T. (1993). The tyrosinase gene codes for an antigen recognized by autologous cytolytic T lymphocytes on HLA-A2 melanomas. *J. Exp. Med.* **178,** 489–495.
5. Coulie, P. G., Brichard, V., Van Pel, A., Wolfel, T., Schneider, J., Traversari, C., Mattei, S., De Plaen, E., Lurquin, C., Szikora, J. P., *et al.* (1994). A new gene coding for a differentiation antigen recognized by autologous cytolytic T lymphocytes on HLA-A2 melanomas [see comments]. *J. Exp. Med.* **180,** 35–42.
6. Kawakami, Y., Eliyahu, S., Delgado, C. H., Robbins, P. F., Sakaguchi, K., Appella, E., Yannelli, J. R., Adema, G. J., Miki, T., and Rosenberg, S. A. (1994). Identification of a human melanoma antigen recognized by tumor-infiltrating lymphocytes associated

with *in vivo* tumor rejection. *Proc. Natl. Acad. Sci. U.S.A.* **91**, 6458–6462.

7. Wang, R. F., Robbins, P. F., Kawakami, Y., Kang, X. Q., and Rosenberg, S. A. (1995). Identification of a gene encoding a melanoma tumor antigen recognized by HLA-A31-restricted tumor-infiltrating lymphocytes [published erratum appears in J. Exp. Med. 1995 Mar 1;181(3):1261]. *J. Exp. Med.* **181**, 799–804.

8. Coulie, P. G., Weynants, P., Lehmann, F., Herman, J., Brichard, V., Wolfel, T., Van Pel, A., De Plaen, E., Brasseur, F., and Boon, T. (1993). Genes coding for tumor antigens recognized by human cytolytic T lymphocytes. *J. Immunother.* **14**, 104–109.

9. Itoh, K., Hayashi, A., Nakao, M., Hoshino, T., Seki, N., and Shichijo, S. (1996). Human tumor rejection antigens MAGE. *J. Biochem.* (*Tokyo*) **119**, 385–390.

10. Brasseur, F., Marchand, M., Vanwijck, R., Herin, M., Lethe, B., Chomez, P., and Boon, T. (1992). Human gene MAGE-1, which codes for a tumor-rejection antigen, is expressed by some breast tumors [letter]. *Int. J. Cancer* **52**, 839–841.

11. Chen, Y. T., Stockert, E., Chen, Y., Garin-Chesa, P., Rettig, W. J., van der Bruggen, P., Boon, T., and Old, L. J. (1994). Identification of the MAGE-1 gene product by monoclonal and polyclonal antibodies. *Proc. Natl. Acad. Sci. USA* **91**, 1004–1008.

12. Rimoldi, D., Romero, P., and Carrel, S. (1993). The human melanoma antigen-encoding gene, MAGE-1, is expressed by other tumour cells of neuroectodermal origin such as glioblastomas and neuroblastomas [letter]. *Int. J. Cancer* **54**, 527–528.

13. Becker, J. C., Gillitzer, R., and Brocker, E. B. (1994). A member of the melanoma antigen-encoding gene (MAGE) family is expressed in human skin during wound healing. *Int. J. Cancer* **58**, 346–348.

14. Celis, E., Tsai, V., Crimi, C., DeMars, R., Wentworth, P. A., Chesnut, R. W., Grey, H. M., Sette, A., and Serra, H. M. (1994). Induction of anti-tumor cytotoxic T lymphocytes in normal humans using primary cultures and synthetic peptide epitopes. *Proc. Natl. Acad. Sci. USA* **91**, 2105–2109.

15. Tanaka, F., Fujie, T., Go, H., Baba, K., Mori, M., Takesako, K., and Akiyoshi, T. (1997). Efficient induction of antitumor cytotoxic T lymphocytes from a healthy donor using HLA-A2-restricted MAGE-3 peptide in vitro. *Cancer Immunol. Immunother.* **44**, 21–26.

16. Traversari, C., van der Bruggen, P., Luescher, I. F., Lurquin, C., Chomez, P., Van Pel, A., De Plaen, E., Amar-Costesec, A., and Boon, T. (1992). A nonapeptide encoded by human gene MAGE-1 is recognized on HLA-A1 by cytolytic T lymphocytes directed against tumor antigen MZ2-E. *J. Exp. Med.* **176**, 1453–1457.

17. van der Bruggen, P., Bastin, J., Gajewski, T., Coulie, P. G., Boel, P., De Smet, C., Traversari, C., Townsend, A., and Boon, T. (1994). A peptide encoded by human gene MAGE-3 and presented by HLA-A2 induces cytolytic T lymphocytes that recognize tumor cells expressing MAGE-3. *Eur. J. Immunol.* **24**, 3038–3043.

18. van der Bruggen, P., Szikora, J. P., Boel, P., Wildmann, C., Somville, M., Sensi, M., and Boon, T. (1994). Autologous cytolytic T lymphocytes recognize a MAGE-1 nonapeptide on melanomas expressing HLA-Cw*1601. *Eur. J. Immunol.* **24**, 2134–2140.

19. Jaeger, E., Bernhard, H., Romero, P., Ringhoffer, M., Arand, M., Karbach, J., Ilsemann, C., Hagedorn, M., and Knuth, A. (1996). Generation of cytotoxic T-cell responses with synthetic melanoma-associated peptides *in vivo*: implications for tumor vaccines with melanoma-associated antigens. *Int. J. Cancer* **66**, 162–169.

20. Hu, X., Chakraborty, N. G., Sporn, J. R., Kurtzman, S. H., Ergin, M. T., and Mukherji, B. (1996). Enhancement of cytolytic T lymphocyte precursor frequency in melanoma patients following immunization with the MAGE-1 peptide loaded antigen presenting cell-based vaccine. *Cancer Res.* **56**, 2479–2483.

21. Rosenberg S. A. (1992). Gene therapy for cancer [clinical conference]. *Jama* **268**, 2416–2419.

22. Sugita, S., Sagawa, K., Mochizuki, M., Shichijo, S., and Itoh, K. (1996). Melanocyte lysis by cytotoxic T lymphocytes recognizing the MART-1 melanoma antigen in HLA-A2 patients with Vogt-Koyanagi-Harada disease. *Int. Immunol.* **8**, 799–803.

23. Lee, T. (1990). Distribution of HLA antigens in North American Caucasians, North American Blacks and Orientals. In The HLA System. J. Lee, ed. (NY: Springer-Verlag), pp. 141.

24. Saper, M. A., Bjorkman, P. J., and Wiley, D. C. (1991). Refined structure of the human histocompatibility antigen HLA-A2 at 2.6 A resolution. *J. Mol. Biol.* **219**, 277–319.

25. Falk, K., Rotzschke, O., Stevanovic, S., Jung, G., and Rammensee, H. G. (1991). Allele-specific motifs revealed by sequencing of self-peptides eluted from MHC molecules. *Nature* **351**, 290–296.

26. Celis, E., Fikes, J., Wentworth, P., Sidney, J., Southwood, S., Maewal, A., Del Guercio, M. F., Sette, A., and Livingston, B. (1994). Identification of potential CTL epitopes of tumor-associated antigen MAGE-1 for five common HLA-A alleles. *Mol. Immunol.* **31**, 1423–1430.

27. Hunt, D. F., Henderson, R. A., Shabanowitz, J., Sakaguchi, K., Michel, H., Sevilir, N., Cox, A. L., Appella, E., and Engelhard, V. H. (1992). Characterization of peptides bound to the class I MHC molecule HLA-A2.1 by mass spectrometry [see comments]. *Science* **255**, 1261–1263.

28. Ruppert, J., Sidney, J., Celis, E., Kubo, R. T., Grey, H. M., and Sette, A. (1993). Prominent role of secondary anchor residues in peptide binding to HLA-A2.1 molecules. *Cell* **74**, 929–937.

29. Fruci, D., Rovero, P., Falasca, G., Chersi, A., Sorrentino, R., Butler, R., Tanigaki, N., and Tosi, R. (1993). Anchor residue motifs of HLA class-I-binding peptides analyzed by the direct binding of synthetic peptides to HLA class I alpha chains. *Hum. Immunol.* **38**, 187–192.

30. Santamaria, P., Lindstrom, A. L., Boyce-Jacino, M. T., Myster, S. H., Barbosa, J. J., Faras, A. J., and Rich, S. S. (1993). HLA class I sequence-based typing [published erratum appears in Hum. Immunol. 1994 Dec;41(4):292]. *Hum. Immunol.* **37**, 39–50.

31. Drijfhout, J. W., Brandt, R. M., J, D. A., Kast, W. M., and Melief, C. J. (1995). Detailed motifs for peptide binding to HLA-A*0201 derived from large random sets of peptides using a cellular binding assay. *Hum. Immunol.* **43**, 1–12.

32. Kubo, R. T., Sette, A., Grey, H. M., Appella, E., Sakaguchi, K., Zhu, N. Z., Arnott, D., Sherman, N., Shabanowitz, J., Michel, H., *et al.* (1994). Definition of specific peptide motifs for four major HLA-A alleles. *J. Immunol.* **152**, 3913–3924.

33. Maeurer, M. J., Storkus, W. J., Kirkwood, J. M., and Lotze, M. T. (1996). New treatment options for patients with melanoma: review of melanoma-derived T-cell epitope-based peptide vaccines. *Melanoma Res.* **6**, 11–24.

34. Cole, D. J., Weil, D. P., Shilyansky, J., Custer, M., Kawakami, Y., Rosenberg, S. A., and Nishimura, M. I. (1995). Characterization of the functional specificity of a cloned T-cell receptor heterodimer recognizing the MART-1 melanoma antigen. *Cancer Res.* **55**, 748–752.

35. Kawakami, Y., Eliyahu, S., Jennings, C., Sakaguchi, K., Kang, X., Southwood, S., Robbins, P. F., Sette, A., Appella, E., and Rosenberg, S. A. (1995). Recognition of multiple epitopes in the human melanoma antigen gp100 by tumor-infiltrating T lympho-

cytes associated with *in vivo* tumor regression. *J. Immunol.*
154, 3961–3968.

36. Kawakami, Y., Eliyahu, S., Sakaguchi, K., Robbins, P. F., Rivol-tini, L., Yannelli, J. R., Appella, E., and Rosenberg, S. A. (1994). Identification of the immunodominant peptides of the MART-1 human melanoma antigen recognized by the majority of HLA-A2-restricted tumor infiltrating lymphocytes. *J. Exp. Med.* **180,** 347–352.

37. van der Burg, S. H., Visseren, M. J., Brandt, R. M., Kast, W. M., and Melief, C. J. (1996). Immunogenicity of peptides bound to MHC class I molecules depends on the MHC-peptide complex stability. *J. Immunol.* **156,** 3308–3314.

38. van der Burg, S. H., Visseren, M. J., Offringa, R., and Melief, C. J. (1997). Do epitopes derived from autoantigens display low affinity for MHC class I? [letter]. *Immunol. Today* **18,** 97–98.

39. Tsang, K. Y., Zaremba, S., Nieroda, C. A., Zhu, M. Z., Hamilton, J. M., and Schlom, J. (1995). Generation of human cytotoxic T cells specific for human carcinoembryonic antigen epitopes from patients immunized with recombinant vaccinia-CEA vaccine [see comments]. *J. Natl. Cancer Inst.* **87,** 982–990.

40. Wei, C., Storozynsky, E., McAdam, A. J., Yeh, K. Y., Tilton, B. R., Willis, R. A., Barth, R. K., Looney, R. J., Lord, E. M., and Frelinger, J. G. (1996). Expression of human prostate-specific antigen (PSA) in a mouse tumor cell line reduces tumorigenicity and elicits PSA-specific cytotoxic T lymphocytes. *Cancer Immunol. Immunother.* **42,** 362–368.

41. Peoples, G. E., Goedegebuure, P. S., Smith, R., Linehan, D. C., Yoshino, I., and Eberlein, T. J. (1995). Breast and ovarian cancer-specific cytotoxic T lymphocytes recognize the same HER2/neu-derived peptide. *Proc. Natl. Acad. Sci. U.S.A.* **92,** 432–436.

42. Peace, D. J., Smith, J. W., Chen, W., You, S. G., Cosand, W. L., Blake, J., and Cheever, M. A. (1994). Lysis of ras oncogene-transformed cells by specific cytotoxic T lymphocytes elicited by primary in vitro immunization with mutated ras peptide. *J. Exp. Med.* **179,** 473–479.

43. Mayordomo, J. I., Loftus, D. J., Sakamoto, H., De Cesare, C. M., Appasamy, P. M., Lotze, M. T., Storkus, W. J., Appella, E., and DeLeo, A. B. (1996). Therapy of murine tumors with p53 wild-type and mutant sequence peptide-based vaccines. *J. Exp. Med.* **183,** 1357–1365.

44. Theobald, M., Biggs, J., Dittmer, D., Levine, A. J., and Sherman, L. A. (1995). Targeting p53 as a general tumor antigen. *Proc. Natl. Acad. Sci. U.S.A.* **92,** 11993–11997.

45. Hodge, J. W. (1996). Carcinoembryonic antigen as a target for cancer vaccines. *Cancer Immunol. Immunother.* **43,** 127–134.

46. Xue, B. H., Zhang, Y., Sosman, J. A., and Peace, D. J. (1997). Induction of human cytotoxic T lymphocytes specific for prostate-specific antigen. *Prostate* **30,** 73–78.

47. Linehan, D. C., Goedegebuure, P. S., Peoples, G. E., Rogers, S. O., and Eberlein, T. J. (1995). Tumor-specific and HLA-A2-restricted cytolysis by tumor-associated lymphocytes in human metastatic breast cancer. *J. Immunol.* **155,** 4486–4491.

48. Disis, M. L., Gralow, J. R., Bernhard, H., Hand, S. L., Rubin, W. D., and Cheever, M. A. (1996). Peptide-based, but not whole protein, vaccines elicit immunity to HER-2/neu, oncogenic self-protein. *J. Immunol.* **156,** 3151–3158.

49. Moudgil, K. D., and Sercarz, E. E. (1994). The T cell repertoire against cryptic self determinants and its involvement in autoimmunity and cancer. *Clin. Immunol. Immunopathol.* **73,** 283–289.

50. Juretic, A., Jurgens-Gobel, J., Schaefer, C., Noppen, C., Willimann, T. E., Kocher, T., Zuber, M., Harder, F., Heberer, M., and Spagnoli, G. C. (1996). Cytotoxic T-lymphocyte responses against mutated p21 ras peptides: an analysis of specific T-cell-receptor gene usage. *Int. J. Cancer* **68,** 471–478.

51. Yokomizo, H., Matsushita, S., Fujisao, S., Murakami, S., Fujita, H., Shirouzu, M., Yokoyama, S., Ogawa, M., and Nishimura, Y. (1997). Augmentation of immune response by an analog of the antigenic peptide in a human T-cell clone recognizing mutated Ras-derived peptides. *Hum. Immunol.* **52,** 22–32.

52. Nijman, H. W., Van der Burg, S. H., Vierboom, M. P., Houbiers, J. G., Kast, W. M., and Melief, C. J. (1994). p53, a potential target for tumor-directed T cells. *Immunol. Lett.* **40,** 171–178.

53. Ropke, M., Hald, J., Guldberg, P., Zeuthen, J., Norgaard, L., Fugger, L., Svejgaard, A., Van der Burg, S., Nijman, H. W., Melief, C. J., and Claesson, M. H. (1996). Spontaneous human squamous cell carcinomas are killed by a human cytotoxic T lymphocyte clone recognizing a wild-type p53-derived peptide. *Proc. Natl. Acad. Sci. U.S.A.* **93,** 14704–14707.

54. Yu, Z., Liu, X., McCarty, T. M., Diamond, D. J., and Ellenhorn, J. D. (1997). The use of transgenic mice to generate high affinity p53 specific cytolytic T cells. *J. Surg. Res.* **69,** 337–343.

55. Theobald, M., Biggs, J., Hernandez, J., Lustgarten, J., Labadie, C., and Sherman, L. A. (1997). Tolerance to p53 by A2.1-restricted cytotoxic T lymphocytes. *J. Exp. Med.* **185,** 833–841.

56. Liu, Y., Liggitt, D., Zhong, W., Tu, G., Gaensler, K., and Debs, R. (1995). Cationic liposome-mediated intravenous gene delivery. *J. Biol. Chem.* **270,** 24864–24870.

57. Parkhurst, M. R., Salgaller, M. L., Southwood, S., Robbins, P. F., Sette, A., Rosenberg, S. A., and Kawakami, Y. (1996). Improved induction of melanoma-reactive CTL with peptides from the melanoma antigen gp100 modified at HLA-A*0201-binding residues. *J. Immunol.* **157,** 2539–2548.

58. Wolff, J. A., Malone, R. W., Williams, P., Chong, W., Acsadi, G., Jani, A., and Felgner, P. L. (1990). Direct gene transfer into mouse muscle *in vivo*. *Science* **247,** 1465–1468.

59. Kumar, V., and Sercarz, E. (1996). Genetic vaccination: the advantages of going naked [comment]. *Nat. Med.* **2,** 857–859.

60. Conry, R. M., LoBuglio, A. F., Kantor, J., Schlom, J., Loechel, F., Moore, S. E., Sumerel, L. A., Barlow, D. L., Abrams, S., and Curiel, D. T. (1994). Immune response to a carcinoembryonic antigen polynucleotide vaccine. *Cancer Res.* **54,** 1164–1168.

61. Conry, R. M., Widera, G., LoBuglio, A. F., Fuller, J. T., Moore, S. E., Barlow, D. L., Turner, J., Yang, N. S., and Curiel, D. T. (1996). Selected strategies to augment polynucleotide immunization. *Gene Ther.* **3,** 67–74.

62. Sato, Y., Roman, M., Tighe, H., Lee, D., Corr, M., Nguyen, M. D., Silverman, G. J., Lotz, M., Carson, D. A., and Raz, E. (1996). Immunostimulatory DNA sequences necessary for effective intradermal gene immunization. *Science* **273,** 352–354.

63. Hartikka, J., Sawdey, M., Cornefert-Jensen, F., Margalith, M., Barnhart, K., Nolasco, M., Vahlsing, H. L., Meek, J., Marquet, M., Hobart, P., Norman, J., and Manthorpe, M. (1996). An improved plasmid DNA expression vector for direct injection into skeletal muscle. *Hum. Gene Ther.* **7,** 1205–1217.

64. Ciernik, I. F., Berzofsky, J. A., and Carbone, D. P. (1996). Induction of cytotoxic T lymphocytes and antitumor immunity with DNA vaccines expressing single T cell epitopes. *J. Immunol.* **156,** 2369–2375.

65. Irvine, K. R., Rao, J. B., Rosenberg, S. A., and Restifo, N. P. (1996). Cytokine enhancement of DNA immunization leads to effective treatment of established pulmonary metastases. *J. Immunol.* **156,** 238–245.

66. Davis, H. L., Whalen, R. G., and Demeneix, B. A. (1993). Direct gene transfer into skeletal muscle *in vivo*: factors affecting efficiency of transfer and stability of expression. *Hum. Gene Ther.* **4,** 151–159.

67. Geissler, M., Gesien, A., Tokushige, K., and Wands, J. R. (1997). Enhancement of cellular and humoral immune responses to hep-

atitis C virus core protein using DNA-based vaccines augmented with cytokine-expressing plasmids. *J. Immunol.* **158**, 1231–1237.

68. Corr, M., Lee, D. J., Carson, D. A., and Tighe, H. (1996). Gene vaccination with naked plasmid DNA: mechanism of CTL priming. *J. Exp. Med.* **184**, 1555–1560.

69. Doe, B., Selby, M., Barnett, S., Baenziger, J., and Walker, C. M. (1996). Induction of cytotoxic T lymphocytes by intramuscular immunization with plasmid DNA is facilitated by bone marrow-derived cells. *Proc. Natl. Acad. Sci. U.S.A.* **93**, 8578–8583.

70. Iwasaki, A., Torres, C. A., Ohashi, P. S., Robinson, H. L., and Barber, B. H. (1997). The dominant role of bone marrow-derived cells in CTL induction following plasmid DNA immunization at different sites. *J. Immunol.* **159**, 11–14.

71. Schirmbeck, R., Bohm, W., and Reimann, J. (1996). DNA vaccination primes MHC class I-restricted, simian virus 40 large tumor antigen-specific CTL in H-2d mice that reject syngeneic tumors. *J. Immunol.* **157**, 3550–3558.

72. Torres, C. A., Iwasaki, A., Barber, B. H., and Robinson, H. L. (1997). Differential dependence on target site tissue for gene gun and intramuscular DNA immunization. *J. Immunol.* **158**, 4529–4532.

73. Cayeux, S., Richter, G., Noffz, G., Dorken, B., and Blankenstein, T. (1997). Influence of gene-modified (IL-7, IL-4, and B7) tumor cells vaccines on tumor antigen presentation. *J. Immunol.* **158**, 2834–2841.

74. Baskar, S., Clements, V. K., Glimcher, L. H., Nabavi, N., and Ostrand-Rosenberg, S. (1996). Rejection of MHC class II-transfected tumor cells requires induction of tumor-encoded B7-1 and/or B7-2 costimulatory molecules. *J. Immunol.* **156**, 3821–3827.

75. Schultze, J., Nadler, L. M., and Gribben, J. G. (1996). B7-mediated costimulation and the immune response. *Blood Rev.* **10**, 111–127.

76. Condon, C., Watkins, S. C., Celluzzi, C. M., Thompson, K., and Falo, L. D., Jr. (1996). DNA-based immunization by in vivo transfection of dendritic cells. *Nat. Med.* **2**, 1122–1128.

77. Steinman, R. M. (1991). The dendritic cell system and its role in immunogenicity. *Annu. Rev. Immunol.* **9**, 271–296.

78. Inaba, K., Metlay, J. P., Crowley, M. T., Witmer-Pack, M., and Steinman, R. M. (1990). Dendritic cells as antigen presenting cells in vivo. *Int. Rev. Immunol.* **6**, 197–206.

79. Macatonia, S. E., Taylor, P. M., Knight, S. C., and Askonas, B. A. (1989). Primary stimulation by dendritic cells induces antiviral proliferative and cytotoxic T cell responses *in vitro. J. Exp. Med.* **169**, 1255–1264.

80. Lotze, M. T. (1997). Getting to the source: dendritic cells as therapeutic reagents for the treatment of patients with cancer [editorial; comment]. *Ann. Surg.* **226**, 1–5.

81. Morse, M. A., Zhou, L. J., Tedder, T. F., Lyerly, H. K., and Smith, C. (1997). Generation of dendritic cells *in vitro* from peripheral blood mononuclear cells with granulocyte-macrophage-colony-stimulating factor, interleukin-4, and tumor necrosis factor-alpha for use in cancer immunotherapy [see comments]. *Ann. Surg.* **226**, 6–16.

82. Lanzavecchia, A. (1996). Mechanisms of antigen uptake for presentation. *Curr. Opin. Immunol.* **8**, 348–354.

83. Shurin, M. R. (1996). Dendritic cells presenting tumor antigen. *Cancer Immunol. Immunother.* **43**, 158–164.

84. Heemels, M. T., and Ploegh, H. (1995). Generation, translocation, and presentation of MHC class I-restricted peptides. *Annu. Rev. Biochem.* **64**, 463–491.

85. Norbury, C. C., Chambers, B. J., Prescott, A. R., Ljunggren, H. G., and Watts, C. (1997). Constitutive macropinocytosis allows TAP-dependent major histocompatibility complex class I presentation of exogenous soluble antigen by bone marrow-derived dendritic cells. *Eur. J. Immunol.* **27**, 280–288.

86. Inaba, K., Metlay, J. P., Crowley, M. T., and Steinman, R. M. (1990). Dendritic cells pulsed with protein antigens in vitro can prime antigen-specific, MHC-restricted T cells in situ [published erratum appears in J. Exp. Med. 1990. Oct. 1;172(4):1275]. *J. Exp. Med.* **172**, 631–640.

87. Liu, L. M., and MacPherson, G. G. (1993). Antigen acquisition by dendritic cells: intestinal dendritic cells acquire antigen administered orally and can prime naive T cells in vivo. *J. Exp. Med.* **177**, 1299–1307.

88. Svensson, M., Stockinger, B., and Wick, M. J. (1997). Bone marrow-derived dendritic cells can process bacteria for MHC-I and MHC-II presentation to T cells. *J. Immunol.* **158**, 4229–4236.

89. Romani, N., Gruner, S., Brang, D., Kampgen, E., Lenz, A., Trockenbacher, B., Konwalinka, G., Fritsch, P. O., Steinman, R. M., and Schuler, G. (1994). Proliferating dendritic cell progenitors in human blood. *J. Exp. Med.* **180**, 83–93.

90. Young, J. W., Szabolcs, P., and Moore, M. A. (1995). Identification of dendritic cell colony-forming units among normal human CD34+ bone marrow progenitors that are expanded by c-kit-ligand and yield pure dendritic cell colonies in the presence of granulocyte/macrophage colony-stimulating factor and tumor necrosis factor alpha. *J. Exp. Med.* **182**, 1111–1119.

91. Inaba, K., Inaba, M., Romani, N., Aya, H., Deguchi, M., Ikehara, S., Muramatsu, S., and Steinman, R. M. (1992). Generation of large numbers of dendritic cells from mouse bone marrow cultures supplemented with granulocyte/macrophage colony-stimulating factor. *J. Exp. Med.* **176**, 1693–1702.

92. McLellan, A. D., Starling, G. C., and Hart, D. N. (1995). Isolation of human blood dendritic cells by discontinuous Nycodenz gradient centrifugation. *J. Immunol. Methods* **184**, 81–89.

93. Hsu, F. J., Benike, C., Fagnoni, F., Liles, T. M., Czerwinski, D., Taidi, B., Engleman, E. G., and Levy, R. (1996). Vaccination of patients with B-cell lymphoma using autologous antigen-pulsed dendritic cells. *Nat. Med.* **2**, 52–58.

94. Jacobsen, S. E., Okkenhaug, C., Myklebust, J., Veiby, O. P., and Lyman, S. D. (1995). The FLT3 ligand potently and directly stimulates the growth and expansion of primitive murine bone marrow progenitor cells in vitro: synergistic interactions with interleukin (IL) 11, IL-12, and other hematopoietic growth factors. *J. Exp. Med.* **181**, 1357–1363.

95. Maraskovsky, E., Brasel, K., Teepe, M., Roux, E. R., Lyman, S. D., Shortman, K., and McKenna, H. J. (1996). Dramatic increase in the numbers of functionally mature dendritic cells in Flt3 ligand-treated mice: multiple dendritic cell subpopulations identified. *J. Exp. Med.* **184**, 1953–1962.

96. Nair, S. K., Boczkowski, D., Snyder, D., and Gilboa, E. (1997). Antigen-presenting cells pulsed with unfractionated tumor-derived peptides are potent tumor vaccines. *Eur. J. Immunol.* **27**, 589–597.

97. Paglia, P., Chiodoni, C., Rodolfo, M., and Colombo, M. P. (1996). Murine dendritic cells loaded in vitro with soluble protein prime cytotoxic T lymphocytes against tumor antigen in vivo [see comments]. *J. Exp. Med.* **183**, 317–322.

98. Celluzzi, C. M., Mayordomo, J. I., Storkus, W. J., Lotze, M. T., and Falo, L. D., Jr. (1996). Peptide-pulsed dendritic cells induce antigen-specific CTL-mediated protective tumor immunity [see comments]. *J. Exp. Med.* **183**, 283–287.

99. Mayordomo, J. I., Zorina, T., Storkus, W. J., Zitvogel, L., Celluzzi, C., Falo, L. D., Melief, C. J., Ildstad, S. T., Kast, W. M., Deleo, A. B., *et al.* (1995). Bone marrow-derived dendritic cells

pulsed with synthetic tumour peptides elicit protective and therapeutic antitumour immunity. *Nat. Med.* **1,** 1297–1302.

100. Garbilovich, D. I., Cunningham, H. T., and Carbone, D. P. (1996). IL-12 and mutant P53 peptide-pulsed dendritic cells for the specific immunotherapy of cancer. *J. Immunother. Emphasis Tumor Immunol.* **19,** 414–418.

101. Murphy, G., Tjoa, B., Ragde, H., Kenny, G., and Boynton, A. (1996). Phase I clinical trial: T-cell therapy for prostate cancer using autologous dendritic cells pulsed with HLA-A0201-specific peptides from prostate-specific membrane antigens. *Prostate* **29,** 371–380.

102. Mukherji, B., Chakraborty, N. G., Yamasaki, S., Okino, T., Yamase, H., Sporn, J. R., Kurtzman, S. K., Ergin, M. T., Ozols, J., Meehan, J., *et al.* (1995). Induction of antigen-specific cytolytic T cells *in situ* in human melanoma by immunization with synthetic peptide-pulsed autologous antigen presenting cells. *Proc. Natl. Acad. Sci. USA* **92,** 8078–8082.

103. Zitvogel, L., Mayordomo, J. I., Tjandrawan, T., DeLeo, A. B., Clarke, M. R., Lotze, M. T., and Storkus, W. J. (1996). Therapy of murine tumors with tumor peptide-pulsed dendritic cells: dependence on T cells, B7 costimulation, and T helper cell 1-associated cytokines [see comments]. *J. Exp. Med.* **183,** 87–97.

104. Nair, S. K., Snyder, D., Rouse, B. T., and Gilboa, E. (1997). Regression of tumors in mice vaccinated with professional antigen-presenting cells pulsed with tumor extracts. *Int. J. Cancer* **70,** 706–715.

105. Alijagic, S., Moller, P., Artuc, M., Jurgovsky, K., Czarnetzki, B. M., and Schadendorf, D. (1995). Dendritic cells generated from peripheral blood transfected with human tyrosinase induce specific T cell activation. *Eur. J. Immunol.* **25,** 3100–3107.

106. Boczkowski, D., Nair, S. K., Snyder, D., and Gilboa, E. (1996). Dendritic cells pulsed with RNA are potent antigen-presenting cells in vitro and in vivo. *J. Exp. Med.* **184,** 465–472.

107. Aicher, A., Westermann, J., Cayeux, S., Willimsky, G., Daemen, K., Blankenstein, T., Uckert, W., Dorken, B., and Pezzutto, A. (1997). Successful retroviral mediated transduction of a reporter gene in human dendritic cells: feasibility of therapy with gene-modified antigen presenting cells. *Exp. Hematol.* **25,** 39–44.

108. Reeves, M. E., Royal, R. E., Lam, J. S., Rosenberg, S. A., and Hwu, P. (1996). Retroviral transduction of human dendritic cells with a tumor-associated antigen gene. *Cancer Res.* **56,** 5672–5677.

109. Bronte, V., Carroll, M. W., Goletz, T. J., Wang, M., Overwijk, W. W., Marincola, F., Rosenberg, S. A., Moss, B., and Restifo, N. P. (1997). Antigen expression by dendritic cells correlates with the therapeutic effectiveness of a model recombinant poxvirus tumor vaccine. *Proc. Natl. Acad. Sci. USA* **94,** 3183–3188.

110. Kim, C. J., Prevette, T., Cormier, J., Overwijk, W., Roden, M., Restifo, N. P., Rosenberg, S. A., and Marincola, F. M. (1997). Dendritic cells infected with poxviruses encoding MART-1/Melan A sensitize T lymphocytes in vitro. *J. Immunother.* **20,** 276–286.

111. Arthur, J. F., Butterfield, L. H., Roth, M. D., Bui, L. A., Kiertscher, S. M., Lau, R., Dubinett, S., Glaspy, J., McBride, W. H., and Economou, J. S. (1997). A comparison of gene transfer methods in human dendritic cells. *Cancer Gene Ther.* **4,** 17–25.

112. Zhai, Y., Yang, J. C., Kawakami, Y., Spiess, P., Wadsworth, S. C., Cardoza, L. M., Couture, L. A., Smith, A. E., and Rosenberg, S. A. (1996). Antigen-specific tumor vaccines. Development and characterization of recombinant adenoviruses encoding MART1 or gp100 for cancer therapy. *J. Immunol.* **156,** 700–710.

113. Ribas, A., Butterfield, L. H., McBride, W. H., Jilani, S. M., Bui, L. A., Vollmer, C. M., Lau, R., Dissette, V. B., Hu, B., Chen, A. Y., Glaspy, J. A., and Economou, J. S. (1997). Genetic immunization for the melanoma antigen MART-1/Melan-A using recombinant adenovirus-transduced murine dendritic cells. *Cancer Res.* **57,** 2865–2869.

114. Brossart, P., Goldrath, A. W., Butz, E. A., Martin, S., and Bevan, M. J. (1997). Virus-mediated delivery of antigenic epitopes into dendritic cells as a means to induce CTL. *J. Immunol.* **158,** 3270–3276.

115. Gong, J., Chen, D., Kashiwaba, M., and Kufe, D. (1997). Induction of antitumor activity by immunization with fusions of dendritic and carcinoma cells. *Nat. Med.* **3,** 558–561.

116. Hart, I., and Colaco, C. (1997). Immunotherapy. Fusion induces tumour rejection [news]. *Nature* **388,** 626–627.

117. Gabrilovich, D. I., Ciernik, I. F., and Carbone, D. P. (1996). Dendritic cells in antitumor immune responses. I. Defective antigen presentation in tumor-bearing hosts. *Cell Immunol.* **170,** 101–110.

118. Tjoa, B., Erickson, S., Barren, R., 3rd, Ragde, H., Kenny, G., Boynton, A., and Murphy, G. (1995). *In vitro* propagated dendritic cells from prostate cancer patients as a component of prostate cancer immunotherapy. *Prostate* **27,** 63–69.

119. Van Waes, C., Monach, P. A., Urban, J. L., Wortzel, R. D., and Schreiber, H. (1996). Immunodominance deters the response to other tumor antigens thereby favoring escape: prevention by vaccination with tumor variants selected with cloned cytolytic T cells *in vitro. Tissue Antigens* **47,** 399–407.

PART VI

Genetically Modified Cells for Immunization

CHAPTER 19

Engineering Cellular Cancer Vaccines: Gene and Protein Transfer Options

MARK L. TYKOCINSKI

Department of Pathology and Laboratory Medicine, University of Pennsylvania, Philadelphia, Pennsylvania 19104

I. INTRODUCTION

Discourse in the field of tumor immunology has been dominated by controversial hypotheses, and for years, the "immune surveillance" hypothesis has loomed especially large. This hypothesis proposes the existence of natural immunity to cancer, envisioning an ongoing duel between an ever-vigilant immune system and tumor cells that arise now and then. In the past, this hypothesis tended to eclipse a far more practical question, Can the immune system be artificially harnessed to cure established tumors? Whether the immune system naturally

kills tumor cells is less important than whether it can be programmed with "vaccines" to kill tumor cells if called upon by the immunotherapist.

Cancer vaccines are geared toward disease cure, not prevention. Thus, they differ fundamentally from the strictly protective vaccines used as standard fare for infectious diseases. The notion of using "vaccines" to elicit curative antitumor immune responses has been around for some time, and despite their as yet unrealized potential, cancer vaccines remain a promising cancer therapeutic option. The documented efficacy of cancer vaccines in a growing number of animal tumor systems has, of late, helped counter the skepticism over such vaccines that was fostered by early clinical disappointments. As is so often the case, poorly conceived clinical trials early on can torpedo a promising field for years. This was the situation in the antitumor monoclonal antibody field, and recent developments in that field provide a compelling case study of how a therapeutic modality can be resurrected after early clinical failures [1]. Like antitumor monoclonal antibodies, cancer vaccines may now be poised to reenter the therapeutic mainstream. Effective clinical translation will demand an approach that deals with scientific details in a comprehensive fashion and avoidance of the reductionist schemes that are used so frequently to simplify experiments in animal models.

The special emphasis in this chapter is on cellular cancer vaccines. There is a long history documenting the use of cells for cancer vaccination, and it has become clearer with time that cellular vaccines, especially when enhanced with gene and protein transfer tools, may have unique advantages for this purpose [2]. In early studies, irradiated, but otherwise unmodified, tumor cells were used as immunogens [3]. In derivative studies, irradiated tumor cells were combined with nonspecific immunostimulants, but clinical outcomes remained equivocal [4]. It became clear that more sophisticated cellular immunogens were called for, and these indeed materialized in the form of an array of different types of ex vivo–engineered cancer vaccine cells.

This chapter deals with two major classes of cellular cancer vaccines, dendritic and tumor cell vaccines, both of which are designed to activate tumor-specific CD8+ cytotoxic T-lymphocyte (CTL) effectors. The focus here is on the cellular engineering tools that are currently available for the ex vivo production of both classes of cancer vaccine cells, especially those tools applicable to engineering cell surfaces. Most studies to date have concentrated upon ex vivo gene transfer approaches, but openness to other cellular engineering strategies is needed. Hence, a special case is made here for ex vivo protein transfer as an ancillary tool for cellular cancer vaccine engineering (see Sections III.D and IV.E).

II. TUMOR PEPTIDE ANTIGENS

The nomenclature for tumor antigens has traditionally been complex. Standard textbook categories, such as "tumor-specific transplantation antigen," were functionally defined years ago on the basis of tumor transplantation experiments. Despite their intuitive elegance, these classical experiments are hard to force fit into an increasingly detailed molecular picture of the immune system. The recognition of major histocompatibility complex (MHC)-associated peptides, recognized by autologous CTLs, as a distinct class of tumor antigens has fundamentally changed this situation and has, in a sense, brought tumor antigens into the mainstream of immunology [5].

MHC class-I-associated peptides, also referred to as CTL epitopes, trigger CD8+ CTLs, whereas MHC class-II-associated peptides interface with CD4+ T lymphocytes. Classically, class-I- and class-II-associated peptides have been thought to derive from endogenous and exogenous proteins, respectively. Class I peptides are produced by proteases located in the cytosol, transported into the endoplasmic reticulum through the action of the TAP transporter, loaded onto class I molecules, and then transported to the cell surface [6]. However, the idea of an exclusive link between MHC class I and endogenous proteins has begun to fade. It is now known that professional antigen-presenting cells (APCs) can capture exogenous antigens by phagocytosis or macropinocytosis for presentation on MHC class I molecules. These internalized antigens can be transferred from the endocytic compartment either into the cytosol, where they can be processed by the usual proteasome- and TAP-dependent class I pathway [7], or to the cell surface directly [8]. These exogenous class I pathways may be especially important for the in vivo priming of CTL responses to antigens that are not synthesized by professional APCs. The connection between MHC class II and exogenous proteins has also begun to unravel in the face of findings that peptides derived from endogenously synthesized proteins can be presented by MHC class II molecules [9]. For a more complete discussion of immune mechanisms, see Chapter 3.

Several human MHC class-I-associating tumor peptide antigens have been analyzed in some detail. Although the first defined tumor antigen recognized by T cells, reported in 1991, was from a murine mastocytoma [10], the most extensively characterized tumor antigens are from melanomas. Two categories of melanoma peptide antigens, both deriving from nonmutated self-proteins shared by tumors from different individuals, can be distinguished: tumor-specific shared antigens (MAGE, BAGE, GAGE) and differentiation- or lineage-specific antigens (MART-1/Melan-A, gp100/Pmel-17, tyrosinase, gp75/trp-1) [11–13]. Some of these melanoma CTL epitopes are expressed in nonmelanoma tumors as well [12]. Outside of the melanoma realm, an intensively studied tumor peptide antigen derives from HER-2/neu, an oncogene protein overexpressed in approximately 30 to 40% of all ovarian and breast carcinomas [14]. Other classes of tumor peptide antigens for CTLs include ones derived from mutated oncogene proteins (e.g., mutated Ras), mutated nononcogenic proteins (e.g., β-catenin), and viral proteins (e.g., E6 and E7 of HPV).

For the moment, the set of well-defined MHC-associating tumor peptide antigens is limited. Several experimental strategies have been used successfully for uncovering new tumor peptide antigens, and these can be expected to yield more antigens in the future. Expression cloning is the strategy that yielded the MAGE and MART antigens of melanoma [15,16], and as a function-oriented screening method, it has the distinct advantage of directly searching from the outset for functional CTL epitopes relevant to patients. Typically, an expression cloning procedure might involve the sequential steps of culturing tumor-infiltrating lymphocytes from patient tumor samples, identifying cellular transfectants that can trigger these tumor-infiltrating lymphocytes, and characterizing the antigenic protein within these transfectants. Clearly, this approach is labor intensive.

A second strategy for identifying tumor peptide antigens involves eluting resident MHC-associated peptides from tumor cell surfaces, fractionating them by high-performance liquid chromatography, and then characterizing the isolated peptides by mass spectrometry and sequencing [12,17,18]. A third strategy involves identifying "candidate proteins" that are uniquely expressed, or are overexpressed, in tumor cells, and then scanning their protein sequences for short peptide stretches containing HLA-specific anchor residues. In this way, candidate CTL epitopes can be identified that might be derived from these tumor-associated proteins. The next steps in this approach entail chemically synthesizing the peptides, testing their abilities to specifically associate with the chosen MHC class I allelic products using MHC/peptide binding assays, and critically evaluating their immunogenic functions and clinical relevance. This candidate protein approach has now been used with success to identify several new tumor peptide antigens [19,20].

For both the peptide elution/mass spectrometry and candidate protein approaches, the major challenge is to efficiently sift through the large number of candidate CTL epitopes that emerge and to identify those that can trigger optimal antitumor responses. CTLs derived from both cancer patients [16] and healthy donors [20] can be used for this functional peptide screening. Only a minority of motif-containing peptides from a given protein are expected to be immunogenic, and even fewer can be expected to generate clinically useful antitumor T-cell responses. Nontumor cells primed with the tumor peptide antigen candidates can be used to functionally screen them, with the proviso that in some cases, native CTLs from patients can recognize the peptide-expressing nontumor cells, but not the peptide-expressing tumor cells themselves [21].

MHC polymorphism adds substantially to the challenge for the tumor immunotherapist. Most peptides studied to date are restricted to HLA-A2.1, the most common human MHC class I haplotype among Caucasians. A random assortment of peptides relevant to some other haplotypes have also been identified to date [12], but a larger collection of peptides relevant to these and other MHC haplotypes will ultimately be needed.

III. DENDRITIC CELL VACCINES

A. General Concept

One class of cellular cancer vaccines consists of professional APCs that have been "primed" with tumor antigens. Such *ex vivo* engineered APCs have been used as cellular immunogens for inducing tumor-directed CD8+ T-cell responses to CTL epitopes *in vivo* [22]. Antigen-primed APCs can also be used *ex vivo*, for example, to expand tumor-antigen-specific tumor-infiltrating lymphocytes [23] and to prime peptide-specific CTLs from peripheral blood [24].

Dendritic cells have emerged as the preferred professional APCs for cellular vaccination, with demonstrated utility for inducing class-I-restricted antitumor immunity [25]. Dendritic cells, first described in 1973 [26] and later shown to be especially potent APCs [27], are now known to be quite heterogeneous with regard to their origins, maturation states, and rates of turnover [28]. They have distinct pathways of differentiation and have been subdivided into myeloid- and lymphoid-derived lineages that can be derived from several tissue compartments. In the clinical setting, a convenient source for myeloid-derived dendritic cells are CD14+ monocytes in peripheral blood, which can be induced to differentiate into "immature" dendritic cells with granulocyte-macrophage colony-stimulating factor (GM-CSF) and IL-4 [29–31]. These cells can be further converted into "mature" dendritic cells by exposure to proinflammatory cytokines such as TNF-α or IL-1β [30]. Whereas immature dendritic cells are efficient at antigen uptake via macropinocytosis, mature dendritic cells are preferentially geared toward migration and efficient T-cell activation [28]. TGF-β, rather than GM-CSF, promotes the differentiation of lymphoid dendritic cells [32]. These cytokines now provide the means for generating sufficient numbers of autologous dendritic cells from leukapharesed cells from cancer patients for clinical trials. Other cytokines, such as stem cell factor (Kit ligand) and Flt3 ligand, promise to substantially increase the yields of cultured dendritic cells even further.

Despite this embrace of dendritic cells by immunotherapists, the clinical use of these cells is complicated by their heterogeneity, with evidence that different dendritic cell lineages may perform different functions. Data suggest a dichotomy between myeloid and lymphoid dendritic cells wherein the two are oppositely oriented towards immune activation and inhibition, respectively. CD8+ lymphoid dendritic cells may exert suppressive effects on both CD4+ and CD8+ T cells [33,34]. Although CD95L (Fas ligand) on the lymphoid dendritic cells has been implicated as an inducer of T-cell apoptosis [34], the CD8 on these cells could also account, at least in part, for the observed inhibitory effects, given its documented *trans* inhibitory potential [35–37].

This dendritic cell heterogeneity provides a strong argument in favor of using *ex vivo* engineered dendritic cells, as opposed to soluble peptide and protein antigens, for vaccination. The *ex vivo* setting permits one to direct antigens to preselected dendritic cell subsets that possess optimal vaccine attributes. Questions to be answered before the therapeutic potential of dendritic cell vaccines can be realized pertain to which dendritic cell lineage is optimal, whether maturation should be induced *ex vivo* before injection of the cells into patients, and how best to generate sufficient numbers of dendritic cells to be reinfused into patients. When cytokines are removed, dendritic cells rapidly revert to a less stimulatory state, posing yet another challenge to their clinical use. To counter this problem, improved methods are being developed for generating dendritic cells from CD14+ blood cells with a stable immunogenic phenotype [38,39]. Moreover, the possibility emerges for further tailoring the immunogenic properties of these dendritic cells with *ex vivo* cellular engineering tools. The engineering of dendritic cells to express high levels of certain cytokines has been suggested [40].

B. Antigen Priming: The Standard Approach

The standard approach for priming APCs involves the simple step of co-incubating these cells with synthetic peptide antigens. A number of mechanisms may account for the peptide loading that is attained by this method, including direct trafficking of free peptides to the endoplasmic reticulum, where MHC class I binding occurs [41], binding of peptides to "peptide-receptive" MHC class I molecules at the cell surface that are created by the binding of β_2-microglobulin present in the culture medium to free class I alpha chains on the cell surface [42], and displacement by the exogenously added peptides of the endogenously loaded peptides that accompany MHC heterodimers when they reach the cell surface. The efficiency of this "peptide-pulsing"

method is idiosyncratic, and generally speaking, the method does not permit fine control of final MHC/peptide complex densities at APC surfaces. However, notwithstanding this limitation, therapeutic APCs prepared by peptide pulsing have been shown to be effective for generating antitumor responses in animal models [22], as well as for inducing CTL responses from peripheral blood lymphocytes of healthy human donors [43,44].

Despite the extensive literature on antigen presentation, key questions pertaining to quantitative aspects of antigen triggering still need to be resolved. Several considerations make the control of antigen levels on therapeutic APCs an imperative. First, studies have tended to underscore the need for high antigen densities at APC surfaces for optimal primary stimulation of CTLs [45,46]. Second, the T cell receptor (TCR) activation threshold model argues for the importance of antigenic levels in dictating not just quantitative, but also qualitative aspects of T-cell responses at the single-cell level. CTL clones require a higher antigen density for triggering lymphokine release as opposed to cytolysis [47,48]. Moreover, according to this model, a given T cell may produce different cytokine arrays when triggered by varying levels of antigen [47,49]. Third, high antigen expression provides one route for skewing the response toward low-avidity CTLs. Such CTLs may be preferable for tumor immunotherapy because they can reject tumor cells expressing high levels of antigen and yet remain tolerant of normal body cells bearing normal levels of the antigen [50]. The need to control antigen levels on therapeutic APCs argues further for the use of *ex vivo* engineered APCs whose antigen load can be controlled.

Driven by the goal of maximizing CTL epitope densities at APC surfaces, some groups have proposed enhancements and/or variations of the peptide-pulsing method. For instance, by adding a preliminary mild acid wash step, it is possible to remove resident peptides that are prebound to MHC molecules and thereby enhance the uptake of exogenous peptides into vacated MHC acceptors [19,21,51,52]. Another approach involves using peptide-coated phagocytic substrates, for example, latex beads, that target bound peptides to the MHC class I pathway [24]. Polycations, such as polyarginine, have also been used to enhance the transport of peptides into cells [53].

As an alternative to loading dendritic cells with well-defined peptides one at a time, it is instead possible to acid-strip class-I-associated peptides from tumor cells and then load the recovered peptides *en masse* onto dendritic cells for immunization purposes [54,55]. Yet another approach for delivering mixtures of tumor-derived class-I-associating peptides to APCs involves the use of heat shock proteins as carriers. After extraction from tumor cells, heat-shock-protein/peptide complexes can be added to macrophages of any MHC haplotype, which can, in turn, internalize the complexes and process the chaperoned peptides for class-I-restricted presentation [56]. Tumor-specific CD8+ CTLs can be primed by this approach, using substantially less peptide than is required for peptide pulsing.

C. Antigen Priming: Gene Transfer

Gene transfer has been invoked as a newer generation antigen-priming tool. Whole tumor proteins can be expressed in APCs, such as dendritic cells, via gene transfer; the expressed proteins are processed, and component peptides are presented via the cells' endogenous MHC class I pathways. Compared to the exogenous loading of dendritic cells with whole proteins [57,58], gene transfer of sequences encoding whole proteins into dendritic cells permits higher levels of protein expression, directs processing preferentially to the class I pathway, and induces specific CTLs with fewer *in vitro* stimulations than does peptide pulsing [59]. Sequence modifications can be introduced at the N terminus of the transfected protein to enhance proteasome-dependent degradation and CTL induction [60]. Another advantage of the gene transfer approach is that it could in principle enable long-term antigen presentation *in vivo;* in contrast, the expression of pulsed peptides and proteins at the cell surface is only transient due to intracellular proteolysis, dissociation from MHC, and MHC turnover at the surface. Transfected dendritic cells have been used successfully in murine [61,62] and human [59,63] tumor studies, where they were shown to stimulate both CD8+ [59,62] and CD4+ [63] T cells. Retroviral [59], adenoviral [62], and poxvirus [64] vectors have been shown to be suitable for dendritic cell gene transfer. Lipofection with cationic liposomes represents another gene transfer modality that has been used in conjunction with dendritic cells [65].

Alternatively, gene transfer can be used to express within APCs artificial minigenes that each encode an endoplasmic reticulum insertion sequence linked in tandem to a multiunit CTL epitope [45,66]. This approach bypasses the vagaries of peptide selection associated with TAP-dependent proteolysis of whole proteins by APCs. Moreover, expression of preselected peptides enables the induction of CTL-mediated immunity to otherwise silent or subdominant CTL epitopes and allows one to avoid otherwise tolerogenic epitopes that sometimes dominate when APCs are pulsed with whole self-proteins [67]. For further discussion of transfected DC vaccines, see Chapter 19.

There are other gene transfer options. Instead of expressing proteins or multiunit CTL epitopes within an APC, it is also possible to express MHC/β_2-microglobulin (β_2m)/peptide antigen fusion proteins that incorporate all three elements in a single polypeptide chain, with the peptide antigen appended to the amino terminus of the MHC heavy chain [68,69]. Tumor-derived RNA can be used instead of recombinant DNA as the gene transfer tool. In this case, bulk RNA is extracted from tumor cells and then introduced *en masse* into dendritic cells; the RNA-encoded protein pool is processed by the endogenous class I pathway [70]. Antisense gene transfer can also be invoked for antigen priming. Antisense TAP-2 modulation of APCs [71] interferes with endogenous class I peptide loading and thereby promotes exogenous peptide engagement.

D. Antigen Priming: MHC Protein Transfer

Our group has developed another option for displaying a uniform array of a particular peptide antigen, and controlling its density, on APC surfaces (Fig. 1). This approach bypasses gene transfer altogether and instead involves the "painting" of APC surfaces with genetically engineered MHC/peptide antigen complexes.

To accomplish this protein transfer feat, glycosyl-phosphatidylinositol (GPI)-modified MHC proteins were produced and used as "protein paints." GPI proteins exhibit the remarkable property of membrane reincorporability. Unlike proteins with transmembrane peptide anchors, proteins with GPI anchors at their carboxyl termini remain in solution as pseudomicelles when isolated from detergent membrane extracts and depleted of detergent. This feature allows GPI protein solutions to be combined with cells without causing cell lysis, and the exogenously added proteins reincorporate into the cell membranes. A native GPI protein, human decay-accelerating factor (DAF), was shown to reincorporate into cell membranes in this fashion back in 1984, and the reincorporated DAF retained complement regulatory activity [72].

Once DAF cDNA was cloned [73,74], its GPI modification signal sequence was identified and used to confer GPI anchors to other proteins through gene chimerization. More specifically, it was shown by our group and others that any protein of interest can be produced in a GPI-modified (and thus, in principle, a "paintable") form by appending to its carboxyl terminus the GPI modification signal sequence from DAF's 3' end [75,76]. Subsequent studies have established the feasibility of the GPI reanchoring method for a diversity of proteins [77,78].

Our studies early on, with CD8α as a model protein, established the principle that artificial GPI modification does not perturb native *trans* signaling functions [36].

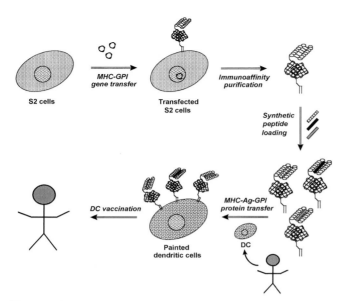

FIGURE 1 *Dendritic cell vaccination: the MHC/Ag/GPI protein transfer approach.* This schematic depicts a multi-step procedure for preparing and using MHC/Ag/GPI-painted dendritic cells for cellular cancer vaccination. The first steps are directed towards the production of recombinant MHC/Ag/GPI complexes. To this end, expression constructs encoding a GPI-modified MHC class I heavy chain of a defined haplotype and the non-polymorphic β_2-m light chain are stably co-transfected into Drosophila S2 Schneider cells. The expressed β_2-m polypeptide, which is an immunoglobulin supergene family (IgSF) domain, associates with the membrane-proximal α_3 IgSF domain of the expressed GPI-modified MHC class I heavy chain. The GPI-modified MHC heterodimer, whose N-terminal antigen-binding pocket is free of endogenous peptides when produced in the S2 cell background, is purified using an immunoaffinity matrix and is then loaded with synthetic peptides corresponding to different tumor CTL epitopes. The next steps of the procedure deal with the therapeutic utilization of the heterotrimeric MHC/Ag/GPI complexes, each consisting of a GPI-modified class I heavy chain, β_2-m light chain, and a tumor peptide antigen bearing the relevant CTL epitopes. These "painted dendritic cells" are then administered as a cellular vaccine back to the patient.

The subsequent goal has been to develop therapeutic applications for GPI protein transfer, especially in the area of cellular vaccines. The use of MHC/GPI proteins for antigen priming of APCs exemplifies this approach and constituted a first example of the use of protein transfer for purposes of APC engineering. In one study, we showed that GPI-reanchored HLA-A2.1 can function as an effective alloantigen when expressed by gene transfer [79]. Specifically, this study demonstrated that (1) HLA-A2.1/GPI can be expressed on the surfaces of class-I-deficient human C1R B cells, and (2) this recombinant protein can both allostimulate and be recognized by allospecific CTLs.

These findings were extended in a second study [80], which directly demonstrated that CTL epitopes can indeed be painted onto APC surfaces using MHC/GPI

proteins as transfer vehicles. The specific findings in this study were as follows: (1) Recombinant HLA-A2.1/GPI can be produced in Drosophila Schneider S2 cells, a cell background in which transfected human MHC class I molecules reach the cell surface without preengaged antigenic peptides. This special feature allows one to produce a homogeneous population of peptide-free GPI-modified MHC class I product, which can then be homogeneously loaded with a defined antigenic peptide. (2) The HLA-A2.1/GPI heavy chain (like its native counterpart) heterodimerizes to β_2m light chain and, following immunopurification from Drosophila S2 cells, can be painted onto APCs (C1R B cells) in finely titrated amounts in a time-dependent fashion; and (3) painted HLA-A2.1/GPI/β_2m/antigenic peptide heterotrimers can present HLA-A2.1–restricted antigenic peptides to CTL clones of appropriate antigenic specificity.

In contrast to our data with gene- and protein-transferred human HLA/GPI protein, mixed results have been reported with murine H-2/GPI proteins expressed by gene transfer [81–84]. There are several possible explanations for this difference. First, it is possible that murine H-2 molecules may, in general, be less amenable to GPI modification than are human HLA molecules. Second, the murine studies uniformly utilized murine Qa-2's GPI modification signal sequence, in some cases even including the Qa-2 α_3 domain [81,82]. The Qa-2 α_3 domain is known to bind poorly to CD8 [82], and further peculiarities of the Qa-2 GPI modification signal sequence could be problematic. Third, these murine studies relied exclusively on gene transfer of the GPI proteins and, in the case of peptide antigens, were confined to endogenously loaded peptides expressed by virally infected cells. The one study that compared endogenously loaded to exogenously added peptides found that the latter were indeed immunostimulatory on an H-2D^b/GPI carrier (incorporating an H-2D^b α_3 domain [84]). Hence, even if there is a deficit in the endogenous loading of certain MHC class I/GPI proteins, this does not preclude their therapeutic use as antigen paints.

IV. TUMOR CELL VACCINES

A. General Concept

Tumor cells are thought to lack immunogenicity as a consequence of deficits in antigenic and/or costimulatory potentials. Poor antigenicity can ensue from the down-regulation of surface MHC expression, alterations in antigen-processing pathways, and selection of antigen-loss variants. Deficient costimulatory activity is presumed to reflect the lack of critical cell surface and soluble factors that are needed for the *trans* activation of T cells.

It was predicted some time ago that tumor cells with intact antigenicity might be converted into active "tumor APCs" by remedying their costimulatory defects. This prediction has in fact been realized with a diverse set of "tumor cell vaccines," each consisting of *ex vivo* engineered tumor cells that can elicit curative antitumor CTL responses *in vivo* (for example, see references [85–92]). There is evidence that engineered tumor cells can act as APCs themselves (Section IV.C), eliciting systemic antitumor T-cell responses that effectively eliminate residual cancer within animals. The animal data have now spawned Phase I clinical trials in which autochthonous, genetically modified cancer cells are being reinjected back into patients as cellular cancer vaccines.

The fact that curative CTLs can be artificially amplified with modified tumor cells as stimulators argues against an absolute state of T-cell "tolerance" in cancer patients. The likelihood instead is that a state of immunologic "ignorance," not tolerance, characterizes a significant component of the T-cell repertoire in cancer, with the T cells remaining amenable to activation by appropriate APC triggering [93], and even if tumor tolerance exists in some fashion, there is evidence that immunity can be generated against antigens for which tolerance is preexisting [94]. Whether the well-documented signaling defects in T cells infiltrating human tumor beds [95] can be reversed remains to be determined.

B. Cytokine Gene Transfer

The primary objective in the tumor vaccine arena is to convey features of *bona fide* APCs to tumor cells. One of the earliest, and perhaps most intensively explored, strategies for achieving this end is enforced cytokine expression via gene transfer. Over the years, a diverse array of cytokine-expressing tumor cell transfectants have been shown to acquire the capacity to elicit systemic antitumor responses [96]. Both tumor cure and prevention have been repeatedly achieved in syngeneic animal tumor models. Gene transfer has been applied to a broad range of cytokines, and IL-12, a cytokine capable of interfacing directly with T cells, has emerged as a particularly promising candidate [97,98].

The *in vivo* effects of cytokine-engineered tumor cells may be substantially more complex than was first thought. This complexity undoubtedly mirrors the functional pleiotropism of the individual cytokines themselves. There is evidence that diverse immunologic mechanisms, including non-T-cell-dependent ones, may be operative when cytokine-transfected tumor cells are administered. For example, GM-CSF tumor cell transfectants may act principally by recruiting monocytes/macrophages and thereby promoting the reprocessing of tumor antigens by host APCs, a process referred to as cross-priming [99]. Such cross-priming presumably calls upon the exogenous MHC class I pathway. There are data to support this mechanism for tumor cell transfectants expressing certain other cytokines, such as IL-3 [100], IL-4, and IL-7 [101]. Moreover, given Flt3-ligand's documented dendritic cell stimulatory activity, the vaccine efficacy of Flt3-ligand- transfected tumor cells [102] could likely turn out to be a consequence of the enhancement of cross-priming. However, notwithstanding this complexity, at least some cytokine-expressing tumor cells may function as tumor APCs [86], and there is no reason to believe that tumor APC function cannot be promoted by more complexly engineering the cells with a more representative set of APC molecules and by focusing on lymphocyte-directed cytokines.

Some cytokines with pleiotropic activities pose mechanistic puzzles when considered in the context of tumor cell vaccines. IL-10, for example, has been invoked by some for tumor cell vaccine production, yet has been implicated by others as a cause of poor tumor cell immunogenicity (Section IV.G). Similarly, IL-4 has been used by several groups for augmenting tumor cell immunogenicity [85,87], whereas others have proposed that antibody blocking of IL-4 may be appropriate for promoting tumor resistance, given this cytokine's immunosuppressive properties [103]. Further complicating the issue are data suggesting that IL-4 expressing tumor cell vaccines may act by promoting eosinophil infiltration by inducing local production of eotaxin, a chemokine with chemoattractant properties for eosinophils [104]. For a fuller discussion of the clinical use of cytokine-transfected tumor vaccines, see Chapters 22 and 23.

C. Costimulator Gene Transfer: B7

Another option is to tailor the protein display at the tumor cell surface. Early efforts along this line have dealt with cell surface proteins that function as costimulators of T-cells. T-cell activation depends on the engagement of both its antigen receptor (i.e., the TCR/CD3 molecular complex) and costimulator receptors [105]. TCR/CD3 engagement in the absence of costimulation is insufficient to induce IL-2 production by T cells and, remarkably, can result in long-lasting anergy [106]. Hence, efficient T-cell activation relies on coordinate *trans* signaling along two parallel molecular axes: the MHC/antigen-to-T-cell receptor axis and the costimulator-to-costimulator receptor axis.

The preponderance of data in the literature relates to the B7-1 (CD80) and B7-2 (CD86) costimulators, which share the capacity to bind to and trigger two counterreceptors on T cells, CD28 and CTLA-4 [105]. Considerable controversy exists as to whether B7-1 and B7-2 are interchangeable or mediate distinct functions through the CD28 and CTLA-4 counterreceptors. Differences in their binding kinetics to CD28 and CTLA-4 could explain some of the observed differences [107]. For a complete discussion of the role of costimulator molecules in immune stimulation, see Chapter 3.

B7 proteins are absent from most tumor cell surfaces. Many groups have shown that B7-1 costimulator neo-expression on a tumor cell's surface by gene transfer can in and of itself substantially enhance its immunogenicity (for example, see references [88,90,108–111]). The results with B7-2 neo-expression on tumor cells have been mixed [110,112], and this could reflect documented effects of cell background on endogenously expressed B7-2's functional activities [113]. B7-1-modified tumor cells, corresponding to diverse tumor types, were successfully used as cancer vaccines to elicit curative systemic antitumor responses.

It has been presumed that modified B7-1+ tumor cells function primarily as APCs to directly prime CTLs in vivo [88,90]. A reasonable scenario envisions the B7-1 on the transfected tumor APC acting to lower TCR activation thresholds [47,49], thereby potentiating responses to CTL epitopes present at limiting concentrations, or even uncovering entirely new CTL epitopes [93,114]. This view of direct antigen presentation by B7-transfected tumor cells has received support from recent murine [101] and human [115] studies, the latter showing that B7-1+ melanoma cells can elicit tumor-specific CTLs from purified CD8+ T cells in vitro.

There is evidence to suggest that transfected B7 may also function, at least in some settings, to promote indirect tumor antigen presentation via a cross-priming mechanism. B7-mediated stimulation of CD4+ T cells [108,111] and natural killer cells [116] could tie into such cross-priming. For instance, one group has speculated that B7 expression leads to increased lysis of the B7-1+ tumors by natural killer cells in vivo, in turn making tumor antigens available for reprocessing by professional host APCs [117]. However, even their data, based upon the use of surrogate tumor antigens in a bone marrow chimera system, show some degree of direct priming of naïve CTLs by the B7-1+ tumor cells themselves and further indicate that B7-1-transfected tumor cells can restimulate and expand already primed CD8+ CTL [117].

The cross-priming issue remains a complex one, with another group showing cross-priming by B7-1+ transfectants only when they are combined with IL-7- and IL-4-expressing transfectants; on their own, the B7-1+ transfectants prime directly [101]. Others have failed to observe any cross-priming by host APCs in their in vivo systems [118]. Moreover, certain quantitative aspects of the cross-priming phenomenon need to be considered in more depth, given the immense protein heterogeneity to be expected within transfectant tumor cell debris. In principle, the two mechanisms, immunogenic tumor cells functioning as APCs and cross-priming of host APCs with immunogenic tumor cell debris, need not be mutually exclusive [101], and there is the possibility that direct and indirect pathways may be operative at different postinjection time frames. One group has suggested that tumor APCs dominate the stage early on, while they are still physically intact [9]; another has alternatively suggested that tumor APCs may serve to reinforce responses initiated by host APCs [119].

D. Costimulator Gene Transfer: Beyond B7

B7 costimulator gene transfer, on its own, may be insufficient for those tumor cells with no "intrinsic" immunogenicity [108]. Such a "nonimmunogenic" phenotype may be a common feature of primary human tumor cells, so there is a pressing need for ways to augment the efficacy of B7+ tumor cell vaccines. Maximizing the immunogenicity of cancer vaccine cells becomes especially important in the face of theoretical analyses and data suggesting that tumor cells with only marginally increased immunogenicity could paradoxically have protumor effects [120].

One possible means for enhancing B7+ tumor cell vaccines stems from insights into the contrasting functions of B7 receptors. A body of data now point to opposing actions for the CD28 and CTLA-4 receptors, with CTLA-4 mediating T-cell inhibition. The notion that CTLA-4 is a T-cell-inhibitory receptor is supported by experiments showing that selective blockade of CTLA-4 (with anti-CTLA-4 antibodies) augments antitumor responses at early stages of tumor growth [121,122]. Based upon this insight, the possibility emerges for coadministering a CTLA-4 blocking reagent in conjunction with B7-1+ tumor cell transfectants, in this way channeling the B7 signal to the activating CD28 receptor and thus potentiating vaccine efficacy.

A second option is to combine B7 with cytokines or other cell surface molecules linked to T-cell activation. On the cytokine side, synergies have in fact been demonstrated when B7-1 costimulation is combined with cytokines such as IL-12 [97,98], IL-4 [123], IL-7 [124], IL-2 [125], and GM-CSF [126]. On the cell surface molecule side, a growing number of costimulators have been shown to function in a cooperative fashion with B7 to synergistically amplify, or change the cytokine output of, T-cell responses. For example, synergies have been demonstrated when B7 is combined with CD48 [127], HSA [128], ICAM-1 [129], VCAM-1 [130], CD72 [131], or LFA-3 (CD58) [132].

4-1BB ligand, a costimulator expressed on several APC types, including dendritic cells, is another costimulator that might be used effectively in combination with B7 proteins. This protein can provide a costimulus to T cells independent of signaling through the CD28 receptor [133], and it may be especially useful for activating the CD8+ T-cell subset [134]. Hence, a B7/4-1BB combination could potentially serve to broaden the repertoire of T cells that are recruited in the course of the antitumor response.

E. Costimulator Protein Transfer

The engineering of tumor cell vaccines must be simplified in order to facilitate their application in clinical settings. Because most primary tumor cells are difficult to transfect, alternatives to gene transfer are needed for surface costimulator expression. Protein transfer, unlike gene transfer, is readily applicable to primary tumor cells that are difficult to transfect and additionally facilitates the simultaneous expression of multiple proteins at the cell surface, making costimulator coexpression feasible. Furthermore, for some solid cancers, such as colon and breast carcinoma, in which large amounts of primary tumor can often be obtained via excisional biopsy, protein transfer offers the opportunity to altogether bypass primary tumor cell culturing and stable gene transfer as a means for engineering immunogenic tumor cells.

The GPI protein transfer technology described in Section III.D in connection with dendritic cell vaccines is similarly well suited for tumor cell vaccine engineering. In this case, it can be used to paint costimulators onto tumor cell surfaces (Fig. 2). In one study [135], our laboratory showed that murine B7-1 and B7-2 costimulators remain functional after GPI modification, indicating that function is not dependent upon native transmembrane anchorage. The costimulatory functions of murine B7-1/GPI and B7-2/GPI were demonstrated in both *in vitro* and *in vivo* cellular immunologic assays. In subsequent work, we have proceeded to show that B7-1/GPI and B7-2/GPI, once actually painted onto tumor cell surfaces, continue to costimulate well. Specifically, both murine B7-1/GPI and B7-2/GPI were produced in a murine glutamine synthetase amplification/expression system and then purified from the chinese hamster ovary cell transfectants using immunoaffinity

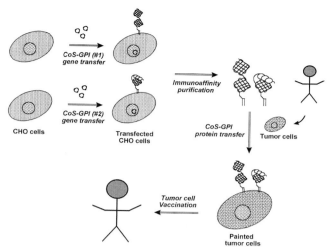

FIGURE 2 *Tumor cell vaccination: the CoS/GPI protein transfer approach.* This schematic depicts a multistep procedure for preparing and using CoS/GPI-painted autologous tumor cells for cellular cancer vaccination. The first steps are directed towards the production of recombinant CoS/GPI complexes, and involve: 1) generating stable CHO cell transfectants that express GPI-modified derivatives of each of the costimulators of interest [shown here as CoS/GPI (#1) and CoS/GPI (#2)]; and 2) immunopurifying the CoS/GPIs from each of the transfectants. These recombinant CoS/GPIs are used clinically for vaccination by: 1) obtaining tumor cells from a patient's excisional biopsy, and amplifying them via culturing if necessary; 2) "painting" the tumor cells *ex vivo* with the appropriate CoS/GPIs; and 3) administering these immunogenic "painted tumor cells" as a cellular vaccine back to the patient.

chromatography. Efficient transfer of these proteins to the cell surface was accomplished for each of these recombinant proteins on different murine tumor lines. That these costimulator/GPIs were appropriately integrated into cell membranes was verified by showing that the reincorporated proteins (1) are phosphoinositol-specific phospholipase D resistant, (2) cannot be stripped off with lipid vesicles, and (3) can be chased into membrane microdomains. After protein transfer, both of these costimulator/GPIs effectively costimulated T cells *ex vivo*.

Similar results have been reported for GPI-modified human B7-1 [136]. In addition, in recent experiments, we have developed a novel approach for epitope-tagging GPI proteins and have applied it specifically to human B7-1. In particular, we have shown that a hexahistidine sequence can be inserted just upstream of the GPI modification signal sequence. This tag permits rapid enrichment of recombinant GPI proteins by metal chelate chromatography. This should simplify the goal of working with an array of costimulator/GPIs in parallel.

When it comes to B7, there may be a special advantage in protein, as opposed to gene, transfer. The function of transfected B7-2 is highly contingent upon cell background, with evidence that B7-2 expressed in murine T cells may preferentially engage CTLA-4 and trigger inhibition [113]. The use of a defined recombinant protein product, such as a B7/GPI protein, circumvents cell-specific idiosyncrasies associated with posttranslational modification processes such as glycosylation [137] and allows better control of surface costimulator densities. Furthermore, GPI derivatization may offer unique advantages for *trans* signaling molecules. Hence, whereas a "tailless" B7-1, mutagenized to eliminate the 16 carboxyl-terminal amino acids, lacked function [138], GPI-reanchored B7 proteins preserve costimulatory function. This difference may be a consequence of their differential abilities to localize to specific sites of cell-to-cell contact that are optimal for T-cell costimulation.

F. Other Cell Surface Engineering Options

Tumor cell surfaces can be engineered in yet other ways. Interactions between APCs and T cells are strengthened by natural adhesins, and selectins can direct cellular homing *in vivo*. By expressing appropriate adhesins and/or selectins on tumor cell surfaces, it should be possible, in principle, to enhance T-cell engagement and modify *in vivo* trafficking properties.

Our group has proposed a general method for artificially modifying the adhesiveness cells by painting onto them "artificial adhesins." We documented the efficacy of a model adhesin/GPI [139], specifically showing that (1) a chimeric M-CSF/GPI can be expressed on the surface of transfectants, and (2) cells with this M-CSF/GPI tethered to their surfaces show markedly enhanced, antibody-inhibitable binding to transfectants bearing the M-CSF receptor. Numerous adaptations of this strategy can be envisioned.

G. Endogenous Gene Inhibition

Another strategy for enhancing the immunogenicity of tumor cells involves the inhibition of endogenous genes within them. The assumption is that certain proteins within tumor cells serve to mask their immunogenic potential; by inhibiting the expression of these proteins, the latent immunogenicity of the tumor cells can be evoked. Antisense gene transfer provides one common means for inhibiting the expression of selected proteins. Ribozymes, triplex-forming polynucleotides, and intrabodies provide yet other means for achieving this goal.

Two classes of proteins have been targeted by antisense RNA expression to produce tumor cell vaccines. One class encompasses proteins implicated in the control of the differentiation states and tumorigenic pheno-

types in tumor cells. The paradigmatic proteins in this class are insulin-like growth factor-1 (IGF-1) and its receptor (IGF-1R). These proteins are known to be expressed in a broad range of tumor types, and IGF-1R signaling has been shown to be obligatory for the establishment and maintenance of the transformed phenotype [140]. We [91,141] and others [142] have shown that interruption of the IGF-1/IGF-1R signaling axis, by targeting either IGF-1 or IGF-1R with antisense RNA, enhances tumor cell immunogenicity. In these experiments, antisense IGF-1 tumor transfectants were effective as tumor cell vaccines for inducing curative and preventive antitumor responses. The mechanisms underlying this phenomenon remain unclear, although it has been reproduced in different tumor types.

A second class of proteins that has been targeted in tumor cell vaccine studies encompasses those with documented anti-inflammatory activities. Transforming growth factor-β (TGF-β), a protein with recognized T-cell inhibitory activity, is expressed by certain tumors, for example, gliomas. It has been shown that antisense inhibition of TGF-β expression in a rat gliosarcoma line (9L) confers upon it vaccine potential [143]. IL-10, known to be produced by a variety of human tumors [144], is another cytokine that has been targeted in this context. Its immunosuppressive activities are directed at APCs, including macrophages [145], dendritic cells [146], and lymphoid cells [147]. These activities include down-regulation of both costimulator expression [145] and MHC class-I-mediated antigen presentation (via TAP transporter inhibition) [147], with additional evidence that IL-10 may convert immature dendritic cells into tolerogenic APCs [148]. In one study [149], an antisense IL-10 retrovirus was used to inhibit IL-10 expression in a murine tumor line (J558L) previously transfected with GM-CSF. Whereas the host GM-CSF transfectant lacked immunogenicity, superimposed antisense IL-10 inhibition elicited the tumor cells' vaccine potential. The investigators proposed that tumor-derived IL-10 somehow prevents GM-CSF-elicited APCs, specifically dendritic cells, from obtaining access to tumor antigens, and in this way interferes with putative cross-priming.

V. CONCLUDING REMARKS

This chapter has focused upon two categories of cellular cancer vaccines, dendritic cell and tumor cell vaccines, along with the gene and protein strategies that can be used to produce them. Cellular cancer vaccines must combine antigenic and costimulatory potentials to optimally trigger CTLs, and the dendritic and tumor cell vaccines arrive at these potentials by different routes.

Dendritic cells come armed with their innate costimulatory potential and thus require priming with tumor antigens. In contrast, tumor cells come armed with their endogenous repertoire of antigens and thus require enhancing of costimulatory potential.

Significant challenges remain on the road to effective cellular cancer vaccines. Looking to the future, some of the fundamental issues are these:

1. *What are the optimal tumor peptide antigens for dendritic cell vaccination?*

Tumors sharing the same histopathologic features can nonetheless demonstrate considerable antigenic heterogeneity [150]. Moreover, there can be significant antigenic diversity even among tumor cells of the same patient. This multilevel (inter- and intrapatient) heterogeneity makes it imperative to more fully molecularly profile tumor cells prior to vaccination and to tailor vaccines accordingly in a patient-specific fashion. The tools for accomplishing this task are steadily evolving. Emerging RNA and protein profiling technologies offer the prospect of generating comprehensive molecular profiles of tumors with relatively rapid turnaround [46,150]. In this way, the molecular complexity of cancer may become more tractable, with improved selection of patients to be included in antigen-specific immunization protocols and with the prospect of "custom" vaccines tailored to tumor phenotypes.

The antigenic diversity of tumors within individual patients demands the use of "multivalent" cancer vaccines, comprising more than one peptide antigen, that evoke heterogeneous CTL responses broad enough to capture within their nets those tumor cells that drift antigenically [151]. Alternatively, antigenic diversity can be tackled by directing repeat vaccination attempts at new peptide antigen targets within the same patient [152]. An enlarged catalogue of effective tumor peptide antigens, corresponding to both immunodominant and subdominant CTL epitopes [44,153] and applicable to a broader set of MHC haplotypes, will be needed to make this feasible. The fact that low-affinity-MHC-binding peptides can in some instances be used to generate effective antitumor CTLs could simplify this accrual process [154].

Given that there are few *bona fide* tumor-specific peptides (that is, peptides absent from normal tissues) that can be used as CTL epitopes in vaccines, it remains necessary to turn to peptides derived from more broadly expressed proteins that happen to be overexpressed in particular tumor types. For such peptides, this leads to a sort of tightrope wherein the cancer vaccine must trigger selected autoreactivity without precipitating full-blown autoimmunity. Much remains to be learned about this "autoimmunity paradox," how it can be exploited,

and how immunologic "tolerance" to normal cellular proteins can be broken. The goal will be to load dendritic cells with antigens in such a way as to promote CTL responses without inducing autoimmune responses. Skewing the responses toward low-avidity CTLs may provide one route for splitting antitumor and autoimmune responses ([50]; Section III.B). In the meantime, the risk for induction of chronic autoimmune diseases appears to be limited [155]. Paradoxically, for some tumors, purposeful induction of autoimmunity has been proposed as the route toward tumor eradication [156].

2. *What are the optimal combinations of costimulators and cytokines to use for tumor cell vaccination?*

As more complete molecular profiles of APCs and T cells become available, a more comprehensive view of what transpires at the APC/T-cell interface should emerge. In turn, a clearer picture of what the APC surface actually looks like should bring new sophistication to the *ex vivo* engineering of tumor cell vaccines. Ideally, one would like to tailor tumor APCs that can selectively target defined T-cell subsets. A further goal may ultimately be not just the differential activation of distinct T-cell subsets, but also distinctive ways of activating the same T-cell. The T-cell activation threshold model (Section III.B) has pointed to the considerable plasticity of T-cell responses at the single-cell level, and tumor APCs could potentially be designed to trigger very specific types of cytokine responses in individual T-cells. In short, increasing control of inputs (therapeutic agents) and outputs (immunologic responses) is expected.

3. *What can be done about MHC- and/or TAP-transporter-deficient tumor cells?*

By definition, tumor cells that lack surface MHC/antigen complexes are not expected to activate, or be lysed by, CTLs. Fortunately, despite the considerable variability in MHC and TAP transporter expression in tumor cells, complete MHC negativity is not the rule [157]. Moreover, tumor cells with subthreshold levels of MHC/antigen can be rescued for use as cellular vaccines. For example, in the case of melanoma, otherwise subthreshold levels of MHC/antigen can trigger CTL clones when sufficient ICAM-1 and LFA-3 are coexpressed on the tumor cell surfaces [46]. Furthermore, once activated, CTL effectors may require fewer antigenic complexes on the tumor targets.

Ultimately, strategies may evolve for increasing MHC expression on tumor cells to make tumor cells into better APCs and/or more optimal CTL targets. Regulatory cytokines or gene transfer could be used to accomplish this goal, with potential proteins to be upregulated including not just MHC heavy and light chains but also proteasome and TAP transporter components of the antigen-processing pathway. In addition, although MHC class I expression on tumor cells has attracted the most attention, there are data in the literature suggesting that optimal tumor cell vaccines should also express MHC class II [9]. There are now early efforts to identify class-II-restricted tumor peptide antigens recognized by CD4+ T cells [18,158]; interestingly, the tyrosinase protein, known to encompass a class-I-restricted melanoma peptide (Section II), is a source for a class-II-restricted one as well [158].

4. *Is it possible to move from* ex vivo *to* in vivo *engineering of cancer vaccine cells?*

Some continue to argue for systemic or local synthetic peptide administration as the simplest of cancer vaccination strategies [159], with some partial clinical responses reported in one study for a naked melanoma peptide vaccine without adjuvant [160]. However, for others, the imperatives of dosing, antigen coexpression, and APC selection make a compelling case for cell-based vaccines as a preferred strategy [2,161]. This view is reinforced by the inordinate complexities associated with the use of synthetic peptides, emulsified in adjuvants, as immunogens [162,163].

Despite the promise offered by cellular cancer vaccine experimentation to date, including the movement to clinical trials [164,165], there remains a pressing need to simplify the cell targeting and engineering steps. With respect to tumor cell vaccines, one possibility is to produce immunogenic tumor cells *in situ* directly within tumor beds. There is evidence that genetically modified, nonmetastatic tumor cells do indeed enter draining nodes, where they would be positioned to elicit systemic antitumor immune responses [119]. However, much remains unknown about the trafficking patterns of tumor cells leaving solid tumors, and it is probable that tumor cell migration from primary tumor differs for different orthotopic sites. Such an approach would be applicable to tumors that are accessible to direct injection methods (e.g., via skin or endoscopic procedures), and cytokine or costimulator expression could be achieved by either gene or protein transfer. Vaccinia virus recombinants have been proposed for directly transfecting tumor cells *in situ* with cytokine genes, as has been shown for intratumoral injection in patients with accessible melanoma lesions (see Chapter 8) [166]. Adenoviral vectors have been used with success for intratumoral delivery of the IL-2 gene in a murine tumor model, although tumor regression in this study was attributed to nonspecific immune effectors [167]. Adenoviral-vector-mediated intratumoral expression of IL-12 was achieved in a murine bladder carcinoma and induced tumor-specific T-cell immunity [168]. Intradermal B7-1 gene transfer was shown to enhance the response to a surrogate MHC class-I-restricted tumor antigen [66].

In vivo engineering of dendritic cell vaccines is a more distant prospect. One might envision directing tumor antigen expression cassettes, such as peptide minigenes (Section II.C), to dendritic cells *in vivo*. The steps would involve first amplifying dendritic cells within patients, for instance by systemic administration of Flt3 ligand [169], and then administering antigen-expressing vectors that home to the amplified dendritic cells. Fundamental advances in gene-targeting technologies would be needed to make this dream a reality. A first step along these lines consists of the recent demonstration that cutaneous injection of naked DNA in mice yields expression of transfected genes in skin-derived dendritic cells, which then localize in draining lymph nodes [170]. An alternative to *in vivo* gene delivery has been proposed that entails the injection of dendritic cell-attracting cytokines in conjunction with tumor antigens to attract dendritic cells to the immunization site [161].

5. *How does the general immune status of cancer patients evolve with time, and is there a role for nonspecific immunopotentiation as an adjunct to cancer vaccination?*

There has been much speculation over the cause of the depressed immune status associated with increasing tumor burden in cancer patients [171]. Potential strategies for dealing with this immunosuppression associated with the tumor-bearing state include blockade of tumor-derived immunosuppressive factors and nonspecific activation of various mononuclear cell effectors. Data with systemically administered anti-CTLA-4 blocking antibodies (Section IV.D) point to the synergistic benefits that may accrue by simultaneously enhancing tumor-cell-mediated costimulation and blocking T-cell inhibitory pathways within the costimulated antitumor T-cells. Even more dramatic is the recent demonstration that systemically administered anti-4-1BB antibodies, functioning as activators of primed T-cells, can lead to the regression of relatively large established tumors within animals [172]. Clearly, cellular cancer vaccine efficacy could be augmented by combining them with anti-CTLA-4, anti-4-1BB, or other antibodies. Other nonspecific adjuvants could be used in combination with vaccines, such as thymomimetic peptides (e.g., thymosin α1), drugs (e.g., levamisole and isoprinosine), or proinflammatory interleukins [103]. Flt3 ligand could also potentially serve as a potent immunologic adjuvant by enhancing dendritic cell development [173].

The growing appreciation for the need to consider the general immunologic status of patients at the time of vaccination has shifted the cancer vaccine field to a new and more promising track, and has created a more realistic mind-set. What emerges now is the concept of using vaccines as adjuvant immunotherapy, to be used in conjunction with standard tumor-debulking procedures (surgical resection, radiotherapy, antitumor monoclonal antibodies, and so on) for the treatment of minimal residual disease. The early cancer vaccine trials, with their inclusion of patients with advanced disease, no longer seem appropriate.

6. *What is the relevance of experimental findings with cell lines and animal tumor models to immunotherapy of primary human cancer?*

As in all of cancer therapeutics, the transition from cell lines and animals to humans is not a trivial one. Susceptibility to cytolysis by T-cell clones can differ substantially between cultured and uncultured tumor cells. In one study, tumor-specific CTLs did not lyse autologous melanoma cell lines but lysed "fresh" autologous tumor cells in an MHC class-I-dependent manner [174]. In contrast, another study has shown that although MAGE-3-peptide-specific CTLs expanded from melanoma patients were highly active against peptide-pulsed or transfected target cells, these CTLs could not recognize MAGE-3-expressing melanoma cells [21]. Hence, the immune responses to long-term cultured tumor cell lines may not be totally indicative of the *in situ* immune status.

A reductionist approach has limited the success of early cancer vaccine trials. Current efforts are dealing more effectively with the complexities of tumor cells and of the immune system. The use of sophisticated cellular cancer vaccines with potentiated efficacy, along with more careful attention to the immunologic status of patients, have added new dimensions to our thinking of the problem. The prospects for cellular cancer vaccines now seem more promising. The real tests for the cancer vaccine concept are now ahead of us in the clinic.

References

1. Vitetta, E. S. (1994). From the basic science of B cells to biological missiles at the bedside. *J. Immunol.* **153**, 1407–1420.
2. Tykocinski, M. L., D. R. Kaplan, and M. E. Medof. (1996). Antigen-presenting cell engineering. The molecular toolbox. *Am. J. Pathol.* **148**, 1–16.
3. Gross, L. (1943). Intradermal immunization of C3H mice against a sarcoma that originated in an animal of the same line. *Cancer Res.* **3**, 326–333.
4. Hoover, H. C., J. S. Brandhorst, L. C. Peters, M. G. Surdyke, Y. Takeshita, J. Madariaga, L. R. Muenz, and M. G. Hanna. (1993). Adjuvant active specific immunotherapy for human colorectal cancer: 6.5-year median follow-up of a phase III prospectively randomized trial. *J. Clin. Oncol.* **11**, 390–399.
5. Boon, T., J.-C. Cerottini, B. Van den Eynde, P. van der Bruggen, and A. Van Pel. (1994). Tumor antigens recognized by T lymphocytes. *Annu. Rev. Immunol.* **12**, 337–365.
6. York, I., and K. L. Rock. (1996). Antigen processing and presentation by the class I major histocompatibility complex. *Annu. Rev. Immunol.* **14**, 369–396.

7. Rock, K. L. (1996). A new foreign policy: MHC class I molecules monitor the outside world. *Immunol. Today* **17**, 131–137.
8. Harding, C. V. (1995) Phagocytic processing of antigens for presentation by MHC molecules. *Trends Cell. Biol.* **5**, 105–109.
9. Armstrong, T. D., V. K. Clements, and S. Ostrand-Rosenberg. (1998). MHC class II-transfected tumor cells directly present antigen to tumor-specific CD4+ T lymphocytes. *J. Immunol.* **160**, 661–666.
10. Van den Eynde, B., B. Lethe, A. van Pel, *et al.* (1991). The gene coding for a major tumor rejection antigen of tumor P815 is identical to the normal gene of syngeneic DBA/2 mice. *J. Exp. Med.* **173**, 1373–1384.
11. Boon, T., and P. Van der Bruggen. (1996). Human tumor antigens recognized by T lymphocytes. *J. Exp. Med.* **183**, 725–729.
12. Slingluff, C. L. (1996). Tumor antigens and tumor vaccines: peptides as immunogens. *Semin. Surg. Oncol.* **12**, 446–453.
13. Robbins, P. F., and Y. Kawakami. (1997). Human tumor antigen recognized by T cells. *Curr. Opin. Immunol.* **8**, 628–636.
14. Peoples, G. E., P. S. Goedegebuure, R. Smith, D. C. Lihehan, I. Yoshino, and T. J. Eberlein. (1995). Breast and ovarian cancer-specific cytotoxic T lymphocytes recognize the same HER2/neu-derived peptide. *Proc. Natl. Acad. Sci. U.S.A.* **92**, 432–436.
15. Traversari, C., P. van der Bruggen, I. F. Luescher, C. Lurquin, P. Chomez, A. Van Pel, E. De Plaen, A. Amar-Costesec, and T. Boon. (1992). A nonapeptide encoded by human gene MAGE-1 is recognized on HLA-A1 by cytolytic T lymphocytes directed against tumor antigen MZ2-E. *J. Exp. Med.* **176**, 1453–1457.
16. Kawakami, Y., S. Eliyahu, C. H. Delgado, P. F. Robbins, L. Rivoltini, S. L. Topalian, T. Miki, S. A. Rosenberg. (1994). Cloning of the gene coding for a shared human melanoma antigen recognized by autologous T cells infiltrating into tumor. *Proc. Natl. Acad. Sci. U.S.A.* **91**, 3515–3519.
17. Castelli, C., W. J. Storkus, M. J. Mauere, D. M. Martin, E. C. Huang, B. M. Pramanik, T. L. Nagabhushan, G. Parmiani, and M. T. Lotze. (1995). Mass spectrometric identification of a naturally processed melanoma peptide recognized by CD8+ cytotoxic T lymphocytes. *J. Exp. Med.* **181**, 363–368.
18. Halder, T., G. Pawelec, A. F. Kirkin, J. Zeuthen, H. E. Meyer, L. Kun, and H. Kalbacher. (1997). Isolation of novel HLA-DR restricted potential tumor-associated antigens from the melanoma cell line FM3. *Cancer Res.* **57**, 3238–3244.
19. van der Bruggen, P., J. Bastin, T. Gajewski, P. Coulie, P. Boel, C. De Smet, C. Traversari, A. Townsend, and T. Boon. (1994). A peptide encoded by human gene MAGE-3 and presented by HLA-A2 induces cytolytic T lymphocytes that recognize tumor cells expressing MAGE-3. *Eur. J. Immunol.* **24**, 3038–3043.
20. Salazar-Onfray, F., T. Nakazawa, V. Chhajlani, M. Petersson, K. Karre, G. Masucci, E. Celis, A. Sette, S. Southwood, E. Appella, and R. Kiessling. (1997). Synthetic peptides derived from the melanocyte-stimulating hormone receptor MC1R can stimulate HLA-A2-restricted cytotoxic T lymphocytes that recognize naturally processed peptides on human melanoma cells. *Cancer Res.* **57**, 4348–4355.
21. Valmori, D., D. Lienard, G. Waanders, D. Rimoldi, J.-C. Cerottini, and P. Romero. (1997). Analysis of MAGE-3-specific cytolytic T lymphocytes in human leukocyte antigen-A2 melanoma patients. *Cancer Res.* **57**, 735–741.
22. Mayordomo, J. I., T. Zorina, W. J. Storkus, L. Zitvogel, C. Celluzzi, L. D. Falo, C. J. Melief, S. T. Ildstad, W. M. Kast, A. B. DeLeo, and M. T. Lotze. (1995). Bone marrow-derived dendritic cells pulsed with synthetic tumor peptides elicit protective and therapeutic antitumor immunity. *Nat. Med.* **1**, 1297–1302.
23. Toso, J. F., C. Oei, F. Oshidari, J. Tartaglia, E. Paoletti, H. K. Lyerly, S. Talib, and K. J. Weinhold. (1996). MAGE-1-specific

precursor cytotoxic T-lymphocytes present among tumor-infiltrating lymphocytes from a patient with breast cancer: characterization and antigen-specific activation. *Cancer Res.* **56,** 16–20.

24. Protti, M. P., M. A. Imro, A. A. Manfredi, G. Consogno, S. Heltai, C. Arcelloni, M. Bellone, P. Dellabona, G. Casorati, and C. Rugarli. (1996). Particulate naturally processed peptides prime a cytotoxic response against human melanoma in vitro. *Cancer Res.* **56,** 1210–1213.

25. Young, J. W., and K. Inaba. (1996). Dendritic cells as adjuvants for class I major histocompatibility complex-restricted antitumor immunity. *J. Exp. Med.* **183,** 7–11.

26. Steinman, R. M., and Z. A. Cohn. (1973). Identification of a novel cell type in peripheral lymphoid organs of mice. I. Morphology, quantitation, tissue distribution. *J. Exp. Med.* **137,** 1142–1162.

27. Steinman, R. M., and K. Inaba. (1985). Stimulation of the primary mixed leukocyte reaction. *CRC Crit. Rev. Immunol.* **5,** 331–338.

28. Cella, M., F. Sallusto, and A. Lanzavecchia. (1997). Origin, maturation and antigen presenting function of dendritic cells. *Curr. Opin. Immunol.* **9,** 10–16.

29. Kasinrerk, W., T. Baumruker, O. Majdic, W. Knapp, and H. Stockinger. (1993). CD1 molecule expression on human monocytes induced by granulocyte-macrophage colony-stimulating factor. *J. Immunol.* **150,** 579–584.

30. Sallusto, F., and A. Lanzavecchia. (1994). Efficient presentation of soluble antigen by cultured human dendritic cells is maintained by granulocyte/macrophage colony-stimulating factor plus interleukin 4 and downregulated by tumor necrosis factor alpha. *J. Exp. Med.* **179,** 1109–1118.

31. Zhou, L. J., and T. F. Tedder. (1996). CD14+ blood monocytes can differentiate into functionally mature CD83+ dendritic cells. *Proc. Natl. Acad. Sci. U.S.A.* **93,** 2588–2592.

32. Saunders, D., K. Lucas, J. Ismaili, L. Wu, E. Maraskovsky, A. Dunn, and K. Shortman. (1996). Dendritic cell development in culture from thymic precursor cells in the absence of granulocyte/macrophage colony-stimulating factor. *J. Exp. Med.* **184,** 2185–2196.

33. Kronin, V., K. Winkel, G. Suss, A. Kelso, W. Heath, J. Kirberg, H. von Boehmer, and K. Shortman. (1996). A subclass of dendritic cells regulates the response of naive CD8 T cells by limiting IL-2 production. *J. Immunol.* **157,** 3819–3827.

34. Suss, G., and K. Shortman. (1996). A subclass of dendritic cells kills CD4 T cells via Fas/Fas-ligand-induced apoptosis. *J. Exp. Med.* **183,** 1789–1796.

35. Kaplan, D. R., J. E. Hambor, and M. L. Tykocinski. (1989). An immunoregulatory function for the CD8 molecule. *Proc. Natl. Acad. Sci. U.S.A.* **86,** 8512–8515.

36. Hambor, J. E., D. R. Kaplan, and M. L. Tykocinski. (1990). CD8 functions as an inhibitory ligand in mediating the immunoregulatory activity of CD8+ cells. *J. Immunol.* **145,** 1646–1652.

37. Sambhara, S. R., and R. G. Miller. (1991). Programmed cell death of T cells signaled by the T cell receptor and the alpha3 domain of class I MHC. *Science* **252,** 1424–1427.

38. Romani, N., D. Reider, M. Heuer, S. Ebner, E. Kampgen, B. Eibl, D. Niederwieser, and G. Schuler. (1996). Generation of mature dendritic cells from human blood. An improved method with special regard to clinical applicability. *J. Immunol. Methods* **196,** 137–151.

39. Bender, A., M. Sapp, G. Schuler, R. M. Steinman, and N. Bhardwaj. (1996). Improved methods for the generation of dendritic cells from nonproliferating progenitors in human blood. *J. Immunol. Methods* **196,** 121–135.

40. Girolomoni, G., and P. Ricciardi-Castagnoli. (1997). Dendritic cells hold promise for immunotherapy. *Immunol. Today* **18,** 102–104.

41. Day, P. M., J. W. Yewdell, A. Porgador, R. N. Germain, and J. R. Bennink. (1997). Direct delivery of exogenous MHC class I molecule-binding oligopeptides to the endoplasmic reticulum of viable cells. *Proc. Natl. Acad. Sci. U.S.A.* **94,** 8064–8069.

42. Rock, K. L., S. Gamble, L. Rothstein, C. Gramm, and B. Benacerraf. (1991). Dissociation of beta-2-microglobulin leads to the accumulation of a substantial pool of inactive class I MHC heavy chains on the cell surface. *Cell* **64,** 611–620.

43. Van Elsas, A., S. H. Van der Burg, C. E. Van der Minne, M. Borghi, J. S. Mourer, C. J. Melief, and P. I. Schrier. (1996). Peptide-pulsed dendritic cells induce tumoricidal cytotoxic T lymphocytes from healthy donors against stably HLA-A*0201-binding peptides from the Melan-A/MART-1 self antigen. *Eur. J. Immunol.* **26,** 1683–1689.

44. Tsai, V., S. Southwood, J. Sidney, K. Sakaguchi, Y. Kawakami, E. Appella, A. Sette, and E. Celis. (1997). Identification of subdominant CTL epitopes of the GP100 melanoma-associated tumor antigen by primary *in vitro* immunization with peptide-pulsed dendritic cells. *J. Immunol.* **158,** 1796–1802.

45. Restifo, N. P., I. Bacik, K. R. Irvine, J. W. Yewdell, B. J. McCabe, R. W. Anderson, L. C. Eisenlohr, S. A. Rosenberg, and J. R. Bennink. (1995). Antigen processing *in vivo* and the elicitation of primary CTL responses. *J. Immunol.* **154,** 4414–4422.

46. Labarriere, N., E. Diez, M.-C. Pandolfino, C. Viret, Y. Guilloux, S. Le Guiner, J.-F. Fonteneau, B. Dreno, and F. Jotereau. (1997). Optimal T cell activation by melanoma cells depends on a minimal level of antigen transcription. *J. Immunol.* **158,** 1238–1245.

47. Valitutti, S., S. Muller, M. Dessing, and A. Lanzavecchia. (1996). Different responses are elicited in cytotoxic T lymphocytes by different levels of T cell receptor occupancy. *J. Exp. Med.* **183,** 1917–1921.

48. Gervois, N., Y. Guilloux, E. Diez, and F. Jotereau. (1996). Suboptimal activation of melanoma infiltrating lymphocytes (TIL) due to low avidity of TCR/MHC-tumor peptide interactions. *J. Exp. Med.* **183,** 2403–2407.

49. Itoh, Y., and R. N. Germain. (1997). Single cell analysis reveals regulated hierarchical T cell antigen receptor signaling thresholds and intraclonal heterogeneity for individual cytokine responses of CD4+ T cells. *J. Exp. Med.* **186,** 757–766.

50. Morgan, D. J., H. T. C. Kreuwel, S. Fleck, H. I. Levitsky, D. M. Pardoll, and L. A. Sherman. (1998). Activation of low avidity CTL specific for a self-epitope results in tumor rejection but not autoimmunity. *J. Immunol.* **160,** 643–651.

51. Storkus, W. J., H. J. I. Zeb, R. D. Salter, *et al.* (1993). Identification of T cell epitopes: rapid isolation of class I-presented peptides from viable cells by mild acid elution. *J. Immunother.* **14,** 94–103.

52. Langlade, P., J.-P. Levraud, P. Kourilsky, and J.-P. Abastado. (1994). Primary cytotoxic T lymphocyte induction using peptide-stripped autologous cells. *Int. Immunol.* **6,** 1759–1766.

53. Buschle, M., W. Schmidt, W. Zauner, K. Mechtler, B. Trska, H. Kirlappos, and M. L. Birnstiel. (1997). Transloading of tumor antigen-derived peptides into antigen-presenting cells. *Proc. Natl. Acad. Sci. USA* **94,** 3256–3261.

54. Franksson, L., M. Petersson, R. Kiessling, and K. Karre. (1993). Immunization against tumor and minor histocompatibility antigens by eluted cellular peptides loaded on antigen processing defective cells. *Eur. J. Immunol.* **23,** 2606–2613.

55. Bellone, M., G. Iezzi, A. Martin-Fontecha, L. Rivolta, A. A. Manfredi, M. P. Protti, M. Freschi, P. Dellabona, G. Casorati, and C. Rugarli. (1997). Rejection of a nonimmunogenic melanoma by vaccination with natural melanoma peptides on engineered antigen-presenting cells. *J. Immunol.* **158,** 783–789.

56. Suto, R., and P. K. Srivastava. (1995). A mechanism for the specific immunogenicity of heat shock protein-chaperoned peptides. *Science* **269**, 1585–1588.

57. Paglia, P., C. Chiodoni, M. Rodolfo, and M. P. Colombo. (1996). Murine dendritic cells loaded *in vitro* with soluble protein prime cytotoxic T lymphocytes against tumor antigen *in vivo*. *J. Exp. Med.* **183**, 317–322.

58. Hsu, F. J., C. Benike, F. Fagnoli, T. M. Liles, D. Czerwinski, B. Taidi, E. G. Engleman, and R. Levy. (1996). Vaccination of patients with B-cell lymphoma using autologous antigen-pulsed dendritic cells. *Nat. Med.* **2**, 52–58.

59. Reeves, M. E., R. E. Royal, J. S. Lam, S. A. Rosenberg, and P. Hwu. (1996). Retroviral transduction of human dendritic cells with a tumor-associated antigen gene. *Cancer Res.* **56**, 5672–5677.

60. Wu, Y., and T. J. Kipps. (1997). Deoxyribonucleic acid vaccines encoding antigens with rapid proteasome-dependent degradation are highly efficient inducers of cytolytic T lymphocytes. *J. Immunol.* **159**, 6037–6043.

61. Flamand, V., T. Sornasse, K. Thielemans, C. Demanet, M. Bakkus, H. Bazin, F. Tielemans, O. Leo, J. Urbain, and M. Moser. (1994). Murine dendritic cells pulsed *in vitro* with tumor antigen induce tumor resistance *in vivo*. *Eur. J. Immunol.* **24**, 605–610.

62. Ribas, A., L. H. Butterfield, W. H. McBride, S. M. Jilani, L. A. Bui, C. M. Vollmer, R. Lau, V. B. Dissette, B. Hu, A. Y. Chen, J. A. Glaspy, and J. S. Economou. (1997). Genetic immunization for the melanoma antigen MART-1/Melan-A using recombinant adenovirus-transduced murine dendritic cells. *Cancer Res.* **57**, 2865–2869.

63. Henderson, R. A., M. T. Nimgaonkar, S. C. Watkins, P. D. Robbins, E. D. Ball, and O. J. Finn. (1996). Human dendritic cells genetically engineered to express high levels of the human epithelial tumor antigen mucin (MUC-2). *Cancer Res.* **56**, 3763–3770.

64. Bronte, V., M. W. Carroll, T. J. Goletz, M. Wang, W. W. Overwijk, F. Marincola, S. A. Rosenberg, B. Moss, and N. P. Restifo. (1997). Antigen expression by dendritic cells correlates with the therapeutic effectiveness of a model recombinant poxvirus tumor vaccine. *Proc. Natl. Acad. Sci. USA* **94**, 3183–3188.

65. Alijagic, S., P. Moller, M. Artuc, K. Jurgovsky, B. M. Czarnetzki, and D. Schadendorf. (1995). Dendritic cells generated from peripheral blood transfected with human tyrosinase induce specific T-cell activation. *Eur. J. Immunol.* **25**, 3100–3107.

66. Corr, M., H. Tighe, D. Lee, J. Dudler, M. Trieu, D. C. Brinson, and D. A. Carson. (1997). Costimulation provided by DNA immunization enhances antitumor immunity. *J. Immunol.* **159**, 4999–5004.

67. Disis, M. L., J. R. Gralow, H. Bernhard, S. L. Hand, W. D. Rubin, and M. A. Cheever. (1996). Peptide-based, but not whole protein, vaccines elicit immunity to HER-2/neu, an oncogenic self-protein. *J. Immunol.* **156**, 3151–3158.

68. Lee, L., L. McHugh, R. K. Ribaudo, S. Kozlowski, D. H. Margulies, and M. G. Mage. (1994). Functional cell surface expression by a recombinant single-chain class I major histocompatibility complex molecule with a *cis*-active beta 2-microglobulin domain. *Eur. J. Immunol.* **24**, 2633–2639.

69. Mottez, E., P. Langlade-Demoyen, H. Gournier, F. Martinon, J. Maryanski, P. Kourilsky, and J.-P. Abastado. (1995). Cells expressing a major histocompatibility complex class I molecule with a single covalently bound peptide are highly immunogenic. *J. Exp. Med.* **181**, 493–502.

70. Boczkowski, D., S. K. Nair, D. Snyder, and E. Gilboa. (1996). Dendritic cells pulsed with RNA are potent antigen-presenting cells *in vitro* and *in vivo*. *J. Exp. Med.* **184**, 465–472.

71. Nair, S. K., D. Snyder, and E. Gilboa. (1996). Cells treated with TAP-2 antisense oligonucleotides are potent antigen-presenting cells *in vitro* and *in vivo*. *J. Immunol.* **156**, 1772–1780.

72. Medof, M. E., T. Kinoshita, and V. Nussenzweig. (1984). Inhibition of complement activation on the surface of cells after incorporation of decay-accelerating factor (DAF) into their membranes. *J. Exp. Med.* **160**, 1558–1578.

73. Medof, M. E., D. M. Lublin, V. M. Holers, D. J. Ayers, R. R. Getty, J. F. Leykam, J. P. Atkinson, and M. L. Tykocinski. (1987). Cloning and characterization of cDNAs encoding the complete sequence of decay-accelerating factor of human complement. *Proc. Natl. Acad. Sci. USA* **84**, 2007–2011.

74. Caras, I. W., M. A. Davitz, L. Rhee, G. Weddell, D. W. Martin, and V. Nussenzweig. (1987). Cloning of decay-accelerating factor suggests novel use of splicing to generate two proteins. *Nature* **325**, 545–549.

75. Tykocinski, M. L., H. K. Shu, D. J. Ayers, E. I. Walter, R. R. Getty, R. K. Groger, C. A. Hauer, and M. E. Medof. (1988). Glycolipid reanchoring of T-lymphocyte surface antigen CD8 using the 3' end sequence of decay-accelerating factor's mRNA. *Proc. Natl. Acad. Sci. USA* **85**, 3555–3559.

76. Caras, I. W., G. N. Weddell, M. A. Davitz, V. Nussenzweig, J. and D. W. Martin. (1987). Signal for attachment of a phospholipid membrane anchor in decay accelerating factor. *Science* **238**, 1280–1283.

77. Medof, M. E., S. Nagarajan, and M. L. Tykocinski. (1996). Cell surface engineering with GPI-anchored proteins. *FASEB J.* **10**, 574–586.

78. Ilangumaran, S., P. J. Robinson, and D. C. Hoessli. (1996). Transfer of exogenous glycosylphosphatidylinositol (GPI)-linked molecules to plasma membranes. *Cell Biol.* **6**, 163–167.

79. Huang, J.-H., N. S. Greenspan, and M. L. Tykocinski. (1994). Alloantigenic recognition of artificial glycosylphosphatidylinositol-anchored HLA-A2.1. *Mol Immunol.* **31**, 1017–1028.

80. Huang, J.-H., R. R. Getty, F. V. Chisari, P. Fowler, N. S. Greenspan, and M. L. Tykocinski. (1994). Protein transfer of preformed MHC-peptide complexes sensitizes target cells to T cell cytolysis. *Immunity.* **1**, 607–613.

81. Mann, D. W., I. Stroynowski, L. Hood, and J. Forman. (1989). An H-2L^d hybrid molecule with a Qa-2, alpha 3 domain and phosphatidylinositol anchor is not recognized by H-2L^d-specific cytotoxic T lymphocytes. *J. Immunol.* **142**, 318–322.

82. Aldrich, C. J., L. C. Lowen, D. Mann, M. Nishimura, L. Hood, I. Stroynowski, and J. Forman. (1991). The Q7 α3 domain alters T cell recognition of class I antigens. *J. Immunol.* **146**, 3082–3090.

83. Zamoyska, R., T. Ong, G. Kwan-Lim, P. Tomlinson, and P. J. Robinson. (1996). Unprimed T cells are inefficiently stimulated by glycosylphosphatidylinositol-linked H-2K^b because of its lipid anchor rather than defects in CD8 binding. *Int. Immunol.* **8**, 551–557.

84. Cariappa, A., D. C. Flyer, C. T. Rollins, D. C. Roopenian, R. A. Flavell, D. Brown, and G. I. Waneck. (1996). Glycosylphosphatidylinositol-anchored H-2D^b molecules are defective in antigen processing and presentation to cytotoxic T lymphocytes. *Eur. J. Immunol.* **26**, 2215–2224.

85. Tepper, R. I., P. K. Pattengale, and P. Leder. (1989). Murine interleukin-4 displays potent anti-tumor activity in vivo. *Cell* **57**, 503–512.

86. Fearon, E. R., D. M. Pardoll, T. Itaya, P. Golumbek, H. I. Levitsky, J. W. Simons, H. Karasuyama, B. Vogelstein, and P. Frost. (1990). Interleukin-2 production by tumor cells bypasses T helper function in the generation of an immune response. *Cell* **60**, 397–403.

87. Golumbek, P. T., A. J. Lazenby, H. I. Levitsky, L. M. Jaffee, H. Karasuyama, M. Baker, and D. M. Pardoll. (1991). Treatment of established renal cancer by tumor cells engineered to secrete interleukin-4. *Science* **254,** 713–716.

88. Chen, L., S. Ashe, W. A. Brady, I. Hellstrom, K. E. Hellstrom, J. A. Ledbetter, P. McGowan, and P. S. Linsley. (1992). Costimulation of antitumor immunity by the B7 counterreceptor for the T-lymphocyte molecules CD28 and CTLA-4. *Cell* **71,** 1093–1102.

89. Lukacs, K. V., D. B. Lowrie, R. W. Stokes, and M. J. Colston. (1993). Tumor cells transfected with a bacterial heat-shock gene lose tumorigenicity and induce protection against tumors. *J. Exp. Med.* **178,** 343–348.

90. Townsend, S. E., and J. P. Allison. (1993). Tumor rejection after direct costimulation of CD8+ T cells by B7-transfected melanoma cells. *Science* **259,** 368–370.

91. Trojan, J., T. R. Johnson, S. D. Rudin, J. Ilan, M. L. Tykocinski, and J. Ilan. (1993). Treatment and prevention of rat glioblastoma by immunogenic C6 cells expressing antisense IGF-1 RNA. *Science* **259,** 94–97.

92. Guo, Y., M. Wu, H. Chen, X. Wang, G. Liu, G. Li, J. Ma, and M.-S. Sy. (1994). Effective tumor vaccine generated by fusion of hepatoma cells with activated B cells. *Science* **263,** 518–520.

93. Melero, I., N. Bach, and L. Chen. (1997). Costimulation, tolerance and ignorance of cytolytic T lymphocytes in immune responses to tumor antigens. *Life Sci.* **69,** 2035–2041.

94. Ridge, J. P., E. J. Fuchs, and P. Matzinger. (1996). Neonatal tolerance revisited: turning on newborn T cells with dendritic cells. *Science* **271, 1723–1726.**

95. Kolenko, V., Q. Wang, M. C. Riedy, J. O'Shea, J. Ritz, M. K. Cathcart, P. Rayman, R. Tubbs, M. Edinger, A. Novick, R. Bukowski, and J. Finke. (1997). Tumor-induced suppression of T lymphocyte proliferation coincides with inhibition of Jak3 expression and IL-2 receptor signaling. Role of soluble products from human renal carcinoma. *J. Immunol.* **159,** 3057–3067.

96. Colombo, M., and G. Forni. (1994). Cytokine gene transfer in tumor inhibition and tumor therapy: where are we now? *Immunol. Today* **15,** 48–51.

97. Gajewski, T. F., J.-C. Renauld, A. Van Pel, and T. Boon. (1995). Costimulation with B7-1, IL-6, and IL-12 is sufficient for primary generation of murine anti-tumor cytolytic T lymphocytes *in vitro.* *J. Immunol.* **154,** 5637–5648.

98. Coughlin, C. M., M. Wysocka, H. L. Kurzawa, W. M. F. Lee, G. Trinchieri, and S. L. Eck. (1995). B7-1 and interleukin 12 synergistically induce effective antitumor immunity. *Cancer Res.* **55,** 4980–4987.

99. Huang, A. Y. C., P. Golumbek, M. Ahmadzadeh, E. Jaffee, D. M. Pardoll, and H. I. Levitsky. (1994). Role of bone marrow-derived cells in presenting MHC class I-restricted tumor antigens. *Science* **264,** 961–965.

100. Pulaski, B., K. Yeh, N. Shastri, K. Maltby, D. Penney, E. Lord, and J. Frelinger. (1996). IL-3 enhances CTL development and class I MHC presentation of exogenous antigen by tumor-infiltrating macrophages. *Proc. Natl. Acad. Sci. U.S.A.* **93,** 3669–3672.

101. Cayeux, S., G. Richter, G. Noffz, B. Dorken, and T. Blankenstein. (1997). Influence of gene-modified (IL-7, IL-4, and B7) tumor cell vaccines on tumor antigen presentation. *J. Immunol.* **158,** 2834–2841.

102. Chen, K., S. Braun, S. Lyman, Y. Fan, C. M. Traycoff, E. A. Wiebke, J. Gaddy, G. Sledge, H. E. Broxmeyer, and K. Cornetta. (1997). Antitumor activity and immunotherapeutic properties of Flt3-ligand in a murine breast cancer model. *Cancer Res.* **57,** 3511–3516.

103. Hadden, J. W. (1994). T-cell adjuvants. *Int. J. Immunopharmacol.* **16,** 703–710.

104. Rothenberg, M. E., A. D. Luster, and P. Leder. (1995). Murine eotaxin: an eosinophil chemoattractant inducible in endothelial cells and in interleukin 4-induced tumor suppression. *Proc. Natl. Acad. Sci. U.S.A.* **92,** 8960–8964.

105. Bluestone, J. (1995). New perspectives of CD28-B7-mediated T cell costimulation. *Immunity.* **2,** 555–559.

106. Mueller, D. L., M. K. Jenkins, and R. H. Schwartz. (1989). Clonal expansion versus functional clonal inactivation: a costimulatory signaling pathway determines the outcome of T cell antigen receptor occupancy. *Annu. Rev. Immunol.* **7,** 445–480.

107. Linsley, P. S., J. L. Greene, W. Brady, J. Bajorath, J. A. Ledbetter, and R. Peach. (1994). Human B7-1 (CD80) and B7-2 (CD86) bind with similar avidities but distinct kinetics to CD28 and CTLA-4 receptors. *Immunity* **1,** 793–801.

108. Chen, L., P. McGowan, S. Ashe, J. Johnston, Y. Li, I. Hellstrom, and K. Hellstrom. (1994). Tumor immunogenicity determines the effect of B7 costimulation on T cell-mediated tumor immunity. *J. Exp. Med.* **179,** 523–532.

109. Baskar, S., S. Ostrand-Rosenberg, N. Nabavi, L. Nadler, G. Freeman, and L. Glimcher. (1993). Constitutive expression of B7 restores immunogenicity of tumor cells expressing truncated major histocompatibility complex class II molecules. *Proc. Natl. Acad. Sci. U.S.A.* **90,** 5687–5690.

110. Hodge, J. W., S. Abrams, J. Schlom, and J. A. Kantor. (1994). Induction of antitumor immunity by recombinant vaccinia viruses expressing B7-1 or B7-2 costimulatory molecules. *Cancer Res.* **54,** 5552–5555.

111. Li, Y., P. McGowan, L. Hellstrom, K. Hellstrom, and L. Chen. (1994). Costimulation of tumor-reactive CD4+ and CD8+ T lymphocytes by B7, a natural ligand for CD28, can be utilized to treat established mouse melanoma. *J. Immunol.* **153,** 421–428.

112. Gajewski, T. F. (1996). B7-1 but not B7-2 efficiently costimulates CD8+ T lymphocytes in the P815 tumor system *in vitro.* *J. Immunol.* **156,** 465–472.

113. Greenfield, E. A., E. Howard, T. Paradis, K. Nguyen, F. Benazzo, P. McLean, P. Hollsberg, G. Davis, D. A. Hafler, A. H. Sharpe, G. J. Freeman, and V. K. Kuchroo. (1997). B7.2 expressed by T cells does not induce CD28-mediated costimulatory activity but retains CTLA4 binding. Implications for induction of antitumor immunity to T cell tumors. *J. Immunol.* **158,** 2025–2034.

114. Johnston, J. V., A. R. Malacko, M. T. Mizuno, P. McGowan, I. Hellstrom, K. E. Hellstrom, H. Marquardt, and L. Chen. (1996). B7-CD28 costimulation unveils the hierarchy of tumor epitopes recognized by major histocompatibility complex class I-restricted CD8+ cytolytic T lymphocytes. *J. Exp. Med.* **183,** 791–800.

115. Yang, S., T. L. Darrow, and H. F. Seigler. (1997). Generation of primary tumor-specific cytotoxic T lymphocytes from autologous and human lymphocyte antigen class I-matched allogeneic peripheral blood lymphocytes by B7 gene-modified melanoma cells. *Cancer Res.* **57,** 1561–1568.

116. Wu, T.-C., A. Y. C. Huang, E. M. Jaffee, H. I. Levitsky, and D. M. Pardoll. (1995). A reassessment of the role of B7-1 expression in tumor rejection. *J. Exp. Med.* **182,** 1415–1421.

117. Huang, A. Y. C., A. T. Bruce, D. M. Pardoll, and H. I. Levitsky. (1996). Does B7-1 expression confer antigen-presenting cell capacity to tumors *in vivo? J. Exp. Med.* **183,** 769–776.

118. Kundig, T. M., M. F. Bachmann, C. DiPaolo, J. J. L. Simard, M. Battegay, H. Lother, A. Gessner, K. Kuhlcke, P. S. Ohashi, H. Hengartner, and R. M. Zinkernagel. (1995). Fibroblasts as efficient antigen-presenting cells in lymphoid organs. *Science* **268,** 1343–1349.

119. Yang, G., M. T. Mizuno, K. E. Hellstrom, and L. Chen. (1997). B7-negative versus B7-positive P815 tumor. Differential requirements for priming of an antitumor immune response in lymph nodes. *J. Immunol.* **158,** 851–858.

120. Prehn, R. T. (1994). Stimulatory effects of immune reactions upon the growths of untransplanted tumors. *Cancer Res.* **54,** 908–914.

121. Leach, D. R., M. F. Krummel, and J. P. Allison. (1996). Enhancement of antitumor immunity by CTLA-4 blockade. *Science* **271,** 1734–1736.

122. Yang, Y.-F., J.-P. Zou, J. Mu, R. Wijesuriya, S. Ono, T. Walunas, J. Bluestone, H. Fujiwara, and T. Hamaoka. (1997). Enhanced induction of antitumor T-cell responses by cytotoxic T lymphocyte-associated molecule-4 blockade: the effect is manifested only at restricted tumor-bearing stages. *Cancer Res.* **57,** 4036–4041.

123. Cayeux, S., C. Beck, B. Dorken, and T. Blankenstein. (1996). Coexpression of interleukin-4 and B7.1 in murine tumor cells leads to improved tumor rejection and vaccine effect compared to single gene transfectants and a classical adjuvant. *Hum. Gene Ther.* **7,** 525–529.

124. Cayeux, S., C. Beck, A. Aicher, B. Dorken, and T. Blankenstein. (1995). Tumor cells cotransfected with interleukin 7 and B7.1 genes induce CD25 and CD28 on tumor infiltrating lymphocytes and are strong vaccines. *Eur. J. Immunol.* **25,** 2325–2331.

125. Salvadori, S., B. Gansbacher, I. Wernick, S. Tirelli, and K. Zier. (1995). B7.1 amplifies the response to interleukin-2-secreting tumor vaccines *in vivo,* but fails to induce a response by naive cells *in vitro. Hum. Gene Ther.* **6,** 1299–1306.

126. Bueler, H., and R. C. Mulligan. (1996). Induction of antigen-specific tumor immunity by genetic and cellular vaccines against MAGE: enhanced tumor protection by coexpression of granulocyte-macrophage colony-stimulating factor and B7-1. *Mol. Med.* **2,** 545–555.

127. Li, Y., K. E. Hellstrom, S. A. Newby, and L. Chen. (1996). Costimulation by CD48 and B7-1 induces immunity against poorly immunogenic tumors. *J. Exp. Med.* **183,** 639–644.

128. Liu, Y., B. Jones, W. Brady, and C. A. J. Janeway. (1992). Costimulation of murine CD4+ T cell growth: cooperation between B7 and heat-stable antigen. *Eur. J. Immunol.* **22,** 2855–2859.

129. Damle, N. K., K. Klussman, P. S. Linsley, A. Aruffo, and J. A. Ledbetter. (1992). Differential regulatory effects of intercellular adhesion molecule-1 on costimulation by the CD28 counter-receptor B7. *J. Immunol.* **149,** 2541–2548.

130. Damle, N. K., K. Klussman, G. Laytze, H. D. Ochs, A. Aruffo, P. S. Linsley, and J. A. Ledbetter. (1993). Costimulation via vascular cell adhesion molecule-1 induces in T cells increased responsiveness to the CD28 counter-receptor B7. *Cell. Immunol.* **148,** 144–156.

131. Kroesen, B.-J., A. Bakker, R. A. W. van Lier, H. T., and L. de Leij. (1995). Bispecific antibody-mediated target cell-specific costimulation of resting T cells via CD5 and CD28. *Cancer Res.* **55,** 4409–4415.

132. Parra, E., A. G. Wingren, G. Hedlund, T. Kalland, and M. Dohlsten. (1997). The role of B7-1 and LFA-3 in costimulation of CD8+ T cells. *J. Immunol.* **158,** 637–642.

133. DeBenedette, M. A., A. Shahinian, T. W. Mak, and T. H. Watts. (1997). Costimulation of CD28- T lymphocytes by 4-1BB ligand. *J. Immunol.* **158,** 551–559.

134. Shuford, W. W., K. Klussman, D. D. Tritchler, D. T. Loo, J. Chalupny, A. W. Siadak, T. J. Brown, J. Emswiler, H. Raecho, C. P. Larsen, T. C. Pearson, J. A. Ledbetter, A. Aruffo, and R. S. Mittler. (1997). 4-1BB costimulatory signals preferentially induce CD8+ T cell proliferation and lead to the amplification *in vivo* of cytotoxic T cell responses. *J. Exp. Med.* **186,** 47–55.

135. Brunschwig, E. B., E. Levine, U. Trefzer, and M. L. Tykocinski. (1995). Glycosylphosphatidylinositol-modified murine B7-1 and B7-2 retain costimulator function. *J Immunol.* **155,** 5498–5505.

136. McHugh, R. S., S. N. Ahmed, Y.-C. Wang, K. W. Sell, and P. Selvaraj. (1995). Construction, purification, and functional incorporation on tumor cells of glycolipid-anchored human B7-1 (CD80). *Proc. Natl. Acad. Sci. U.S.A.* **92,** 8059–8063.

137. Hollsberg, P., C. Scholz, D. E. Anderson, E. A. Greenfield, V. K. Kuchroo, G. J. Freeman, and D. A. Hafler. (1997). Expression of a hypoglycosylated form of CD86 (B7-2) on human T cells with altered binding properties to CD28 and CTLA-4. *J. Immunol.* **159,** 4799–4805.

138. Doty, R. T., and E. A. Clark. (1996). Subcellular localization of CD80 receptors is dependent on an intact cytoplasmic tail and is required for CD28-dependent T cell costimulation. *J. Immunol.* **157,** 3270–3279.

139. Weber, M. C., R. K. Groger, and M. L. Tykocinski. (1994). A glycosylphosphatidylinositol-anchored cytokine can function as an artificial cellular adhesin. *Exp. Cell. Res.* **145,** 1646–1652.

140. Sell, C., G. Dumenil, C. Deveaud, M. Miura, D. Coppola, T. DeAngelis, R. Rubin, A. Efstratiadis, and R. Baserga. (1994). Effect of a null mutation of the type 1 IGF receptor gene on growth and transformation of mouse embryo fibroblasts. *Mol. Cell. Biol.* **14,** 3604–3612.

141. Trojan, J., T. R. Johnson, S. D. Rudin, B. K. Blossey, K. M. Kelley, A. Shevelev, F. W. Abdul-Karim, D. D. Anthony, M. L. Tykocinski, J. Ilan, and J. Ilan. (1994). Gene therapy of murine teratocarcinoma: separate functions for IGF-I and IGF-II in immunogenicity and differentiation. *Proc. Natl. Acad. Sci. U.S.A.* **91,** 6088–6092.

142. Resnicoff, M., C. Sell, M. Rubini, D. Coppola, D. Ambrose, R. Baserga, and R. Rubin. (1994). Rat glioblastoma cells expressing an antisense RNA to the insulin-like growth factor-1 (IGF-I) receptor are nontumorigenic and induce regression of wild-type tumors. *Cancer Res.* **54,** 2218–2222.

143. Fakhrai, H., O. Dorigo, D. L. Shawler, H. Lin, D. Mercola, K. L. Black, I. Royston, and R. E. Sobol. (1996). Eradication of established intracranial rat gliomas by transforming growth factor beta antisense gene therapy. *Proc. Natl. Acad. Sci. U.S.A.* **93,** 2909–2914.

144. Lattime, E. C., M. J. Mastrangelo, O. Baserga, W. Li, and D. Berd. (1995). Expression of cytokine mRNA in human melanoma tissues. *Cancer Immunol. Immunother.* **41,** 151–156.

145. Ding, L., P. S. Linsley, L.-Y. Huang, R. N. Germain, and E. M. Shevach. (1993). IL-10 inhibits macrophage costimulatory activity by selectively inhibiting the up-regulation of B7 expression. *J. Immunol.* **151,** 1224–1234.

146. Caux, C., C. Massacrier, B. Vanbervliet, C. Barthelemy, Y. J. Liu, and J. Banchereau. (1994). Interleukin 10 inhibits T cell alloreaction induced by human dendritic cells. *Int. Immunol.* **6,** 1177–1185.

147. Salazar-Onfray, F., J. Charo, M. Petersson, S. Freland, G. Noffz, Z. Qin, T. Blankenstein, H.-G. Ljunggren, and R. Kiessling. (1997). Down-regulation of the expression and function of the transporter associated with antigen processing in murine cell lines expressing IL-10. *J. Immunol.* **159,** 3195–3202.

148. Steinbrink, K., M. Wolfl, H. Jonuleit, J. Knop, and A. H. Enk. (1997). Induction of tolerance by IL-10-treated dendritic cells. *J. Immunol.* **159,** 4772–4780.

149. Qin, Z., G. Noffz, M. Mohaupt, and T. Blankenstein. (1997). Interleukin-10 prevents dendritic cell accumulation and vaccina-

tion with granulocyte-macrophage colony-stimulating factor gene-modified tumor cells. *J. Immunol.* **159,** 770–776.

150. de Vries, T. J., A. Fourkour, T. Wobbes, G. Verkroost, D. J. Ruiter, and G. N. P. van Muijen. (1997). Heterogeneous expression of immunotherapy candidate proteins gp100, MART-1, and tyrosinase in human melanoma cell lines and in human melanocytic lesions. *Cancer Res.* **57,** 3223–3229.

151. Tanaka, Y., and S. S. Tevethia. (1988). *In vitro* selection of SV40 T antigen epitope loss variants by site-specific cytotoxic T lymphocyte clones. *J. Immunol.* **140,** 4348.

152. Robbins, P. F., M. el-Garnil, Y. F. Li, S. L. Topalian, L. Rivoltini, K. Sakaguchi, E. Appella, Y. Kawakami, and S. A. Rosenberg. (1995). Cloning of a new gene encoding an antigen recognized by melanoma-specific HLA-A24-restricted tumor-infiltrating lymphocytes. *J. Immunol.* **154,** 5944–5950.

153. Van Waes, C., P. A. Monach, J. L. Urban, R. D. Wortzel, and H. Schreiber. (1996). Immunodominance deters the response to other tumor antigens thereby favoring escape: prevention by vaccination with tumor variants selected with cloned cytolytic T cells in vitro. *Tissue Antigens* **47,** 399–407.

154. Apostolopoulos, V., V. Karanikas, J. S. Haurum, and I. F. C. McKenzie. (1997). Induction of HLA-A2-restricted CTLs to the mucin 1 human breast cancer antigen. *J. Immunol.* **159,** 5211–5218.

155. Speiser, D. E., R. Miranda, A. Zakarian, M. F. Bachmann, K. McKall-Faienza, B. Odermatt, D. Hanahan, R. M. Zinkernagel, and P. S. Ohashi. (1997). Self antigens expressed by solid tumors do not efficiently stimulate naive or activated T cells: implications for immunotherapy. *J. Exp. Med.* **186,** 645–653.

156. Fong, L., C. L. Ruegg, D. Brockstedt, E. G. Engleman, and R. Laus. (1997). Induction of tissue-specific autoimmune prostatitis with prostatic acid phosphatase immunization. Implications for immunotherapy of prostate cancer. *J. Immunol.* **159,** 3113–3117.

157. Elliott, B. E., D. A. Carlow, A. Rodricks, and A. Wade. (1989). Perspectives on the role of MHC antigens in normal and malignant cell development. *Adv. Cancer Res.* **53,** 181–245.

158. Topalian, S. L., L. Rivoltini, M. Mancini, *et al.* (1994). Human CD4+ T cells specifically recognize a shared melanoma-associated antigen encoded by the tyrosinase gene. *Proc. Natl. Acad. Sci. U.S.A.* **91,** 9461–9465.

159. Porgador, A., H. F. Staats, B. Faiola, E. Gilboa, and T. J. Palker. (1997). Intranasal immunization with CTL epitope peptides from HIV-1 or ovalbumin and the mucosal adjuvant cholera toxin induces peptide-specific CTLs and protection against tumor development *in vivo*. *J. Immunol.* **158,** 834–841.

160. Marchand, M., P. Weymants, E. Rankin, *et al.* (1995). Tumor regression responses in melanoma patients treated with a peptide encoded by gene MAGE-3. *Int. J. Cancer* **63,** 883–885.

161. Mayordomo, J. I., T. Zorina, W. J. Storkus, L. Zitvogel, M. D. Garcia-Prats, A. B. DeLeo, and M. T. Lotze. (1997). Bone marrow-derived dendritic cells serve as potent adjuvants for peptide-based antitumor vaccines. *Stem Cells* **15,** 94–103.

162. Aichele, P., K. Brduscha Riem, R. M. Zinkernagel, *et al.* (1995). T cell priming versus T cell tolerance induced by synthetic peptides. *J. Exp. Med.* **182,** 261–266.

163. Toes, R. E. M., R. J. J. Blom, R. Offringa, W. M. Kast, and C. J. M. Melief. (1996). Enhanced tumor outgrowth after peptide vaccination: functional deletion of tumor-specific CTL induced by peptide vaccination can lead to the inability to reject tumors. *J. Immunol.* **156,** 3911–3918.

164. Hsu, F. J., C. Benike, F. Fagnoni, D. Czerwinski, T. Liles, B. Taidi, E. Engleman, and R. Levy. (1996). A clinical trial of antigen-pulsed dendritic cells in the treatment of patients with B-cell lymphoma. *Proc. Am. Soc. Clin. Oncol.* **15,** A1288.

165. Stingl, G., E. B. Brocker, R. Mertelsmann, K. Wolff, S. Schreiber, E. Kampgen, A. Schneeberger, J. Trcka, U. Brennscheidt, H. Veelken, M. L. Birnstiel, K. Zatloukal, G. Maass, E. Wagner, M. Buschle, E. R. Kempe, H. A. Weber, and T. Voigt. (1997). Phase I study to the immunotherapy of metastatic malignant melanoma by a cancer vaccine consisting of autologous cancer cells transfected with the human IL-2 gene. *J. Mol. Med.* **75,** 297–299.

166. Lattime, E. C., S. S. Lee, L. C. Eisenlohr, and M. J. Mastrangelo. (1996). *In situ* cytokine gene transfection using vaccinia virus vectors. *Semin. Oncol.* **23,** 88–100.

167. Levraud, J.-P., M.-T. Duffour, L. Cordier, M. Perricaudet, H. Haddada, and P. Kourilsky. (1997). IL-2 gene delivery within an established murine tumor causes its regression without proliferation of preexisting anti-tumor-specific CTL. *J. Immunol.* **158,** 3335–3343.

168. Chen, L., D. Chen, E. Block, M. O'Donnell, D. W. Kufe, and S. K. Clinton. (1997). Eradication of murine bladder carcinoma by intratumor injection of a bicistronic adenoviral vector carrying cDNAs for the IL-12 heterodimer and its inhibition by the IL-12 p40 subunit homodimer. *J. Immunol.* **159,** 351–359.

169. Maraskovsky, E., K. Brasel, M. Teepe, E. R. Roux, S. D. Lyman, K. Shortman, and H. J. McKenna. (1996). Dramatic increase in the numbers of functionally mature dendritic cells in Flt3 ligand-treated mice: multiple dendritic cell subpopulations identified. *J. Exp. Med.* **184,** 1953–1962.

170. Condon, C., S. C. Watkins, C. M. Celluzzi, K. Thompson, and L. D. J. Falo. (1996). DNA-based immunization by *in vitro* transfection of dendritic cells. *Nat. Med.* **2,** 1122–1128.

171. Fujiwara, H., and T. Hamaoka. (1995). Regulatory mechanisms of anti-tumor T cell responses in the tumor-bearing mice. *Immunol. Rev.* **14,** 271–291.

172. Melero, I., W. W. Shuford, S. A. Newby, A. Aruffo, J. A. Ledbetter, K. E. Hellstrom, R. S. Mittler, and L. Chen. (1997). Monoclonal antibodies against the 4-1BB T-cell activation molecule eradicate established tumors. *Nature Med.* **3,** 1–4.

173. Lynch, D. H., A. Andreasen, E. Maraskovsky, J. Whitmore, R. E. Miller, J., and C. L. Schuh. (1997). Flt3 ligand induces tumor regression and antitumor immune responses *in vivo*. *Nat. Med.* **3,** 625–631.

174. Dufour, E., G. Carcelain, C. Gaudin, C. Flament, M.-F. Avril, and F. Faure. (1997). Diversity of the cytotoxic melanoma-specific immune response. Some CTL clones recognize autologous fresh tumor cells and not tumor cell lines. *J. Immunol.* **158,** 3787–3795.

CHAPTER 20

Cancer Gene Therapy by Direct Transfer of Plasmid DNA in Cationic Lipids

EVAN M. HERSH

Arizona Cancer Center, Tucson, Arizona 85724

I. INTRODUCTION

Several major approaches to cancer gene therapy are being investigated in preclinical models and in clinical systems. The first of these includes various approaches to gene-modified tumor cells as vaccines [1]. The tumor cells may be modified *in vitro* and then reinjected as a vaccine (the *ex vivo, in vivo* approach), or the genes in the appropriate vector may be transferred directly intratumorally *in vivo* or may be transferred *in vivo* by other methods that target the tumor. The objective of this approach is to introduce genes, such as cytokine genes, into the tumor cells and thus make them more immunogenic (see Chapter 8). A second, related approach is the use of DNA vaccines, whereby the genes for tumor antigens in the appropriate vector are injected, usually into striated muscle, where they are expressed and induce an antitumor immune response (see Chapter 17) [2].

Another approach to gene therapy is the administration of prodrug activating enzyme genes, again into the tumor, followed by systemic administration of the prodrug (see Chapter 10) [3]. The prodrug is then metabolized, releasing the active cytotoxic drug at a high concentration within the tumor. Other approaches include introduction of genes into the tumor that express antitumor toxins (see Chapter 25) [4], the use of gene-modified effector T cells (see Chapter 24) [5], whereby either their potential to secrete effector molecules or their binding affinities have been modified, introduction of wild-type tumor suppressor genes (see Chapter 14) [6], introduction of anti-oncogene molecules including antisense molecules (see Chapter 14) [7] or ribozymes (see Chapter 9) [8], and the use of hematopoietic stem cells, gene-modified to express the multidrug resistance gene for use in bone marrow transplantation protocols (see Chapters 11–13) [9].

Over the last 5 years there has been a steady increase in the number and diversity of clinical trials of gene therapy in cancer as well as other diseases. As of June 1997 approximately 1,500 patients with cancer had been entered onto gene therapy trials. The majority of these were patients with solid tumors [10]. Approximately one quarter of the patients were entered onto trials with genes delivered with cationic lipids. Approximately 150 protocols for cancer gene therapy had been registered and were reported in a recent compendium [11]. About 40% were in the category of cytokines/immunotherapy, and about one quarter of those involved the delivery of plasmid DNA complexed with cationic lipids. A few protocols also reported the delivery of tumor suppressor genes or anti-oncogene antisense molecules with cationic lipids. The majority of these studies were done using intratumoral injection of plasmid DNA/lipid complexes. All of these studies involved small numbers of patients and only patients with advanced disease refractory to conventional therapy. Thus, these clinical studies are to be considered pilot and phase I types of investigations.

A variety of vectors have been investigated for the delivery of gene therapy. The most extensive studies have been done in vitro and in animal model systems, and their utility in the clinic is yet to be proven. Many reviews have compared the advantages and disadvantages of these vectors. Retroviral vectors can only transduce dividing cells and require in vitro selection via antibiotic resistance to develop a population with a reasonable number of cells expressing the gene of interest [12]. Therefore, they are most useful for the in vitro generation of stably transduced populations of cells for subsequent in vivo administration. Similar constraints apply to adeno-associated virus vectors [13]. Adenovirus is a powerful vector that can transduce a variety of cells, both in vitro and in vivo [14]. One hundred percent of most solid tumor cells exposed at a multiplicity of infection (MOI) greater than 10 to 100 will be transduced and express the gene of interest. This expression is transient, usually lasting 1 to 2 weeks. A major disadvantage of the adenovirus vector is that it and the cells it transduces are highly immunogenic. Therefore, repeated administrations may be limited by the development of neutralizing antibody. A potentially detrimental inflammatory response may also be induced. Finally, adenovirus may be cytotoxic to the transduced cells, although mutant strains have been developed that do not induce cytotoxicity. A variety of other viral vectors as well as modified retroviruses and adenoviruses now under development may indeed overcome many of these limitations. (Further in-depth discussions of adenovirus strategies can be found in Chapters 15 and 16.)

Particle-mediated gene delivery with the gene gun is fairly efficient in vitro, with about a 20% transfection efficiency and with expression lasting about 1 to 2 weeks [15]. These in vitro-transfected cells may be transferred and express the gene in vivo. If the tumor can be exposed to the gene gun in situ, it may be transfected in vivo as well [16]. The effects are likely to be limited, however, because of the limited penetration of the particles into the tumor.

Gene delivery with cationic lipids is currently the most clinically developed approach to gene therapy. With cationic lipid and liposome delivery systems, efficiency of gene transfer is usually low [17]. However, genes can be transferred in vitro or directly in vivo. Gene expression has been documented to last up to several weeks or longer after direct, intratumoral injection [18]. The cationic lipid vector is nonimmunogenic and induces neither an inflammatory nor an autoimmune response [19]. Thus, repeated delivery over many months is possible. With genes encapsulated into liposomes, even intravenous delivery is possible, and liposomes targeting to specific tissues are being developed (see Chapter 7) [20]. Thus, at the moment the cationic lipid and liposomal delivery vectors are among the most promising for immediate clinical development.

This review will focus on cancer gene therapy with plasmid DNA administered in cationic lipid complexes. Both preclinical models and the current status of clinical trials will be discussed.

II. GENE-MODIFIED TUMOR VACCINES

As noted earlier, gene therapy for cancer may be used to deliver prodrugs, toxin genes, gene-modified effector cells, or genetic material that modifies tumor suppressor gene function or oncogene function. Gene therapy to date has primarily been the use of gene-modified tumor cells as vaccines. The tumor cells, gene-modified in vitro or in vivo, are engineered to be more immunogenic. Their tumor antigens may be upregulated by the introduction of the interferon gamma gene [21], as can their MHC class I expression. They may be engineered to express costimulatory molecules such as B-7.1, which are necessary to make them better antigen-presenting cells [22]. They may be transfected or transduced with a variety of cytokine genes of diverse function, which include granulocyte-macrophage colony-stimulating factor (GM-CSF) to call forth an antigen-presenting cell response [23], or with genes for cytokines, which will drive T-cell proliferative responses such as IL-2, IL-4, IL-7, IL-12, and others [24–26]. Finally, tumor cells may be engineered to down-regulate constitutively produced immunosuppressive molecules such as transforming growth factor (TGF-β) [27]. All of these approaches and others have been demonstrated to be effective in various animal model systems and have been reviewed extensively [28–32].

A potentially important advance in the use of gene-modified tumor cells as vaccines is the use of two genes concurrently. Investigators of this approach hypothesize that the simultaneous introduction of two genes may be more effective than introducing one gene. The two genes should stimulate different components of the antitumor immune response such as antigen presentation plus lymphocyte proliferation or MHC up-regulation plus lymphocyte proliferation. Combinations reported to show increased therapeutic activity in animal models include IL-4 plus IL-12 [33], B7 plus IL-12 (gene or protein) [34], and interferon gamma plus IL-2 [35]. In animals with pulmonary metastases of B16 melanoma established 3 days prior to treatment, weekly intraperitoneal injections of 2×10^6 inactivated tumor cells secreting both IL-2 and IL-6 had 60% survival at 90 days compared to 10 and 30% for each gene administered separately and 0% for untreated controls [36].

A major obstacle to gene-modified tumor cells as vaccines is the array of mechanisms by which tumors evade host control or antitumor host defenses fail, either generally or in a tumor-specific fashion. Many tumor cells are poor antigen-presenting cells. For instance, in melanoma half of metastatic tumors lose MHC class I expression [37]. In murine B16 melanoma, tumor-infiltrating lymphocytes (TILs) from class-I-negative tumors do not lyse tumor cells and have no therapeutic activity compared to those from class-I-positive tumors [38]. This suggests that up-regulation of class I by interferon gamma may be a therapeutic strategy to overcome this problem [39]. Solid tumor cells usually do not express the costimulatory molecule B7.1, which is necessary for effective antigen presentation. Transfection of B7.1 into murine melanoma results in increased immunogenicity, loss of tumorigenicity, and resistance to rechallenge with wild-type tumor [40].

IL-10, a negative regulatory cytokine [41], promotes Th2 cell activity and down-regulates the Th1 response necessary for the generation of cytotoxic T lymphocytes (CTLs). Melanoma cells produce increased amounts of IL-10 [42,43]. Melanoma patients also have elevated levels of circulating IL-10 [44]. Thus, the down-regulation of IL-10 production is a potential target for gene therapy. Many tumors also secrete TGF-β, which inhibits T-cell proliferation and the generation of lymphokine-activated killer (LAK) cells and CTLs [45]. Our group has shown that murine breast cancer cells

transduced with antisense to TGF-β have reduced TGF-β production, have reduced tumorigenicity, and induce resistance to subsequent wild-type tumor cell challenge [46]. Tumor cells also secrete SPARC (secreted protein acidic and rich in cysteine), which is a counteradhesive lipoprotein molecule that promotes matrix metalloproteinase action and angiogenesis [47]. Human melanoma cells transduced with antisense to SPARC lose their tumorigenicity [48].

Tumors may also evade host control through the Fas, Fas ligand system. This system mediates lymphocyte cytotoxicity. Activated lymphocytes expressing Fas ligand interact with Fas on target tumor cells inducing apoptosis [49]. Interestingly, tumor cells may also express Fas ligand, and its up-regulation appears to be increased by chemotherapy [50]. Fas-ligand-expressing human melanoma cells induce apoptosis in Fas-bearing lymphoma cells [51]. Fas ligand has been found on human melanoma, hepatoma, colon polyps, and colon cancer cells [52–56]. The assumption is that Fas-ligand-expressing tumor cells can induce apoptosis in attacking lymphocytes. These findings suggest that the Fas ligand system is an attractive target for gene therapy. Antibody, antisense, or ribozymes directed to inactivate Fas ligand might promote more effective CTL activity.

Cancer patients may also have generalized or tumor-specific immunodeficiency. For example, tumor-bearing mice and patients have a structural abnormality in signal-transducing zeta chains of CD3 and CD16 [57]. This is associated with a decreased lymphocyte cytokine response. Survival in melanoma patients correlates with the level of T-cell receptor (TCR) zeta chain [58]. Another possible mechanism in host failure in cancer patients is poor dendritic cell function. In breast cancer patients poor antigen presentation by dendritic cells can be overcome by culturing dendritic cell precursors in GM-CSF [59]. Several cytokines, including IL-2 and IL-12, can also be immunorestorative in cancer patients when administered systemically. Their coadministration with gene therapy should be considered as an approach to overcoming general immunodeficiency. For example, low-dose IL-2 administered daily over several months has been shown to be immunorestorative in severely immune-deficient AIDS patients [60].

III. GENE DELIVERY WITH CATIONIC LIPIDS

Gene delivery with cationic lipids has moved quickly from preclinical studies to clinical trials. The plasmids containing the desired expression cassettes can be manufactured easily and reproducibly and are stable. The cationic lipids are manufactured synthetically and are also stable. Plasmid DNA and the selected cationic lipids readily form stable complexes of about 100 nm in diameter. This means that a true pharmaceutical can be produced that would fulfill all of the usual regulatory requirements for a biological drug. Furthermore, these complexes may be administered *in vivo* by a variety of routes such as intravenous, intraarterial, intramuscular, intratumoral, or intracavitary, usually without any toxicity. Finally, these complexes are nonimmunogenic and therefore may be administered repeatedly without fear of allergic or anaphylactic reactions. There are also important limitations to this approach, which should be mentioned. For example, at present we have only a very limited ability to target genes of interest to specific tissues with cationic lipids. When given intravenously the complexes may be cleared rapidly by the reticuloendothelial system (RES). This situation may be remedied by liposomally encapsulated plasmids with targeting ligand in the lipid bilayer [61]. Because of our current inability to target specific tumors *in vivo,* the use of tissue-specific promoters within the expression cassette has been suggested [62]. Thus, the systemically administered DNA lipid complexes would be distributed to many tissues but activated only in a specific tissue. There needs to be a great deal more emphasis on the pharmacokinetics of systemically administered gene therapy, which will be critical for the clinical development of this field.

An issue in the use of cationic lipid DNA complexes to transfer genes in order to produce gene-modified tumor cells for vaccines is the specificity of the induced biological activity. Some studies have indicated that cationic lipids themselves, without DNA, can have an immunostimulatory effect. Also, there appears to be some activity of irrelevant DNA as an immunostimulant. Thus, both lipid alone and nonspecific DNA might augment tumor cell immunogenicity or have a nonspecific adjuvant effect. Specific effects might then be attributed to the specific gene, which is actually acting nonspecifically.

Another issue is the whether the use of cationic lipids actually promotes gene transfer and expression. Most studies do indeed show that the cationic lipids are necessary, but gene transfer with naked plasmid DNA has also been effective. One study showed that both lipofectamine and DC cholesterol/1,2-dioleoyl-sn-glycero-3-phosphoethanolamine (DOPE) may actually inhibit gene expression after intratumoral injection of the complexes [63]. Our group has generated data suggesting that free plasmid DNA may be more effective in gene transfer than the same complexed with $(\pm)$-*N*-(2-hydroxyethyl)-*N,N*-dimethyl-2,3-bis(tetradecyloxy)-1-propanaminium bromide (DMRIE)/DOPE for the intratumoral transfer and expression of the IL-2 gene in

UM449 human melanoma implanted in the SCID mouse [64].

A variety of lipids have been used to deliver plasmid DNA by a variety of routes. Generally, the cationic lipid is coadministered with DOPE, a neutral lipid. They are combined by mixing or vortexing with DNA over a range of DNA-to-lipid ratios. Higher lipid concentrations improve transfection but may lead to precipitation of the suspension. Recent modifications have improved complex stability and the potential for intravenous administration. These include condensing the DNA with polyamines [65] and stabilizing the liposome membrane with phospholipid/peg complexes [66]. This reduces RES clearance and prolongs circulating $t_{1/2}$, permitting more time for intratumoral penetration. To achieve the full potential of these complexes, it may be necessary to target such liposomes via monoclonal antibody or other ligands to tumor cell surface antigens [67]. Otherwise, they may still localize predominantly in the liver and the lung.

A broad spectrum of genes have been delivered *in vivo* with cationic lipids in animal models. These include the marker genes, β-galactosidase (lacZ) and chloramphenicol amino-transferase (CAT), cytokine genes such as IL-2, allogenic MHC molecules, and immunostimulatory molecules such as HSP65 and the prodrug-activating enzyme herpes simplex virus thymidine kinase (HSV-TK) gene. Felgner and co-workers first described the delivery of plasmid DNA in lipids or liposomes in the early 1980s [68]. Since then, investigators have focused on various lipid preparations, their stability and transfection efficiency, gene delivery by different routes and to different sites, the pharmacokinetics of locally and intravenously delivered DNA lipid complexes, and issues of safety, toxicity, and efficacy. These studies have shown that plasmid DNA delivered with cationic lipids *in vivo* is practical, moderately efficient, easily adapted to various clinical circumstances, and of limited toxicity. Limited biological activity and efficacy have also been demonstrated.

The study of the pharmacokinetics of cationic lipid DNA complexes administered either locally or systemically will be very important for clinical trials. Several studies have compared the pharmacokinetics of naked DNA to DNA lipid complexes. Intravenously administered naked plasmid DNA degrades rapidly and is inefficient for gene transfer [69]. Intravenous administration of marker genes in lipid complexes often leads to high levels of gene expression in the lung and liver, particularly after repeated dosing [70]. Uptake in the lung is mainly in the vascular endothelial cells, whereas in the liver it is mainly in Kupfer's cells. Repeated i.v. administration of the genes for GM-CSF or G-CSF in lipid complexes resulted in effective circulating levels of the growth factors sufficient to induce leukocytosis [71].

When the *CAT* gene in DOTMA/DOPE (Lipofectin, Gibco) was given intravenously, there was effective uptake and retention in B16 melanoma nodules [72]. When the *HLA-B7* gene was administered with DMRIE/DOPE intravenously the $t_{1/2}$ was 5 minutes and the DNA was found in multiple organs at 28 days and in muscle only at 6 months. There was no toxicity. The DNA was detected at these later times, but the gene product was not [73].

Recently, a multicompartment model for studying gene transfer has been developed. It examines the kinetics of the gene, mRNA, and protein product, the fate of the DNA, and the rate of DNA uptake, transcription, translation, and posttranslational processing [74]. Each of six compartments had a finite $t_{1/2}$ whereby first-order kinetics could result from a summation of each step.

Other routes of administration have also been investigated. Thus, both *lacZ* and *HLA-B7* have been instilled intra-arterially. Uptake was rapid, and gene expression persisted in all layers of the blood vessel for 6 weeks [75]. Local installation by lung aerosolization has also resulted in high levels of gene expression in the proximal tracheobronchial tree, associated, however, with a substantial inflammatory response [76]. All of the reported therapeutic studies in animal models have used the local or regionally delivery route. We feel that intravenous delivery awaits the further development and refinement of stable DNA lipid complexes, targeted delivery through monoclonal antibodies or other ligands, inclusion of tissue-specific promoters, or induction of local gene activation via such mechanisms as local hyperthermia.

IV. *IN VITRO* STUDIES USING HUMAN CELLS

Cationic lipid plasmid DNA complexes are fairly efficient in delivering genes into human cells and expressing the gene product *in vitro*. These have given results comparable to those obtained in particle-mediated transfer with the gene gun and with electroporation but are not as efficient as adenovirus or retrovirus vectors. We have studied delivery of plasmid DNA containing the IL-2 gene under the control of the CMV promoter via the cationic lipid complex DMRIE/DOPE in fresh tumor cells and tumor cell lines [77]. We observed 10 to 100 times more IL-2 production (1,000 to 10,000 IU/10^6 cells/24 h compared to fresh tumor cells. IL-2 expression persisted for up to 4 weeks in culture. The IL-2 produced was active as measured by its ability to stimulate lymphocyte proliferation and generate LAK cells. Cationic lipids and other polycationic substances (polybrene and protamine) also promote more efficient ade-

novirus transduction of human tumor cells *in vitro* [78]. β-galactosidase expression was increased to 2- to 10-fold in UM449 melanoma cells by the concurrent administration of adenovirus and cationic lipid.

V. THERAPEUTIC STUDIES IN ANIMAL MODELS

Gene therapy of cancer in animal models using plasmid DNA lipid complexes was first described by Plautz and co-workers in 1992 [79]. H_2K^d mice bearing the syngeneic CT26 colon tumor or the syngeneic MCA106 sarcoma were given a gene for the allogeneic H_2K^s class I MHC antigen intratumorally after preimmunization with the allogeneic MHC protein. Slowing of tumor growth and actual tumor regression were noted. Repeated doses were better than a single dose. The treatment was ineffective in nude mice, suggesting that the effect was immunologic. Tumor-specific CTLs were generated. The investigators hypothesized that when the tumor expressed the allogeneic MHC antigen, a cytokine cascade was induced, which triggered the development of effective specific antitumor immunity. This study provided the preclinical basis for the initial clinical trials subsequently conducted by Nabel and co-workers.

In another study Parker and co-workers administered the IL-2 gene in DMRIE/DOPE intratumorally in B16 melanoma [80]. Tumor regression was induced not only by the IL-2 plasmid cationic lipid complex, but also by the cationic lipid alone and the cationic lipid complexed to an irrelevant DNA. However, when the B16 melanoma cells were transfected *in vitro* with cationic lipid and then administered *in vivo*, only the specific IL-2-transfected cells reduced tumor growth and prolonged survival. This occurred in spite of the fact that only about 10% of the cells were transfected. In addition, when the *in vitro*–transfected cells were given intravenously they yielded far fewer lung colonies than nontransfected cells. Parker and co-workers have also studied IL-2 plasmid DNA lipid complexes administered intratumorally in the Renca murine renal cell carcinoma model [81]. Intratumoral injection of established tumors resulted in complete regression in 80% of the animals, which were then resistant to subsequent challenge with wild-type tumor. Spleen cells were capable of transferring the resistance, and cytotoxic T cells were also demonstrated *in vitro*. There was an IL-2 dose-response curve, and higher doses of DNA were more effective than lower doses, as was a more frequent dosing schedule.

IL-2 plasmid DNA lipid complexes injected intratumorally in 0.4-cm-diameter nodules of human squamous cell lung cancer growing in SCID mice have also been studied. Complete regression of tumor was noted, which was reversed by anti-asialo-GM1. This indicated that the mechanism of action was natural killer (NK) cell activation [82]. This study has important implications for gene therapy in the face of the general immune deficiency seen in cancer patients. Another gene studied in this context is HSV-TK, the prodrug-activating enzyme gene for ganciclovir. Intraperitoneal injection of HSV-TK plasmid DNA in cationic lipids in nude mice bearing the human pancreatic cell lines TSN-1 [83] or human ovarian cancer cells [84], followed by treatment with ganciclovir at 50 mg/kg every 12 hours for 8 days, showed major benefit. About half the animals had reduced mortality and prolonged survival compared to none of the controls. In a mouse meningiomatosis model HSV-TK plasmid DNA in an HVJ liposome (Sendai virus coat-protein plus pspc cholesterol) delivered intrathecally 2 days after tumor inoculation led to 80% versus 0% survival in the treated versus the control animals at 30 days [85]. Additional HSV-TK studies are discussed in Chapter 10. In another model human glioma was inoculated intracerebrally in nude mice followed 7 days later by intracerebral DNA expressing the IFN-β gene in cationic lipid [86]. At a dose of 0.6 μg of DNA there was a 90% versus 0% survival in the treated versus the control animals at 80 days.

Transfer of immune activating genes also seems to be quite effective. For example, intratumoral *Mycobacterium leprae* HSP-65 DNA in cationic lipid has been effective against the J774 reticulum cell sarcoma inoculated intraperitoneally into normal syngeneic or SCID mice [87]. After tumor inoculation, mice received treatment twice a week for 2 weeks, resulting in a tumor weight reduction from 5.5 to 0.8 g in normal mice and 6.5 to 2 g in SCID mice, even when treatment was initiated as late as 14 days after tumor cell inoculation. Naked DNA was not effective. The normal but not the SCID mice were resistant to subsequent challenge with wild-type tumor, suggesting the generation of immunologic memory. This same group has recently extended these observations, showing very similar results in mice bearing a syngeneic mesothelioma inoculated and treated intraperitoneally [88].

Some studies have showed a nonspecific effect. Thus, human ovarian cancer growing intraperitoneally in SCID mice was treated with nonspecific DNA in DC cholesterol/DOPE [89]. Treatment with 50 μg of DNA three times a week intraperitoneally resulted in an increased survival, from 40 days in the controls to 70 days in the treated animals. The mechanism probably relates to an inflammatory reaction and activation of NK cells and macrophages.

In summary, local or regional therapy with plasmid DNA in cationic lipids has shown major antitumor effects in murine studies. There were effects both in immu-

nologically intact mice and in immunodeficient SCID mice. In general, the effects related to a specific immunologically active plasmid DNA such as that expressing the IL-2 gene. Naked DNA and irrelevant DNA generally were not effective. High doses of DNA and more intense schedules of treatment generally were more effective. These animal studies have formed a reasonable basis for the initiation of clinical trials of gene transfer with cationic lipids in humans.

VI. CLINICAL TRIALS

Gene therapy of cancer is being studied in variety of clinical trials throughout the developed world. As noted earlier, approximately 1,500 cancer patients have been treated with gene therapy, of whom about 1,300 had solid tumors [10], and about 25% of these were treated under protocols involving plasmid DNA complexed with cationic lipids. Of the clinical trials for cancer gene therapy [11], 144 protocols included 64 involving cytokines/immunotherapy. Of these 17 utilized the delivery of plasmid DNA with cationic lipids, of which 11 utilized the HLA-B7 gene and 6 the IL-2 gene. A variety of tumor types are under treatment including breast, colon, lung, ovarian, prostate, and renal cell carcinoma and malignant melanoma. Six of these protocols utilize the *ex vivo/in vivo* approach; 11 utilize direct intratumoral transfer of the DNA lipid complex. Three also involve the subsequent collection of TILs from the treated tumor nodule, their *in vitro* expansion, and their subsequent reinfusion as TIL cell therapy along with systemic IL-2. Of 15 protocols involving the transfer of tumor suppressor genes or anti-oncogene, antisense molecules, two involve transfer with cationic lipids, one antisense to IGF-1 for the treatment of brain tumors and the other the E-1A gene for the treatment of breast cancer. The majority of these clinical trials are yet to be reported in detail.

Gary Nabel and co-workers at the University of Michigan were the initial pioneers of clinical trials using direct gene transfer to tumors using cationic lipids [90]. Their work was based on their previously reported animal model studies described earlier [79]. They used plasmid DNA containing the *HLA-B7* gene under the control of the RSV promoter mixed with DMRIE/DOPE in 5 HLA-B7$-$ patients with metastatic malignant melanoma. Tumor nodules were injected with 2 μg of DNA weekly for three treatments. There was no major toxicity. Transferred DNA, mRNA, and expressed protein were detected in all five patients. Two patients developed CTLs to their autologous tumor. The patients also developed CTLs to HLA-B7. One patient developed a partial remission, including regression of noninjected pulmonary masses. One of these patients underwent catheter delivery of *HLA-B7* DNA in cationic lipids into a pulmonary artery feeding a lung tumor nodule [91]. DNA was not detected in the peripheral circulation, and there was no toxicity. The investigators were unable to document gene transfer because of the patient's general medical condition.

Based on these observations the American gene therapy company Vical, Inc., sponsored three phase I studies of HLA-B7/β_2-microglobulin (β_2m) in DMRIE/DOPE as intratumoral gene therapy for patients with advanced malignant disease. The HLA-B7/β_2m plasmid (Allovectin 7) is shown in Fig. 1 and the lipid DMRIE/DOPE in Fig. 2. The three protocols were identical in design. Patients with malignant melanoma were studied at the Arizona Cancer Center [92]. Groups of 3 patients received a single dose of 10, 50, or 250 μg of DNA once intratumorally or 10 μg twice or three times, all into a single tumor nodule. A retrocaval mass undergoing injection is shown in Fig. 3. There were 14 patients evaluable for response. The maximally tolerated dose was not reached. The only toxicity observed related to the mechanics of needle placement or of posttreatment needle biopsy of the treated tumor. Five patients had local hemorrhage, one of whom required overnight hospitalization for a subcapsular hepatic hemorrhage associated with a drop in hemoglobin. Two patients experienced minor pneumothorax not requiring hospitalization, and 4 patients had local pain. DNA was detected in 64%, mRNA in 29%, HLA-B7 protein in 79% and HLA-B7 CTLs in 21% after treatment. Seven of the 14 patients showed at least 25% regression of the injected tumor nodule; none had major regression of an uninjected nodule. The response of a pulmonary nodule is shown in Fig. 4. There was no correlation between the DNA dose or schedule and response. This suggested that the lowest dose was both safe and as effective as

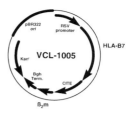

FIGURE 2   Map of the Allovectin-7® plasmid containing the MHC class-1 *HLA-B7* gene and the β_2-microglobulin gene.

higher-dose, more frequent administrations. However, one could argue that the number of patients was too small and the dose range of DNA too narrow to draw this conclusion firmly. One of these patients achieved complete remission after several more doses of 10 μg of DNA intratumorally [93]. This patient's only site of disease was a 5-cm retrocaval mass. She has remained in remission for 3½ years, but has recently developed acute myelomonocytic leukemia. The median survival of the 14 evaluable patients was 8.1 months. Three of the original 14 remain alive at 3 years, one in complete remission, one in a stable partial remission after subsequent radiotherapy and chemotherapy to a lung mass, and one with progressive disease.

At the Mayo clinic 15 HLA-B7− patients with hepatic metastases of colon cancer were treated with the same protocol [94]. In posttreatment biopsies, transferred DNA was detected in 93%, specific mRNA in 33%, HLA B7 protein in 63%, and an infiltrate of CD8+ T cells in 100%. CTLs to HLA-B7 were detected in 53%. There was no major toxicity, but there were no responses, and only 6 patients manifested stable disease after treatment. The investigators speculated that the lack of response related to the large tumor burden and the small fraction of tumor transfected by the procedure. At the University of Chicago, 14 HLA-B7− patients with renal cell carcinoma were treated under the same protocol [95]. Gene transfer was detected in 78% and T-cell infiltration of the tumor in posttreatment biopsies or an immune response to HLA-B7 in 93%. Gene transfer was well tolerated. There were no responses, and only 6 patients showed stable disease after treatment. Again, the lack of response was attributed to the large tumor burden. These three phase I studies were summarized in a subsequent brief report [96]. As noted earlier, responses were only seen in patients with melanoma. Of interest, 98% of the patients in the studies had some evidence of gene uptake and expression and there was no major toxicity except for the mechanical effects of intratumoral injection or needle biopsy.

Several other preliminary studies have been done in malignant melanoma. Nabel and his group treated an additional 10 HLA-B7− patients [97]. Gene transfer

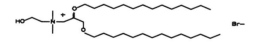

DMRIE: (±)-*N*-(2-hydroxyethyl)-*N*,*N*-dimethyl-2,3-bis(tetradecyloxy)-1-propanaminium bromide (or DiMyristoyl Rosenthal Inhibitor Ether)

DOPE: 1,2-dioleoyl-sn-glycero-3-phosphoethanolamine

FIGURE 1   Structure of DMRIE and DOPE, the cationic and neutral lipids utilized in forming the cationic lipid complexes with Allovectin and Leuvectin.

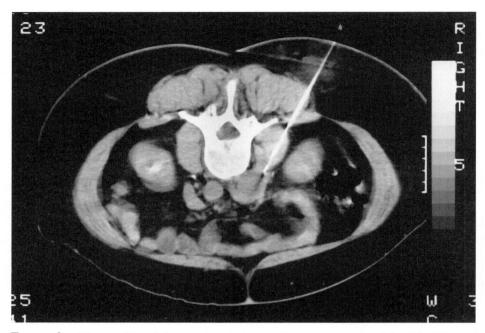

FIGURE 3   CT scan of the abdomen taken during the direct gene transfer of the *HLA-B7* gene. The needle and fluid track of the therapeutic agent can be seen inserted into a retrocaval mass. The patient eventually achieved a complete remission.

and expression were detected in all patients. Six of seven posttreatment biopsies showed TILs, and CTLs to autologous tumor were detected among TILs but not peripheral blood lymphocytes in 2 patients. Of great interest, this group demonstrated a change in the TCR Vβ usage of TILs after intratumoral *HLA-B7* treatment [98]. This was seen in injected and uninjected tumor nodules. After treatment there was an increased diversity of usage and a decline in the previously observed predominance of Vβ13. This suggested generalized immunologic activation. Two of these 10 patients showed regression of the injected nodule, and one showed regression of noninjected nodules constituting a partial remission. TILs were obtained from this patient's injected tumor nodule, expanded *in vitro* and used as TIL therapy. The patient then entered a complete remission.

Seven patients with metastatic melanoma, predominately skin and nodal disease, were treated at the British Columbia Cancer Agency [99]. These investigators speculated that HLA-B7+ patients might also respond to the treatment because their tumors may have downregulated MHC expression and its up- regulation by gene transfer might cause the tumor to become immunogenic. Three of the patients were HLA-B7+, and four were HLA-B7−. Patients received 6 intratumoral injections of 5 or 10 μg of *HLA-B7* DNA in DMRIE/DOPE over 8 weeks. Only local toxicity was observed. There were 3 responses, 2 in patients who were HLA-B7+.

The data on these four studies have been compiled and reviewed [100]. Overall, local responses were seen in 36% of 36 patients; distant responses of uninjected nodules were seen in 19%. The *HLA-B7* gene was de-

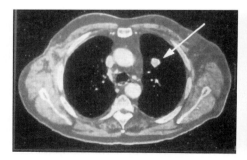
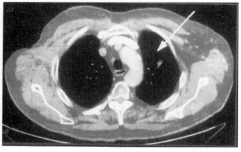

FIGURE 4   Before and after scans of a patient treated with Allovectin-7® for a lung metastasis of malignant melanoma. Complete disappearance of the lesion is evident.

tected in 64%, protein in 79%, and increased T-cell infiltration in 100% of the study patients. These preliminary trials are summarized in Table 1.

After these encouraging results, a phase II study was done in HLA-B7− patients treated with 10 μg of DNA intratumorally into a single tumor nodule on days 1, 14, 42, and 56 [101]. Thirty-eight patients were treated. Thirteen progressed rapidly and were inevaluable because they did not complete the four intratumoral injections. Among the 25 evaluable patients there was no grade 3 or 4 toxicity, and again the only toxicity was mechanical toxicity due to needle placement. Six patients had greater than 50% regression of the injected tumor nodule and 1 had 30% regression of an injected nodule. Two patients had regression of noninjected nodules constituting a partial remission. Of the 11 patients with cutaneous or nodal disease, 5 showed regression greater than 50%, whereas only 1 of the 14 with visceral metastases showed regression of the injected nodule greater than 50%. This suggested that tumor burden may have a strong influence on response, with better responses being seen in patients with a limited tumor burden. This observation may form the basis of future studies.

Heo and co-workers treated a total of 9 HLA-B7− patients with various malignancies with HLA-B7/β_2m in lipofectamine at 10, 20, or 50 μg of DNA every 2 weeks [102]. One hundred percent of patients showed DNA and mRNA in posttreatment biopsies, and 60% showed HLA-B7 protein. There was no toxicity. One partial remission and one mixed response were reported in 2 patients with non-small-cell lung cancer.

Gleich and co-workers treated 9 HLA-B7− patients with squamous cell carcinoma of the head and neck with

direct intratumoral injection of 10 μg of HLA-B7/β_2m DNA in DMRIE/DOPE [103]. One cycle of therapy was 2 doses at 14-day intervals. Four of 9 patients began to respond in the first cycle and proceeded to receive the second cycle; the other 5 progressed during the first cycle and were taken off the study. There was no significant toxicity. Of 4 responders, one remains completely free of disease and alive at 17 months. Of interest, apoptosis was observed in the tumors of the patients with regressing disease.

Nineteen patients were treated with intratumoral injections of plasmid DNA expressing HLA-A2, HLA-B13, or a murine H$_2$K gene in DC cholesterol/DOPE by Hui and co-workers [104]. Cutaneous nodules of advanced cancer were treated with 4 weekly injections. Two complete and 4 partial regressions of 8 nodules injected with HLA-A2 were observed in patients with cervical and ovarian cancer.

We have also initiated studies of the delivery of plasmid DNA containing the IL-2 gene under the control of the CMV promoter complexed DMRIE/DOPE intratumorally in patients with various malignancies. The structure of the IL-2 plasmid (Leuvectin) is shown in Fig. 5. Twenty-three patients were treated intratumorally for 6 weeks with 10, 30, 100, or 300 μg of DNA [105]. Patients had renal cell or colon cancer, melanoma, and sarcoma. Sites injected included subcutaneous nodules, lymph nodes, lung, liver, pancreas, and retroperitoneal tumor masses. Toxicities involved local symptomatology, but with this gene also mild flulike symptoms, mild myalgias, and minimal arthralgias. We attributed the latter to the systemic release of small amounts of IL-2, although IL-2 was not detected in the circulation.

TABLE 1 Clinical Trials of Gene Therapy for Cancer Using Plasmid DNA in Cationic Lipids

First author	Genes	Lipid	Number of patients	Diagnoses	Responses (local or general)	Reference
Nabel	HLA-B7	DMRIE/DOPE	5	Melanoma	1	91
Stopeck	HLA-B7/β2M	DMRIE/DOPE	14	Melanoma	7	92
Rubin	HLA-B7/β2M	DMRIE/DOPE	15	Colon cancer	0	94
Vogelzang	HLA-B7/β2M	DMRIE/DOPE	14	Renal cancer	0	95
Nabel	HLA-B7/β2M	DMRIE/DOPE	10	Melanoma	2	97
Silver	HLA-B7/β2M	DMRIE/DOPE	7	Melanoma	3	99
Hersh	HLA-B7/β2M	DMRIE/DOPE	25	Melanoma	7	101
Heo	HLA-B7/β2M	Lipofectamine	9	Non-small-cell lung cancer	2	102
Gleich	HLA-B7/β2M	DMRIE/DOPE	9	Head and neck cancer	4	103
Hui	HLA-A2	DC chol/DOPE	19	Cervical, ovarian cancer	6	104
Stopeck	IL-2	DMRIE/DOPE	23	Various	5	105
Rubin	IL-2	DMRIE/DOPE	25	Various	(2)[a]	106

[a] Study still undergoing definitive analysis.

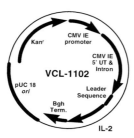

FIGURE 5   Map of the Leuvectin® plasmid containing the IL-2 gene under the control of the CMV promoter.

DNA was detected in 75%, IL-2 by immunohistochemistry in 60%, and CD8+ T-cell infiltration by immunohistochemistry in 50% of the patients' posttreatment biopsies. Tumor regression ranging from 31 to 71% of the injected tumor nodules was seen in 2 melanoma patients and one patient each with pancreatic cancer, sarcoma, and basal cell carcinoma. There was no regression on noninjected nodules. More recently we have extended the phase I study to up to 1,500 μg of DNA [106]. Tolerance was good, with minimal flulike symptoms. An additional 3 partial remissions of distant, noninjected metastases were seen in 25 patients, including 2 renal cell carcinomas and one melanoma.

VII. DISCUSSION

These preclinical and clinical studies of gene therapy of cancer with plasmid DNA complexed with cationic lipids have established several facts. Gene transfer *in vivo* using this methodology is safe, nontoxic, nonimmunogenic and nonallergenic, and can be give repeatedly. Toxicity has been limited for the most part to the mechanical or traumatic effects of needle insertion into tumors or tissue or to the posttreatment biopsies required by the experimental protocols. This doesn't mean that if the gene encodes a potentially toxic substance there will be no toxicity. This is actually hinted at by our observation of mild systemic symptoms after administration of the IL-2 gene intratumorally. In addition, in these early clinical studies there has been clear-cut evidence of gene transfer, persistence of the gene, gene expression, and biological activity, such as T-cell infiltration, immune responses to the transferred gene, development of specific antitumor CTLs, and development of apoptosis in the tumor. A small fraction of the patients have shown partial or complete remission; a larger fraction have shown measurable regression only of the injected tumor nodules. These responses are interesting because only 1 to 10% of the cells in the injected tumor are transfected. This suggests that the antitumor effect

is indeed due to the activation of host defense mechanisms.

These safety and biological activity data are encouraging, but they are only the first steps along the pathway to a truly effective cancer gene therapy using plasmid DNA in cationic lipids. Intratumoral injections are very time-consuming, labor-intensive, and costly, particularly when utilizing computed tomography (CT) or ultrasound guidance. In addition, therapy must be administered intratumorally by a physician, often an interventive radiologist. The efficiency of gene transfer is very low, and a very small percentage of tumor cells are transfected with the genes. We speculate that if a larger fraction of the tumor could be transfected, a larger fraction of the tumor would become immunogenic and a greater response would be observed. This might be achieved through improved expression vectors, the use of tissue-specific promoters, or local activation of transferred genes by such mechanisms as hyperthermia. Targeted gene delivery via the intravenous route would also be a step forward if it could be developed technically. The limited access to tumor sites that can be injected is another major obstacle to the potential effectiveness of this approach. Also, tumor nodules under 1 cm in diameter cannot be injected, and if the tumor is over 5 cm in diameter it is unlikely that sufficient DNA can be delivered to change the intratumoral hosttumor interactions.

In terms of the reported clinical studies there has been a lack of optimization of dose, schedule, and duration of treatment for the intratumoral route. We have not explored the number of injections per nodule, the volume to be delivered per injection, the concurrent injection of multiple nodules, and the potential for the use of two genes injected into the same or different nodules. Such studies will be clinically difficult because of the general lack of toxicity of the intratumoral injections. Consequently, biological and surrogate end points must be developed in order to decide upon the relative efficacy of these dose and schedule response studies.

Other factors that have not been addressed are the immunosuppressive factors produced by the tumor and host defense failure in the advanced cancer patients. Some of the approaches that should be considered would be abrogation of tumor produced factors that lead to evasion of host control by the tumor (such as TGF-β) and systemic immunomodulation given concurrently with these gene therapy approaches.

In this light, our next studies include one study of HLA B7 in patients with limited metastatic disease that is mainly cutaneous or nodal in location. Patients with limited disease should be more immunocompetent and therefore mount a more vigorous antitumor response. Another study will utilize a combination of gene therapy interspersed between courses of cytoreductive chemotherapy. Based on our prior observations, a reduced tumor burden should lead to better responses. Combination of the two genes *HLA-B7* and IL-2 will be explored based on the animal studies showing better efficacy of two-gene therapy. Finally, combination of gene therapy with systemic immunomodulation will be studied. Our first approach in this regard is *HLA-B7* plus systemic low-dose IL-2. Hopefully, these manipulations and modifications will lead to more effective human cancer gene therapy.

References

1. Lotze, M. T. (1996) Cytokine gene therapy of cancer. *Cancer J. Sci. Am.* **2,** 63–72.
2. McDonnell, W. M., Askari, F. K. (1996) Molecular medicine (DNA vaccines). *N. Engl. J. Med.* **334,** 40–45.
3. Freeman, S. M., Whartenby, K. A., Freeman, J. L., *et al.* (1996) *In situ* use of suicide genes for cancer therapy. *Semin. Oncol.* **23,** 31–45.
4. Ying W, Marineau D, Beitz J, Lappi DA, Baird A. (1994) Anti-B-16-F10 melanoma activity of a basic fibroblast growth factor-saporin mitotoxin. *Cancer* **74, 848,** 853.
5. Hwu, P., Rosenberg, S. A. (1994) The genetic modification of T cells for cancer therapy: an overview of laboratory and clinical trials. *Cancer Detect. Prev.* **18,** 43–50.
6. Beaudry, G. A., Bertelsen, A. H., Sherman, M. I. (1996) Therapeutic targeting of the p53 tumor suppressor gene. *Curr. Opin. Biotechnol.* **7,** 592–600.
7. Tonkinson, J. L., Stein, C. A. (1996) Antisense oligodeoxynucleotides as clinical therapeutic agents. *Cancer Invest.* **14,** 54–65.
8. Sullivan, S. M. (1994) Development of ribozymes for gene therapy. *J. Invest. Dermatol.* **103,** 85S–89S.
9. Licht, T., Pastan, I., Gottesman, M. M., *et al.* (1996) The multidrug resistance gene in gene therapy of cancer and hematopoietic disorders. *Ann. Hematol.* **72,** 184–93.
10. Marcel, T., Grausz, J. D. (1997) The TMC Worldwide Gene Therapy Enrollment Report, End 1996. *Hum. Gene Ther.* **8,** 775–800.
11. Anonymous. Section III: gene therapy clinical trials worldwide (1995) http/www.appleton-lange.com/genetherapy/trials. In Sobol RE, Scanlon KJ (eds), *The Internet Book of Gene Therapy/ Cancer Therapeutics.* Appleton and Lange, New Haven, Connecticut, pp. 283–296.
12. Uckert, W., Walther, W. (1994) Retrovirus-mediated gene transfer in cancer therapy. *Pharmacol. Ther.* **63,** 323–347.
13. Kotin, R. M. (1994) Review: prospects for the use of adeno-associated virus as a vector for human gene therapy. *Hum. Gene Ther.* **5,** 793–801.
14. Shaughnessy, E., Lu, D., Chatterjee, S., *et al.* (1996) Parvoviral vectors for gene therapy of cancer. *Semin. Oncol.* **23,** 159–171.
15. Mahvi, D. M., Burkholder, J. K., Turner, J., *et al.* (1996) Particle-mediated gene transfer of granulocyte-macrophage colony-stimulating factor CDNA to tumor cells: implications for a clinically relevant tumor vaccine. *Human Gene Therapy* **7**(13), 1535–1543.
16. Yang, N. S., Sun, W. H. (1995) Gene gun and other non-viral approaches for cancer gene therapy. *Nat. Med.* **1,** 481–483.
17. Lasic, D. P. (1997) Mechanisms of transfection. In *Liposomes in Gene Delivery.* CRC Press, New York, pp. 189–198.
18. Puyal, C., Milhaud, P., Bienvenue, A., Philipott, J. R. (1995) A new cationic liposome encapsulating genetic material, a potential delivery system for polynucleotides. *Eur. J. Biochem.* **228,** 697–703.
19. Lee, E. R., Marshall, J., Siegel, C. S., Jiang, C., Yew, N. S. (1996) Detailed analysis of structures and formulations of cationic lipids for efficient gene transfer to the lung. *Hum. Gene Ther.* **7,** 1701–1717.
20. Cheng, P. W. (1996) Receptor ligand-facilitated gene transfer: enhancement of liposome-mediated gene transfer and expression by transferrin. *Hum. Gene Ther.* **7,** 275–282.
21. Warner, S. J., Friedman, G. B., Libby, P. (1989) Regulation of major histocompatability gene expression in human vascular smooth muscle cells. *Atherosclerosis* **9,** 279–288.
22. Townsend, S. E., Allison, J. P. (1993) Tumor rejection after direct costimulation of CD8+ T cells by B7-transfected melanoma cells. *Science* **259,** 368–370.
23. Dranoff, G., Jaffee, E., Lazenby, A., *et al.* (1993) Vaccination with irradiated tumor cells engineered to secrete murine granulocyte-macrophage colony-stimulating factor stimulates potent, specific, and long-lasting anti-tumor immunity. *Proc. Natl. Acad. Sci. U.S.A.* **90,** 3539–3543.
24. Parker, S. E., Khatibi, S., Margalith, M., *et al.* (1996) Plasmid DNA gene therapy: studies with the human interleukin-2 gene in tumor cells *in vitro* and in the murine B16 melanoma model *in vivo. Cancer Gene Ther.,* **3**(3), 175–185.
25. Gansbacher, B., Rosenthal, F. M., Zier, K. (1993) Retroviral vector-mediated cytokine-gene transfer into tumor cells. *Cancer Invest.* **11,** 345–354.
26. Klingemann, H. G., Dougherty, G. J. (1996) Site-specific delivery of cytokines in cancer. *Mol. Med. Today* **2,** 154–159.
27. Palladino, M. A., Morris, R. A., Starnes, H. F., Levinson, A. D. (1990) The transforming growth factor betas: a new family of immunoregulatory molecules. *Ann. N.Y. Acad. Sci.* **593,** 181–187.
28. Fujiwara, T., Grimm, E. A., Roth, J. A. (1994) Gene therapeutics and gene therapy for cancer *Curr. Opin. Oncol.* **6,** 96–105.
29. Porgador, A., Feldman, M., Eisenbach, L. (1994) Immunotherapy of tumor metastasis via gene therapy. *Natl. Immunol.* **13,** 113–130.
30. Ettinghausen, S., Rosenberg, S. A. (1995) Immunotherapy and gene therapy of cancer. *Adv. Surg.* **8,** 223–25
31. Gilboa, E., Lyerly, H. K. (1991) Specific active immunotherapy of cancer using genetically modified tumor vaccines. *Biol. Ther. Cancer Updates* **6,** 1–15
32. Miller, A., McBride, W. H., Hunt, K., Economou, J. S. (1994) Cytokine-mediated gene therapy for cancer. *Ann. Surg. Oncol.* **1**(5), 436–450.
33. Hollingsworth, S. J., Darling, D., Gaken, J., *et al.* (1996) Abstract 1: the effect of combined expression of interleukin-2 and interleukin-4 on the tumorigenicity and treatment of B16F10 melanoma. *Br. J. Cancer.* **74**(1), 6–15.
34. Kato, K., Yamada, K., Wakimoto, H. (1995) Abstract P-30. Combination gene therapy with B7 and IL-12 for lung metastasis of mouse lung carcinoma. *Cancer Gene Ther.* **2**(4), 316.
35. Abdel-Wahab, Z., Dar, M., Osanto, S., *et al.* (1997) Eradication of melanoma pulmonary metastases by immunotherapy with tumor cells engineered to secrete interleukin-2 or gamma interferon. *Cancer Gene Ther.* **4**(1), 33–41.
36. Cao, X., Chen, G., Zhang, W., *et al.* (1996) Enhanced efficacy of combination of IL-2 gene and IL-6 gene-transfected tumor cells in the treatment of established metastatic tumors. *Gene Ther.* **3**(5), 421–426.
37. Van Duinen, S. G., Ruiter, D. J., Broecker, E. B., *et al.* (1988) Level of HLA antigens in locoregional metastases and clinical

course of the disease in patients with melanoma. *Cancer Res.* **48**, 1019–1025.

38. Weber, J. S., Rosenberg, S. A. (1990) Effects of murine tumor class I Major histocompatibility complex expression on antitumor activity of tumor-infiltrating lymphocytes. *J. Natl. Cancer Inst.* **82**(9), 755–761.

39. Weber, J. S., Rosenberg, S. A. (1988) Modulation of murine tumor major histocompatibility antigens by cytokines *in vivo* and *in vitro*. *Cancer Res.* **48**, 818–824.

40. Guinan, E. C., Gribben, J. G., Boussiotis, V. A., *et al.* (1994) Pivotal role of the B7:CD28 pathway in transplantation tolerance and tumor immunity. *Blood* **84**(10), 3261–3282.

41. Howard, M., O'Garra, A. (1992) Biological properties of interleukin 10. *Immunol. Today* **13**, 198–200.

42. Fuchs, A. C., Granowitz, E. V., Shapiro, L., *et al.* (1996) Clinical, hematologic, and immunologic effects of interleukin-10 in humans. *J. Clin. Immunol.* **16**(5), 291–304.

43. Sato, T., McKue, P., Kazuhiro, M., *et al.* (1996) Interleukin 10 production by human melanoma. *Clin. Cancer Res.* **2**, 1383–1390.

44. Sato, T., Inoue, G., Takao, C., Clark, C., Medley, E., Bloome, M. J., Berd, D. (1996) Elevated serum IL-10 in metastatic melanoma. *Proc. Am. Soc. Clin. Oncol.* **15**, 274.

45. Massague, J. (1990) The transforming growth factor-β family. *Annu. Rev. Cell. Biol.* **6**, 597–641.

46. Park, J. A., Kurt, R. A., Wang, E., Schluter, S. F., Hersh, E. M., Akporiaye, E. T. (1997) Expression of an antisense TGF-β1 transgene reduces tumorigenicity of EMT 6 mammary tumor cells. *Cancer Gene Ther.* **4**, 42–50.

47. Sage, E. H. (1997) Terms of attachment: SPARC and tumorigenesis. *Nat. Med.* **3**(2), 144–146.

48. Ledda, M. F., Adris, S., Bravo, A. I., *et al.* (1997) Suppression of SPARC expression by antisense RNA abrogates the tumorigenicity of human melanoma cells. *Nat. Med.* **3**(2), 171–176.

49. Nagata, S., Goldstein, P. (1995) The Fas death factor. *Science* **267**, 1449–1456.

50. Strand, S., Hofmann, W. J., Hug, H., *et al.* (1996) Lymphocyte apoptosis induced by CD95 (APO-1/Fas) ligand-expressing tumor cells–a mechanism of immune evasion? *Nat. Med.* **2**(12), 1361–1366.

51. Seino, K.-I., Kayagaki, N., Okumura, K., *et al.* (1997) Antitumor effect of locally produced CD95 ligand. *Nat. Med.* **3**(2), 165–170.

52. Hahne, M., Rimoldi, D., Schroter, M., *et al.* (1996) Melanoma cell expression of Fas (Apo-/CD95) ligand: implications for tumor immune escape. *Science* **274**, 1363–1366.

53. Galle, P. R., *et al.* (1995) Involvement of the CD95 (APO-1/Fas) receptor and ligand in liver damage. *J. Exp. Med.* **182**, 1223–1230.

54. Dean, M., *et al.* (1996) Genetic restriction of HIV-1 infection and progression to AIDS by a deletion allele of the CCR5 structural gene. *Science* **273**, 1856–1862.

55. Nagata, S. (1996) Fas ligand and immune evasion. *Nat. Med.* **2**(12), 1306–1307.

56. Griffith, T., Brunner, T., Fletcher, S., Green, D., Ferguson, T. (1995) Fas ligand-induced apoptosis as a mechanism of immune privilege. *Science* **270**, 1189–1192.

57. Kono, K., Ressing, M., Brandt, R. M. P., *et al.* (1996) Decreased expression of signal-transducing chain in peripheral T cells and natural killer cells in patients with cervical cancer. *Clin. Cancer Res.* **2**, 1825–1828.

58. Zea, M. in A. H., Curti, B. D., Longo, C. D., Alvord, W. G., Srobl, S. L., Mizoguchi, H., Creekmore, S. P., O'Shea, J. J., Powers, G. C., Urba, W. J., Ochoa, A. C. (1995) Alterations in T cell receptor and signal transduction molecules in melanoma patients. *Clin. Cancer Res.* **1**, 1327–1335.

59. Gabrilovich, D. I., Corak, J., Ciernik, I. F., *et al.* (1997) Decreased antigen presentation by dendritic cells in patients with breast cancer. *Clin. Cancer Res.* **3**, 483–490.

60. Jacobson, E. L., Pilaro, F., Smith, K. A. (1996) Rational interleukin-2 therapy for HIV positive individuals. Daily low doses enhance immune function without toxicity. *Proc. Natl. Acad. Sci. USA* **93**, 10405–10410.

61. Cristiano, R. J., Curiel, D. T. (1996) Strategies to accomplish gene delivery via the receptor-mediated endocytosis pathway. *Cancer Gene Ther.* **3**, 49–57.

62. Hart, I. R. (1996) Tissue specific promoters in targeting systemically delivered gene therapy. *Semin. Oncol.* **23**, 154–158.

63. Yang, J. P., Huang, L. (1996) Direct gene transfer to mouse melanoma by intratumor injection of free DNA. *Gene Ther.* **3**, 542–548.

64. Clark, P. R., Ferrari, M., Felgner, P. L., *et al.* (1998) Development of an intra-tumoral DNA injection assay system in SCID mice: serum IL-2 and plasmid efflux detected following intra-tumoral injection of naked plasmid DNA. *Cancer Gene Ther.* In press.

65. Labat-Moleur, F., Steffan, A. M., Brisson, C., *et al.* (1996) An electron microscopy study into the mechanism of gene transfer with lipopolyamines. *Gene Ther.* **3**, 1010–1017.

66. Hong, K., Zheng, W., Baker, A., *et al.* (1997) Stabilization of cationic liposome-plasmid DNA complexes by polyamines and poly (ethylene glycol)-phospholipid conjugates for efficient *in vivo* gene therapy. *FEBS Lett.* **400**, 233–237.

67. Gregoriadis, G. (1995) Engineering liposomes for drug delivery: progress and problems. *Trends Biotechnol.* **13**, 527–537.

68. Felgner, P. L., Zaugg, R. H., Norman, J. A. (1995) Synthetic recombinant DNA delivery for cancer therapeutics. *Cancer Gene Ther.* **2**, 61–65.

69. Lew, D., Parker, S. E., Latimer, T., *et al.* (1995) Cancer gene therapy using plasmid DNA: pharmacokinetic study of DNA following injection in mice. *Hum. Gene Ther.* **6**, 553–564.

70. Stewart, M. J., Plautz, G. E., Del Buono, L., *et al.* (1992) Gene transfer *in vivo* with DNA-liposome complexes: safety and acute toxicity in mice. *Hum. Gene Ther.* **3**, 267–275.

71. Liu, Y., Liggitt, D., Zhong, W., *et al.* (1995) Cationic liposome-mediated intravenous gene delivery. *J. Biol. Chem.* **270**, 24864–24870.

72. Zhu, N., Liggitt, D., Liu, Y., *et al.* (1993) Systemic gene expression after intravenous DNA delivery into adult mice. *Science* **261**, 209–211.

73. Parker, S. E., Ducharme, S., Norman, J., *et al.* (1997) Tissue distribution of the cytofectin component of a plasmid-DNA/cationic lipid complex following intravenous administration in mice. *Hum. Gene Ther.* **8**, 393–401.

74. Ledley, T. S., Ledley, F. D. (1994) Multicompartment, numerical model of cellular events in the pharmacokinetics of gene therapies. *Hum. Gene Ther.* **5**, 679–691.

75. Nabel, E. G., Plautz, G., Nabel, G. J. (1992) Transduction of a foreign histocompatibility gene into the arterial wall induces vasculitis. *Proc. Natl. Acad. Sci. U.S.A.* **89**, 5157–5161.

76. Scheule, R. K., St. George, J. A., Bagley, R. G., *et al.* (1997) Basis of pulmonary toxicity associated with cationic lipid-mediated gene transfer to the mammalian lung. *Hum. Gene Ther.* **8**, 689–707.

77. Hersh, E. M., Stopeck, A. T., Warnecke, J., *et al.* (1996) Abstract P-28: *in vitro* studies of tumor cell transfection with plasmid DNA in a cationic lipid vector. *Cancer Gene Ther.* **3**(6), S18.

78. Clark, P. R., Stopeck, A. T., Wang, Q., *et al.* (1996) Abstract P-48: Enhanced adenoviral transduction and transgene expression *in vitro* using polycations and cationic lipids. *Cancer Gene Ther.* **3**(6), S23.

79. Plautz, G. E., Yang, Z. Y., Wu, B.-Y., *et al.* (1993) Immunotherapy of Malignancy by *in vivo* gene transfer into tumors. *Proc. Natl. Acad. Sci. U.S.A.* **90**, 4645–4649.

80. Parker, S. E., Khatibi, S., Margalith, M., *et al.* (1996) Plasmid DNA gene therapy: studies with the human interleukin-2 gene in tumor cells *in vitro* and in the murine B16 melanoma model *in vivo*. *Cancer Gene Ther.* **3**, 175–185.

81. Parker, S. E., Anderson, D., Khatibi, M., *et al.* (1996) Gene therapy of cancer using plasmid DNA delivery of the interleukin-2 gene. *Cancer Gene Ther.* **3**, S16.

82. Egilmez, N. K., Cuenca, R., Yokota, S. J., *et al.* (1996) *In vivo* cytokine gene therapy of human tumor xenografts in SCID mice by liposome-mediated DNA delivery. *Gene Ther.* **3**, 607–614.

83. Aoki, K., Yoshida, T., Matsumoto, N., *et al.* (1997) Gene therapy for peritoneal dissemination of pancreatic cancer by liposome-mediated transfer of herpes simplex virus thymidine kinase gene. *Hum. Gene Ther.* **8**, 1105–1113.

84. Kikuchi, A., Sugaya, S., Tanaka, K. (1996) Gene therapy for peritonitis carcinomatosa using cationic liposome-mediated gene transfer. (Meeting abstract). *Proc. Annu. Meet. Am. Assoc. Cancer Res.* **37**, A2389.

85. Mabuchi, E., Shimizu, K., Miyao, T., *et al.* (1997) Gene delivery by HVJ-liposome in the experimental gene therapy of murine glioma. *Gene Ther.* **4**, 768–772.

86. Yagi, K., Hayashi, Y., Ishida, N., *et al.* (1994) Interferon-β endogenously produced by intratumoral injection of cationic liposome-encapsulated gene: cytocidal effect on glioma transplanted into nude mouse brain. *Biochem. Mol. Biol. Int.* **32**, 167–171.

87. Lukacs, K. V., Nakakes, A., Atkins, C. J., *et al.* (1997) *In vivo* gene therapy of malignant tumors with heat shock protein-65 gene. *Gene Ther.* **4**, 346–350.

88. Lukacs, K. V., Steel, R. M., Oakley, R. E., Pardo, D. E., Porter, C. D., Sorgi, F. (1997) Cancer Gene Therapy with a heat shock protein gene. *Cancer Gene Ther.* **4**, S51.

89. Hofland, H., Huang, L. (1995) Inhibition of human ovarian carcinoma cell proliferation liposome-plasmid DNA complex. *Biochem. Biophys. Res. Commun.* **207**, 492–496.

90. Nabel, G. J., Nabel, E. G., Yang, Z. Y., *et al.* (1993) Direct gene transfer with DNA-liposome complexes in melanoma: expression, biological activity and lack of toxicity in humans. *Proc. Natl. Acad. Sci. U.S.A.* **90**, 11307–11311.

91. Nabel, E. G., Yang, Z., Muller, D., *et al.* (1994) Safety and toxicity of catheter gene delivery to the pulmonary vasculature in a patient with metastatic melanoma. *Hum. Gene Ther.* **5**, 1089–1094.

92. Stopeck, A. T., Hersh, E. M., Akporiaye, E. T., *et al.* (1997) Phase I study of direct gene transfer of an allogeneic histocompatibility antigen, HLA-B7, in patients with metastatic melanoma. *J. Clin. Oncol.* **15**, 341–349.

93. Hersh, E., Stopeck, A., Harris, D., *et al.* (1996) Long-term follow-up and retreatment studies on patients with metastatic malignant melanoma (MM) treated in a phase I/II study of direct intratu-moral injection of the HLA-B7/β₂m gene (Allovectin-7) in a cationic lipid vector. *Proc. Am. Soc. Clin. Oncol.* **15**, 235.

94. Rubin, G., Galanis, E., Pitot, H. C., *et al.* (1997) Phase I study of immunotherapy of hepatic metastases of colorectal carcinoma by direct gene transfer of an allogeneic histocompatibility antigen, HLA-B7. *Gene Ther.* **4**, 419–425.

95. Vogelzang, N. J., Sudakoff, G., McKay, S., *et al.* (1995) A Phase I study of intra-lesional (IL) gene therapy in metastatic renal cell cancer (RCC). *Proc. Am. Soc. Clin. Oncol.* Los Angeles, CA. **14**, A641.

96. Vogelzang, N. J., Sudakoff, G., Hersh, E. M., *et al.* (1996) Clinical experience in phase I and phase II testing of direct intratumoral administration with Allovectin-7: a gene-based immunotherapeutic agent. *Proc. Am. Soc. Clin. Oncol.* **15**, 235.

97. Nabel, G. J., Gordon, D., Bishop, D. K., *et al.* (1996) Immune response in human melanoma after transfer of an allogeneic class I major histocompatibility complex gene with DNA-liposome complexes. *Proc. Natl. Acad. Sci. U.S.A.* **93**, 15388.

98. DeBruyne, L. A., Chang, A. E., Cameron, M. J., *et al.* (1996) Direct transfer of foreign MHC gene into human melanoma alters T cell receptor Vβ usage by tumor infiltrating lymphocytes. *Cancer Immunol. Immunother.* **43**, 49–58.

99. Silver, H. K. B., Klasa, R. J., Bally, M. B., *et al.* (1996) Phase I gene therapy study of HLA-B7 transduction by direct injection in malignant melanoma. *Proc. Am. Assoc. Cancer Res.* Washington, DC, p. 342.

100. Hersh, E. M., Nabel, G., Silver, H., *et al.* (1996) Intratumoral injection of plasmid DNA in cationic lipid vectors for cancer gene therapy. *Cancer Gene Ther.* **3**, S11.

101. Hersh, E. M., Stopeck, A. T., Silver, H. K. B., *et al.* (1997) Phase II study of intratumoral injection of HLA-B7/β₂m plasmid DNA-cationic lipid complex (Allovectin-7) therapy for metastatic malignant melanoma (MMM). *Cancer Gene Ther.* **4**, S48.

102. Heo, D. S., Yoon, S. J., Kim, W. S., *et al.* (1996) Locoregional response and increased NK activity after intratumoral gene therapy with HLA-B7/β₂-microglobulin gene. *Cancer Gene Ther.* **3**, S11.

103. Gleich, L. L., Gluckman, J. L., Armstrong, S., *et al.* (1997) Alloantigen gene therapy for squamous cell carcinoma of the head and neck. *Cancer Gene Ther.* **4**, S48.

104. Hui, K. M., Ang, P. T., Huang, L., *et al.* (1997) Phase I study of immunotherapy of cutaneous metastases of human carcinoma using allogeneic and xenogeneic MHC DNA-liposome complexes. *Gene Ther.* **4**, 783–790.

105. Stopeck, A., Hersh, E, Warneke, J. (1996) Results of a Phase I study of direct gene transfer of interleukin-2 (IL-2) formulated with cationic lipid vector, Leuvectin, in patients with metastatic solid tumors. *Proc. Am. Soc. Clin. Oncol.* **15**, 235.

106. Rubin, J., Galanis, E., Burch, P., *et al.* (1997) Phase I/II trial of the IL-2 DNA/DMRIE/DOPE lipid complex as an immunotherapeutic agent by direct gene transfer in patients with advanced melanoma, renal cell carcinoma, and sarcoma. *Cancer Gene Ther.* **4**, S49.

CHAPTER 21

Ex Vivo and *in Vivo* Gene Therapy Strategies for Prostate Cancer: Emerging Clinical Pharmacology

JONATHAN W. SIMONS, ANGELO A. BACCALA, AND HO YEONG LIM
NIH SPORE in Prostate Cancer, The Johns Hopkins Oncology Center and Brady Urological Institute, The Johns Hopkins University School of Medicine, Baltimore, Maryland 21287

I. INTRODUCTION

A pressing need exists for new therapeutic treatments for advanced prostate cancer (PCA). Prostate cancer is the second most common cause of death from cancer in the United States, and no systemic therapy is curative for metastatic disease. Cytoreductive gene therapy research is a particularly compelling approach to experimental therapeutics for PCA. The prostate is a unique, accessory organ. It is not required for fertility or potency and expresses over 500 unique gene products in expressed sequence-tagged libraries as potential therapeutic targets. Prostate cancer thus offers potentially unique antigens, particularly for the generation of autoimmune antitumor immune responses following the use of *ex vivo* cytokine-gene-transduced tumor vaccines. Prostate-organ-unique genes also possess tissue-specific promoters and enhancers that permit po-

333

tential prostate cancer specificity in transcription of suicide genes for *in vivo* cytoreductive gene therapy applications. These tissue-specific promoters and enhancers also allow the creation of novel, target-specific oncolytic viruses for *in vivo* gene therapy applications.

Among all mammals, only *homo sapiens* are afflicted with a high incidence of spontaneous prostate cancer. Thus, integral to both *ex vivo* and *in vivo* gene therapy research is evaluation of the basic molecular mechanisms of the antineoplastic action of gene therapy in patients with prostate cancer. Consequently, phase I and II trials in prostate cancer gene therapy research can be a unique opportunity for rigorous, basic pharmacological research, as well as vital steps in therapeutic development of a concept.

II. BACKGROUND

A. Human Prostate Cancer as an Ideal Disease for Gene Therapy Research

Prostate cancer (PCA) is the most commonly diagnosed cancer and second leading cause of cancer death in North American men [1]. An American now dies on average every 14 minutes around the clock from this neoplasm. Paradoxically, the outlook for earlier treatment of the disease has never been brighter. With advances in diagnosis by prostate-specific antigen (PSA) screening, earlier clinical detection of PCA has increased [1]. Yet, despite significant advances in diagnosis, surgery, and radiation therapy in the past decade, complete cure can be expected only when the disease has been localized within the prostate gland. A large conceptual advance in the systemic treatment of PCA has not been forthcoming since 1941, when Charles Huggins reported on the androgen dependence of PCA. For patients with advanced, metastatic, or recurrent PCA, androgen ablation, by castration [2] or medical castration using antiandrogens [3] or luteinizing-hormone-releasing hormone (LHRH) analogs [4], is still the mainstay of therapy. Although androgen ablation has resulted in significant palliation for hundreds of thousands of patients and has achieved clinical responses in over 70% of them, disease progressions from androgen-refractory tumors are inevitable [5].

Given its incidence and the absence of active systemic therapies including chemotherapy currently available, advanced prostate cancer is a compelling target for cytoreductive gene therapy research. Several reasons exist for this. First, the societal need for new treatment for PCA is great: as many patients carry a PCA diagnosis as carry an HIV diagnosis. Second, hormone-refractory PCA is refractory in general to conventional and efficacious cytotoxic agents, which have increased disease-free survival in breast cancer and colon cancer in adjuvant therapy settings. Although a subset of PCA patients might ultimately benefit from newer forms of chemotherapy, it seems unlikely that continued exploration of dose-intensive combinations of currently available cytotoxic drugs in clinical trials will yield large incremental improvements in survival for the majority of PCA patients. The clinical resistance of PCA to many S-phase-dependent cytotoxic drugs may be due in part to its intrinsic low proliferative index [6–8]. In general, the lethal phenotype of this neoplasm has acquired defects in programmed cell death pathways rather than genomic damage that accelerates proliferation. The clinical doubling time of PCA can be over 150 days, as only a very small population of the cells (<5%) may be actively replicating at any moment [7,8]. The development of new therapeutic techniques, such as cytoreductive gene therapy approaches, that do not require a high proliferative rate has a strong rationale in treating PCA. Indeed, it is the phenotypic resistance of PCA *in vivo* to conventional cytotoxic agents, established in clinical trials from 1975 to 1995, that makes PCA such a logical focus for antineoplastic strategies that do not work like cytotoxic drugs but can kill with selective toxicity. As described later, cytoreductive gene therapy strategies, which can kill tumor cells selectively but independent of PCA growth fraction, have thus emerged as an attractive research strategy for clinical translation.

III. EMERGENT STRATEGIES FOR CLINICAL TRANSLATION IN PROSTATE CANCER

To date, two approaches have undergone preclinical evaluation and early clinical evaluation in gene therapy for PCA. As of 10 March 1998, 10 clinical trials have received NIH RAC and FDA approval in the field (Table 1). Corrective gene therapy (gene replacement) and cytoreductive gene therapy have both been expanding areas for preclinical research. Corrective gene therapy approaches to prostate cancer have not emerged yet as an active clinical research area. The genes to correct are still being identified. For example, the hereditary prostate cancer (HPC)-1 gene, located by linkage studies on human chromosome 1, and other hereditary prostate cancer loci have not had cDNAs identified yet.

A. Corrective Gene Therapy for Prostate Cancer

Prostate cancer is not a monogenic disease like cystic fibrosis, yet targets already clearly exist for corrective

TABLE 1   NIH RAC-Approved Human Gene Therapy Trials in Prostate Cancer

Vector Rx cDNA	Principal investigator	Modality	Review date
Retroviral: MFG-GM-CSF	Simons, Johns Hopkins	*Ex vivo;* autologous PCA vaccines	8/3/94
Retroviral: IL-2 + gamma IFN	Gansbacher, MSK	*Ex vivo;* allogeneic PCA vaccine	5/14/95
Retroviral: anti-sense *myc*	Steiner, Vanderbilt	*In vivo;* intraprostatic injection	9/30/95
Vaccinia: PSA cDNA	Chen, Bethesda Naval	*In vivo;* intradermal injection	9/22/95
Adenovirus 5: RSV-TKr	Scardino, Baylor	*In vivo;* intraprostatic injection	1/29/96
Vaccinia: PSA cDNA	Kufe and Eder, Harvard	*In vivo;* intradermal injections	9/18/96
Vaccinia: PSA cDNA	Sanda, U. Michigan	*In vivo;* Intradermal injections	5/13/97
Retroviral: GM-CSF	Simons, Johns Hopkins	*Ex vivo;* allogeneic PCA vaccines	9/9/97
Adenovirus 5: p53 wild-type cDNA	Belldegrun, UCLA	*In vivo;* intratumoral injection	9/17/97
Adenovirus 5: p53 wild-type cDNA	Logothetis, MDA	*In vivo;* intratumoral injection	11/6/97

Source: http://www.nih.gov/od/orda/protocol.htm.

strategies. In the realm of known tumor suppressor genes, p53 has been the target for preclinical studies of gene replacement in prostate cancer. Depending on the population being studied, perhaps 25% of U.S. men with metastatic PCA have tumors with mutant p53 genes [9]. Wild-type p53 transfection can decrease the tumorgenicity of some p53-mutated prostate cancer cell lines [10,11]. For example, NIH RAC approval has been granted for phase I studies at UCLA and the M. D. Anderson Cancer Center to deliver wild-type p53 in an adenoviral vectors to primary PCA tumors using a replication-defective adenoviral vector [12]. (See Chapter 14 for a further discussion of the use of adeno-p53 as therapy.)

The inactivation of the retinoblastoma (Rb) gene may also be important in some cases of prostate cancer progression [13]. Allelic loss of the Rb gene has been observed in 27% of prostate tumors by PCR methods [14]. Retrovirus-mediated introduction of wild-type Rb can suppress the tumorigenicity of DU145 cells that have nonfunctional truncated Rb protein [14,15]. Thus, wild-type Rb transfection potential candidate for corrective gene therapy of prostate cancer.

In prostate cancer gene therapy, the glutathione-*S*-transferase (GST) π-gene is theoretically a very attractive target gene for corrective gene therapy. The promoter of the GST π-gene is located on chromosome 11q13. Recent research on the GST π-gene showed that methylation of this region was detected in every one of 30 prostate cancer DNAs examined; on the other hand, no methylation was detected in any normal or hyperplastic prostate tissue [15]. This genomic inactivation in PCA makes it the most common genomic alteration in PCA and may occur as early as PINs (prostatic intraepithelial neoplasia). The GST π-gene plays a key role in detoxifying potential carcinogens, so the inactivation

of the GST π-gene could result in an increase in the susceptibility of prostate tissue to DNA damage and the accumulation of mutations in the DNA of the stem cells of the prostate epithelium. Conceptually, the reintroduction of a GST π-gene that detoxifies potential carcinogens to the prostate epithelial cells could serve as a cancer prevention strategy employing corrective gene therapy.

Cell adhesion molecules (CAMs) are also under investigation as candidate therapeutic genes for potential delivery *in vivo,* as they have important roles in regulating cell growth and differentiation. E-cadherin protein levels have been found to be reduced or absent in half of PCAs [16]. Inactivation of E-cadherin has a strong correlation with metastatic and/or invasive potential of PCA [17]. Loss of E-cadherin is a powerful predictor of poor outcomes in prostate cancer [18]. C-CAM acts as a tumor suppressor in PCA. Adenoviral C-cam transfection into a tumorigenic PCA cell line (PC-3) appears to reduce growth rate *in vitro* and decrease tumor take rate and tumor growth *in vivo* [19,20].

Anti-oncogene therapy includes complementary or antisense oligonucleotides to target oncogenes and ribozymes. By the annealing of antisense sequences to specific oncogenes, this strategy results in blocking the transcription or translation of oncogenes eliminating expression of oncogenic proteins. A retroviral vector carrying antisense sequences to c-*myc* is being developed for intraprostatic injection to block PCA growth [21]. The *ras* genes are a potential target for antisense strategy, as they are frequent mutated oncogenes in human cancer development. Antisense DNA oligonucleotides to *ras* mRNA has been shown to block the production of *ras* mRNA and reduce the growth of human lung cancer *in vitro* and *in vivo* [22–24]. In contrast to other tumors, several studies have found low frequencies (2–5%) of *ras* gene mutations in prostate cancer [25,26], although *ras* mutations occur at a higher frequency (25%) in prostate tissue from Japanese men [26]. Therefore, *ras* anti-oncogene therapy may have some limitations for cancer gene therapy of PCA. (See Chapters 15 and 9 for discussions of antisense and ribozyme strategies respectively.)

The study of corrective gene therapy for prostate cancer is predicted on the assumption that higher and higher efficiency vectors will become available to achieve nearly 100% gene transfer efficiencies in prostate epithelial stem cells. Correcting tumor cells of genomic lesions with systemic administration of a vector has not been achieved so far at high efficiencies. Of particular importance will be verification that the self-renewing stem cell population of the prostate epithelium is "corrected" with many of these approaches [27].

B. Cytoreductive Gene Therapy for Prostate Cancer

Many efforts have centered on generating either locoregional or systemic antineoplastic effects using vectors designed to kill PCA cells. *Ex vivo* approaches have centered upon the generation of PCA tumor vaccines that express immunostimulatory molecules after transduction [28]. *Ex vivo* gene therapy research has burgeoned in large measure as a consequence of cloning, and characterization of cytokines has generated a number of candidate immunostimulatory genes that can be used in gene therapy strategies. Improvements in vector efficiency have allowed investigators to express cytokines or other immunologically important genes in tumor cells and evaluate these as new vaccines. A central concept driving these applications of gene transfer is that T-cell immune responses are essential in antitumor immunity [28]. Nearly every strategy of immunotherapy for PCA employing genetic engineering of tumor cells is aimed at the enhanced stimulation of T-cell immune responses to tumor-associated antigens (Fig. 1) [29–31]. (See Chapter 3 for a detailed discussion of antitumor immune mechanisms.)

In vivo gene therapy has generally concentrated upon suicide gene therapy strategies. Replication-defective vectors have been evaluated for delivery of suicide genes. Adenoviral vectors have been the principal gene transfer system studied and published to date. The suicide gene that has been studied clinically is the thymidine kinase gene [32]. Preclinically, the array of suicide genes under study includes cytosine deaminase (CDA), purnine nucleotide phosphorylase (PNP), and more recently the diphtheria toxin A chain (DT-A) cDNA. A second approach involves the generation of specific, on-

A Strategy of *Ex Vivo* Gene Therapy for Prostate Cancer

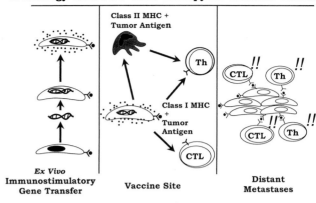

FIGURE 1

colytic vectors for PCA cells. As discussed later, these replication-competent viruses kill by replication in PCA cells. The vectors are genetically engineered to be restricted to prostate cells. Genes critical to vector replication and PCA cell killing are placed under the transcriptional control of prostate-specific gene regulatory elements [33]. For example, the prostate specifice antigen (PSA) gene enhancer and promoter can be used to regulate adenoviral replication [33] (Fig. 2).

C. Vector Development for Human Trials in Prostate Cancer

Critical to human gene therapy as applied to cancer pharmacology is the basic and applied science of vector development. Numerous methods are currently available for delivering nucleotide sequences into human prostate cancer cells. Many have been tested either *ex*

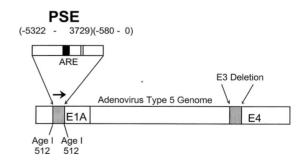

FIGURE 2

vivo or *in vivo* in animal xenografts or now in humans. The general principles of vector design and delivery methods are described elsewhere (see Section II.). With respect to basic research in cancer gene therapy applied to human PCA, Table 2 describes vector systems currently employed and either in clinical trials currently or near final translation to phase I trials in PCA. Some of the favorable and limiting properties of these vectors with respect to PCA gene therapy applications are noted.

Most experience in man in the field of PCA gene therapy has to date been with viral vectors. Early PCA gene therapy trials have used retroviral vectors,

vaccinia-based vectors, or adenoviral-based vectors as gene delivery systems. With retroviral vectors, high infectivity and expression can be achieved into prostate cancer cells *ex vivo*. This has allowed clinical application in phase I and phase II trials using cytokine-transduced autologous and allogeneic PCA vaccines using the MFG retroviral vector, for example [30,31]. Retroviruses are able to stably integrate desired genes into the chromosomal DNA of the target cell, making them favorable systems for autologous tumor cell vaccine generation. The absence of a requirement of a coselectable gene for transduction (neomycin resistance marker gene) in some high-efficiency retroviral vectors permits them to

TABLE 2   Candidate Vectors for Human Prostate Cancer Gene Therapy Trials

Vector	Potential insert size	Advantages	Disadvantages
Liposomes		Completely synthetic No limitation on size and type of nucleic acid	No targeting Inefficiency
Retrovirus	4–5 kb	Integration Requirement of cell division for transduction	Low transduction efficiency Packaging cell line required No targeting Replication competence requirement
Adenovirus	30 kb	High transduction efficiency Infection of many cell types Infection does not require cell division Active as oncolytic agent *in vivo*	No integration Packaging cell line required Immunogenicity Replication competence No targeting
Adeno-associated virus	5 kb	Integration No viral genes Infection does not require cell division	No targeting Packaging cell line required Immunogenicity
Vaccinia virus	25 kb	Large insert size Active as oncolytic agent *in vivo*	Immunogenicity Toxicity Efficiency No targeting
Herpes simplex virus	40–50 kb	Neuronal tropism Large insert size Latency expression	No targeting Packaging cell line Toxicity
Avian Pox virus	5–10 kb	Transduction does not require cell division Large insert size	Immunogenicity Toxicity Efficiency No targeting
Baculovirus		Expression of protein at high levels Liver-directed gene transfer	Immunogenicity Toxicity Safety Efficiency No targeting
Mechanical administration: colloidal gold gene gun		No limitation on size of nucleic acid	No targeting Possible requirement for surgical procedure Inefficiency
Protein/DNA complex		No limitation on size and type of nucleic acid Cell-specific targeting	No integration Safety/toxicity Inefficiency *in vivo* intravenously Immunogenicity

be employed for propagation of autologous tumor vaccine cells *ex vivo* without loss to drug selection [28,30]. In the case of *ex vivo* gene transfer of immunostimulatory genes for PCA vaccines, increases in expression of the integrated therapeutic gene by nearly 300-fold can be achieved with a single transduction with clinical-grade vector. This can be readily achieved with either autologous or allogeneic human PCA cells *ex vivo*. As described in Fig. 3, vaccines can be generated from primary cultures of autologous tumors resected at surgery. A nontrivial aspect in vaccine development is optimizing primary culture conditions for expansion of vaccine cells after transduction. Adeno-associated virus (AAV) based vectors can be used instead of retroviral vectors for the generation of IL-2 cytokine-transduced autologous vaccines from primary prostate tumors (Robertson, personal communication). Although they can have cytopathic effects at high MOIs, Ad5 vectors carrying the granulocyte-macrophage colony-stimulating factor (GM-CSF) gene can also be used to achieve high levels of GM-CSF gene expression within 96 hours of transduction, without establishment of a long-term primary culture of PCA cells. (Jungles, Cohen, and Simons, unpublished observations). Thus, many options for *ex vivo* generation of genetically modified autologous and allogeneic PCA tumor vaccines exist if early clinical trials show bioactivity of the central concept.

Retroviral vectors have some disadvantages for *ex vivo* gene-transduced tumor vaccines. One disadvantage is emotional. Even with an expanding data set showing their safety, retroviral vectors and now lentiviral vectors carry the words "retrovirus-based gene transfer" in their FDA IND applications. This necessarily generates regulatory and patient concerns regarding replication-competent retrovirus generation that other vector systems do not. In terms of cancer gene therapy pharmacology for PCA, Moloney-based retroviral vectors are currently not capable of carrying large cDNAs or poly-

cistronic constructs, and are restricted to 6 kb. Currently, retroviral vectors can only transduce actively dividing cells [28], which makes it potentially difficult to achieve high levels of gene transfer in slow-growing tumors. Lentiviral vectors may circumvent this, as preliminary data suggest that these retroviral vectors could transduce PCA cells in the G_0 phase of the cell cycle (Verma, personal communication). Treatment of patients with PCA with autologous vaccines will always be limited to total achievable dose of vaccine cells. Thus, improvements in primary culture conditions are needed to expand vaccine yields if long schedules of vaccinations or boosting after primary vaccinations are to be studied in phase II/III studies.

Multiple approaches to *ex vivo* gene transfer for PCA vaccines, including the use of AAV vectors or plasmid vectors reliant upon synthetic gene transfer systems like liposomes are all under investigation currently. Essential to these and further improvements in *ex vivo* gene transfer is the clear identification of clinically active strategies employing immunostimulatory gene-transduced PCA vaccines.

Sanda and colleagues and other groups have explored the use of vaccinia vectors for cytoreductive immunotherapy for PCA. Vaccinia vectors have potentially large cDNA insert capacities. These vectors replicate within the cytoplasm of target cells, but gene expression with these vectors is transient. Vaccinia vectors expressing the CEA gene have been shown to be safe in colorectal carcinoma, and those expressing the PSA gene cDNA are nearing phase I trial completion at the University of Michigan (Sanda, personal communication). PSA as a unique antigen is used as the immunizing peptide in the Michigan study. Constructs elaborating several immune costimulatory molecules such as CTLA-4 or cytokines inducing enhanced peptide antigen presentation or T-cell activation are also under intensive preclinical study. In general, peptide motifs for the HLA A2 molecule have been used for clinical grade virus. A2 is chosen because it is the most common HLA class I molecule in the North American population (approximately 50% of Caucasians and 34% of African-Americans). With the identification of new prostate-unique peptide antigen cDNAs in the human genome project, vaccinia vectors allow the possible expression of a polyvalent set of prostate-unique peptides from the identified cDNAs. These vaccinia vectors may allow the induction of antiprostate epithelial cell immunity. They can be given intradermally as a vaccination, which appears particularly attractive in a strategy of adjuvant therapy following radical prostatectomy when the total autologous tumor is under 1 g so that it is currently technically difficult currently to generate large yields of

Ex Vivo Gene Therapy for Prostate Cancer

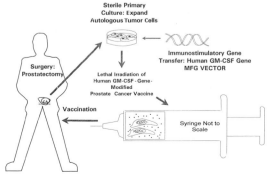

FIGURE 3

autologous PCA vaccine with *ex vivo* gene transfer. These vectors can also be given intratumorally.

One significant potential limitation is the fact that most patients have preexisting immunity to vaccinia vectors from smallpox vaccinations. How significant the attenuation is of the immune response with multiple vaccinations with vaccinia vectors remains to be seen in the phase II trials. (For a further discussion of the use of vaccinia vectors see Chapter 8 and 16.)

An alternative approach exists with avian pox virus vectors expressing immunostimulatory cytokines such as GM-CSF, IL-2 and tumor necrosis factor (TNF-α). One strategy is to use direct intratumoral injection prior to surgery or radiation therapy. The autologous tumor *in situ* serves as the source of autologous antigen without the requirement of *ex vivo* expansion of primary tumor cells. Avian pox virus vectors have apparently high tropism for human prostate epithelial cells and allow presentation of potential autologous antigens in the context of a highly immunogenic pox vector and paracrine expression of cytokines (T. Ratliffe, personal communication). As humans are not generally exposed to canary pox virus, neutralizing immune responses to pox virus proteins at the treatment site may allow less early attenuation of the early effects of prostate-tumor-associated antigen presentation at the treatment site.

Adenoviral vectors show excellent tropism for PCA both for *ex vivo* and *in vivo* applications in patients with PCA (Cohen and Simons, unpublished observations). As adenoviral vectors remain episomal, they can be diluted out in vaccine cell propagation over an extensive period of time and may be of particular value for autologous tumor vaccines when no extensive cell propagation *in vitro* is required after transduction and vaccination will occur shortly thereafter. Both replication-defective and replication-competent adenoviral vectors are undergoing clinical as well as preclinical evaluation in PCA gene therapy research.

An expanding human experience in PCA research is in the use of replication-defective Ad5 vectors. The E1 region of adenovirus is responsible for viral replication. After placing the desired therapeutic gene into the deleted E1 region, replication-deficient adenoviral vectors carrying suicide genes can be produced at high titer ($>10^{12}$ pfu/mL) with packaging cell line 293 cells (human embryonic kidney cell line containing adenovirus E1 DNA) [32,34–36]. A particular advantage to the use of adenoviral vectors for prostate cancer clinical trials is the safety profile of even wild-type adenoviral infections. High doses of replication-defective or replication-competent Ad5 vectors are likely to compel management of systemic symptoms of coldlike symptoms such as fever associated with wild-type Ad5 infections in up-

per respiratory infections (Scardino, personal communication).

A large area of opportunity for molecular pharmacology exists with Ad5 vectors in PCA gene therapy research. These vectors can transfer potentially polycistronic therapeutic DNA sequences that are larger than current retroviral vector constructs. The adenoviral genome also allows potential placement of antineoplastic genes for PCA in the E1, E3, and E4 regions, making it highly attractive for the use of tissue-specific enhancer/promoter constructs driving cytotoxic or immunostimulatory gene products. For example, the ORF 4 region of Ad5 vectors could be used to express immunostimulatory molecules such as GM-CSF and the E1 region could be used to carry a suicide gene. The early human data with GM-CSF by our group in renal cell carcinoma and PCA suggest that GM-CSF activation of dendritic cells and other antigen presentation cells such as macrophages could serve as a potent adjuvant around directly cytotoxic strategies of PCA gene therapy.

As described earlier, the majority of PCA metastases in humans do not have a high S-phase fraction. As adenoviral vectors are capable of entering prostate cancer cells through a receptor-mediated/integrin protein endocytosis, they can transfer antineoplastic genes into target cells at any stage of the cell cycle.

A current major limitation in Ad5 vector use for PCA gene therapy is immunogenecity [37]. Statistically, over 90% of potential PCA patients for treatment will have neutralizing antibody levels against Ad5 detectable prior to therapy. Treatment with Ad5 vectors is already reported to induce antiadenoviral antibodies and potent T-cell responses [37]. Currently, the prospects for long-term, intermittent therapy with these vectors in sensitized individuals appear limited. For phase II trials of *in vivo* gene therapy with Ad5 vectors, an active dose in a tight administration schedule (2–4 weeks) is likely to be employed. Alternatively, a brief controlled period of immunosuppression will be required to extend schedule and dose. Efforts to create a less immunogeneic set of adenoviral vectors is also underway in the biotechnology industry. In anticipation of potential clinical activity in phase II trials with the use of these vectors, many studies are underway attempting to block or circumvent neutralizing immune responses.

Several other DNA-based viral vectors are being investigated for gene transfer. AAVs need a helper virus for replication, and these viruses cause no illness to human beings [38,39]. They have the property of mediating gene transfer via stable integration of the treatment gene into the host chromosome [39,40], so gene expression with these vectors may be durable.

Vaccinia viruses have a large DNA insert capacity and do not need packaging cell lines for replication [42].

This allows *in vivo* use without culture of vaccine cells. These vectors replicate within the cytoplasm of target cells, but gene expression with these vectors is transient.

Synthetic gene transfer systems remain an intensive area for pharmaceutical research. Liposome-mediated gene transfer has been easy to achieve at low efficiencies in *ex vivo* studies of PCA. Significant gene transfer efficiencies *in vivo* administration remain challenging. Nonetheless, many new ligand/binding strategies are under investigation.

D. Preclinical Pharmacology of Cytoreductive Gene Therapy for Prostate Cancer

1. IMMUNOLOGIC APPROACHES

An alternative approach to direct *ex vivo* gene transfer is the use of autologous or HLA-matched, genetically engineered fibroblasts elaborating the cytokine of interest, admixed with irradiated, autologous unpassaged tumor [43]. These "bystander" fibroblasts provide the paracrine source of cytokine surrounding autologous tumor antigens. (For a further discussion of fibroblast transfectants see Chapter 23.) Alternatively, allogeneic, cytokine-gene-transduced tumor vaccines are being evaluated in patients who have minimal residual metastatic cancer following surgery. It is hoped that many tumor-associated antigens are retained in these cell lines. The limitless supply of genetically modified allogeneic tumor cells allows testing of repetitive vaccinations and large cumulative doses of treatment, whereas autologous cells are always in finite supply, depending on the size of the resected autologous tumor. In PCA, for instance, T2b cancers can rarely be expected to yield more than 3 g of tumor for *ex vivo* gene transfer even with high gene transfer efficiency vectors [30].

Several potentially interesting therapeutic genes in *ex vivo* gene therapy for genitourinary cancers are showing preliminary evidence of activity and novel properties of immune activation. A comparison involving over 20 immunostimulatory genes found that GM-CSF conferred the most potent induction of long-lasting, specific tumor immunity, even in the most poorly immunogenic tumor models [44]. The preclinical efficacy of GM-CSF-gene-transduced tumor vaccines has been shown in preclinical models of renal cell carcinoma and prostate carcinoma among others. We have shown in preclinical models of hormone refractory and poorly immunogenic models of PCA that GM-CSF-gene-transduced, irradiated tumor vaccines can eradicate small, preestablished tumors [30,31]. The MFG retroviral vector was used [45]. Rats with anaplastic, hormone-refractory PCA

treated with MFG-GM-CSF-gene-transduced vaccines could be rendered disease-free for greater than 100 days, with animals on control arms dying in 21 days. The therapeutic action of GM-CSF after gene transfer involves the paracrine (local) spread of the cytokine at high concentrations, which triggers antigen-presenting cell (APC) ingress and activation. These APCs include macrophages and dendritic cells—the most potent APCs for helper T cells known. Systemic, antitumor immune responses are catalyzed by the APCs priming of the T-cell arm of the immune response. Both CD-4+ and CD-8+ T-cells are activated and tumor cells are selectively destroyed upon T-cell recognition of tumor-associated antigens at distant sites [44]. As discussed later, early clinical trial data suggests that this approach is bioactive, and gene transfer can be achieved at high efficiency for the generation of autologous, cytokine-transduced PCA vaccines.

2. CYTOTOXIC APPROACHES: SUICIDE VECTORS, APOPTOTIC VECTORS, ONCOLYTIC VECTORS

Suicide gene therapy is one strategy to transfect virus-directed apoptotic cDNAs or enzymes into tumor cells selectively and administer prodrugs systemically [46–53]. A gene encoding a nonmammalian enzyme in tumor cells can convert a nontoxic prodrug into a potent toxic metabolite. Normal mammalian cells lack herpes simplex virus thymidine kinase (HSV-TK); this enzyme converts nontoxic ganciclovir (GCV) into a phosphorylated compound, ganciclovir triphosphate, that is toxic to cancer cells through its high affinity for DNA polymerase. Therefore, ganciclovir triphosphate can terminate DNA synthesis and kill cancer cells in the S phase of the cell cycle. This approach is attractive because of the "bystander," which produces cell death in untransfected nearby cells not transfected with an enzyme cDNA. The mechanisms of this unique antitumor effect by suicide gene therapy include the transfer of toxic metabolites via gap junctions, the endocytosis of apoptotic vesicles, and the up-regulation of an immune response. This bystander effect is able to amplify the low efficiency of actual gene transfer *in vivo* into measurable shrinkage of solid tumors in rodent models. Adenovirus-mediated HSV-TK genes in conjunction with GCV in PCA have been shown to produce growth suppression *in vitro* and significant regression of tumor growth and of spontaneous metastasis in several animal models. Remarkably little is known about the relative contribution of the immune responses versus gap junctions to suicide gene therapy in PCA. A phase I trial of *in vivo* gene therapy for PCA using an adenoviral RSV promote driving an HSV-TK vector nearing completion at Baylor University is evaluating this in part with the use of biopsies of treated tumors (Scardino, personal communication). An early clinical impression is that secondary immune responses to apoptotic PCA cells may be generated, and systemic T-cell immune responses to PCA antigens may be generated from intralesional treatment.

Other suicide substrate genes, for example, cytosine deaminase (CDA) and purine nucleoside phosphorylase (PNP), are also being investigated currently both preclinically and clinically. CDA, which is present in bacteria and fungi, not in mammalian cells, deaminates the nontoxic antifungal drug 5-fluorocytosine (5-FC) to the cytotoxic chemotherapeutic agent 5-fluorouracil (5-FU). In addition, vectors designed for CDA/5-FC produce a significant preclinical activity because tumor cells expressing CDA produce 5-FU, which diffuses into the nontransfecting neighboring cells. 5-FU is a membrane-permeant toxin and thus is readily diffusible from one cell to another. Recent report shows significant tumor regression *in vivo* in colorectal cancers when only 2% of the tumor mass contains CDA-expressing cells. PNP in eukaryotic cells converts the prodrug 6-methylpurine-deoxyriboside to the membrane-permeant toxin 6-methylpurine. This PNP/prodrug stategy also has a obviously significant bystander effect. CDA has yet to be fully evaluated in preclinical or clinical trials of PCA. (See Chapters 11 and 12 for further discussion of cytosine deaminase and other drug-resistance mechanisms.)

Unfortunately, HSV-TK, CDA, and other cytotoxic prodrug-activating enzyme genes may have some limitations for clinical application to PCA, which are self-evident to medical oncologists. In metastatic PCA, only a very small population (4–10%) of the cells are in the S-phase fraction, so human prostate cancer may often to be resistant to HSV-TK or CDA/5-FC gene therapy. (PCA) has an intrinsic resistance to many S-phase-dependent chemotherapeutic agents [6]. Enthusiasm for new forms of immunotherapy for PCA is based in large measure on the absence of a PCA cell to be in S phase to be killed by a T-cell. However, to date, no combination of cytotoxic drugs has convincing been shown to generate improved disease-free or improved overall survival in men with hormone-refractory prostate cancer. Phenotypic resistance to S-phase killing is the clinical and molecular hallmark of metastatic PCA cells, at least in the schedules and doses of drugs evaluated so far. Hence, some investigators are skeptical that gene therapy approaches poised to protect the bone marrow stem cells from cytotoxic drugs will have a strong rationale for PCA medical oncology in the near future.

Many efforts have concentrated preclinically on over-expression of wild-type p53 in PCA cells as in other solid tumors. In general, adenoviral vectors have been employed for these studies. As in other tumor systems, overexpression of p53 can induce PCA cell apoptosis [46–48].

Gene therapy aiming at transferring the multidrug resistance (MDR)-1 gene or other drug metabolite protection genes into bone marrow stem cells has been to protect bone marrow stem cells during chemotherapy with doxorubicin, actinomycin D, vinblastine, etoposide, and taxol. This strategy allows delivery of higher doses of chemotherapeutic agents in the treatment of drug-sensitive tumor with less bone marrow toxicity. Several human gene therapy clinical trials currently underway are transferring the MDR-1 gene into bone marrow stem cells to diminish the toxicity of chemotherapeutic agents in tumors such as breast cancer for which the clinical activity of cytotoxic drugs is not disputed as it is in PCA. Perhaps a subset of patients with prostate cancer who have tumors that respond to cytotoxic drugs will be identified. Thus, this strategy of bone marrow rescue using the MDR-1 gene has some limitations to apply to the general treatment of PCA but may become relevant when the chemotherapy-sensitive tumors can be phenotyped in advance of treatment by new molecular diagnostics (see Chapters 11, 12).

Antiangiogenesis gene therapy is also emerging as potentially new area for ultimate clinical translation in PCA. Angiogenesis appears essential for the development and progression of malignancy. Angiogenic activators for neovasculature of prostate tumors include basic fibroblast growth factor (FGF), acidic FGF, and vascular endothelial growth factor (VEGF). Many more genes are being scrutinized for their role in PCA angiogenesis as potential targets for gene therapy approaches. Recently, many studies show that angiogenesis inhibitors retard prostate tumor growth in several systems. Prolonged delivery of the angiogenesis inhibitor platelet factor 4 using retroviral and adenoviral vectors selectively inhibits endothelial cell proliferation *in vitro* and results in hypovascular tumors *in vivo*. In addition, this antiangiogenic strategy inhibits tumor angiogenesis and prolongs survival of animal models. The early 21st century can be forecast to include several antiangiogenesis strategies being evaluated by gene therapy clinical research in PCA.

3. PROSTATE-CANCER-SPECIFIC CYTOTOXIC AND ONCOLYTIC VECTORS

Selective expression of cytotoxic genes regulated by PSA enhancer/promoter sequences has been proposed for cytoreductive *in vivo* gene therapy of PCA. As the prostate is an accessory organ and is not required for a male fertility or sexual potency, selective and effective eradication of all PSA-expressing cells would be a desirable therapeutic outcome in men diagnosed with PCA. Some have proposed that upon identification of the hereditary prostate cancer (HPC)-1 gene, prophylactic ablation of the gland if achieved with acceptable toxicities in men carrying the allele offers a rational primary prevention approach to reducing mortality from this neoplasm. Furthermore, serum PSA itself is useful in monitoring progression of the disease and rapidly testing *in vivo* therapeutic response to PSA-expressing-cell killing. Thus, studies of the potential selective toxicity of PSA gene enhancer/promoter sequences for cytoreductive gene therapy of PCA or the preneoplastic prostate gland have both a compelling molecular and clinical rationale.

Many prostate-specific promoters might be employed. The PSA gene is not the only choice but is an excellent choice for molecular proof of the principle. PSA is tightly regulated and is identified even in EST libraries as being a prostate-specific mRNA. Clinically, over 95% of metastatic prostate cancers express PSA, allowing monitoring of disease activity or response to therapy. Other prostate-specific genes (e.g., *DD-3, HGK-2,* and *PCSF*), are also being identified and may be considered in future vector designs. Schuur and Henderson identified and published the first tissue-specific enhancer for the prostate derived from the 5′ flank of the human PSA gene, which is located between −5,824 and −3,738 relative to the PSA cap site [54]. The resulting prostate-specific enhancer (PSE) + PSA enhancer/promoter constructs contain two AREs, 5′ to the TATA box. Cytotoxic adenoviral vectors were chosen to test the specificity of "lead compound" vector constructs using the PSE to express genes cytotoxic to PSA-expressing PCA cells. The rationale for initially using Ad5 vectors included preparation at high titer, absent potential for insertional mutagenesis, a high level of recombinant protein expression in a large number of nonproliferating cells, a favorable human safety profile in phase I studies to date, and impressive tropism followed by cytopathic effects in human CAP cells (Rodriguez, Lim, and Simons, unpublished data).

Two approaches out of many novel vectors considered have been evaluated. For example, we created a replication-defective, novel suicide gene vector for PCA: PSE-DT-A Ela-deleted/E3-deleted virus. The choice of DT-A as a cytotoxin was based on our screening studies showing that DT was the most potent cytotoxin for gene therapy research that does *not* require any PCA cell replication for induction of apoptosis and whose cDNA could be easily genetically engineered into Ad5 [55]. First, Dr. Rodriguez genetically constructed a DT-A-resistant 293 packaging cell line by generating a DT-resistant Ad5 vector-packaging cell line by transfecting the DT-resistant, mutant EF-2 gene cDNA from Chinese hamster ovary (CHO) cells into L-293 cells. The Ad5 PSE-DT-A virus has impressive potential antineoplastic effects preclinically against PSA-expressing tumors. However, low-production titers require further

technical improvements in the 293 packaging cell line before it is feasible to consider large-scale clinical translation.

Oncolytic vectors that are attenuated to replicate selectively in PSA-expressing cells and lyse can generate impressive antineoplastic activity *in vitro* and *in vivo*. For example, we created a replication-competent, attenuated cytotoxic Ad5 vector for PSA-gene-expressing PCA cells. Minimal enhancer/promoter constructs derived from the 5′ flank of the human PSA gene (PSE) were inserted into adenovirus type 5 DNA so as to drive the E1A gene, thereby creating a PSE-containing virus, CN706 [33]. E1A protein was expressed at 100-fold higher levels in CN706-infected human-PSA-producing LNCaP cells than in CN706-infected DU145 cells. The selective cytotoxicity for PSA gene expression was evaluated *in vitro*. By titer of CN706 compared to wild-type Ad5, production of CN706 was 20- to 3,000-fold higher in LNCaP cells than in several human tumor cell lines that do not produce PSA. Compared to wild-type Ad5, CN706 has a selective cytotoxicity for PSA+ tumor cells of 3,000 : 1 to 20 : 1 *in vitro*. Most FDA-approved chemotherapy drugs have clinical specificities described by therapeutic indices between 2 : 1 and 6 : 1. (Elliot *et al.*, Cancer Res. 54:4412–4418, 1994). In PSA-expressing LNCaP cells, the oncolytic yield of CN706 was dependent on exogenous androgen (R1881). *In vivo*, CN706 destroyed via oncolytic infections large LNCaP tumors (1×10^9 cells) and abolished PSA production in *nu/nu* mouse xenograft models with a single intratumoral injection of 5×10^8 pfu. With clinical-grade CN706, dose-response studies show a clear dose dependence with a single intratumoral dose (Simons and Henderson, unpublished data). Schedule studies show that four daily injections rather than one can reduce by potentially two orders of magnitude the cumulative dose required to eradicate large (10^9 LNCaP cells) PSA-expressing tumors (Bacalla and Simons, unpublished data). Ongoing studies suggest a profound and non-immune-mediated bystander effect from microinfarctions surrounding apoptotic (TdT-positive) LNCaP cells (Lim, Rodriguez, and Simons, manuscript in preparation). These and other studies on schedule, dose, and mechanism of action were permitted in FDA concurrence with the design of the first phase I study for CN706. Regulatory agencies, however, require proof of safety and bioactivity via phase I dose escalation by intratumoral injection of CN706 in recurrent T-3 tumors following radiotherapy. The same clinical development path is being followed with the Onynx 015 oncolytic adenoviral vector in p53-mutated head and neck cancers (see Chapter 15). In advance of, and in parallel to, the intralesional phase I dose-escalation trial scheduled to begin after NIH RAC approval, CN706 is being evaluated critically as an intravenous antineoplastic for CAP. Oncolytic adenoviral vectors will be limited in the duration of their clinical use due to neutralizing B-cell and T-cell immune responses to Ad5 viruses. Approaches to transient immunosuppression or the use of second generation, less immunogeneic adenoviral vectors are issues requiring additional research if intravenous, oncolytic, PSE-regulated Ad5 vectors are to be used repeatedly in patients with advanced, metastatic PCA.

IV. CURRENT HUMAN GENE THERAPY TRIALS IN PROSTATE CANCER

As described in Table 1, the majority of human gene therapy trials in PCA are cytoreductive in their approaches. Most are still accruing at this writing, and have not had data mature for publication. The projected completion and reporting times suggest that 1999 will be a year in which at least four human gene therapy trials in PCA will have undergone peer review.

The first phase I trial in human gene therapy for PCA (Simons, Johns Hopkins) has completed accrual, and is undergoing analysis. The first NIH RAC- and FDA-approved trial for human gene therapy of PCA employed *ex vivo* gene transfer and the use of GM-CSF-gene-transduced autologous tumor vaccines. The MFG vector was employed as in the preclinical studies, which showed efficacy in preestablished tumors and defined some parameters for GM-CSF gene expression levels and vaccine schedules. In this respect, the preclinical basic gene therapy research had an inherent translational research bias toward human gene therapy clinical trial evaluation.

Patients with metastatic PCA found at prostatectomy were enrolled (Simons, Johns Hopkins). Two dose levels of vaccine were evaluated (1×10^7 and 5×10^7 cells) secreting GM-CSF at 150–1,500 ng/million cells/24 h. The schedule was vaccination every 2 weeks until dose-limiting toxicities or exhaustion of vaccine. In preclinical models, secretion by vaccine cells of >35 ng GM-CSF/million vaccine cells/24 h conferred antitumor activity in PCA models. In human renal cell carcinoma, clinical activity (7-month partial response) was demonstrated by Simons and colleagues in the first phase I study using the identical MFG vector system at the dose level of 4×10^7 autologous vaccine cells secreting 150 ng of GM-CSF/million cells/24 h [56]. In this phase I trial, the *ex vivo* gene therapy trial evaluated the strategy in patients with metastatic PCA (T2–3/N + M0) found incidentally at radical prostatectomy. Autologous PCA vaccines were generated following MFG-GM-CSF transduction as originally reported by our group. In a

small dose-escalation study, 8 patients were treated as outpatients with irradiated, autologous PCA vaccine cells after *ex vivo* human GM-CSF gene transfer using the MFG retrovirus vector. The major limitation of this treatment approach was *in vitro* expansion (3/11) cases; *ex vivo* GM-CSF gene transfer was never a technical problem. The treatment was 100% successful in 8/8 established cultures to specification of GM-CSF secretion. No NCI CTC dose-limiting toxicities were observed among vaccinated patients. Toxicities were restricted to pruritis and erythema at the vaccination sites. These resolved within 7 days following vaccination. No recombinant retrovirus was detected in clinical vaccines or in patients following treatment. Over 42 fully evaluable vaccination PCA cycles confirmed the outpatient safety of this early approach to *ex vivo* gene therapy for PCA.

Posttreatment biopsies of intradermal vaccination sites displayed distinctive inflammatory infiltrate, composed of macrophages, dendritic cells, eosinophils, and T cells similar to those observed in preclinical models of efficacy. They also closely resembled those seen in our first renal cell carcinoma trial, which used MFG-GM-CSF gene-transduced vaccines in patients randomized to the plus GM-CSF gene transfer arm of the trial [56]. Antigen-presenting cells, dendritic cells, and macrophages were clearly recruited by vaccination with GM-CSF-gene-transduced, irradiated, autologous PCA vaccines at both dose levels studied in the phase I trial. The cellular infiltrates appeared to be more intense at the high vaccine cell dose level, but the trial was not statistically powered. It was designed only to define a safe, outpatient phase II dose range based on previous experience with renal cell carcinoma.

A striking recruitment of intradermal eosinophils surrounding APCs at the intradermal vaccination sites was also noted. These consistent molecular studies of vaccination sites suggest that the required antigen-presenting cells were recruited to vaccination sites by high levels of paracrine expression of GM-CSF from autologous vaccine cells.

Unfortunately, quantitative human antitumor immune response assays lag behind the clinical development of immunotherapy for PCA. The potency of one schedule of immunotherapy versus another is not yet rigorously testable in patients with cancer. The specific, relevant, immunodominant, PCA-associated antigens in the autologous vaccine that are presented to T-cells by GM-CSF-activated APCs are also unknown. Peptide antigen discovery techniques are being used in archived patient T cells before and after vaccination in this trial to try elucidate some of these tumor-associated peptides.

Nevertheless, systemic treatment effects can be detected if not well quantitated. Delayed type hypersensitivity (DTH) reactivity assessment to intradermal challenge with autologous, irradiated, untransduced tumor cells has been used in multiple tumor vaccine strategies to confirm generation of CD4 cells that recognize antigens on the cancer vaccine and the patient's primary tumor. A dose dependence was discovered using autologous tumor DTH reactivity in the first clinical trial employing MFG-GM-CSF-gene-transduced renal cell carcinoma vaccines [56]. In the PCA phase I study using MFG-GM-CSF-gene-transduced prostate cancer vaccines, every patient displayed DTH reactivity to untransduced, autologous PCA target cells after vaccination. Interestingly, the largest biopsy confirmed that DTH reactivity after vaccination was observed in a patient who had a documented past history of prostatitis.

The median serum PSA level before surgery was 28.85 (with a range of 6.7–75), and the median PSA level at first vaccination was 0.65 (with a range of 0.1–30.4). By ultrasensitive serum PSA, 7/8 patients progressed after surgery and vaccination: average follow-up is now at greater than 20 months. Exhaustion of the supply of vaccine cells limited evaluation of multiple vaccinations over several months. Total vaccine yield, not *ex vivo* gene transfer efficiency, is the major technical hindrance for large phase II studies. This study demonstrated the feasibility, outpatient safety, and bioactivity of *ex vivo* GM-CSF-gene-transduced PCA vaccines for phase II trials powered to estimate efficacy.

A significant limitation of current autologous vaccine strategies is that when primary cultures of gene-transduced tumor cells are used, vaccine supply simply runs out well before questions of cumulative dose or long schedules of vaccination can be explored potentially in phase III studies. In addition, many prostate tumors are under 2 g at radical prostatectomy, further limiting total yields of vaccine cells. Technical improvements in autologous vaccine-expansion cell culture may circumvent some of these technical limitations.

The use of GM-CSF-gene-transduced prostate-cancer-derived allogeneic vaccines allows a limitless supply of vaccine for phase II studies. The cost is the loss of potentially immunodominant autologous-tumor-associated antigens. Fortunately, at least, HLA matching may not be required with GM-CSF-gene-transduced allogeneic tumor vaccines. After vaccination, the GM-CSF-activated antigen-presenting cells in the patient process tumor-associated peptides and present them to the autologous T cells of the patient. The first phase II clinical question in PCA gene therapy is being conducted currently at Johns Hopkins in patients with serum-PSA-detected recurrences of PCA following anatomic prostatectomy. These patients have had no other therapy and have low-tumor-burden disease with negative bone scans as a final entry criterion. In this trial, sterile, unpassaged, autologous tumor is not available

from most patients who underwent surgery where tumor cell processing as a clinical research effort was not taking place. In the phase II allogeneic trial, two MFG-GM-CSF-gene-transduced allogeneic PCA cell lines are given simultaneously within the phase II dose range defined by the phase I autologous PCA trial. The high level of GM-CSF gene expression chosen in this allogeneic trial was also designed in large measure based on the data generated from the autologous trials in renal cell carcinoma and PCA that preceded it. The PC-3 (bone-metastasis-derived) and the PSA-expressing Ln-CAP (lymph-node-metastasis-derived) cell lines are being used as potential PCA-associated antigen sources. This trial is powered to estimate efficacy.

The second NIH RAC- and FDA-approved clinical trial in gene therapy for PCA employs replication-defective Ad5 vectors carrying the TK gene. It is designed to evaluate dose escalation using direct intratumoral injection of the suicide gene construct under ultrasound guidance (Scardino, Baylor). No prostate-specific promoter construct is used; rather, the trial explores the safety of dose-escalation TK and GCV effects for intralesional therapy. Thus, this vector at the level of nucleotide sequence has no intrinsic PCA cell specificity for potential intravenous use. The vector employed is AD5 with the RSV promoter driving the TK gene. Other phase I trials continue to accrue and mature as described in Table 1. The projected completion and reporting times suggest that 1999 will be a year in which at least four human gene therapy trials in PCA will have undergone peer review.

In parallel, entirely new avenues for oncolytic gene therapy (CN706, Simons, Johns Hopkins), osteocalcin-promoter-driven suicide gene therapy for PCA bone metastases (OC-TK, Gardener and Chung, University of Virginia), and p53 Ad5 vectors as proof of principle to preprostatectomy specimens (Belldegrun, UCLA), are all in translation to phase I testing in 1998.

V. FUTURE DIRECTIONS

If the past 5 years of preclinical research activity are predictive, many gene therapy clinical trial reports will be available by the new millenium to stimulate a new generation of urologists, radiation oncologists, and medical oncologists. Entirely novel concepts for the treatment of PCA continue to emerge from basic gene therapy research. Vector improvements and advances in basic understanding of the human antitumor immune responses can be predicted to impact heavily on the field. Immunotherapies applying gene transfer techniques and new immunologic concepts are continuing in their development. The more potent and long-lasting generation of cytolytic T cells against PCA using *ex vivo* and *in vivo* gene transfer systems is one obvious area for further development. Of note, even the first small phase I human gene therapy trial of 8 patients with PCA consistently demonstrated activation of both antigen presentation at vaccination sites and systemic generation of CD4 T cells that could recognize antigens expressed by the patient's primary tumor. New technologies may allow exploration of autologous and allogeneic approaches to genetically modified vaccines and *in situ* autovaccination using *in vivo* gene transfer.

A huge pragmatic hurdle to be overcome is the generation of systemic antineoplastic responses using purely cytotoxic vectors, which look interesting when given intralesionally. Clearly tissue-specific promoters allow multiple iterations of vectors that could be prostate specific if given intravenously. Less clear, however, is how well the adenoviral-vector-based approaches to suicide gene approaches, p53 vectors, or PSA-specific oncolytic vectors will work systemically. Evaluation in combination with antiandrogens and other cytotoxic or cytostatic drugs will also be critical. Each approach will require innovative clinical trial designs with molecular intermediate end points of therapeutic action. A rigorous attention to pharmacokinetics of vector clearance in clinical trials will be critical as well. In the 4 years since the first NIH RAC approval of a human gene therapy trial in PCA, cytoreductive treatment of PCA using recombinant DNA technology has advanced beyond proof-of-principle trials to approved phase II questions. The field can be anticipated to be a vibrant area of basic and clinical pharmacology research for the foreseeable future. The final measure, however, will be to meet an as yet unmet social need: prolongation of PCA-free survival in men with metastatic disease through the use of the vast therapeutic potential in the code of the DNA molecule.

Acknowledgments

The authors wish to acknowledge the insightful and scholarly comments of Ms. Kimberly Cordwell and Mr. Thomas Boswell in the preparation of this manuscript.

References

1. Wingo, P. A., Troy, T., Bolden, S. Cancer statistics, 1996. *CA Cancer J. Clin.* **46,** 113–125, 1996.
2. Huggins, C., Hodges, C. Studies on prostate cancer. I. The effect of castration, of estrogen and of androgen injection on serum phosphatases in metastatic carcinoma of the prostate. *Cancer Res.* **1,** 293–297, 1941.
3. Redding, T. W., Schally, A. V. Investigation of the combination of the agonist D-Trp-6-LH-RH and the anti-androgen flutamide in the treatment of Dunning R3327H prostate cancer model. *Prostate* **6,** 219–232, 1985.
4. Schally, A. V., Comaru-Schally, A. M., Redding, T. W. Antitumor effects of analogues of hypothalamic hormones in endocrine-dependent cancers. *Proc. Soc. Exp. Biol. Med.* **175,** 259–281, 1984.
5. Scher, H. I. Prostate cancer: improving the therapeutic index. *Semin. Oncol.* **21,** 688–693, 1994.
6. Coffey, D. S., Issacs, J. T. Theory—prostate tumor biology and cell kinetics. *Urology* **17,** 40–53, 1981.
7. Berges, R. R., Vulkanovic, J., Epstein, J. I., *et al.* Implication of cell kinetic changes during the progression of human prostatic cancer. *Clin. Can. Res.* **1,** 473–480, 1995.
8. Shackney, S. E., McCormick, G. W., Cuchural, G. J. Growth rate patterns of solid tumors and their relation to responsiveness to therapy. *Ann. Intern. Med.* **89**(1), 107–121, 1978.
9. Brooks, J. D., Bova, G. S., Ewing, C. M., Piantadosi, S., Carter, B., Robinson, J. C., Epstein, J. I., Isaacs, W. B. An uncertain role for p53 gene alterations in human prostate cancers. *Cancer Res.* **56,** 3814–3822, 1996.
10. Issacs, W. B., Carter, B. S., Ewing, C. M. Wild-type p53 suppresses growth of human prostate cancer cells containing mutant p53 alleles. *Cancer Res.* **51,** 4716–4720, 1991.
11. Asgari, K., Sesterhenn, I. A., McLeod, D. G., *et al.* Inhibition of the growth of pre-established subcutaneous tumor nodules of human prostate cancer cells by single injection of the recombinant adenovirus p53 expression vectors. *Int. J. Cancer* **71,** 377–382, 1997.
12. Taneja, S. S., Pang, S., Cohan, P., Belldegrun, A. Gene therapy: principles and potential. *Cancer Surv.* **23,** 247–266, 1995.
13. Bookstein, R., Rio, P., Madreperla, S. A., *et al.* Promoter deletion and loss of retinoblastoma gene expression in human prostate carcinoma. *Proc. Natl. Acad. Sci. U.S.A.* **87,** 7762–7766, 1990.
14. Bookstein, R., Shew, J.-Y., Chen, P.-L., *et al.* Suppression of tumorigenicity of human prostate carcinoma cells by replacing a mutated RB gene. *Science* **247,** 712–715, 1990.
15. Lee, W.-H., Morton, R. A., Epstein, J. I., Isaacs, W. B., Bova, G. S., Nelson, W. G. Cytidine methylation of regulatory sequences near the p-class glutathione-*S*-transferase gene accompanies human prostate cancer carcinogenesis. *Proc. Natl. Acad. Sci. U.S.A.* **91,** 11733–11737, 1994.
16. Umbas, R., Isaacs, W. B., Bringuier, P. P., *et al.* Decreased E-cadherin expression is associated with poor prognosis in patients with prostate cancer. *Cancer Res.* **54,** 3929–3933, 1994.
17. Bussemakers, M. J., van Moorselaar, R. J., Giroldi, L. A., *et al.* Decreased expression of E-cadherin in the progression of rat prostatic cancer. *Cancer Res.* **52,** 2916–2922, 1992.
18. Umbas, R., Isaacs, W. B., Bringuier, P. P., *et al.* Decreased E-cadherin expression is associated with poor prognosis in patients with prostate cancer. *Cancer Res.* **54,** 3929–3933, 1994.
19. Hsieh, J. T., Luo, W., Song, W., *et al.* Tumor suppressive role of an androgen-regulated epithelial cell adhesion molecule (C-CAM) in prostate carcinoma cell revealed by sense and antisense approaches. *Cancer Res.* **55,** 190–197, 1995.
20. Kleinerman, D. I., Zhang, W. W., Lin, S. H., *et al.* Application of a tumor suppressor (C-CAM1)-expressing recombinant adenovirus in androgen-independent human prostate cancer therapy: a preclinical study. *Cancer Res.* **55,** 2831–2836, 1995.
21. Steiner, M. S., Holt, J. T. Gene therapy for the treatment of advanced prostate cancer by *in vivo* transduction with prostate targeted retroviral vectors expressing antisense c-*myc* RNA. RAC report 9509-123, 1995.
22. Mukhopadhyay, T., Tainsky, M., Cavender, A. C., *et al.* Specific inhibition of K-*ras* expression and tumorigenicity of lung cancer cells by antisense RNA. *Cancer Res.* **51,** 1744–1748, 1991.

23. Zhang, Y., Mukhopadhyay, T., Donenhower, L. A., *et al.* Retroviral vector-mediated transduction of K-*ras* antisense RNA into human lung cancer cells inhibits expression of the malignant phenotype. *Hum. Gene Ther.* **4,** 451–460, 1993.

24. Georges, R. N., Mukhopadhyay, T., Zhang, Y., *et al.* Prevention of orthotopic human lung cancer growth by intratracheal instillation of a retroviral antisense K-*ras* construct. *Cancer Res.* **53,** 1743–1746, 1993.

25. Carter, B. S., Epstein, J. I., Issacs, W. B. *ras* gene mutations in human prostate cancer. *Cancer Res.* **50,** 6830–6832, 1991.

26. Moul, J. W., Friedrichs, P. A., Lance, R. S., *et al.* Infrequent *ras* oncogene mutations in human prostate cancer. *Prostate* **20,** 327–338, 1992.

27. Mulligan, R. The basic science of gene therapy. **260,** 926–932, 1993.

28. Sweatt, W. H., Simons, J. W. Gene therapy for genitourinary cancer. In *Principles and Practice of Genitourinary Oncology,* Appleton, pp. 121–130, 1997.

29. Vieweg, J., Rosenthal, F., Bannerji, R., *et al.* Immunotherapy for prostate cancer in the Dunning rat model: use of cytokine gene modified tumor vaccines. *Cancer Res.* **54,** 1760–1765, 1994.

30. Sanda, M., Ayyagari, S., Jaffee, E., *et al.* Demonstration of a rational strategy for human prostate cancer gene therapy. *J. Urol.* **151,** 622–628, 1994.

31. Carducci, M. A., Ayyagari, S. R., Sanda, M. G., *et al.* Gene therapy for human prostate cancer: translational research in the hormone refractory Dunning prostate cancer model. *Cancer* **75**(1), 2014–2019, 1995.

32. Hall, S. J., Mutchnik, S. E., Chen, S.-H., *et al.* Adenovirus-mediated herpes simplex virus thymidine kinase gene and ganciclovir therapy leads to systemic activity against spontaneous and induced metastasis in an orthotopic mouse model of prostate cancer. *Int. J. Cancer* **70,** 183–187, 1997.

33. Rodriguez, R., Schuur, E. R., Lim, H. Y., Henderson, G. A., Simons, J. W., Henderson, D. R. Prostate attenuated replication competent adenovirus (ARCA) CN706: a selective cytotoxic for prostate-specific antigen-positive prostate cancer cells. *Cancer Res.* **57,** 2559–2563, 1997.

34. Ghosh-Choudhury, G., Haj-Ahmad, Y., Brinkley, P., *et al.* Human adenovirus cloning vectors based on infectious bacterial plasmid. *Gene* **50,** 161–171, 1986.

35. Berkner, K. L. Development of adenovirus vectors for the expression of heterologous genes. *Biotechniques* **6,** 616–629, 1988.

36. Rosenfeld, M. A., Yoshimura, K., Trapnell, B. C., *et al.* In vivo transfer of the human cystic fibrosis transmembrane conductance regulator gene to the airway epithelium. *Cell* **68,** 143–155, 1992.

37. Yang, Y., Nunes, F., Berencsi, K., *et al.* Cellular immunity to viral antigens limits E1-deleted adenoviruses for gene therapy. *Proc. Natl. Acad. Sci. U.S.A.* **91,** 4407–4441, 1994.

38. Berns, K. I., Giraud, C. Adenovirus and adeno-associated virus as vectors for gene therapy. *Ann. N.Y. Acad. Sci.* **772,** 95–104, 1995.

39. McKeon, C., Samulski, R. J. NIDDK workshop on AAV vectors: gene transfer into quiescent cells. *Hum. Gene Ther.* **7,** 1615–1619, 1996.

40. Kotin, R. M., Siniscalco, M., Samulski, R. J., *et al.* Site-specific integration by adeno-associated virus. *Proc. Natl. Acad. Sci. U.S.A.* **87,** 2211–2215, 1990.

41. Sutter, G., Moss, B. Non-replicating vaccina virus vector efficiently expresses recombinant genes. *Proc. Natl. Acad. Sci. U.S.A.* **89,** 10847–10851, 1992.

42. Moss, B., Flexner, C. Vaccinia virus expression vectors. *Annu. Rev. Immun.* **5,** 305–324, 1987.

43. Levitsky, H., Lazenby, A., Hayashi, R., *et al.* In vivo priming of 2 distinct antitumor effector populations: the role of MHC class-I expression. *J. Exp. Med.* **179,** 1215–1224, 1994.

44. Dranoff, G., Jaffee, E., Lazenby, A., *et al.* Vaccination with irradiated tumor cells engineered to secrete murine granulocyte-macrophage colony-stimulating factor stimulates potent, specific, and long-lasting antitumor immunity. *Proc. Natl. Acad. Sci. U.S.A.* **90,** 3539–3543, 1993.

45. Danos, O., Mulligan, R. C. Safe and efficient generation of recombinant retroviruses with amphotropic and ecotropic host ranges. *Proc. Natl. Acad. Sci. U.S.A.* **85,** 6460–6464, 1988.

46. Yang, C., Cirielli, C., Capogrossi, M. C., *et al.* Adenovirus-mediated wild-type p53 expression induces apoptosis and suppresses tumorigenesis of prostatic tumor cells. *Cancer Res.* **55,** 4210–4213, 1995.

47. Ko, S. C., Gotoh, A., Thalmann, G. N., *et al.* Molecular therapy with recombinant p53 adenovirus in an androgen-independent, metastatic human prostate cancer model. *Hum. Gene Ther.* **7,** 1683–1691, 1996.

48. Asgari, K., Sesterhenn, I. A., McLeod, D. G., *et al.* Inhibition of the growth of pre-established subcutaneous tumor nodules of human prostate cancer cells by single injection of the recombinant adenovirus p53 expression vectors. *Int. J. Cancer* **71,** 377–382, 1997.

49. Hall, S. J., Mutchnik, S. E., Chen, S. H., *et al.* Adenovirus-mediated herpes simplex virus thymidine kinase gene and ganciclovir therapy leads to systemic activity against spontaneous and induced metastasis in an orthotopic mouse model of prostate cancer. *Int. J. Cancer* **70,** 183–187, 1997.

50. Eastham, J. A., Chen, S. H., Sehgal, I., *et al.* Prostate cancer gene therapy: herpes simplex virus thymidine kinase gene transduction followed by ganciclovir in mouse and human prostate cancer models. *Hum. Gene Ther.* **7,** 515–523, 1996.

51. Scardino, P. T., Thompson, T. C., Woo, S. L. C. Phase I study of adenoviral vector delivery of the HSV-TK gene and the intravenous administration of ganciclovir in men with local recurrence of prostate cancer after radiation therapy. *RAC Report* 9601-144, 1996.

52. Huber, B. E., Austin, E. A., Richards, C. A., *et al.* Metabolism of 5-fluorocytosine to 5-fluorouracil in human colorectal tumor cells transduced with the cytosine deaminase gene: significant antitumor effects when only a small percentage of tumor cells express cytosine deaminase. *Proc. Natl. Acad. Sci. U.S.A.* **91,** 8302–8306, 1994.

53. Hughes, B. W., Wells, A. H., Bebok, Z., *et al.* Bystander killing of melanoma cells using the human tyrosinase promoter to express the *Escherichia coli* purine nucleoside phosphorylase gene. *Cancer Res.* **55,** 3339–3345, 1995.

54. Schuur, E. R, Henderson, G. A., Kemetee, L. A., Henderson, D. Prostate-specific antigen expression is regulated by an upstream enhancer. *J. Biol. Chem.* **271**(12), 7043–7051, 1996.

55. Rodriguez, R., Lim, H. Y., *et al. The Prostate,* in press, 1998.

56. Simons, J. W., Jaffee, E. M., Weber, C. E., *et al.* Bioactivity of autologous irradiated renal cell carcinoma vaccines generated by *ex vivo* granulocyte-macrophage colony-stimulating factor gene transfer. *Cancer Res.* **57,** 1537–1546, 1997.

PART VII

Genetically Modified Effector Cells for Immune Based Immunotherapy

CHAPTER 22

Applications of Gene Transfer in the Adoptive Immunotherapy of Cancer

ALFRED E. CHANG AND A. P. SALAS

Division of Surgical Oncology, University of Michigan Medical School, Ann Arbor, Michigan 48109

I. INTRODUCTION

The rapid strides made in recombinant gene technology have been the impetus for equally dramatic developments in our understanding of various aspects of cancer carcinogenesis, growth factors, cellular proliferation, differentiation, apoptosis, and metastasis. Novel therapeutic approaches that take advantage of this new knowledge involve biological alterations of the host cell at a genetic level. Many of these approaches are discussed in greater detail elsewhere in this book. This chapter focuses on the applications of gene transfer to the adoptive immunotherapy of malignancy. This form of cellular therapy refers to the infusion of tumor-reactive immune cells into the tumor-bearing host to mediate, directly or indirectly, regression of established tumor. This review is divided into three areas, each one involving different methods of genetic transfer to generate immune cells into the tumor-bearing host for subsequent adoptive transfer: (1) the use of gene-modified tumors to serve as immunogens to generate effector T cells, (2) genetic manipulation of T cells to enhance antitumor reactivity, and (3) genetic modulation of DC.

The feasibility of adoptive immunotherapy for cancer is predicated on two fundamental observations derived from animal models. The first is that tumor cells express antigens that are qualitatively or quantitatively different from normal cells and can elicit an immune response within the syngeneic host. The second is that the immune rejection of established tumors can be mediated by the adoptive transfer of appropriately sensitized lymphoid cells.

In 1943, Gross [1] was the first to recognize that inbred mice could be immunized against a tumor that was developed in a mouse of the same inbred strain, thus documenting the existence of tumor-associated antigens. Over the years, it has become apparent that individual tumors vary greatly in the nature of their "immunogenicity." The immunogenicity of a tumor has a direct influence on the ability to develop cellular antitumor immune responses.

The transfer of immunity to a naive host by the use of cells was first described by Landsteiner and Chase [2] in 1942. They reported that hypersensitivity to simple compounds could be transferred to normal rats by the transfer of peritoneal exudate cells of sensitized donor animals. In 1954, Billingham et al. [3] documented the ability to transfer skin allograft immunity to a normal murine host by the use of regional lymph node cells from animals that had rejected primary skin allografts. These investigators developed the term *adoptive immunity* to describe the acquisition of immunity in a normal subject as a result of the transference, not of preformed antibody, but of immunologically competent cells. In 1955, Mitchison [4] was the first to report about the adoptive immunotherapy of tumors in a rodent model. In this study, the adoptive transfer of lymph node cells from mice that rejected tumor allografts conferred accelerated rejection of the same tumor allografts in naive hosts. However, more germane to clinical therapy is the ability to transfer immunity to autologous tumors (i.e., syngeneic tumors in inbred rodent strains) using lymphocytes. Borberg et al. [5] was the first to clearly show that the infusion of syngeneic immune cells from hyperimmunized donor animals was capable of mediating the regression of established tumor. During the ensuing years, several other investigators have documented the ability to successfully treat established syngeneic tumors by the adoptive transfer of immune "effector" cells [6,7].

One of the major obstacles in extrapolating the concepts developed in the animal models to the clinical treatment of cancers was the inability to generate adequate numbers of immune cells for therapy. Methods to grow or expand lymphoid cells while retaining their immunologic reactivity were limited. However, in 1976, the discovery of interleukin-2 (IL-2) as a T-cell growth factor made it possible to culture activated T cells in large quantities [8]. In 1981, Cheever et al. [9] showed that tumor-reactive T cells could be expanded in IL-2 and still maintain their therapeutic efficacy in adoptive immunotherapy of tumors in mice.

The first successful clinical application of cellular therapy in man was reported by Rosenberg et al. [10] in 1985. These investigators generated large quantities of IL-2-activated peripheral blood lymphocytes (approximately 1 to 2 × 10^{11} cells/patient), which were infused along with the concomitant administration of IL-2. These lymphokine-activated killer (LAK) cells were nonspecifically cytolytic to tumor cells by *in vitro* measurements. In these early clinical trials, significant tumor burdens regressed in a subset of patients and the feasibility of generating large numbers of cells for clinical therapy was achieved. Based on subsequent animal studies [11,12], TILs were found to be an alternative population of cells that were more potent than LAK cells in mediating tumor regression. Rosenberg and co-workers were the first to report significant clinical responses using TILs in the treatment of patients with advanced melanoma [13,14].

Despite the progress in this field, cellular therapy is still in its infancy. The isolation and expansion of potent immune effector cells derived from the tumor-bearing host remains a formidable task. The use of genetic approaches to alter the host-tumor immune response offers potential opportunities to develop more potent cellular reagents (Table 1).

TABLE 1   Genetic Approaches to
Adoptive Immunotherapy

1. Generation of effector cells using gene-modified tumors
   a. TILs isolated from gene-modified tumors
   b. Lymph node cells draining gene-modified tumors
2. Gene modification of effector cells
   a. Marking studies of effector cells
   b. Cytokine gene transfer into effector cells
   c. Modification with signal transduction genes
   d. Redirection of effector cells with chimeric receptor genes
   e. Genetic engineering of effector cells from naive cells (i.e., resting lymphoid cells or stem cells)
3. Genetic modification of DCs
   a. Transfer of DCs with genes encoding tumor-associated antigens

II. USE OF GENE-MODIFIED TUMORS TO GENERATE ANTITUMOR-REACTIVE T CELLS

Genetic modification of tumors to secrete or express immunomodulatory peptides has been found to significantly alter the biology of the tumor cells when these are inoculated into the syngeneic host. A majority of these studies have involved the use of various cytokine genes introduced into tumor cells. A further discussion of such studies can be found in Chapter 24. Many of the observed changes are related to an enhanced cellular immune response to tumor-associated antigens expressed on the parental tumor. In selected animal models, regression of established tumors by the inoculation of genetically modified tumor cells that secrete IL-4 or interferon gamma (IFN-γ) administered as a tumor vaccination has been observed [15–17]. These studies have served as a rationale to initiate vaccination trials in humans for the therapy of cancers. Based on these early animal studies, investigators have used gene-modified tumors to develop cellular reagents for adoptive immunotherapy. The generation of TILs and vaccine-primed lymph node cells using gene-modified tumors as immunogens is described in this section.

A. TILs Derived from Gene-Modified Tumors

TILs represent lymphoid cells derived from tumors that are disaggregated *ex vivo* and cultured in IL-2 [11]. A significant impediment to generating therapeutic TILs resides in the inherent immunogenicity of the tumor from which the TILs are derived. In animal models using poorly immunogenic tumors, the therapeutic efficacy of the TILs is limited [11]. These observations suggest that TILs represent a heterogeneous population of cells and that a significant portion of these cells are not appropriately sensitized within "poorly" or nonimmunogenic tumors. Because human tumors are postulated to be nonimmunogenic based on their spontaneous origins and ability to escape the host immune system, methods to enhance the isolation of therapeutic TILs from these tumors could have significant clinical applications.

Tumors genetically engineered to produce certain cytokines have been found to contain TILs with enhanced *in vitro* and *in vivo* antitumor reactivity. Using the poorly immunogenic methylcholanthrene (MCA)-induced 101 murine fibrosarcoma, Restifo *et al.* [18] showed that therapeutic TILs could be generated if these tumor cells were genetically engineered to secrete

IFN-γ. Transduction of MCA 101 tumor cells to secrete IFN-γ resulted in up-regulated expression of major histocompatibility complex (MHC) class I molecules. TILs derived from IFN-γ-secreting tumors mediated regression of established parental tumor metastases compared with TILs derived from wild-type tumor, which were ineffective. In additional studies, they found that TILs from the transduced tumors were capable of presenting viral antigen to sensitized T cells whereas TILs from the parental tumor could not. These studies showed that tumors can be genetically altered with the IFN-γ gene to become "nonprofessional" antigen-presenting cells and that such tumors appeared to be a more reliable source for therapeutically effective TILs.

In a model of established lung metastases, Marincola *et al.* [19] evaluated the immunologic response of the host to an inoculation of syngeneic tumor modified to secrete tumor necrosis factor alpha (TNF-α). TNF-α can stimulate T-cell proliferation and cytotoxic T lymphocyte (CTL) activity. In these studies, the poorly immunogenic MCA 102 sarcoma was modified to secrete TNF-α. The tumorigenicity of the TNF-α-secreting tumor cells was not different from that of wild-type tumor in normal animals. However, TNF-α-secreting tumor cells inoculated subcutaneously regressed in the presence of established wild-type pulmonary metastases, in contrast to inocula of wild-type tumor cells, which grew progressively. In this setting, TILs that were generated from TNF-α-secreting tumors mediated significant antitumor reactivity in adoptive transfer studies compared with wild-type tumor cells, which could not. These observations are important in documenting that therapeutic TILs can be generated from poorly immunogenic tumors modified to secrete TNF-α; moreover, this was accomplished in hosts bearing significant tumor burden.

IL-7 is another cytokine that has been shown to enhance recovery of TILs from tumors. When the gene for murine IL-7 was retrovirally transferred into an immunogenic murine fibrosarcoma, the tumorigenicity of the tumor was significantly diminished [20]. Moreover by flow cytometry, the IL-7-secreting tumor showed a greater than fivefold increase in infiltrating T cells compared with wild-type tumor. These infiltrating cells were primarily CD8+ cells that mediated enhanced *in vitro* cytotoxicity against wild-type tumor compared with T cells isolated from control tumors transfected with a neomycin-resistance gene. These studies indicated that IL-7 secretion by the gene-modified tumors can promote the recruitment or induction of cytolytic T cells to the site of that same tumor.

A novel variation to alter TIL reactivity has been developed at the University of Michigan. In earlier animal studies, tumors treated by the *in vivo* transfer of an allogeneic MHC class I gene complexed with liposomes resulted in expression of the class I molecules by tumor cells [21]. This also resulted in tumor regression and induction of T cells that were reactive not only to transfected tumor cells, but also to unmodified cancer cells. We have also been able to achieve gene expression in advanced melanoma patients treated by the *in vivo* inoculation of tumor with DNA/liposome complexes containing a foreign MHC class I gene [22]. Based upon these initial observations, we are currently evaluating the immune reactivity of TILs derived from tumors modified by direct *in vivo* gene transfer utilizing an allogeneic class I MHC gene, *HLA B7*, in patients with stage IV melanoma. We have confirmed, in *in vitro* assays, an enhanced reactivity of TIL derived from patients inoculated with the foreign class I gene (Fig. 1). We postulate that the allogeneic response induced by the expression of *HLA B7* results in elaboration of other cytokines which enhance TIL reactivity to tumor-associated antigens. The advantage of this approach is that it does not employ a viral vector and does not require the establishment of a cultured tumor line to accomplish gene transfer.

B. Lymph Node Cells Sensitized with Gene-Modified Tumor Vaccines

Our laboratory has had a long-standing interest in the use of tumor-draining lymph nodes (TDLNs) or vaccine-primed lymph nodes (VPLNs) as a source of T cells for adoptive immunotherapy in murine models and in human trials [23–31]. We have shown that lymphoid cells derived from TDLNs or VPLNs by themselves do not possess antitumor reactivity, but require secondary activation *in vitro* to gain functional antitumor activity. We have called these TDLN or VPLN cells, "pre-

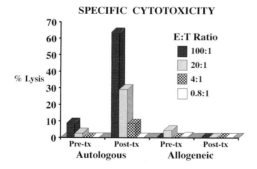

FIGURE 1 *In vitro* cytolytic activity of TILs derived from a melanoma nodule before (pretreatment) and after (posttreatment) intralesional *HLA B7*/liposome inoculations given 3 times every 2 weeks. Antitumor reactivity was assessed against autologous and allogeneic melanoma targets at various effector/target (E/T) ratios. Tumor-specific TIL cytolytic reactivity was enhanced after *HLA B7* inoculation.

effector" cells. Secondary *in vitro* activation may be accomplished by coculture with irradiated tumor and IL-2 [23,24,26] or the sequential activation with anti-CD3 monoclonal antibody (MAb) and expansion in IL-2 [25,27]. Systemic micrometastases from weakly immunogenic murine tumors such as MCA 205 can be successfully treated using TDLN cells derived from either method of secondary activation. Poorly immunogenic tumors such as the B16-BL6 melanoma, however, fail to sensitize the TDLN cells to become pre-effector cells. To overcome this problem, we have previously shown that the admixture of the B16-BL6 tumor with the potent immunologic adjuvant, *Corynebacterium parvum* primes draining lymph nodes to develop pre-effector cells [28]. These cells, upon secondary *in vitro* activation, mediated regression of established metastases in murine models. This reinforces the premise that poorly immunogenic tumors can be genetically modified to be more immunogenic.

Another method that we have explored to sensitize T cells within the draining lymph node has included the transfection of tumor with an allogeneic MHC class I gene [29]. The B16-BL6 melanoma (H-2^b) was transfected *in vivo* with an allogeneic MHC class I(H-2^d) gene using lipofection techniques. Cells from the TDLN were sequentially cultured in an anti-CD3 mAb and IL-2. When adoptively transferred into animals with wild-type pulmonary metastases, these activated TDLN cells showed significant antitumor activity compared to parental TDLN cells, which had minimal therapeutic effect.

We have found that sensitization of TDLN using tumor cells that have been genetically modified to secrete cytokines has been useful in generating T cells reactive to poorly immunogenic tumors. B16-BL6 tumor that has been transfected with the murine IL-4 gene was used to sensitize the regional TDLN. The pre-effector cells sensitized by the IL-4-secreting tumor were effective in mediating regression of preexisting lung micrometastases in a tumor-bearing animal. The *in vivo* antitumor reactivity of these lymphocytes was comparable to that of lymphocytes sensitized by wild-type tumor admixed with *C. parvum*. Another cytokine that we found enhanced the sensitization of the TDLN cells was granulocyte-macrophage colony-stimulating factor (GM-CSF) [30]. GM-CSF is a potent stimulator of macrophages and DCs, which are important antigen-presenting cells involved in the induction of immune responses. (See Chapter 19 for a further discussion of DCs.) Using the B16-BL6 tumor, GM-CSF-secreting tumors were associated with a significant influx of tissue macrophages within the tumor and TDLN. Activated TDLN cells from mice inoculated with GM-CSF-secreting tumors mediated significant regression of established tumor in adoptive immunotherapy compared with parental TDLN cells, which had no activity. More important, TDLN cells primed with GM-CSF-secreting tumors were more effective in adoptive immunotherapy than were those sensitized by parental tumor admixed with *C. parvum* or tumor cells transduced to secrete other cytokines (e.g., IL-2, IFN-γ, IL-4) (Fig. 2) [31]. Based upon these observations, we are currently conducting a phase I trial to evaluate autologous tumor cells transduced to secrete GM-CSF as a vaccine to sensitize lymph node cells for use in the adoptive immunotherapy of melanoma patients (Fig. 3).

Other investigators have utilized similar models to evaluate the sensitization of TDLN cells. Shiloni *et al.* [32] used the poorly immunogenic MCA 102 sarcoma that was modified with the gene encoding for IFN-γ and reported an increased expression of MHC class I molecules on the transduced tumor cells. These cells were inoculated in the flanks of animals to induce an immune response in the TDLNs, which were excised several days later. After secondary *in vitro* activation, these TDLN cells were adoptively transferred into animals with systemic micrometastases and mediated enhanced antitumor efficacy compared with lymph node cells draining wild-type tumor.

In summary, genetic modification of tumor cells to secrete cytokines or express immunomodulatory proteins has shown promise in enhancing the antitumor reactivity of TILs and TDLN cells in animal models. These observations have established the rationale for evaluating these approaches in clinical studies, which our laboratory is currently pursuing.

III. GENETIC MANIPULATION OF T CELLS TO ENHANCE ANTITUMOR REACTIVITY

The administration of immunocompetent effector T cells and IL-2 has been documented to have therapeutic efficacy in a small percentage of patients with metastatic tumors [13]. However, many of these clinical responses have been of limited duration, so several methods to enhance the therapeutic efficacy of such effector cells are being investigated, including direct gene modification (Fig. 4).

A. TIL-Marking Studies

The first human trial involving gene transfer was reported by Rosenberg et al [33] and addressed the issues of safety of infusion of retrovirally transduced TILs infused into patients. The study involved the infusion of TILs transfected with an antibiotic-resistance (neomycin phosphotransferase) gene, which served as a marker to identify adoptively transferred TILs. Ten patients with metastatic melanoma received gene-marked TILs. No toxicities associated with gene transfer were observed. The *in vitro* cytolytic capacity of the TILs was not altered by the genetic transfer. In some of the patients treated with the gene-marked TILs, tumor responses were observed. Infectious retrovirus has not been found in any of the patients. Furthermore, gene-modified TILs were detected using the polymerase chain reaction (PCR) method to detect the neomycin-resistance gene in tumor deposits for up to 64 days, and in the circulation for up to 189 days after infusion.

B. Cytokine Gene Transfection of Effector Cells

In studies using [111]indium-labeled TIL, Griffith *et al.* [34] documented preferential localization of TILs to sites of tumor when compared with similarly labeled peripheral blood lymphocytes in patients with melanoma. Because of the selective trafficking of TILs to sites of tumor, these cells represent potential reagents to deliver immunoregulatory proteins to tumors for therapy. TNF-α is a potent cytokine with antitumor properties when delivered locally by intratumoral inoculation or regional perfusion [35,36]. However, systemic

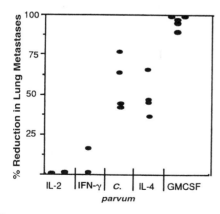

FIGURE 2 Comparison of the adjuvant effect of various cytokines (IL-2, IFN-γ, IL-4, and GM-CSF) elaborated at the site of tumor inoculation in priming pre-effector cells in the tumor-draining lymph nodes. *C. parvum* was also included as a bacterial adjuvant admixed with tumor cells. The antitumor reactivity of the tumor-draining lymph node cells was assessed in adoptive immunotherapy experiments. Each dot represents a group of animals from separate experiments and the percent reduction of pulmonary metastases recorded for each group. GM-CSF was the most potent adjuvant compared to the other agents studied.

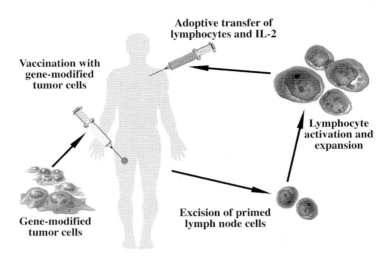

FIGURE 3   Schema of a clinical protocol being performed at the University of Michigan. Melanoma patients with stage IV disease have tumor harvested for transduction with a retrovirus encoding GM-CSF. The gene-modified tumor cells are inoculated intradermally in the thigh, and the vaccine-primed lymph nodes are harvested one week later for *ex vivo* activation and expansion. These cells are subsequently transferred back to the patient intravenously along with the concomitant administration of IL-2.

administration of TNF-α is considerably toxic. The local production by TILs of TNF-α, which selectively migrates to tumor sites, may enhance the therapeutic effectiveness of the TILs [37–39]. In 16 of 16 attempts, TNF-α gene insertion was successful, with an estimated 28 to 93% cells transduced within each culture. However, efficiency of gene expression and protein production in genetically modified TILs remains problematic, with evidence for decreased mRNA transcripts compared with transfected tumor cell lines [39]. Current methods

being explored to enhance transcription efficiency include the use of different promoters and modification of the TNF-α cDNA transmembranous portion to contain the signal peptide from IFN-γ, which appears to enhance transgene production. The addition of retinoic acid to TNF-α-transduced TIL cultures also has been shown to increase mRNA transcripts for TNF-α, with a corresponding increase in protein production [38].

Another approach to enhance the antitumor reactivity of TILs is to genetically modify the cells to secrete IL-2. In a murine B16-F10 melanoma model, Nakamura *et al.* [40] modified TILs to secrete IL-2 using an adenoviral vector, with IL-2 production of 3,000 IU/10^6 cells/24 h. The secretion of IL-2 by TILs could potentially enhance their survival after *in vivo* transfer in animals. These IL-2-secreting TILs were shown to have enhanced therapeutic efficacy in the treatment of pulmonary metastatic compared to nontransduced TILs. In addition, the administration of IL-2-secreting TIL prolonged the survival of mice with established intracranial tumor.

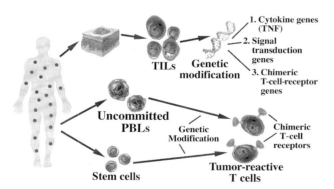

FIGURE 4   Schema of clinical approach to genetically modify T cells to enhance antitumor reactivity. One approach is to alter cytolytic T cells (i.e., TILs) to be more potent effector cells by elaborating cytokines, increasing their capacity to proliferate in response to tumor antigen, or redirecting them to recognize specific tumor antigens. These approaches are reviewed in this chapter. Another approach is to genetically modify uncommitted lymphoid cells or stem cells to become tumor-reactive T cells. This approach will be reviewed in the Chapter 24.

C. Modification of Effector Cells with Signal Transduction Genes

Another strategy of genetic modification of effector cells, reported by Finn *et al.* [41] has been directed toward more efficient signal transduction in T cells. The activation of protein kinase C (PKC) is believed to be part of the final common pathway resulting from T-cell-receptor activation. Cytotoxic murine T-cell clones with previously well-defined antigen-specific cytotoxic activity were transduced with the PKC-γ. This resulted in constitutive expression of PKC-γ. The modified T cells were enabled to propagate indefinitely in culture, without the need for periodic antigen stimulation, while maintaining their antigen specificity.

Further studies reported by Chen et al. [42] involving tumor-specific T-cell lines reactive to the FBL-3 tumor showed that PKC-γ-transduced CD8+ cells maintained specific antitumor cytotoxicity in vitro and that trans-duced CD4+ cells maintained proliferative activity in response to tumor [42]. These cells have shown a continued expression of IL-2 receptor in culture, indicative of their activated state, even without periodic antigen stimulation. In addition, bulk T-cell cultures transduced with the PKC-γ gene were shown to have in vivo antitumor efficacy against systemic metastases in adoptive immunotherapy experiments.

D. Redirection of Effector Cells with Chimeric Receptor Genes

A novel approach to redirect effector cell function involves the genetic modification of these cells with chimeric receptor genes constructed from the variable domains obtained from MAbs linked with the constant regions of the TCR chains. The rationale for this approach is to combine the antitumor specificity of many available MAbs with signal transduction molecules of effector cells. This would bypass the need for MHC identity necessary in conventional antigen recognition by T cells. Eshhar et al. [43] described the construction of a chimeric gene consisting of a single-chain Fv domain (V_L linked to V_H) from Sp6, and anti-TNP (trinitrophenyl) MAb, linked to the signal transducing subunit of the TCR/CD3 complex of the γ-signal-transducing sub-unit of the immunoglobulin/Fc receptor complex. Retro-viral transduction of this chimeric gene into a cytolytic T-cell hybridoma line resulted in gene expression. In vitro, these modified T cells were cytolytic and released IL-2 specifically against TNP-labeled cells, but not unla-beled cells.

These studies were extended to determine if human TIL could be redirected to recognize tumor targets. A gene encoding a single-chain Fv from MOv18, an antibody that binds a folate-binding protein expressed on most ovarian cell lines, was linked to the FcR-γ gene to form a chimeric gene [44,45]. TILs were transduced with this chimeric gene and evaluated for immunological function. The MOv-γ-modified TILs showed specific ly-sis of an ovarian carcinoma cell line but not nonovarian

cell lines. Specific cytokine release by MOV-γ TILs was also shown against ovarian but not nonovarian targets. MOv-γ TILs have also been found to be effective in vivo against peritoneal ovarian tumors expressing MOv 18 and against sarcoma tumor cells that were transduced with the MOv18 gene. In a direct comparison, MOv-γ TILs showed greater therapeutic efficacy than that of conventional TILs or mock-transduced TILs. In murine models, cytolytic T-cell lines have been modified with chimeric receptor genes to recognize human Neu/Her2 (also known as Her-2/neu, ERBB2) antigen-expressing tumors. Stancovski et al. [46] generated cytotoxic murine T-cell hybridomas that specifically recognized Neu/Her2-expressed cells by production of IL-2 and lysis of target cells. In an in vivo model, Moritz et al. [47] showed that the adoptive transfer of murine CTLs genetically modified to recognize ERBB-2 retarded the growth of ERBB-2 transformed cells inoculated in athymic nude mice. This latter study showed the feasibility of using redirected effector cells in clinical therapy.

E. Genetic Engineering of Immunocompetent Effector Cells from Naive Lymphoid Cells

The ability of retrieving tumor-reactive effector T cells from the tumor-bearing host has been technically difficult for a variety of reasons: (1) the relatively low frequency of tumor-reactive lymphoid cells present in vivo, (2) the weak immunogenicity of human tumors, and (3) the presence of tumor-induced immune suppression mechanisms that can inhibit T-cell sensitization. The difficulties in expanding mature effector lymphoid cells while maintaining their therapeutic efficacy can also hamper the application of this approach to clinical therapy. Recently, several tumor-associated antigens have been identified and cloned from human tumors such as melanoma, breast, and ovarian cancers [48–54]. These antigens have been identified by their binding to the TCR-αβ complex. The subsequent cloning of functional TCR genes capable of recognizing tumor-associated antigens offers a potential opportunity to genetically modify alternate cells that have not been previously exposed to tumor antigen to become competent in recognizing tumor.

Cole et al. [55] transfected the cDNA for the TCR-α and -β chains of an HLA A2–restricted, melanoma-reactive T-cell clone into the human Jurkat T-cell line. The transfected Jurkat line was able to mediate recognition of the melanoma antigen, melanoma tumor-associated antigen (MART-1), when presented by antigen-presenting cells. This represents the first report of a naive cellular construct genetically engineered to mediate functional tumor antigen recognition. This technology allows the potential development of novel cellular reagents that may be useful in adoptive immunotherapy.

IV. GENETIC MODULATION OF DENDRITIC CELLS

DCs are highly potent antigen-presenting cells. These cells process antigens and present them to lymphoid cells in association with MHC molecules. In addition, they express a large number of costimulatory molecules on their surface. DCs play a central role in the induction of immune responses and can be isolated from the peripheral blood [56,57]. However, the techniques available to isolate DCs are relatively cumbersome and result in low yields. The combination of GM-CSF and IL-4 was found by Romani *et al.* [58] to be effective in the generation of functional DCs from mobilized stem cells isolated from the peripheral blood of patients undergoing chemotherapy. These studies lend support to the notion that DCs are derived from uncommitted hematopoietic stem cells.

One of the major interests in DCs is their ability to present tumor antigen to resting lymphoid cells, which results in immunocompetent T and B cells. Several animal studies have demonstrated that the *in vitro* "pulsing" of DCs with tumor antigen in the form of whole tumor cells, tumor lysates, or tumor peptides will generate DCs capable of priming naive T cells [59,60]. Moreover, the adoptive transfer of these pulsed DCs by either intravenous or intradermal inoculation can result in regression of established tumors. Hsu *et al.* [61] reported that the administration of human DCs pulsed *ex vivo* with tumor-specific lymphoma idiotype could mediate the regression of recurrent lymphoma. Impressively, some of these patients developed durable complete tumor responses to their therapy.

Methods to enhance the therapeutic efficacy of DCs include genetic modification (Fig. 5). Liposomal transfection, retroviral gene transfer, and the gene gun have all been used to successfully modify DC. Genes encoding for tumor antigens have been successfully transferred into DCs to enhance their ability sensitize lymphoid cells [62,63]. Reeves *et al.* [63] retrovirally transduced human DCs with the melanoma tumor-associated antigen gene *MART-1*. These DCs were able to generate specific antitumor CTLs and stimulate greater levels of cytokine release by MART-1-specific TILs. In an *in vivo* murine model, Ribas *et al.* [64] demonstrated induction of immunity with *MART-1*-transduced murine DCs that were capable of successfully treating established tumors expressing the

DENDRITIC CELL-BASED VACCINE

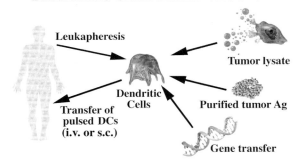

FIGURE 5 Schema of the clinical use of DC-based vaccines. Patients are leukapheresed to obtain DCs from the peripheral circulation. The DCs are "pulsed" with tumor lysate or purified tumor antigen. Alternatively, the DCs can be transfected with genes encoding tumor antigens. The genetic material can be in the form of DNA or RNA.

MART-1 antigen. For a further discussion of DCs, see Chapter 18.

V. SUMMARY

Adoptive cellular therapy remains a powerful method of eradicating established tumor in experimental models. The use of gene transfer techniques has resulted in more effective methods to generate tumor-specific T cells. Another area of tremendous interest is the adoptive transfer of DCs manipulated to present tumor antigen to resting T cells. Gene transfer techniques may offer more optimal ways to generate therapeutic DCs. The application of these methods in clinical studies has been extremely important in identifying new directions to explore in the experimental setting. The ability to genetically alter the host–tumor interaction has dramatically increased our understanding of the immune response to tumor cells. It is anticipated that this increased understanding will lead to more effective therapies of malignancy.

References

1. Gross, L. (1943). Intradermal immunization of C3H mice against a sarcoma that originated in an animal of the same line. *Cancer Res.* **3**, 326–333.
2. Landsteiner, K., and Chase, M. W. (1942). Experiments on transfer of cutaneous sensitivity to simple compounds. *Proc. Soc. Exp. Biol. Med.* **49**, 688–690.
3. Billingham, R. E., Brent, L., and Medawar, P. B. (1954). Quantitative studies on tissue transplantation immunity, vol. II: the origin, strength and duration of actively and adoptively acquired immunity. *Proc. R. Soc. Biol.* **143**, 58–80.
4. Mitchison, N. A. (1955). Studies on the immunological response to foreign tumor transplants in the mouse, vol. I: the role of lymph

node cells in conferring immunity by adoptive transfer. *J. Exp. Med.* **102**, 57–177.

5. Borberg, H., Oettgen, H. F., Choudry, K., *et al.* (1972). Inhibition of established transplants of chemically induced sarcomas in syngeneic mice by lymphocytes from immunized donors. *Int. J. Cancer* **10**, 539–547.

6. Rosenberg, S. A., and Terry, W. D. (1977). Passive immunotherapy of cancer in animals and man. *Adv. Cancer Res.* **25**, 323–388.

7. Greenberg, P. D. (1990). Adoptive T cell therapy of tumors: mechanisms operative in the recognition and elimination of tumor cells. *Adv. Immunol.* **49**, 281–355.

8. Morgan, D. A., Ruscetti, F. W., and Gallo, R. C. (1976). Selective *in vitro* growth of T lymphocytes from normal human bone marrows. *Science* **193**, 1007–1008.

9. Cheever, M. A., Greenberg, P. D., and Fefer, A. (1981). Specific adoptive therapy of established leukemia with syngeneic lymphocytes sequentially immunized *in vivo* and *in vitro* and nonspecifically expanded by culture with interleukin 2. *J. Immunol.* **126**, 1318–1322.

10. Rosenberg, S. A., Lotze, M. T., Muul, L. M., *et al.* (1985). Observations on the systemic administration of autologous lymphokine-activated killer cells and recombinant interleukin-2 in patients with metastatic cancer. *N. Engl. J. Med.* **313**, 1485–1492.

11. Rosenberg, S. A., Spiess, P., and Lafreniere, R. (1986). A new approach to the adoptive immunotherapy of cancer with tumor-infiltrating lymphocytes. *Science* **233**, 1318–1321.

12. Spiess, P. J., Yang, J. C., and Rosenberg, S. A. (1987). *In vivo* antitumor activity of tumor-infiltrating lymphocytes expanded in recombinant interleukin-2. *J. Natl. Cancer Inst.* **79**, 1067–1075.

13. Rosenberg, S. A., Packard, B. S., Aebersold, P. M., *et al.* (1988). Use of tumor-infiltrating lymphocytes and interleukin-2 in the immunotherapy of patients with metastatic melanoma. *N. Engl. J. Med.* **319**, 1676–1680.

14. Aebersold, P., Hyatt, C., Johnson, S., *et al.* (1991). Lysis of autologous melanoma cells by tumor-infiltrating lymphocytes association with clinical response. *J. Natl. Cancer Inst.* **83**, 932–937.

15. Gansbacher, B., Bannerji, R., Daniels, B., *et al.* (1990). Retroviral vector-mediated γ-interferon gene transfer into tumor cells generates potent and long lasting antitumor immunity. *Cancer Res.* **50**, 7820–7825.

16. Golumbek, P. T., Lazenby, A. J., Levitsky, H. I., *et al.* (1991). Treatment of established renal cancer by tumor cells engineered to secrete interleukin-2. *Science* **254**, 713–716.

17. Tepper, R. I., Coffman, R. I., and Leder, P. (1992). An eosinophil dependent mechanism for the antitumor effect of interleukin-2. *Science* **257**, 548–551.

18. Restifo, N. P., Speiss, P. J., Karp, S. E., *et al.* (1992). A nonimmunogenic sarcoma transduced with the cDNA for interferon gamma elicit CD8+ cells against the wild-type tumor: correlation with antigen presentation capability. *J. Exp. Med.* **175**, 1423–1431.

19. Marincola, F. M., Ettinghausen, S., Cohen, P. A., *et al.* (1994). Treatment of established lung metastases with tumor-infiltrating lymphocytes derived from a poorly immunogenic tumor engineered to secrete human TNF-alpha. *J. Immunol.* **152**, 3500–3513.

20. McBride, W. H., Thacker, J. D., Comora, S., *et al.* (1992). Genetic modification of a murine fibrosarcoma to produce interleukin-7 stimulates host cell infiltration and tumor immunity. *Cancer Res.* **52**, 3931–3937.

21. Plautz, G. E., Yang, Z., Wu, B., *et al.* (1993). Immunotherapy of malignancy by *in vivo* gene transfer into tumors. *Proc. Natl. Acad. Sci.* **90**, 4645–4649.

22. Nabel, G. J., Gordon, D., Bishop, D. K., *et al.* (1996). Immune response in human melanoma after transfer of an allogenoic class I major histocompatibility complex gene with DNA-liposome complexes. *Proc. Natl. Acad. Sci.* **93**, 15388–15393.

23. Chou, T., Chang, A. E., Shu, S. (1988). Generation of therapeutic T lymphocytes from tumor-bearing mice by *in vitro* sensitization: culture requirements and characterization of immunologic specificity. *J. Immunol.* **140**, 2453–2461.

24. Shu, S., Chou, T., Sakai, K. (1989). Lymphocytes generated by *in vivo* priming and *in vitro* sensitization demonstrate therapeutic efficacy against a murine that lacks apparent immunogenicity. *J. Immunol.* **143**, 740–748.

25. Yoshizawa, H., Chang, A. E., Shu, S. (1991). Specific adoptive immunotherapy mediated by tumor-draining lymph node cells sequentially activated with anti-CD3 and IL-2. *J. Immunol.* **147**, 729–737.

26. Chang, A. E., Yoshizawa, H., Sakai, K., *et al.* (1993). Clinical observations on adoptive immunotherapy with vaccine-primed T lymphocytes secondarily sensitized to tumor *in vitro*. *Cancer Res.* **53**, 1043–1050.

27. Chang, A. E., Atsushi, A., Cameron, M., *et al.* (1997). Adoptive immunotherapy with vaccine-primed lymph node cells secondarily activated with anti-CD3 and Interleukin-2. *J. Clin. Oncol.* **15**, 796–807.

28. Geiger, J. D., Wagner, P. D., Cameron, M. J., *et al.* Generation of T cells reactive to the poorly immunogenic B16-BL6 melanoma with efficacy in the treatment of spontaneous metastases. *J. Immunother.* **13**, 153–165.

29. Wahl, W. L., Strome, S. E., Nabel, G. J., *et al.* (1995). Generation of therapeutic T lymphocytes after *in vivo* tumor transfection with an allogeneic class I major histocompatibility complex gene. *J. Immunother.* **17**, 1–11.

30. Arca, M. J., Krauss, J. C., Aruga, A., *et al.* (1996). Therapeutic efficacy of T cells derived from lymph nodes draining immunogenic tumor transduced to secrete GM-CSF. *Cancer Gene Ther.* **3**, 39–47.

31. Arca, M., Krauss, J., Strome, S., *et al.* (1996). Diverse manifestations of tumorigenicity and immunogenicity displayed by the poorly immunogenic B16-BL6 melanoma transduced with cytokine genes. *Cancer Immunol. Immunother.* **42**, 237–245.

32. Shiloni, E., Karp, S. E., Custer, M. C., *et al.* (1993). Retroviral transduction of interferon-gamma cDNA into a nonimmunogenic murine fibrosarcoma: generation of T cells in draining lymph nodes capable of treating established parental metastatic tumor. *Cancer Immunol. Immunother.* **37**, 286–292.

33. Rosenberg, S. A., Aerbersold, P., Cornetta, K., *et al.* (1990). Gene transfer into humans—immunotherapy of patients with advanced melanoma, using tumor-infiltrating lymphocytes modified by retroviral gene transduction. *N. Engl. J. Med.* **323**, 570–578.

34. Griffith, K. D., Read, E. J., Carrasquillo, J. A., *et al.* (1989). *In vivo* distribution of adoptively transferred Indium-111 labeled tumor infiltrating lymphocytes and peripheral blood lymphocytes in patients with metastatic melanoma. *J. Natl. Cancer Inst.* **81**, 1709–1717.

35. Thom, A. K., Fraker, D. L., Taubenberger, J. K., *et al.* (1992). Effective regional therapy of experimental cancer with paralesional administration of tumor necrosis factor-α and interferon-γ. *Surg. Oncol.* **1**, 291–298.

36. Lienard, D., Ewalenko, P., Delmotte, J. J., *et al.* (1992). High-dose recombinant tumor necrosis factor alpha in combination with interferon gamma and melphalan in isolation perfusion of limbs for melanoma and sarcoma. *J. Clin. Oncol.* **10**, 52–60.

37. Hwu, P., Yannelli, J., Kriegler, M., *et al.* (1993). Functional and molecular characterization of tumor-infiltrating lymphocytes transduced with tumor necrosis factor-α cDNA for the gene therapy of cancer in humans. *J. Immunol.* **150**, 4104–4115.

38. Treisman, J., Hwu, P., Yannelli, J. R., *et al.* (1994). Upregulation of tumor necrosis factor-alpha production by retroviral transduced tumor-infiltrating lymphocytes using *trans*-retinoic acid. *Cell. Immunol.* **156,** 448–457.

39. Hwu, P., Rosenberg, S. A. (1994). The genetic modification of T cells for cancer therapy: an overview of laboratory and clinical trials. *Cancer Detect. Prev.* **18,** 43–50.

40. Nakamura, Y., Wakimoto, H., Abe, J., *et al.* (1994). Adoptive immunotherapy with murine tumor specific T lymphocytes engineered to secrete interleukin-2. *Cancer Res.* **54,** 5757–5760.

41. Finn, O. J., Persons, D. A., Bendt, K. M., *et al.* (1991). Retroviral transduction of protein kinase C-gamma into cytotoxic T lymphocyte clones leads to immortalization with retention of specific function. *J. Immunol.* **146,** 1099–1103.

42. Chen, W., Schweins, E., Chen, X., *et al.* (1994). Retroviral transduction of protein kinase C-gamma into tumor-specific T cells allows antigen-independent long-term growth in IL-2 with retention of functional specificity in vitro and ability to mediate tumor therapy *in vivo. J. Immunol.* **153,** 3630–3638.

43. Eshhar, Z., Waks, T., Gross, G., *et al.* (1993). Specific activation and targeting of cytotoxic lymphocytes through chimeric single chain consisting of antibody domains and the γ or ζ subunits of the immunoglobin and T cell receptors. *Proc. Natl. Acad. Sci. USA* **90,** 720–724.

44. Hwu, P., Shafer, G. E., Treisman, J., *et al.* (1993). Lysis of ovarian cancer cells by human lymphocytes redirected with a chimeric gene composed of an antibody variable region and the Fc receptor γ chain. *J. Exp. Med.* **178,** 361–366.

45. Hwu, P., Yang, T. C., Cowherd, R., *et al.* (1995). *In vivo* antitumor activity of T cells redirected with chimeric antibody/T cells receptor genes. *Cancer Res.* **55,** 3369–3373.

46. Stancovski, I., Schindler, D. G., Waks, T. (1993). Targeting of T lymphocytes to Neu/Her2-expressing cells using chimeric single chain Fv receptors. *J. Immunol.* **151,** 6577–6582.

47. Moritz, D., Wels, W., Mattern, J., *et al.* (1994). Cytotoxic T lymphocytes with a grafted recognition specificity for ERBB2-expressing tumor cells. *Proc. Natl. Acad. Sci. USA* **91,** 4318–4322.

48. Traversari, C., Van der Bruggen, P., Luescher, I. F., et al. (1992). A nonapeptide encoded by human gene MAGE-1 is recognized on HLA-A1 by cytolytic T lymphocytes directed against tumor antigen MZ2-E. *J. Exp. Med.* **176,** 1453–1457.

49. Zakut, R., Topalian, S. L., Kawakami, Y., *et al.* (1993). Differential expression of MAGE-1, -2 and -3 messenger RNA in transformed and normal human cell lines. *Cancer Res.* **53,** 5–8.

50. Topalian, S. L., Rivoltini, L., Mancini, M., *et al.* (1994). Human CD4+ T cells specifically recognize a shared melanoma associated antigen encoded by the tyrosinase gene. *Proc. Natl. Acad. Sci. USA* **91,** 9461–9465.

51. Rivoltini, L., Kawakami, Y., Sakaguchi, K., *et al.* (1995). Induction of tumor-reactive CTL from peripheral blood and tumor-infiltrating lymphocytes of melanoma patients by *in vitro* stimulation with an immunodominant peptide of the human melanoma antigen MART-1. *J. Immunol.* **154,** 2257–2265.

52. Jerome, K. R., Barnd, D. L., Bendt, K. M., *et al.* (1991). Cytotoxic T lymphocytes derived from patients with breast adenocarcinoma recognize an epitope present on the protein core of a mucin molecule preferentially expressed by malignant cells. *Cancer Res.* **51,** 2908–2916.

53. Yoshino, I., Peoples, G. E., Goedegebuure, P. S., *et al.* (1994). Association of Her2/neu expression with sensitivity to tumor-specific CTL in human ovarian cancer. *J. Immunol.* **152,** 2392–2400.

54. Yoshino, I., Goedegebuure, P. S., Peoples, G. E., *et al.* (1994). HER2/neu-derived peptides are shared antigens among human non-small cell lung cancer and ovarian cancer. *Cancer Res.* **54,** 3387–3390.

55. Cole, D. J., Weil, D. P., Shilyanski, J., *et al.* (1995). Characterization of the functional specificity of a cloned T cell receptor heterodimer recognizing the MART-1 melanoma antigen. *Cancer Res.* **55,** 748–752.

56. Mehta-Damani, A., Markowicz, S., Engleman, E. (1994). Generation of antigen-specific CD8+ CTLs from naive precursors. *J. Immunol.* **153,** 996–1003.

57. Mehta-Damani, A., Markowicz, S., Engleman, E. (1995). Generation of antigen-specific CD4+ T cell lines from naive precursors. *Eur. J. Immunol.* **25,** 1206–1211.

58. Romani, N., Gruner, S., Brang, D., *et al.* (1994). Proliferating dendritic cell progenitors in human blood. *J. Exp. Med.* **180,** 83–93.

59. Zitvogel, L., Mayordomo, J., Tjandrawan, T., *et al.* (1996). Therapy of murine tumors with tumor peptide-pulsed dendritic cells: Dependence on T cells, B7 costimulation, and T helper cell 1-associated cytokines. *J. Exp. Med.* **183,** 87–97.

60. Coveney, E., Wheatley, G., Lyerly, H. K. (1997). Active immunization using dendritic cells mixed with tumor cells inhibits the growth of primary breast cancer. *Surgery* **122,** 226–234.

61. Hsu, F., Benike, C., Fagnoni, F., *et al.* (1996). Vaccination of patients with B-cell lymphoma using autologous antigen-pulsed dendritic cells. *Nat. Med.* **2,** 52–58.

62. Henderson, R., Nimgaonkar, M., Watkins, S., *et al.* (1996). Human dendritic cells genetically engineered to express high levels of the human epithelial tumor antigen mucin (MUC-1). *Cancer Res.* **56,** 3763–3770.

63. Reeves, M., Royal, R., Lam, J., *et al.* (1996). Retroviral transduction of human dendritic cells with a tumor-associated antigen gene. *Cancer Res.* **56,** 5672–5677.

64. Ribas, A., Butterfield, L., McBride, W., *et al.* (1997). Genetic immunization for the melanoma antigen MART-1/Melan-A using recombinant adenovirus-transduced murine dendritic cells. *Cancer Res.* **57,** 2865–2869.

CHAPTER 23

Cytokine Gene Therapy of Cancer

JEAN-MARIE PÉRON, MICHAEL R. SHURIN, AND MICHAEL T. LOTZE

University of Pittsburgh Cancer Institute, 300 Kaufmann Building, Department of Surgery, University of Pittsburgh Medical Center, Pittsburgh, Pennsylvania 15261

I. INTRODUCTION

Clinical observations including the occurrence of spontaneous regression and a prolonged latency period between removal of the primary tumor and the appearance of metastases in some patients with cancer suggest the existence of host immune responses to several malignancies. Unfortunately these mechanisms often fail to fully protect the host. In the past decade, advances in tumor immunology have led to a better understanding of the molecular basis of tumor antigen processing, presentation, and recognition, as well as the role of cytokines in the regulation of the antitumor immune response. Means by which the tumor evades immune elimination are now more readily identified. Cytokines regulate immune responses and direct the maturation, activation, and migration of inflammatory cells, but their use systemically is hampered by serious side effects [1,2] and their effectiveness is decreased by their rapid degradation or elimination [3,4]. These observations led to the development of alternative means of cytokine delivery by transfecting the cytokine gene into tumor or carrier cells that will express the cytokine at the primary tumor site, mimicking paracrine cytokine release *in vivo* and enhancing the induction of tumor-specific immune responses without many of the troublesome systemic side effects. Several cancer gene therapy approaches involving cytokines are explored next.

II. MECHANISM OF CYTOKINE-INDUCED TUMOR IMMUNITY

Active cancer immunotherapy can be defined as cancer treatments that act primarily through the intermediaries of natural host defense mechanisms. Cytokine gene therapy can be directed toward making cellular immune elements more effective at controlling tumor cells or making tumor cells more easily recognized by the immune system. T cells stimulated by tumor cells often fail to secrete cytokines, without which proliferation

359

and differentiation of effector T cells proceed poorly. Furthermore, the survival of immune cells is dependent on their local elaboration. Thus came the initial notion to express cytokine genes in the tumor in order to bypass the inability of tumor to initiate a protective immune response.

The immediate goal of immunotherapy is to induce an effective immune response that primarily involves a cellular immune response [cytotoxic CD8+ T lymphocytes (CTL) and T-helper (Th)1 cells]. Indeed it was originally shown that the rejection of transplantable tumors after immunization procedures in murine systems is primarily mediated by T cells [5–7]. Moreover, the adoptive transfer of bulk populations of effector T cells derived from either tumor tissue or peripheral blood has been used to protect and even treat relevant tumor-bearing, syngeneic animals [8–10]. These results demonstrated the important role of T cells and in particular CTLs.

To participate in an adaptive immune response circulating naïve T cells must be induced to proliferate and differentiate into cells capable of responding to tumor cells when they encount the specific antigen (Ag) on the surface of a professional Ag-presenting cell, of which the most important are dendritic cells (DCs) [11]. DCs also express the specialized costimulatory molecules that enable them to activate naïve T cells [12]. This priming occurs in the secondary lymphoid organs, where the APCs migrate to present the Ag after it has been taken up at the tumor site and processed [13]. Three different functional classes of effector T cells derive from this priming, leave the lymphoid organs, and reenter the bloodstream so that they can migrate to the tumor site: the CD8 cytotoxic T cells that recognize Ags presented by major histocompatibility complex (MHC) class I molecules on the cell surface, the CD4 Th1 cells that activates macrophages and other resident cells at the site of inflammation, and the CD4 Th2 cells, which activate specific B cells to produce antibodies (Abs). CD8 T cells mediate their killing through two predominant pathways, a membrane attack complex initiated with the formation of pores in target cell membranes by perforin, allowing delivery of apoptosis-inducing granzymes, and a nonsecretory one initiated by receptor-mediated triggering of apoptosis by "death factor," including tumor necrosis factor (TNF), Fas ligand and TRAIL (TNF-related apoptosis-inducing ligand) [14–19]. At the same time immunologic memory is generated. These memory T cells are long lived and will respond to Ag with an accelerated response. For a further discussion of immune effector mechanisms and targets, see Chapter 3.

III. GENETIC MODIFICATION OF TUMOR CELLS FOR CYTOKINE DELIVERY

Cancer-bearing hosts are often immunologically impaired with regard to cell-mediated immunity: absent or low expression of MHC class I and/or class II molecules, lack of costimulatory signals, defective antigen-processing machinery in the antigen-presenting cell (APC), loss of signal-transducing molecules in the T cells, and induction of suppressor cells have all been identified. The rationale for transfecting cytokine genes into tumor cells is to render them vulnerable to immune attack, enhancing the ability of the tumor to initiate a protective immune response. Moreover, cytokine expression by a tumor itself may affect the ability of the host to generate and maintain an effective immunologic response.

The direct modification of tumor cells using cytokine genes to increase host immunity has been studied intensively in animal models over the last 8 years (see reviews in references [20–24]). The introduction of cytokine genes directly into tumor cells by transfection (or infection in the case of viral vectors) followed by inoculation of the genetically engineered cells into animal hosts has been termed the tumor-cytokine transplantation assay [25,26]. This assay, which originally studied the role of interleukin (IL)-2 and IL-4, helped demonstrate that a number of cytokine genes reduced tumorigenicity by stimulating localized inflammatory and/or immune responses. These included TNF-α, G-CSF, GM-CSF, interferon gamma, IL-1, IL-2, IL-4, IL-6, IL-7, IL-12, and IL-18. The assay also demonstrated the lack of tumor killing or immunity induced by IL-5 and the inhibition of tumor immuniginicity by transforming growth factor (TGF)-β [27–29]. One report revealed the augmentation of metastatic tumor growth by TNF-α expression [30]. As most of these studies showed, in vitro growth of tumor cells is not affected by the transfected cytokine's action, suggesting that the inhibition of tumor growth in vivo results from the stimulation of host immune effector cells by the activity of the secreted cytokine. Among the most important features emphasized by all of these studies is that the local sustained release of cytokines produces local inflammatory or immune-mediated effects without significant evidence of systemic effects or toxicity. These cytokines have different mechanisms of action, and it is important to distinguish localized tumor killing (which is still beneficial to the host) and the desired generation of systemic, T-cell-mediated antitumor immunity (Table 1). This in turn allows for an effective antitumor response at a distal site.

TABLE 1   Effector Cells Associated with an Antitumor Response Induced by Different Cytokines

Cytokine	Major effector cells	Other possible effector cells	Reference
G-CSF	Neutrophils	Macrophages, eosinophils, CD8+ T cells	42 45
GM-CSF	CD4+ T cells, CD8+ T cells		31
IFN-γ	CD8+ T cells	NK cells	112 36
IFN-α	CD8+ T cells, NK cells		106
TNF-α	CD4+ T cells, CD8+ T cells	Macrophages	113 21
IL-1	ND	ND	114
IL-2	CD8+ T cells	NK cells, neutrophils, eosinophils, macrophages	115 112 116
IL-4	Eosinophils	Macrophages, CD8+ T cells	25, 42
IL-6	CD4+ T cells, CD8+ T cells	NK cells, macrophages	34 33
IL-7	CD4+ T cells, CD8+ T cells, macrophages		36–38, 117
IL-12	CD4+ T cells, CD8+ T cells, NK cells		39
IL-18	NK cells	CD4+ T cells, CD8+ T cells	40

* ND, not described.

Several cytokines have been demonstrated to enhance T-cell-mediated tumor immunity. They include GM-CSF [31], IL-2 [32], IL-6 [33–35], IL-7 [36–38], IL-12 [39], and IL-18 [40]. Eosinophils have been shown to be important in IL-4- and IL-2-mediated tumor killing [26,41]. Neutrophils play a role in the antitumor activity of G-CSF [42] and macrophages in the antitumor activity of G-CSF [42], IFN-γ [43], TNF-α [21], IL-2 [44], IL-4 [25], and IL-7 [38]. IL-4 has recently been shown to be more effective than IL-2, IL-18, CSF, or GM-CSF in increasing the influx of dendritic cells in the tumor [45]. When injected in the ear pinna GM-CSF-transfected tumor cells stimulate whereas IL-4-transfected tumor cells inhibit DC-mediated priming of delayed type hypersensitivity reaction by 2,4-dinitro-1-fluorobenzene, proving that transduced cytokines differentially influence not only DC recruitment at the tumor site but also DC function in nearby tissues. It is noteworthy that localized killing of tumor cells by

cytokines or direct irradiation may generate nonviable tumor cells and an associated inflammatory cell influx (including APCs). This may indeed stimulate the recognition of an immunogenic tumor. Some cytokines may also have a direct action on the endothelium. In fact, IFN-γ, IFN-α, IL-1, and IL-4 have been shown to induce or up-regulate the expression of various adhesion molecules on endothelial cells [46,47]. This action on endothelial cells could also allow transmigration of specific inflammatory cells into tissue sites. It is still not clear from these models which cytokine is the most potent for generating antitumor responses, and it is quite possible that different cytokines generate different types of responses against histologically different tumors.

Based on this preclinical data phase I/II clinical trials using cytokine gene-modified human tumors have been initiated (Table 2). Clinical vaccine trials can be divided into two general types depending on the cell type transfected, autologous tumor cells or allogeneic tumor cell lines, both having advantages and disadvantages. The use of autologous tumor cells require that the patient's tumor (primary or metastatic site) be surgically removed followed by introduction of the gene *ex vivo*. Indisputably, this strategy has the advantage that the patient's own tumor cells have the greatest chance to vaccinate against the spectrum of relevant tumor antigens both shared and unique to the individual. The major disadvantage relates to the labor intensity associated with individualized vaccines and the variability of gene transfer in cultures derived from primary tumors. The use of allogeneic vaccines decreases these two parameters (a single standardized transduced cell line is used for the vaccine), but for this technique to be successful it is critical that the transduced cells share antigens with the patient's tumor. Most studies use irradiated tumor cells over nonirradiated ones to prevent having the live vaccine tumor cells continue to multiply *in vivo*. Not only does the antigen dose increase, so does the introduced gene product in nonirradiated cells. Both can cause potential risks. Irradiation causes growth arrest, allowing a more accurate determination of vaccine inoculum and better comparison of the relative efficacy between different transfectants. These clinical trials utilize escalating cell dosage challenges of either irradiated or nonirradiated tumor cells genetically engineered to express a given cytokine. The modified tumor cells are then injected directly into the tumor. They generally obtain information on toxicity from cytokine production, effects on the patient's humoral and cellular immune response, risk assessment of the gene modification by replication-defective retroviral constructs, and impact on the growth of established metastatic nodules at the injection site and other distant sites. For a further

TABLE 2   Current Cytokine Gene Therapy Protocols for Cancer

Institution	Patient population	Cytokine gene transduced	Method of transfer	Cell transduced
National Institutes of Health, Bethesda, MD	Advanced cancer	IL-2	Retrovirus	Tumor cell
San Diego Regional Cancer Center, San Diego, CA	Brain tumor	IL-2	Retrovirus	Tumor cell, fibroblasts
San Diego Regional Cancer Center, San Diego, CA	Colon cancer	IL-2	Retrovirus	Tumor cell, fibroblasts
Institute of Basic Medical Sciences, Beijing	Lung	IL-2	Retrovirus	Direct gene transfer
University of Miami and University Veterans Administration Hospital, Miami, FL	Small cell lung cancer	IL-2	Lipofection	Tumor cell
Memorial Sloan-Kettering Cancer Center, New York, NY	Melanoma	IL-2	Retrovirus	Tumor cell
University of California Medical Center, Los Angeles, CA	Melanoma	IL-2	Retrovirus	Tumor cell
University Hospital, Leiden	Melanoma	IL-2	Transfection method	Tumor cell
University of Illinois at Chicago, Chicago, IL	Melanoma	IL-2	Retrovirus	Tumor cell
St. Jude Children's Research Hospital, Memphis, TN	Neuroblastoma	IL-2	Retrovirus	Tumor cell
Memorial Sloan-Kettering Cancer Center, New York, NY	Renal cell carcinoma	IL-2	Retrovirus	Tumor cell
University of Pittsburgh, Pittsburgh, PA	Advanced cancer	IL-4	Retrovirus	Tumor cell, fibroblast
University of Michigan, Ann Arbor, MI	Melanoma	IL-4	Retrovirus	Tumor cell
Universitatsklinikum Rudolf Virchow, Berlin	Colon cancer, renal cell carcinoma, melanoma, lymphoma	IL-7	Electroporation and ballistic transfer	Tumor cell, CIK
Duke University, Durham, NC	Melanoma	IFN-γ	Retrovirus	Tumor cell
NIH, Bethesda, MD	Advanced cancer	TNF	Retrovirus	TIL
NIH, Bethesda, MD	Advanced cancer	TNF	Retrovirus	Tumor cell
Johns Hopkins Oncology Center, Baltimore, MD	Renal cell	GM-CSF	Retrovirus	Tumor cell
University of Pittsburgh, Pittsburgh, PA	Advanced cancer	IL-12	Retrovirus	Fibroblast

discussion of the clinical use of cytokine-transfected vaccines, see Chapter 21.

IV. GENETIC MODIFICATION OF LYMPHOCYTES FOR CYTOKINE DELIVERY

Lymphocytes are attractive vehicles for gene transfer because of their ability to proliferate in response to stimuli, circulate and migrate throughout the host and express recombinant proteins. Many human solid tumors are infiltrated by T cells (TILs), [48], which are believed to reflect the ability of the patient's immune system to develop a local antitumor response. In fact, in many tumors the presence of TILs has been considered to be a favorable prognostic indicator. Cultured TILs adoptively transferred to patients with metastatic melanoma, along with the administration of IL-2, induce clinical responses in approximately a third of the treated

patients [49]. These promising results encouraged us to use these TILs in cytokine-based gene therapy protocols. TNF was initially chosen because, although the injection of recombinant TNF can mediate the necrosis and regression of a variety of established murine cancers, humans can tolerate only 2% of the dose required to mediate an antitumor effect [50]. TNF-transfected TILs would presumably allow an increase of local concentrations of this cytokine at the tumor site. Unfortunately it was difficult to achieve consistently high levels of TNF-α production in lymphocytes following retrovirus-mediated gene transfer, probably because of posttranscriptional control of TNF expression and secretion.

Schmidt-Wolf et al. [51] have initiated a pilot study using genetically modified cytokine-induced killer cells (CIKs) transfected with the IL-7 gene. IL-7 promotes tumor rejection via expansion of CD8+ T cells, generation of CTLs, and generation of lymphokine-activated killer (LAK) cell activity by CD8+ and natural killer (NK) cells [37]. In other tumor types, CD4+ T cells and macrophages seem to be of importance to enhance tumor inhibition via a delayed type hypersensitivity reaction [36]. CIK cells are highly proliferating lymphocytes generated by culturing peripheral blood lymphocytes (PBLs) in the presence of IFN-γ, IL-2, IL-1, and anti-CD3. In this pilot study, CIK cells and/or tumor cells were transfected by either electroporation or ballistic transfer and injected i.v. in a dose (amount of cells)-escalation protocol to patients with end-stage metastatic renal cancer, melanoma, colon carcinoma, and malignant lymphoma.

V. GENETIC MODIFICATION OF ENDOTHELIAL CELLS AND FIBROBLASTS FOR CYTOKINE DELIVERY

The endothelium is a metabolically active cell type that produces proteins that act on perivascular cells in a paracrine fashion. Thus, endothelial cells function not only as a vascular framework for the intravascular delivery and removal of nutrients and metabolites from tissues, but as a source of cytokines that influence the growth and differentiation of neighboring vascular and parenchymal cells in response to disease and during development. In fact, all normal cells are believed to be within 6 to 8 cells from a blood vessel. In a tumor setting, in addition to perfusing the tumor, neovascularization also has a paracrine effect. Tumor cells can be stimulated by growth factors and other cytokines produced by vascular endothelial cells [52,53]. These two cell compartments within a tumor, endothelial cells and tumor cells, can stimulate each other. Unfortunately, the endothelium of solid tumors has been difficult to isolate and culture, but it seems reasonable to assume that biochemical differences may exist between endothelium from malignant and from nonmalignant tissues.

Zwiebel et al. [54] were the first to successfully infect endothelial cells (rabbit aortic endothelial cells) using Moloney murine leukemia virus–derived vectors achieving a high level of transgene expression. Endothelial cells are attractive as vehicles for the delivery of therapeutic recombinant molecules in vivo because they maintain their capacity to proliferate in vivo and offer a potential renewable and expandable source of therapeutic gene product at sites of tumor angiogenesis. Indeed, endothelial cells can undergo genetic manipulations and, after i.v. injection, migrate into and survive within an angiogenic site [55].

Transfected endothelial cells can also be stably engrafted to growing gliomas [56]. Johnston et al. [57], showed that brain microvascular endothelial cells trans-

fected with the mammalian expression vector pBCMG-hygro-FGF containing the fibroblast growth factor gene using the polycationic transfecting reagent lipofectamine could maintain their characteristic in vitro morphology, growth pattern, and endothelial phenotype, that these cells survived following implantation within rat brain for at least 21 days, and that they expressed the transfected gene for this period.

Su et al. [58] evaluated retroviral-mediated cytokine transduction of endothelial cells in breast cancer. Human umbilical vein endothelial cells were transduced with retroviral vectors containing cDNA encoding full-length rhIL-2 or rhIL-1α genes placed under the transcriptional control of the cytomegalovirus immediate early gene promoter. Cells expressing 19 ng/10^6 cells/day of IL-1α or 120 units/10^6 cells/day of IL-2 were used for the tumor implant experiment. Highly metastatic human breast cancer cells (MDA-MB-435) were injected with the genetically manipulated endothelial cells expressing IL-1α or IL-2 into the mammary pad of nude mice. Cytokine-expressing endothelial cells not only inhibited the tumorigenesis of MDA-MB-435 cells, they also abrogated the formation of metastasis.

Ojeifo et al. [59] studied the expression of IL-1α and IL-2 in retroviral-transduced human and murine endothelial cells. Expression ranging from 1.5 ng/10^6 cells/24 h to 246.7 ng/10^6 cells/24 h was achieved in the human and murine endothelial cells, the highest level of expression being in the human IL-2-transfected cells. Cytokine production was sufficient for the activation of potent cytotoxic lymphocyte activity. The cytokine-transduced endothelial cells retained normal endothelial cell features, including uptake of acetylated low-density lipoprotein and expression of von Willebrand factor.

An antitumor effect using endothelial-cell-based gene delivery was also demonstrated in a 9L glioma model. Immortalized rat brain endothelial cells that express the lacZ reporter gene and the gene for murine IL-2 were co-inoculated intracranially with 9L glioma cells to Fisher rats. The tumor growth was significantly inhibited by the genetically modified endothelial cells and rats receiving these cells showed enhanced survival. The endothelial cells were immortalized by transfection with adenovirus 2 E1A gene under transcriptional control of the SV40 promoter, then transfected in the presence of lipofectamine with a bovine papilloma virus expression vector containing a murine IL-2 cDNA clone.

Autologous fibroblasts were selected for gene delivery due to technical difficulties in establishing primary tumor cell lines from the vast majority of cancers. In contrast, dermal fibroblasts can be more easily obtained and cultured from virtually all individuals. Fibroblasts in culture can be infected by retroviral vectors selected for antibiotic resistance and expanded to sufficient numbers for therapeutic purposes [60]. This procedure has several theoretical advantages compared with other *ex vivo* approaches: It is independent of tranfection efficiency, avoids the delivery of infectious viral particles to the host, and does not include potential immunogens other than the nominal tumor antigen. Moreover, cytokine-expressing fibroblasts may be implanted into a tumor bed or inoculated together with tumor cells to immunize the patient against the tumor. Few data are available on whether or not coexpression of cytokines and tumor-associated antigens by the same immunizing cells is required for augmentation of host immunity.

This approach was pioneered by Bubenik *et al.* [61,62] using a retroviral expression construct in which human IL-2 cDNA had been inserted. Mertelsmann *et al.* [63] were the first to be given approval for gene therapy in Germany. They studied the use of autologous tumor cells plus IL-2-gene-transfected allogeneic fibroblasts as a vaccine in patients with advanced malignant tumors that were refractory of standard treatment regimens [64]. For this study a clone of a permanent human fibroblast line (KMST-6) that stably secreted 5290 IU IL-2/10^6 cells/24 h was obtained by cationic lipofection with an expression construct for human IL-2. Fifteen patients received 3 or 4 injections of irradiated transfected fibroblasts and autologous tumor cells. In three cases CTLs with cytotoxic activity against the autologous tumor were detected at the vaccination site. There was no objective tumor response, but this question was not specifically addressed in this phase I study of patients with high tumor burden and impaired T cell function. Further analysis of the data from this study also showed that in one patient the same CTL clone had infiltrated the tumor, circulated in the peripheral blood, and was amplified at the vaccination site [65].

Kim and Cohen [66] have also retrovirally cotransfected fibroblasts with the gene encoding IL-2 and with genomic DNA from the poorly immunogenic mouse B16 melanoma. The fibroblasts were then used as a therapeutic vaccine in mice with established melanoma, allowing the immunogenic-melanoma-bearing mice to survive longer than mice untreated or injected with irradiated melanoma cells.

Our group's experience in cancer gene therapy using transfected fibroblasts began with the use of IL-4 genetically modified fibroblasts together with autologous tumor cells to elicit an immune response [67]. The rationale for this protocol was that we and others had demonstrated that murine tumor cells transfected with the IL-4 gene regressed after a brief period of growth and that protective immunity was acquired, suggesting that constant local production of IL-4 is associated with antitumor immunity [25,42]. The mechanism underlying

the antitumor effect, although not clearly defined, involves macrophage and lymphocyte infiltration. IL-4 may therefore play a critical role in the homing of these cells to sites of tumor by inducing the expression of vascular cell adhesion molecule (VCAM) by the tumor microvascular endothelium. Eighteen patients with metastatic and/or unresectable melanoma, renal cell carcinoma, breast cancer, or colon cancer were studied. Three patients had a partial response, and one is still alive without tumor ([68]; Lotze *et al.*, manuscript in preparation).

Our interests also include IL-12, which serves as a bridge between the adoptive and the innate immune responses. It is one of the key cytokines involved in the development of antitumor immunity, facilitating a number of immune mechanisms including activation and generation of CTL, stimulation of NK cell activity, increase in macrophage function, and promotion of Th1 cell generation (see references [69,70] for reviews). IL-12 is a disulfide-linked heterodimeric cytokine composed of a 35-kDa light chain (p35) and a 40-kDa heavy chain (p40) [71,72]. Unlike most other cytokines, simultaneous transfection of mammalian cells with two different genes is necessary for the production of biologically active IL-12 [73,74]. In this initial study NIH3T3 cells were stably transfected to express 100–240 units/10^6 cells/48 h with three plasmids encoding the genes for p35, p40, and neomycin phosphotransferase using standard calcium phosphate techniques. We showed that fibroblasts stably expressing IL-12 could suppress tumor growth and induce antitumor immunity to a murine melanoma in an establishment model [75]. In a subsequent study, the antitumor effects of local, high-level IL-12 expression were examined using a retroviral vector that can express both IL-12 subunits (p35 and p40) and the neomycin phosphotransferase gene from a polycistronic message utilizing internal ribosome entry site sequences [76]. Injection of these engineered fibroblasts at the site of an established (day 8) MCA207 sarcoma could efficiently eliminate or suppress tumor growth in a dose dependent manner requiring >150 ng/kg/dose of bioactive IL-12. Seven-day-established lung metastases of less immunogenic tumors (MCA102, MC38) could also be significantly reduced.

These encouraging results led to our current clinical protocol using direct injection of resident tumors with autologous fibroblasts transduced with a retroviral vector encoding both subunits of human IL-12 [76,77]. An MFG-based retroviral vector was used to generate a recombinant retrovirus in which a long terminal repeat (LTR)-driven polycistronic transcript encodes both subunits of human IL-12: hp35 and hp40 cDNAs are linked and coexpressed using the internal ribosome entry site

(IRES) from the encephalomyocarditis virus (Fig. 1). This clinical trial was designed to evaluate local and systemic toxicity as well as local and systemic immuno-modulatory effects of peritumoral injections of these genetically engineered fibroblasts. Thirty-three patients with metastatic melanoma, breast cancer, head and neck cancer, or cutaneous T-cell lymphoma have been treated so far. The protocol is ongoing, but lesional and regional responses in patients with melanoma, head and neck carcinoma, and breast cancer have been observed [78]. Phase II studies in patients with melanoma and head and neck cancer are being initiated at the University of Pittsburgh Cancer Institute.

VI. FUTURE DIRECTIONS: APC TRANSFECTION

Processing and presentation of tumor antigen by APCs is required in the induction of effective antitumor-reactive T cells [79]. In fact, presentation of antigen to T cells in the absence of a second signal (costimulation) may induce T-cell apoptosis or antigen-specific tolerance [12]. DCs are the most effective APCs and are found in all tissues and organs in the body [11]. They are critical for the induction of primary, cell-mediated immune responses due to their ability to acquire antigen in the peripheral tissues and process, transport, and present it to naive or memory-antigen-specific T cells in secondary lymphoid organs [13,80,81].

DCs have been shown to express a variety of cytokines including IL-1α, IL-1β, IL-6, IL-8, IL-10, IL-12, IL-15, IFN-α, TNF-α and GM-CSF [82–85]. It is now clear that cytokines are crucially involved in the regulation of tumor Ag presentation by APCs and that primary and secondary immune responses are differentially modulated by these cytokines. It is likely that DCs *in situ* reside in an "immature" state, being relatively inefficient at inducing primary immune response, and that they acquire this activity only after stimulation with the appropriate cytokines [86].

The DC's effectiveness in presenting tumor Ag is potently modulated by cytokines, indicating that the local cytokine environment is crucial for the generation and elicitation of tumor immune responses. Because of difficulties in obtaining sufficient amounts of pure DCs, their use as immunotherapeutic agents has been impractical. They constitute only 0.05 to 0.3% of peripheral blood mononuclear cells [87]. However, methods have been developed recently to generate sizable numbers of pure DCs in both humans and rodents. Human, mouse, and rat DCs can be generated *in vitro* from hematopoietic progenitors in media containing GM-CSF in all species with the addition of IL-4 or TNF in the murine species or TNF and SCF in humans [13,80,81]. Studies are ongoing to identify the most efficient cytokine cocktail to be used. DCs can also be generated from human peripheral blood cells cultured with GM-CSF and IL-4 [88,89]. Human DC cell lines have also been established from histiocytic lymphoma cells or immortalized blood mononuclear cells [90,91].

The ready availability of cultured DCs and their capacity to stimulate potent and Ag-specific antitumor immunity make them an attractive target for gene therapy. Expression of cytokines by gene-modified DCs should significantly enhance the initial T-cell response to tumor antigen presentation or maintain them effectively in the periphery. The initial difficulties of transfecting DCs have been overcome and a number of studies in the past year have shown efficient transfection of DCs (Tables 3 and 4). Almost all of these initial studies evaluated the transfection of tumor antigens or reporter genes. Using a number of different vectors, our group is currently investigating the transfection of cytokine genes including IL-12.

Numerous studies have shown successful retroviral mediated transduction of human DC [92–99]. All used tumor Ag or reporter genes. Of these, five evaluated retroviral transduction of CD34 precursors from bone marrow, cord blood, or peripheral blood cells, obtaining a transfection efficiency ranging from 11 to 28% [92–94,96,98]. The rationale for transfecting these DC precursors is that as nonreplicating terminally differential cells, mature DCs are poor candidates for retroviral gene modification. However, CD34 precursors actively proliferate and can be matured *in vitro* into DCs expressing the wanted gene. Szabolcs *et al.* [96] and Reeves *et al.* [94] infected the CD34 precursors by coculturing them with producer cells. This coculture system is interesting for *in vitro* studies, but it is incompatible with current guidelines for therapeutic uses of retroviral vectors. Henderson *et al.* [93], Reeves *et al.* [94], and Bello-Fernandez *et al.* [98] added an additional step: At the beginning of the infection, they centrifuged the infected CD34 precursors aiming to enhance infection efficiency. To further enhance the transfection efficiency they included multiple infection cycles (from 2 to 3 separate

FIGURE 1 IL-12 retroviral vector (TFG-hIL-12-Neo). MFG-based retroviral plasmid where human p40-IL-12, p35-IL-12 and NeoR are subcloned. Both subunits and the selectable marker NeoR are promoted by the 5'LTR. Utilization of intervening IRES sequences allows the transcription of a single polycistronic message.

TABLE 3   Human DC Transfection

Reference	Transfection means	Cell transduced (DC or precursor)	Centrifugation step during transduction	Coculture with producer cells	Multiple transfections	Efficiency
92	Retrovirus	CD34 precursor	No	No	No	17%
118	Liposome	DC (PBMC cultured in GM/IL4)	No	No	No	ND
93	Retrovirus	CD34 precursor	Yes	No	Yes(2)	13.5%
94	Retrovirus	CD34 precursor	Yes	Yes	Yes(3)	22–28%
97	Adenovirus, liposome electroporation, CaPO$_4$ precipitation	DC (PBMC cultured in GM/IL4)	No	No	No	<30% at 100:1 MOI >95% at 1000:1 MOI
96	Retrovirus	CD34 precursor	No	Yes	No	11.5–21.2%
98	Retrovirus	CD34 precursor	Yes	No	Yes(3)	14%
95	Retrovirus	DC (PBMC cultured in GM/IL4)	No	No	Yes(3)	51%
119	Gene gun	DC (PBMC cultured in GM/IL4)	No	No	No	5–10%

* ND, not determined.

days). In fact, Bello-Fernandez *et al.* [98] compared one infection to three infections on three consecutive days and showed that the latter regimen resulted in a two- to threefold increase in transduction efficiency. They also found that the presence of Flt3-ligand in the culture medium enhanced transfection efficiency. This effect was not mediated by an increase in the number of cycling cells facilitating retroviral infection, suggesting that other mechanisms, including possibly an up-regulation of the number or increase in the affinity of retroviral receptors on progenitor cells, likely contribute to the effect of Flt3-ligand in enhancing retroviral infection. This is consistent with previously published data show-

ing that Flt3-ligand increases the capacity of retroviruses to transduce hematopoietic stem cells [99,100]. Only one group studied the retroviral transduction of DCs obtained from PBMCs cultured in medium supplemented with GM-CSF and IL-4 and found a transfection efficiency of 51%, which is peculiarly high for a retroviral construct in view of the fact that peripheral blood precursors of DCs are nonproliferating [95,101].

Only one paper compared gene transfer methods in human dendritic cells, unfortunately leaving out retroviral vectors [97]. Human DCs were prepared by culturing the adherent fraction of peripheral blood mononuclear cells (PBMCs) in GM-CSF and IL-4. Adenoviral trans-

TABLE 4   Murine DC Transduction

Reference	Transduction means	Centrifugation step during transduction	Coculture with producer cells	Multiple transfections	Efficiency
102	Retrovirus, gene gun, endocytosis	No	No	No	Retrovirus: 8–10% Gene gun: ND Endocytosis: 34–49%
103	Adenovirus	No	No	No	24% at MOI 10^2 33% at MOI 10^3
106	Gene gun	No	No	No	5–10%
104	DOTAP (naked DNA)	No	No	No	ND
105	Adenovirus	No	No	No	ND
107	Adenovirus	No	No	No	90% at MOI 10^2
108	Retrovirus	No	Yes	No	41–72%
109	Retrovirus	Yes	No	Yes(3)	20–50%

* ND, not determined; MOI, multiplicity of infection.

duction was compared to liposomes, electroporation, and CaPO$_4$ precipitation, and was clearly the most efficient method: Transgene expression was up to 1000-fold more efficient. Cells were transduced in suspension with adenoviral vectors encoding reporter genes, IL-2, or IL-7 at various multiplicities of infection (MOIs) (MOI = 1–10,000) at 37°C in 5% CO$_2$ for 2 hours. They also compared DC and tumor cell transfection and showed that DCs consistently produced 2 to 4 orders of magnitude fewer genes than did melanoma cells at similar MOIs. Moreover the transfection efficiency was MOI dependent, reaching more than 95% at 1000:1 MOI, whereas at 100:1 MOI, less than 30% of DCs were transduced. The cytopathic effect of a virus load of 1000:1 that is observed in melanoma cells was not observed in DCs. DCs transduced with IL-2 produced high levels of IL-2 over the course of 7 days, approximately 500 ng/10^6 cells/24 h, and the level of IL-2 transgene expression was appreciable for up to 2 weeks in some experiments.

Tüting et al. [99] transfected monocyte-derived dendritic cells with plasmid DNA encoding human melanoma antigens associated with IL-12 and TNF-α by particle bombardment. In these studies, transduced DCs expressed a normal phenotype and maintained their functional properties, stimulating allogeneic T cells with efficiency similar to that of nontransduced DCs. Furthermore, the transduction procedures did not have negative effects on the viability of these cultures.

Eight studies evaluated murine DC transfection (Table 3) [102–109]. Basically the same protocol was used to generate murine DCs. Bone marrow cells were cultured for a week in media containing GM-CSF and IL-4 as previously described [110]. This allows for a high proportion of nonproliferating DCs at the end of the 6-day culture. Zitvogel et al. [102] compared three transduction techniques: retroviral, particle bombardment, and endocytic delivery of particulate DNA. In this latter technique retroviruses and uncoated gold beads were mixed together immediately before employing the standard retrovirus infection protocol. Surprisingly, DCs transfected with the latter technique exhibited a higher transduction efficiency when compared with retrovirus alone (49 vs. 10%) and higher expression levels of IL-12 when compared with retroviruses alone or particle bombardment, respectively (25 ng/mL/10^6 cells vs. 2 ng/mL/10^6 cells vs. 0.3 ng/mL/10^6 cells). It was then shown that genetically engineered DCs that secrete IL-12 were enhanced in their efficacy in peptide-based therapy. Wan et al. [107] used an adenoviral approach to transduce murine DCs. They were harvested after 6 days of culture, and recombinant viruses encoding LacZ or the Ag exchange PymT were added at an MOI of 100 plaque-forming units (pfu)/cell. This resulted in a trans-

fection efficiency of more than 90%. Importantly, a mixed lymphocyte reaction assay demonstrated that Ad-infected DCs were as potent as noninfected DCs in stimulating the proliferation of allogeneic T cells, indicating that the Ag-presenting function of DCs was not impaired by the adenoviral transfection. Furthermore, no hepatic lesions or serum enzyme increases were detected in mice infected with adenoviral-infected DCs, in contrast to findings of dose-dependent hepatic inflammation and degeneration associated with increases of liver enzymes after direct i.v. injection of adenovirus.

The adenoviral approach was also used by Ribas et al. [105] and Brossart et al. [103]. The latter group had a transfection efficiency of 33% at MOI 10^3 and 24% at MOI 10^2, and showed that they could infect an immortalized bone marrow DC cell line as well as fresh DCs could. Repeated injections of virus-infected DCs induced only low titers of neutralizing antibodies compared to direct injection of the same adenovirus. Moreover, the presence of neutralizing antibodies specific for the virus (animal already in contact with the adenoviral vector) did not affect the usefulness of infected DCs (transduction of antigenic epitopes), as repeated applications of DCs boosted the CTL response. Tüting et al. [111] obtained a 5 to 10% transfection efficiency using the "gene gun" approach and detected significant levels of transgenes within 8 hours, with maximal levels occurring 16–24 hours after transfection. The transfection efficiency using this approach is significantly lower than with the viral approaches, but what percentage of DCs need to be transfected to obtain a clinical effect is still unknown, and although "more is often better," there are actually no data that support this. Moreover they showed using a tumor Ag exchange that enough DCs were transfected to stimulate peptide-specific CTLs and promote protective antitumor immunity in vivo.

Two groups studied retroviral transduction of bone-marrow-derived DCs [108,109]. They both transfected bone marrow precursors in the first few days of culture, while they were still proliferating and in the differentiating process. Specht et al. [108] cocultured bone marrow precursors and an irradiated-fibroblast-derived, ecotropic-packaging cell line encoding the lacZ gene in the MFG retroviral backbone. Transduction efficiencies between 41 and 72% were consistently observed by this method. Nishioka et al. [109] studied the retroviral transfection of a reporter and IL-12 gene. Retroviral centrifugation was performed three times, on days 2, 3, and 4 in bone marrow culture with GM-CSF and IL-4. Twenty to 50% transfection efficiency was achieved using DFG-hCD80-Neo retroviral supernatant produced by BOSC23 ecotropic producer cells (more than 2 × 10^6 pfu). Transfection with the IL-12 gene resulted in expression of 4-80 ng/10^6 cells/48 h IL-12 for at least 5 days. When IL-12-transduced DCs were injected into established MCA207 tumors, tumor growth was suppressed. Manickan et al. [104] transfected murine DCs with naked plasmid DNA encoding proteins of herpes simplex virus in the presence of DOTAP in a nontumor model. They did not report the transfection efficiency but were able to induce significant enhancement of resistance to viral challenge. For a further discussion of the use of transfected DCs as vaccines see Chapter 18.

VII. CONCLUSION

The interactions between tumor cells and the immune system are complex and Darwinian, played out over several years of coevolution in the host. Cytokines secreted either by the tumor or by neighboring cells play a crucial role in their development. The immune manipulations currently evaluated and described in this chapter are in the beginning of their development, but the fact that at least some human tumors respond to these immunologic maneuvers is encouraging for future developments in this field. Considerable work in this area should open ways to a better treatment of patients with cancer and to a more fundamental understanding of antitumor immunity. At the present time, primarily for ethical reasons, most clinical trials include only patients with disseminated disease. Patients with minimal tumor burden, but at high risk of recurrence, are expected to especially benefit from such treatment approaches.

References

1. Lotze, M. T., Matory, Y. L., Rayner, A. A., Ettinghausen, S. E., Vetto, J. T., Seipp, C. A., Rosenberg, S. A. Clinical effects and toxicity of interleukin-2 in patients with cancer. *Cancer* 1986; **58,** 2764–2772.
2. Trehu, E. G., Isner, J. M., Mier, J. W., Karp, D. D., Atkins, M. B. Possible myocardial toxicity associated with interleukin-4 therapy. *J. Immunother.* 1993; **14,** 348–351.
3. Rosenberg, S. A., Lotze, M. T., Yang, J. C., Aebersold, P. M., Linehan, W. M., Seipp, C. A., White, D. E. Experience with the use of high-dose interleukin-2 in the treatment of 652 cancer patients. *Ann. Surg.* 1989; **210,** 474–484.
4. Lotze, M. T., Matory, Y. L., Ettinghausen, S. E., Rayner, A. A., Sharrow, S. O., Seipp, C. A., Custer, M. C., Rosenberg, S. A. In vivo administration of purified human interleukin 2. II. Half life, immunologic effects, and expansion of peripheral lymphoid cells in vivo with recombinant IL 2. *J. Immunol.* 1985; **135,** 2865–2875.
5. Foley, E. J. Antigenic properties of methylcholanthrene-induced tumors in mice of the strain of origin. *Cancer Res.* 1953; **13,** 835–837.
6. Prehn, R. T., Main, J. M. Immunity to methylcholanthrene-induced sarcomas. *J. Natl. Cancer Inst.* 1957; **18,** 769–778.
7. Borberg, H., Oettgen, H. F., Choudry, K., Beattie, E. J., Jr. Inhibition of established transplants of chemically induced sarcomas in syngeneic mice by lymphocytes from immunized donors. *Int. J. Cancer* 1972; **10**(3), 539–547.
8. Kast, W. M., Offringa, R., Petres, P. J., Voordouw, A. C., Meleon, R. H., van der Eb, A. J., Melief, C. J. M. Eradication of adenovirus E1-induced tumors by E1A-specific cytotoxic T lymphocytes. *Cell* 1989; **59,** 603–614.
9. Melief, C. J. M. Tumor eradication by adoptive transfer of cytotoxic lymphocytes. *Adv. Cancer Res.* 1991; **58,** 143–175.
10. Rosenberg, S. A., Spiess, P., Lafreniere, R. A. A new approach to the adoptive immunotherapy of cancer with tumor-infiltrating lymphocytes. *Science* 1986; **233,** 1318–1321.
11. Caux, C., Liu, Y. J., Bancherau, J. Recent advances in the study of dendritic cells and follicular dendritic cells. *Immunol. Today* 1995; **16,** 2–4.
12. Chen, L., Linsley, P. S., Hellstrom, K. E. Costimulation of T-cells for tumor immunity. *Immunol. Today* 1993; **14,** 483–486.
13. Steinman, R. M. The dendritic cells system and its role in immunogenicity. *Annu. Rev. Immunol.* 1991; **9,** 271–296.
14. Berke, G. The binding and lysis of target cells by cytotoxic lymphocytes: molecular and cellular aspects. *Annu. Rev. Immunol.* 1994; **12,** 735–773.
15. Wiley, S. R., Schooley, K., Smolak, P. J., Din, W. S., Huang, C. P., Nicholl, J. K., Sutherland, G. R., Smith, T. D., Rauch, C., Smith, C. A., *et al.* Identification and characterization of a new member of the TNF family that induces apoptosis. *Immunity* 1995; **3,** 673–682.
16. Fraser, A., Evan, G. A license to kill. *Cell* 1996; **85,** 781–784.
17. Pitti, R. M., Marsters, S. A., Ruppert, S., Donahue, C. J., Moore, A., Ashkenazi, A. Induction of apoptosis by Apo-2 ligand, a new member of the tumor necrosis factor cytokine family. *J. Biol. Chem.* 1996; **271,** 12687–12690.
18. Marsters, S. A., Sheridan, J. P., Donahue, C. J., Pitti, R. M., Gray, C. L., Goddard, A. D., Bauer, K. D., Ashkenazi, A. Apo-3, a new member of the tumor necrosis factor receptor family, contains a death domain and activates apoptosis and NF-kappa B. *Curr. Biol.* 1996; **6,** 1669–1676.
19. Nagata, S. Apoptosis by death factor. *Cell* 1997; **88,** 355–365.
20. Russel, S. J., Flemming, C., Eccle, S. A., Johnson, C., Collins, M. K. L. Transfer in and expression of the human IL-2 gene in a transplantable rat sarcoma. In *Cellular Immunology and Immunotherapy of Cancer,* Lotze, M. T., Finn, O. J., Eds. AJ Liss and Co., New York, 1990; 275–278.
21. Blankenstein, T., Rowley, S., Schreiber, H. Cytokines and cancer: experimental systems. *Curr. Opin. Immunol.* 1991; **3,** 694–698.
22. Pardoll, D. M. Paracrine cytokine adjuvants in cancer immunotherapy. *Annu. Rev. Immunol.* 1995; **13,** 399–415.
23. Colombo, M., Forni, G. Cytokine gene transfer in tumor inhibition and tumor therapy: where are we now? *Immunol. Today* 1994; **15,** 48–51.
24. Tahara, H., Zitvogel, L., Storkus, W. J., Robbins, P. D., Lotze, M. T. Murine models of cancer cytokine gene therapy using Interleukin-12. *Ann. N.Y. Acad. Sci.* 1996; **795,** 275–283.
25. Tepper, R. I., Pattengale, P. K., Leder, P. Murine interleukin-4 displays potent antitumor activity *in vivo. Cell* 1989; **57,** 503–512.
26. Tepper, R. I. The tumor-cytokine transplantation assay and the antitumor activity of interleukin-4. *Bone Marrow Transplant* 1992; 9(Suppl 1):177–181.
27. Krueger-Krasagakes, S., Li, W., Richter, G., Diamantstein, T., Blankestein, T. Eosinophils infiltrating interleukin-5 gene-transfected tumors do not suppress tumor growth. *Eur. J. Immunol.* 1993; **23,** 992–995.

28. Torre-Amione, G., Beauchamp, R. D., Koeppen, H., Park, B. H., Schreiber, H., Moses, H. L., Rowley, D. A. A highly immunogenic tumor transfected with a murine transforming growth factor type β1 cDNA escapes immune surveillance. *Proc. Natl. Acad. Sci. U.S.A.* 1990; **87**, 1486–1490.

29. Chang, H. L., Gillet, N., Figari, I., Lopez, A. R., Palladino, M. A., Derynck, R. Increased transforming growth factor β expression inhibits cell proliferation *in vitro*, yet increases tumorigenicity and tumor growth of MethA Sarcoma cells. *Cancer Res.* 1993; **53**, 4391–4398.

30. Qin, Z., Kruger-Krasagakes, S., Kunzendorf, U., Hock, H., Diamantstein, T., Blankenstein, T. Expression of tumor necrosis factory by different tumor cell lines results either in tumor suppression or augmented metastasis. *J. Exp. Med.* 1993; **178**, 355–360.

31. Dranoff, G., Jaffee, E., Lazenby, A., Golumbek, P., Levitsky, H., Brose, K., Jackson, V., Hamada, H., Pardoll, D., Mulligan, R. C. Vaccination with irradiated tumor cells engineered to secrete murine granulocyte-macrophage colony-stimulating factor stimulates potent, specific, and long-lasting anti-tumor immunity. *Proc. Natl. Acad. Sci. U.S.A.* 1993; **90**, 3539–3543.

32. Cavallo, P., Giovarelli, M., Gulino, A., Vacca, A., Stoppacciaro, A., Modesti, A., Forni, G. Role of neutrophils and CD4 T lymphocytes in the primary and memory response to nonimmunogenic murine mammary adenocarcinoma made immunogenic by IL-2 gene. *J. Immunol.* 1992; **149**, 3627–3635.

33. Mullen, C. A., Coale, M., Levy, A. T., Stetler-Stevenson, W. G., Liotta, L. A., Brandt, S., Blaese, R. M. Fibrosarcoma cells transduced with the IL-6 gene exhibit reduced tumorigenicity, increased immunogenicity and decreased metastatic potential. *Cancer Res.* 1992; **52**, 6020–6024.

34. Porgador, A., Tzehoval, E., Katz, A., Vadai, E., Revel, M., Feldman, M., Eisenbach, L. Interleukin-6 gene transfection into Lewis lung carcinoma tumor cells suppresses the malignant phenotype and confers immunotherapeutic competence against parental metastatic cells. *Cancer Res.* 1992; **52**, 3679–3686.

35. Sun, W. H., Kreisle, R. A., Phillips, A. W., Ershler, W. B. *In vitro* and *in vivo* characteristics of interleukin 6-transfected B16 melanoma cells. *Cancer Res.* 1992; **52**, 5412–5415.

36. Hock, H., Dorsch, M., Diamantstein, T., Blankenstein, T. Interleukin 7 induces CD4+ T-cell-dependent tumor rejection. *J. Exp. Med.* 1991; **174**, 1291–1298.

37. Jicha, D. L., Mule, J. J., Rosenberg, S. A. Interleukin 7 generates antitumor cytotoxic T lymphocytes against murine sarcomas with efficacy in cellular adoptive immunotherapy. *J. Exp. Med.* 1991; **174**, 1511–1515.

38. McBride, W. H., Thacker, J. D., Comora, S., Economou, J. S., Kelley, D., Hogge, D., Dubinett, S. M., Dougherty, G. J. Genetic modification of a murine fibrosarcoma to produce interleukin 7 stimulates host cell infiltration and tumor immunity. *Cancer Res.* 1992; **52**, 3931–3937.

39. Tahara, H., Zitvogel, L., Storkus, W. J., Zeh, H. J., McKinney, T. G., Schreiber, Gubler, U., Robbins, P. D., Lotze, M. T. Effective eradication of established murine tumors with interleukin-12 gene therapy using a polycystronic retroviral vector. *J. Immunol.* 1995; **154**, 6466–6474.

40. Osaki, T., Péron, J.-M., Cai, Q., Okamura, H., Robbins, P. D., Kurimoto, M., Lotze, M. T., Tahara, H. Interferon-γ-inducing factor/interleukin 18 administration mediates interleukin 12 and interferon-γ independent anti-tumor effects. *J. Immunol.* 1998; **160**, 1742–1749.

41. Huland, E., Huland, H. Treated patients: evidence of toxic Tumor-associated eosinophilia in interleukin-2-eosinophil degranu-

lation on bladder cancer cells. *J. Cancer Res. Clin. Oncol.* 1992; **118**, 463–467.

42. Colombo, M. P., Ferrari, G., Stoppacciaro, A., *et al.* Granulocyte colony-stimulating factor gene transfer suppresses tumorigenicity of a murine adenocarcinoma *in vivo*. *J. Exp. Med.* 1991; **173**, 889–897.

43. Giovarelli, M., Cofano, F., Vecchi, A., Forni, M., Landolfo, S., Forni, G. Interferon-activated tumor inhibition in vivo. Small amounts of interferon-gamma inhibit tumor growth by eliciting host systemic immunoreactivity. *Int. J. Cancer* 1986; **37**, 141–148.

44. Forni, G., Giovarelli, M., Santoni, A., Modesti, A., Forni, M. Interleukin 2 ativated tumor inhibition in vivo depends on the systemic involvement of host immunoreactivity. *J. Immunol.* 1987; **138**, 4033–4041.

45. Stoppacciaro, A., Paglia, P., Lombardi, L., Parmiani, G., Baroni, C., Colombo, M. P. Genetic modification of a carcinoma with the IL-4 gene increases the influx of dendritic cells relative to other cytokines. *Eur. J. Immunol.* 1997; **27**, 2375–2382.

46. Bevilacqua, M. P., Prober, J. S., Wheeler, M. E., Cotran, R. S., Gimbrone, M. J. Interleukin 1 acts on cultured human vascular endothelium to increase the adhesion of polymorphonuclear leukocytes, monocytes, and related leukocyte cell lines. *J. Clin. Invest.* 1985; **76**, 2003–2009.

47. Thornill, M. H., Wellicome, S. M., Mahiouz, D. L., Lanchbury, J. S. S., Kyan-Aung, U., Haskard, D. O. Tumor necrosis factor combines with IL-4 or IFN-γ to selectively enhance endothelial cell adhesiveness for T-cells. The contribution of vascular cell adhesion molecule-1-dependent and independent binding mechanism. *J. Immunol.* 1991; **146**, 592–598.

48. Miescher, S., Whiteside, T. L., Moretta, L., von Fliedner, V. Clonal and frequency analyses of tumor-infiltrating T lymphocytes from human solid tumors. *J. Immunol.* 1987; **138**, 4004–4011.

49. Rosenberg, S. A., Packard, B. S., Aebersold, P. M., *et al.* Use of tumor-infiltrating lymphocytes and interleukin-2 in the immunotherapy of patients with metastatic melanoma: a preliminary report. *N. Engl. J. Med.* 1988; **319**, 1676–1680.

50. Rosenberg, S. A. Immunotherapy and gene therapy of cancer. *Cancer Res.* 1991; **51**, 5074–5079.

51. Schmidt-Wolf, I. G. H., Huhn, D., Neubauer, A., Wittig, B. Interleukin-7 gene transfer in patients with metastatic colon carcinoma, renal cell carcinoma, melanoma or with lymphoma. *Hum. Gene Ther.* 1994; **5**, 1161–1168.

52. Rak, J., Filmus, J., Finkenzeller, G., Grugel, S., Marme, D., Kerbel, R. S. Oncogenes as inducers of tumor angiogenesis. *Cancer Metastasis Rev.* 1995; **14**, 263–277.

53. Folkman, J. Tumor angiogenesis and tissue factor. *Nature Med.* 1996; **2**, 167–168.

54. Zwiebel, J. A., Freeman, S. M., Kantoff, P. W., Cornetta, K., Ryan, U. S., Anderson, W. F. High-level recombinant gene expression in rabbit endothelial cells transduced by retroviral vectors. *Science* 1989; **243**, 220–222.

55. Ojeifo, J. O., Forough, R., Paik, S., Maciag, T., Zwiebel, J. A. Angiogenesis-directed implantation of genetically modified endothelial cells in mice. *Cancer Res.* 1995; **55**, 2240–2244.

56. Lal, B., Indurti, R. R., Couraud, P. O., Goldstein, G. W., Laterra, J. Endothelial cell implantation and survival within experimental gliomas. *Proc. Natl. Acad. Sci. U.S.A.* 1994; **91**, 9695–9699.

57. Johnston, P., Nam, M., Hossain, M. A., Indurti, R. R., Mankowski, J. L., Wilson, M. A., Laterra, J. Delivery of human fibroblast growth factor-1 gene to brain by modified rat brain endothelial cells. *J. Neurochem.* 1996; **67**, 1643–1652.

58. Su, N., Ojeifo, J. O., MacPherson, A., Zwiebel, J. A. Breast cancer gene therapy: transgenic immunotherapy. *Breast Cancer Res. Treat.* 1994; **31**, 349–356.

59. Ojeifo, J. O., Su, N., Ryan, U. S., Verma, U. N., Mazumder, A., Zwiebel, J. A. Towards endothelial-cell-directed cancer immuno-therapy—*in vitro* expression of human recombinant cytokine genes by human and mouse primary endothelial cells. *Cytokines Mol. Ther.* 1996; **2**, 89–101.

60. Elder, E. M., Lotze, M. T., Whiteside, T. L. Successful culture and selection of cytokine gene-modified human dermal fibroblasts for the biologic therapy of patients with cancer. *Hum. Gene Ther.* 1996; **7**, 479–487.

61. Bubenik, J., Voitenok, N. N., Kieler, J., Prassolov, V. S., Chuma-kov, P. M., Bubenikova, D., Simova, J., Jandlova, T. Local admin-istration of cells containing an inserted IL-2 gene and producing IL-2 inhibits growth of human tumours in nu/nu mice. *Immunol. Lett.* 1988; **19**, 279–282.

62. Bubenik, J., Simova, J., Jandlova, T. Immunotherapy of cancer using local administration of lymphoid cells transformed by IL-2 cDNA and constitutively producing IL-2. *Immunol. Lett.* 1990; **23**, 287–292.

63. Mertelsman, R., Lindemann, A., Boehm, T., Brennscheidt, U., Franke, B., Kulmburg, P., Lahn, M., Mackensen, A., Rosenthal, F. M., Vellken, H. Pilot study for the evaluation of T-cell medi-ated tumor immunotherapy by cytokine gene transfer in patients with malignants tumors. *J. Mol. Med.* 1995; **73**, 205–206.

64. Veelken, H., Mackensen, A., Lahn, M., Kohler, G., Becker, D., Franke, B., Brennscheidt, U., Kulmburg, P., Rosenthal, F. M., Keller, H., Hasse, J., Schultze-Seemann, W., Farthmann, E. H., Mertelsmann, R., Lindemann, A. A phase-I clinical study of autologous tumor cells plus interleukin-2-gene-transfected allo-geneic fibroblasts as a vaccine in patients with cancer. *Int. J. Cancer* 1997; **70**, 269–277.

65. Mackensen, A., Veelken, H., Lahn, M., Wittnebel, S., Becker, D., Kohler, G., Kulmburg, P., Brennscheidt, U., Rosenthal, F., Fanke, B., Mertelsmann, R., Lindemann, A. Introduction of tumor-specific cytotoxic T lymphocytes by immunization with autologous tumor cells and interleukin-2 gene transfected fibro-blasts. *J. Mol. Med.* 1997; **75**, 290–296.

66. Kim, T. S., Cohen, E. P. Interleukin-2-secreting mouse fibroblasts transfected with genomic DNA from murine melanoma cells prolong the survival of mice with melanoma. *Cancer Res.* 1994; **54**, 2531–2535.

67. Lotze, M. T., Rubin, J. T., Carty, S., Edington, H., Ferson, P., Landreneau, R., Pippin, B., Posner, M., Rosenfelder, D., Watson, C. Gene therapy of cancer: a pilot study of IL-4-gene-modified fibroblasts admixed with autologous tumor to elicit an immune response. *Hum. Gene Ther.* 1994; **5**, 41–55.

68. Suminami, Y., Elder, E. M., Lotze, M. T., Whiteside, T. L. *In situ* IL-4 gene expression in cancer patients treated with a genetically-modified tumor vaccine. *J. Immunother.* 1995; **17**, 238–248.

69. Shurin, M. R., Esche, C., Peron, J. M., Lotze, M. T. Antitumor activities of IL-12 and mechanisms of action. *Chem. Immunol.* 1997; **68**, 153–174.

70. Storkus, W. J., Tahara, H., Lotze, M. T. Interleukin 12. In *The Cytokine Handbook, Third Edition;* Thomson A. Ed. Academic Press, Ltd., London, in press.

71. Kobayashi, M., Fitz, L., Ryan, M., Hewick, R. M., Clark, S. C., Chan, S., Loudon, R., Sherman, F., Perussia, B., Trinchieri, G. Identification and purification of natural killer cell stimulatory factor, a cytokine with multiple biologic effects on human lym-phocytes. *J. Exp. Med.* 1989; **170**, 827–845.

72. Stern, A. S., Magram, J., Presky, D. H. Interleukin-12, an integral cytokine in the immune response. *Life Sci.* 1996; **58**, 639–654.

73. Wolf, S. F., Temple, P. A., Kobayashi, M., Trinchieri, G., Perus-sia, B. Cloning of cDNA for NKSF, a heterodimeric cytokine

with multiple biologic effects on T and NK cells. *J. Immunol.* 1992; **146**, 3074–3081.

74. Gubler, U., Chua, A. O., Schoenchaut, D. S., Dwyer, C. M., McGately, M. K. Coexpression of two distinct genes is required to generate secreted bioactive cytotoxic lymphocyte maturation factor. *Proc. Natl. Acad. Sci. U.S.A.* 1991; **88**, 4143–4147.

75. Tahara, H., Zeh, H. J., Storkus, W. J., Pappo, I., Watkins, S. C., Gubler, U., Wolf, S. F., Pobbins, P. D., Lotze, M. T. Fibroblasts genetically engineered to secrete interleukin-12 can suppress tumor growth and induce antitumor immunity to a murine mela-noma *in vivo. Cancer Res.* 1994; **54**, 182–189.

76. Tahara, H., Lotze, M. T., Robbins, P. D., Storkus, W. J., Zitvogel, L. Clinical protocol: IL-12 gene therapy using direct injections of tumors with genetically engineered autologous fibroblasts. *Hum. Gene Ther.* 1995; **6**, 1607–1624.

77. Zitvogel, L., Tahara, H., Cai, Q., Storkus, W. J., Muller, G., Wolf, S. F., Gately, M., Robbins, P. D., Lotze, M. T. Construction and characterization of retroviral vectors expressing biologically active human interleukin-12. *Hum. Gene Ther.* 1994; **5**, 1493–1506.

78. Tahara, H., Zitvogel, L., Storkus, W. J., Elder, E. M., Kinzler, D., Whiteside, T. L., Robbins, P. D., Lotze, M. T. Antitumor effects in patients with melanoma, head and neck and breast cancer in a phase I/II clinical trial of Interleukin-12 gene therapy. *Proc. Am. Soc. Clin. Oncol.* 1997; **16**, 438a.

79. Huang, A. Y. C., Golumbek, P., Ahmadzadeh, M., Jaffee, E., Pardoll, D., Levitsky, H. Role of bone marrow-derived cells in presenting MHC class I-restricted tumor antigens. *Science* 1994; **264**, 961–965.

80. Stingl, G., Bergstresser, P. R. Dendritic cells: a major story un-folds. *Immunol. Today* 1995; **16**, 330–333.

81. Shurin, M. R. Dendritic cells presenting tumor antigen. *Cancer Immunol. Immunother.* 1996; **43**, 158–164.

82. Tiemessen, C. T., Chalekoff, S., Morris, L., Becker, Y., Martin, D. J. Characterization of human blood dendritic cells: cytokine profiles. *Adv. Exp. Med. Biol.* 1995; **378**, 85–87.

83. Macatonia, S. E., Hosken, N. A., Litton, M., Vieira, P., Hsieh, C. S., Culpepper, J. A., Wysocka, M., Trinchieri, G., Murphy, K. M,. O'Garra, A. Dendritic cells produce interleukin-12 and direct the development of Th1 cells from naive CD4+ T-cells. *J. Immunol.* 1995; **154**, 5071–5079.

84. Ghanekar, S., Zheng, L., Logar, A., Navratil, J., Borowski, L., Gupta, P., Rinaldoc. Cytokine expression by human peripheral blood dendritic cells stimulated *in vitro* with HIV-1 and herpes simplex virus. *J. Immunol.* 1996; **157**, 4028–4036.

85. Cumberbatch, M., Dearman, R. J., Kimber, I. Constitutive and inducible expression of interleukin-6 by Langerhans cells and lymph node dendritic cells. *Immunology* 1996; **87**, 513–518.

86. Streilein, J. W., Grammer, S. F., Yoshikawa, T., Demidem, A., Vermeer, M. Functional dichotomy between Langerhans cells that present antigen to naive and to memory/effector T lympho-cytes. *Immunol. Rev.* 1990; **117**, 159–183.

87. Thomas, R., Lipsky, P. E. Human peripheral blood dendritic cell subsets: isolation and characterization of precursor and mature antigen-presenting cells. *J. Immunol.* 1994; **153**, 4016–4028.

88. O'Doherty, U., Steinman, R. M., Peng, M., Cameron, P. U., Gezelter, S., Kopeloff, I., Swiggard, W. J., Pope, M., Bhardwaj, N. Dendritic cells freshly isolated from human blood express CD4 and mature into typical immunostimulatory dendritic cells after culture in monocyte-conditioned medium. *J. Exp. Med.* 1993; **178**, 1067–1076.

89. Strunk, D., Nemeth, P., Elbe, A., Rappersberger, C., Stingl, G. Generation of Langerhans cells from CD34+ peripheral blood stem cells. *J. Invest. Dermatol.* 1994; **102**, 525.

90. Nunez, R. Review of the characteristics of six cell lines with Langerhans cell phenotype. *Adv. Exp. Med. Biol.* 1995; **378,** 35–37.

91. Paglia, P., Girolomoni, G., Robbiati, F., Granucci, F., Ricciardi-Castagnoli, P. Immortalized dendritic cell line fully competent in antigen presentation initiates primary T-cell responses *in vivo*. *J. Exp. Med.* 1993; **178,** 1893–1901.

92. Cassel, A., Cottler-Fox, M., Doren, S., Dunbar, C. E. Retroviral-mediated gene transfer into CD34-enriched human peripheral blood stem cells. *Exp. Hematol.* 1993; **21,** 585–591.

93. Henderson, R. A., Nimgaonkar, M. T., Watkins, S., Robbins, P. D., Ball, E. D., Finn, O. J. Human dendritic cells genetically engineered to express high levels of the human epithelial tumor antigen mucin (MUC-1). *Cancer Res.* 1996; **56,** 3763–3770.

94. Reeves, M. E., Royal, R. E., Lam, J. S., Rosenberg, S. A., Hwu, P. Retroviral transduction of human dendritic cells with a tumor-associated antigen gene. *Cancer Res.* 1996; **56,** 5672–5677.

95. Aicher, A., Westermann, J., Cayeux, S., Willimsky, G., Daemen, K., Blankenstien, T., Uckert, W., Dörken, B., Pezzutto, A. Successful retroviral mediated transduction of a reporter gene in human dendritic cells: feasibility of therapy with gene-modified antigen presenting cells. *Exp. Hematol.* 1997; **25,** 39–44.

96. Szabolcs, P., Gallardo, H. F., Ciococ, D. H., Sadelain, M., Young, J. W. Retrovirally transduced human dendritic cells express a normal phenotype and potent T-cell stimulatory capacity. *Blood* 1997; **90,** 2160–2167.

97. Arthur, J. F., Butterfield, L. H., Roth, M. D., Bui, L. A., Kiertscher, S. M., Lau, R., Dubinett, S., Glaspy, J., McBride, W. H., Economou, J. S., A comparison of gene transfer methods in human dendritic cells. *Cancer Gene Ther.* 1997; **4,** 17–25.

98. Bello-Fernandez, C., Matyash, M., Strobl, H., Pickl, W. F., Majdic, O., Lyman, S. D., Knapp, W. Efficient retrovirus-mediated gene transfer of dendritic cells generated from CD34+ cord blood cells under serum-free conditions. *Hum. Gene Ther.* 1997; **8,** 1651–1658.

99. Dao, M. A., Hannum, C. H., Kohn, D. B., Nolta, J. A. Flt3-ligand preserves the ability of human CD34+ progenitors to sustain long-term hematopoiesis in immune-deficient mice after *ex vivo* retroviral-mediated transduction. *Blood* 1997; **89,** 446–456.

100. Elwood, N. J., Zogos, H., Willson, T., Begley, C. G. Retroviral transduction of human progenitor cells: use of granulocyte colony-stimulating factor plus stem cell factor to mobilize progenitor cells in vivo and stimulation by Flt3/Flk2 ligand *in vitro*. *Blood* 1996; **88,** 4452–4462.

101. Bender, A., Sapp, M., Schuler, G., Steinman, R. M. Improved methods for the generation of dendritic cells from nonproliferating progenitors in human blood. *J. Immunol. Methods* 1996; **196,** 121.

102. Zitvogel, L., Couderc, B., Mayordomo, J. L., Robbins, P. D., Lotze, M. T., Storkus, W. J. IL-12 engineered dendritic cells serve as effective tumor vaccine adjuvants *in vivo*. *Ann. N.Y. Acad. Sci.* 1996; **795,** 284–293.

103. Brossart, P., Goldrath, A. W., Butz, E. A., Martin, S., Bevan, M. J. Virus-mediated delivery of antigenic epitopes into dendritic cells as a means to induce CTL. *J. Immunol.* 1997; **158,** 3270–3276.

104. Manickan, E., Kanangat, S., Rouse, R. J. D., Yu, Z., Rouse, B. T. Enhancement of immune response to naked DNA vaccine by immunization with transfected dendritic cells. *J. Leukocyte Biol.* 1997; **61,** 125–132.

105. Ribas, A., Butterfield, L. H., McBride, W. H., Jilani, S. M., Bui, L. A., Vollmer, C. M., Lau, R., Dissetten, V. B., Hu, B., Chen, A. Y., Glaspy, J. A., Economou, J. S. Genetic immunization for the melanoma antigen MART-1/Melan-A using recombinant

106. Tüting, T., Gambotto, A., Baar, J., Davis, I. D., Storkus, W. J., Zavodny, P. J., Narula, S., Tahara, H., Robbins, P. D., Lotze, M. T. Interferon-α gene therapy for cancer—retroviral transduction of fibroblasts and particle-mediated transfection of tumor cells are both effective strategies for gene delivery in murine tumor models. *Gene Ther.* 1997; **4,** 1053–1060.

107. Wan, Y., Bramson, J., Carter, R., Graham, F., Gauldie, J. Dendritic cells transduced with an adenoviral vector encoding a model tumor-associated antigen for tumor vaccination. *Hum. Gene Ther.* 1997; **8,** 1355–1363.

108. Specht, J. M., Wang, G., Do, M. T., Lam, J. S., Royal, R. E., Reeves, M. E., Rosenberg SA, Hwu P. Dendritic cells retrovirally transduced with a model antigen gene are therapeutically effective against established pulmonary metastases. *J. Exp. Med.* 1997; **186,** 1213–1221.

109. Nishioka, Y., Shurin, M., Robbins, P. D., Storkus, W. J., Lotze, M. T., Tahara, H. Effective tumor immunotherapy using bone marrow-derived dendritic cells genetically engineered to express interleukin-12. *J. Immunother.* 1997; **20,** 419.

110. Mayordomo, J. I., Zorina, T., Storkus, W. J., Zitvogel, L., Celuzzi, C., Falo, L. D., Melief, C. J., Ilstad, S. T., Kast, W. M., Deleo, A. B., Lotze, M. T. Bone marrow-derived dendritic cells pulsed with synthetic tumour peptides elicit protective and therapeutic antitumour immunity. *Nat. Med.* 1995; **1,** 1297–1302.

111. Tüting, T., Storkus, W. J., Lotze, M. T. Gene-based strategies for the immunotherapy of cancer. *J. Mol. Med.* 1997; **75,** 478–491.

112. Gansbacher, B., Banner, J. R., Daniels, B., Zier, K., Cronin, K., Gilboa, E. Retroviral vector-mediated gamma-interferon gene transfer into tumor cells generates potent and long lasting antitumor immunity. *Cancer Res.* 1990; **50,** 7820–7825.

113. Asher, A. L., Mule, J. J., Kasid, A., Restifo, N. P., Salo, J. C., Reichert, C. M., Jaffe, G., Fendly, B., Kriegler, M., Rosenberg, S. A. Murine tumor cells transduced with the gene for tumor necrosis factor-alpha. Evidence for paracrine immune effects of tumor necrosis factor against tumors. *J. Immunol.* 1991; **146,** 3227–3234.

114. Douvdevani, A., Huleihel, M., Zoller, M., Segal, S., Apte, R. N. Reduced tumorigenicity of fibrosarcomas which constitutively generate IL-1 alpha either spontaneously or following IL-1 alpha gene transfer. *Int. J. Cancer* 1992; **51,** 822–830.

115. Fearon, E. R., Pardoll, D. M., Itaya, T., Golumbek, P., Levitsky, H. I., Simons, J. W., Karasuyama, H., Vogelstein, B., Frost, P. Interleukin-2 production by tumor cells bypasses T helper function in the generation of an antitumor response. *Cell* 1990; **60,** 397–403.

116. Karp, S. E., Farber, A., Salo, J. C., Hwu, P., Jaffe, G., Asher, A. L., Shiloni, E., Restifo, N. P., Mule, J. J., Rosenber, S. A. Cytokine secretion but genetically modified nonimmunogenic murine fibrosarcoma. Tumor inhibition by IL-2 but not tumor necrosis factor. *J. Immunol.* 1993; **150,** 896–908.

117. Aoki, T., Tashiro, K., Miyatake, S., Kinashi, T., Nakano, T., Oda, Y., Kikuchi, H., Honjo, T. Expression of murine interleukin 7 in a murine glioma cell line results in reduced tumorigenicity in vivo. *Proc. Natl. Acad. Sci. U.S.A.* 1992; **89,** 3850–3854.

118. Alijagic, S., Moller, P., Artuc, M., Jurgovsky, K., Czarnetzki, B. M., Schadendorf. Dendritic cells generated from peripheral blood transfected with human tyrosinase induce specific T-cell activation. *Eur. J. Immunol.* 1995; **25,** 3100–3107.

119. Tüting, T., Wilson, C. C., Martin, D. M., Kasamon, Y. L., Rowles, J., Ma, D. I., Slingluff, C. L., Jr, Wagner, S. N., van der Bruggen, P., Baar, J., Lotze, M. T., Storkus, W. J. Autologous human monocyte-derived dendritic cells genetically-modified to express melanoma antigens elicit primary cytotoxic T-cell responses *in vitro*: enhancement by cotransfection of genes encoding the Th1-biasing cytokines IL-12 and IFN-α. *J. Immunol.*, in press.

CHAPTER 24

The Use of Genetically Modified Hematopoietic Stem Cells for Cancer Therapy

LEE G. WILKE AND JAMES J. MULÉ

Department of Surgery, University of Michigan, Ann Arbor, Michigan 48109

I. INTRODUCTION

Hematopoietic stem cells (HSCs) are rare cells found within the bone marrow (BM), peripheral blood (PB), and cord blood that are self-renewing and capable of differentiating into progeny that multiply to become the erythroid, lymphoid, and myeloid components of the blood. HSCs have been characterized by their expression of the CD34 and Thy antigens, and by their lack of mature lymphoid and myeloid lineage markers [1,2]. These primitive progenitor cells are attractive targets for gene therapy, which is defined as the introduction of new genetic material into somatic cells, because of their potential to self-renew and generate a large population of progeny cells with new genetic material. As evidenced by a number of recent reviews and consensus conferences, gene transfer to HSCs has rapidly become a focus of research into the mechanisms and/or therapeutic options for a variety of inherited, infectious, and neoplastic diseases [3–7].

How can this new technology be applied to the potential treatment of a neoplasm? Originally, the transfer of genetic material into HSC was conceptualized as a means to correct an inherited disease with a monogenic defect, such as adenosine deaminase deficiency or Gaucher's Disease [3]. However, because a number of blood-borne and solid cancers are currently being treated with bone marrow transplants (BMTs) or peripheral blood stem cell transplants (PBSCTs), the use of genetically modified HSCs has entered the armamentarium of the clinician treating cancer. The potential use of gene-modified HSCs has branched out in several directions, as it has been applied to cancer therapy. First, the introduction of gene-modified HSCs into BMTs and/or PBSCTs is being developed as a means to investigate the source of long-term hematopoietic reconstitution in the patient receiving marrow-ablative chemotherapy. It is also conceptualized as a means to determine the source of relapse in the cancer patient receiving a BMT or PBSCT. Second, the introduction of HSCs modified with a drug-resistance gene is being investigated as a means to select those cells that express the introduced drug-resistance gene and to protect the reinfused marrow from destruction by repeated chemotherapy treatments. An additional strategy might be the use of this drug-resistance selection to increase the frequency of gene-modified cells that also contain a therapeutic transgene with antitumor activity. Finally, HSCs have been proposed as targets for the introduction of chimeric receptor gene(s) with reactivity to tumor-associated antigens, which would potentially offer the patient receiving a gene-modified HSC transplant a long-term antitumor immune response that is biased to recognize autologous tumor.

In this chapter the advantages and disadvantages of HSCs as recipients of new genetic material will be discussed. This discussion will center on the use of retroviral vectors, as these are the most widely used vehicles of gene transfer into HSCs. Next, the chapter will focus on the preclinical studies that uncovered experimental techniques and difficulties of genetically manipulating HSCs. The final portion of this chapter will focus on the status of the three applications of genetically manipulated HSCs to the therapy of cancer highlighted earlier.

II. HUMAN HEMATOPOIETIC STEM CELLS AS VEHICLES OF GENE TRANSFER

The ideal gene transfer vehicle for HSCs would have the following properties: (1) integrate into chromosomes and be transmitted to progeny, (2) integrate into quiescent cells, (3) be free of replication-competent viral particles, and (4) integrate at a "safe" location in the host [4]. Reports of successful gene integration into HSCs using adeno-associated vectors and a molecular conjugate vector containing steel factor (SF) have appeared in the literature; however, retroviral constructs remain the primary vectors used to genetically modify or transduce HSCs [8,9]. Retroviruses are not the optimal

gene transfer vehicles, however, and vector development research continues to search for a more efficient vector for HSC transduction.

Retroviruses are RNA viruses of approximately 10,000 base pairs of DNA. For gene delivery, the structural proteins, or gag, pol, and env proteins, of the RNA virus are replaced with the gene of interest [10]. The packaging cell line provides these missing structural proteins in trans to enable the retrovirus to be infectious but replication incompetent in the host [6]. Retroviruses satisfy two of the desired requirements for delivery: chromosomal integration and potential for replication incompetence. The production of a replication-competent virus remains a major safety concern when discussing retroviruses. To date there have been no known deleterious effects from the use of retrovirally modified cells in human clinical trials, but T-cell lymphomas did develop in rhesus monkeys after they had received BMTs with CD34 cells exposed to a replication-competent retrovirus [11,12]. Stringent regulatory practices remain in effect to prevent the potential contamination of vector preparations with replication-competent retroviruses [13–15].

As delineated in Table 1, one of the major disadvantages of retroviruses as the gene delivery mechanism for HSCs is their requirement for actively dividing cells for insertion. Hematopoietic stem cells are primarily nondividing cells. The murine leukemia virus (MLV)-based vectors have been used the most extensively as gene transfer vehicles for HSCs, but their efficiency of transduction remains low due to the requirement for active mitosis in the host cell population [13,15]. To improve the transduction efficiency into HSCs, most investigators use a variety of different cytokines, prior to and/or during exposure to the retrovirus, to induce

TABLE 1 Retroviral Gene Transfer to Human Hematopoietic Stem Cells: Advantages and Disadvantages

Advantages	Disadvantages
Genetic integration into stem cells enables transmission to multilineage progeny in vivo.	Retroviruses require cells in mitoses = a rare event for "true" stem cells.
Large-scale expansion of transduced stem cells is not required ex vivo.	Gene integration site unknown in host stem cell chromosome = risk of "insertional oncogenesis".
High transduction efficiency not necessary due to in vivo expansion, self-renewal.	Poor in vitro assay systems to determine transfer efficacy into stem cell.
Increasing number of hematologic and solid cancers treatable (curable) with hematopoietic stem cell transplants.	Need for myelosuppression with associated risk of infection.

these target cells to proliferate and become more susceptible to retroviral infection [16–25]. Whether this *ex vivo* manipulation of HSC-containing populations actually results in adequate cycling of HSCs or induces their terminal differentiation and loss of repopulating capacity is not currently known and is an area where continued research into the biology of hematopoiesis is necessary to optimize gene delivery to HSCs [26]. Theoretically, if only a small number of true HSCs are transduced with the transgene of interest, a large number of progeny should be apparent in the host once reconstitution occurs *in vivo*. This expansion property of the HSC also obviates the need for large-scale expansion of genetically modified cells *ex vivo*, a process currently required for genetic manipulation of committed lymphocytes [27].

A second potential disadvantage to the use of retroviruses in HSCs is the risk of "insertional mutagenesis" [4,13]. The insertion site for retroviral gene transfer is not defined, so there is the risk that retroviral transduction will cause activation of genes involved in growth control or inactivation of tumor suppressor genes [28]. This potential disadvantage has, again, not been borne out in human trials to date. However, as previously noted, T-cell lymphomas have appeared in rhesus monkeys that received HSCs that had been exposed to a replication-competent murine-leukemia-based retrovirus prior to transplant [11,12].

A disadvantage to the use of HSCs as the host cells for retroviral genetic manipulation is the lack of *in vitro* assay systems to determine the transfer efficacy into a true HSC. The colony-forming unit (CFU) assays in methylcellulose and the long-term marrow culture (LTMC) assays do not definitively study the long-lived pluripotent HSC [13,15]. Transplantation assays into immunocompromised mice exist and attempt to replicate the reconstitution properties of the true human HSC, but they are time-consuming, requiring up to 9 months to achieve results, and may not directly predict the success or failure of transfer into a human host [15]. Similar to the need for further research in the areas of vector development and *ex vivo* HSC proliferation, more investigation into the exact microenvironment necessary to study pluripotent HSC maintenance and differentiation will need to be done to better ease the transition of this strategy into the clinical arena.

III. PRECLINICAL STUDIES OF GENE TRANSFER INTO HEMATOPOIETIC STEM CELLS

Successful retroviral transduction of murine HSCs has been reported, using as examples marker genes, the human adenosine deaminase (huADA) gene, and

a human multidrug resistance (MDR) gene [29–33]. The murine studies uncovered several experimental techniques that improved HSC transduction efficiency and permitted long-term gene expression in progeny cell lineages. These methodologies included pretreatment of the mice with 5-fluorouracil (5-FU) prior to bone marrow retrieval to increase the number of stem cells in cell cycle [31]. These findings were utilized in subsequent human HSC work. It is now common practice to give patients chemotherapy and/or cytokines prior to obtaining their peripheral blood for stem cell transduction in the hope of improving retroviral gene insertion by increasing the absolute numbers of HSCs and by altering their cell cycle status in the mobilized peripheral blood (MPB) [34].

The findings that the murine HSCs were more effectively transduced on a stroma of retroviral producer cells have been expanded to human HSC studies [29–31]. Multiple groups continue to use cocultivation to maximize human HSC gene transduction [16,18,20–22]. The risk of retroviral producer cell contamination in a HSC sample being used for transplantation and cancer therapy has, however, led others to identify alternate stromal transduction protocols. Moritz *et al.*, using a neomycin (*neo*) marker gene, transduced human cord blood using four different protocols [18]. The most efficient transduction protocol, measured by CFU growth in a neomycin analog (G418), utilized coculture; the least efficient used supernatant exposure (45% versus 11%). Those protocols that resulted in intermediate transduction efficiencies employed autologous bone marrow stroma or a murine-derived stromal cell line that expressed human SF (32% and 19%) (Table 2). More recent studies have shown improved transduction of human HSCs on a fragment of the fibronectin molecule [35,36]. These improvements, however, are in comparison to transduction with a retroviral vector on a bovine serum albumin (BSA)-coated plate. In CFU assays, the number of G418 resistant colonies grown from BM transduced with a *neo* marker gene increased from approximately 2% on BSA to 18% on a fibronectin coated plate in two studies [35,36]. The transduction efficiency on fibronectin does not necessarily differ significantly from results obtained by other groups who utilize supernatant exposure alone, although direct comparative studies have not been performed (Table 2, Cassel *et al.* [17], and Lu *et al.* [19]). The use of a fibronectin molecule, however, does avoid the risk of producer cell contamination and the need for BM aspiration if autologous marrow is to be used as the stomal monolayer for transduction protocols.

Another technical contribution from the murine studies was the introduction of cytokines into the culture media during exposure of the HSC to the retrovirus and during a prestimulation phase prior to transduction. As

TABLE 2 Preclinical Studies of Gene Transfer into Human Progenitor Cells

Progenitor cell source	CD34 isolation	Retroviral vector	Infection protocol	Transduction efficiency	Murine experiments	Ref.
BM, normal donors	No	N2	1. 24-hour preincubation of BM in 15% 5637 human bladder carcinoma-conditioned media, IL-6, and LIF. 2. 24-hour cocultivation of BM with producer cells.	CFU assay: 5–20% growth in G418 PCR for neo: approx. 20%+	5 bg/nu/xid mice (BMT with transduced human BM): 0.1% BM and spleen cells neo+ by PCR at 4 months	16
BM and MPB from normal and multiple myeloma patients	Yes	LNL6 G1Na.4	1. 72-hour exposure of progenitor cells to vector supernatant in media with IL-3, IL-6, and SCF.	CFU assay: 17% growth in G418 (MPB & BM) PCR for neo: 20%+ (MPB), 27%+ (BM)	NA	17
MPB breast cancer patients	Yes	LNL6	1. 24-hour prestimulation in media with IFG-1, IL-3, IL-6, SCF, GMCSF, and erythropoietin. 2. 12-hour vector supernatant exposure.	CFU assay: 3–30% growth in G418 PCR for neo: 67–100%+ RT-PCR for neo: 8%+	NA	19
BM, normal donors MPB, breast cancer patients	Yes	MSCV-HSA.NEO	1. 48-hour prestimulation of BM or MPB in media with IL-3, IL-6, SF. 2. 48-hour cocultivation of producer cells with BM or MPB.	CFU assay: 11–12% growth in G418 (unsorted) 70–100% growth in G418 (HSA sorted)	NA	22
BM, normal donors	Yes	LGsFH	1. 4 "x" vector supernatant exposure on autologous BM stroma in media with IL-3, IL-6, and SCF.	FACS: 10% HSA + PCR for GC gene: 38–50%+ (unsorted) 100%+ (HSA sorted)	NA	23
BM and MPB, normal donors	Yes	LN	1. 72-hour vector supernatant exposure on allogenic BM stroma.	CFU assay: 35% growth in G418	bnx mice transplanted with transduced CD34+ cells in conjunction with BM stroma engineered to produce human IL-3: 3/24 mice found to have identical proviral fragment sites in T and myeloid cells via PCR	24

MPB, normal donors	Yes	L(mCD4tr)SN	1. 3 to 6-day vector supernatant exposure of CD34+ cells on fibronectin coated flasks in media with IL-1B, IL-3, IL-6, and SCF.	FACS: 7%+ for mCD4 CFU assay: 8–40% growth in G418 after FACS sort for mCD4	NA	25
MPB, cancer patients	Yes	HaMDR/A	1. 24 to 48-hour preincubation with IL-3, IL-6, and SCF. 2. 24-hour transduction on fibronection-coated plates with two viral supernatant changes.	PCR for MDR: 64.7%+ BFU-E 77.3%+ CFU-GM CFU assay: 19–26% BFU-E taxol resistant 20–48% CFU-GM taxol resistant	NA	38
Cord blood	No	TK NEO	1. 48-hour prestimulation of cord blood cells in media with IL-6 and SCF. 2. Four infection protocols. a. coculture b. vector supernatant exposure on S1^4-h220[a] c. vector supernatant exposure on allogeneic BM d. vector supernatant exposure	CFU assay: 45% growth in G418 (coculture) 32% growth in G418 (allo BM) 19% growth in G418 (S1^4-h220)	NA	18
Cord blood	Yes	MFG-mCD2	1. 4-day coultivation of producer cells with cord blood CD34+ cells in media with IL-3 and SCF.	FACS: 40%+ for murine CD2	SCID-hu Thymic constructs: 4–9% thymocytes mCD2+ via FACS in 4/9 SCID chimera (5–10 weeks)	21
Fetal liver cells	Yes	LNL6	1. 72-hour cocultivation of producer cells with fetal liver cells in media with IL-3, IL-6, and SCF.	PCR for neo: 10%+	SCID-hu Thymic liver constructs: 3% thymocytes + for neo by PCR (6 weeks) 2% CD4+ and CD8+ cells + for neo by PCR (4 weeks)	20

Note. BFU-E, burst-forming unit—erythroid; BM, bone marrow; BMT, bone marrow transplant; CFU, colony-forming unit; CFU-GM, colony forming unit—granulocyte macrophage; HSA, heat stable antigen; LIF, leukemia inhibitory factor; MPB, mobilized peripheral blood; NA, not applicable; neo, neophophotransferase; PCR, polymerase chain reaction; RT-PCR, reverse transcription–polymerase chain reaction; SCF, stem cell factor; SCID, severe combined immunodeficient (mice).

[a] S1^4-h220 is a murine-derived, genetically modified stromal cell line that expresses membrane bound steel factor (SF).

previously noted, efficient retroviral transduction requires the recipient cells to be in active cell cycle. The most commonly used cytokines for human HSC transduction include IL-3, IL-6, and stem cell factor (SCF). Though protocols vary considerably, as seen in Table 2, most studies utilize cytokines prior to and/or during retroviral vector exposure to attain transduction efficiencies in the 10 to 30% range as determined by CFU assay. One recent report indicated, however, that cytokine stimulation after transduction decreased the efficiency of gene expression, implying that HSC growth and differentiation could "turn off" the introduced gene of interest [37]. Though this work focused on the potential effects of cytokines after transduction, it is possible that the cytokine protocols currently in use before and during transduction may be affecting the HSC transduction efficiencies. As more insight into the biology of HSC and its microenvironment is obtained, more efficient, timely transduction protocols for HSCs could be designed.

An important finding from the murine studies was the fact that progeny development was not affected by genetic manipulation of HSCs [29,30]. Several human studies have used genes encoding for a membrane-bound protein to identify HSCs that have been transduced and, in some protocols, have sorted by flow cytometry for those marked cells to obtain a purer population of gene-modified HSCs. Examples of these surface-expressed marker genes include the heat-stable antigen (HSA), the murine CD4 antigen (mCD4), the murine CD2 antigen (mCD2), and the truncated nerve growth factor receptor (NGFR) [21–23,25,38]. These studies utilized multiple sources of human HSCs–bone marrow, mobilized peripheral blood, and cord blood–indicating the broader feasibility of transduction. An important advantage of these protocols is the ability to perform fluorescence-activated cell sorting (FACS) on the transduced population and obtain more rapid assessment of the HSC transduction efficiency than is possible with growth in selection media. Champseix et al. transduced CD34+ cells from cord blood via a cocultivation protocol with a retroviral vector containing the mCD2 cDNA [21]. The transduced cells were then injected into human fetal thymic pieces, which were then implanted into severe combined immunodeficient mice (SCID). Using this modified in vivo system, these investigators obtained 4 to 9% of the resulting progeny thymocytes with the mCD2 marker. These results support the hypothesis that development of progeny cells from the transduced human HSCs is not altered. Another study described in Table 2 expands this approach to illustrate that transduction of a pluripotent HSC does not alter development of distinct progeny lineages. Nolta et al. have described infection of CD34+ cells from BM and

MPB with a neomycin-resistant gene marker vector followed by injection into beige/nude mice [24]. Three of 24 of the transplanted mice were found to have identical proviral fragment sites in their progeny T and myeloid cells, confirming that a pluripotent HSC had been transduced with its ability to differentiate into separate lineages left intact.

In addition to marker genes, it has been shown that a functional gene, human MDR, can be inserted into murine HSC and enrichment for transduced progeny bone marrow-derived lineage cells observed following taxol therapy [30]. These results have been duplicated in human in vitro studies [39]. Using the CFU assay, 20 to 48% of progeny granulocyte-macrophage colonies from transduced MPB CD34+ cells were found to be relatively taxol resistant.

Despite the technical advancements described earlier and the variety of transgenes capable of being introduced into human HSCs, those studies that have examined the in vivo development of the infected HSCs have shown a markedly low transduction efficiency, which translates into a small number of progeny cells with the transgene. As delineated in Table 2, those investigators who introduced the transduced HSCs into immunocompromised mice with or without a human fetal hematopoietic microenvironment, found that only 0.1 to 9% of the BM, spleen, or thymic cells contained the transgene of interest [16,20,21]. Similar results have been obtained in large-animal studies. Table 3 describes several of the primate studies performed to date with retrovirally transduced HSCs. Like the in vitro preclinical studies, the infection protocols for the primate studies have been varied, including coculture on producer cells, engineered stroma, simple exposure to vector supernatant, or exposure on long-term marrow culture (LTMC), with or without cytokine support [40–44]. The results from these studies have, however, been similar in that the long-term appearance of transduced progeny was minimal. Bienzle et al. reported the largest long-term detection of a transgene, with 5% of marrow cells from 2 dogs maintaining G418 resistance in CFU assay after transplantation with BM exposed to the N2 retroviral vector 2 years earlier [42].

The large-animal models, in spite of the low numbers of resultant progeny that contain the transgene of interest, have provided several important results, which led many investigators to begin human HSC gene therapy: first, that primate HSC could be transduced with a variety of retroviral genes including the standard neo marker transgene and potentially functional genes, the murine ADA and human glucocerebrosidase gene; second, that the transduction protocols, if a non-replication-competent retrovirus was used, did not lead to any long-term side effects or significant mortality;

TABLE 3   Primate Studies Involving Progenitor Cell Gene Transduction

Progenitor cell source	CD34 isolation	Retroviral vector	Infection protocol	*In vitro* transduction efficiency	*In vivo* outcome	Ref.
Canine BM	No	N2	Long-term marrow culture (LTMC) was established from BM and exposed to vector supernatant 3× over 21 days.	PCR: 70%+ for neo on day 21 of LTMC	4 dogs underwent marrow ablation prior to infusion of transduced BM (1 death occurred), 3 dogs had no marrow ablation prior to transduced marrow infusion. 1. 0.1–10% BM + for neo via PCR at 3 months 2. 0.1–1% BM + for neo via PCR at 10–21 months 3. 0.1–1% T cells + for neo via PCR at 12–19 months	40
Canine BM	No	N2	LTMC was established from BM and exposed to vector supernatant 3× over 21 days.	CFU assay: 44% growth in G418	18 dogs transplanted with transduced marrow without marrow ablation 1. Maximum G418 resistant marrow cells in CFU assay was 10–30% at 3 months. 2. 2 dogs maintained 5% G418 resistant marrow cells in CFU assay at 2 years.	42
Canine MPB	No	LN	24-hour coculture of MPB on producer cells followed by a 10-day incubation of the MPB in LTMC with daily vector supernatant changes	NA	3 dogs received transduced MPB and untransduced BM after total body irradiation. 1. Week 6, 1 dog with PCR + neo in peripheral granulocytes/lymphocytes 2. Week 20, 1 dog 2% CFU-GM G418 resistant	34
Rhesus Monkey BM	Yes	PGK-mu ADA	96-hour cocultivation of CD34+ cells on S1⁴-h220 stroma with IL-6 and SCF and vector supernatant	NA	3 monkeys transplanted after marrow ablation 1. 2% peripheral blood cells (T cells and granulocytes) + for muADA via PCR at 1 year	41
Rhesus monkey BM	Yes, with Thy1+ cells	LG4	7 day exposure of CD34+/Thy1+ cells to vector supernatant in media with IL-6 and SCF	NA	4 monkeys were transplanted after marrow ablation (1 death occurred). 1. 1–13% of B cells were PCR + for the GC gene at days 48–117 2. 1.2% of CD2+ T cells were GC + via PCR at >300 days 3. No long-term gene transfer occurred in the granulocyte population	43
Rhesus monkey BM	No	N2	3-day coculture of BM on producer cells followed by 3-day coculture on fresh producer cells	NA	3 monkeys were transplanted with autologous BM that had been exposed to high titer producer cells. 1. PCR + for neo in BM on days 20–99 (approximately 1% of total cells)	44

Note. ADA, adenosine deaminase; GC, glucocerebrosidase gene; PGKpr, human phosphoglycerate kinase promoter.

TABLE 4 Clinical Trials Involving Gene Marking of Human Hematopoietic Stem Cells

Patient population	Completed vs. ongoing	Protocol	CD34 selection	Results vs desired outcomes	Institution	Ref.
Pediatric AML and pediatric neuroblastoma	Completed	One third of autologous BM was exposed to either LNL6 or G1N retroviral vectors for 6 hours without cytokine supplementation and reinfused after ablative therapy with two-thirds unmanipulated BM.	No	1. 15 of 18 evaluable patients had 0–29% G418-resistant colonies in CFU assay at 1 month. 2. 5 of 5 evaluable patients had 0–15% G418 resistant colonies in CFU assay at 1 year. 3. 1 patient had T and B cells + by PCR for neo at 18 months. 4. 4 AML patients relapsed: 2 had leukemic blasts that were G418 resistant in CFU assay. 5. 4 neuroblastoma patients relapsed: all 4 had an estimated .05–1% of neuroblasts + for neo via limiting cycle PCR.	St Jude's Children's Hospital, Memphis, TN	45,46, 51, 54
Adult multiple myeloma and breast cancer	Completed	One third of BM and MPB exposed to LNL6 or G1Na.40 vector supernatant over a 72-hour period in media with IL-3, SCF, (and IL-6 Breast cancer patients only) Transduced and untransduced BM and MPB was reinfused into each patient. The BM and MPB for each patient were exposed to distinct vectors for differentiation of the progeny posttransplant.	Yes	1. Both the MPB and BM were shown to contribute to patient engraftment based on differential recognition of the two neo vector sequences. 2. 3 (2 multiple myeloma and 1 breast cancer) of 9 evaluable patients had PB cells positive for the neo gene via PCR at 18 months post-transplant.	National Institutes of Health, Bethesda, MD	49
Adult CML	Completed	MPB was collected immediately following chemotherapy, and 30% of cells were exposed to LNL6 retroviral vector supernatant for 6 hours. Transduced and untransduced MPB were reinfused following ablative therapy.	Yes	2 patients enrolled in study: 1. Patient 1 had return of blast crisis by day 159 and the neo gene was detected in leukemic cells by PCR. 2. Patient 2 also relapsed and had Ph+ cells with the neo gene by PCR. 3. Patient 1 had normal and leukemic cells + for neo via RT-PCR >280 days posttransplant.	M.D. Anderson Cancer Center, Houston, TX	48
Adult AML or ALL	Completed	5–19% of BM obtained during second clinical remission (AML), or first clinical remission (ALL) was exposed to the G1N retroviral vector supernatant for 4 hours, and reinfused with untransduced marrow following ablative therapy.	No	1. Patient had BM and PB + for neo by PCR at 1 year. 2. Neither of the 2 relapsed patients (AML) had neo gene-marked leukemic blasts.	Indiana School of Medicine, Indianapolis, IN	47

Patient population	Status	Protocol	Gene marking	Objective	Institution	Ref.
Pediatric AML in first clinical remission	Ongoing	One third of BM exposed to LNL6, a second third of BM exposed to G1N for 6 hours without cytokine support; each set of BM would then be exposed to a separate *ex vivo* purging regimen.	No	With relapse, assess the presence of a neo-specific marker in the leukemic blasts to determine the efficacy of the two purging regimens.	St Jude's Children's Hospital, Memphis, TN	51,57
Breast cancer or malignant lymphoma	Ongoing	A portion of MPB to be transduced via a 5-day retroviral vector (LN) supernatant exposure in media with IL-1, IL-3, IL-6, and SCF. Transduced and untransduced MPB to be reinfused post ablative therapy.	Yes	Study designed to determine the ability of MPB to contribute to long-term hematopoiesis.	Fred Hutchinson Cancer Center, Seattle, WA	34
Metastatic breast cancer or lymphoma	Ongoing	One third of harvested BM and one-half of any 2 MPB harvests will be enriched for CD34 cells, preincubated for 42 hours in media with IL-3 and IL-6 and then incubated for 6 hours in LNL6 or G1Na retroviral vector supernatant.	Yes	Differential gene marking of BM and MPB to determine the relative contributions of each to hematopoietic reconstitution.	USC Comprehensive Cancer Center, Los Angeles, CA	52
Multiple myeloma	Ongoing	10–30% of BM will be placed in LTMC and exposed to GTk1.SvNa.7 (LN derivative with herpes simplex thymidine kinase gene) retroviral vector supernatant 3 times in 21 days.	No	Study designed to determine the contribution to relapse of the gene marked autograft.	University of Toronto, Toronto, Canada	55
Patients receiving ABMT for leukemias or solid tumors	Ongoing	Two-thirds of BM will undergo CD34+ cell selection. One-half of the selected population will be exposed to LNL6 and the other half exposed to G1Na for 6 hours. The transduced cells will then be randomized to either cytokine exposure (IL-3, IL-6, and SCF for 5 to 7 days) or no cytokine exposure.	Yes	Study designed to investigate the *ex vivo* exposure of CD34+ BM cells to growth factors and the time to hematopoietic engraftment in patients receiving ablative therapy.	St Jude's Children's Hospital, Memphis, TN	53
Relapsed follicular non-Hodgkin's lymphoma	Ongoing	A portion of the MPB and BM was exposed to either LNL6 or G1Na retroviral vector supernatant without exposure to cytokines or stroma. Both transduced and nontransduced MPB and BM were reinfused into patients after ablative therapy.	Yes	2 patients currently enrolled in the study: 1. At day 45 both patients have neo+ BM cells by PCR (1.5 and 4.3%). 2. The goal of the study is to determine the contribution of contaminated BM or MPB to relapse.	M.D. Anderson Cancer Center, Houston, TX	56

Note. ABMT, autologous bone marrow transplant; ALL, acute lymphoblastic leukemia; AML, acute monocytic leukemia; CML, chronic monocytic leukemia.

finally, that HSCs could be transduced and progeny cells of different lineages develop with each possessing the transgene of interest, thus indicating that a primitive, pluripotent HSC had incorporated the retroviral vector into its DNA [41,43].

IV. APPLICATIONS OF GENETICALLY MANIPULATED HEMATOPOIETIC STEM CELLS TO THE THERAPY OF HUMAN CANCER

A. Gene Marking of Human Hematopoietic Stem Cells

The first use of retrovirally transduced HSCs was not as a therapy for human cancer but rather as a tool for understanding the biology of BMTs and PBSCTs. BMTs and more recently PBSCTs have been used to restore hematopoiesis in patients who have received ablative chemotherapy for cancer, including a variety of blood-borne leukemias as well as solid tumors such as breast cancer and neuroblastoma [34,45–49]. Table 4 delineates those completed and ongoing clinical trials that utilize(d) a retrovirally inserted transgene to mark the HSC and study the biology of BMTs and PBSCTs in cancer patients. The first question investigators sought to answer was whether the reinfused autologous bone marrow and/or the peripheral blood stem cells provided the source of long-term hematopoietic cells or whether the patient retained a small population of functioning HSCs that could reconstitute their hematopoietic system despite the myeloablative therapy. Of those studies that have been completed, it can be concluded that human HSCs can be "marked" with a retroviral vector containing a neomycin-resistant gene [45–49]. Brenner et al. revealed that 5 of 5 evaluable pediatric patients with acute monocytic leukemia (AML) or neuroblastoma had from 0 to approximately 15% of cells in their BM that were G418 resistant in CFU assay at one year posttransplant. These investigators thus concluded that harvested autologous bone marrow does contribute to hematopoietic recovery after ablative chemotherapy [45]. Cornetta et al. also demonstrated that in an adult leukemic patient treated with a BMT in which a portion of the HSC had been exposed to a retroviral vector with the neomycin resistant gene, both BM and PB cells contained the transgene 1 year posttransplant [47]. Deisseroth et al. found that one of their adult chronic myelogenous leukemia (CML) patients, who had received a PBSCT, had cells in the peripheral blood that were positive via reverse transcription–polymerase chain reaction (RT-PCR) for the neomycin gene 280 days post-

transplant. This finding led to the conclusion that reinfused, retrovirally exposed mobilized peripheral blood, as well as BM, could contribute to the patient's hematopoietic recovery [48]. Dunbar et al. reported the long-term presence of a marker gene in the circulating granulocytes, T and B cells (0.01–1% of cells) of 2 patients with myeloma and 1 patient with breast cancer 18 months after a combined transplant with marked bone marrow and mobilized peripheral blood [4,49]. By using two different vectors with the neomycin gene, the investigators were able to discriminate the source of the engrafted cells after the transplant. Using a transduction protocol similar to that of Brenner et al., in which the HSCs were exposed to vector supernatant for 6 hours without cytokine support, Dunbar et al. were unable to demonstrate gene marking in adult patients with breast cancer or myeloma who received a BMT and PBSCT [50]. Further study into the conditions necessary for optimal HSC gene transduction for individual cancers at different stages will need to be performed to resolve the discrepancies in results from different laboratories. Despite the differences in outcomes, it has been possible to use the technology of gene marking to address a clinically relevant question about the restoration of human hematopoiesis [51]. As shown in Table 4, several ongoing trials are seeking to further address the question of the source of hematopoietic recovery after myeloablation [34,52–53].

A second question of interest to those investigators focusing on the biology of BMTs and PBSCTs in the cancer setting is that of the origin of relapse after transplantation for leukemia. Brenner et al. and Deisseroth et al. both discovered that in some of their patients who relapsed from pediatric AML, pediatric neuroblastoma, and adult CML, the leukemic blasts contained the neomycin gene. The authors thus concluded that the reinfused autologous BM or MPB was a contributor to the cancer recurrence [46,48,51,54]. Cornetta et al. reported that 2 of 4 patients with AML, treated with BMT, in which a portion of the graft was exposed to a retroviral vector with the neomycin-resistant gene, relapsed. However, neither patient had evidence of vector-marked leukemic blasts [47]. Clinical studies are ongoing to further clarify this question concerning the source of relapse in patients with varied forms of cancer at different clinical stages, including multiple myeloma and non-Hodgkin's lymphoma [55,56].

A third area of interest is the role "purging" of the BMT and/or PBSCT with cytokines or chemotherapeutic agents prior to reinfusion into the patient will play in preventing disease recurrence. Brenner et al. have initiated studies using two distinct retroviral vectors with the neomycin-resistant gene in pediatric AML patients to address this question [57]. As improvements in vector design and HSC transduction occur, more clinical studies can be undertaken to further elucidate the biology of BMTs and PBSCTs, and those *ex vivo* and *in vivo* therapies that will benefit the patient with cancer.

B. Drug Resistance Genes in Hematopoietic Stem Cells

In addition to its role in the study of BMTs and PBSCTs, gene therapy with HSCs has been introduced as a means to potentially improve existing transplant protocols. The dose of chemotherapeutic agents used to treat malignancies is limited to some extent by myelosuppression. If a patient's bone marrow were tolerant to the antineoplastic drugs, increasing doses of the agents could be used to improve disease eradication. The MDR-1 gene encodes a 170-kDa P-glycoprotein that functions as an adenosine triphosphate–dependant efflux pump for lipophilic compounds [58]. Those chemotherapeutic agents that are "removed" by cells with the MDR-1 pump include the anthracyclines, the vinca alkaloids, the epipodophyllotoxins, actinomycin D, and taxol [58]. As previously noted, murine HSCs have been transduced with the human MDR-1 gene via a retroviral vector and selection with taxol has been successful [30]. The ability to select for *MDR-1*-containing cells in humans would conceivably permit dose escalation of the antineoplastic agents as well as decreased intervals between chemotherapy administration due to improved hematopoietic recovery [58]. Several human clinical trials, with primarily breast and/or ovarian cancer patients, are ongoing to answer the question whether human HSCs from BM and/or MPB can be transduced without toxicity to the HSC and in adequate quantities to permit growth selection with taxol [59–62]. Preliminary results from one group indicated that the CD34+ cells from the BM and MPB of ovarian and breast cancer patients respectively could be transduced with the MDR-1 gene by exposing the cells to the retroviral supernatant on a stromal monolayer with cytokine supplementation in the culture media. The transgene was identified in the BM of 5 of 8 evaluable patients using a solution DNA PCR assay [63].

As results from the ongoing trials with MDR-1-transduced HSC become available and indicate the feasibility of BM selection, a second use for the drug-resistance genes may arise. Retroviral constructs containing the MDR-1 gene and a second therapeutic gene could be introduced into the HSCs and those cells with the transgene selected for using taxol [58]. This strategy would then increase the frequency of gene-modified cells with a potentially therapeutic antitumor gene (discussed later).

Without the selection process, the antitumor gene might only be present in a small number of HSCs due to low transduction efficiency with current retroviral vectors. With selection, however, the number of cells with the transgene DNA would increase and thereby improve therapeutic potential after the BMT and/or PBSCT. For a fuller discussion of drug resistance mechanisms and stem cell transfection see Chapter 11.

C. Chimeric Receptor Genes in Hematopoietic Stem Cells

"Purging" autologous marrow or peripheral blood prior to reinfusion after myeloablative therapy is one way to address the problem of disease recurrence. Another way of approaching this problem in the cancer patient might be to provide the patient receiving a gene-modified HSC transplant with an immune system that is biased to recognize autologous tumor. Eshhar *et al.* developed a chimeric single-chain receptor that consists of the F_V domain (scFv) of an antibody linked with the γ or ζ chains, the signal-transducing subunits of the immunoglobulin receptor and the T-cell receptor (TCR) [64]. The scFv domains, which are a combination of the antibody's heavy and light variable regions (V_H and V_L), appear to possess the specificity and affinity of the intact Fab' fragment [64]. Originally a chimeric gene with scFv from an antitrinitophenyl (anti-TNP) antibody was found to be expressed as a functional surface receptor in a murine cytolytic T-cell line and to specifically secrete IL-2 upon exposure to the TNP antigen [64]. T-cell activation via this receptor is not major histocompatibility complex (MHC) restricted because of antibody recognition as opposed to peptide recognition with the TCR complex. Studies utilizing the chimeric receptor were expanded to investigate antitumor antibodies. Stancoviski *et al.* introduced a chimeric gene with an anti-HER-2/neu antibody into a murine CTL line and demonstrated specific antigen recognition and lysis of cells overexpressing Neu/HER-2 [65]. Moritz *et al.* similarly utilized an anti-HER-2/neu scFv linked to the ζ chain of the TCR complex and transduced a murine CTL line. Target cells that overexpress HER-2/neu were lysed *in vitro* by the transduced CTLs, and the growth of HER-2/neu-transformed NIH3T3 cells in nude mice was retarded, though not prevented, by CTLs transduced by the chimeric receptor and subsequently injected into the mice [66]. Moving beyond the stable cell lines, Hwu *et al.* transduced CD8+ human CTL with a chimeric receptor designed to recognize a defined human ovarian carcinoma antigen [67]. Again, the transduced cells recognized target cells with the ovarian cancer antigen and secreted granulocyte-macrophage col-ony-stimulating factor (GM-CSF) upon incubation with the specific antigen [67].

The next step toward applying this technology to the therapy of human cancer involves transduction of human HSCs with a chimeric receptor. Figure 1 depicts a model therapeutic strategy for patients receiving a chimeric receptor gene-modified HSC transplant. The progeny lymphocytes and myeloid cells (monocytes and neutrophils) from the transduced HSC would poten-tially have the ability to recognize and kill in a non-MHC-restricted fashion those cells expressing the spe-cific tumor antigen incorporated into the chimeric gene. Because the HSCs are self-renewing and pluripotent, they would offer the cancer patient receiving a BMT and/or PBSCT a long-term immune system with antitu-mor function. We have recently introduced into human CD34+ cells, isolated from mobilized peripheral blood, the gene encoding for a chimeric receptor to HER-2/neu (68). Following selection, under G418, 81% of the colonies derived from the transduced HSC (11%) con-tained the transgene. We are now planning a phase I clinical trial of chimeric receptor gene-modified HSCs into advanced breast cancer patients receiving a PBSCT. Because of the concern that recognition of "normal" levels of HER-2/neu on normal tissues, as opposed to "overexpressed" levels on tumor might lead to toxicity, vector constructs will also contain suicide cytosine de-aminase (CD) or thymidine kinase (TK) genes to elimi-nate expressing effector progeny cells if adverse target-ing occurs.

Another strategy based on redirecting the immune response to recognize tumor-associated antigens after BMT or PBSCT is the introduction of genes encoding "classic" TCRs. Unlike the chimeric receptor approach, this strategy would be limited to progeny T cells only, would be MHC restricted, and would be more specific to tumor cells (as opposed to normal tissue). Recently, TCRs for a cytotoxic T-cell line (CTL)-defined peptide expressed by *HLA*A2* on melanoma has been cloned (69). The prohibitive limitation of this approach, how-ever, may be the difficulty in successfully expressing two independent genes that encode the separate alpha and beta chains of the TCR in HSC target cells and their T-lymphoid progeny.

V. CONCLUSIONS

Genetic modification of human HSCs for cancer ther-apy remains in its infancy but already has shown great potential as a tool to enhance the disease-free survival of patients undergoing a BMT and/or PBSCT for blood-borne or solid tumors. Areas of ongoing and future focus include novel vectors and improvement on exist-ing retroviral vectors for HSC transduction, *ex vivo* pro-liferation of HSCs and the cytokines necessary to sup-port survival of the self-renewing, pluripotent HSCs, and *in vivo* models to study transduced HSC. As human clinical trials are completed, more information concern-ing the biology of BMT and/or PBSCT, the feasibility of HSC transduction with multidrug resistance genes

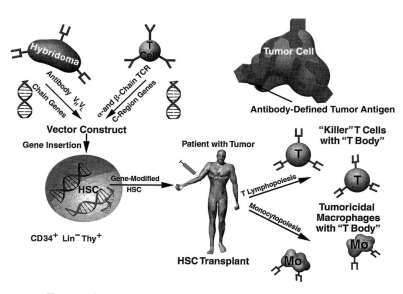

FIGURE 1   HSC Gene Transfer: "Chimeric" TCR Approach

and their ability to provide bone marrow protection will become available. These data will provide information on the safety of this technology and offer a stepping stone towards the use of dual drug-resistance and therapeutic genes including those encoding receptor molecules that recognize tumor-associated antigens.

References

1. Baum, C. M., Weissman I. L., Tsukamoto, A. S., *et al.* (1992). Isolation of a candidate human hematopoietic stem-cell population. *Proc. Natl. Acad. Sci. USA* **89**, 2804–2808.
2. Morrison, S. J., Uchida, N., and Weissman, I. L. (1995). The biology of hematopoietic stem cells. *Annu. Rev. Cell. Dev. Biol.* **11**, 35–71.
3. Roth, J. A., and Cristiano, R. J. (1997). Gene therapy for cancer: what have we done and where are we going? *J. Natl. Cancer Inst.* **89**, 21–39.
4. Dunbar, C. E. (1996). Gene transfer to hematopoietic stem cells: implications for gene therapy of human disease. *Annu. Rev. Med.* **47**, 11–20.
5. Brenner, M. K. (1996). Gene transfer to hematopoietic cells. *N. Engl. J. Med.* **335**, 337–339.
6. Kerr, W. G., and Mule, J. J. (1994). Gene therapy: current status and future prospects. *J. Leukoc. Biol.* **56**, 210–214.
7. Bagnis, C., Chabannon, C., and Mannoni, P. (1996). Gene transfer into haematopoietic cells: a challenge for gene therapy-European concerted action, workshop. *Gene Ther.* **3**, 362–264.
8. Schwarzenberger, P., Spence, S. E., Gooya, J. M., *et al.* (1996). Targeted gene transfer to human hematopoietic progenitor cell lines through the c-kit receptor. *Blood* **87**, 472–478.
9. Fisher-Adams, G., Wong, K. K., Podsakoff, G., *et al.* (1996). Integration of adeno-associated virus vectors in CD34+ human hematopoietic progenitor cells after transduction. *Blood* **88**, 492–504.
10. Geraghty, P. J., and Chang, A. E. (1995). Basic principles associated with gene therapy of cancer. *Surg. Oncol.* **4**, 125–137.
11. Donahue, R. E., Kessler, S. W., Bodine, D., *et al.* (1992). Helper virus induced T cell lymphoma in nonhuman primates after retroviral gene transfer. *J. Exp. Med.* **176**, 1125–1135.
12. Purcell, D. F., Broscius, C. M., Vanin, E. F., *et al.* (1996). An array of murine leukemia virus-related elements is transmitted and expressed in a primate recipient of retroviral gene transfer. *J. Virol.* **70**, 887–897.
13. Kohn, D. B. (1997). Gene therapy for haematopoietic and lymphoid disorders. *Clin. Exp. Immunol.* **107**, 54–57.
14. Mastrangelo, M. J., Berd, D., Nathan, F. E., *et al.* (1996). Gene therapy for human cancer: an essay for clinicians. *Semin. Oncol.* **23**, 4–21.
15. Medin, J. A., and Karlsson, S. (1997). Viral vectors for gene therapy of hematopoietic cells. *Immunotechnology* **3**, 3–19.
16. Dick, J. E., Kamel-Ried, S., Murdoch, B., *et al.* (1991). Gene transfer into normal human hematopoietic cells using *in vitro* and *in vivo* assays. *Blood* **78**, 624–634.
17. Cassel, A., Cottler-Fox, M., Doren, S., *et al.* (1993). Retroviral-mediated gene transfer into CD34-enriched human peripheral blood stem cells. *Exp. Hematol.* **21**, 585–591.
18. Moritz, T., Keller, D. C., and Williams, D. A. (1993). Human cord blood cells as targets for gene transfer: potential use in genetic therapies of severe combined immunodeficiency disease. *J. Exp. Med.* **178**, 529–536.
19. Lu, M., Maruyama, M., Zhang, N., *et al.* (1994). High efficiency retroviral-mediated gene transduction into CD34+ cells purified from peripheral blood of breast cancer patients primed with chemotherapy and granulocyte-macrophage colony-stimulating factor. *Hum. Gene Ther.* **5**, 203–208.
20. Akkina, R. K., Rosenblatt, J. D., Campbell, A. G., *et al.* (1994). Modeling human lymphoid precursor cell gene therapy in the SCID-hu mouse. *Blood* **84**, 1393–1398.
21. Champseix, C., Marechal, V., Khazaal, I., *et al.* (1996). A cell surface marker gene transferred with a retroviral vector into CD34+ cord blood cells is expressed by their T-cell progeny in the SCID-hu thymus. *Blood* **88**, 107–113.
22. Conneally, E., Bardy, P., Eaves, C. J., *et al.* (1996). Rapid and efficient selection of human hematopoietic cells expressing murine heat-stable antigen as an indicator of retroviral-mediated gene transfer. *Blood* **87**, 456–464.
23. Medin, J. A., Migita, M., Pawliuk, R., *et al.* (1996). A bicistronic therapeutic retroviral vector enables sorting of transduced CD34+ cells and corrects the enzyme deficiency in cells from Gaucher patients. *Blood* **87**, 1754–1762.
24. Nolta, J. A., Dao, M. A., Wells, S., *et al.* (1996). Transduction of pluripotent human hematopoietic stem cells demonstrated by clonal analysis after engraftment in immune-deficient mice. *Proc. Natl. Acad. Sci. USA* **93**, 2414–2419.
25. Bauer, T. R., and Hickstein, D. D. (1997). Transduction of human hematopoietic cells and cell lines using a retroviral vector containing a modified murine CD4 reporter gene. *Hum. Gene Ther.* **8**, 243–252.
26. Emerson, S. G. (1996). *Ex vivo* expansion of hematopoietic precursors, progenitors, and stem cells: the next generation of cellular therapeutics. *Blood* **87**, 3082–3088.
27. Arca, M. J., Mule, J. J., and Chang, A. E. (1996). Genetic approaches to adoptive cellular therapy of malignancy. *Semin. Oncol.* **23**, 108–117.
28. Herrmann, F. (1995). Cancer gene therapy: principles, problems, and perspectives. *J. Mol. Med.* **73**, 157–163.
29. Wilson, J. M., Danos, O., Grossman, M., *et al.* (1990). Expression of human adenosine deaminase in mice reconstituted with retrovirus-transduced hematopoietic stem cells. *Proc. Natl. Acad. Sci. USA* **87**, 439–443.
30. Sorrentino, B. P., Brandt, S. J., Bodine, D., *et al.* (1992). Selection of drug-resistant bone marrow cells *in vivo* after retroviral transfer of human MDR1. *Science* **257**, 99–103.
31. Bodine, D. M., McDonagh, K. T., Seidel, N. E., *et al.* (1991). Survival and retrovirus infection of murine hematopoietic stem cells *in vitro*: effects of 5-FU and method of infection. *Exp. Hematol.* **19**, 206–212.
32. Fraser, C. C., Eaves, C. J., Szilvassy, S. J., *et al.* (1990). Expansion *in vitro* of retrovirally marked totipotent hematopoietic stem cells. *Blood* **76**, 1071–1076.
33. Sorrentino, B. P., McDonagh, K. T., Woods, D., *et al.* (1995). Expression of retroviral vectors containing the human multidrug resistance 1 cDNA in hematopoietic cells of transplanted mice. *Blood* **86**, 491–501.
34. Schuening, F., Miller, A. D., Torok-Storb, B., *et al.* (1994). Study on contribution of genetically marked peripheral blood repopulating cells to hematopoietic reconstitution after transplantation. *Hum. Gene Ther.* **5**, 1523–1534.
35. Moritz, T., Patel, V. P., and Williams, D. A. (1994). Bone marrow extracellular matrix molecules improve gene transfer into human hematopoietic cells via retroviral vectors. *J. Clin. Invest.* **93**, 1451–1457.
36. Traycoff, C. M., Srour, E. F., Dutt, P., *et al.* (1997). The 30/35 kDa chymotryptic fragment of fibronectin enhances retroviral-mediated gene transfer in purified chronic myelogenous leukemia bone marrow progenitors. *Leukemia* **11**, 159–167.

37. Lu, M., Zhang, N., Maruyama, M., *et al.* (1996). Retrovirus-mediated gene expression in hematopoietic cells correlates inversely with growth factor stimulation. *Hum. Gene Ther.* **7,** 2263–2271.

38. Leonard, J. P., May, C., Gallardo, H., *et al.* (1996). Retroviral transduction of human hematopoietic progenitor cells using a vector encoding a cell surface marker (LNGFR) to optimize transgene expression and characterize transduced cell populations. *Blood* **88,** 433a.

39. Ward, M., Pioli, P., Ayello, J., *et al.* (1996). Retroviral transfer and expression of the human multiple drug resistance (MDR) gene in peripheral blood progenitor cells. *Clin. Cancer Res.* **2,** 873–876.

40. Carter, R. F., Abrams-Ogg, A. C. G., Dick, J. E., *et al.* (1992). Autologous transplantation of canine long-term marrow culture cells genetically marked by retroviral vectors. *Blood* **79,** 356–364.

41. Bodine, D. M., Mortiz, T., Donahue, R. E., *et al.* (1993). Long-term *in vivo* expression of a murine adenosine deaminase gene in rhesus monkey hematopoietic cells of multiple lineages after retroviral mediated gene transfer into CD34+ bone marrow cells. *Blood* **82,** 1975–1980.

42. Bienzle, D., Abrams-Ogg, A. C. G., Kruth, S. A., *et al.* (1994). Gene transfer into hematopoietic stem cells: long-term maintenance of *in vitro* activated progenitors without marrow ablation. *Proc. Natl. Acad. Sci. USA* **91,** 350–354.

43. Donahue, R. E., Byrne, E. R., Thomas, T. E., *et al.* (1996). Transplantation and gene transfer of the human glucocerebrosidase gene into immunoselected primate CD34+Thy-1+ cells. *Blood* **88,** 4166–4172.

44. Bodine, D. M., McDonagh, K. T., Brandt, S. J., *et al.* (1990). Development of a high titer retrovirus producer cell line capable of gene transfer into rhesus monkey hematopoietic stem cells. *Proc. Natl. Acad. Sci. USA* **87,** 3738–3742.

45. Brenner, M. K., Rill, D. R., Holladay, M. S., *et al.* (1993). Gene marking to determine whether autologous marrow infusion restores long-term haemopoiesis in cancer patients. *Lancet* **342,** 1134–1137.

46. Rill, D. R., Santana, V. M., Roberts, W. M., *et al.* (1994). Direct demonstration that autologous bone marrow transplantation for solid tumors can return a multiplicity of tumorigenic cells. *Blood* **84,** 380–383.

47. Cornetta, K., Srour, E. F., Moore, A., *et al.* (1996). Retroviral gene transfer in autologous bone marrow transplantation for adult acute leukemia. *Hum. Gene Ther.* **7,** 1323–1329.

48. Deisseroth, A. B., Zu, Z., Claxton, D., *et al.* (1994). Genetic marking shows that Ph+ cells present in autologous transplants of chronic myelogenous leukemia (CML) contribute to relapse after autologous bone marrow in CML. *Blood* **83,** 3068–3076.

49. Dunbar, C. E., Cottler-Fox, M., O'Shaughnessy, J. A., *et al.* (1995). Retrovirally marked CD34-enriched peripheral blood and bone marrow cells contribute to long-term engraftment after autologous transplantation. *Blood* **85,** 3048–3057.

50. Emmons, R. V., Doren, S., Zujewski, J., *et al.* (1997). Retroviral gene transduction of adult peripheral blood or marrow-derived CD34+ cells for six hours without growth factors or on autologous stroma does not improve marking efficiency assessed in vivo. *Blood* **89,** 4040–4046.

51. Heslop, H. E., Rooney, C. M., Rill, D. R., *et al.* (1996). Use of gene marking in bone marrow transplantation. *Cancer Detect. Prev.* **20,** 108–113.

52. Douer, D., Levine, A., Anderson, A. F., *et al.* (1996). Clinical protocol: high-dose chemotherapy and autologous bone marrow plus peripheral blood stem cell transplantation for patients with lymphoma or metastatic breast cancer: use of marker genes to investigate hematopoietic reconstitution in adults. *Hum. Gene Ther.* **7,** 669–684.

53. Heslop, H. E., Brenner, M. A., Krance, R. A., *et al.* (1996). Clinical protocol: use of double marking with retroviral vectors to determine rate of reconstitution of untreated and cytokine expanded CD34+ selected marrow cells in patients undergoing autologous bone marrow transplantation. *Hum. Gene Ther.* **7,** 655–667.

54. Brenner, M. K., Rill, D. R., Moen, R. C., *et al.* (1993), Gene-marking to trace origin of relapse after autologous bone-marrow transplantation. *Lancet* **341,** 85–86.

55. Stewart, A. K., Dube, I. D., Kamel-Reid, S., *et al.* (1995). Clinical protocol: a phase I study of autologous bone marrow transplantation with stem cell gene marking in multiple myeloma. *Hum. Gene Ther.* **6,** 107–119.

56. Bachier, C., Giles, R. E., Ellerson, D., *et al.* (1996). Retroviral gene marking in relapsed follicular non-Hodgkin's lymphoma (FNHL) to determine the origin of relapse following autologous bone marrow (ABMT) and peripheral stem cell transplant (PSCT) (meeting abstract). *Proc. Annu. Meet. Am. Assoc. Cancer Res.* **37,** A1398.

57. Brenner, M. A., Krance, R., Heslop, H. E., *et al.* (1994). Clinical protocol: assessment of the efficacy of purging by using gene marked autologous marrow transplantation for children with AML in first complete remission. *Hum. Gene Ther.* **5,** 481–499.

58. Koç, O. N., Allay, J. A., Lee, K., *et al.* (1996). Transfer of drug resistance genes into hematopoietic progenitors to improve chemotherapy tolerance. *Semin. Oncol.* **23,** 46–65.

59. Hesdorffer, C., Antman, K., Bank, A., *et al.* (1994). Clinical protocol: human MDR gene transfer in patients with advanced cancer. *Hum. Gene Ther.* **5,** 1151–1160.

60. Deisseroth, A. B., Kavanagh, J., and Champlin, R. Clinical protocol: use of safety-modified retroviruses to introduce chemotherapy resistance sequences into normal hematopoietic cells for chemoprotection during the therapy of ovarian cancer: a pilot trial. *Hum. Gene Ther.* **5,** 1507–1522.

61. O'Shaughnessy, J. A., Cowan, K. H., Nienhuis, A. W., *et al.* (1994). Clinical protocol: retroviral mediated transfer of the human multidrug resistance gene (MDR-1) into hematopoietic stem cells during autologous transplantation after intensive chemotherapy for metastatic breast cancer. *Hum. Gene Ther.* **5,** 891–911.

62. Deisseroth, A. B., Holmes, F., Hortobagyi, G., *et al.* (1996). Clinical protocol: use of safety modified retroviruses to introduce chemotherapy resistance sequences into normal hematopoietic cells for chemoprotection during the therapy of breast cancer: a pilot trial. *Hum. Gene Ther.* **7,** 401–416.

63. Hanania, E. G., Giles, R. E., Kavanagh, J., *et al.* (1996). Results of MDR-1 vector modification trial indicate that granulocyte/macrophage colony-forming unit cells do not contribute to post-transplant hematopoietic recovery following intensive systemic therapy. *Proc. Natl. Acad. Sci. USA* **93,** 15346–15351.

64. Eshhar, Z., Waks, T., and Gross, G., *et al.* (1993). Specific activation and targeting of cytotoxic lymphocytes through chimeric single chains consisting of antibody-binding domains and the g or z subunits of the immunoglobulin and T-cell receptors. *Proc. Natl. Acad. Sci. USA* **90,** 720–724.

65. Stancovski, I., Schindler, D. G., Waks, T. *et al.* (1993). Targeting of T lymphocytes to Neu/HER2-expressing cells using chimeric single chain Fv receptors. *J. Immunol.* **151,** 6577–6582.

66. Moritz, K., Wels, W., Mattern, J., *et al.* (1994). Cytotoxic T lymphocytes with a grafted recognition specificity for ERBB2-expressing tumor cells. *Proc. Natl. Acad. Sci. USA* **91,** 4318–4322.

67. Hwu, P., Shafer, G. E., Treisman, J., *et al.* (1993). Lysis of ovarian cancer cells by human lymphocytes redirected with a chimeric gene composed of an antibody variable region and the Fc receptor g chain. *J. Exp. Med.* **178,** 361–366.

68. Wilke, L. G., Reynolds, C. M., McDonagh, K. T., *et al.* (1997) Engineering immune effector cells to target the HER-2/neu breast cancer antigen. (meeting abstract). SSO *Annual Cancer Symposium.* **51,** P4.

69. Cole, D. J., Weil, D. P., Shilyansky, J., *et al.* (1995) Characterization of the functional specificity of a cloned T-cell receptor heterodimer recognizing the MART-1 melanoma antigen. *Cancer Res.* **55,** 748.

CHAPTER 25

Novel Artificial Tumor-Specific Killer Cell and Intrabody Approaches for Cancer Therapy

SI.-YI CHEN

Department of Cancer Biology, Comprehensive Cancer Center, Wake Forest University School of Medicine, Winston-Salem, North Carolina 27157

I. INTRODUCTION

With advances in molecular biology and genetics, increasing numbers of oncoproteins or tumor-associated antigens, either displayed on the cell surface or located inside a cell, have been defined that can be used as targets for cancer therapy. Two fundamental approaches for tumor immunotherapy have been explored so far: the antibody-directed targeting of toxic and cytolytic activity to tumor cells, and the augmentation of host cellular immune responses to tumor cells [1–6]. Monoclonal antibodies that recognize cell surface markers on tumor cells provide a potential means for selective targeting of the tumor cells [4–7]. However, a major obstacle in the use of antibody-mediated therapy is the low percentage of an injected dose of antibodies that are delivered to the tumor site [8]. Clinical studies have indicated that considerably <0.001% of the in-

389

jected antibodies actually bind to and accumulate in the tumor, despite the use of high-avidity antibody to tumor antigens [9–11]. Meanwhile, considerable efforts have been made to stimulate or modify immune cells, such as tumor-infiltrating lymphocytes (TILs), by transduction of cytokine genes or by specific antigen stimulation, to increase the effectiveness of the cellular immune response [1–3,12–20]. However, a major limitation in the use of adoptive cellular immunotherapy lies in the difficulty in obtaining tumor-specific cytotoxic cells, which may be a major reason for the limited clinical success of using adoptive cellular immunotherapy for the treatment of cancer so far [12–20]. It is known that only small percentages of TILs or LAK (lymphokine-activated killer) cells, which are a heterogeneous population consisting mainly of CD4+ and CD8+ lymphocytes [16–20], are reactive against autologous tumors. Considerable fractions of reinfused cells do not have antitumor activities. To improve the efficacy of cancer immunotherapy, it is crucial to generate potent and antigen-defined tumor-specific cytotoxic cells. In our recent studies, we have developed several novel genetic approaches with unique features for cancer therapy. One approach is to generate a new class of tumor-specific killer cells by genetically modifying lymphocytes to produce immunotoxins that target molecules on the surface of tumor cells [21]. As a result, the transduced lymphocytes expressing immunotoxins function not only as a vehicle for delivering antibody/toxins to tumor tissues, but also as a producer of targeted toxins within the tumor tissues. Thus, this artificial tumor-specific killer cell approach may have implication for cancer therapy. In addition, a novel intrabody approach was designed to target the molecules inside malignant cells to specifically block target protein functions [22,23]. In this article, we briefly illustrate the design of these approaches and their potential for cancer gene therapy.

II. ARTIFICIAL TUMOR-SPECIFIC KILLER CELLS

A. Design of a New Class of Tumor-Specific Killer Cells

In our recent study a new class of tumor-specific killer cells that combine the specificity of antibody, extreme potency of toxin molecules, and effector cell properties of lymphocytes (Fig. 1) was designed and generated [21]. This new class of tumor-specific killer cells, which are able to produce and secrete targeted toxin proteins, was designed based on the knowledge of protein-trafficking and cell-killing mechanism of toxins. Toxin molecules, such as *Pseudomonas* exotoxin A (PEA),

block cellular protein synthesis and thus kill the cell by inactivating the elongation factor-2 (EF-2) in the cytosol [24–28], indicating that toxin molecules have to be located in the cytosol in order to kill a cell. Accordingly, it may be feasible that a mammalian cell can be genetically modified to produce and secrete targeted toxins by using a leader signal sequence [29]. Subsequently, newly synthesized targeted toxins are translocated into the lumen of the endoplasmic reticulum (ER) cotranslationally and then secreted out of the cells. The toxin-expressing cells should remain viable because the interaction of the newly synthesized fusion toxins with the EF-2 in the cytosol is blocked by the membrane lipid bilayer of the ER and secretory vesicles. The secreted targeted toxins selectively bind and destroy target cells after being internalized and released into the cytosol [4–6]. Due to the lack of antigens on the cell surface, the secreted toxins are unable to kill the toxin-expressing cells.

The feasibility of this strategy was demonstrated by generating a killer cell recognizing HER2, a transmembrane protein of the epidermal growth factor receptor family [30]. Malignant cells arise as a result of a stepwise progression of genetic events that include the overexpression of proteins or oncoproteins in the signaling pathways of cells. One such overexpressed protein that has been implicated in malignant transformation of cancer is the oncoprotein HER2 (p185^{erbB2}) [30–34]. *In vitro* studies demonstrate that malignant transformation of NIH3T3 cells can be induced by overexpression of HER2 [31]. In animal model studies, transgenic mice carrying HER2 develop a variety of tumors such as mammary carcinoma [32]. Amplification and/or overexpression of HER2 occurs in a variety of human carcinomas, including malignancies of the breast, ovary, gastrointestinal tract, and prostate [33–39]. Moreover, in breast carcinomas there is a direct correlation between the extent of amplification and overexpression of HER2 and a patient's prognosis [38,39]. A single-chain antibody (sFv) comprising the heavy chain variable region linked to the light chain variable region [40,41], derived from a monoclonal anti-HER2 23e, has been shown to exhibit high-affinity binding to the extracellular domain of HER2. When linked to a bacterial toxin, the recombinant proteins of the anti-HER2 sFv23e demonstrated potent selective cytotoxicity to targeted tumor cells *in vitro* and *in vivo* [40,41]. In this study, the sFv23e gene and a truncated gene *(PE40)* encoding domains II (translocation) and III (catalytic) (253 to 613 amino acids) of PEA [24] were linked with a leader signal sequence under the control of the cytomegalovirus (CMV) promoter. This vector with the Neo selection marker pCMV-sFv23e-PE40 was transduced into lymphocytes on which HER2 surface expression is undetectable.

Artificial Cytotoxic Cells

**Targeted Cells
(Breast cancer cells)**

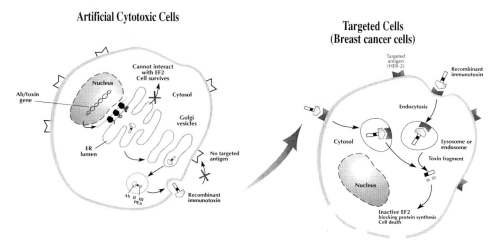

FIGURE 1 Schematic representation of generation and cytotoxic mechanism of artificial tumor-specific killer cells. Tumor-specific killer cells (left panel) were generated by transducing an anti-HER2/toxin expression vector into human lymphocytes. As a result, the genetically modified cells would express and secrete targeted antibody/toxins, which recognize and specifically kill the targeted HER2-overexpressing cells (right panel). These killer cells would function as targeted toxin producers as well as carriers, which may combine the advantages of antibody-directed and cell-mediated immunotherapy.

B. Potent Antitumor Activity

The transduced killer cells were generated and shown to produce and secrete the anti-HER2/toxin. The killer cells were demonstrated to have potent and selective cytotoxicity to various tumor cells overexpressing HER2 in cell culture [21]. To further evaluate potential *in vivo* application, human LAK cells derived from peripheral blood lymphocytes (PBLs) were transduced with the anti-HER2/PE40 gene by using a transient retroviral transfection procedure [21]. Over 40% of transduced LAK cells were expressing the toxin proteins after G418 selection, as detected by immunofluorescent staining. The mice with SKBR3 (HER-2-overexpressing breast cancer line) xenografts were administered with the transduced or mock-transduced LAK cells via tailvein injection weekly for 2 weeks. One group of SKBR3 xenograft mice were injected daily for 2 weeks with the purified anti-HER2 recombinant immunotoxin (e23sFv/PE38KDEL), a modified form of e23sFv/PE40 with 6- to 10-fold increase in cytotoxicity, in the amount (0.15 μg) comparable to the estimated e23sFv/PE40 produced by the transduced LAK cells (0.15 μg/10^6 cells/24 h). The administration of the transduced cells almost completely inhibited the growth of the tumor xenografts, but no apparent inhibition of tumor growth was found in the mice directly administered that amount of the immunotoxin.

We then examined the ability of adoptively transduced LAK cells to infiltrate tumor tissues. The SKBR3 xenograft mice were administered transduced or mock-transduced LAK cells (i.v.) and killed at different times. The toxin-expressing cells were detected in the tumor tissues one hour after injection and still observed one week later. Significant toxin ADP-ribosylation activity was found only in the tumor samples 1 week after injection. In addition, no histological abnormality of normal tissues (liver, lung, kidney, and heart) was observed under microscopy after H&E staining, although toxin-expressing cells were detected in these normal tissues and no animal died due to the toxicity of the transduced LAK cells. In contrast, in the mice administered the immunotoxin, significant toxin activities in blood and tumor were detected only shortly after administration and 12 hours later dropped to trace levels. These results suggest that e23sFv/PE40 continuously produced by the transduced cells inside the tumor may accumulate locally to exert selective antitumor activities, whereas the administered immunotoxins were quickly cleared from blood circulation and poorly penetrated into tumor tissues. Thus, the cell-based approach appears to have more potent and selective cytotoxicity than immunotoxin administration. By utilizing toxin molecules, among the most potent molecules capable of killing mammalian cells [6], these cells may have the most potent cytotoxicity to targeted cells delivered in an MHC-independent manner.

C. Potential Application for Cancer Therapy

One of the potential applications of this approach for cancer therapy is to transduce TILs or other lymphocytes with an antibody/toxin gene, because TILs can proliferate rapidly *in vitro* and recirculate in and localize at tumor sites after reinfusion [14–20]. The tumor-homing characteristic of transduced TILs may allow them to function not only as a vehicle for delivering antibody/toxins to tumor tissues, but also as a producer of targeted toxins within the tumor tissues. Thus, the major problem facing antibody-directed toxin therapy, such as the limited accessibility of circulating immunotoxins to solid tumor cells, could be overcome by using this approach. In addition, a transduced lymphocyte is capable of destroying many tumor cells. Given the low transduction efficiency of current vector systems, this feature may have a decisive advantage over other gene-based strategies, which are only able to destroy the gene-transferred tumor cells. It is especially appealing that tumor micrometastases, which are almost impossible to detect, can be recognized and destroyed by the transduced killer cells when these are reinfused back to patients, thereby functioning as immune surveillance.

Given the ability of antibodies to target antigens on tumor cells, this approach can be used for the treatment of many types of tumors. For example, in a recent study (unpublished data), killer cells specific for B-lineage leukemia/lymphomas were generated by transducing the cells with an antibody/toxin fusion gene. A monoclonal antibody, Lym-1, which recognizes a polymorphic variant of the HLA-DR antigen present on the cell surface of normal and malignant B cells and is remarkably B-cell specific, was used [42,43]. Radioimaging studies with ^{123}I-labeled Lym-1 have demonstrated selective localization to the sites of lymphomas, as well as the ability to detect tumor sites [42,43]. There has been no evidence of binding to T cells or other normal cells and tissues as demonstrated by immunochemistry of tissue sections and radioimaging [42,43]. The single-chain antibody (sFv) gene derived from Lym-1 was linked to a leader signal sequence and then fused with PE40. The specific killer cells, which were able to produce and secrete the sFv-Lym/toxin fusion protein, were generated by transducing the sFv-Lym/toxin gene into human T lymphocytes. The selective cytotoxicity of the transduced killer cells to the B-cell leukemia/lymphoma cells was observed *in vitro*. Thus, this new class of killer cells with defined specificity may have broad application for the treatment of cancers as well as other diseases [44].

In addition to the systemic application, this toxin-expressing-cell approach can also be used locally. The glial neoplasms constitute about 40 to 60% of all childhood primary CNS tumors [45]. The amplification/overexpression of the epidermal growth factor receptor (EGFR) gene was found in about 25 to 50% of pediatric glial neoplasms [46]. Thus, the overexpressed EGFR has been used as a target for brain tumor therapy [47]. In our recent study (manuscript in preparation), a plasmid DNA expressing TGF-PE40 fusion toxins was transfected into normal neurons or tumor cells. It was found that the TGF-PE40 fusion toxins were produced and secreted from the neurons, in which the EGFR expression is at an undetectable level. In contrast, the tumor cells were effectively killed after transfection. When the TGF-PE40 plasmid DNAs were also transfected into tumor xenografts in nude mouse models, the tumor growth was effectively inhibited. This study illustrates the local application of the artificial killer cell approach.

III. INTRABODY FOR CANCER GENE THERAPY

Recent advances in antibody engineering have allowed antibody genes to be manipulated and antibody molecules to be reshaped [48,49]. By harvesting the genetic information of the immune system in the form of rearranged immunoglobulin genes, engineered antibodies with high affinity and binding specificity can be created. This technological advance, combined with the wealth of information that has been obtained on intracellular protein trafficking signals [50], has allowed engineered antibodies to be directed to and function in different subcellular compartments. The marriage between these two disciplines has created a new intracellular antibody technique to block a target protein inside a cell [22,23], which may have profound implications for basic research as well as for gene therapy [51].

Intracellular antibodies, hereafter called "intrabodies," that are synthesized by the target cell and targeted to specific cellular compartments, represent a recent innovation. Intracellular expression of immunoglobulin heavy and light chains had been previously shown in mammalian cells other than B cells [52–54]. However, the potent intracellular binding activity of engineered antibodies was not demonstrated until very recently. Several recent studies from our and other groups have provided clear evidence that engineered antibodies, especially single-chain antibodies (sFvs), can be expressed and folded to inactivate target proteins in the ER, cytoplasm, and nucleus [22,23,55–58]. With the possibility of unlimited targets, the intrabody approach can be widely used to inactivate the target proteins for functional analysis as well as for therapeutic application (Fig. 2).

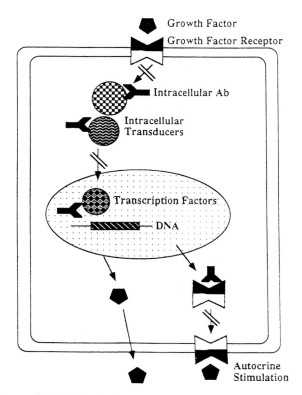

FIGURE 2   The intrabody approach can be used to block the malignant growth of cancer cells. Intracellular antibodies can be targeted to the secretory pathway to block the auto or paracrine stimulation, targeted to the cytosol to block the intracellular signal transduction, or targeted to the nucleus to block the transcriptional factors of the malignant cells.

A. Intrabodies to Block Oncoproteins in Secretory Pathway

In our previous studies, we have demonstrated that an anti-HIV-1 gp120 single-chain antibody targeted to the ER can effectively block targeted viral protein transport and function. Subsequently, the intrabody approach has been successfully used to inactivate oncoproteins. Recently, the use of anti-interleukin-2 receptor (IL-2R) intrabodies to efficiently block the cell surface expression of IL-2R was described [56]. IL-2R, a growth factor receptor, is overexpressed in a variety of T- and B-cell leukemias, most notably in HTLV-1-associated adult T-cell leukemia, which may contribute to the malignant growth of the cells. The high-affinity IL-2 receptor is a heterotrimer composed of α-, β-, and γ-chains [59]. For the study, the anti-Tac hybridoma [60] was used as the source of RNA encoding the monoclonal antibody. It was found that the ER-resident intrabody was effective at blocking the cell surface transport of IL-2R by binding and retaining the receptor chain in the ER, indicating that the sFvTac intrabodies will provide a

valuable tool for examining the role of IL-2R in T-cell activation, IL-2 signal transduction, and the deregulated growth of leukemia cells.

Intracellular expression of sFv intrabodies directed against the extracellular domain of HER-2 has been demonstrated to revert HER-2 transformation [57,58,61]. The HER-2 gene encodes a 185-kDa transmembrane glycoprotein that is a member of the subclass I epidermal growth factor receptor–related tyrosine kinases [30–36]. Amplification and/or overexpression of HER-2 is observed in tumors arising at many sites, including breast and ovary, where it correlates with an unfavorable patient prognosis [34–39]. Beerli et al. [57,61] reported on the biological effects of anti-HER-2 sFv intrabodies that bind to the extracellular domain of HER-2 and that were targeted to the lumen of the ER. When the anti-HER-2 sFv intrabodies were expressed in HER-2-transformed NIH/3T3 cells that express an oncogenically activated form of the receptor with constitutive and ligand-independent kinase activity, the sFv intrabodies bound to the receptor and prevented its transit through the ER. This resulted in the functional and stable inactivation of the oncoprotein as demonstrated by a decrease in the intensity of phosphotyrosine-containing proteins in these cells and reversion of the transformed phenotype.

In a related study, Deshane et al. [58] reported on the biological effects of anti-HER-2 sFv intrabodies that bind to the extracellular domain of HER-2 and that were targeted to the lumen of the ER. Using the HER-2-overexpressing ovarian carcinoma cell line SKOV3, these investigators demonstrated that transfection of these ER-directed anti-HER-2 sFvs resulted in inhibition of both HER-2 surface expression and cellular proliferation, and a marked reduction in survival of neoplastic clones. In addition, the biological effects of an ER-directed anti–human epidermal growth factor receptor (EGFR) sFv intrabody (scFv225R) were evaluated in EGFR-transformed NIH/3T3 cells [57]. The inhibitory effects of the intrabody on EGFR protein expression, EGFR-induced activation of cellular phosphotyrosine content, and EGF-dependent colony formation in soft agar were observed.

B. Intrabodies to Block Oncoproteins in Cytosol and Nucleus

Expression of sFv intrabodies against the p21^{ras} proto-oncoprotein in the cytoplasm has been reported [53,62,63]. In studies by Biocca et al. [62,63], microinjection of mRNA encoding an anti-Ras sFv intrabody into the cytoplasm of Xenopus oocytes was used to test the feasibility of perturbing the function of the cellular p21^{ras} protein. Xenopus oocytes could be induced to mature meiotically *in vitro* by incubation with insulin, and p21^{ras} had been shown to be involved in this process. Oocytes microinjected with mRNA encoding the cytoplasmically expressed anti-Ras sFv intrabodies demonstrated the inhibition of the insulin-induced meiotic maturation. Werge *et al.* [64] performed transient cotransfection studies in Jurkat T cells with an anti-Ras sFv intrabody expresser and a CAT reporter gene under the transcriptional control of several copies of the binding site for the transcription factor NF-AT. These investigators demonstrated that the anti-Ras sFv intrabody interfered with NF-AT activation upon direct activation of the T-cell antigen receptor.

IV. FUTURE DIRECTION

Recent studies from our and other groups illustrate that intrabodies provide a simple and powerful new approach to inactivate oncoproteins, indicating that intracellular antibodies can be used as a new class of molecules for cancer therapy as well as for research. One obvious requirement for this strategy to be successful for clinical application is that intrabody genes should be efficiently transduced into cancer cells. Currently, several vector systems such as retroviral vectors, adeno-associated virus (AAV) vectors, or adenovirus vectors can be used to transduce the gene into target cells. In addition, intrabodies can be used as a research tool to analyze a target protein function. As such, this technology may complement traditional "gene knock-out" studies in transgenic animals and may have the additional ability to "knock out" more than one protein target simultaneously. Further insight into the mechanism(s) by which cellular oncoproteins alter cell growth and promote neoplastic transformation can be expected to be made through this technology.

The new class of tumor-specific killer cells can be generated by transducing lymphocytes to produce and secrete an antibody/toxin fusion protein. This approach combines the specificity of antibody, the extreme potency of toxin molecules, and the effector cell properties of lymphocytes. These tumor-specific killer cells function not only as a vehicle for delivering targeted toxins to tumor tissues, but also as a producer of targeted toxins within the tumor tissues, in addition to their own antitumor activities. The anti-HER-2 killer cells were demonstrated to have selective and potent antitumor activities superior to administered immunotoxins *in vivo*, which may be translated into a better treatment with enhanced antitumor activity and decreased nonspecific toxicity. Toward the ultimate goal of translating this novel strategy to the practice of cancer treatment, we are currently improving and testing the transduced killer cells expressing the anti-HER2/toxin in animal models to determine their safety and antitumor activity.

References

1. Rosenberg, S. A., Packard, B. S., Aerbersold, P. M., *et al.* (1988). Use of tumor-infiltrating lymphocytes and interleukin-2 in the immunotherapy of patients with metastatic melanoma. *N. Engl. J. Med.* **319**, 1676–1680.

2. Rosenberg, S. A., Aerbersold, P., Cornetta, K., *et al.* (1990). Gene transfer into humans–immunotherapy of patients with advanced melanoma, using tumor-infiltrating lymphocytes modified by retroviral gene transduction. *N. Engl. J. Med.* **323**, 570–578.

3. Rosenberg, S. A. Immunotherapy and gene therapy of cancer. (1991). *Cancer Res.* **51**(suppl), 5074s–5079s.

4. Vitetta, E. S., Fulton, R. J., May, R. D., Till, M., and Uhr, J. W. (1987). Redesigning major poisons to create antitumor reagents. *Science* **238**, 1098–1104.

5. Vitetta, E. S., and Uhr, J. W. (1994). Monoclonal antibodies as agonists: an expanded role for their use in cancer therapy. *Cancer Res.* **54**, 5301–5309.

6. Pastan, I., and FitzGerald, D. J. P. (1992). Recombinant toxins for cancer treatment. *Science* **254**, 1173–1177.

7. Waldmann, Y. (1991). Monoclonal antibodies in diagnosis and therapy. *Science* **252**, 1657–1662.

8. Jain, R. K. (1989). Delivery of novel therapeutic agents in tumors: physiological barriers and strategies. *J. Natl. Cancer Inst.* **81**, 570–576.

9. Shockley, T. R., Lin, K., Sung, C., Nagy, J. A., Tompkins, R. G., Derick, R. L., Dvorak, H. F., and Yarmush, M. L. (1992). A quantitative analysis of tumor-specific monoclonal antibody uptake by human melanoma xenografts: effects of antibody immunological properties and tumor antigen expression levels. *Cancer Res.* **52**, 357–366.

10. Sung, C., Shockley, T. R., Morrison, P. F., Dvorak, H. F., Yarmush, M. L., and Dedrick, R. L. (1992). Predicted and observed effects of antibody affinity and antigen density on monoclonal antibody uptake in solid tumors. *Cancer Res.* **52**, 377–384.

11. Baxter, L. T., Yuan, F., and Jain, R. K. (1992). Pharmacokinetic analysis of the perivascular distribution of bifunctional antibodies and haptens: comparison with experimental data. *Cancer Res.* **52**, 5838–5844.

12. Matsumura, T., Sussman, J. J., Krinock, R. A., Chang, A. E., and Shu, S. (1994). Characteristics and *in vivo* homing of long-term T-cell lines and clones derived from tumor-draining lymph nodes. *Cancer Res.* **54**, 2744–2750.

13. Rosenberg, S. A., Yannelli, J. R., Yang, J. C., *et al.* (1994). Treatment of patients with metastatic melanoma with autologous tumor-infiltrating lymphocytes and IL-2. *J. Natl. Cancer Inst.* **86**, 1159–1166.

14. Bolhuis, R. L. H., Strum, E., Braakman, E. (1991). T cell targeting in cancer therapy. *Cancer Immunol. Immunother.* **34**, 1–8.

15. Goverman, J., Gomez, S. M., Segesman, K. D., *et al.* (1990). Chimeric immunoglobulin-T cell receptor proteins form functional receptors: implications for T cell receptor complex formation and activation. *Cell* **60**, 929–939.

16. Griffith, K. D., Read, E. L., Carrasquill, J. A., *et al.* (1989). *In vivo* distribution of adoptively transferred Indium-111-labeled tumor infiltrating lymphocytes and peripheral blood lymphocytes in patients with metastatic melanoma. *J. Natl. Cancer Inst.* **81**, 1709–1717.

17. Eshhar, Z., Waks, T., Gross, G., and Schindler, D. (1993). Specific activation and targeting of cytotoxic lymphocytes through chimeric single chains consisting of antibody-binding domains and the gamma or zeta subunits of the immunoglobulin and T-cell receptors. *Proc. Natl. Acad. Sci. USA* **90**, 720–724.

18. Gross, G., Waks, T., and Eshhar, Z. (1989). Expression of immunoglobulin-T-cell receptor chimeric molecules as functional receptors with antibody-type specificity. *Proc. Natl. Acad. Sci. USA* **86,** 10024–10028.

19. Topalian, S., Solomon, D., and Rosenberg, S. A. (1989). Tumor-specific cytolysis by lymphocytes infiltrating human melanomas. *J. Immunol.* **142,** 3714–3725.

20. Whiteside, T. L., and Parmiani, G. (1994). Tumor-infiltrating lymphocytes: their phenotype, functions and clinical use. *Cancer Immunol. Immunother.* **39,** 15–21.

21. Chen, S.-Y., Yang, A., Chen, J., Kute, T., King, R., Cong, Y., Yao, C., and Huang, X. (1997). Potent anti-tumor activities of a new class of tumor-specific cytotoxic cells. *Nature* **385,** 78–80.

22. Marasco, W. A., Haseltine, W. A., and Chen, S. Y. (1993). Design, intracellular expression, and activity of a human anti-HIV-1 gp120 single chain antibody. *Proc. Natl. Acad. Sci. USA* **90,** 7889–7893.

23. Chen, S. Y., Bagley, J., and Marasco, W. A. (1994). Intracellular antibodies as a new class of molecules for gene therapy. *Hum. Gene Ther.* **5,** 595–601.

24. Siegall, C. B., Chaudhary, V., FitzGerald, D. J., *et al.* (1989). Functional analysis of domains II, Ib and III of *Pseudomonas* exotoxin. *J. Biol. Chem.* **264,** 4256–4261.

25. Allured, V. S., Collier, R. J., Carroll, S. F., and Mckay, D. B. (1986). Structure of exotoxin A of *Pseudomonas aeruginosa* at 3.0-angstrom resolution. *Proc. Natl. Acad. Sci. USA* **83,** 1320–1324.

26. Gary, G. L., Smith, D. H., Baldridge, J. S., et al. (1984). Cloning, nucleotide sequence, and expression of *E. coli* of the exotoxin A structural gene of *Pseudomonas aeruginosa. Proc. Natl. Acad. Sci. USA* **81,** 2645–2649.

27. Collier, R. J., and Kandel, J. (1971). Structure and activity of diphtheria toxin. I. Thiol-dependent dissociation of a fraction of toxin into enzymically active and inactive fragments. *J. Biol. Chem.* **246,** 1485–1491.

28. Iglewski, B. H., Sadoff, J. B., Bjorn, M. J., and Maxwell, E. S. (1978). *Pseudomonas* exoenzyme S: an adenosine diphosphate ribosyltransferase from toxin A. *Proc. Natl. Acad. Sci. USA* **75,** 3211–3215.

29. Walter, P., and Lingappa, V. R. (1986). Mechanism of protein translocation across the endoplasmic reticulum membrane. *Annu. Rev. Cell Biol.* **2,** 499–516.

30. Bargmann, C. I., Huang, M. C., and Weinberg, R. A. (1986). The *neu* oncogene encodes an epidermal growth factor receptor-related protein. *Nature* **319,** 226–230.

31. Hudziak, R. M., Schlessinger, J., and Ullrich, A. (1987). Increased expression of the putative growth factor receptor p185HER2 causes transformation and tumorigenesis of NIH3T3 cell. *Proc. Natl. Acad. Sci. USA* **84,** 7159–7163.

32. Muller, W. J., Sinn, E., Pattengale, P. K., *et al.* (1988). Single-step induction of mammary adenocarcinoma in transgenic mice bearing the activated c-*neu* oncogene. *Cell* **54,** 105–115.

33. Van de Vijver, M. J., Peterse, J. L., Mooi, W. J., *et al.* (1988). Neu-protein overexpression breast cancer. *N. Engl. J. Med.* **319,** 1239–1245.

34. Slamon, D. J., Godolphin, W., Jones, L. A., *et al.* (1989). Studies of the HER2/neu proto-oncogene in human breast and ovarian cancer. *Science* **244,** 707–712.

35. Walker, R. A., Gullick, W. J., and Varley, J. M. (1989). An evaluation of immunoreactivity for c-erbB-2 protein as a marker of poor short-term prognosis in breast cancer. *Br. J. Cancer* **60,** 426–429.

36. Berchuck, A., Kamel, A. Whitaker, R., *et al.* (1990). Overexpression of HER2/neu is associated with poor survival in advanced epithelial ovarian cancer. *Cancer Res.* **50,** 4087–4091.

37. Wen, D., Peles, E., Cupples, R., Suggs, S. V., Bacus, S. S., Luo, Y., Trail, G., Hu, S., Silbiger, S. M., Levy, R. B., Koski, S. M., Lu, H. S., and Yarden, Y. (1992). Neu differentiation factors: a transmembrane glycoprotein containing an EGF domain and an immunoglobulin homology unit. *Cell* **69,** 559–572.

38. Muss, H. B, Thor, A. D., Berry, D. A., *et al.* (1994). c-erbB-2 expression and response to adjuvant therapy in women with node-positive early breast cancer. *N. Engl. J. Med.* **330,** 1260–1265.

39. Hynes, N. E. (1993). Amplification and overexpression of the erbB-2 gene in human tumors: its involvement in tumor development, significance as a prognostic factor, and potential as a target for cancer therapy. *Cancer Biol.* **4,** 19–26.

40. Kasprzyk, P. G., Song, S. U., Di Fiore, P. G., and King, C. R. (1992). Therapy of an animal model of human gastric cancer using a combination of anti-erbB2 monoclonal antibodies. *Cancer Res.* **52,** 2771–2776.

41. Batra, J. K., Kasprzyk, P. G., Bird, R. E., *et al.* (1992). Recombinant anti-erbB2 immunotoxins containing *Pseudomonas* exotoxin. *Proc. Natl. Acad. Sci. USA* **89,** 5876–5871.

42. Epstein, A. L., Marder, R. J., Winter, J. N., Stathopoulos, E., Chen, F.-M., Parker, J. W., and Taylor, C. R. (1987). Two new monoclonal antibodies, Lym-1 and Lym-2, reactive with human B-lymphocytes and derived tumors, with immunodiagnostic and immunotherapeutic potential. *Cancer Res.* **47,** 830–839.

43. DeNardo, S. J., DeNardo, G. L., O'Grady, L. F., Hu, E., Systma, V. M., Mills, R. L., Levy, N. B., Macey, D. J., Miller, C. H., Epstein, A. L. (1988). Treatment of B-cell malignancies with [131]I Lym-1 monoclonal antibodies. *Int. J. Cancer* **3,** 96–104.

44. Yang, A., and S.-Y. Chen (1997). A new class of antigen-specific cytotoxic cells. *Nature Biotechnology* **15,** 46–51.

45. Young, J. L., and Miller, R. W. (1975). Incidence of malignant tumors in U.S. children. *J. Pediatr.* **86,** 254–258

46. Humphrey, P. A., Wong, A. J., Vogelstein, B., *et al.* (1988). Amplification and expression of EGFR in human glioma xenografts. *Cancer Res.* **48,** 2231–2238.

47. Theuer, C. P., FitzGerald, D., and Pastan, I. (1992). A recombinant form of *Pseudomonas* exotoxin directed at the EGFR that is cytotoxic without requiring proteolytic processing. *J. Biol. Chem.* **267,** 16872–16877.

48. Riechmann, L., Clark, M., Waldmann, H., *et al.* (1988). Reshaping human antibodies for therapy. *Nature* **332,** 323–327.

49. Winter, G., and Milstein, C. (1991). Man-made antibodies. *Nature* **349,** 293–299.

50. Pugsley, A. P. (1989) In *Protein Targeting,* Academic Press, San Diego, CA.

51. Marasco, W. A. (1997). Intrabodies: turning the humoral immune system outside in for intracellular immunization. *Gene Ther.* **4,** 11–15.

52. Biocca, S., Neuberger, M. S., and Cattaneo, A. (1990). Expression and targeting of intracellular antibodies in mammalian cells. *EMBO J.* **9,** 101–108.

53. Werge, T. M., Biocca, S., and Cattaneo, A. (1990). Intracellular immunization: cloning and intracellular expression of a monoclonal antibody to the p21ras protein. *FEBS Lett.* **274,** 193–198.

54. Piccioli, P., Ruberti, F., Biocca, S., *et al.* (1991). Neuroantibodies: molecular cloning of a monoclonal antibody against substance P for expression in the central nervous system. *Proc. Natl. Acad. Sci. USA* **88,** 5611–5615.

55. Mhashikar, A. M., Bagley, J., Chen, S. Y., *et al.* (1995). Inhibition of HIV-1 infection using cytoplasmic anti-Tat antibodies. *EMBO J.* **14,** 1542–1551.

56. Richardson, J. H., Sodroski, J. G., Waldmann, T. A., *et al.* (1995). Phenotypic knockout of the high-affinity human IL2 receptor by

intracellular single-chain antibodies against the α subunit of the receptor. *Proc. Natl. Acad. Sci. USA* **92,** 3137–3141.

57. Beerli, R. R., Weis, W., and Hynes, N. E. (1994). Autocrine inhibition of the epidermal growth factor receptor by intracellular expression of a single-chain antibody. *Biochem. Biophys. Res. Commun.* **204,** 666–672.

58. Deshane, J., Loechel, F., Conry, R. M., *et al.* (1994). Intracellular sFv directed against erb B-2 down-regulates surface erb-B2 and exhibits a selective active proliferative effect in erb B2 over-expressing cancer cell lines. *Gene Therapy* **1,** 332–337.

59. Waldmann, T. A., Pastan, I. H., Gansow, O. A., *et al.* (1992). The multichain interleukin-2 receptor: a target for immunotherapy. *Ann. Intern. Med.* **116,** 148–160.

60. Uchiyama, T., Broder, S., and Waldmann, T. A. (1981). A mono-clonal antibody (anti-Tac) reactive with activated and functionally mature human T cells. I. Production of anti-Tac monoclonal anti-body and distribution of Tac (+) cells. *J. Immunol.* **126,** 1393–1397.

61. Beerli, R. R., Weis, W., and Hynes, N. E. (1994). Intracellular expression of single chain antibodies reverts erbB-2 transforma-tion. *J. Biol. Chem.* **289,** 23931–23936.

62. Biocca, S., Pierandrei-Amaldi, P., and Cattaneo, A. (1993). Intra-cellular expression of anti-p21ras single chain Fv fragments inhib-its meiotic maturation of *Xenopus* oocytes. *Biochem. Biophys. Res. Commun.* **197,** 422–427.

63. Biocca, S., Pierandrei-Amaldi, P., Campioni, N., *et al.* (1994). Intracellular immunization with cytosolic recombinant antibodies. *Biotechnology* **12,** 396–399.

64. Werge, T. M., Baldari, C. T., and Telford, J. L. (1994). Intracellular single chain Fv antibody inhibits Ras activity in T-cell antigen receptor stimulated Jurkat cells. *FEBS Lett.* **351,** 393-396.

Index

X